Study Guide with Student Solutions Manual

College Algebra

SEVENTH EDITION

Richard N. Aufmann

Vernon C. Barker

Richard D. Nation

Prepared by

Joy St. John Johnson
Oakwood University, Calhoun Community College

Christine S. Verity

Australia • Brazil • Japan • Korea • Mexico • Singapore • Spain • United Kingdom • United States

For product information and technology assistance, contact us at
Cengage Learning Customer & Sales Support,
1-800-354-9706

For permission to use material from this text or product, submit all requests online at **www.cengage.com/permissions**
Further permissions questions can be emailed to
permissionrequest@cengage.com

ISBN-13: 978-0-538-75764-5
ISBN-10: 0-538-75764-7

Brooks/Cole
20 Davis Drive
Belmont, CA 94002-3098
USA

Cengage Learning is a leading provider ofustomized learning solutions with office locations aroid the globe, including Singapore, the United Kingdom, ustralia, Mexico, Brazil, and Japan. Locate your lod office at:
www.cengage.com/global

Cengage Learning products are representd in Canada by Nelson Education, Ltd.

To learn more about Brooks/Cole, visit
www.cengage.com/brookscole

Purchase any of our products at your locacollege store or at our preferred online store
www.cengagebrain.com

Printed in the United States of America
1 2 3 4 5 6 7 14 13 12 11 10

Table of Contents

Study Guide

Student Solutions Manual

Study Guide

Preface

It has been suggested that the three most-desired traits an employer looks for in a potential employee are communication skills, critical thinking ability, and teamwork know-how. I don't pretend to be writing a book to prepare you for the job market, but these three traits do interest me and also seem crucial to success in math. In order to improve communication about math, I have expanded on the definitions of terms and sometimes tried to reframe problems in slightly different language in order to stretch comprehension. I believe that all the thinking-type questions of "why" and "what if" that I've included here encourage critical thinking. I also intend that my intermediate-style of problem solving will entrain brains a little further into the problem-solving mindset. Of course, reading a book is usually a solitary endeavor, but I've tried to let you know that I'm on your team with my tips about common pitfalls and important points. Don't forget, too, that your fellow students and, importantly, your instructor are team members you can work together with for everyone's benefit.

Most of all I have enjoyed writing this study guide. Although it by no means replaces the textbook, I have in mind that it will get you from reading text to solving problems on your own a little more easily. For those headed to a calculus series, I hope you will be better prepared.

This study guide is organized in chapters, sections, and subsections -- exactly like the text which it accompanies. The test-prep problems follow the recipe for test-prep as the textbook lays it out, but my problems are included with the content instead of at the end of the chapter. (By the time you face the same material at the end of the chapter in the textbook, you will be readier for test prep having seen it with its original material already.) The many suggestions are easy to spot: they are **For Fun** and **Questions to Ask Your Teacher** items. I've created a game at the start and end of each chapter -- something I'm freer to do than the *serious* textbook -- and I hope these help with communication and critical thinking.

This study guide gives you more worked-out examples and problems in hopes that you will find yourself able to bite off a little bit more before you are left to handle problems on your own. All main ideas and theories have been restated here: repetition, repetition, repetition! This book is formatted much more informally than a textbook in hopes of providing a welcoming atmosphere for exploration by budding algebraists.

I'm grateful to have had this opportunity to write a little book about math. Thanks go to Tom Ash, my Brooks/Cole sales rep, for suggesting me for the job and to Stefanie Beeck, the assistant editor for Math at Cengage Learning, who encouraged me and stuck with me through this process. My accuracy reviewer, Dr. Charles Heuer professor emeritus of Concordia College, was a gem to work with, and I can only thank the heavens he was looking over my shoulder. Diane Leisher, a former colleague from Alabama A&M University, took a look and gave me valuable feedback. I owe a debt of gratitude to my writing partner Khris Chess who bravely took on proofing and editing a book about *math* of all things! I wish for her, everyone else who helped me, and all my intrepid students bright futures and dreams fulfilled.

Chapter P: Preliminary Concepts

Word Search

```
L W I S P Z X T Z C J X V C I
H A F M Q R N E R C E I O E N
L O C G A A O O D X J E R L F
L A I O T G T D P N F Q A B I
A H N S R A I O U F I Z D A N
E N N O R P N N I C W L I I I
R O I E I E I C A S T A C R T
C X M X N T I C I R Q I A A Y
I U C T W E A X E E Y M L V U
N Y D O N K A R C R Z O G V J
R S V T F A C T O R I N G H H
R O T A N I M O N E D Y G Q C
L A V R E T N I L D I L F R B
I N T E G E R Y S E T O M V T
H E V Y M U O G X E L P M O C
```

AXIS	IMAGINARY	PRODUCT
COEFFICIENT	INDEX	RADICAL
COMPLEX	INFINITY	RATIONAL
CONSTANT	INTEGER	REAL
DENOMINATOR	INTERVAL	RECIPROCAL
EXPONENT	NUMERATOR	SET
FACTORING	POLYNOMIAL	VARIABLE

P.1: The Real Number System

The Real Numbers are the numbers we need to get along in our everyday lives -- from just plain counting things, to representing numbers too big to count easily, to balancing the checkbook, to building houses, and more. We also need to be able to communicate about these numbers in order to get along in our everyday lives, so now we will talk a little about the ways we communicate about and classify these numbers.

There are a lot of basic concepts in this section. Now is not the time to panic. Read and enjoy as much as you can!

Sets

In order to communicate about numbers, one of the first things we need to do is to distinguish between different kinds of numbers. So if we collect numbers together based on different attributes, we can call those collections "sets." Then we apply reasoning and set operations to them.

✤ **composite number**
A composite number is a positive integer that can be written as a product of other numbers besides one and itself. For example, 21 can be written as a product of 3 and 7, or 1, 3, 7, and 21 are its factors. In other words, a composite number is *not* a prime number. By the way, being able to represent numbers as a product is extremely useful. (When we *factor* polynomials later, for instance, we will use some of the same ideas.)

✤ **negative integers**
The counting or natural numbers all have opposites, created simply by putting a negative sign in front of them. These would be the negative integers. The negative integers with the counting numbers (positive integers) and the number zero make up the set of integers.

✓ **Questions to Ask Your Teacher**
Sometimes a special symbol is used to symbolize the set of integers, ***Z***. Ask if your teacher uses this notation.

★ **For Fun**
If you drew a number line with zero in the middle, placed some positive and negative numbers on it, could you come up with a workable meaning for the negative sign?

✤ **positive integers**
Of course, positive integers are the counting numbers, right? (They're also called the natural numbers.)

★ **For Fun**
Is zero a negative integer or a positive integer?

★ **For Fun**
What is the difference between the natural numbers and the whole numbers? Does that difference make the names of these sets sensible?

✤ **empty set, null set**
The empty set is actually the very subtle idea of a collection of absolutely nothing. (How could it be a collection, then?)

★ **For Fun**
Can you see a relation between the idea of the empty, or null, set and the idea of zero?

✤ **finite set**
One way to describe a finite set, or a collection of objects, is that it is conceivable that all the members of the set could be listed ... even if it took a lifetime or more to do so. For example, an easy finite set to list might look like $\{a,b,c\}$; on the other hand, the set of all integers between zero and two hundred million is a set that you can conceive of listing, but who would want to? The elements of a finite set can be counted, and importantly, it is conceivable that the counting would not go on forever and ever.

✤ **infinite set**
An infinite set contains an infinite number of elements. Now, infinity is a hard concept to come to terms with, and that's what makes it different from finiteness. Counting until infinity is inconceivable. An infinite set is inconceivably large. Infinity is an idea you have to relax into and let your mind accept!

★ **For Fun**
But could an infinite set be *countable*? Don't answer too quickly.

✣ **set builder notation**

Set builder notation is another method to communicate about the members in a set. Since it's a notation, the notation holds all the information: $\{x | 0 < x \leq 5\}$, for example. $\{x | 0 < x \leq 5\}$ means "the set of all x's such that x is greater than zero and less than or equal to five." (Just a matter of translating from *math-ese* into English.)

✓ **Questions to Ask Your Teacher**

Maybe now would be a good time to go over the number line and greater than and less than if you have problems sorting out those symbols.

★ **For Fun**

Do you think that the set $\{x | 0 < x \leq 5\}$ is finite or infinite? Again, don't answer too quickly!

Sets: Test Prep	
Specify those numbers which are integers, rational numbers, irrational numbers, prime, and reals: $-\sqrt{3},\ 0,\ 3\frac{4}{5},\ 0.\overline{3},\ \pi,\ -1,\ -\frac{100}{\sqrt{7}}$	By inspection, we can sort these numbers accordingly. • integers: 0, -1 • rationals: 0, $3\frac{4}{5}$, $0.\overline{3}$, -1 • irrationals: $-\sqrt{3}$, π, $-\frac{100}{\sqrt{7}}$ • primes: none • reals: all
Try Them	
Put the elements above into order from smallest to largest.	$-\frac{100}{\sqrt{7}},\ -\sqrt{3},\ -1,\ 0,\ 0.\overline{3},\ \pi,\ 3\frac{4}{5}$
Find the exact decimal representation of $-\frac{4}{7}$	$0.\overline{571428}$

✓ **For Fun**
Where do the Whole Numbers fit in this Venn diagram?

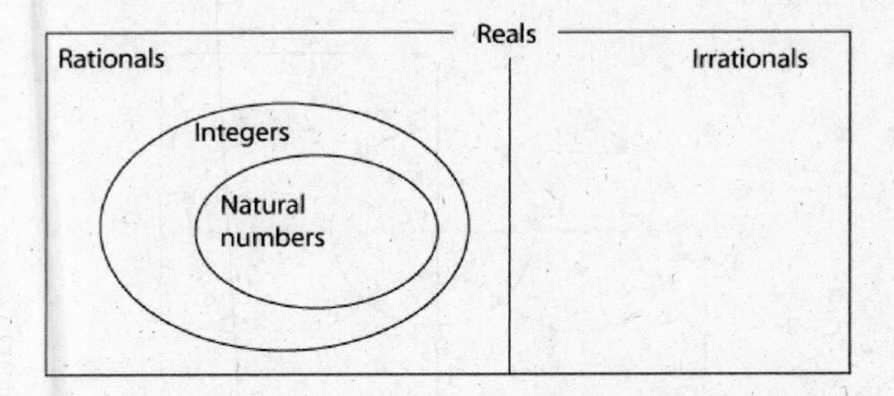

Set Builder Notation: Test Prep	
Write a description of this set in English: $\{x \mid 0 < x\}$	Noting that x's are strictly greater than zero, we can say, "The set of all x such that x is a positive real number."
Try Them	
List the smallest three elements of the set $\{y \mid y = 2x, \quad x \in naturals\}$	2, 4, 6
List the largest three elements of $\{x \mid x < 0, \quad x \in Z\}$	-1, -2, -3

Union and Intersection of Sets

Just like there are well-defined operations for numbers like addition and multiplication, sets have operations that apply especially to them. The union of two or more sets is the operation that *includes* all the members of those sets in one new set. The intersection is a more exclusive operation; it admits for members only those items that were already in all of the sets under consideration.

★ **For Fun**

Can you see the different sets, unions of sets, and intersections of sets in this Venn diagram?

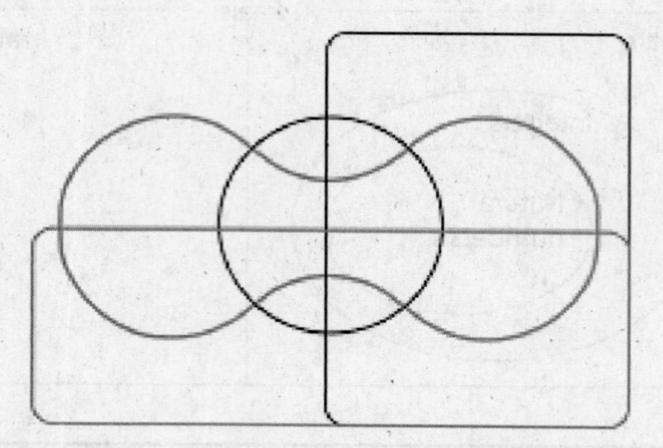

Union and Intersection of Sets: Test Prep	
Find $A \cap B$ given A= {p, e, a, c, h} and B= {p, e, a, r}	By inspection, we see that the elements common to A and B are {p, e, a}
Try Them	
What is the set formed by $\{0\} \cup \mathbf{Z}$	**W**, the set of whole numbers.
Find $A \cup B$ given A and B above.	{p, r, e, a, c, h}

Interval Notation

Dense sets that cannot be listed, like the rational numbers (fractions), require different symbols for describing them. One such tool is set-builder notation; another is interval notation.

✤ **interval notation**

Interval notation is a special kind of set notation. In interval notation, the set $\{x|0 < x \leq 5\}$ would be written $(0,5]$. The parenthesis means that zero is *not* included in the set; the square bracket means five is included. And everything between those numbers is included, too.

✤ **open interval**

Intervals that include neither of the end-points are open intervals: $(0,5)$, for example.

✤ **closed interval**

Both end-points are included in a closed interval: $[-3.5,10]$, for example.

★ **For Fun**

Could you write $[-3.5,10]$ in set-builder notation? Could you signify this closed interval on a number line?

✤ **half-open interval**

You could just as well call this a half-closed interval, couldn't you? $(0,5]$ is one.

✤ **infinity symbol: ∞**

In interval notation, "all real numbers" is symbolized by the open interval $(-\infty,\infty)$, "from negative infinity to positive infinity." Please notice that infinity, ∞, is *not* a number, but an idea, and so can never be an included end-point. (Could you ever actually reach infinity in order to include it?)

✤ **negative infinity symbol: $-\infty$**

The number line goes forever in both directions. Negative infinity is where the arrowhead on the negative side of the number line is headed!

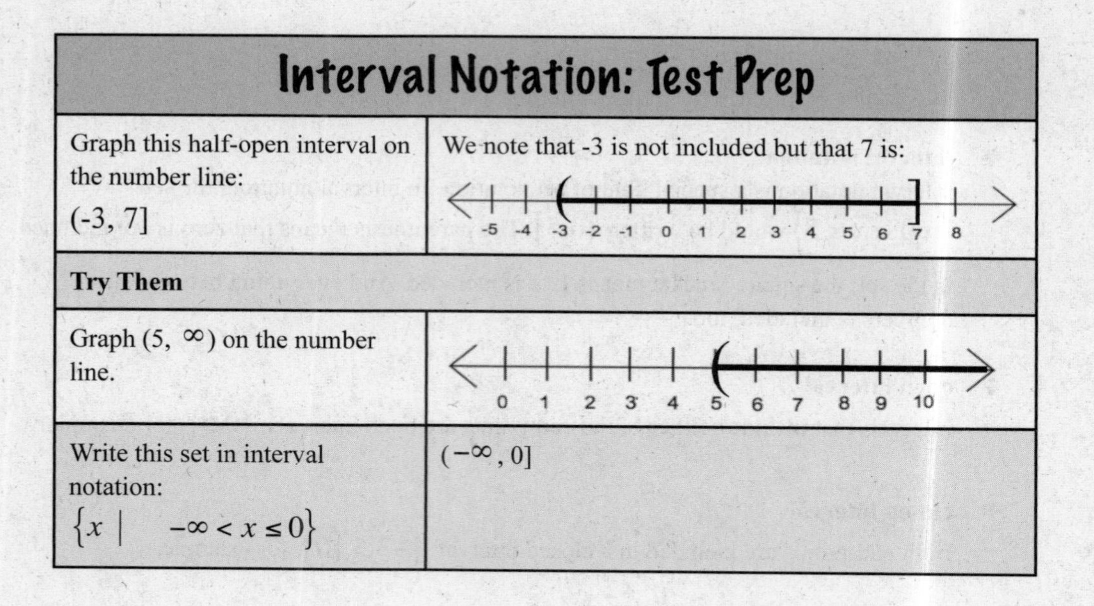

Interval Notation: Test Prep	
Graph this half-open interval on the number line: (-3, 7]	We note that -3 is not included but that 7 is: -5 -4 -3 -2 -1 0 1 2 3 4 5 6 7 8
Try Them	
Graph (5, ∞) on the number line.	0 1 2 3 4 5 6 7 8 9 10
Write this set in interval notation: $\{x \mid \quad -\infty < x \le 0\}$	$(-\infty, 0]$

Absolute Value and Distance

Absolute Value is one of the easiest operations you learned in grade school. You know: just change it to a positive. It turns out, though, that the idea of absolute value is tied up with what it means to be a certain *distance* away. When it comes to numbers, the distance between them is very important, indeed. See if you can think differently about this easy operation.

- **coordinate axis**

  An axis is a fixed line; the coordinates are the fixed values on that line. By the way, it's better to be consistent with distances between coordinates!

- **real number line**

  The real number line is a coordinate axis that includes all real numbers: $(-\infty,\infty)$.

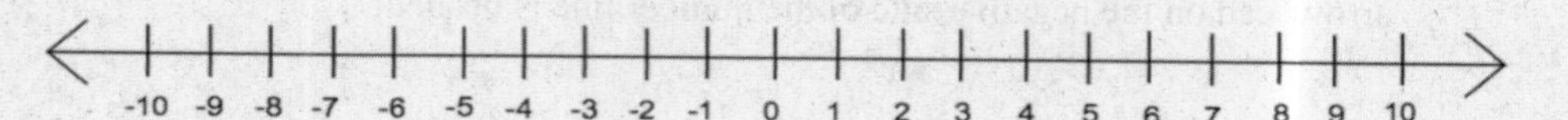

✤ **coordinate**

A coordinate is a member of a group of numbers that specifies a position, like the numbers on the number line above!

✤ **origin**

Traditionally this is at zero, the point that means you've gone nowhere and are perhaps ready to embark! The origin is like the "You Are Here" symbol on the map at the mall.

✤ **absolute value**

Absolute value tells you have far you've gone, the distance you have travelled from the origin. And distance is always a positive notion, isn't it? Make sure you understand the formal definition of absolute value given on page 8 of your text book; the functional notation is important and understanding it will help tremendously in subsequent sections and math classes.

★ **For Fun**

Can you imagine a meaning for negative distance? How could you go negative three miles from where you are?

✤ **exponential notation**

Exponential notation is short hand for multiplication. 3^4 means $3 \cdot 3 \cdot 3 \cdot 3$.

✤ **base**

In the expression 3^4, the 3 is the base. In the expression x^4, what is the base?

✤ **exponent**

The presence of an exponent is what exponential notation is all about, right? In the expression 3^4, the 4 is the exponent! The exponent tells you how many times to multiply the base by itself.

Absolute Value: Test Prep

Use the definition of absolute value to evaluate: 1. $\lvert 5\rvert$ 2. $\lvert 0\rvert$ 3. $\left\lvert -\frac{1}{7}\right\rvert$	1. Since $5 > 0$, $\lvert 5\rvert = 5$ 2. Since $0 = 0$, $\lvert 0\rvert = 0$ 3. Since $-\frac{1}{7} < 0$, $\left\lvert -\frac{1}{7}\right\rvert = -\left(-\frac{1}{7}\right) = \frac{1}{7}$
Try Them	
Given $x = -3$, evaluate $\lvert 2x+1\rvert$	5
Evaluate $-\lvert -5-1\rvert + \left\lvert -\frac{3}{4}\right\rvert$	$-5\frac{1}{4}$ or $-\frac{21}{4}$

Distance: Test Prep

Find the distance between -5 and 14.	$d(-5,14)$ $= \lvert -5-14\rvert$ $= \lvert -19\rvert$ $= 19$
Try Them	
Use absolute value notation to describe the distance between x and -9.	$\lvert x+9\rvert$
Use absolute value notation to describe that the distance between z and 2 is not more than 7.	$\lvert z-2\rvert \le 7$

Exponential Notation: Test Prep	
Evaluate $\left(-4^3\right)\left(-2\right)^2$	$\left(-4^3\right)\left(-2\right)^2$ $=-64\cdot 4$ $=-256$
Try Them	
Given $a=3$ and $b=-2$, evaluate $\dfrac{b^2}{-a^2}$	$-\dfrac{4}{9}$
Given $x=-1$ and $y=5$, evaluate $-x^2+y$	4

Order of Operations Agreement

We have algebraic operations -- addition, subtraction, multiplication, division -- and now exponents, and we have algebraic expressions that use these operations between their parts. In order for the meaning in our mathematics to remain consistent and useful, we have to agree on how these symbols combine and what instruction is intended by the author. So we have the convention of "order of operations":

1. **Groups First**
   Use the following three steps to simplify any expressions inside of grouping symbols like brackets and parentheses, then use them to continue to simplify the expression as far as possible.
2. **Exponents**
   Simplify exponential expressions.
3. **Multiplication and Division**
   Carry out multiplication from left to right as those operations appear in the expression. Take as many steps as you need here to get the correct result!
4. **Addition and Subtraction**
   Carry out addition and subtraction from left to right as those operations appear in the expression.

Take your time; don't be fooled by attractive, easy steps if they are not the next correct step as dictated by the order of operations agreement. And please, realize that just because you know an acronym does not mean you cannot be "fooled" into violating the agreement. Practice, and practice taking your time.

Another very important aspect of using the order of operations agreement: know how to input expressions in your calculator so that the order is observed when you evaluate in this manner, too.

✤ **variable term**
Variables are what algebra is all about, I like to think. It requires a lot of sophisticated thought to allow a letter to stand in for several possible choices for the idea of a number. Layers of abstraction. At any rate, a product that contains a letter standing in for a number is a variable term. For example, $-3x$, $12xy$, x^2, and so forth.

★ **For Fun**
Can you think of anything else besides letters to use as variables?

✤ **constant term**
A constant term does not have a variable in it. Constant terms have the appearance of ... [drum roll, please] ... numbers! For example, -17, 2, and the like.

✤ **numerical coefficient**
In a variable term, the product under question always has a constant, or numerical, factor. It's almost always written at the beginning of the term. For example, the numerical coefficient of $-3x$ is -3. It's the constant factor at the front.

★ **For Fun**
See how good you are at specifying the coefficients of these numbers: $17x^2y$, $-y^3$, xyz, $\frac{x}{3}$. The answers are, respectively: 17, -1, 1, and $\frac{1}{3}$.

✓ **Questions to Ask Your Teacher**
If you're not clear about the last coefficient from the list just above, you might ask your teacher to go over fractions, and multiplication, and show a few more examples. The whole class might benefit from that clarification.

- **variable part**
  Put simply: the part made up of letters, not the numerical coefficient, that's the variable part. Name the variable parts from the For Fun above. Easy.

- **evaluate**
  To eVALUate means to find the VALUE of. By the way, values are numbers. So keep in mind that the answers to "evaluate" questions will be numbers; no variables left over! To evaluate an algebraic expression, you must replace each variable with its specified numerical value ... then simplify *carefully* according to the proper order of operations.

Order of Operations Agreement: Test Prep	
Evaluate $-2\cdot 3^2-\lvert -5\rvert$	$-2\cdot 3^2-\lvert -5\rvert$ $=-2\cdot 9-5$ $=-18-5$ $=-23$
Try Them	
Evaluate $7-3(3-5)^2$	-5
Evaluate $-5+[9-(-1)]^2\div 4$	20

Simplifying Variable Expressions

Simplifying means to make something simpler or easier to do or understand. In math something is usually "simpler" if parenthesis are removed and operations are carried out wherever possible, like terms are combined, and, if applicable, ratios are reduced. In order to know what operations you can perform, and when and how, you have to know the operations, how to communicate about them, and how they are defined for use with the real numbers. Here follow some words about the operations, but be sure to thoroughly understand the rules for how these operations can be used with the real numbers; that is, what the properties of real numbers are. These properties, and the properties of equality, are discussed on pages 12 through 14 in your text.

- **addition**

  Addition is just a special word for "counting up all the units in order to find their total." It's a good thing we know how to add pretty quickly using addition tables and shortcuts (or even, gasp, calculators). Addition is where arithmetic starts, and so it's the basis for a lot of math. That means *counting* is the basis for a lot of math -- Does that make sense?

- **sum**

  A sum is the total you arrive at after performing an addition.

- **terms**

  In an algebraic expression, like $3x + 2y + z$, three *terms* are separated by the plus sign. Please note that neither the numerical coefficients nor the variable parts make up a term in themselves!

- **multiplication**

  Multiplication is simply a shortcut for addition (see? I told you addition formed the basis of a lot of stuff; here's one already!). 3 times 2 means 3 plus 3, or 2 plus 2 plus 2; that is, $3 \cdot 2 = 3 + 3 = 2 + 2 + 2 = 6$.

- **product**

  Like a sum is the result of addition, a product is the result of multiplication. In the above example, 6 is the product.

- **factor**

  In the above example, the factors are 3 and 2. Notice that if $3 \cdot 2 = 6$, then it's equally true that $6 = 3 \cdot 2$, and we've *factored* 6.

- **additive inverse**

  The additive inverse is the number you can add to any number and obtain a sum of zero. Thus, $a + (-a) = 0$. Another word for "additive inverse" is "opposite."

★ **For Fun**
By the way, zero is called the additive identity -- it is the number you can add to any number and not change the *identity* of that number: $a + 0 = a$. Can you perform a little algebra on this equation to turn it back into the equation $a + (-a) = 0$?

✤ **subtraction**
And subtraction, again in terms of addition, is the addition of the opposite! For example: $10 - 3 = 10 + (-3) = 7$.

✓ **Questions to Ask Your Teacher**
Now is also a good time to review the difference between addition and subtraction using a number line. What does the minus sign really mean?

✤ **difference**
A difference is the result of subtraction. But it wouldn't hurt to also hang onto the idea of what a difference means in English; you'll see how that meaning holds very soon in this course!

✤ **multiplicative inverse**
The multiplicative inverse is the number that you can *multiply* any number by and find the product to be one. Thus, $b \cdot \frac{1}{b} = 1$. Another word for "multiplicative inverse" is the reciprocal (keep reading).

✤ **reciprocal**
Same as the multiplicative inverse, you find the reciprocal by "turning a number upside down."

✓ **For Fun**
Find the reciprocals of $\frac{1}{3}, -\frac{8}{9}, 15$. Make sure when you multiply the reciprocal you specify with the original number that the product is 1.

✤ **division**

As soon as you write something like $\frac{1}{b}$ you're already talking about division. For instance, $b \cdot \frac{1}{b}$ means $b \div b$. What I'm saying is that division is multiplication by the reciprocal; maybe you've heard that before? In other words, we really know division as multiplication, just like subtraction is really defined as addition of the opposite. Note that you know $32 \div 8 = 4$ not because you really know or understand how many times 8 goes into 32; what you do know is $32 = 8 \cdot 4$.

✤ **quotient**

The special name we give to the result of a division operation is the quotient.

✤ **numerator**

Every division question can be written in the form of a fraction: $\frac{32}{8}$, for instance. The top part of a fraction is called the numerator. In a division problem, whether it's formatted as a fraction or not, the "numerator" can also be called the dividend.

★ **For Fun**

Is there a relation between this dividend and the dividend that is a positive return on an investment?

✤ **denominator**

The denominator is the bottom of a fraction, also called the divisor.

★ **For Fun**

Okay, this is important: what number can you never divide by?

On September 21, 1997, a divide by zero error on board the USS Yorktown (CG-48) Remote Data Base Manager brought down all the machines on the network, causing the ship's propulsion system to fail. "Sunk by Windows NT". *Wired News.* 1998-07-24.

Properties of Real Numbers: Test Prep

Fill in the blank and name the property used: $-a + \_\_\_\_ = 0$	$-a + a = 0$, Additive Inverse
Try Them	
Name all the properties of real numbers used in $2u + 4w = 2(2w + u)$	Distributive Property of Multiplication Over Division and Commutative Property of Addition.
Simplify the variable expression: $5 - 7[2 - (x+1)] - (x+10)$	$6x - 12$

The commutative and associative laws do not hold for subtraction or division:

a – b is not equal to b – a
a / b is not equal to b / a
a – (b – c) is not equal to (a – b) – c
a / (b / c) is not equal to (a / b) / c

Try some examples with numbers and you will see that they do not work.

Properties of Equality: Test Prep	
Name the property of equality used If $a(b+c) = ab + ac$, then $ab + ac = a(b+c)$.	Since the left- and right-hand sides have traded places, this is the Symmetric Property of Equality.
Try Them	
Name the Property of equality used: $2a = 2a$	Reflexive Property of Equality.
Name the Property of equality used: If $x = 3$, then $x + 5 = 8$	Substitution Property of Equality.

P.2: Integer and Rational Number Exponents

Remembering from last section that an exponent is shorthand for multiplication -- that is, $b^n = b \cdot b \cdot b \cdot \ldots \cdot b$ with b repeated n times in the product -- we are now ready to explore all sorts of interesting uses and properties of exponents. Be sure you have reviewed adding, subtracting, multiplying, and dividing fractions.

Properties of Exponents

1. For any non-zero real number b, $b^0 = 1$.
2. If $b \neq 0$ and n is a natural number, then $b^{-n} = \frac{1}{b^n}$ and $\frac{1}{b^{-n}} = b^n$.
3. $b^n \cdot b^m = b^{n+m}$
4. $\frac{b^n}{b^m} = b^{n-m}$
5. $\left(b^m\right)^n = b^{mn}$
6. $\left(a^m b^n\right)^p = a^{mp} b^{np}$,
7. $\left(\frac{a^m}{b^n}\right)^p = \frac{a^{mp}}{b^{np}}$, $b \neq 0$

Know them, love them, practice them until they are no longer a stumbling block or a mystery. Each and every one of them can be easily derived from the basic definition of an exponential: $b^n = b \cdot b \cdot b \cdot \ldots \cdot b$ where b is repeated n times in the product.

✓ **For Fun**

You know what I'm going to suggest: prove a couple of these properties for yourself. Or you can more easily do a few examples with real numbers and small exponents!

Zero and Negative Exponents: Test Prep

Evaluate $-\frac{1}{2^{-8}}$	$-\frac{1}{2^{-8}}$ $= -2^8$ $= -256$
Try Them	
Given non-zero variables, evaluate $(abc)^0$	1
Given non-zero variables, simplify $\frac{4^{-2}}{x^0}$	$\frac{1}{16}$

Properties of Exponents: Test Prep

Use the quotient rule of exponents to simplify $\frac{a^2}{a^{-10}}$	$\frac{a^2}{a^{-10}}$ $= a^{2-(-10)}$ $= a^{12}$
Try Them	
Simplify $3^2 \cdot 3$	27
Simplify $\left(\frac{2}{x^2}\right)^5$	$\frac{32}{x^{10}}$

Operations with Rational Exponents: Test Prep	
Given non-zero variables, simplify $\left(\frac{x^0y^{-10}}{3xy^2}\right)^{-2}$	$\left(\frac{x^0y^{-10}}{3xy^2}\right)^{-2}$ $=\left(\frac{3xy^2}{x^0y^{-10}}\right)^{2}$ $=\left(\frac{3xy^2y^{10}}{1}\right)^{2}$ $=\left(3xy^{12}\right)^{2}$ $=9x^2y^{24}$
Try Them	
Simplify $\left(a\cdot a^4\cdot b^3\right)^3$	$a^{15}b^9$
Simplify $\left[\frac{1}{(-2)^3}\right]^{-\frac{5}{3}}$	-32

Scientific Notation

By now you know that mathematicians like to come up with shortcuts and shorthand for various notations -- exponential notation itself is one example. Scientific notation is another shorthand especially used to specify extremely large and extremely small numbers.

✤ **scientific notation**

In scientific notation, all numbers are written in the form $a \times 10^n$. Some examples:

decimal notation	scientific notation	on the calculator
2000	2×10^3	2 E3
-6,720,000	-6.72×10^6	-6.72 E6
0.000 012	1.2×10^{-5}	1.2 E-5

The exponential part of scientific notation obeys all the properties of scientific notation that you have become so familiar with!

★ **For Fun**
Who "invented" scientific notation: mathematicians or other scientists? What disciplines use it often?

★ **For Fun**
What is the difference between scientific notation and engineering notation? (How much do you know about the metric system?)

Scientific Notation: Test Prep	
Rewrite in scientific notation: 23,400,000	Moving the (understood) decimal point 7 places to the left so that $0 < 2.34 \le 10$, write 2.34×10^7. And check your answer!
Try Them	
Rewrite in decimal notation: -5.6×10^{-3}	-0.0056
Rewrite in scientific notation: $\frac{3}{1000}$	3.0×10^{-3}

Rational Exponents and Radicals

Okay, this is a serious heads-up for any students going on to Calculus: you have to "get" this stuff! (And if you're not necessarily planning on Calculus, you need it for this course anyway.)

1. If n is an even positive integer and $b \geq 0$, then $b^{\frac{1}{n}}$ is the nonnegative real number such that $\left(b^{\frac{1}{n}}\right)^n = b$.

2. If n is an odd positive integer, then $b^{\frac{1}{n}}$ is the real number such that $\left(b^{\frac{1}{n}}\right)^n = b$.

Both these cases basically say " $b^{\frac{1}{n}}$ is the nth root of b."

In other words, you can write $\sqrt{2} = 2^{\frac{1}{2}}$, or $\sqrt[4]{16} = 16^{\frac{1}{4}} = 2$, and so on and so forth! Now all these roots are just *exponential expressions*, and you can use all the exponential rules to handle them. That's a really, really good thing.

Simplify Radical Expressions

Once you know how to handle radical expressions in their exponential form, a lot of the work of simplifying the radical can become easier ... especially if you have reviewed all your fractional operations.

Here come some properties of radicals. I recommend that you translate the second item in particular into its rational exponent version and just see how much sense it makes.

1. If n is a positive integer and b is a real number such that $b^{\frac{1}{n}}$ is a real number, then $b^{\frac{1}{n}} = \sqrt[n]{b}$.

2. For all positive integers n, all integers m, and all real numbers b such that $\sqrt[n]{b}$ is a real number, $\sqrt[n]{b^m} = \left(\sqrt[n]{b}\right)^m = b^{\frac{m}{n}}$.

3. If n is an even natural number and b is a real number, then $\sqrt[n]{b^n} = |b|$.

4. If n is an odd natural number and b is a real number, then $\sqrt[n]{b^n} = b$.

✤ **radicals**

Radicals are *roots*, like square roots, cube roots, 10th roots, you name it. "Radical" also refers to the symbol: $\sqrt{}$.

✤ **radicand**

All the information *under* the radical symbol is the radicand. For example, $1-x$ is the radicand of the expression $\sqrt{1-x}$.

✤ **index**

The little integer nestled in the pocket in front of the radical symbol is the index: $\sqrt[9]{22}$, whose index is 9. The index tells you the denominator of the corresponding rational exponent version: $22^{\frac{1}{9}}$ in this case. So if $\sqrt{2} = 2^{\frac{1}{2}}$, what is the "invisible" index? Square roots are understood to have an index of 2.

✣ **principal square root, square root**

The principal square root is the square root you get if you plug the radical into your calculator. Technically we know that $\sqrt{4} = \pm 2$, but nobody says that when you ask "what is the square root of 4?" The principal square root is the *positive* square root. If you actually want *both* roots, you must specify $\pm\sqrt{4} = \pm 2$, or if you want the negative root: $-\sqrt{4} = -2$.

✣ **simplest form of a radical**

Here are the requirements for changing any radical into its simplest form:

1. The radicand contains only powers less than the index.

2. The index of the radical is as small as possible.

3. The denominator has been rationalized; that is, no radicals appear in the denominator.

4. No fractions occur under the radical sign (in the radicand).

✣ **like radicals**

You can add and subtract radicals just like you do with variable terms if they are *like* terms (next section, coming attractions!); that is, the variable parts are the same. Like radicals' radical parts are the *same*. Of course, you might have to *simplify* the radicals before you can tell if they are "like."

For example, $\sqrt{2} + \sqrt{200} = \sqrt{2} + 10\sqrt{2} = 11\sqrt{2}$

✤ **rationalize the denominator**

Step 3 above in the requirements for the simplest form of a radical makes this topic necessary. Unfortunately not all radical expressions come automatically with with no radicals in the denominator. The underlying idea involves multiplying the top and bottom of the ratio by something that will cause the denominator to become a perfect root of the radical in question. For example:

$$\frac{1}{3-\sqrt{5}} = \frac{1}{3-\sqrt{5}} \cdot \frac{3+\sqrt{5}}{3+\sqrt{5}} = \frac{1\left(3+\sqrt{5}\right)}{\left(3-\sqrt{5}\right)\left(3+\sqrt{5}\right)}$$

$$= \frac{3+\sqrt{5}}{9+3\sqrt{5}-3\sqrt{5}-\sqrt{25}} = \frac{3+\sqrt{5}}{9-5}$$

$$= \frac{3+\sqrt{5}}{4}$$

★ **For Fun**

Which property of real numbers are you using if you multiply the top and the bottom by the same value?

Properties of Radicals: Test Prep	
Rewrite the radical using rational exponents: $\sqrt[3]{\frac{2^3}{x^2}}$	$\sqrt[3]{\frac{2^3}{x^2}} = \frac{\sqrt[3]{2^3}}{\sqrt[3]{x^2}} = \frac{\left(2^3\right)^{\frac{1}{3}}}{\left(x^2\right)^{\frac{1}{3}}} = \frac{2}{x^{\frac{2}{3}}}$

Properties of Radicals: Test Prep

Try Them	
Rewrite the radical using rational exponents: $\sqrt[4]{a^2bc^3}$	$a^{\frac{1}{2}}b^{\frac{1}{4}}c^{\frac{3}{4}}$
Rewrite in radical notation: $(3x)^{\frac{2}{5}}$	$\sqrt[5]{9x^2}$

Simplifying Radicals: Test Prep

Simplify the radical expression: $\sqrt{24x^2y}$	$\sqrt{24x^2y}$ $= \sqrt{4 \cdot 6 \cdot x^2y}$ $= \sqrt{4}\sqrt{x^2}\sqrt{6y}$ $= 2x\sqrt{6y}$
Try Them	
Simplify the radical expression: $2\sqrt{27} - 5\sqrt{3}$	$\sqrt{3}$
Simplify the rational exponent expression: $\left(2x^{\frac{1}{3}}y\right)\left(5x^{\frac{2}{3}}y^{\frac{1}{3}}\right)$	$10xy^{\frac{4}{3}}$

Rationalizing Denominators: Test Prep	
Simplify the radical expression: $\frac{3}{\sqrt{x}-1}$	$\frac{3}{\sqrt{x}-1} \cdot \frac{\sqrt{x}+1}{\sqrt{x}+1} = \frac{3(\sqrt{x}+1)}{x-1} = \frac{3\sqrt{x}+3}{x-1}$
Try Them	
Simplify the radical expression: $\frac{4}{\sqrt{32x}}$	$\frac{\sqrt{2x}}{2x}$
Simplify the rational exponent expression and rewrite in simplified radical form: $\frac{x^{\frac{1}{3}}y^{\frac{5}{6}}}{x^{\frac{2}{3}}y^{\frac{1}{6}}}$	$\frac{\sqrt[3]{x^2y^2}}{x}$

P.3: Polynomials

Polynomials, like real numbers, are important mathematical objects that we can use to solve many problems. They are especially nice because, like real numbers, they have very specific properties that make them the next least complicated (yes, least complicated!) mathematical object to study. Calculus students adore polynomials for how nicely behaved and easy to analyze they are.

Properties of Polynomials: Test Prep	
Determine the a) degree, b) leading coefficient, and c) rewrite the trinomial in descending order of degree: $-x-x^2+3$	a) 2 b) -1 c) $-x^2-x+3$
Try Them	
Determine the degree of the polynomial: $3xy-5xy^2+11xyz$	3
Give an example of a fourth degree trinomial.	example: x^4-x+1 Answers will vary.

Operations on Polynomials

Now that we've reviewed a great deal about operations (adding, subtracting, etc.) with real numbers, it's time to learn about a new object, the polynomial, and how to perform operations on it. So first comes all the vocabulary for polynomials.

✤ **degree of a monomial**
First of all, a monomial is a single polynomial term. In order to be a term in a polynomial, the degree of the variable(s) has to be a whole number. (First, remember that the whole numbers are the counting numbers plus zero. Second, remember that radicals translate into rational exponents. Third, remember your negative exponent rules.) And the degree of a variable is its exponent. That's the degree of the monomial.

★ **For Fun**

Don't forget that $x = x^1$, so that the degree of x is 1. Having said that, I ask you, what is the degree of a constant?

★ **For Fun**

Now can you figure out what the degree of a term like $-3x^2yz^5$ is? First of all, do you agree that it's a monomial by definition?

✤ **polynomial**
A polynomial is a collection of one to many monomials that are added and subtracted from each other. Remember terms are separated by addition and subtraction, and so these *monomials* are *terms* in the polynomial. Also, it stands repeating from above, the degree of all these terms must be a whole number.

Some polynomial terms: -3, $13x^3$, $-2.2y$

Some *not* polynomial terms: $-\sqrt{42}$, $\frac{8}{x}$, $12x^{-2}$

✤ **degree of a polynomial**
The degree of the polynomial is the highest degree of one of the monomials that makes it up. Easy.

✤ **like terms**

Like terms have variable parts that match exactly.

Like terms: $-x, \quad 2x, \quad \frac{1}{3}x$

Not like terms: $-xy, \quad 2x^2, \quad \frac{1}{3}y^2$

✤ **binomial**

A binomial is a polynomial with exactly two terms: for example, $x + y$.

✤ **trinomial**

Naturally, that means that a polynomial with exactly three terms is a trinomial. Polynomials with more than three terms don't get special names based on how many terms they have.

✤ **constant polynomial**

If you realized that the degree of a constant term is zero, and you know that zero is a whole number, then you know a constant term is a monomial. As such, it makes a perfectly good constant polynomial: for example, -12.5. That's a very simple and pretty boring polynomial, but it might be very important in the right context!

✤ **leading coefficient**

The leading coefficient is the numerical coefficient of the term with the highest degree (the term that gives its degree to the entire polynomial).

✤ **constant term**

The constant term is easy to recognize -- it's the term with no variable. The degree of the constant term is zero because, for example, $2x^0 = 2 \cdot 1 = 2$. Remember the zero-exponent rule?

✤ **FOIL method**

FOIL is an acronym with stands for First, Outer, Inside, Last and, as such, gives an instruction about how to multiply two binomials: $(a+b)(c+d) = ac + ad + bc + bd$. Be sure to include the +/- signs! When you multiply polynomials, you're building a new polynomial whose terms are separated by ... plus or minus!

✤ **evaluate a polynomial**

To evaluate a polynomial, plug in the given values for the variables, follow the order of operations agreement to the letter, and, keeping in mind the *value* part of "evaluate," end up with a *number value*.

Operations on Polynomials: Test Prep

Simplify $(x^3 - x^2 + 3x) - (2x^2 - x + 10)$	$(x^3 - x^2 + 3x) - (2x^2 - x + 10)$ $= x^3 - x^2 + 3x - 2x^2 + x - 10$ $= x^2 - 3x^2 + 4x - 10$
Try Them	
Simplify and determine the degree of the resulting polynomial: $(2x-1)(x^2 + x + 5)$	$2x^3 + x^2 + 9x - 5$, degree 3
Multiply and simplify: $(5x+2)^2 - (6x-1)^2$	$-11x^2 + 32x + 3$

Special Products: Test Prep	
Simplify using FOIL and using a special product rule; show the results are identical: $(5x-3)(5x+3)$	1. $(5x-3)(5x+3)=25x^2+15x-15x-9=25x^2-9$ 2. $(5x-3)(5x+3)=(5x)^2-(3)^2=25x^2-9$ 3. $25x^2-9=25x^2-9$, identically!
Try Them	
Use a special product rule to simplify: $(x+2y)^2$	$x^2+4xy+4y^2$
Simplify: $(3a-2)^2$	$9a^2-12a+4$

Every polynomial in one variable can be written in the form

$$a_n x^n + a_{n-1}x^{n-1} + \cdots + a_2 x^2 + a_1 x + a_0 x$$

This form is sometimes taken as the definition of a polynomial in one variable.
Can you identify the degree, leading coefficient, and constant term?

Applications of Polynomials

A line happens to be a polynomial (for example: the graph of $y = 3x + 1$ is a line; can you see that both of its terms, *3x* and *1*, satisfy the requirements for polynomial terms?). Everybody knows a straight line is the shortest distance between two points, right? Well, if a line (a polynomial) fails to describe how to get from one place to another, lots of times a higher-order polynomial might be able to do the job instead. (What is the degree of a linear polynomial? What do I mean by a higher-order polynomial? We've factored some of them, yes?) Of course, there are more complicated mathematical expressions that can be used to describe behaviors of systems, but beginning with lines and other polynomials will get you very far. Once you've found a nice polynomial, you'll need to evaluate it -- see the previous section!

★ **For Fun**

Here are some obscure polynomial degree terms you can amaze your friends with:

Degree	Name	Example
$-\infty$	zero	0
0	(non-zero) constant	1
1	linear	$x + 1$
2	quadratic	$x^2 + 1$
3	cubic	$x^3 + 1$
4	quartic (or biquadratic)	$x^4 + 1$
5	quintic	$x^5 + 1$
6	sextic (or hexic)	$x^6 + 1$
7	septic (or heptic)	$x^7 + 1$
8	octic	$x^8 + 1$
9	nonic	$x^9 + 1$

NOW IS THE TIME TO TRY THE MID-CHAPTER QUIZ, PAGE 39, IN YOUR TEXT.

P.4: Factoring

In order to use mathematical ideas like The Zero-Factor Principle or The Factor Theorem, it's necessary to be able to find the *factors* first. Factoring is one of the first and most fundamental activities one undertakes in algebra. Although factoring in and of itself might be intellectually rewarding, there are very good reasons or applications for it, and they are coming!

✤ **factoring**
Factoring is generally recognized as the act of writing a polynomial as a product of polynomials, for example: $2x^3 - 2x^2 - 12x = 2x(x+2)(x-3)$. But remember composite numbers? Finding their prime factorization is factoring also, and many of the reasons for both kinds of factoring overlap.

★ **For Fun**
The factored polynomial above: Can you find the degree of each of its factors?

✤ **factoring over the integers**
Factoring over the integers makes life simpler for us algebra students. It means that you will be expected to learn how to recognize the factors of $(x^2 - 4) = (x+2)(x-2)$, but not necessarily how to factor this: $(x^2 - 2) = (x + \sqrt{2})(x - \sqrt{2})$. In the second case, we had to factor over the irrational numbers, which is much, much harder to do intuitively! Can you see the difference?

Greatest Common Factor

Finding the greatest common factor (GCF) of a polynomial is like using the distributive property in reverse. Instead of the usual $a(b+c) = ab + ac$, look at it the other way: $ab + ac = a(b+c)$. The *a* is the GCF of the expression on the left-hand side of the equation, and on the right-hand side it has been "factored out."

★ **For Fun**
What property of equality says that if $x = y$ then $y = x$? (Check out page 14 of your text if you cannot remember.)

Notice that in order to *check your factoring* all you are doing is applying the distributive property in reverse (or in the way you are used to thinking of it).

- **greatest common factor**
  The first step in factoring is to find the greatest common factor. To find the greatest common factor, look first for a common numerical factor for *all* the coefficients in the polynomial, then find the lowest power of the variable(s) common to *all* the terms. Use the distributive property to rewrite the polynomial with that GCF "factored out." Then check your factoring.

Greatest Common Factor: Test Prep	
Use the distributive property to factor out the GCF: $6a^2z - 2az + 14az^2$	$6a^2z - 2az + 14az^2 = 2az(3a - 1 + 7z)$ Check it!
Try Them	
Factor out the GCF: $5x - 15$	$5(x-3)$
Factor out the GCF: $2x^2 - 10x$	$2x(x-5)$

Factoring Trinomials

After finding and factoring out the GCF, the next experiment is in factoring a trinomial. That's almost as far as this chapter goes, but in later chapters factoring larger and larger polynomials comes into play. So get good at the basics so that you can concentrate on the more difficult techniques later! There are basically two kinds of trinomials to learn to factor: $x^2 + bx + c$ or $ax^2 + bx + c$. In one case the *leading coefficient* is one, and these are usually the easiest to factor, if they can be factored. More techniques, trials, and practice are needed to become proficient at factoring trinomials of the form $ax^2 + bx + c$.

✤ **nonfactorable over the integers**
Sometimes called a prime polynomial, but that's really misleading, so a trinomial is called "nonfactorable over the integers" if you cannot factor it by the means given in this section. This definition actually holds for polynomials of any degree, but you need more methods than those given in this section! So $\left(x^2-2\right)=\left(x+\sqrt{2}\right)\left(x-\sqrt{2}\right)$ is nonfactorable over the integers (but you may notice that it is *not* strictly "prime" -- however that applies to polynomials -- since I did factor it).

Factoring Quadratic Trinomials: Test Prep	
Factor: $b^2-7b+12$	Noting that $-3-4=-7$; whereas, $(-3)(-4)=12$, we factor the trinomial so: $(b-3)(b-4)$. And check it!
Try Them	
Factor: $5x^2+14x-3$	$(5x-1)(x+3)$
Factor the quadratic binomial: $4s^2-9$	$(2s-3)(2s+3)$

Special Factoring

More shortcuts! Again: this takes practice to recognize when you can use these shortcuts. Nice enough, though, if you cannot remember any shortcut that seems to apply, all the other methods of this section will still work.

✓ **Questions to Ask Your Teacher**
Make sure you understand how thoroughly your teacher expects you to know, recognize, and use each of these shortcuts. Don't be penalized for using other methods if you can help it.

Factoring the Difference of Two Perfect Squares:

$$a^2 - b^2 = (a+b)(a-b)$$

✤ **perfect square**

A perfect square is a product of exactly two identical factors. $81 = 9 \cdot 9$ is a numerical perfect square. $x^4 = x^2 \cdot x^2$ is the perfect square of variables. And so on. Are you able to recognize a perfect square when you see it? You need to be able to find the perfect squares in a polynomial like $4x^2 - 81$ in order to be able to use the *Difference of Squares* factoring shortcut.

✤ **square root of a perfect square**

Likewise 9 is the square root of a perfect square, namely 81. Now you need to be able to find the square roots of the terms in $4x^2 - 81$ in order to factor it by this shortcut.

Factoring a Perfect Square Trinomial:

$$a^2 + 2ab + b^2 = (a+b)^2$$

$$a^2 - 2ab + b^2 = (a-b)^2$$

✤ **perfect square trinomial**

A perfect square trinomial is a trinomial that is in the form of the left-hand side of one of the two formulae above. These are harder to recognize than the difference of squares. Some examples are $x^2 - 10x + 25$, $a^2 + 4a + 4$, $4x^2 - 4x + 1$.

Factoring a Sum/Difference of Cubes Binomial:

$$a^3 + b^3 = (a+b)(a^2 - ab + b^2)$$

$$a^3 - b^3 = (a-b)(a^2 + ab + b^2)$$

✤ **perfect cube**

Similarly, you need to recognize a perfect cube, a number or variable expression which is the product of three identical factors, for example 27 or $8x^3$.

✤ **cube root of a perfect cube**

Likewise the 2x is the cube root of the perfect cube $8x^3$.

Special Factoring Formulas: Test Prep	
Factor using the sum of cubes formula: $27a^3 + 8$	$27a^3 + 8 = (3a)^3 + (2)^3 = (3a+2)(9a^2 - 6a + 4)$
Try Them	
Factor using a special factoring formula: $y^2 - 6y + 9$	$(y-3)^2$
Factor: $4x^2 + 81$	Nonfactorable over the integers.

✤ **quadratic in form**

Now everything you've learned about factoring quadratic expressions can be applied to any polynomial expression that is *quadratic in form* even though it may not actually be quadratic:

$x^4 + 7x^2 + 12$

$16x^4 + 8x^2 - 35$

$x^4 - 625$

$x^2 + 6\sqrt{x} + 5$

Whereas these are important forms and become more and more important as you progress in algebra, they're hard to recognize. All I can say is "keep your eyes and mind open, look for patterns, and practice!"

Quadratic in Form: Test Prep	
Translate into quadratic form, $au^2 + bu + c$, factor, and translate back into original variable(s): $z^4 - 5z^2 + 6$	$z^4 - 5z^2 + 6$ $= u^2 - 5u + 6$ $= (u-3)(u-2)$ $= (z^2-3)(z^2-2)$
Try Them	
Factor: $y^4 + y^2 - 2$	$(y^2+2)(y^2-1)$
Factor: $a^2b^2 - ab - 6$	$(ab+2)(ab-3)$

Factor by Grouping

This method of factoring is one of my favorites. It requires that the polynomial be written in four terms, exactly the opportune four terms!

✤ **factored by grouping**

To factor by grouping, first you see the four terms as two groups of two (*grouping*), then factor the GCF out of each of those groups, then see still another GCF in the remaining two terms. (You have to remember that *terms* are separated by + or -, no multiplication involved.)

✓ **Questions to Ask Your Teacher**

There's a way to use factoring by grouping to factor trinomials, too. See if you can get your teacher to show you that. It's fun and very useful!

Grouping: Test Prep	
Factor by grouping: $3x + 6a - ax - 2a^2$	$3x + 6a - ax - 2a^2$ $= 3(x + 2a) - a(x + 2a)$ $= (x + 2a)(3 - a)$
Try Them	
Factor: $x^3 + x^2 - x - 1$	$(x - 1)(x + 1)^2$
Factor: $x^2 - 4xy + 4y^2 - 16$	$(x - 2y + 4)(x - 2y - 4)$

General Factoring

When it comes to factoring, you need to have as many tools as possible on hand to bring to bear on the problem. Review and practice all the methods of this section; look over a synopsis of them on page 47 of your textbook.

General Factoring: Test Prep	
Factor completely over the integers: $x^4 + x^2 - 20$	$x^4 + x^2 - 20 = (x^2 + 5)(x^2 - 4)$ $= (x^2 + 5)(x + 2)(x - 2)$
Try Them	
Factor: $4x^4 - 2x^3 - 6x^2$	$2x^2(x + 1)(2x - 3)$
Factor: $2a^4 - 16a$	$2a(a - 2)(a^2 + 2a + 4)$

P.5: Rational Expressions

Rational expressions are ratios of polynomials, for example: $\frac{x^2 - 2x + 1}{x^2 - 4}$. Here's the definition:

✤ **rational expression**
A rational expression is a fraction in which the numerator and denominator are polynomials.

✤ **domain of a rational expression**
The domain of a rational expression is any set of numbers that can be used for replacement for the variable(s). Keep in mind that division by zero is not allowed!

★ **For Fun**
Find the domains of these rational expressions:

1. $\frac{x+2}{x-5}$
2. $\frac{a^2+9}{a^2-64}$
3. $\frac{y-5}{y}$
4. $\frac{x^2-4}{15}$

Simplifying Rational Expressions

Like simplifying fractions (also known as reducing fractions), you can *factor* the numerator and denominator of a rational expression and cancel common factors.

✤ **simplify a rational expression**
Use the "equivalent expressions property" to eliminate factors common to both the numerator and denominator of the rational expression. Please be sure to factor first and cancel common factors. You can only cancel terms if the numerator and denominator are monomials (and even then, you're frequently canceling factors of the monomials -- think about it).

✓ **Questions to Ask Your Teacher**
Have your teacher review canceling factors if you are unclear about this!

Simplify Rational Expressions: Test Prep	
Factor and simplify: $\frac{x^2+2x-3}{x^2-1}$	$\frac{x^2+2x-3}{x^2-1}=\frac{(x+3)(x-1)}{(x-1)(x+1)}=\frac{x+3}{x+1}$
Try Them	
Describe the domain of the rational expression: $\frac{2x^2-x}{x^2+1}$	all real numbers
Describe the domain: $\frac{1}{x^2-16}$	$x \neq 4,-4$

Operations on Rational Expressions

Just like any other numbers, rational expressions can be added, subtracted, multiplied, and divided by each other. Since they are *fractions*, once again, you need to know how to do all these operations on fractions.

✤ **common denominator**
When you add or subtract fractions, you must write all expressions involved on a common denominator, right? Same thing with rational expressions:

$$\frac{4x-2}{3x+12}-\frac{x-2}{x+4}=\frac{4x-2}{3(x+4)}-\frac{x-2}{x+4}\cdot\frac{3}{3}=\frac{4x-2-3(x-2)}{3(x+4)}=\frac{4x-2-3x+6}{3(x+4)}=\frac{x+4}{3(x+4)}=\frac{1}{3}$$

Be sure to simplify (reduce) your rational expression at the end, too.

✓ **Questions to Ask Your Teacher**
Review with your teacher the methods to find the least common denominator for fractions; see how this applies to rational expressions, too.

Determining the LCD of Rational Expressions

The least common denominator (LCD) of a rational expression is found in fundamentally the same way that the LCD is found for a group of fractions.

Consider again this example:

$$\frac{4x-2}{3x+12}-\frac{x-2}{x+4}=\frac{4x-2}{3(x+4)}-\frac{x-2}{x+4}\cdot\frac{3}{3}=\frac{4x-2-3(x-2)}{3(x+4)}=\frac{4x-2-3x+6}{3(x+4)}=\frac{x+4}{3(x+4)}=\frac{1}{3}$$

1. **Factor all the denominators.**
   In my example above, the two denominators factored thus $3(x+4)$ and $(x+4)$.

2. **Find the least common multiple of all those factors.**
   In the list of factors of the denominators, the least common multiple is the product of each factor in the list raised to the highest power that it occurs there. It's $3(x+4)$ in my example above.

3. **Multiply any rational expressions that are not already over the LCD by any factors missing from their denominators** -- top and bottom because you can multiply by the number 1 and not change anything.

4. **Now combine all the numerators on their common denominator** -- you're combining apples and apples now, instead of apples and oranges!

Operations on Rational Expressions: Test Prep	
Add: $\frac{2}{x+3}+\frac{x}{x+5}$	The LCD is $(x+3)(x+5)$, so $\frac{2}{x+3}+\frac{x}{x+5}$ $=\frac{2}{x+3}\cdot\frac{x+5}{x+5}+\frac{x}{x+5}\cdot\frac{x+3}{x+3}$ $=\frac{2x+10+x^2+3x}{(x+3)(x+5)}$ $=\frac{x^2+5x+10}{x^2+8x+15}$
Try Them	
Factor, simplify, then multiply: $\frac{x^2-2x+1}{x^2-4}\cdot\frac{x^2+4x+4}{x^2+x-2}$	$\frac{x-1}{x-2}$
Divide: $\frac{x^2-x-6}{x+4}\div\frac{x^2-9}{x^2+3x-4}$	$\frac{x^2+x-2}{x+3}$

Complex Fractions

✤ **complex fraction**
A complex fraction is a fraction whose numerator or denominator, or possibly both, also contain a fraction.

Complex fractions entail their own special challenges for simplifying. There are basically two methods:

Method One

1. Find the LCD of all the fractions internal to the complex fraction.
   (For example, $\frac{\frac{1}{3}+\frac{3}{x}}{2}$ has $3 \cdot x \cdot 1 = 3x$ as its overall LCD.)

2. Multiply top and bottom of the complex fraction by this LCD.
   (Continuing the example: $\frac{\frac{1}{3}+\frac{3}{x}}{2} \cdot \frac{3x}{3x} = \frac{x+9}{6x}$. *Be sure you use the distributive property to multiply the LCD correctly!*)

3. Simplify, if needed.
   (The above example cannot be simplified, can it?)

Method Two

1. Simplify the numerator and denominator into a single fraction; that is, a fraction on a common denominator.
   (For example, $\frac{\frac{1}{3}+\frac{3}{x}}{2} = \frac{\frac{x+9}{3x}}{2}$.)

2. Multiply the numerator by the reciprocal of the denominator.
   (For example, $\frac{\frac{x+9}{3x}}{2} = \frac{x+9}{3x} \cdot \frac{1}{2}$.)

3. Multiply the fractions; simplify is necessary.
   (Continuing the example: $\frac{x+9}{3x} \cdot \frac{1}{2} = \frac{x+9}{6x}$.)

Complex Fractions: Test Prep	
Simplify: $\dfrac{1}{\dfrac{1}{S}+\dfrac{2}{R}}$	$\dfrac{1}{\dfrac{1}{S}+\dfrac{2}{R}}\cdot\dfrac{SR}{SR}=\dfrac{SR}{R+2S}$ The LCD of the entire complex fraction is *SR*.
Try Them	
Simplify: $\dfrac{\dfrac{y}{x}+1}{\dfrac{x}{y}-\dfrac{y}{x}}$	$\dfrac{y}{x-y}$
Simplify: $\dfrac{4-\dfrac{1}{x}}{\dfrac{1}{x}-2}$	$\dfrac{4x-1}{1-2x}$

Applications of Rational Expressions

Rational expressions are one type of "more complicated" expression you might find used to describe a system. For example, capacitances in electronic circuits are summed up by their reciprocals: $\dfrac{1}{c_1}+\dfrac{1}{c_2}+\dfrac{1}{c_3}$. Can you simplify this expression so that it shares a common denominator? Then, of course, you might have to evaluate it for certain capacitances, or the capacitances might turn out to be described by polynomials! Who knows?

P.6: Complex Numbers

Like negative numbers and irrational numbers, as mathematics progressed new numbers were required. Imaginary numbers, which express the square roots of negative numbers, were a necessary evolution of the number system. They are not real numbers as have been all the numbers we've considered so far this chapter.

Complex numbers are numbers that have both real parts and imaginary parts.

★ **For Fun**
Can you illustrate the relationship between real numbers, imaginary numbers, and complex numbers in a Venn diagram?

Introduction to Complex Numbers

Complex numbers are numbers that have both real and imaginary parts.

✤ **imaginary unit**
i is the imaginary unit and is equal to the square root of -1.

✤ **imaginary number**
An expression that is a multiple of i, the square root of -1, is an imaginary number. So $\sqrt{-3} = i\sqrt{3}$ is an imaginary number. Notice the degree of the i is one, so you see that imaginary numbers are linear multiples of i.

★ **For Fun**
What do I mean by a "linear multiple"?

✤ **complex number**
A complex number has a real part and an imaginary part, even if either part has zero magnitude. Even zero is a complex number.

✤ **real part**
For example, in the complex number $3-2i$, the real part is "3". The real part is the term that is not a multiple of i.

✤ **complex part**

In the above example, the imaginary part is $-2i$. The imaginary part is the term that involves i.

✤ **standard form**

Complex numbers written in standard form appear like $a+bi$, where a is the real part and bi is the imaginary part. Sometimes it is more lovely to write the complex number in the form $a+ib$; as in the case: $5-i\sqrt{5}$ where the i should not appear as though it might be underneath the radical symbol.

Complex Numbers: Test Prep	
Write the complex number in standard form: $\sqrt{81}-\sqrt{-1}$	By definition of square root and the imaginary unit, $\sqrt{81}-\sqrt{-1}=9-i$
Try Them	
Simplify and write in standard form: $\sqrt{-28}$	$2i\sqrt{7}$ (the result is purely imaginary)
Write in standard form: $\sqrt{24}-\sqrt{-6}$	$2\sqrt{6}-i\sqrt{6}$

Addition and Subtraction of Complex Numbers

If you think of real parts to be *like* real parts and likewise for imaginary parts, then adding and subtracting complex numbers is exactly like combining like parts: $4+3i-5+i=-1+4i$. Keep the sum or difference in standard form!

Multiplication of Complex Numbers

Proceed to multiply complex numbers just like you would polynomials, by the FOIL method, for instance. The difference is that $i^2=-1$, which is a *real* number. So after the

multiplication and simplification, you'll have another real term to combine according to the addition/subtraction algorithm above. For example:
$(2-i)(3+4i)=6+8i-3i-4i^2=6+5i-4(-1)=6+5i+4=10+5i$. Again the answer is given in standard form.

Division of Complex Numbers

Division of complex numbers is not exactly division at all. It's really more of a process like rationalizing denominators was. With complex numbers the idea is to turn the denominator into an entirely real number (you will need the complex conjugate for that) and then write the result in standard form. You do need to understand division by a real number still, but that's just fractions, right?

- **complex conjugates, conjugates**
  The complex conjugate of a complex number $a+bi$ is simply $a-bi$. Did you catch that? All you have to do is change the sign of the imaginary part. So that if you multiplied $(a+bi)(a-bi)$ you'd get a^2+b^2. (Try it! It's just like the difference of squares, but that $i^2=-1$ changes it into the *sum* of squares!

So here's how division works, for example:

$$\frac{3}{2-i}$$
$$=\frac{3}{2-i}\cdot\frac{2+i}{2+i}$$
$$=\frac{6+3i}{4+1}$$
$$=\frac{6+3i}{5}$$
$$=\frac{6}{5}+\frac{3}{5}i$$

Notice how division by a monomial (5 here) works so that the answer may be written in standard form.

Operations on Complex Numbers: Test Prep

Simplify and write the result in standard form: $\sqrt{-10}\cdot\sqrt{-5}$	$\sqrt{-10}\cdot\sqrt{-5} = i\cdot\sqrt{10}\cdot i\cdot\sqrt{5} = -1\cdot\sqrt{50}$ $= -5\sqrt{2}$ (the result is purely real)
Try Them	
Add the complex numbers: $(3+4i)+(4-5i)$	$7-i$
Divide the complex numbers, write the result in standard form: $\dfrac{2}{3-i}$	$\dfrac{3}{5}+\dfrac{1}{5}i$

Powers of i

If you investigate raising i to different powers, you can readily discover a pattern:

$i^0 = 1$	$i^3 = -i$	$i^6 = -1$
$i^1 = i$	$i^4 = 1$	$i^7 =$ ________
$i^2 = -1$	$i^5 = i$	$i^8 =$ ________

This repeat every four powers, yes? So we devise the Powers of i observation:

If n is a positive integer, then $i^n = i^r$, where r is the remainder of the division of n by 4. For example: $i^{14} = i^2 = -1$ because the remainder of dividing 14 by 4 is 2.

Powers of i: Test Prep	
Evaluate the power of i: $-i^{40}$.	Because 40/4 leaves a remainder of zero: $-i^{40} = -i^0 = -1$
Try Them	
Evaluate and write the result in standard form: $\frac{-4 \pm \sqrt{4^2 - 4 \cdot 1 \cdot 6}}{2 \cdot 1}$.	$-2 \pm i\sqrt{2}$
Simplify: $(2-i)^2$.	$3-4i$

$$(a+bi)\left(\frac{a}{a^2+b^2} - \frac{b}{a^2+b^2}i\right) = \frac{(a+bi)(a-bi)}{a^2+b^2} = 1$$

NOW TRY THESE:

REVIEW EXERCISES: TEXTBOOK PAGE 70

PRACTICE TEST: TEXTBOOK PAGE 73.

Final Fun: Matching

FIND THE BEST MATCH. GOOD LUCK.

	$a^2 + 2ab + b^2$
EMPTY SET	$b^0, b \neq 0$
INFINITY	$\sqrt{-1}$
INTERVAL NOTATION	$x + 5\sqrt{x} + 6$
ADDITIVE IDENTITY	$a \times 10^n$
SQUARE ROOT OF 3	ZERO
i	$[-1,8)$
PERFECT SQUARE TRINOMIAL	18x
QUADRATIC IN FORM	∞
GCF of {2x, 6, 18}	2
LCM of {2x, 6, 18}	$3^{\frac{1}{2}}$
MULTIPLICATIVE IDENTITY	5
DEGREE OF A QUINTIC	$\varnothing$
SCIENTIFIC NOTATION	

Chapter 1: Equations and Inequalities

Word Search

```
W E Q U A T I O N O T R H N S
N R C X L N V V U E A L Y O O
Y O S Q U A R E S E A O P I L
F R I Q P N C S N N B R O T U
S A Z T U I E I O N E E T A T
I D U U C M L I T L A Z E I I
T I Z K V I T P N I V I N R O
A C I T A R D A U Q R E U A N
S A R I O C M A I S Q C S V E
Q L S P K S H N R U G G E C S
L M O T N I O J A T I E E U R
W R N M E D S L X V N V L B E
P W A P P L I C A T I O N I V
L A N O I T A R O X Q N C C N
I O D C Y F O R M U L A P V I
```

APPLICATION
CONTRADICTION
CRITICAL
CUBIC
DISCRIMINANT
EQUATION
FORMULA
HYPOTENUSE
INEQUALITY
INVERSE
JOINT
LEGS
LINEAR
PROPORTIONAL
QUADRATIC
RADICAL
RATIONAL
SATISFY
SET
SOLUTION
SOLVE
SQUARE
VARIATION
ZERO

1.1: Linear and Absolute Value Equations

From simplifying and otherwise manipulating expressions, in this chapter we progress to manipulating equations. Now we are working with mathematically complete thoughts, and we can begin to solve problems.

Linear Equations

There is a lot of very specific and meaningful information in the phrase "linear equation." First of all, the word equation lets you know that there must be an equals sign present, for example: $3 = 2x + 1$. What does the word "linear" tell you? Remember the degrees of polynomials discussed last chapter? "Linear" tells you the degree of the equation (not just the statement, but the whole equation) is *one*.

★ **For Fun**
Can you tell that the equation $3 = 2x + 1$ is linear?

✤ **equation**
In order for a mathematical expression to be called an equation, let's keep in mind that it has to start out with an equals sign in it. That makes an equation a complete mathematical sentence: the equals sign is the "verb." And this sentence or statement tells us about the absolute equivalence of the two side of the equals sign -- the subject and the predicate, if you will.

✤ **satisfy**
To satisfy means something like "meeting the desires of," and satisfying an equation meets the particular and only desire of math -- the truth. To satisfy an equation is to find the values, usually of variables, that make the equation true.

★ **For Fun**
Does the value $x = 1$ *satisfy* the equation $3 = 2x + 1$?

✤ **solve**
Solving an equation is the process of satisfying the equation: that is, finding all the values of the variables that make the equation true. That set of values that satisfy the equation is called the solution set.

★ **For Fun**

Does the set $\{x \mid x = 1\}$ represent the *solution set* of the equation $3 = 2x + 1$?

✤ **equivalent equations**

Equivalent equations have exactly the same members in their solution sets. This is how you discover the basic rules of algebra that we can use to find solutions -- Addition, Subtraction, Multiplication, and Division Rules of equality.

★ **For Fun**

See if you can figure out how my statement of the following properties is the same (and/or different?) as the statements of these properties in your textbook, page 76.

Addition and Subtraction Property of Equality

These are equivalent equations:

$$a = b$$

$$a + c = b + c$$

$$a - c = b - c$$

Multiplication and Division Property of Equality

These are equivalent equations:

$$a = b$$

$$a \cdot c = b \cdot c$$

$$a \div c = b \div c$$

✤ **linear equation**

So a linear equation is an equation of the form $ax + b = c$ where x is the variable and a, b, and c are constants, and $a \neq 0$.

★ **For Fun**

Can you tell why the equation $ax + b = c$ would no longer be "linear" if you allowed a to be equal to zero?

Linear or First-Degree Equations: Test Prep

Solve and check your solution: $5x + 7 = x - 1$	$5x + 7 = x - 1$ $5x + 7 - x - 7 = x - 1 - x - 7$ $4x = -8$ $x = -2$ (*answer*) *check:* $5(-2) + 7 \stackrel{?}{=} (-2) - 1$ $-10 + 7 \stackrel{?}{=} -3$ $-3 = -3$
Try Them	
Solve: $6(5s - 11) - 12(2s + 5) = 0$	$s = 21$
Solve: $2(3 - 4x) = 6 - 7(2x + 4)$	$x = -\dfrac{14}{3}$

Related Coming Attractions

(in other words: some of the topics you're preparing for now)

- Graphing Linear Equations
- Linear Inequalities
- Graphing Linear Inequalities
- Systems of Linear Equations

Clearing Fractions: Test Prep	
Solve: $\frac{12+x}{-4}=\frac{5x-7}{3}+2$	$\frac{12+x}{-4}=\frac{5x-7}{3}+2 \quad (LCD=-12)$ $(-12)\left(\frac{12+x}{-4}\right)=\left(\frac{5x-7}{3}+2\right)(-12)$ $3(12+x)=-4(5x-7)-24$ $36+3x=-20x+28-24$ $23x=-32$ $x=-\frac{32}{23}$
Try Them	
Solve and check your solution: $\frac{x+3}{4}=6$	$x=21$
Solve and check your solution: $x-\frac{2}{3}=\frac{x}{6}$	$x=\frac{4}{5}$

Contradictions, Conditional Equations, and Identities

Now that we have equations, we start to organize them into different categories so that we know what we can, and cannot, accomplish *on* them and *with* them. Some might be for eliminating, some for solving, some for using as tools.

✤ **contradiction**
In English, a contradiction is a statement that can never be true because the parts are inconsistent with each other -- they exclude the possibility of truth for the whole statement because they cannot both be true. In math, all it takes to be a contradiction is to exclude the possibility of truth; for example: $2+2=5$ or $a+10=a$.

✤ **conditional equation**
A conditional equation is an equation that is *true* under just the right circumstances. For instance, $3=2x+1$ is true only when $x=1$. $x=1$ is the *condition* that makes $3=2x+1$ a true statement.

✤ **identity**
An identity is a statement that is always true, no conditions. Of course something like $2=2$ is an identity, but although this has value, it's pretty elementary. The identities that are really fruitful are the ones we turn into properties and put to use in, say, simplifying expressions: $a(b+c)=ab+ac$, the Distributive Property. This equation is always true, no matter what *a, b,* and *c* are.

★ **For Fun**
Can you come up with mathematical examples of contradictions, conditional equations, and identities on your own? Can you come up with examples *in English*, too?

Absolute Value Equations

As the title suggests, absolute value equations are equations that contain absolute value expressions in them. Here is a reminder of how absolute value is defined:

Absolute Value

$$|x|=\begin{cases} x, & x\geq 0 \\ -x, & x<0 \end{cases}$$

Remember, if you will, that absolute value has everything to do with measuring distances, so an equation that includes an absolute value is frequently a question of some sort about how far away things are from each other. For example, the equation $|x-1|=2$ is a statement about how far away *x* is from 1; in fact, it is a statement that *x* is exactly 2 away from 1.

★ **For Fun**

Could you draw $|x-1|=2$ on the number line? Can you find the *conditions* that *satisfy* this equation, can you find the member(s) of the *solution set*?

★ **For Fun**

Would you classify the absolute value equation $|x|=-1$ as a contradiction, a conditional equation, or an identity?

✓ **Questions to Ask Your Teacher**

What happens if you try to use an expression of higher order than linear in an absolute value equation?

★ **For Fun**

Investigate the truth of this equation for different values of x: $\sqrt{x^2}=|x|$.
Contradiction, conditional, or identity?

Linear Absolute Value Equations: Test Prep

Solve: $\|3x+2\|=3$	$\|3x+2\|=3$ $3x+2=3 \quad or \quad 3x+2=-3$ $3x=1 \quad or \quad 3x=-5$ $x=\frac{1}{3} \quad or \quad x=-\frac{5}{3}$ $x=\frac{1}{3},-\frac{5}{3}$
Try Them	
Solve: $\|4x-1\|=-12$	no solution
Solve: $3\|x-5\|-16=2$	$x=11,-1$

Applications of Linear Equations

Applications of linear equations are frequent; anywhere in the real world that we can describe a behavior as following a line, we find linear equations applicable. Anywhere we are interested in distances, we might find absolute value equations useful too.

Pointers for working with word problems

1. Read the problem.
2. Read the problem carefully. (Yes, read the problem again!) And read the problem critically -- what is the problem asking you for and what is it giving you to help you get there; what equations can you construct right away; what formulas do you need?
3. Pick (preferably meaningful) variables -- like A for area or S for sum.
4. Are there any relationships between the variables? Can you write the relations in equations?
5. If it's even remotely helpful, Draw A Picture! Sketch A Diagram! Make A Table!
6. Can you write an equation that will give you the solution?
7. Check your answer. Don't simply check your answer in the equation(s) that you came up with, but really check your answer to see if it answers the question and if it makes sense!

1.2: Formulas and Applications

Since we now know how to evaluate all sorts of expressions -- plug in values, right? -- we can use them when they are in equations to actually get results. Formulas are equations with all sorts of variables waiting to be put to use.

Formulas

Formulas are like identities: $PV = nRT$, a statement of fact. Then they are like conditional equations if you need to find what *P* is when *V, n, R,* and *T* are given, for example. It takes someone a lot hours of experimentation to come up with some of these formulas. Just so that we can put them to use!

✣ **formula**
A formula is an *equation* that states how variables are related to each other under certain (usually physical) situations. Formulas range from the extremely simple, the area of a rectangle, $A = lw$; to completely inscrutable, the focal length,

$$\frac{1}{f} = (n-1)\left(\frac{1}{r_1} - \frac{1}{r_2}\right) + \frac{(n-1)^2}{n}\frac{t_c}{r_1 r_2}.$$

Evaluating a formula means plugging the appropriate value(s) into corresponding variable(s). Usually you wind up with one variable left over -- the one you were looking for. Sometimes it's necessary to solve the formula for a particular variable -- again, usually the one you're looking for! Right about now you should be hoping for simpler formulas (not like the focal length one above) and practicing using algebra to isolate the variable in question -- that is, *solve* for it.

Formulas: Test Prep	
Solve the formula for *h*: $A = 2\pi rh + 2\pi r^2$	$A = 2\pi rh + 2\pi r^2$ $A - 2\pi r^2 = 2\pi rh$ $\frac{A - 2\pi r^2}{2\pi r} = h$

Formulas: Test Prep	
Try Them	
Solve for m: $y - y_1 = m(x - x_1)$	$m = \dfrac{y - y_1}{x - x_1}$
Solve for d: $40 = a + 21d$	$d = \dfrac{40 - a}{21}$

Applications

There are geometric formulas (area of a circle, volume of a rectangular solid, length of the hypotenuse of a right triangle), business formulas (compound interest, sales taxes, cost functions), physics formulas (motion, heat exchange, nuclear decay), chemical formulas (mixtures, stoichiometry, titration), and many others (like coin collection, task time, and so forth). **Be sure you understand what each variable stands for**. Whether you name the variables and research your own formula or one is given to you, take the time to understand all the pieces of the puzzle. The careful work and understanding will pay off.

Some frequently used formulas, for example:

- area of a triangle: $A = \frac{1}{2}bh$
- volume of a pyramid: $V = \frac{1}{3}bh$
- compound interest: $P = C\left(1 + \frac{r}{n}\right)^{nt}$
- quadratic formula: $x = \frac{-b \pm \sqrt{b^2 - 4ac}}{2a}$

Applications of Formulas: Test Prep

Freda has 75 coins. Some are quarters, the rest are dimes. Altogether she has \$12.90. How many of each type of coin does she have?	Let x be the number of quarters and y be the number of dimes. $x + y = 75$ $y = 75 - x$ Then $0.25x + 0.10y = 12.90$ $25x + 10y = 1290$ $25x + 10(75 - x) = 1290$ $15x + 750 = 1290$ $15x = 540$ $x = 36$ and $y = 75 - 36 = 39$ So: 36 quarters and 39 dimes.
Try Them	
The width of a rectangle is 3 yards more than twice the length. If the perimeter of the rectangle is 222 yards, find the length and width.	Length = 36 yd.; width = 75 yd.
A bicyclist traveling at 20 mph overtakes a in-line skater who is traveling at 10 mph and had a 0.5 hour head start. How far from the starting point did the bicyclist over take the in-line skater?	10 mi.

1.3: Quadratic Equations

Quadratic polynomials are polynomials of degree two. Linear polynomials are polynomials of degree one, and you now know that a linear equation is like a linear polynomial put into an equation. So, you know what quadratic equations are by extension. And since they're *equations* now, we can set about solving them! That higher degree, though, suddenly admits more solutions and more ways of finding those solutions.

Solving Quadratic Equations by Factoring

The first method is the safest, if your quadratic is factorable. Remember first to put the equation into *standard quadratic form*, then factor the quadratic, and use the Zero Product Principle to set each factor equal to zero. Now you have linear equations to solve, and that's easy. It also yields the solution(s) to your quadratic equation.

The Zero Product Principle

If $A \cdot B = 0$, then either $A = 0$ or $B = 0$.

★ **For Fun**

Does the Zero Product Principle remind you of any of the properties of real numbers you studied?

✤ **quadratic equation**

A quadratic polynomial is a polynomial of degree 2. A quadratic equation is an *equation* containing a quadratic polynomial(s). For example: $3x^2 + x = 1$.

✤ **standard quadratic equation**

The standard form for a quadratic equation is $ax^2 + bx + c = 0$ where a cannot be equal to zero. We could convert the above example into standard form by subtracting 1 from both sides of the equation: $3x^2 + x - 1 = 0$.

- **double solution, double root**
Sometimes, when considering solutions restricted to the set of real numbers, a quadratic equation may have as few as no solutions and as many as two solutions. When a quadratic equation has only one solution, it is called a *double* solution -- it's sorta like the equation actually has two solutions but they just happen to be the same numbers. (Later, when we get to talk about *multiplicity* of solutions, this might make even more sense.)

Solving Quadratic Equations by Taking Square Roots

Another method for solving some quadratic equations is by taking the square root, in particular the square root of x^2 if x is the variable under consideration. Since $\sqrt{x^2} = |x|$ and $|x|$ equals either x or $-x$, we usually say $\sqrt{x^2} = \pm x$ and can use this to solve quadratic equations, especially when x^2 can be written in terms of constants only. For example,

$$x^2 - 49 = 0$$
$$x^2 = 49$$
$$\sqrt{x^2} = \pm\sqrt{49}$$
$$x = \pm 7$$

Two solutions to our quadratic equation, $x^2 - 49 = 0$.

✤ **square root procedure**

The square root procedure involves

1. isolating the x^2 on one side of the equation
2. taking the square root of both sides, **being sure to take "both" square roots of the constant**
3. and simplifying the square root if possible.

Another example might look like

$$x^2 - 24 = 0$$

$$x^2 = 24$$

$$\sqrt{x^2} = \pm\sqrt{24}$$

$x = \pm 2\sqrt{6}$ Once again notice the *two* solutions implied by the plus/minus symbol.

This is an illustration of the square root of a complex number:

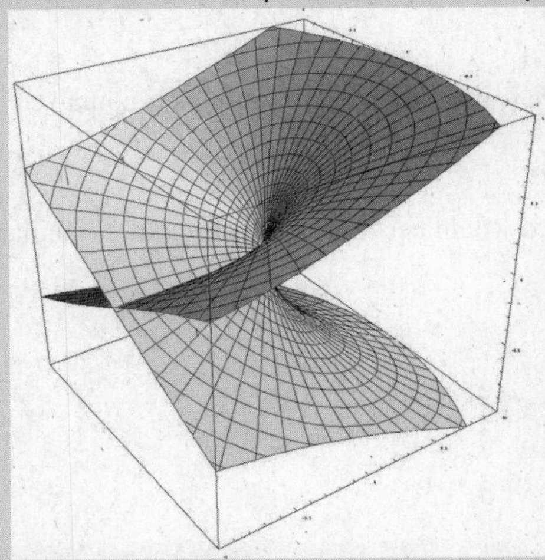

1. Aren't you glad we're not doing that?
2. Isn't it lovely? Doesn't math make lovely pictures sometimes?

Solving Quadratic Equations by Completing the Square

It's a natural progression to use the square root technique from above combined with our knowledge of factoring from last chapter to develop a technique to solve quadratic equations that depends on what is called "completing the square." If you can write $81 = 9^2$, then you have the concept required for completing the square.

First recall...

Factoring a Perfect Square Trinomial:

$$a^2 + 2ab + b^2 = (a+b)^2$$

$$a^2 - 2ab + b^2 = (a-b)^2$$

✤ **completing the square**
Completing the square is the method of rewriting a trinomial so that it is a perfect square trinomial.

For example, $x^2 - 6x$ would become a perfect square trinomial if I added 9 to it: $x^2 - 6x + 9 = (x-3)^2$.

★ **For Fun**
Check it!

<u>Completing the square is used in mathematics in these places, for example:</u>

1. Solving quadratic equations (you're doing this now)
2. Graphing quadratic equations (next chapter)
3. Evaluating some integrals (that's Calculus, in case you're headed there)
4. Finding Laplace Transforms (and that's beyond Calculus)

Completing the square is considered a basic operation in algebra.

Combining this with what we know about solving by taking square roots from above (also remembering the Addition and Subtraction Property of Equality), lets you solve more quadratic equations.

For example:

$$x^2 - 2x = 2$$
$$x^2 - 2x + 1 = 2 + 1$$
$$(x-1)^2 = 3$$
$$\sqrt{(x-1)^2} = \pm\sqrt{3}$$
$$x - 1 = \pm\sqrt{3}$$
$$x = 1 \pm \sqrt{3}$$

Now, the BIG question is, "How did I know to add 1 to both sides of the equation?" Here's how:

1. Find the coefficient of the linear term, then
   (that's -2 in my example)
2. divide that coefficient by 2,
   ($-2 \div 2 = -1$ *for me)*
3. and square that result -- that will be the number to add **to both sides of the equation**
   ($(-1)^2 = 1$ *for my example)*

★ **For Fun**
Could you ever wind up having to *subtract* the same number from both sides of the equation?

✓ **Questions to Ask Your Teacher**
Make sure you know *how* your teacher will expect you to solve a given quadratic equation. (We are about to develop yet another method!)

Since we are developing several methods to solve quadratic equations, get into the very, very good habit of *reading the instructions* given in the question. In other words, be prepared to use the method specified by your teacher in the problem.

Solving Quadratic Equations by Using the Quadratic Formula

By completing the square on the standard form of the quadratic equation, $ax^2 + bx + c = 0$, you can derive a very general solution to quadratic equations. We name it the quadratic formula (see? more formulas!).

Given $ax^2 + bx + c = 0$, where $a \neq 0$, then the solution(s) are

$$x = \frac{-b \pm \sqrt{b^2 - 4ac}}{2a}$$

★ **For Fun**

Can you see at least *two* reasons why a cannot be equal to zero in the above note?

Here are some considerations I would like you to take to heart when solving quadratic equations, by any method:

1. Solve them by factoring if you can -- the method of factoring is slightly less error-prone than the quadratic formula.

2. Get into the habit, when using the quadratic formula, of actually writing the variable you are solving for, the "$x =$" in the above notation because
   - it will help you remember that you are building the *solution set* of allowed variables, and
   - it will help you make far fewer errors when your variables become more complicated (yes, that's a Coming Attractions bulletin).

3. Remember that the plus/minus symbol means $x = \dfrac{-b \pm \sqrt{b^2 - 4ac}}{2a}$ specifies **two** values.

Quadratic Equations: Test Prep

Solve by completing the square: $x^2 - 6x + 7 = 0$	$x^2 - 6x + 7 = 0$ $x^2 - 6x + 9 = -7 + 9 \quad \left[\left(\frac{-6}{2}\right)^2 = 9\right]$ $(x-3)^2 = 2$ $x - 3 = \pm\sqrt{2}$ $x = 3 \pm \sqrt{2}$
Try Them	
Solve by factoring: $3x^2 - 5x = 2$	$x = 2, -\frac{1}{3}$
Solve by using the quadratic formula: $x^2 - x + 2 = 0$	$x = \frac{1 \pm i\sqrt{7}}{2}$

The Discriminant of a Quadratic Equation

Learning to be discriminating about your discriminant can help you discover information *about* the solution(s) to your quadratic equation ... before you ever have to solve it.

✤ **discriminant**
The discriminant of a quadratic equation is the small portion of the quadratic formula that resides under the radical, namely, $b^2 - 4ac$.

The Discriminant and the Solutions of a Quadratic Equation

- If $b^2 - 4ac > 0$, then there are two different real solutions
- If $b^2 - 4ac = 0$, then you have a *double solution*
- If $b^2 - 4ac < 0$, then there are no *real* solutions. (It so happens that there are exactly two *complex* solutions!)

The Discriminant: Test Prep	
Determine the discriminant and state the number of real roots: $8x^2 = 3x - 2$	$8x^2 = 3x - 2$ $8x^2 - 3x + 2 = 0$ $b^2 - 4ac = (-3)^2 - 4(8)(2)$ $= 9 - 64 = -55 < 0$ no real roots
Try Them	
Determine the discriminant and state the number of real solutions: $9x^2 - 12x + 4 = 0$	*0*, one real solution
Determine the discriminant and state the number of real roots: $x^2 - 2x = 2$	12, two real roots

Applications of Quadratic Equations

Any system that can be described by a quadratic equation is a good candidate for exercising your quadratic-solving muscles. (Anything that involves an area, which is measured in *square* units, might be a candidate for the methods of quadratic equations whose dimensions are also squared -- right?)

Simple motion in gravity is also a quadratic equation: $s(t) = -16t^2 + v_0 t + s_0$. s is called a position equation, and the variable is t, which denotes *time*. The position changes as time passes. Go ahead and get used to the notations v_0 and s_0. The subscript *zero* usually means *initial*, and *initial* means the starting condition -- in other words the condition when t was equal to zero.

The Pythagorean Theorem, frequently written $a^2 + b^2 = c^2$, couldn't look much more quadratic, could it? So expect some applications having to do with the sides of right triangles. Here are some definitions to go with those applications:

✤ **right triangle**

A right triangle is a triangle that contains one $90°$ angle.

★ **For Fun**

Is it possible for a triangle to contain more than one $90°$ angle?

✤ **hypotenuse**

The longest side of a right triangle is called the hypotenuse. It also happens to be the side that is opposite, or across from, the $90°$ angle.

★ **For Fun**

Where does the word "hypotenuse" come from? Who coined it?

✤ **legs**

The other two sides of the triangle (not the hypotenuse) are called the legs. (There are *three* sides to a triangle, right?)

Pythagorean Theorem: Test Prep

Problem	Solution
An LCD television screen measures 72 in. on the diagonal, and its aspect ratio is 16 to 9. Find the width and height of the screen. Approximate your answers to the nearest tenth of an inch.	$(16x)^2 + (9x)^2 = 72^2$ $256x^2 + 81x^2 = 5184$ $x^2 = \frac{5184}{256+81} = \frac{5184}{337}$ $x = \sqrt{\frac{5184}{337}}$ $width = 16x = 16\sqrt{\frac{5184}{337}} \approx 62.8in.$ $height = 9x = 9\sqrt{\frac{5184}{337}} \approx 35.3in.$

Pythagorean Theorem: Test Prep

Try Them	
The length of each side of an equilateral triangle is 30 cm. Find the exact altitude of the triangle.	$15\sqrt{3}$
If the legs of a right triangle measure 8 in. and 17 in., find the length of the hypotenuse.	$\sqrt{353}$ in.

★ **For Fun**

I've heard it said that Pythagoras' daughter was admitted into his mathematical and philosophical society. Apparently Pythagoras didn't practice sexual discrimination. What other off-beat tidbits can you find about Pythagoras? There are quite a few out there even though we actually know very little about him.

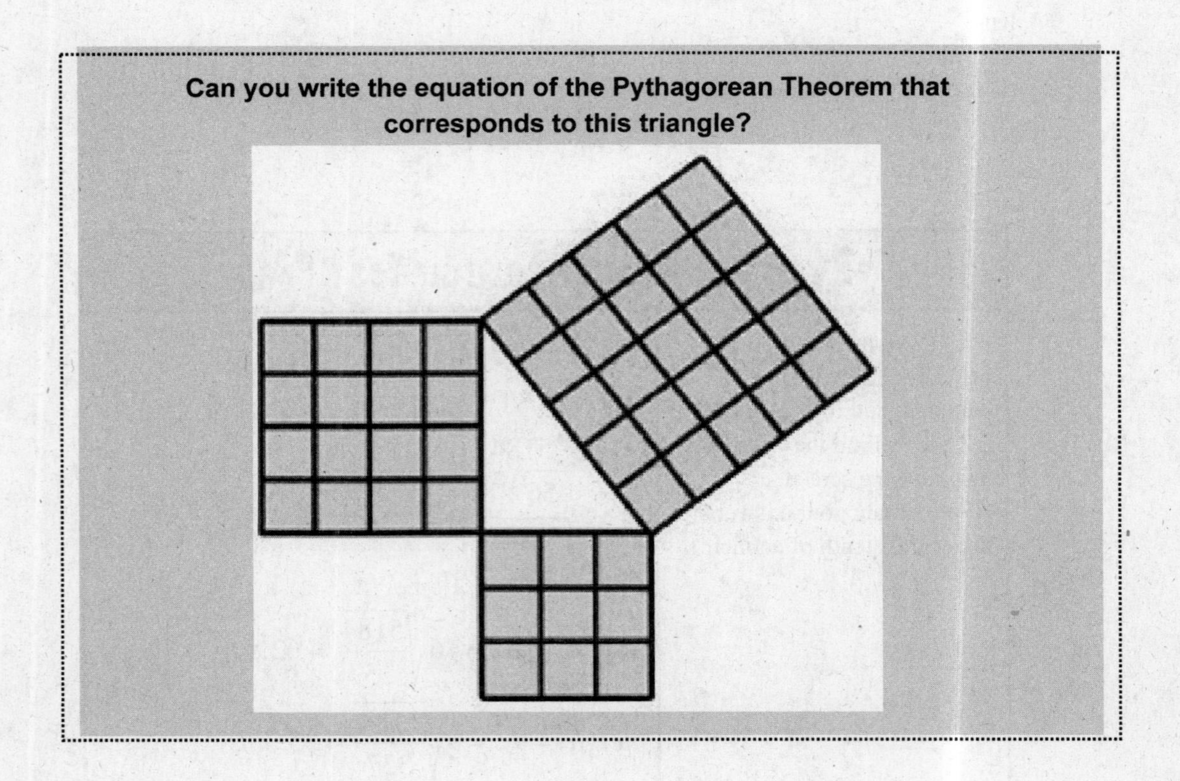

Applications of Quadratics: Test Prep

The height of a rock t seconds after being dropped from a 180 ft. tower is given by $h(t) = -16t^2 + 180$ How long does it take the rock to hit the ground, to the nearest hundredth of a second?	Height is zero at the ground, so $0 = -16t^2 + 180$ $\frac{-180}{-16} = t^2$ $t \approx 3.35\text{ sec}.$
Try Them	
If the production cost of x number of pencil erasers is modeled by the equation $C(x) = 0.03x^2 + 2.5x + 150$ how many widgets can be manufactured for \$4000?	319 erasers
The velocity in cm. per sec. of a fluid in a pipe goes according to the equation $v(r) = 16 - r^2$ where r is the radius of the pipe in cm. Find the radius of the pipe if the velocity of the fluid is found to be 7 cm. per sec.	3 cm.

NOW IS THE TIME TO TRY THE MID-CHAPTER QUIZ, PAGE 109, IN YOUR TEXT.

1.4: Other Types of Equations

All the reasons we've investigated linear, absolute value, and quadratic equations now broaden out and include, well frankly, any kind of equation. Why not? In many ways, progress in algebra means progress in kinds of equations you can use to describe systems and methods you command to solve said equations.

Polynomial Equations

Certainly we already know that polynomials don't come in only *linear* and *quadratic* varieties. As you progress through this textbook, you will learn more and more about solving higher-order polynomials equations. For now, try what you do know: factoring.

- **third-degree equation, cubic equation**
  A third-degree, or cubic, equation refers to a polynomial equation involving cubic polynomial(s). Inspect and see if you might be able to factor it -- try grouping, try finding a GCF. Those are the tools you have now. Maybe you're excited to hear that you'll learn more tools for cubics (and bigger!) soon.

Polynomial Equations: Test Prep	
Solve $2x^3 - x^2 - 4x + 2 = 0$	factor by grouping: $2x^3 - x^2 - 4x + 2 = 0$ $x^2(2x-1) - 2(2x-1) = 0$ $(2x-1)(x^2-2) = 0$ $2x - 1 = 0 \quad or \quad x^2 - 2 = 0$ $2x = 1 \quad or \quad x^2 = 2$ $x = \frac{1}{2} \quad or \quad x = \pm\sqrt{2}$ $x = \frac{1}{2}, \sqrt{2}, -\sqrt{2}$

Polynomial Equations: Test Prep	
Try Them	
Solve: $x^3 = 4x$	$x = 0,2,-2$
Solve: $16x^4 - 1 = 0$	$x = \pm\frac{1}{2}, \pm\frac{1}{2}i$

Rational Equations

You know what a rational expression is -- a ratio of polynomials, right? Put one or more rational expressions into an equations, and, voila, you have a rational equation! Add the tools for solving these to your repertoire.

✤ **rational equation**
A rational equation is an equation that involves one or more rational expressions, which are ratios of polynomials.

★ **For Fun**
If the degree of an expression like *2xy* is two, what is the degree, do you think, of an expression like $(x+1)(x-3)$ or $(x+1)^3$? Now what do you think you might be able to call the *degree* of a rational expression like, for instance: $\frac{x^2+1}{x-3}$?

★ **For Fun**
Remember how to find the domain of a rational expression? This domain question applies just as readily, if not more so, to rational equations. What is the domain of $\frac{x^2+1}{x-3}$? Would it be possible under any condition for $x = 3$ to be a solution to a rational equation like, say: $\frac{x^2+1}{x-3} = \frac{1}{2}$?

Rational Equations: Test Prep

Solve and check your answer: $\frac{x+2}{x-3}+2=\frac{10}{2x-6}$	$\frac{x+2}{x-3}+2=\frac{10}{2x-6}$ $LCD = 2(x-3)$ $2(x-3)\left(\frac{x+2}{x-3}+2\right)=\left(\frac{10}{2x-6}\right)2(x-3)$ $2x+4+4(x-3)=10$ $6x=18$ $x=3$ *answer?* *check*: $\frac{3+2}{3-3}+2\overset{?}{=}\frac{10}{2(3)-6}$ Division by zero is not defined; therefore, there is no solution.
Try Them	
Solve: $\frac{x}{x-1}=\frac{x+3}{x+1}$	$x=3$
Solve: $3x-\frac{x+2}{x+3}=\frac{4x+13}{x+3}$	$x=\frac{5}{3}$

Radical Equations

The Power Principle makes it possible for you to raise both sides of an equation to a power and not upset the truth value of the equation, just like the Addition/Subtraction or Multiplication/Division Principles:

If $P=Q$, then $P^n=Q^n$.

★ **For Fun**
Find examples of this Power Principle for all types of n: positive and negative integers, radicals, etc.

★ **For Fun**
These principles always require that items like P and Q above must be "algebraic objects." Do you understand what an algebraic object is? (If not, maybe check with your teacher?) Can you think of something that is *not* an algebraic object that renders something like the Power Principle meaningless?

✣ **extraneous solution**
Like rational equations, the domain of a radical equation is important! Think about the domain of your radical equation before you solve it, and always check you answer(s)! An extraneous solution is a solution that is NOT a solution ... if you get my drift.

(And, I might add, the same sort of thing can happen with rational equations -- check your answers!)

★ **For Fun**
Can you figure out why extraneous solutions arise when you raise both sides of an equation to an *even* power? (Hint: What happens to a negative number when you square it?)

Radical Equations: Test Prep

Solve and check your answer: $\sqrt{2x-3}=7$	$\sqrt{2x-3}=7$ $2x-3=49$ $2x=52$ $x=26$ *answer*? *check*: $\sqrt{2(26)-3}\stackrel{?}{=}7$ $\sqrt{52-3}\stackrel{?}{=}7$ $\sqrt{49}=7$

Radical Equations: Test Prep	
Try Them	
Solve: $\sqrt{2x}+\sqrt{x+1}=7$	$x=8$
Solve: $\sqrt{2x-7}-5=x+4$	No real solution.

Rational Exponent Equations

If you can solve radical equations, then you can solve rational exponent equations ... because radicals can be written as rational exponents. In fact, using the Power Principle is easier in my opinion because to me it's easier to see what power to raise both sides to:

$$x^{\frac{1}{3}}=-2$$

$$\left(x^{\frac{1}{3}}\right)^3=(-2)^3$$

$$x=-8$$

See how the one-third exponent immediately dictates raising both sides to the third power (cubing) in order to isolate x (that is, find the member of the solution set)?

But, be careful. There's no way your square root could be a negative number, for instance. So is there any solution to ...

$$x^{\frac{1}{2}}=-2$$

$$\left(x^{\frac{1}{2}}\right)^2=(-2)^2$$

$$x=4$$

Check it!

$$4^{\frac{1}{2}}\stackrel{?}{=}-2$$

Remember: unless you are told explicitly $\pm4^{\frac{1}{2}}$, $4^{\frac{1}{2}}$ means the *principal square root*, which is positive 2.

★ **For Fun**

Does $x^{\frac{1}{2}} = -2$ remind you of *complex* numbers from last chapter?

Just as the Power Principle gives rise to extraneous solutions with radicals, it can give rise to the same problems with rational exponent equations (they're really the same thing, after all) ... when you raise both sides of the equation to an even power.

Equations With a Rational Exponent: Test Prep	
Solve: $2x^{\frac{3}{5}} + 7 = 23$	$2x^{\frac{3}{5}} + 7 = 23$ $2x^{\frac{3}{5}} = 16$ $x^{\frac{3}{5}} = 8$ $x = 8^{\frac{5}{3}} = \left(8^{\frac{1}{3}}\right)^5 = 2^5 = 32$ $x = 32$
Try Them	
Solve: $x^{\frac{2}{3}} = 25$	$x = 125$
Solve: $(2x)^{\frac{3}{2}} + 5 = -3$	No solution

Equations That Are Quadratic in Form

No sense in wasting all those skills we have for solving quadratic equations (factoring, taking square roots,etc.). If you can recognize when an equation that is not actually quadratic is, nonetheless, quadratic in form, you can probably make headway solving it.

✤ **quadratic in form**

Here are some examples of equations that are quadratic in form:

$x^4 + 7x^2 + 12 = 0$

$16x^4 + 8x^2 = 35$

$x^4 - 625 = 0$

$x + 6\sqrt{x} + 5 = 0$

Can you recognize that these are quadratic in form even though they are not actually quadratic? Could you perhaps use factoring to solve these? The quadratic formula? Can you imagine a strategy?

★ **For Fun**

How many solutions does an equation like the third one in the above list maybe have? It's not quadratic, after all.

★ **For Fun**

If you used the quadratic formula to try to solve the last equation in the above list, what would go on the left-hand side of the formula? In other words, what are you solving for with the quadratic formula ... because it's not *x*!

Equations That Are Quadratic in Form: Test Prep	
Solve: $x - 8\sqrt{x} + 15 = 0$	$x - 8\sqrt{x} + 15 = 0 \quad \left[let \quad u = \sqrt{x}\right]$ $u^2 - 8u + 15 = 0$ $(u-5)(u-3) = 0$ $u = \sqrt{x} = 5 \quad or \quad u = \sqrt{x} = 3$ $x = 25 \quad or \quad x = 9$ $x = 25, 9$
Try Them	
Solve using the quadratic formula: $2x^{\frac{2}{5}} - x^{\frac{1}{5}} - 6 = 0$	$x = -\frac{243}{32}, 32$
Solve: $x^4 - 3x^2 + 2 = 0$	$x = \pm 1, \pm\sqrt{2}$

Applications of Other Types of Equations

Some of the further applications of other types of equations explored in your textbook include motion problems, work problems, and distance problems. There are a lot of systems to describe, investigate, and solve. These are some of the reasons for studying algebra and building up a healthy set of tools for working out these solutions.

Applications of Other Types of Equations: Test Prep	
If the time T in seconds between the instant you drop a stone into a well and the moment you hear its impact at the surface of the water is given by $T = \frac{\sqrt{s}}{4} + \frac{s}{1100}$ where s is measured in feet, find the depth to the foot of the water level if it takes 4 seconds to hear the impact.	$4(4400) = \left(\frac{\sqrt{s}}{4} + \frac{s}{1100}\right)(4400)$ $17600 = 1100\sqrt{s} + 4s$ $4s + 1100\sqrt{s} - 17600 = 0$ $\sqrt{s} = \frac{-1100 \pm \sqrt{1100^2 - 4(4)(-17600)}}{2(4)}$ $= \frac{-1100 + \sqrt{1491600}}{8} \quad [s > 0]$ $s \approx 230\,ft.$
Try Them	
A ferry travels into a 24 mile deep bay at 10 mph and returns at 12 mph. The entire trip takes 5.5 hr. At what rate is the tide running out of the bay?	4 mph
Given the right triangle: 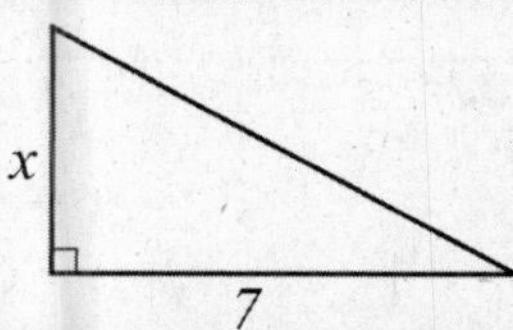 where the ratio of the length of the hypotenuse to 7 is 4, find x. Round to two decimal places.	$x \approx 27.11$

1.5: Inequalities

For many of the sorts of equations we have studied already in this chapter, we can also investigate their behavior and solutions in inequalities instead. We will look at linear and other polynomial inequalities, rational inequalities, absolute value inequalities, etc. with the goal of increasing the tools we have to solve not only equations but also inequalities!

Properties of Inequalities

The next types of "equations" you can learn how to manipulate and solve are not equations. They are inequalities -- mathematical sentences whose verb is not the equals sign but is instead "is greater than," "is less than," "is greater than or equal to," or "is less than or equal to." The symbols are well-known; they are, respectively, $>$, $<$, $\geq$, and $\leq$.

✤ **solution set**
A solution set is the set of all values of the variable that make the *inequality* true. Now this can admit a *lot* of values. For example, $x > 3$ is an infinitely large set of numbers; whereas, $x = 3$ is one number and one number only.

★ **For Fun**
Do you remember how to write $x > 3$ or other similar inequalities in set builder notation? in interval notation? graphed on a number line? (Now would be an especially good time to review all these.)

✤ **equivalent inequalities**
Like equations, equalities can be manipulated in order to find the solution set -- that is, isolate the variable on one side of the *in*equality. These are some of the useful ways you can construct equivalent inequalities:

Properties of Inequalities

1. **Addition/Subtraction Property:**

If $a > b$, then $a + c > b + c$ or $a - c > b - c$. Likewise if $a < b$.

2. Multiplication/Division Property:

i. If $c > 0$ and $a > b$, then $a \cdot c > b \cdot c$ or $a \div c > b \div c$. Likewise if $a < b$.

ii. If $c < 0$ and $a > b$, then $a \cdot c < b \cdot c$ or $a \div c < b \div c$. Likewise if $a < b$.

[Did you catch the difference between 2i. and 2ii.?]

★ **For Fun**

Try these properties out on equalities that you know to be true, like this:

$$3 < 5$$
$$3 - 2 \overset{?}{<} 5 - 2$$
$$1 < 3$$

or

$$3 < 5$$
$$3 \cdot (-2) \overset{?}{>} 5 \cdot (-2)$$
$$-6 > -10$$

Linear Inequalities: Test Prep	
Solve: $-8x < 2x - 15$	$-8x < 2x - 15$ $-10x < -15$ $10x > 15$ $x > \frac{15}{10}$ $x > \frac{3}{2}$

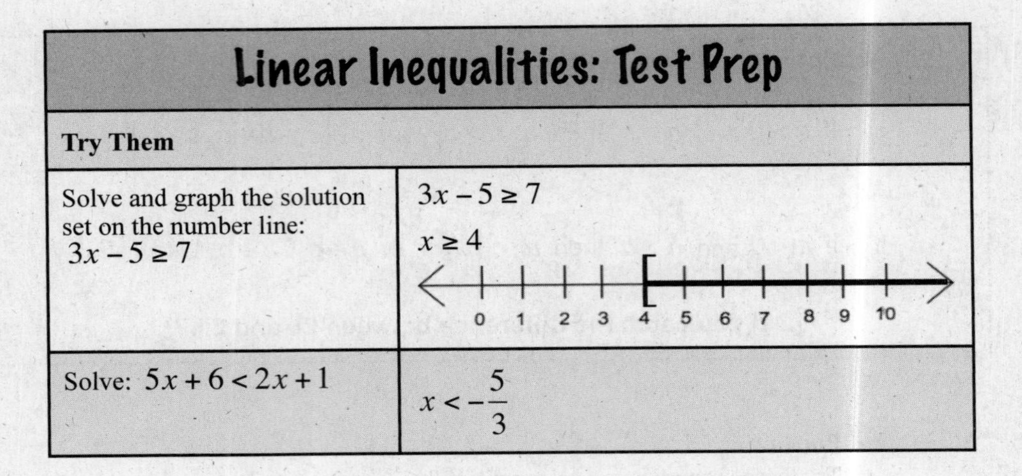

Linear Inequalities: Test Prep	
Try Them	
Solve and graph the solution set on the number line: $3x-5 \ge 7$	$3x-5 \ge 7$ $x \ge 4$ 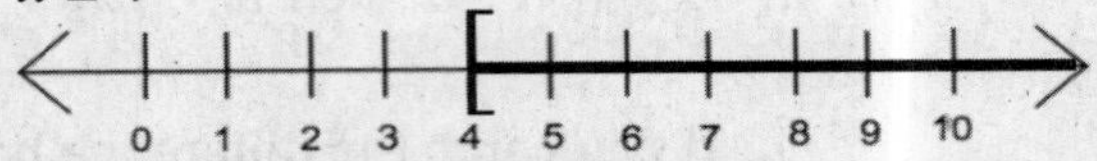
Solve: $5x+6<2x+1$	$x<-\frac{5}{3}$

Compound Inequalities

It isn't really possible to have a compound equation beyond something like $a=b=c$ for the very reason that equality admits of only one possibility. In this case that possibility is that $a=c$.

★ **For Fun**
Can you create an interesting compound equality?

But because inequalities have infinite solutions, it's possible to write something interesting like $-1<3x+2<14$. After all, between -1 and 14 exist an infinity of numbers! And all the *x's* that define those numbers can be discovered -- the solution set.

✤ **compound inequality**
A compound inequality is a combination of inequalities that further defines the solution set of all of the inequalities.

There are two ways to combine inequalities:

1. AND
   And signifies the intersection $\cap$ of the solution sets -- don't forget they are *sets* after all and subject to set operations. For example:

   $x-5>-2$ AND $2x>10$
   $x>3$ AND $x>5$
   $\{x \mid x>3\} \cap \{x \mid x>5\} = \{x \mid x>3\}$

   (Try graphing these sets on a number line if you have problems figuring out the result of the intersection.)

2. OR
   Or signifies the union $\cup$ of the solution sets. For example:

   $-3x>30$ OR $x+1>17$
   $x<-10$ OR $x>16$
   $\{x \mid x<-10\} \cup \{x \mid x>16\} = \{x \mid x<-10 \quad or \quad x>16\}$

   (See if you can graph those on the number line? Can you think of a better way to describe the solution set?)

★ **For Fun**
Computers think with these ideas of AND and OR. Have you ever done a database search using these terms? Some search engines allow them -- look under "Advanced Search Options" or something similar.

★ **For Fun**
Do you believe the compound inequality $-1<3x+2<14$ is an AND or an OR case? Can you write this inequality as two separate inequalities with the correct set AND/OR connector?

Compound Inequalities: Test Prep	
Solve and express the solution in interval notation:	$-7<2-3x\le 11$ $-9<-3x\le 9$ $3>x\ge -3$ $[-3,3)$

Compound Inequalities: Test Prep

Try Them	
Solve and graph the solution on a number line: $3x-2\geq 7 \quad OR \quad 3x-2<-7$	$x\geq 3 \quad OR \quad x<-\dfrac{5}{3}$
Solve: $3\leq 2x+2<8$	$\dfrac{1}{2}\leq x<3$

Absolute Value Inequalities

At this point, combining an absolute value expression with the notion of an inequality becomes interesting. Remember that absolute value has to do with distance. In an inequality absolute value can deal with ranges of distances, like ...

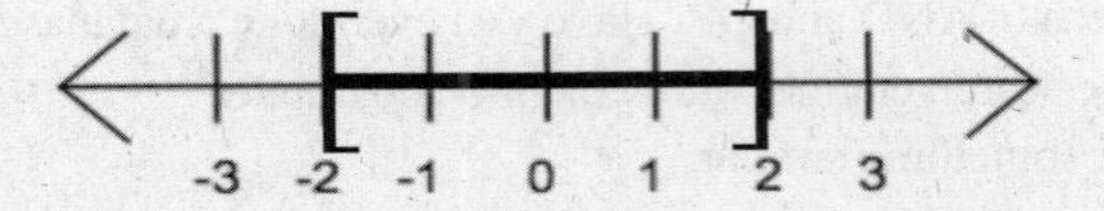

This graph is the set of all numbers that are 2 or less away from zero, right?

Learn these few simple "translation" rules for absolute value inequalities:

Properties of Absolute Value Inequalities

1. If $|E|<k$, then $-k<E<k$
2. If $|E|>k$, then $E>k$ OR $E<-k$

These properties hold for any *E* that is a algebraic expression and any *k* that is a real, non-negative number. They also hold if you replace the < with ≤ and likewise for >.

★ **For Fun**
If k has to be non-negative, can it be zero? Why does it have to be non-negative?

★ **For Fun**
For the set graphed on the number line above, could you use the "translations" above to write an absolute value inequality that matched that set?

★ **For Fun**
Could you graph the set described by the absolute value inequality $|x-1| \le 10$, for instance?

Absolute Value Inequalities: Test Prep	
Solve and graph the solution: $|5-2x| \le 9$	$|5-2x| \le 9$ $-9 \le 5-2x \le 9$ $-14 \le -2x \le 4$ $7 \ge x \ge -2$ 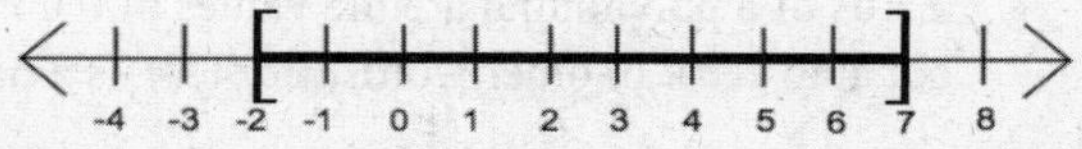
Try Them	
Solve and express the answer in interval notation: $|3x+5| < 7$	$\left(-4, \frac{2}{3}\right)$
Solve: $|2x+1| \ge 2$	$x \ge \frac{1}{2}$ *OR* $x \le -\frac{3}{2}$

Polynomial Inequalities

The next algebraic object you can try out in an inequality is a polynomial. As you might expect, since higher-order polynomials can have more solutions than linear polynomials, the solution sets for polynomial inequalities can get more complicated. Learn the method and try it out. I offer some reminders:

1. Don't forget how to factor.
2. Use the number line.
3. Like the standard form for a quadratic equation, get one side of the inequality equal to zero.

> For all values of x between two consecutive real zeros, all values of the polynomial are positive or all values of the polynomial are negative.
> ("Sign Property of Polynomials" page 127 of your textbook)

- **zero of a polynomial**
  Zeros of a polynomial are the values of the variable that cause the polynomial to be equal to zero. In other words, if $p(x)$ is a polynomial, then the zeros of $p(x)$ are the solution set of the particular case where $p(x) = 0$.

- **critical values of the inequality**
  Real zeros of a polynomial in an inequality are the critical values of that inequality.

- **test value**
  A test value is any value you care to plug into an algebraic expression to find its resultant value. In the case of polynomial inequalities, you will need to pick test values that occur between the critical values of the polynomial in question. (By the way, picking good test values is a skill you will use extensively in calculus, too.)

Solving a Polynomial Inequality

1. If necessary, rearrange the inequality (using properties of equivalent inequalities) so that one side of the inequality is your polynomial and the other is zero.

2. Find all the real zeros of your polynomial -- those are the critical values. Label them on a number line and list the factors in a column either above or below the number line.

3. In between the critical values, pick a test value. Find the sign of each factor for each test value. Determine the result of multiplication of these signs together (as the factors would be multiplied in order to build the original polynomial). Because of the Sign Property of Polynomials, everywhere in each interval the sign of the polynomial is that result.

4. Refer back to the polynomial inequality in step one in order to find the intervals that satisfy, then define the solution set of the inequality accordingly.

Polynomial Inequalities: Test Prep

Solve: $x^2 - x - 2 < 0$	$x^2 - x - 2 < 0$ $(x-2)(x+1) < 0$

$(x-2)$	$-$	$-$	$+$	*factors*
$(x+1)$	$-$	$+$	$+$	
	$x=-2$	$x=0$	$x=3$	*test*
	(critical values: -1, 2)			
$x^2 - x - 2$	$+$	$-$	$+$	*polynomial*

The polynomial is negative between -1 and 2, so the solution set is $-1 < x < 2$

Polynomial Inequalities: Test Prep	
Try Them	
Graph the solution of $9 \le x^2$ on a number line.	-4 -3 -2 -1 0 1 2 3 4
Solve the inequality $x^2+6>5x$ and express the answer in interval notation.	$(-\infty,2)\cup(3,\infty)$

Rational Inequalities

Predictably we will next look at inequalities with rational expressions. We're investigating methods to find the solution set for these, too.

- **rational inequality**
  Of course, a rational inequality is an inequality with a rational expression(s) included. For example: $\frac{x+1}{x-3} \ge \frac{2x}{x+5}$. Remember that a rational expression is a ratio of two polynomials.

- **critical value of a rational expression**
  The critical values of a rational expression are all the zeros of both the numerators and denominators.

Solving a Rational Inequality

1. If necessary, rearrange the inequality (using properties of equivalent inequalities) so that one side of the inequality is zero. Also if necessary, rewrite the other side so that it is a single rational expression -- this may involve all the manipulation of rational expressions that we talked about in the last chapter.

2. Find all the critical values of your rational expression. Label them on a number line and list the factors of both the numerator and denominator in a column either above or below the number line.

3. In between the critical values, pick a test value. Find the sign of each factor for each test value. Determine the result of multiplication and/or division of these signs together (as the factors would be multiplied in order to build the original polynomial).
4. Refer back to the rational inequality in step one in order to find the intervals that satisfy, then define the solution set of the inequality accordingly.

Rational Inequalities: Test Prep

Solve: $\dfrac{x^2-11}{x+1} \le 1$

Express your answer in interval notation.

$$\frac{x^2-11}{x+1} \le 1$$

$$\frac{x^2-11}{x+1} - 1 \le 0$$

$$\frac{x^2-11-(x+1)}{x+1} \le 0$$

$$\frac{x^2-x-12}{x+1} \le 0$$

$$\frac{(x-4)(x+3)}{x+1} \le 0$$

factors				
$(x-4)$	−	−	−	+
$(x+3)$	−	+	+	+
$(x+1)$	−	−	+	+
test	$x=-5$	$x=-2$	$x=0$	$x=5$
critical values	−3	−1	4	
ratio $\dfrac{x^2-x-12}{x+1}$	−	+	−	+

The ratio is negative when x is less than -3 and between -1 and 4; -1 is not in the domain of the ratio, so the solution is $(-\infty,-3]\cup(-1,4]$

Rational Inequalities: Test Prep	
Try Them	
Solve: $\frac{4}{x-1} \geq \frac{3}{x-4}$	$1 < x < 4 \quad AND \quad 13 \leq x < \infty$
Solve: $\frac{5x-9}{x-5} \leq 3$	$[-3,5)$

Applications of Inequalities

As well as 3-D video games, statistics, geometry, and business are a few of the applications of inequalities. Careful reading with a firm basis of technique are the way to succeed with application problems. Be sure to name your variables meaningfully. And practice a lot!

Applications of Inequalities: Test Prep	
The sum of three consecutive odd integers is between 63 and 81. Find all possible sets of integers that satisfy these conditions.	3 consecutive odd integers could be expressed $x, x+2, x+4$. $63 < x + x + 2 + x + 4 < 81$ $63 < 3x + 6 < 81$ $57 < 3x < 75$ $19 < x < 25$ So x is either 21 or 23, and the two possible sets of integers are $\{21,23,25\} \quad or \quad \{23,25,27\}$

Applications of Inequalities: Test Prep

Try Them	
The average cost of producing x coin purses in dollars is $\overline{C} = \frac{1.25x + 2500}{x}$ How many coin purses must be produced to bring the average cost below $3.00?	more than 1429
A model rocket is launched from the top of a cliff 80 ft. high. $s(t) = -16t^2 + 64t + 80$ described the rocket's height in feet after t seconds. When will the rocket's height exceed the height of the cliff?	$0 < t < 4$ sec.

Here are some recent article titles from
The Journal of Inequalities and Applications:

- Refinements, Generalizations, and Applications of Jordan's Inequality and Related Problems
- Several Matrix Euclidean Norm Inequalities Involving Kantorovich Inequality
- Mixed Variational-Like Inequality for Fuzzy Mappings in Reflexive Banach Spaces

Really! This is real math!

1.6: Variation and Applications

All those proportion problems you worked in math before college fall under the heading of variations: relations between variables that *vary* with each other in one way or another. Here come some general remarks, pointers, examples, and encouragements about these sorts of problems.

Direct Variation

Variation has to do with equations that relate variables -- much like the formulas we have already talked about. Direct variation deals with equations of the form: $y = kx^n$.

✤ **variation**

In a lot of ways a variation might seem like a formula; for example, $A = \pi r^2$ is a variation that relates the area of a circle to its radius. In other words, as the radius varies, so does the area.

✤ **vary directly**

If a relationship between variables, x and y for instance, can be written like $y = kx$, where k is a constant then you can say that y varies directly with x.

✤ **directly proportional**

$y = kx$ can also be expressed as "y is directly proportional to x."

✤ **constant of proportionality, variation constant**

And k gets the special label of constant of proportionality or variation constant when it plays this role in a variation equation.

✤ **vary directly as the nth power**

In my example above, $A = \pi r^2$, A varies directly the with second power of r.

★ **For Fun**

What is the constant of proportionality in $A = \pi r^2$?

Direct Variation: Test Prep

Given $y = kx$, find k if y is 10 when x is 2.	$10 = k \cdot 2$ $\frac{10}{2} = k$ $k = 5$
Try Them	
If y varies directly as x, and the constant of variation is $\frac{5}{3}$, what is y when $x = 9$?	15
If y varies directly as x, and $y = 8$ when $x = 12$, find k, write an equation that expresses this variation, and find y when $x = 5$	$k = \frac{2}{3}$ $y = \frac{2}{3}x$ $\frac{10}{3}$

Direct Variation as the nth Power: Test Prep

Given that weight varies with height according to the equation $w = kh^3$, and a man of 6 ft and 201 lb. wishes to meet a woman of similar proportions, how much would such a woman weigh, to the nearest pound, if she were 5.5 ft tall?	$201 = k \cdot 6^3$ $k = \frac{201}{216}$ *so* $w = \frac{201}{216} \cdot 5.5^3$ $\sim 155lb.$

Direct Variation as the nth Power: Test Prep	
Try Them	
The volume of a sphere is described by $V = \frac{4}{3}\pi r^3$. Approximate the volume of a sphere of radius 2.71 cm. to the nearest tenth of a cm.	83.4 c.c.
The distance that an object falls varies directly as the square of the time of the fall. If an object falls 256 ft. in 4 seconds, how long will it take to fall 64 ft?	2 sec.

Inverse Variation

Inverse variation has to do with relationships of the general form: $y = \frac{k}{x^n}$.

✤ **vary inversely, inversely proportional**
If a relationship between variables, x and y for instance, can be written like this -- $y = \frac{k}{x}$, where k is a constant -- then you can say that y varies inversely with x or y is inversely proportional to x.

✤ **vary inversely as the nth power**

More generally you can say that y varies inversely with the nth power of x if $y = \frac{k}{x^n}$.

★ **For Fun**
Usually the n in these sorts of equations means n is a member of the natural numbers (the counting numbers). Does that restriction make sense here and in the direct variation equation?

Inverse Variation: Test Prep	
The water temperature of he Pacific Ocean varies inversely as the water's depth. At a depth of 1000m, the temperature is 4.4 degrees Celsius. What is the water temperature at a depth of 4000m?	$T = \frac{k}{d}$ $4.4 = \frac{k}{1000}$ *so* $k = 4400$ *and* $T = \frac{4400}{4000} = \frac{11}{10}$ $T \approx 1.1°C$
Try Them	
If the number of potential buyers of a house varies inversely with the price and a house priced at $100,000 attracts 250 potential buyers, find the variation constant for this market.	25,000,000
The time required to do a job varies inversely as the number of people working it. If it takes 100 days for 8 workers to complete a job, how long would it take 12 workers?	$66\frac{2}{3}\ days$

Inverse Variation as the nth Power: Test Prep	
Suppose y varies inversely with the square of x, and $y = 4$ when $x = 5$. Write the equation for this relationship and find y when $x = 2$.	$y = \frac{k}{x^2}$ *and* $4 = \frac{k}{25}$ *so* $k = 100$ *and* $y = \frac{100}{x^2}$ *then* $y = \frac{100}{2^2} = \frac{100}{4} = 25$
Try Them	
The number of units of an item varies inversely with the square of the price, and 150 units sold when the price was $2. How many, to the nearest unit, would sell if the price were $1.50?	267
If y varies inversely as the square root of x, and $y = 17$ when $x = 49$, find the equation of variation.	$y = \frac{119}{\sqrt{x}}$

Joint and Combined Variations

Joint and combined variations mix up more variables -- again a lot like some of the formulas you might have already seen. For example: $A = l \times w$, the area of a rectangle, which varies not with just the length or just the width of the rectangle, but both of them.

✤ **vary jointly**

If you can relate a variable z to more than one other variable like this -- $z = kxy$, where k is a constant again -- then z varies jointly with x and y.

★ **For Fun**

This is the kind of equation my area of a rectangle falls under, right? Can you tell what the variation constant is for $A = l \times w$?

✤ **combined variation**

Combined variation means your variation may depend on more than one variable and the dependance may be direct or inverse or some of both. For example: $y = k\frac{xz}{w}$, where y varies directly with x and z and inversely with w.

Joint Variation: Test Prep

y varies jointly as m and the square of n and inversely as p, and $y = 15$ when $m = 2$, $n = 1$, and $p = 6$. Find y when $m = 3$, $n = 4$, and $p = 10$.	$y = \frac{kmn^2}{p}$ *so* $15 = \frac{k \cdot 2 \cdot 1^2}{6} = \frac{k}{3}$ $k = 45$ *and* $y = \frac{45mn^2}{p}$ *then* $y = \frac{45 \cdot 3 \cdot 4^2}{10} = 216$

Joint Variation: Test Prep

Try Them	
Simple interest varies jointly with the principal, rate and time in years: $I = prt$. Find the interest earned if $p = \$6000,\ r = 3.5\%$ after ten years.	\$2100
The pressure of a gas depends on temperature and volume thus: $P = k\frac{T}{V}$. If $P = 2$ when $T = 298$ and $V = 9$, write the gas law.	$P = \frac{9}{149} \cdot \frac{T}{V}$

NOW TRY THESE:
REVIEW EXERCISES: TEXTBOOK PAGE 148
PRACTICE TEST: TEXTBOOK PAGE 151
CUMULATIVE REVIEW: TEXTBOOK PAGE 152

Final Fun: Matching

FIND THE BEST MATCH. GOOD LUCK.

$|x|$

DISCRIMINANT

$x = \dfrac{-b \pm \sqrt{b^2 - 4ac}}{2a}$

STANDARD QUADRATIC

$x^2 + 10x + 25$

VARIATION

CONTRADICTION

$a = b$

RADICAL EQUATION

$X^2 + Y^2 = R^2$

$C = \pi d$

$\dfrac{x+1}{x-3} \geq \dfrac{2x}{x+5}$

QUADRATIC IN FORM

PYTHAGOREAN THEOREM

$x + 3 = x$

CIRCUMFERENCE OF A CIRCLE

$a - c = b - c$

$\sqrt{x-1} = 17$

$x^2 + 6\sqrt{x} + 5 = 0$

PERFECT SQUARE TRINOMIAL

$\begin{cases} x, & x \geq 0 \\ -x, & x < 0 \end{cases}$

$z = kxy$

QUADRATIC FORMULA

$b^2 - 4ac$

RATIONAL INEQUALITY

$ax^2 + bx + c = 0$

Chapter 2: Functions and Graphs

Word Search

```
D T N I O P D I M E D C D P N
N E N O M C E D T F O I E E O
A T T N I P I A O M M T C R I
I N N E O T N R P W A A R P S
S A S L R I A O C X I R E E S
E T S U D M S L I L N D A N E
T S C R I I I S S D E A S D R
R N O E T D S N E N P U I I P
A O N I N V A T A R A Q N C M
C C O U O T Z R A T G R G U O
I N T E R C E P T N I E T L C
N O I T A L E R R O C O R A O
I N C R E A S I N G L E N R Q
N O I T C E L F E R A E N I L
S N O I T C N U F O R I G I N
```

AXIS
CARTESIAN
CENTER
CIRCLE
COMPOSITION
COMPRESSION
CONSTANT
COORDINATE
CORRELATION
DECREASING
DETERMINATION
DISTANCE
DOMAIN
FUNCTIONS
INCREASING
INTERCEPT
LINEAR
MIDPOINT
ODD
ORDINATE
ORIGIN
PERPENDICULAR
QUADRATIC
RADIUS
REFLECTION
REGRESSION
SLOPE
TRANSLATION

2.1: Two-Dimensional Coordinate System and Graphs

Since we're handling variables quite adequately now, let's learn how to handle two at a time and how to communicate about them in a graph.

Cartesian Coordinate Systems

In order to communicate in 2-dimensional geometry and algebra at the same time, we need to agree on a *frame of reference*. In honor of René Descartes, who did much of the early work of combining these two topics, our agreed-upon frame of reference is called *Cartesian*. The Cartesian coordinate system breaks a 2-dimensional surface, such as the blackboard or a piece of notebook paper, into 4 areas that everyone around the globe can navigate and communicate about.

✤ **coordinate**
A coordinate is a number that is associated with a position. A coordinate usually means that a scale is present so that the position has meaning. In algebra that scale is the number line.

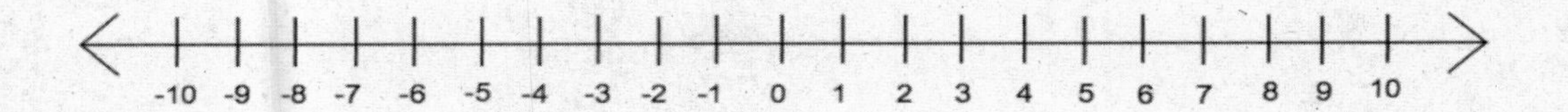

✤ **coordinate plane**
A coordinate plane is 2-dimensional. So instead of one scale, two scales are needed. Each scale is actually just a number line.

✤ **ordered pair**
An ordered pair of numbers, usually (x, y), is simply the coordinates required to specify a position in 2 dimensions. The order of the coordinates in an ordered pair is a *rigorous convention*!

✓ **Questions to Ask Your Teacher**
Make sure you understand when your instructor wants your answers in the form of an ordered pair. And then be sure to include all the elements of the notation: parentheses and comma.

✣ **coordinates**
See ordered pair. In this class coordinates will almost always come in ordered pairs. That means you will be working in two dimensions.

✣ **x-coordinate, abscissa**
Customarily the first coordinate in an ordered pair is the x-coordinate. The first coordinate, and remember that it's just a number, tells you the position along the scale, or number line, that refers to x-values. Abscissa is a slightly old-fashioned word (but still perfectly meaningful) for the x-coordinate. As your textbook points out, *abscissa* is related to the word *scissors*.

★ **For Fun**
Can you find any other words in English that are related to *abscissa* or *scissors*?

✣ ***x*-axis**
The x-axis is one of the number lines (scales) required to find positions on a coordinate plane. Usually the x-axis runs from left to right, from negative values to positive values. The arrowhead always points towards the positive end of the spectrum of x-values (or in the order in which values are increasing).

⟶ x

✓ **Questions to Ask Your Teacher**
How does your instructor want you to label your x-axis? (Or, for that matter, all your *axes*, which is the plural of *axis* even though it looks like the plural of a tool for chopping wood. Context is everything.)

✣ **y-coordinate, ordinate**
Like the x-coordinate, the y-coordinate is a number. This time it is the second number in an ordered pair, and it refers to the position on the scale of y-values.

★ **For Fun**
Look up *ordinate*. Find other meanings and see if they make any sense to you in the context of a 2-dimensional co-*ordinate* plane.

✤ ***y*-axis**

The scale of y-values is called the y-axis. Usually it runs from bottom to top, and negative values are at the bottom while positive values are at the top. It's another number line.

✤ **origin**

If you have two number lines running perpendicular to each other on a 2-dimensional plane, such as your paper or the blackboard, they *will* meet at a point (does that make sense?). We agree that they will meet at the position on each scale corresponding to a value of zero (or nothing). The ordered pair that describes this position on the coordinate plane is (0, 0). This becomes the starting position against which all coordinates are measured: the origin. (The origin is like the "You Are Here" point on the maps at the mall. From "Here" you can get anywhere... given meaningful instructions!)

Here is a short list of possible coordinate systems in 2-dimensions:

Cartesian coordinate system
Polar coordinate system
Parabolic coordinate system
Bipolar coordinates
Hyperbolic coordinates
Elliptic coordinates

Descartes' version is simple. We'll stick with it until trigonometry!

✤ **quadrant**

The two crossing axes break a 2-dimensional surface up into four areas. Each of these areas is one quadrant. The quadrants are always numbered using Roman numerals (see picture).

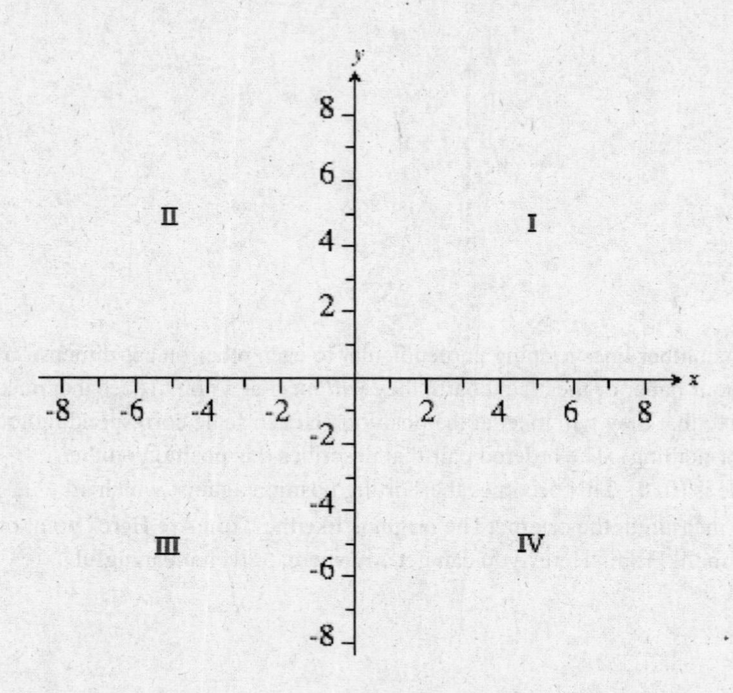

★ **For Fun**

Look up the derivation of the word *quadrant*. Think about it – obviously *quadrant* and *quadratic* are related words. Why?

✤ **Cartesian coordinate system**

The x-axis, y-axis, and quadrants all together are frequently called a Cartesian coordinate system. A standard 2-dimensional coordinate system is the same thing.

★ **For Fun**

Cartesian is named after René Descartes. Can you find out what the French words "*des cartes*" mean? What else is René Descartes known for?

✤ **plot a point**

To plot is to place a point or points on the Cartesian coordinate plane. An actual *plot* is the picture that results from plotting.

Distance and Midpoint Formulas

If we are navigating around a 2-dimensional surface, we might want to know how far we have to go to get from one point to another or where the halfway point lies. The distance and mid-point formulae (Latin plural of formula) enable us to completely specify these quantities.

The distance formula follows:

Given two points, (x_1, y_1), (x_2, y_2), the distance between them is

$$d = \sqrt{(x_2 - x_1)^2 + (y_2 - y_1)^2}$$

✓ **Questions to Ask Your Teacher**
Since the distance formula involves a square root and the mid-point coordinates involve division (that is, fractions!), you might make sure when your instructor wants an exact answer like $d = \sqrt{10}$ or an approximation like $d \cong 3.1623$. In the case of approximations, be sure you understand all about rounding.

★ **For Fun**
Do you readily recognize a similarity between the distance formula and the Pythagorean formula? Can you draw a picture that illustrates this relationship? Is it ever possible to have a negative value under the square root in the distance formula? Why not? Ask yourself, Why is distance always the positive square root?

❖ **midpoint**
A midpoint is the point exactly in the middle of two other points, or halfway between them.
Given two points, (x_1, y_1), (x_2, y_2), the midpoint between them is found at

$$\left(\frac{x_1 + x_2}{2}, \frac{y_1 + y_2}{2}\right)$$

★ **For Fun**
Think about it: How is a midpoint like the average, or mean, of the two end points?

Distance Formula: Test Prep

Find the distance between the points A(2, 1) and B(4, -6).	$x_1 = 2,\ \ y_1 = 1,\ \ x_2 = 4,\ \ y_2 = -6$: $d = \sqrt{(x_2 - x_1)^2 + (y_2 - y_1)^2}$ $= \sqrt{(4-2)^2 + (-6-1)^2}$ $= \sqrt{(2)^2 + (-7)^2}$ $= \sqrt{4 + 49}$ $d = \sqrt{53}$
Try Them	
Find the distance between the points P(-3, -1) and Q(0, 2).	$d = \sqrt{18} = 3\sqrt{2}$
Find the distance between the points P_1(-1, 16) and P_2(4, 4).	$d = 13$

Midpoint Formula: Test Prep

Find the midpoint of the line segment connecting the two points P_1(-6, 2) and P_2(1, -3).	$\left(\frac{x_1 + x_2}{2}, \frac{y_1 + y_2}{2}\right)$ $= \left(\frac{-6+1}{2}, \frac{2+(-3)}{2}\right)$ $= \left(\frac{-5}{2}, \frac{-1}{2}\right)$

Midpoint Formula: Test Prep	
Try Them	
Find the midpoint of the line segment connecting the two points A(5, -3) and B(-3, 4).	$\left(2, \frac{1}{2}\right)$
Find the midpoint of the line segment connecting the two points P $\left(-1, \frac{2}{3}\right)$ and Q(2, 0).	$\left(\frac{1}{2}, \frac{1}{3}\right)$

Graph of an Equation

An equation is a mathematical proposition that frequently relates variables. In 2-dimensions, most equations will have 2 variables, for example:

$$y = 3x - 1$$
$$x^2 + y^2 = 16$$
$$y = |x + 3|$$

In order to "graph an equation," you need a coordinate system, and you need to understand what a graph of an equation means. Then you need to learn the methods to transfer your understanding onto your coordinate system.

✣ **graph of an equation**
The graph of an equation is a set of points that have been plotted on a Cartesian coordinate all of which satisfy the conditions of the equation -- that means the points are in the solution set of the equation. Each ordered pair, (x, y), is a set of values that make the equation true. A graph of an equation is a *picture* of the solution set of the equation!

Graph of an Equation: Test Prep

Graph $y + x = 3$

$y = 3 - x$

x	y
−2	5
−1	4
0	3
1	2
2	1
3	0

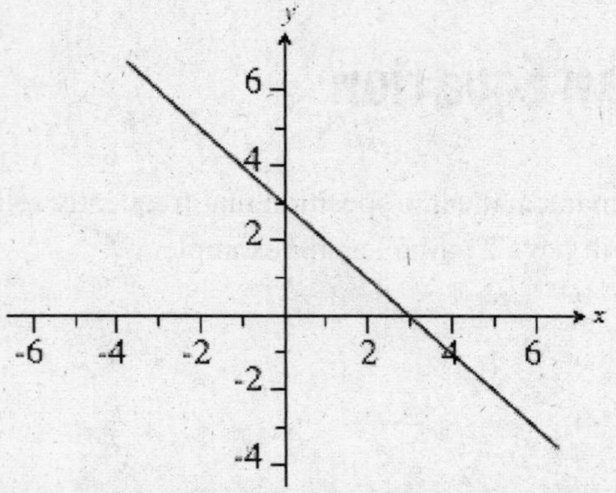

Try Them

Graph $y = -x^2$

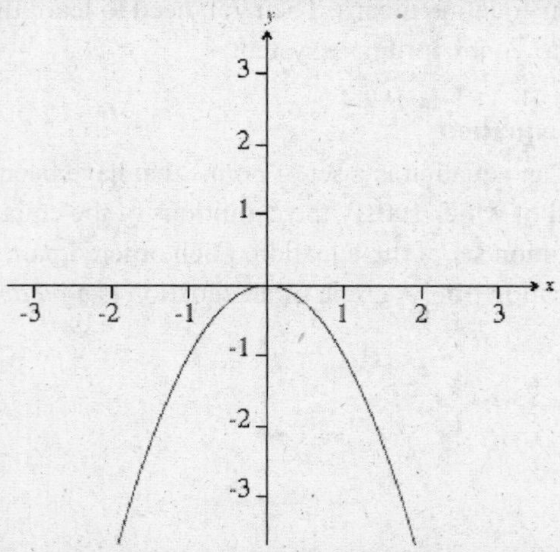

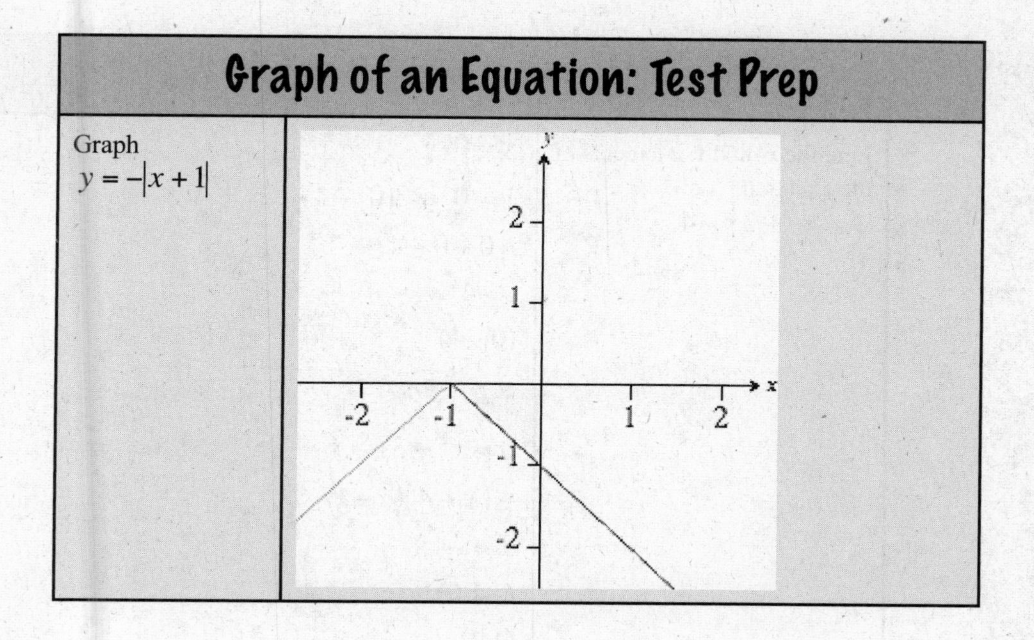

Intercepts

In mathematics an intercept means a meeting point of two curves. (Is that anything like an interception in American football?) "Intercept" frequently refer to the meeting point between the graph of an equation and an axis on a Cartesian coordinate system. (However, later you will discover that *intercept* can also mean where the graphs of two equations meet.)

- **intercept**
  An intercept is any place where two curves meet – like an intersection. For instance, the origin is the intercept of the x-axis and the y-axis.

- ***x*-intercept**
  An x-intercept is the intersection of a graph of an equation with the x-axis. If a point lies *on* the x-axis, what must its y-coordinate be? All ordered pair representations of x-intercepts will have the form: $(x_0,0)$. By the way, x_0 is just a way to say we are talking about an actual number on the x-axis instead of the variable x.

- ***y*-intercept**
  Likewise, all ordered pair representations of y-intercepts will have the form: $(0,y_0)$. If the point is plotted on the y-axis, what must its x-coordinate be?

x-Intercepts and y-Intercepts: Test Prep

Find the x- and y-intercepts of the equation $y = x^2 + 3x - 4$	$x = 0$: $y = 0^2 + 3(0) - 4$ $= 0 + 0 - 4$ $= -4$ $(0,-4)$ $y = 0$: $0 = x^2 + 3x - 4$ $= (x+4)(x-1)$ $x = -4, 1$ $(-4,0)$ $(1,0)$
Try Them	
Find the x- and y-intercepts of the equation $y - x = 9$	(0, 9) and (-9, 0)
Find the x- and y-intercepts of the equation $x^2 = y^2 - 4$	(0, 2), (0, -2), and no x-intercepts

Circles, Their Equations, and Their Graphs

We all have a good idea of what shape a circle is, but can you think of a good definition of that shape? The information contained in an equation that describes a circle depends on precise ideas about just what a circle is. Fortunately, because we know what a circle looks like, we have an advantage when it comes time to graph the equation that describes one -- provided we recognize it and the information it contains.

✤ **circle**

A circle is a 2-dimensional round curve. In fact, it is the set of all points that are exactly the same distance from a central point. (The central point, however, is not part of the circle, is it?)

★ **For Fun**
If the definition for a circle is so "tied up" with the notion of distance, would you expect the distance formula to be involved? See if you can spot it ... or ask your instructor how they are related.

✤ **radius**
The radius of a circle is the length of one-half of its diameter, and a diameter is the length of a line that cuts a circle exactly in two. All points on the circle (on its circumference) are exactly one radius from the center of the circle. (That is, the "distance" mentioned in the definition for a circle above gets the special name of "radius.")

✤ **center**
Of course, the center of a circle is the point which lies exactly in the middle of the area defined by the circumference. Beware: the center of the circle is not part of the circle itself, which is the set of points that make the equation for a circle true. The center is certainly a convenient way to understand more about where the circle is situated on a Cartesian coordinate system.

✤ **standard form of the equation of a circle**
The standard form of the equation of a circle looks like $(x-h)^2+(y-k)^2=r^2$

where the center is situated at the point whose coordinates are (h, k) and the radius is given by r.

✤ **general form of the equation of a circle**
The general form of the equation for a circle usually looks something like

$x^2+y^2+Ax+By+C=0$. When presented with the general form, you can no

longer find the center coordinates or the radius by inspection.

Review the method Completing the Square; you'll need it in order to coax the general form of the equation for a circle into the standard form.

Here's an example:

$$x^2 + y^2 - 8x + 12y - 48 = 0$$

$$x^2 - 8x \qquad + y^2 + 12y \qquad = 48 \qquad \left[\left(\frac{-8}{2}\right)^2 = 16\right] \left[\left(\frac{12}{2}\right)^2 = 36\right]$$

$$x^2 - 8x + 16 + y^2 + 12y + 36 = 48 + 16 + 36$$

$$(x-4)^2 + (y+6)^2 = 100$$

Can you pick out the center and radius now?

Equation of a Circle: Test Prep	
Find the radius and center of the circle described by this equation: $x^2 + y^2 + 6x + 8y - 11 = 0$	$x^2 + y^2 + 6x + 8y - 11 = 0$ $(x^2 + 6x \quad) + (y^2 + 8y \quad) = 11$ $(x^2 + 6x + 9) + (y^2 + 8y + 16) = 11 + 9 + 16$ $(x+3)^2 + (y+4)^2 = 36$ $(h,k) = (-3,-4)$ $r = \sqrt{36} = 6$
Try Them	
Identify the center and radius of the circle described by this equation: $(x-2)^2 + (y+1)^2 = 50$	$(h,k) = (2,-1)$ $r = \sqrt{50} = 5\sqrt{2}$
Sketch the graph of this equation: $x^2 + y^2 - 6x + 2y - 15 = 0$	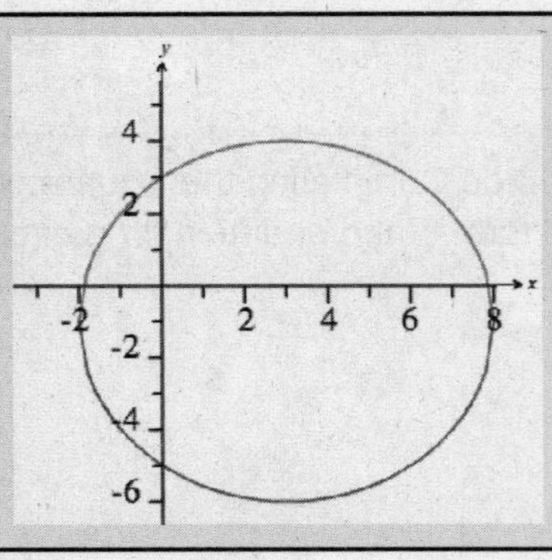

2.2: Introduction to Functions

Functions are a cornerstone not only of algebra but of calculus and beyond. Work on a good foundation now. It'll pay off.

Relations

So the ordered pair is a clear way to relate two coordinates, and we're in the 2-dimensional coordinate system now, so we'll start investigating what ordered pairs mean in terms of what a relation is.

✤ **relation**
Simply put, any collection or set of ordered pairs forms a relationship. It specifies, if you will, how the x's are related to the y's.

For example:
$\{(0,3),\ (2,-1),\ (-5,3),\ (0,-4.79),\ (3,0.25),\ (2,-2),\ (9,9)\}$,
or another less implicit example:
$\{(x,y) \mid \quad y = -2x + 3\}$

Remember that an ordered pair carries with it the idea that you put an x into a relation "machine" and the machine spits out the y that goes with it.

Functions

Functions are special kinds of relations. We're still putting a coordinate in (x) and getting a second coordinate out (y).

[Try diligently to keep in mind that x and y are by no means the only letters we can use for these variables, so you'll see them frequently referred to as the first and second coordinate ... and also as any number of other letter-variable-stand-ins. This requires some abstract thinking, but if you work at it, it will help your algebra (to say the least).]

✤ **function**

A function is actually just a special sort of relation -- a subset of relations, if you will. A function is also a set of ordered pairs but with this important qualification: that no two ordered pairs have the same first coordinate while having different second coordinates.

Function

A function is a set of ordered pairs in which no two ordered pairs have the same first coordinate and different seconds coordinates.

(page 167 of your textbook)

If you read this "math-ese" carefully, you might notice that it would be okay to have the two identical ordered-pairs, (0,3) and (0,3) in a function, but, honestly, what does it add to a set to repeat a member?

The important point in this definition is that you cannot have ordered pairs like (4, 2) and (4, -2) in a function (though it's perfectly okay for them to be in a relation). If two ordered pairs have the same first coordinate, then their second coordinate mustn't be different; that is, if they do *not* have the same first coordinate, everything is okay, and if they do have the same first coordinate, they better be identical in first and second coordinate (and, shrug, then you simply have repetition of a point in your set).

★ **For Fun**

Keep this seemingly convoluted definition in mind for when we look at graphs of functions. Begin to consider what this restriction might mean graphically.

✤ **domain**

All the first coordinates (usually x's) make up the domain. Frequently when asked about the domain of a function, we are really trying to get at what first coordinates will result in *real* second coordinates coming out the other end of the function-machine.

✤ **range**

And those coordinates that come out of the function machine: they are the range, the output of the function.

★ **For Fun**

Can you come up with a definition of a function that doesn't even involve numbers, real or otherwise?

✤ **independent variable**
Variables that correspond to values in the domain are called independent variables.

✤ **dependent variable**
Variables that correspond to values in the range are dependent variables.

★ **For Fun**
Try to imagine why the coordinates are called *dependent* and *independent* like this. Keep it in mind for graphing time.

Definition of a Function: Test Prep	
Determine whether the equation states y as a function of x: $x+(y-2)^2=3$	$x+(y-2)^2=3$ $(y-2)^2=3-x$ $y-2=\pm\sqrt{3-x}$ $y=2\pm\sqrt{3-x}$ Two possible values of y for one x-value means this is not a function.
Try Them	
Determine whether the equation states y as a function of x: $x+4y=14$	yes
Determine whether the equation states y as a function of x: $x^2+y=-1$	yes

Function Notation

Functions, like variables, generally are named with letters; the most common is *f*. *f* would be the name of the function, and it does nothing until you give it something to work on: $f(x)$. Since the x is what you put into the *f*-machine, the x-values are members of the domain of *f*. And the values that come out of $f(x)$, namely $f(x)$, make up the range of *f*. And y often stands in for $f(x)$, a la $y = f(x)$.

As soon as you see someone call an expression by $f(x)$ or $g(x)$ or some other similar name, one thing you know is that it is a function. Not only a relation.

✤ **piecewise-defined functions**
We've already met a piecewise-defined function -- the absolute value function.

$$|x| = \begin{cases} x, & x \geq 0 \\ -x, & x < 0 \end{cases}$$

A piecewise-defined function is a function whose definition is broken up into "pieces, " usually based on different parts of the function's domain.

★ **For Fun**
Can you *see* the domain in the definition above? Can you see (and name) the *x*-value where one piece of the definition changes into the other?

Evaluate a Function: Test Prep	
Given $Q(x) = -3$, find a. $Q(0)$ b. $Q\left(\frac{1}{4}\right)$ c. $Q(2)$	a. $Q(0) = -3$ b. $Q\left(\frac{1}{4}\right) = -3$ c. $Q(2) = -3$ Note that this function is -3 all the time, regardless of the value of *x*. That's a *constant* function!
Try Them	
Given $T(x) = \frac{1}{5}x - 4$, find a. $T(0)$ b. $T(5)$ c. $T(-5)$	a. $T(0) = -4$ b. $T(5) = -3$ c. $T(-5) = -5$

Evaluate a Function: Test Prep

Profit is given by the equation $P = px$, where p is the price and x is the number of units sold. Find the profit when p is \$63.50 and x is 226.	\$14,351

Piecewise-defined Functions: Test Prep

Given $f(x) = \begin{cases} x^2+1 & x \geq 1 \\ x & 0 \leq x < 1 \\ -x^2 & x < 0 \end{cases}$ evaluate a. $f(0)$ b. $f(10)$ c. $f(-5)$	a. $f(0)=0$ (choose the middle definition of the function because $x = 0$) b. $f(10)=10^2+1=101$ (choose the first definition of the function because $x > 1$) c. $f(-5)=-(-5)^2-25$ (choose the last definition of the function because $x < 0$)
Try Them	
Use $\lvert x \rvert = \begin{cases} x & x \geq 0 \\ -x & x < 0 \end{cases}$ to evaluate $\lvert 5 \rvert$ and $\lvert -5 \rvert$.	5, 5
Given $f(x) = \begin{cases} 3x-1 & x \geq 0 \\ 3-x^2 & x < 0 \end{cases}$ find a. $f(0)$ b. $f(-2)$	a. $f(0)=-1$ b. $f(-2)=-1$

★ **For Fun**
Try graphing, by plotting points, any of these piecewise-defined functions.

Domain and Range of a Function: Test Prep

Find the domain of $f(x) = \sqrt{3-x}$	$3 - x \geq 0$ $-x \geq -3$ $x \leq 3$
Try Them	
Find the domain of $f(x) = \dfrac{x}{x^2 - 1}$	$x \neq 1, -1$
Given the graph of *f* below, find the domain and range of *f*. 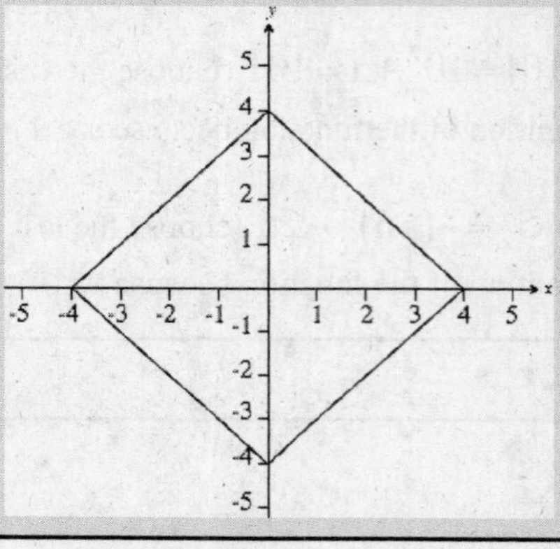	domain: $[-4,4]$ range: $[-4,4]$

Graph of Functions

We already know that functions are nothing more than sets of ordered pairs -- mind you, the set might very well be infinite (and therefore hard to write in a set notation). But what are these ordered pairs?

You know when you evaluate a function that you replace the variable with the value given:

$$g(x) = 2x^2 - x$$

$$g(-2) = 2(-2)^2 - (-2) = 8 + 2 = 10$$

In this example, we have constructed the ordered pair (-2, 10).

More generally, given a function $f(x)$ and an input value *a*, then we get the ordered pair $(a,\ f(a))$.

And now that we have ordered pairs, we can plot them on our Cartesian coordinate system. At some point, we may then connect the dots so that we have a *picture* of the entire solution set of the function. (Before you connect the dots, though, be confident that you do have a very good idea of what the rest of the solution set behaves -- what picture it makes.)

✤ **graph of a function**
The graph of a function is a plot of all the ordered pairs that belong to that function. For any x in the domain of f, the ordered pairs look like (x, $f(x)$). Now we have a real motivation for graphing functions. The picture will be much easier to get information out of than a description in, for instance, set builder notation, which might look something like ... $\{(x,\ f(x)) \mid \ x \in domain(f)\}$ or some other such nearly meaningless jargon. A graph is worth a thousand set notations!

✤ **increasing**
As your eye follows the graph on the page from left to right, if the curve of the graph is going up, then the graph (and by extension the function) is increasing in that region of the domain.

Sometimes a graph of a function will have arrowheads at the ends of the curve that depicts the function, like this:

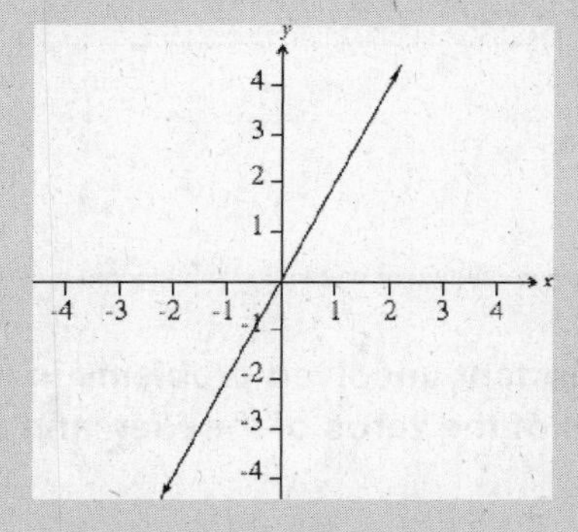

It's super important that you are still able to see that this graph is increasing as you look at it from left to right ... even though the left-hand end of the graph has an arrow pointing down which sometimes makes your eye see it as *decreasing*, but it's not. Do not let your eye fool your mind.

✤ **decreasing**
Similarly, if the graph is going down as your eye follows its curve from left to right, then the graph/function is decreasing in that part of the domain.

✤ **constant**
If the graph is flat, then the function is not changing -- neither increasing nor decreasing -- and it's constant. A constant function looks like, well, a constant. For instance: $f(x) = 3$. (If you're looking at a piecewise-defined function, the function maybe be constant for only part of its domain. Can you read the domain information right out of a piecewise-defined function's description?)

The Zero of a Function

A zero of a function is a value in the domain of the function that results in an output of zero. You'll be working with zeros a lot in the next chapter and beyond.

Meanwhile, (c, 0) would be the point on a graph of a function where $f(c) = 0$. As long as c is a real number that point will be an intercept. Which axis? Check back under in section 1 of this chapter under *Intercepts* if you're not sure.

One of the most important unsolved problems in mathematics concerns the location of the zeros of the Riemann zeta function.

Graph a Function: Test Prep

Use a T-table to graph
$y = 2 - x$

x	y
-1	3
0	2
1	1

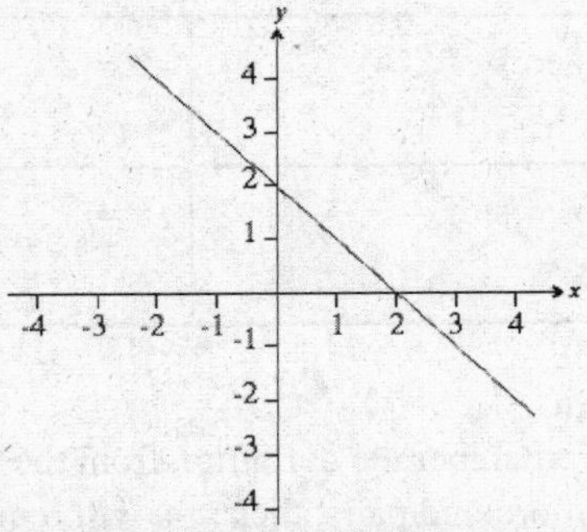

Try Them

Use a T-table to graph
$y = \sqrt{x+2}$

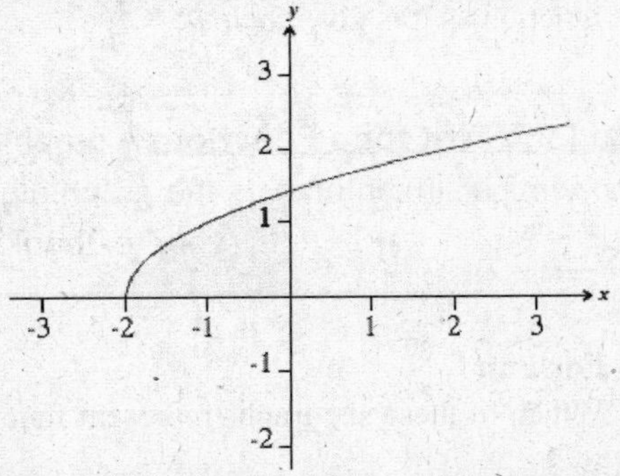

Use a T-table to graph
$y = -|x| + 3$

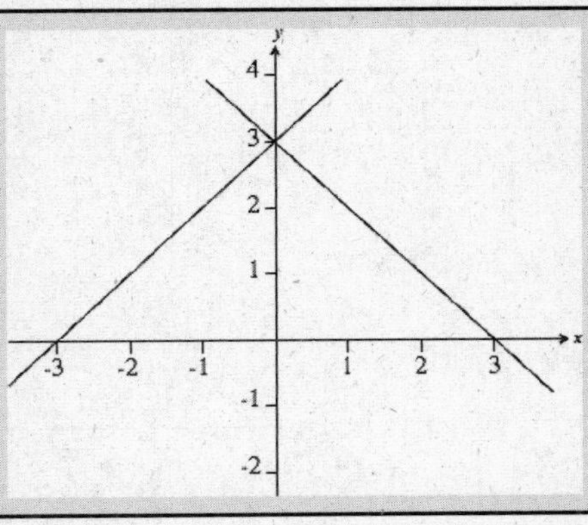

Zero of a Function: Test Prep	
Find all the real zeros of $f(x)=x^2-144$	$0=x^2-144$ $0=(x-12)(x+12)$ $x=12,-12$
Try Them	
Find the zeros of $f(x)=3x-6$	$x=2$
Find the zeros of $y=\|x-2\|-5$	$x=7,-3$

★ **For Fun**
If you remember the definition from the beginning of this chapter for a function in terms of ordered pairs, then you will remember that a set of ordered pairs is not a function if any two ordered pairs match in the first coordinate while their seconds coordinates differ. See if you can make that idea sync up with the visual test for a function as it is given below:

Vertical Line Test for a Function: A graph is the graph of a function if and only if no vertical line intersects the graph at more than one point. (pg. 174 of your textbook)

★ **For Fun**
Which of these six graphs represent functions?

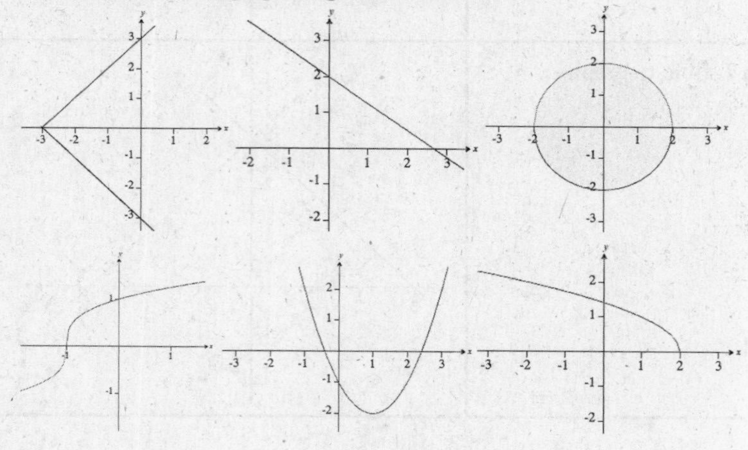

Greatest Integer Function (Floor Function)

A fun function to explore in order to strengthen your understanding of functional notation and graphs!

✤ **greatest integer function, floor function, step function**

Denoted variously by $\lfloor x \rfloor$, int(x), etc., the floor function puts out the greatest integer that is still less than or equal to x. For instance,

$$\left\lfloor \frac{1}{2} \right\rfloor = 0$$

$$\lfloor -3.5 \rfloor = -4$$

$$\lfloor 2 \rfloor = 2$$

And a graph of a floor function generally looks something like this:

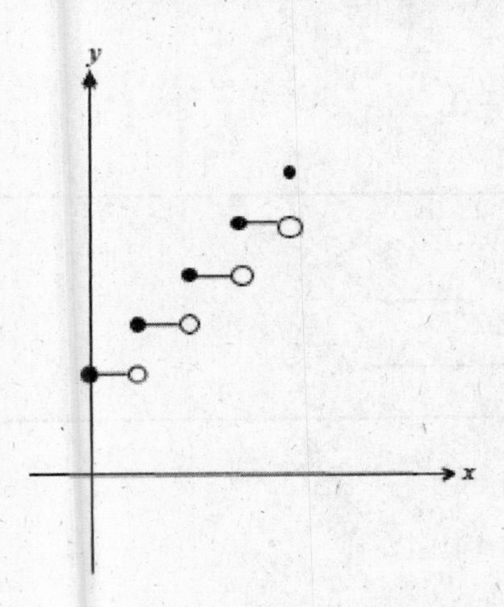

Can you make your understanding of the definition of a floor function match the graph above? Can you also see why this function is sometimes called the step function?

★ **For Fun**

When a computer *truncates* a number, it does something similar to this greatest integer function. Look up *truncation* online; see if truncation is the same as the floor function, or similar, or not at all?

Greatest Integer Function (Floor Function): Test Prep	
Evaluate a. int(3.89) b. $\lfloor -4.1 \rfloor$ c. $\lfloor -13 \rfloor$	a. int(3.89) = 3 b. $\lfloor -4.1 \rfloor = -5$ c. $\lfloor -13 \rfloor = -13$
Try Them	
Evaluate a. $\lfloor -5 \rfloor$ b. $\lfloor 10.2 \rfloor$ c. $\left\lfloor \frac{3}{4} \right\rfloor$	a. $\lfloor -5 \rfloor = -5$ b. $\lfloor 10.2 \rfloor = 10$ c. $\left\lfloor \frac{3}{4} \right\rfloor = 0$
Sketch $y = \frac{1}{2}\lfloor x \rfloor$	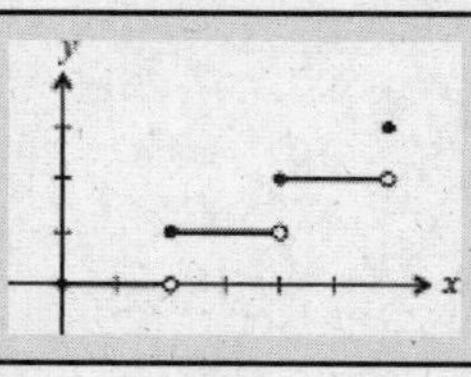

Applications of Functions

Now we can add the challenge of writing the *function* that satisfies the requirements of the application. Then solving the problem! Start thinking "functions!"

2.3: Linear Functions

Lines are easy. Plus we use them in so many ways -- the shortest distance between two points, huh? Lines are basic, basic, basic to algebra, and since they're pretty easy, now is the best of possible times to learn your way around a line from all directions. (If you're headed to calculus, you'll be covered up with the ideas of lines there, too, so might as well master lines now.)

Slopes of Lines

The slope of a line can tell us a lot. It has universal and real-world meaning -- everybody knows it's easier to coast downhill than pedal uphill, right?

✤ **slope**

Aside from saying that the slope of a line is the ratio of the change in y to the change in x, like so

$$m = \frac{change(y)}{change(x)} = \frac{\Delta y}{\Delta x} = \frac{y_2 - y_1}{x_2 - x_1}$$

in algebra, slope also means all the things it means in English. The algebra is arranged so that the number, m, tells you 1) direction (are you going up or down?) and 2) steepness (shallow or sharp?). The graph of a line is increasing from left to right if the slope is positive, and decreasing if it's negative. Steepness is harder to quantify and is really only a comparative measure. In general, a slope greater in absolute value than 1 will be getting pretty steep. A slope less than 1 in absolute value will appear pretty shallow. Experience helps with this recognition.

★ **For Fun**

Consider: What would a slope of exactly one be like? How 'bout a slope of zero? What is the absolutely steepest line you could envision?

The equation for a **horizontal** line has the form: $y = b$

The equation for a **vertical** line has the form: $x = a$

where a and b are constants.

★ **For Fun**

Which of the two cases above is actually a function? (One is definitely not.)

Slope of a Line: Test Prep	
Find the slope of a line that passes through the points (-2, -4) and (2, 5).	$\frac{5-(-4)}{2-(-2)}=\frac{5+4}{2+2}=\frac{9}{4}$
Try Them	
Find the slope of y from the graph: 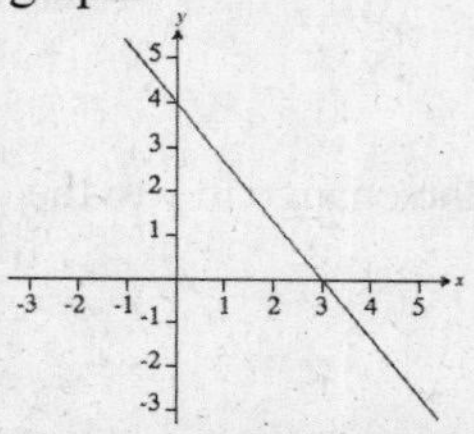	$m=-\frac{4}{3}$
Find the slope of a line that passes through the points (4, 3) and (10, 3).	$m=0$

Slope-Intercept Form

The slope-intercept form of an equation for a line is $y = mx + b$. The x and y are the variables in two dimensions. m is the slope, and the point (0, b) is the y-intercept point for the line. (Does that intercept sync up with what you know about y-intercepts already?)

This equation is surely familiar. It's used frequently in graphing lines. For example, the equation $y = -\frac{1}{2}x + 3$ is in slope-intercept form. To graph this equation, first plot the y-intercept: (0, 3). Then use the slope to plot one (or two?) more points:

$$-\frac{1}{2}=\frac{-1}{2}=\frac{change(y)}{change(x)}.$$

In other words, *from the point you already have* on the line, go -1 in the y-direction then +2 in the x-direction. Follow the process on this graph:

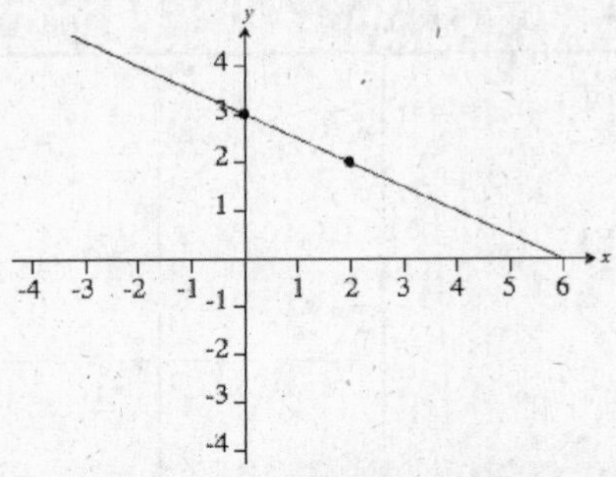

(Now is a good time to check our work. Did the line you just graphed actually turn out to have negative slope -- is it decreasing? Is it pretty shallow since the slope is less in absolute value than 1?)

Slope-Intercept Form of the Equation of a Line: Test Prep

Given a slope of $m = -\frac{2}{3}$ and a y-intercept of (0, 2), write the equation of the corresponding line.	$m = -\frac{2}{3}$ and b is 2, so the equation can be written in slope-intercept form: $y = -\frac{2}{3}x + 2$
Try Them	
Sketch the graph of $y = \frac{3}{4}x - 2$	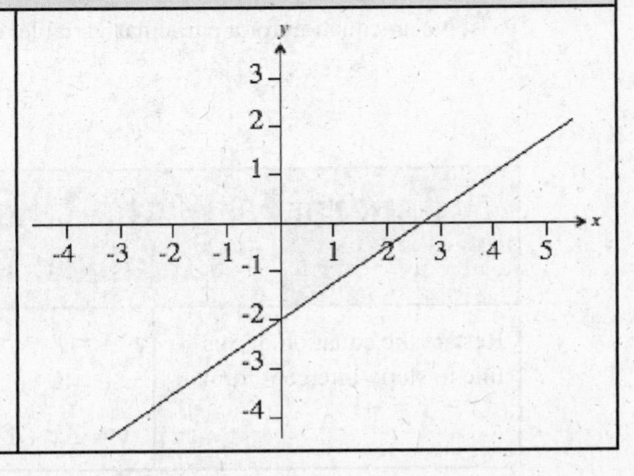

Slope-Intercept Form of the Equation of a Line: Test Prep

Sketch the graph of $3x + 4y = 12$	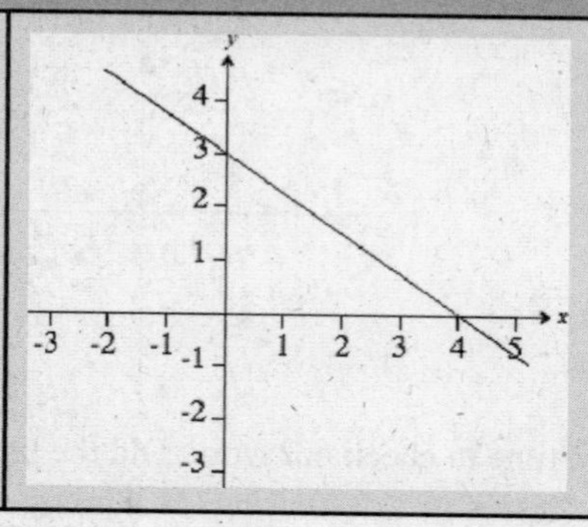

✤ **general form of a linear equation in two variables**

The general form for a linear equation is

$$Ax + By = C$$

where A and B and C and real constants and neither A nor B are zero. (Why not?)

In many cases, you might want to change the general form into the point-slope form for the purpose of graphing the line.

★ **For Fun**

Could you convert the general form into the slope-intercept form? What are the correspondences between the constants in the two forms? (If you forget how to solve an equation for a particular variable, in this case y, see Section 1.2 again.)

General Form of a Linear Equation in Two Variables: Test Prep

Restate the equation of this line in slope-intercept form: $3x - y = -1$	$3x - y = -1$ $-y = -3x - 1$ $y = 3x + 1$

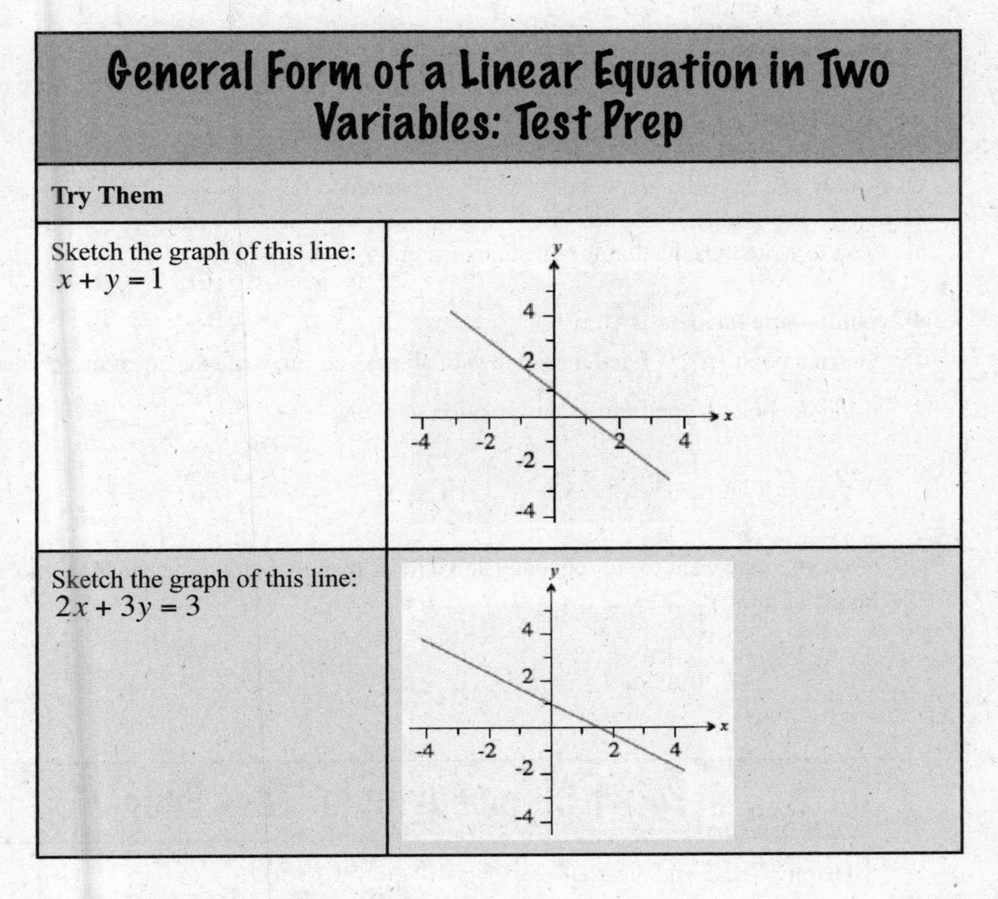

General Form of a Linear Equation in Two Variables: Test Prep

Try Them

Sketch the graph of this line:
$x + y = 1$

Sketch the graph of this line:
$2x + 3y = 3$

✓ **Questions to Ask Your Teacher**
There is another method that is sometimes popular for graphing a line whose equation is in standard form. This method completely skips solving the equation for y in order to put it into slope-intercept form and is sometimes called "the intercept method." If you're interested, you could ask your teacher to show you that method. (It's also illustrated in the alternative method to Example 3 in your text, page 190.)

Finding the Equation of a Line

Given an equation in either the general form or the slope-intercept form for a line in 2 dimensions, you should be able to graph it now. But what happens if you just have *information* about the line instead of its equation? What if someone tells you a point and the slope? Well, graphing that would be easy, right?

★ **For Fun**
Prove it. Graph the line that passes through the point (-1, 3) and has a slope of $m = 2$.

Okay, now you have the graph, but what if you wanted to tell someone on the phone, as succinctly as possible, about this line? Or even more realistically, what if your instructor asks you to write the equation for this line on a quiz?

✤ **point-slope form**
Given a point (x_1, y_1) and the slope *m* of a line, you can write the equation for the line using the point-slope form of the equation of a line:

$$y - y_1 = m(x - x_1)$$

(And you may want to subsequently solve for *y*, thereby putting the equation into slope-intercept form.)

Point-Slope Formula: Test Prep	
Given $m = 3.5$ and the point (-1, 2.5), write the equation for the line with this slope passing through this point, in slope-intercept form.	$y - 2.5 = 3.5(x + 1)$ $y = 3.5x + 3.5 + 2.5$ $y = 3.5x + 6$ *or* $y = \frac{7}{2}x + 6$
Try Them	
Given a line that passes through the points (-1, 4) and (-2, -3), write the equation for this line in slope-intercept form.	$y = 7x + 11$

Point-Slope Formula: Test Prep	
Given a line that passes through the points $\left(\frac{1}{3},\frac{2}{5}\right)$ and $\left(-\frac{1}{3},\frac{1}{5}\right)$, write the equation for this line in slope-intercept form.	$y=\frac{3}{10}x+\frac{3}{10}$.

In two dimensions, you need two pieces of information about a line in order to graph it or to write its equation. Are two points sufficient to graph the line? Sure -- plot two points and draw the line that passes through them. But how would you arrive at an equation for a line given only two points on the line?

1. Find the slope using the slope formula ("Slope of Lines" above).
2. Use that slope and *one* of the points in the point-slope form.
3. Voilà!

Parallel and Perpendicular Lines

And now: a brief discussion of a couple of the special ways lines can be related to each other. Do you already have a clear idea in English what parallel (as in "parallel lines never meet") and perpendicular (meeting at ninety degrees) mean?

★ **For Fun**
The word *perpendicular* has a very interesting derivation. You wanna look it up? Might look up *parallel* too, while you're at it.

✤ **parallel**
Two lines are parallel if their slopes are equal.

★ **For Fun**
Hmm, could a line be parallel to itself?

✤ **perpendicular**
If their slopes are negative reciprocals of each other, then two lines are perpendicular.

★ **For Fun**
Fill in the blanks:

m	*negative reciprocal*
$\frac{1}{2}$	-2
3	________
$-\frac{5}{3}$	________
-1	________

Parallel Lines: Test Prep	
Pick the parallel lines: $y=2x+1$ $x=-\frac{1}{2}y+10$ $y=-\frac{1}{2}x+1$ $y-2x=14$ $4x-2y=5$	$y=2x+1$ $y-2x=14 \quad [y=2x+14]$ $4x-2y=5 \quad [y=2x-\frac{5}{2}]$ They all have a slope of 2.
Try Them	
Write an equation of a line parallel to $y=-2x+3$ that passes through the point (1, 3).	$y=-2x+5$
Write an example of an equation of a line parallel to $3x-y=4$	example: $y=3x+1$

Perpendicular Lines: Test Prep	
Find the equation of a line perpendicular to the line $y = -\frac{1}{3}x + 1$ and passing through the point (3, 0).	$y - 0 = 3(x - 3)$ $y = 3x - 9$
Try Them	
Find a line perpendicular to $y = \frac{2}{5}x$	example: $y = -\frac{5}{2}x + 114$
Sketch a graph of the line perpendicular to $y = -x + 5$ and passing through the origin.	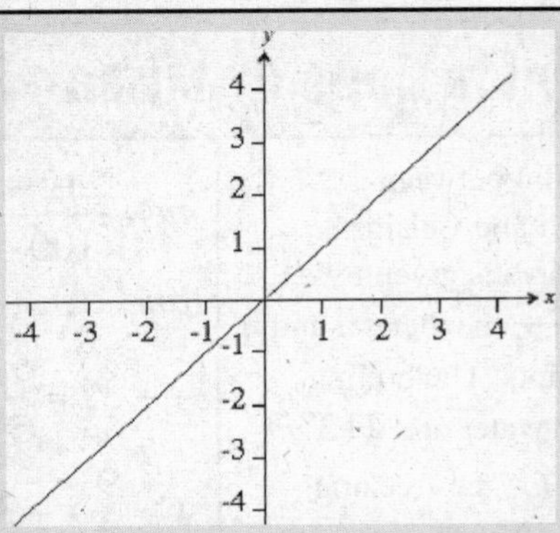

Applications of Linear Functions

As I said at the beginning of this section, we use linear equations to describe lots of systems. Now we'll see a few of these occasions and use these equations to answer questions and draw graphs illustrating the system at hand.

- **cost function**
  A function that models the cost of producing and selling x units: $C(x)$.

- **revenue function**
  A function that describes the amount of money brought in by selling x units: $R(x) = xp$, where p is the price of the unit.

✤ **profit function**
A function that models the profit from selling *x* units. This is usually the difference between the cost $C(x)$ and the revenue $R(x)$: $R(x) - C(x) = P(x)$. (That makes sense, doesn't it?)

✤ **break-even point**
The break-even point is exactly where cost and revenue are equal: $C(x) = R(x)$. You can set two linear equations equal to each other and use this equation to solve or graph the system.

★ **For Fun**
What is the profit at the break-even point?

Applications of Linear Functions: Test Prep

The relation between Fahrenheit and Celsius temperatures is given by $F = mC + n$ where *m* and *n* are constants. The boiling points for water are $212°F$ and $100°C$; the freezing points are $32°F$ and $0°C$. Find *m* and *n*.	$m = \frac{212-32}{100-0} = \frac{9}{5}$ *and* $y - 32 = \frac{9}{5}(x-0)$ $y = \frac{9}{5}x + 32$ *so* $m = \frac{9}{5}$ $n = 32°$
Try Them	
Given a supply $S = \frac{1}{3}p$ and demand $D = -p + 200$, find the value of *p* at the break-even point.	$\frac{1}{3}p = -p + 200$ $\frac{4}{3}p = 200$ $p = 150$

Applications of Linear Functions: Test Prep

<table>
<tr>
<td>Consider this table relating age and number of prescriptions.

age	*prescriptions*
30	1
40	3
50	6
55	8

Write a linear equation based on the points at 40 and 50 years old. Use that equation to estimate the number of prescriptions for age 60.</td>
<td>$y = \frac{3}{10}x - 9$
9 *prescriptions*</td>
</tr>
</table>

NOW IS THE TIME TO TRY THE MID-CHAPTER QUIZ, PAGE 200, IN YOUR TEXT.

★ **For Fun**

Here's a piecewise-defined linear function. You might try graphing this! Good luck.

$$f(x) = \begin{cases} -x - 3 & \text{if} \quad x \le -3 \\ x + 3 & \text{if} \quad -3 < x \le 0 \\ -2x + 3 & \text{if} \quad 0 \le x < 3 \\ x - 3 & \text{if} \quad x \ge 3 \end{cases}$$

2.4: Quadratic Functions

The graphs of quadratic functions are parabolas (a symmetrical open plane curve formed by the intersection of a cone with a plane parallel to its side.... Okay, it's a big U-shaped curve!). The parabola is next degree up from linear and the next more complicated polynomial function. Quadratics functions, whose graphs are parabolas, also come with lots of applications. Here we go.

✤ **quadratic function**

A quadratic function is a *function* of degree 2:

$f(x) = ax^2 + bx + c$ where a is not zero.

★ **For Fun**

Beware, all that is degree two is not a function. Is $y^2 = x + 1$ a quadratic function? You may have to go right back to the definition of a function to find out that, although it *is* quadratic, it's not a function.

✤ **vertex of a parabola**

The vertex of a parabola is the *turning point* of the parabola. It's the lowest point on a parabola that opens up and the highest on one that opens down:

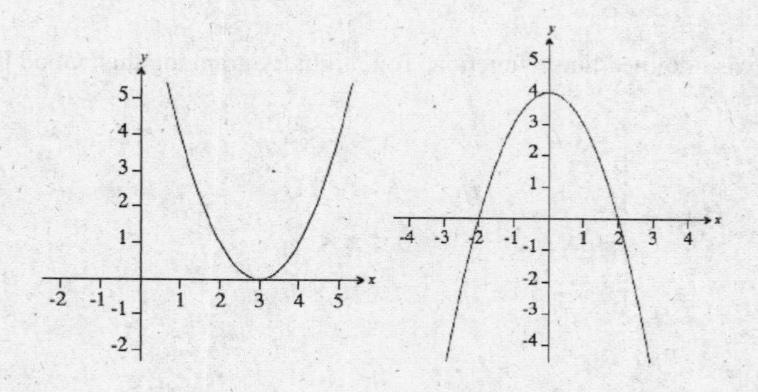

✤ **symmetric with respect to a line**
Symmetry with respect to a line means that the item is made up of exactly similar parts facing each other from opposite sides of an axis; the line is that axis of symmetry.

✤ **axis of symmetry**
The line that divides a symmetric object into its two similar parts is its axis of symmetry. (See above, okay?)

★ **For Fun**
Can you describe the axes of symmetry of the above two parabolas? Each have an axis that is a vertical line, so answer in both cases in the form of an equation for a vertical line. (Vertical lines were discussed in the last subsection.)

✤ **standard form of a quadratic function**
The standard form of a quadratic function is, as follows:

$f(x) = a(x-h)^2 + k$, where a is not zero.

This form is comparable to the standard form of an equation for a circle, and likewise it can be mined for information about the graph of the parabola. (Get ready to complete the square!)

1. Given a quadratic function in standard form, the vertex of the parabola is situated at the point (h, k).

2. If a is positive, then the parabola opens up, and if a is negative, the parabola opens down.

3. The axis of symmetry of the parabola is at the vertical line $x = h$.

Do you think that's enough information to be able to graph a good parabola from an equation in standard form?

By the way, don't forget you can still plot points, too. Don't ever forget you have that skill. It's an excellent way to make sure that your parabola has the right shape.

Quadratic Function: Test Prep	
Find the standard form of this quadratic function: $f(x)=x^2-4x-1$	$f(x)=x^2-4x-1$ $=x^2-4x+4-4-1$ $=(x-2)^2-5$ $f(x)=(x-2)^2-5$
Try Them	
Identify the constants *a*, *b*, and *c* in this quadratic function: $f(x)=-x^2+\frac{x}{3}-12$	$a=-1$ $b=\frac{1}{3}$ $c=-12$
Sketch the parabola: $f(x)=-(x-2)^2+6$	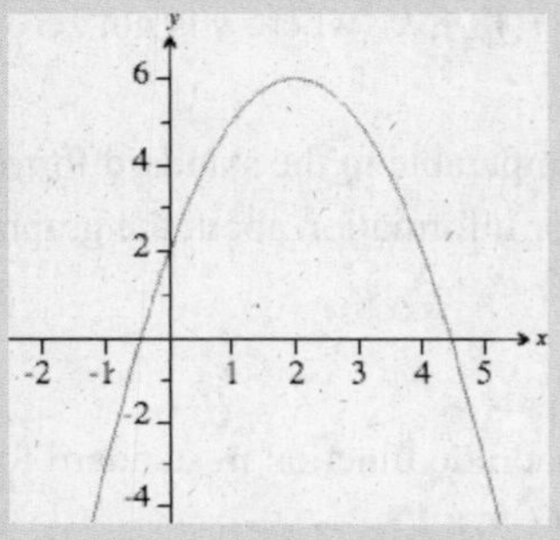

Vertex of a Parabola

If your quadratic function is in standard form, $f(x)=a(x-h)^2+k$, then it's easy to pick out the coordinates of the vertex, namely (*h*, *k*). But even if the quadratic function is given in the form $f(x)=ax^2+bx+c$, you can find the vertex coordinates with those *a*, *b*, and *c*. The vertex is situated at $\left(-\frac{b}{2a}, \frac{4ac-b^2}{4a}\right)$, or $\left(-\frac{b}{2a}, f\left(-\frac{b}{2a}\right)\right)$.

From the function, the sign of *a* will still tell you if the parabola opens up or down, and by extension, you can figure out if the vertex is the highest point (maximum) or lowest point (minimum) of the parabola.

That may be enough, but if you want to graph the parabola, it's generally much easier if you have the function in the standard form: $f(x) = a(x-h)^2 + k$. So here's an example of using completing the square to rewrite a quadratic function into the standard form. This time, instead of adding the same thing to both sides of the equation, I'm going to *add zero* to one side. (I can add zero any time I want and not change the value of an expression, right? Zero, if you recall, is the additive identity.)

$$f(x) = x^2 - 10x + 22 \qquad \left[\left(\frac{-10}{2}\right)^2 = 25\right]$$

$$= x^2 - 10x + 25 - 25 + 22$$

$$= x^2 - 10x + 25 - 3$$

$$= (x-5)^2 - 3$$

$$so \quad f(x) = (x-5)^2 - 3$$

And now you can see that the vertex is at (5, -3), etc.

★ **For Fun**

Could you go ahead and graph that parabola? How 'bout this: could you find the *x*- and *y*-intercepts, if they exist? ... before you graph it? ... and verify them from your graph? Can you tell what the range is? ... before you graph it?

Parabola: Test Prep	
Find the standard form of the equation of this parabola and and its vertex: $f(x) = -x^2 - 2x + 7$	$f(x) = -x^2 - 2x + 7$ $= -(x^2 + 2x + 1 - 1) + 7$ $= -(x^2 + 2x + 1) + 1 + 7$ $f(x) = -(x+1)^2 + 8$ $vertex \quad (-1,8)$

Parabola: Test Prep	
Try Them	
Find the vertex of this parabola using the vertex formula: $f(x) = -x^2 - 4$	(0, -4)
Sketch the parabola $y = -2x^2 + 4$	

Maximum and Minimum of a Quadratic Function

Think of a parabola; for example:

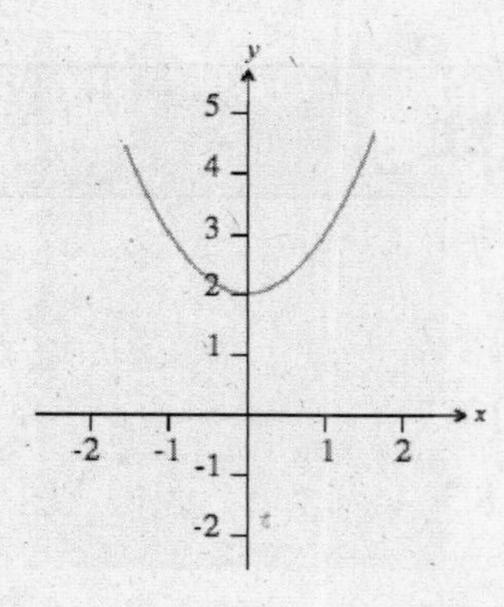

You can see from the graph that there are no coordinates on this curve with a y-value less than 2. In other words, since the vertex is at the point (0, 2) and the parabola opens up, the vertex is at the lowest point of the parabola. That means that the value 2 is the smallest value that this quadratic function will ever put out. If you think about it, that tells you everything you need to know about the *range* of this particular quadratic. Seems like its range is [2, ∞). What do you think?

✤ **minimum value**
From a quadratic function in standard form, the k is a minimum value of the function, or of its range, if the parameter a is positive.

✤ **maximum value**
Similarly, k is a maximum value if a is negative.

★ **For Fun**
What do you have if a is zero? (This is a trick question.)

Minimum or Maximum of a Quadratic Function: Test Prep

Find the range of $f(x)=-(x-2)^2-5$	vertex: (2, -5) opens DOWN therefore, range is $(-\infty,-5]$.
Try Them	
Find the range of $f(x)=x^2+7x+2$	$\left[-\frac{41}{4},\infty\right)$
Find the range of $f(x)=-2x^2-4x+15$	$(-\infty,17]$

There are lots of kinds of functions. So far we've looked at linear and quadratic functions. These are both algebraic functions. Here are some examples of functions that are not algebraic. Some you'll see later, most not!

- Hyperbolic functions
- Bessel functions
- Airy function
- Transcendental functions
- Theta function
- Hermite polynomials

Applications of Quadratic Functions

Areas (*square* inches, *square* centimeters, *square* light-years) and trajectories (baseballs, rockets, thrown snowballs) are natural systems to be described by quadratic functions (they travel in parabolas). There are others, too (business, engineering, etc.). Now you can try out your expertise with quadratic functions on some applications.

Applications of Quadratic Functions: Test Prep

Say the height h in feet of a grand slam baseball after t seconds is approximated by $h(t) = -16t^2 + 89t + 4$. When will the ball be caught be a fan 7 feet up in the stands, to the nearest tenth of a second?	$7 = -16t^2 + 89t + 4$ $0 = -16t^2 + 89t - 3$ $t = \frac{-89 \pm \sqrt{89^2 - 4 \cdot 16 \cdot 3}}{-32}$ approximately 5.5 seconds (disregarding the smaller solution because it will most likely be the second time the ball passes through 7 feet if its path is a parabola, right?)
Try Them	
Given two rafter of 7.5 feet and roof-frame height of 1 foot, what will be the width of the ceiling, to the nearest foot? 	15 feet
A dog breeder has 1200 meters of fence to enclose a rectangular pen with another fence dividing it in the middle as in the diagram: 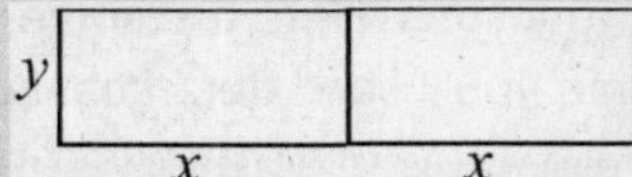 The breeder wishes to use all the fence and enclose the largest possible area. What must x and y be?	$x = 150\,ft$ $y = 200\,ft$

2.5: Properties of Graphs

Now that we are graphing quite a bit, we can try to get some basic understanding about how graphs work and how to manipulate them. Or, perhaps, learn how manipulating the function changes the graph in predictable ways. Knowing this will become a tremendous time-saver, and, it goes without saying, increasing your comprehension of the meaning of displaying 2-dimensional systems as graphs is going to really pay off.

Symmetry

We investigated the idea of symmetry with respect to a line in the last section. That had to do with parabolas, but of course parabolas are not the only things that exhibit symmetry. In this section, we'll look into some special symmetries with an eye toward developing an aid for graphing. (Remember the graph is a picture of the solution set of the equation.)

- **symmetric with respect to the y-axis**
  Symmetry with respect to (w.r.t.) the *y*-axis means that a graph includes all the similar points on both sides of the *y*-axis. In other words, if you folded the graph on the *y*-axis all points would coincide with their similar points on the other side of the axis.

  Without a graph, you can test the equation of a function to see if it's symmetrical w.r.t. the *y*-axis by replacing *x* with -*x* and finding, after simplification, that the equation remains unaffected. In symbols: $f(-x) = f(x)$. Can you see that mathematical expression says that if you replace *x* with -*x* it doesn't make a difference?

  Visually, here's an example of symmetry w.r.t. the y-axis:

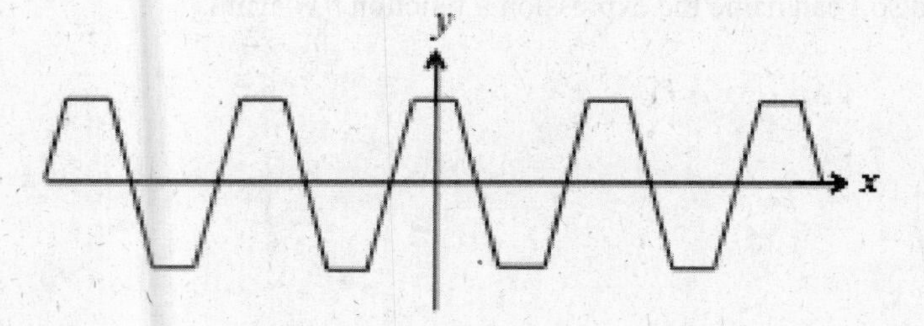

- **symmetric with respect to the *x*-axis**
  Symmetry w.r.t. the *x*-axis is analogous to the above, but, of course, you would be folding the graph on the *x*-axis.

  To test a function for this kind of symmetry, it's best to first rewrite every $f(x)$ with *y*; that is, $f(x) = y$. Then replace *y* with -*y*, simplify, and see if the equation is unaffected. In other words: $y = -y$. [Think about it: if a expression is symmetrical w.r.t. the *x*-axis, is it a function? If not, would I call it *f(x)* then?]

  An example:

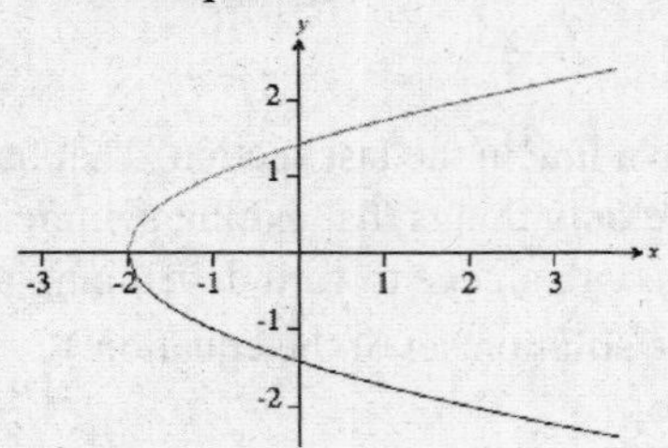

- **symmetric with respect to a point**
  Symmetry w.r.t. an arbitrary point starts to sound much more technical: To be symmetrical about a point *Q*, there must exist *for every point P* its similar point P' so that *Q* is the midpoint of the line from *P* to P'.

  One special point, which is common to test for symmetry about, is the origin (0, 0). In order to test an equation for symmetry about the origin, first you replace *both x* with *-x* and *y* with *-y*, simplify, and see if the equation remains unaffected. Symbolically: $-f(-x) = f(x)$. [Symmetry about the origin does not preclude function-hood (see graph below), and so I can name the expression a function *f(x)* again.]

  Example:

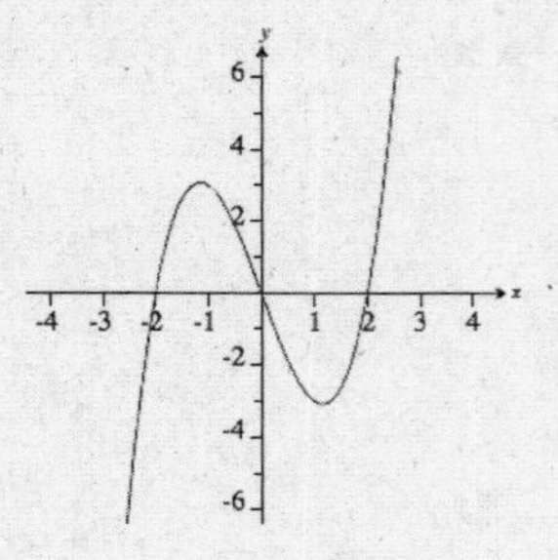

Symmetry of a Graph with Respect to ...: Test Prep	
Test the relation $xy = 2$ for symmetry w.r.t. ... *a.* x-axis *b.* y-axis *c.* origin	a. $x(-y) = 2$ $-xy = 2$ NO b. $(-x)y = 2$ $-xy = 2$ NO c. $(-x)(-y) = 2$ $xy = 2$ YES
Try Them	
Test the relation $x^2 + y^2 = 4$ for symmetry w.r.t. ... *a.* x-axis *b.* y-axis *c.* origin	a. YES b. YES c. YES
Test the relation $xy = \|x\|$ for symmetry w.r.t. ... *a.* x-axis *b.* y-axis *c.* origin	a. NO b. NO c. YES

Even and Odd Functions

Like integers, some functions are even, some are odd. Unlike integers, some functions are neither even nor odd. This does have meaning for the graph of a function, but it's also an idea that becomes more and more important as you progress through trigonometry and calculus.

✤ **even function**

Like a function that is symmetrical w.r.t. the *y*-axis, an even function exhibits the behavior that replacing *x* with -*x* leaves the function unchanged.

$f(-x) = f(x)$

✤ **odd function**

Odd functions are harder to recognize. They behave like this: $f(-x) = -f(x)$. If you replace *x* with -*x*, then you get the function's *opposite* after simplifying. For example:

$f(x) = x^3$

$f(-x) = (-x)^3$

$= -x^3$

$= -f(x)$

Did you catch that? It takes some practice to see that one sometimes.

Even and Odd Functions: Test Prep	
Determine if this function is even or odd or neither: $f(x) = 3x + 1$	$f(-x) = 3(-x) + 1 = -3x + 1 = -(3x - 1)$ $f(-x) \neq f(x)$ $f(-x) \neq -f(x)$ Neither.
Try Them	
Determine if this function is even or odd or neither: $f(x) = \lvert x \rvert - 3$	Even
Determine if this function is even or odd or neither: $f(x) = 2x^3 - 4x$	Odd

Translations of Graphs

Now we can learn how to move graphs around the 2-dimensional plane. No, better still we'll learn to recognize when the function is a basic graph we already know but that has been moved horizontally or vertically or both to a new situation on the plane.

Vertical Translations

Assume that *c* is a positive constant, then

1. If $y = f(x) + c$, then *y* is *f(x)* moved UP *c* units.

2. If $y = f(x) - c$, then *y* is *f(x)* moved DOWN *c* units.

For example:

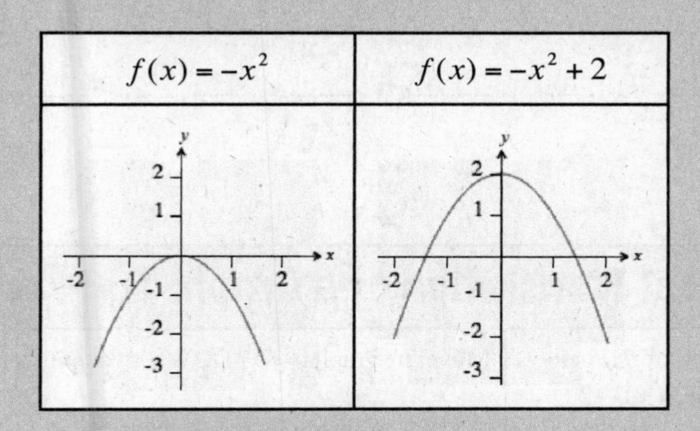

Horizontal Translations

Again assume that c is a positive constant, then

1. If $y = f(x+c)$, then y is $f(x)$ moved LEFT c units.

2. If $y = f(x-c)$, then y is $f(x)$ moved RIGHT c units.

For example:

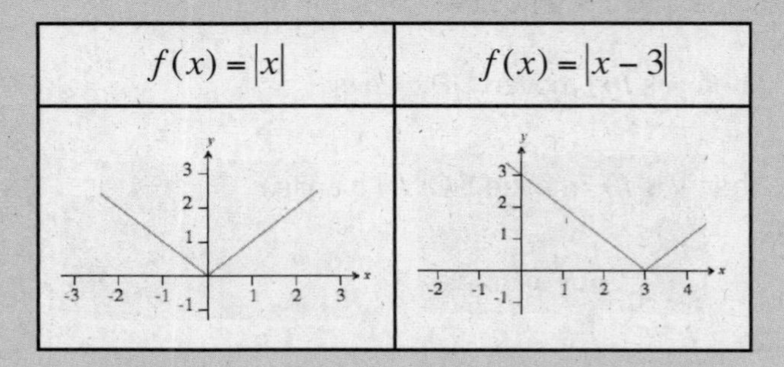

Vertical Translation of a Graph: Test Prep

Given the graph of $f(x)$ below, graph $g(x) = f(x) - 2$

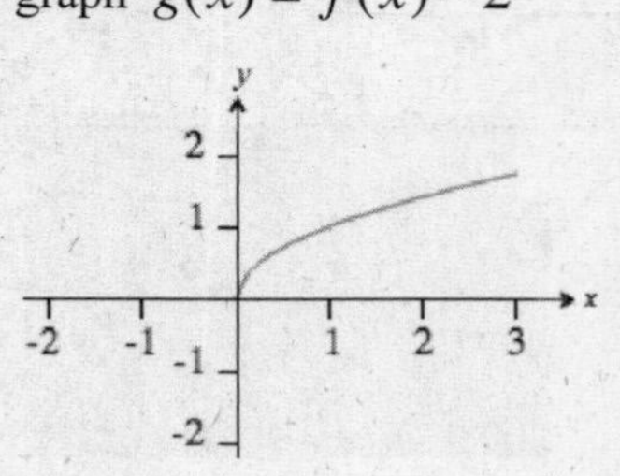

Move the graph of $f(x)$ DOWN two units:

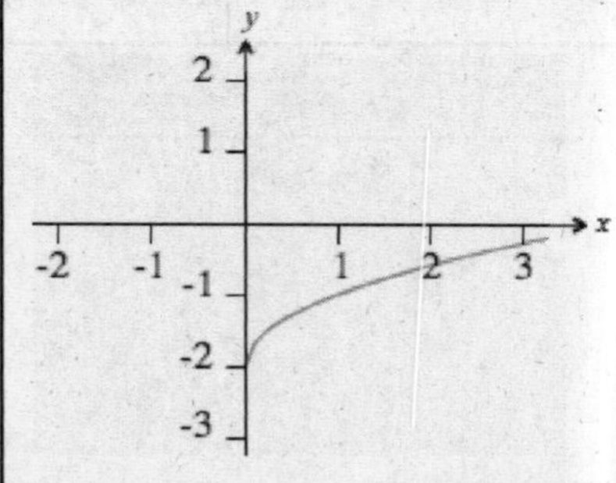

Vertical Translation of a Graph: Test Prep

Try Them

Given the graph of *f(x)* below, graph $f(x)+3$ 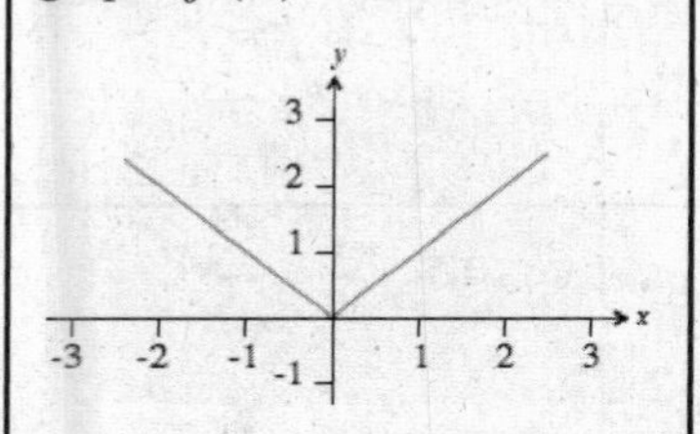	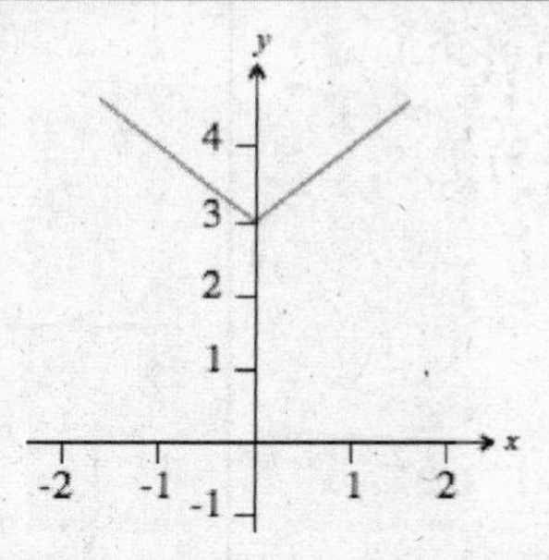
Given $f(x)=x^2$, describe this translation: $g(x)=x^2-10$	$g(x)=x^2-10$ is the graph of *f(x)* moved DOWN ten units.

Translation:

What does *translation* mean in English? Do any of its several possible meanings ring true for what we want to signify when we talk about translating graphs?

Horizontal Translation of a Graph: Test Prep

Given the graph of *f(x)* below, graph $g(x)=f(x+2)$. 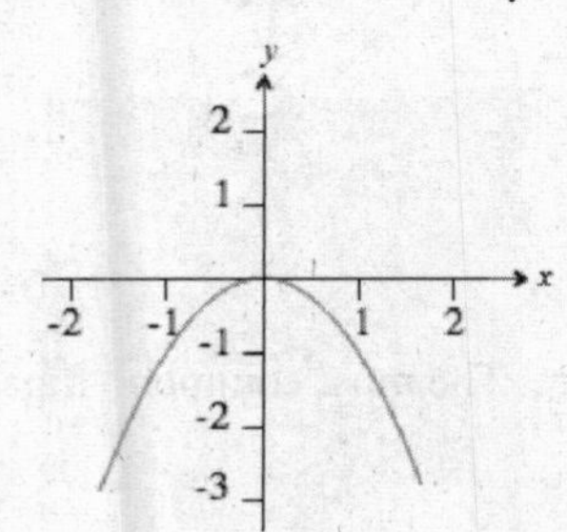	Move the graph of *f(x)* LEFT two units: 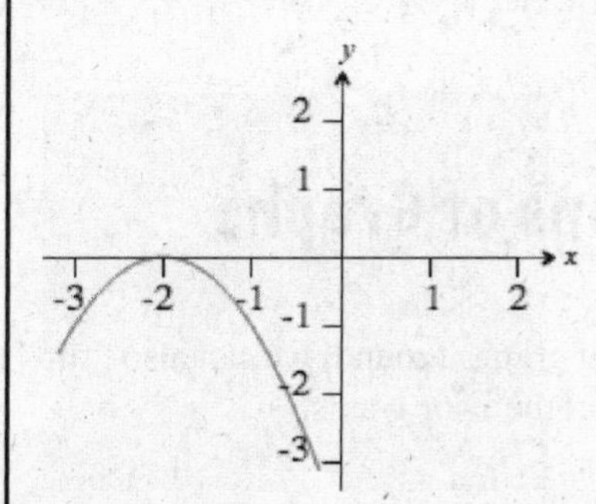

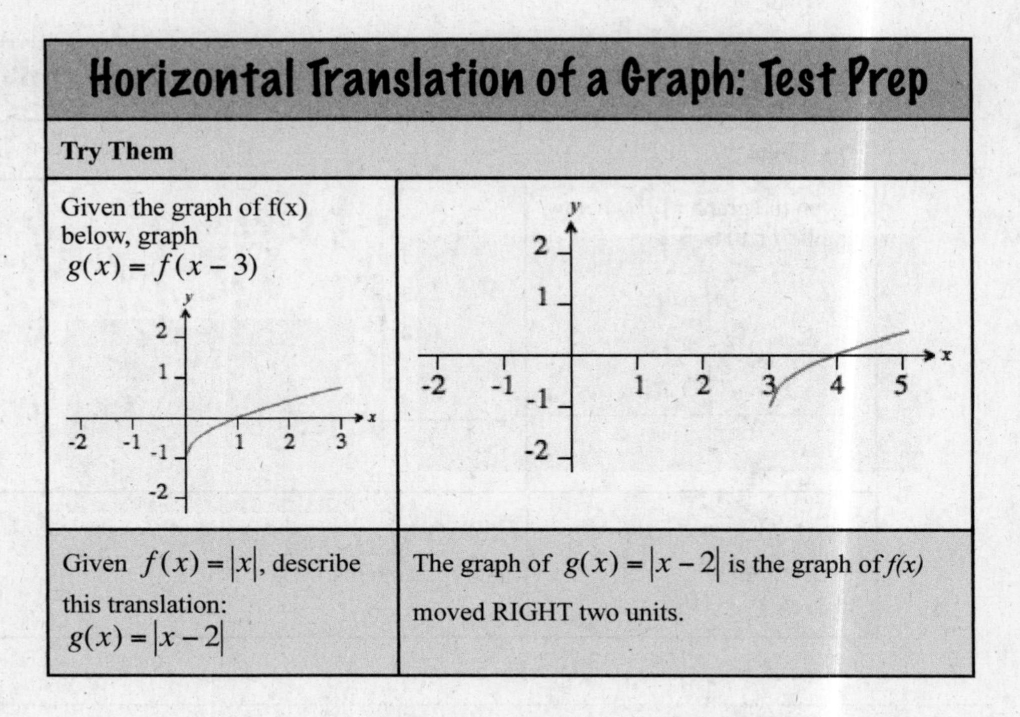

Horizontal Translation of a Graph: Test Prep

Try Them	
Given the graph of f(x) below, graph $g(x) = f(x-3)$	
Given $f(x) = \lvert x \rvert$, describe this translation: $g(x) = \lvert x-2 \rvert$	The graph of $g(x) = \lvert x-2 \rvert$ is the graph of *f(x)* moved RIGHT two units.

★ **For Fun**

Okay, now try doing both to one graph. If this is a graph of *f(x)*, sketch the graph of $y = f(x-2) + 3$

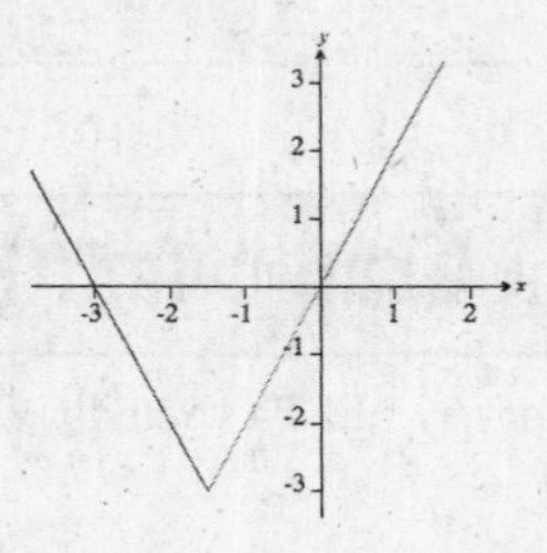

Reflections of Graphs

Besides moving graphs around, we can also "flip" them. The most common "flips" are reflections about the *x*- or *y*-axis.

<u>Reflections</u>

1. If $y = -f(x)$, then *y* is *f(x)* reflected about the *x*-axis.

For example:

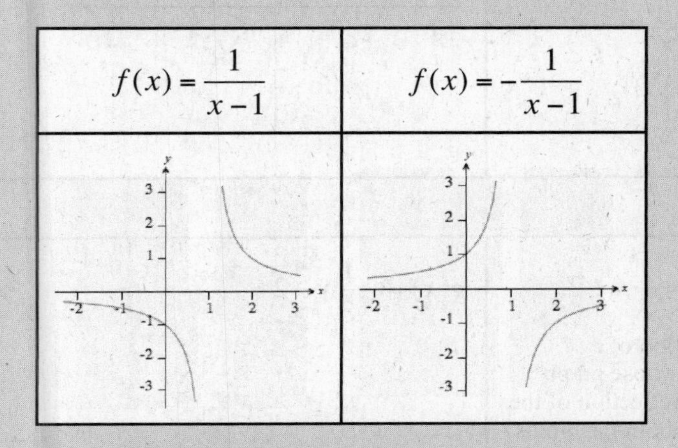

2. If $y = f(-x)$, then *y* is *f(x)* reflected about the *y*-axis.

For example:

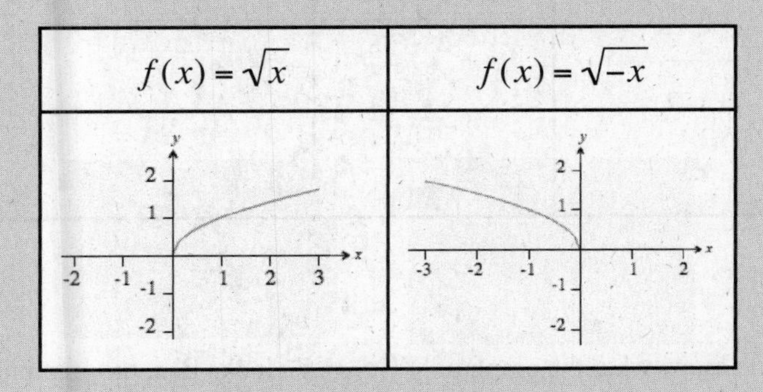

Reflections of a Graph: Test Prep

Given the graph of *f(x)* below, graph $g(x) = f(-x)$.	*g(x)* is a reflection of *f(x)* about the *y*-axis, thus:
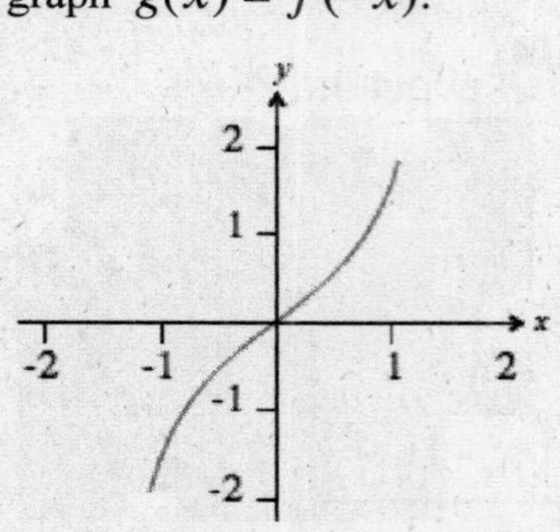	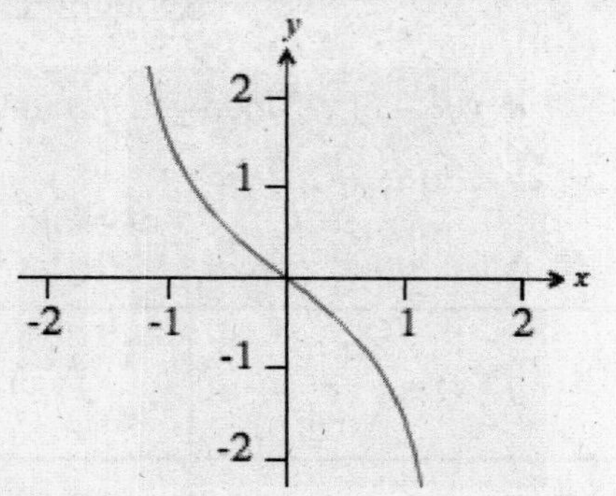
Try Them	
Given $f(x) = \frac{1}{3}x + 2$, write an equation of a function *g(x)* whose graph would be the reflection of the graph of *f(x)* about the *x*-axis.	$g(x) = -\frac{1}{3}x - 2$
Given the graph of *f(x)* below, graph $g(x) = -f(x)$. 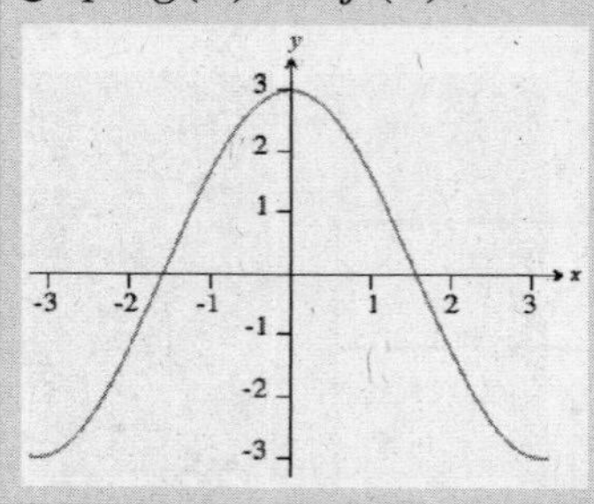	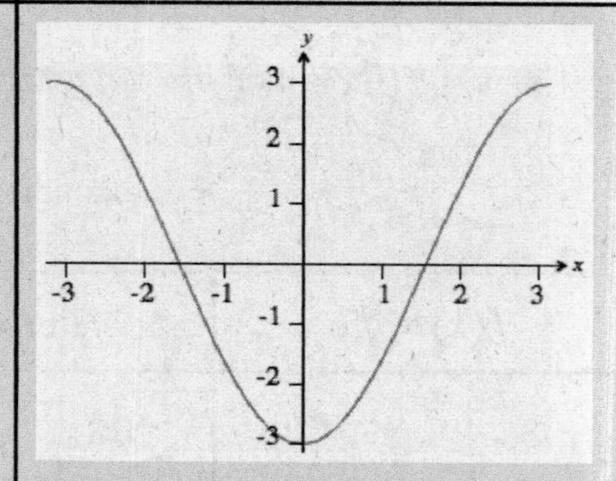

★ **For Fun**

Given that you know what the graph of $f(x) = x^2$ looks like, try graphing this combination of manipulations: $f(x) = -(x+3)^2$.

Compressing and Stretching of Graphs

Compression and stretching of graphs is more subtle. But you'll know it when you see it! Basically, what is involved here is a *multiplicative* factor working on the function, as opposed to adding and subtracting in translations. (However, it's somewhat different from the multiplication by -1 that causes reflections.)

Vertical Stretching/Compression

Let c be a positive constant, and if $y = c \cdot f(x)$ then

1. if $c > 1$, y is $f(x)$ stretched vertically away from the x-axis by a factor of c
2. if $0 < c < 1$, y is $f(x)$ compressed towards the x-axis by a factor of c

For example:

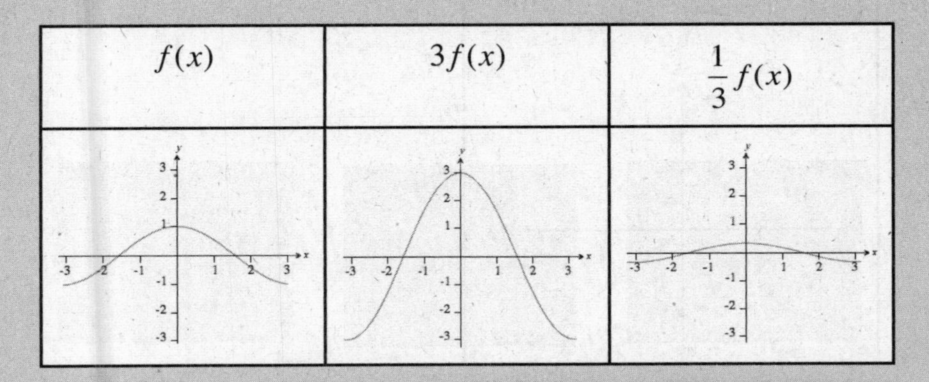

★ **For Fun**
Think about it: why is $c = 1$ not included in the above possibilities?

★ **For Fun**
What would happen if we allowed c to be negative? (Hint: see Reflections above.)

Vertical Stretching and Compressing of Graphs: Test Prep

Given the graph of *f(x)* below, graph $g(x) = 3f(x)$

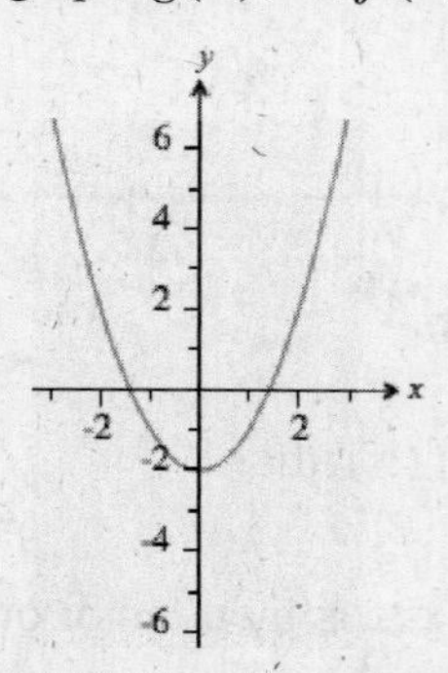

Every point will be 3 times bigger -- the graph will "rise" 3 times faster than *f(x)*. This is a vertical stretching.

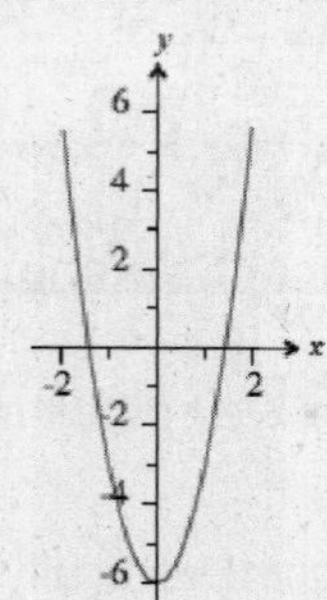

Try Them

Given the graph of *f(x)* below, graph $g(x) = \frac{1}{3}f(x)$.

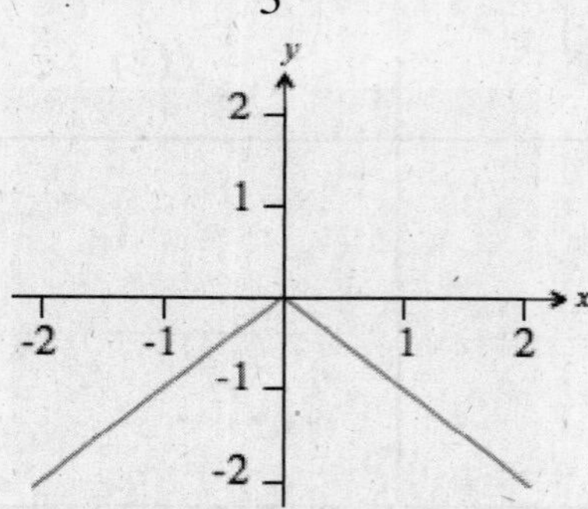

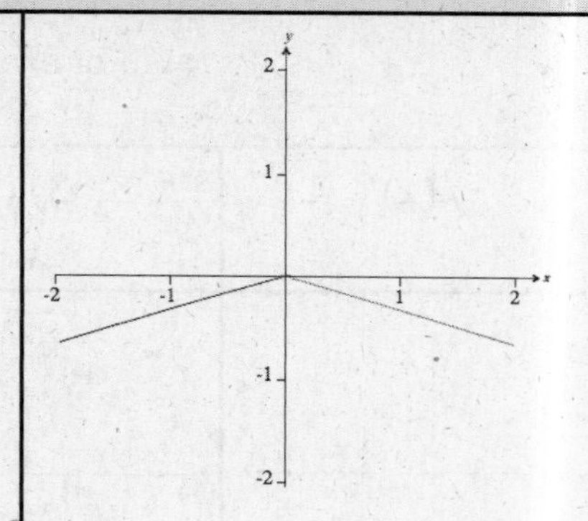

Given the graph of *f(x)* below, graph $g(x) = 2f(x) + 1$.

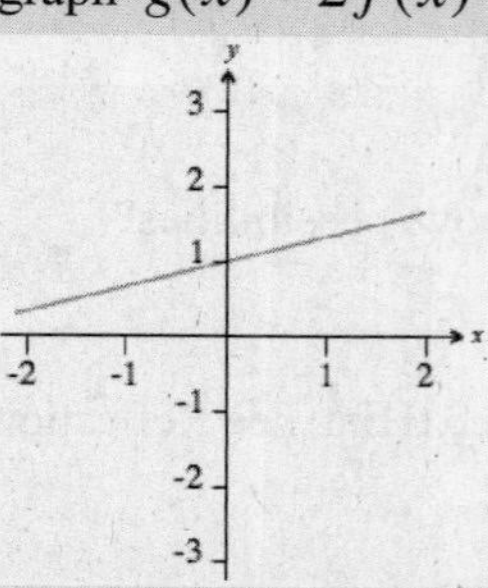

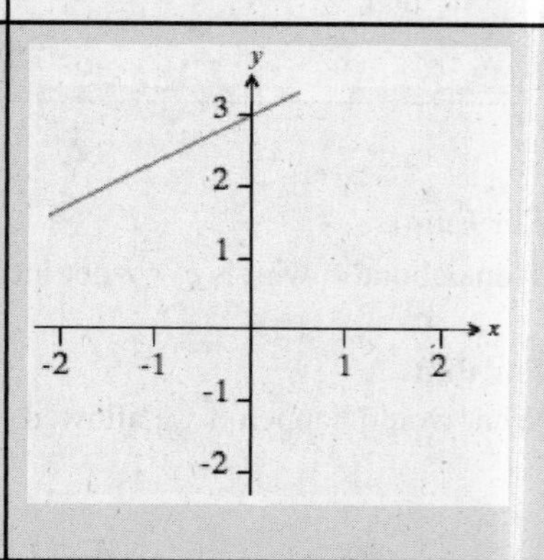

Horizontal Stretching/Compression

Let *c* be a positive constant, and if $y = f(c \cdot x)$ then

1. if $c > 1$, *y* is *f(x)* compressed horizontally toward the *y*-axis by a factor of $\frac{1}{c}$.

   (Some people might see this as the same as a vertical stretching away from the *x*-axis.)

2. if $0 < c < 1$, *y* is *f(x)* stretched horizontally away from the *y*-axis by a factor of $\frac{1}{c}$.

   (Some people will see this one as the same as compressing vertically towards the *x*-axis.)

For example:

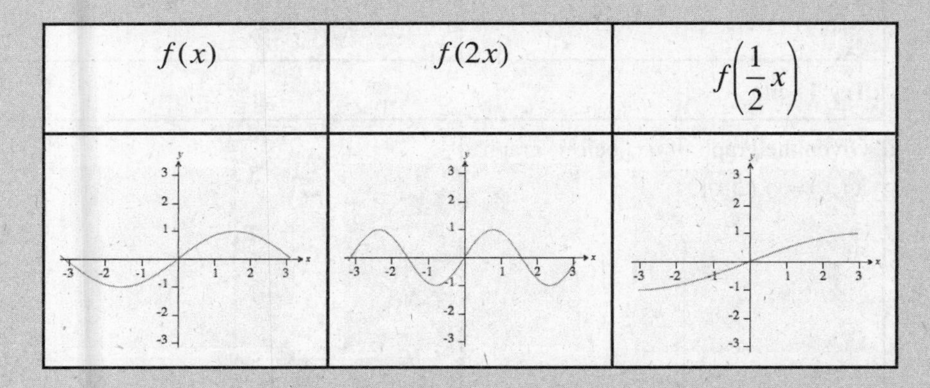

★ **For Fun**

Combine what you know about the graphs of parabolas and all these modifications on functions to see if you can match up all the translations, reflections, and compressions / stretchings in this equation: $f(x) = -2(x-4)^2 - 5$

You might try sketching this parabola and see if you can also visually find all the elements from this section that apply.

Horizontal Stretching and Compressing of Graphs: Test Prep

Given the graph of $f(x)$ below, graph $g(x) = f\left(\frac{1}{4}x\right)$.

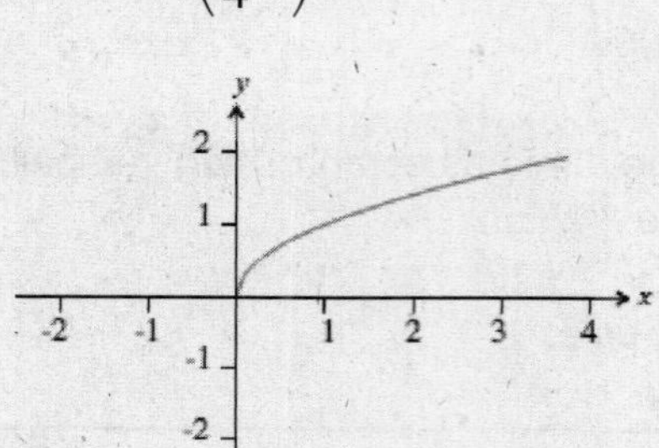

This graph is stretch horizontally away from the y-axis by a factor of four:

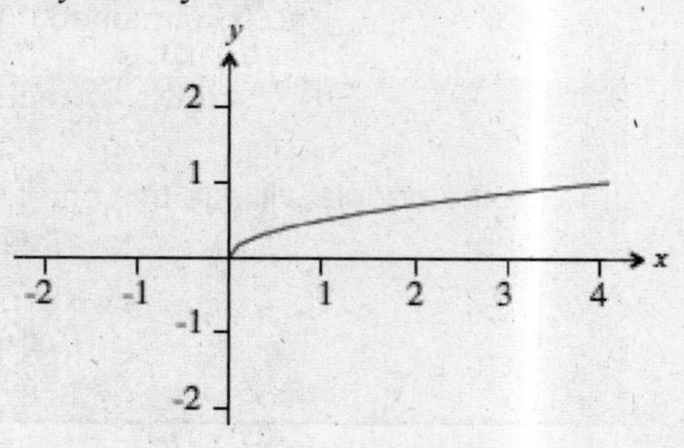

Try Them

Given the graph of $f(x)$ below, graph $g(x) = f(5x)$.

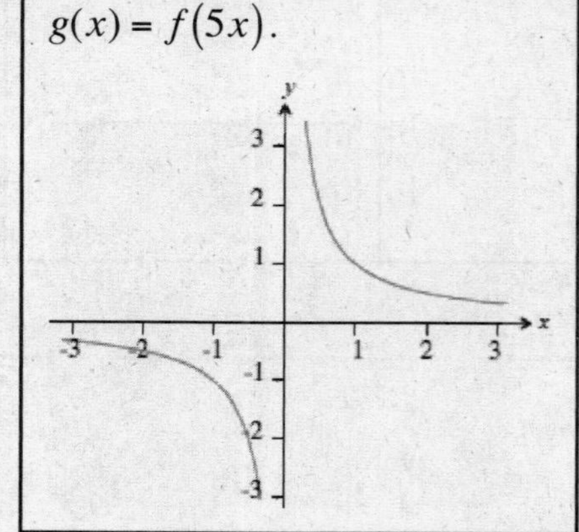

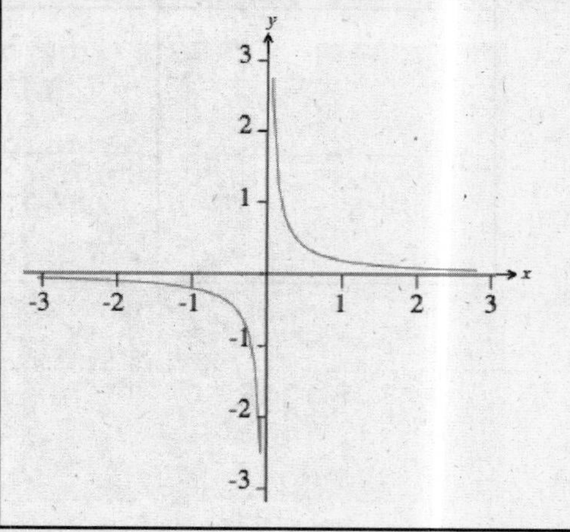

Horizontal Stretching and Compressing of Graphs: Test Prep

Consider $f(x) = x^2$ and use what you know about translations and stretching and compression to graph $f(x) = 3(x-1)^2 - 2$

(Does this match what you expect from what you already know about parabolas?)

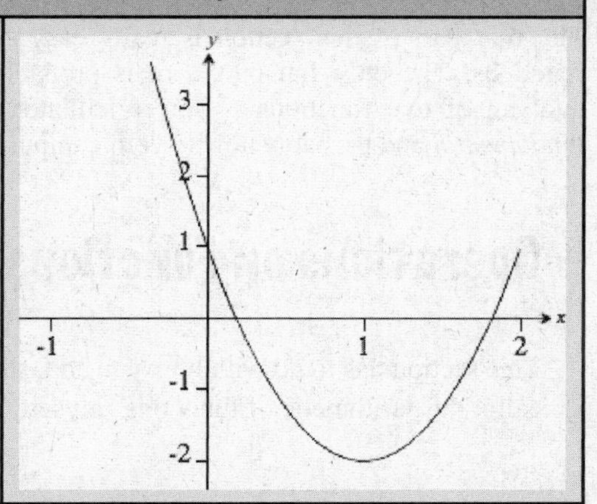

2.6: Algebra of Functions

The first part of this section is really easy -- the sum, difference, product, and quotient of functions. The only fun part here is probably the reminder about *domains* involved in the quotient of two functions -- you're still not allowed to divide by zero. The second part is *fundamental* and probably new to you: composition of functions.

Operations on Functions

This section has to do with how you add, subtract, multiply, and divide functions. I like to call it the "arithmetic of functions" myself. It's straightforward:

The Arithmetic of Functions

1. **Sum**: $(f+g)(x) = f(x) + g(x)$

   for example:
   $Given \quad f(x) = 3x \quad g(x) = 5x - 1$
   $(f+g)(x) = f(x) + g(x) = 3x + 5x - 1 = 8x - 1$

2. ***Difference***: $(f-g)(x) = f(x) - g(x)$

   for example:
   $Given \quad f(x) = 3x \quad g(x) = 5x - 1$
   $(f-g)(x) = f(x) - g(x) = 3x - (5x - 1) = -2x + 1$

3. ***Product***: $(f \cdot g)(x) = (f(x))(g(x))$

   for example:
   $Given \quad f(x) = 3x \quad g(x) = 5x - 1$
   $(f \cdot g)(x) = (f(x))(g(x)) = 3x(5x - 1) = 15x^2 - 3x$

4. *Quotient*: $\left(\frac{f}{g}\right)(x) = \frac{f(x)}{g(x)}$

for example:

Given $f(x) = 3x \quad g(x) = 5x - 1$

$$\left(\frac{f}{g}\right)(x) = \frac{f(x)}{g(x)} = \frac{3x}{5x-1}, \qquad x \neq \frac{1}{5}$$

Operations on Functions: Test Prep

Given $f(x) = x^2$ and $g(x) = x^2 - 1$, find a. $\left(\frac{f}{g}\right)(x)$ b. state its domain	a. $\left(\frac{f}{g}\right)(x) = \frac{f(x)}{g(x)} = \frac{x^2}{x^2-1}$ b. $x^2 - 1 = 0$ $(x+1)(x-1) = 0$ $x = -1, 1$ Since the denominator cannot be zero, the domain is $(-\infty, -1) \cup (-1, 1) \cup (1, \infty)$.
Try Them	
Given $f(x) = x + 1$ and $g(x) = \frac{x}{2}$, find a. $(f+g)(x)$ b. $(f \cdot g)(x)$ c. $(f \cdot g)(4)$	a. $\frac{3}{2}x + 1$ b. $\frac{x}{2}(x+1) = \frac{1}{2}x^2 + \frac{1}{2}x$ c. 10

Operations on Functions: Test Prep	
Given $f(x) = x$ and $g(x) = x^2 - x - 6$, find a. $\left(\frac{3f}{g}\right)(x)$ b. state its domain	a. $\frac{3x}{x^2 - x - 6}$ b. $x \neq 3, -2$

Difference Quotient

The difference quotient is another formula. It pops up lots of places -- calculating averages, slopes, etc. -- and so we practice on it a bit. By the way, knowing about the difference quotient is excellent preparation for the next section (and calculus even).

✤ **difference quotient**
This is the formula that is called the difference quotient (can you see a difference -- subtraction -- and a quotient -- division?)

$$\frac{f(x+h) - f(x)}{h}$$

By the way, can you see why h cannot be equal to zero in this formula? (That's algebra thinking.)

The difference quotient is the slope of a secant line.
A secant line of a curve is a line that intersects two points on the curve.
Secant comes from the Latin *secare*, to cut.
Secant functions are explored in trigonometry; the slope of the secant line (the difference quotient) comes around again in calculus.

How do you find $f(x+h)$?

Let $f(x) = x^2 + 1$.

Find these:

1. $f(3) = 3^2 + 1 =$ ________________
2. $f(-1) = (-1)^2 + 1 =$ ____________
3. $f(a) =$ ______________________
4. $f(b+1) =$ ______________________

Did you get $f(b+1) = (b+1)^2 + 1 = b^2 + 2b + 1 + 1 = b^2 + 2b + 2$ for the last one?

★ **For Fun**

Look at the picture below. Could you pick out the coordinates of the two points where the curve and the line intersect? Could you use the slope formula to write an expression for the slope of that line from those coordinates?

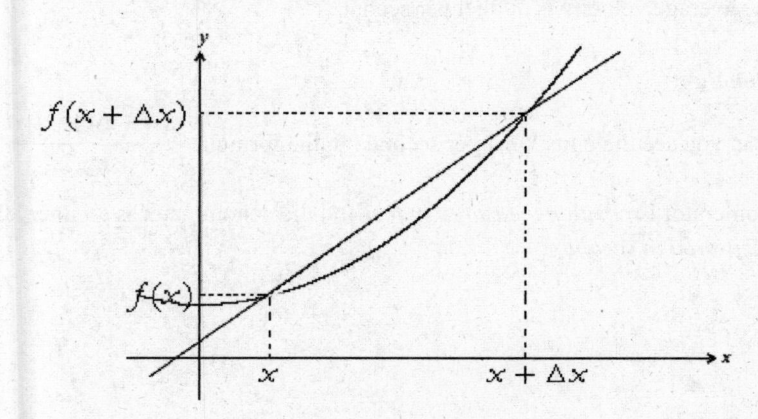

✤ **average velocity**

Average velocity is one of those quantities that can be calculated with the difference quotient.

The formula is the same, but the variables and function have difference names:

$$\frac{s(a+\Delta t)-s(a)}{\Delta t}$$

where *s* is the position function (position is changing if a velocity is involved, right?), and *a* is the point in time when the beginning velocity is measured, and Δt is simply how long time runs.

So, for instance, say $s(t)=6t^2$ measured in feet with *t* is measured in seconds. Let's find the average velocity for $2 \le t \le 3$:

Time is starting at 2 (seconds) so $a=2$. Time ends at 3 (seconds), so $\Delta t = 3-2=1$.

$$\frac{s(a+\Delta t)-s(a)}{\Delta t}=\frac{s(2+1)-s(2)}{1}$$
$$= s(3)-s(2)$$
$$= 6(3)^2-6(2)^2$$
$$= 6\cdot 9-6\cdot 4$$
$$= 54-24$$
$$= 30$$

and the average velocity is 30 feet per second.

★ **For Fun**

Can you see the units "feet per second" in the formula $\frac{s(a+\Delta t)-s(a)}{\Delta t}$ if the numerator is *position measured in feet* and the denominator is an interval of *time measured in seconds*?

Difference Quotient: Test Prep	
Given $f(x)=x^2$ and $x=1$, and $h=2$, find and simplify the difference quotient.	$\frac{f(x+h)-f(h)}{h}=\frac{f(1+2)-f(2)}{2}$ $=\frac{3^2-2^2}{2}$ $=\frac{5}{2}$
Try Them	
Find the difference quotient for $g(x)=x^3$.	$3x^2+3xh+h^2$
Given $s(t)=-6t$, find the average velocity over the time interval [1, 2].	-6

Composition of Functions

The composition of functions has to do with putting an x-value through two (or more) functions to see what you get out. That means that you take the result from one function and put that result into another function. Symbolically this is what that looks like:

$$f(g(x))$$

Do you recognize what's going on there? You're putting the result from running x through the g-function *into* the f-function. It's like any system that you move into another situation; the idea is very universal -- it's like a human moving to a new city and seeing how he/she does.

Algebraically, you may find you need to practice this a bit, but it can definitely become very easy. Here's an example using the two functions that are defined like this:

$$f(x)=x+1 \quad and \quad g(x)=x^2$$

Let's find out what the result is if you run an x through g first and then through f, like so:

$$f(g(x)) = f(x^2) = x^2 + 1$$

Did you catch that? Everywhere there is an x in $f(x)$, you have to insert $g(x)$. That means wherever there is an x in $f(x)$, you have to insert x^2. And so you get the final result: $x^2 + 1$.

Now let's try $g(f(x))$ using the very same g- and f-functions.

$$g(f(x)) = g(x+1) = (x+1)^2 = x^2 + 2x + 1$$

Everywhere there was an x in $g(x)$, we had to insert $(x+1)$. Then we simplified. *Are you catching how that works?* It's no big deal really -- if you can plug ,say, a into f and get $f(a)$, then you already have all the skills you need to do composition of functions.

While we're here, it's important to notice that $f(g(x))$ was *not* equal to $g(f(x))$. In general, the composition of two functions is not *commutative*; that is, the order of the nesting makes a difference. (Later on we'll see a special case where the composition *is* commutative, and that case will have very special meaning.)

★ **For Fun**

You can try the "redundant" compositions of functions from my example f- and g-functions: $f(f(x))$ and $g(g(x))$. See what you get. I think that's all the possible combinations of those two functions ... unless you want to try nesting them more than 2 deep. Can you imagine that?

There is a special *operator* that we use to denote the composition of functions. Now that you know how to "compose" two functions, all you have to know is what this operator means. Here's exactly how it translates into what you already know:

$(f \circ g)(x) = f(g(x))$ or vice versa $(g \circ f)(x) = g(f(x))$. First of all, can you see how the order of the nesting of the functions is specified by this $\circ$ operator?

Second of all, and this is important because students' eyes are sometimes fooled by this, you must be able to keep in mind that $f(x)$ is *functional notation* and there is absolutely *nothing* about multiplying some "*f*" by some (x). There is also *no* multiplication going on between some $(f \circ g)$ and some (x). $(f \circ g)$ is the name of some new function created by composing *f* and *g* -- it's just the name of a function. You still have to give it an *x* to work on before it can actually do anything. And the work $(f \circ g)$ will do on its *x* is not that it multiplies itself by *x* -- it does whatever the definition of the composition turns out to be.

It's like this: $\sqrt{}$ can't actually do anything until you put something inside it. Well, $(f \circ g)$ is just like that, and trying to multiply this $(f \circ g)$ times anything makes as much sense as this: $\sqrt{} \times 14$. (Please tell me that doesn't make sense to you, okay?)

See if you can dissect these compositions of functions -- in other words, can you see a function inside another function? (Like $(x+1)^2$ is $x+1$ inside a squaring function.)

$\sqrt{x+1}$

$(x-2)^3$

$\sqrt[3]{212x}$

$(3x)^5$

$\left(4\sqrt{x^5}\right)^3$

✓ **Questions to Ask Your Teacher**
How many layers of function-nesting can you spot in that last one? You might want to discuss this and other examples carefully and thoroughly in class. Make sure your "eye" is getting properly trained to spot these.

By the way, when you learned in the last subsection to do $f(x+h)$ for the difference quotient, that was a composition of functions, wasn't it? (Can you see that $x + h$ is a function inside of *f*?)

Composition of Functions: Test Prep

Given $f(x)=2x-1$ and $g(x)=x^2$, find and simplify a. $(f\circ g)(x)$ b. $(g\circ f)(x)$	$(f\circ g)(x)=f(g(x))$ a. $=f(x^2)$ $=2x^2-1$ $(g\circ f)(x)=g(f(x))$ b. $=g(2x-1)$ $=(2x-1)^2$ $=4x^2-4x+1$
Try Them	
Given $f(x)=\frac{1}{3}x^2$, find $(f\circ f)(3)$.	3
Given t the temperature in Fahrenheit, the temperature in Celsius is given by $C(t)=\frac{5}{9}(t-32)$ and the temperature Kelvin is given by $K(t)=C(t)+273.15$. Convert $70°F$ to Kelvin. Round to the nearest hundredth.	$294.26°K$

2.7: Modeling Data Using Regression

After we learn so much about linear and quadratic functions and their graphs, how about trying to use these functions to *approximate* data that might fall roughly on a line or in a U-shape? Regression is a way to find functions that approximate graphs of data that you already have. (What we're doing is finding an equation from a graph instead of graphing from an equation like we have been doing throughout this chapter.)

Linear Regression Models

Just like a graph sometimes displays information in a way that minds can appreciate much more than a string a numbers, regression analysis allows us to construct curves on the cartesian coordinate system that our minds can grasp more meaningfully than a collection of dots.

- **regression analysis**
  Regression analysis is the process we use to find a function (in this section, a line) that best describes the data (points on a 2-dimensional graph).

- **line of best fit, least-squares regression line**
  The result of a linear regression analysis will be the line of best fit or the least squares regression line.

Although the equations for calculating the line of best fit are somewhat labor intensive and perhaps scary looking, this is something that you will likely have to learn to do in an introductory statistics course, or even sometimes in an introductory physics course. Don't let that scare you. The methods are actually straightforward -- but a little bit beyond the scope, at this point, of precalculus algebra. Nonetheless, I assure you that it is definitely do-able with basic algebra skills... and time.

Having said that, here we will leave the analysis to a computer package. So get out your calculator guide and look up regression. Also Microsoft Excel and other spreadsheet applications can compute regression lines given *(x, y)* data points. Your textbook, page 239, also gives some steps for some TI calculators.

Linear Regression: Test Prep

Question	Answer
For which set of data is the correlation coefficient closer to 1? 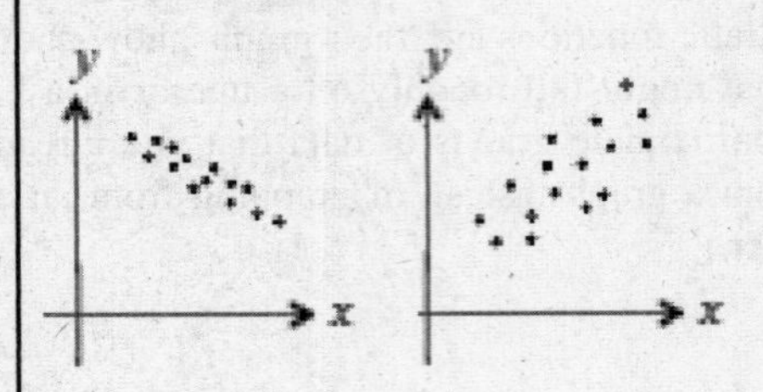	The set on the left because the points would cluster more closely to their regression line.
Try Them	
The data show the distance in feet that a ball travels for various swing speeds in mph of a bat. (see table below) a. Find the linear regression equation for these data. Keep 10 decimal places. b. What is the expected distance a ball will travel when the bat speed is 58 mph? Round to the nearest mph.	a. $y = 3.1410344282x + 65.09359606$ b. 263 ft.
Find the linear regression line for these ordered pairs: (72, 5), (75, 7), (78, 8), (81, 12), (84, 15). Round to nearest hundredths.	$y = 0.83x - 55.60$

bat speed (mph)	distance (ft)
40	200
45	213
50	242
60	275
70	297
75	326
80	335

Correlation Coefficient

Now we introduce a way to quantify how good your "best-fit" line is.

✤ **linear correlation coefficient r**

Again, find out how to get the correlation out of your calculator when you give it data to find a best-fit line. The correlation coefficient, usually denoted by r, indicates how well that best-fit line fits your points. The correlation coefficient takes on values between -1 and 1. The closer it is to zero, the worse the fit. Both 1 and -1 are indicators of a good fit. Can you see those trends in these graphs and the values for their correlation coefficients?

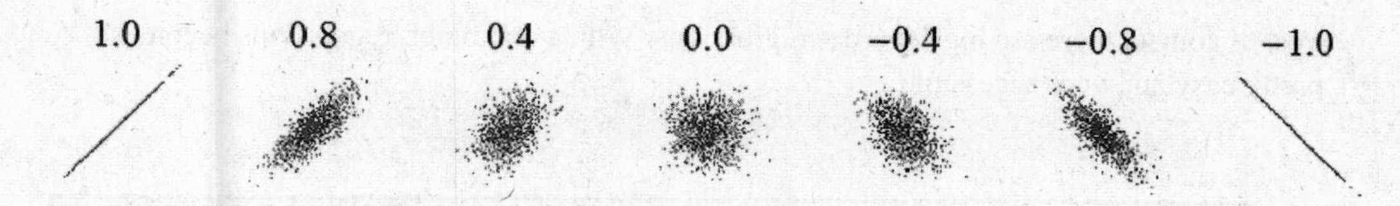

✤ **coefficient of determination r^2**

The square of the correlation coefficient is the coefficient of determination (see below).

> The coefficient of determination r^2 measures the proportion of the variation in the dependent variable that is explained by the regression equation. (page 242 in your text)

For example, if $r \cong 0.929$, then $r^2 \cong 0.86$ or something like 86%. That means that about 86% of the relation of the data to the best fit line is due to the way that your independent and dependent variables interact. It also means that there is a leftover approximate 15% that depends on something else.

★ **For Fun**

If a value closer to zero for the correlation coefficient means "not a good fit," what do you imagine that a coefficient of determination near zero would mean?

Quadratic Regression Models

Well, not everything fits a linear regression profile well. You can try a quadratic (power 2) regression instead; see how that goes. Instructions for quadratic regression for some TI calculators can be found in your textbook. Check out your calculator or spreadsheet application (like Excel).

This method will also furnish you with a coefficient of determination to aid you in deciding how much your independent and dependent variables really depend on each other. At least you can use it to decide if quadratic regression is a better tool than linear regression for your system.

Yes, of course, there are higher-order regressions. With a computer, regressions becomes pretty easy and very accessible.

Quadratic Regression: Test Prep

The data show the oxygen consumption, in milliliters per minute, of a bird flying level at various speeds in kilometers per hour.

speed	consumption
20	32
25	27
28	22
35	21
42	26
50	34

a. Find a quadratic model for these data. Round to 5 decimal places.

b. Use your model to find the speed to the nearest kph at which the bird has minimum oxygen consumption.

a. Enter the data in the table into your calculator or a spreadsheet application. Then find the quadratic regression model:

$$y = 0.05208x^2 - 3.56026x + 82.32999$$

b. You can use the vertex formula to find the minimum or maximum of a quadratic function. In this case it will be a minimum:

$$x = -\frac{b}{2a}$$

$$= -\frac{-3.56026}{2(0.05208)}$$

$$\approx 34$$

Minimum consumption occurs at approximately 34 kph.

Quadratic Regression: Test Prep

Try Them

Find the quadratic regression equation for these data:

year	AIDS cases
1999	41356
2000	41267
2001	40833
2002	41289
2003	43171

Round to 4 decimal places.

$y = 345.1429x^2 - 1705.6571x + 42903.6000$
where x is years since 1998.

Find the quadratic regression equation for these data:

(-1, 6), (0, -1), (1, -3),
(2, -1.5), (3, 5), (4, 10).

Round a, b, and c to two decimal places.

$y = 1.68x^2 - 3.91x - 0.23$

NOW TRY THESE:
REVIEW EXERCISES: PAGE 253,
PRACTICE TEST: PAGE 257.

STAY UP ON YOUR GAME -- DO THE CUMULATIVE
REVIEW EXERCISES: PAGE 258

Final Fun: Matching

FIND THE BEST MATCH. GOOD LUCK.

ORDERED PAIR	LINE
FUNCTIONAL NOTATION	DISTANCE
$x^2 + y^2 - 8x + 12y - 48 = 0$	MIDPOINT
QUADRANT	$h(x)$
$(0,b)$	III
$3x + 2y = 12$	Y-INTERCEPT
$\left(-\frac{b}{2a}, \frac{4ac - b^2}{4a}\right)$	$(3,-2.1)$
$f(g(x))$	SLOPE
DIFFERENCE QUOTIENT	COMPOSITION
$\left(\frac{x_1 + x_2}{2}, \frac{y_1 + y_2}{2}\right)$	CIRCLE
$d = \sqrt{(x_2 - x_1)^2 + (y_2 - y_1)^2}$	$f(x) = -11$
QUOTIENT OF FUNCTIONS	$\lfloor x \rfloor$
CONSTANT FUNCTION	$\frac{x^2}{x-1}$
$m = \frac{y_2 - y_1}{x_2 - x_1}$	$\frac{f(x+h) - f(x)}{h}$
FLOOR FUNCTION	VERTEX FORMULA

Chapter 3: Polynomial and Rational Functions

Word Search

```
D S D N U O B X M E B R L V R
Q I N C D N N D U U L H C Q E
U M V S I L F B L O M W P C D
O U T I U B M G T R Q I O N N
T M H G D O U T I E Y E X I I
I I E V F E U C P Z F H Q A A
E N O G D G N N L F P R S X M
N I R T Q G C D I D A M U W E
T M E F I I Q C C T O N D O R
U Z M S T I I H I O N N C J H
L U Z R P E L O T I A O X L S
Y G A N N O N H Y L T E C D L
Z U A T L A I M O N Y L O P K
Q J C M L A S Y M P T O T E X
K R O S I V I D F A C T O R S
```

ASYMPTOTE
BOUNDS
COEFFICIENT
CONTINUOUS
CUBIC
DIVIDEND
DIVISOR
FACTORS
MAXIMUM
MINIMUM
MULTIPLICITY
POLYNOMIAL
QUARTIC
QUOTIENT
RATIONAL
REMAINDER
SIGN
SMOOTH
THEOREM
ZERO

3.1: Remainder Theorem and Factor Theorem

In Section 3.1, we need to get some basic skills in place. These are skills we'll use throughout the rest of the chapter (and you'll need them heavily in calculus, too). These are techniques for the most part. You can learn a technique -- practice it -- and we can see how it's used soon enough.

Division of Polynomials

Let's cover some basic vocabulary about division. These are all words you've heard before.

- **zeros**
  A zero of a function is an *x*-value that causes the value of the function to become zero. In other words, if $x = c$ is a zero of $f(x)$, then $f(c)=0$.

- **remainder**
  Just like in grade school when you first learned about division, the remainder is still exactly the same item: what is left over when you carry out your division. If the remainder is zero, then the divisor divided the dividend exactly.

  ★ **For Fun**

  When you write an improper fraction like $\frac{31}{9}$ as a mixed number, there's a remainder, right? What do you do with it in the mixed number? Once you're taking pre-calculus algebra, you no longer write the remainder like "R4." The pattern is more-or-less like the way you write the remainder for this mixed number.

- **quotient**
  The quotient is the answer to a division problem which had no remainder.

- **dividend**
  The dividend is the numerator when the division problem is written in fractional form; it's inside the bracket when the quotient is written out for long division. The dividend is the expression being divided *into*.

✤ **divisor**
The divisor is the denominator, the bottom of the fraction, outside the bracket in long division. It is the expression to be divided into the dividend.

★ **For Fun**
If $\frac{31}{9} = 3\frac{4}{9}$, can you name the quotient, divisor, dividend, and remainder?

Synthetic Division

First we'll discuss the long division of polynomials you've probably seen before. This method of division of polynomials is lengthy, and so you'll learn a new method: synthetic division.

Regardless of the method you use to divide polynomials, I assure you, wholeheartedly, that you need to practice these methods. Don't be fooled into thinking they look simple and sensible and so you know them. You won't remember the first step later, and then you're sunk. Plus they are simple and straightforward, so a little practice will go a long way and the speed you gain will pay off at test time.

✤ **long division**
Long division refers to the process (algorithm) you learned in grade school to carry out division of big numbers. Here we use it to divide polynomials. The algorithm is basically the same, but it looks somewhat more confusing:

$$\begin{array}{r} x + 12 \\ x - 3 \overline{) x^2 + 9x - 16} \\ \underline{x^2 - 3x } \\ 12x - 16 \\ \underline{12x - 36} \\ 20 \end{array}$$

The process starts when you ask yourself, "How many times does x go into x^2?" The answer to that question is x, and you write that on the top of the bracket *in the x place value column*:

$$\begin{array}{r} x \\ x - 3 \overline{) x^2 + 9x - 16} \end{array}$$

Then multiply that x by the divisor and write the result underneath the dividend, respecting the place value of the terms in the polynomial.

$$\begin{array}{r} x \\ x - 3 \overline{) x^2 + 9x - 16} \\ \underline{x^2 - 3x} \end{array}$$

Then subtract that result, bring down the next term from the dividend, and ask yourself if x goes into that result.

$$\begin{array}{r} x \\ x - 3 \overline{) x^2 + 9x - 16} \\ \underline{x^2 - 3x} \\ 12x - 16 \end{array}$$

In our example, x goes into *12x* twelve times. Write that on top of the bracket; include the operation of either addition or subtraction when you do because we're building a new polynomial up there.

$$\begin{array}{r} x + 12 \\ x - 3 \overline{) x^2 + 9x - 16} \\ \underline{x^2 - 3x} \\ 12x - 16 \end{array}$$

Multiply the new term in the quotient on top of the bracket by the divisor and write the result underneath the bottom row, respecting place value. Subtract and thereby find your remainder:

$$\begin{array}{r} x + 12 \\ x - 3 \overline{) x^2 + 9x - 16} \\ \underline{x^2 - 3x} \quad \\ 12x - 16 \\ \underline{12x - 36} \\ 20 \end{array}$$

So the division problem with its answer is

$$\left(x^2+9x-16\right)\div\left(x-3\right)=x+12+\frac{20}{x-3}$$

✤ **synthetic division**

Synthetic division is long division of polynomials written in a sort of shorthand.

> First and foremost, in order to use synthetic division, the divisor must be able to be written in the form *x - c*. If the divisor cannot be written in this form, for example x^2-1 or $2x+3$, then you need to use good ol' long division.

Let's try the synthetic division for $\left(2x^3-x^2+3x-1\right)\div\left(x-3\right)$. First make a frame, with the value of x that makes the divisor zero on the outside (that's c of x - c). Across the top row, write the coefficients of the dividend. Copy the first coefficient to the result row under the frame:

$$\begin{array}{r|rrrr} 3 & 2 & -1 & 3 & -1 \\ & & & & \\ \hline & 2 & & & \end{array}$$

Multiply c (here it's 3) by that first coefficient that you copied into the bottom row. Write that result under the next place in the internal second row, add the result to the second coefficient in the top row, and write that result in the bottom, result row:

$$\begin{array}{r|rrrr} 3 & 2 & -1 & 3 & -1 \\ & & 6 & & \\ \hline & 2 & 5 & & \end{array}$$

Now multiply that number by c again (here, it's 3 times 5) and do the same thing with that result as you did in the last step, but this time everything is one more step to the right:

$$\begin{array}{r|rrrr} 3 & 2 & -1 & 3 & -1 \\ & & 6 & 15 & \\ \hline & 2 & 5 & 18 & \end{array}$$

Continue exactly like that until you run out of positions:

$$\begin{array}{r|rrrr} 3 & 2 & -1 & 3 & -1 \\ & & 6 & 15 & 54 \\ \hline & 2 & 5 & 18 & 53 \end{array}$$

The coefficients of the result and the remainder are in that bottom, result row. The quotient will be exactly one degree less than the dividend:

Now we can see the result: $\left(2x^3 - x^2 + 3x - 1\right) \div (x - 3) = 2x^2 + 5x + 18 + \dfrac{53}{x - 3}$

★ **For Fun**

Now try this division $\left(x^2 + 9x - 16\right) \div (x - 3)$ by synthetic division. We already arrived at answer by long division for it. Pay very careful attention to all the ways that long division and synthetic division produce the same results, step-by-step. Also, please do make sure you get the right answer when you do it by a different method!

✤ **fractional form**

Fractional form includes writing the division in the form of numerator over denominator and the remainder over the divisor, thus:

$$\frac{2x^3 - x^2 + 3x - 1}{x - 3} = 2x^2 + 5x + 18 + \frac{53}{x - 3}$$

Synthetic Division: Test Prep

Do the following division by both long division and synthetic division:

$$\frac{6x^4 - 2x^3 - 3x^2 - x}{x - 5}$$

- long division:

$x^2 - 1 = (x - 1)(x + 1) = 0$

$x = 1 \, and$

$$
\begin{array}{r|rrrrr}
 & 6x^3 & +28x^2 & +137x & +684 & \\
x-5 & 6x^4 & -2x^3 & -3x^2 & -x & +0 \\
 & -(6x^4 & -30x^3) & & & \\
 & & 28x^3 & -3x^2 & & \\
 & & -(28x^3 & -140x^2) & & \\
 & & & 137x^2 & -x & \\
 & & & -(137x^2 & -685x) & \\
 & & & & 684x & +0 \\
 & & & & -(684x & -3420) \\
 & & & & & 3420
\end{array}
$$

- synthetic division:

$$
\begin{array}{r|rrrrr}
5 & 6 & -2 & -3 & -1 & 0 \\
 & & 30 & 140 & 685 & 3420 \\
\hline
 & 6 & 28 & 137 & 684 & 3420
\end{array}
$$

- answer:

$$6x^3 + 28x^2 + 137x + 684 + \frac{3420}{x - 5}$$

Try Them

Use synthetic division to carry out $\left(-x^3 + 3x^2 + 5x + 30\right) \div (x - 8)$	$-x^2 - 5x - 35 + \dfrac{-250}{x - 8}$
Divide: $\dfrac{x^5 + 3x^4 - 2x^3 - 7x^2 - x + 4}{x^2 + 1}$	$x^3 + 3x^2 - 3x - 10 + \dfrac{2x + 14}{x^2 + 1}$

Remainder Theorem

The Remainder Theorem is a nifty little item we'll use a lot:

The Remainder Theorem

Given

$$\frac{P(x)}{x-c} = Q(x) + \frac{R(x)}{x-c}$$

then $P(c) = R(x)$.

That means that if a polynomial *P(x)* is divided by *x - c*, then the remainder of that division *is equal to P(c)*.

For example, if $\frac{2x^3 - x^2 + 3x - 1}{x-3} = 2x^2 + 5x + 18 + \frac{53}{x-3}$ (from our last example),

and if we name the numerator like this $P(x) = 2x^3 - x^2 + 3x - 1$, then $P(3) = 53$. Did you catch that one? Try calculating $P(3)$ the old-fashioned way. Is it 53?

If *x* gets to be anything remotely difficult, using synthetic division and finding the remainder that way becomes must quicker and easier ... and less error-prone.

The Remainder Theorem: Test Prep

Use synthetic division and the Remainder Theorem to find *P(-2)* if $P(x) = 4x^3 - 5x^2 + 2x - 10$	-2 | 4 -5 2 -10 -8 26 -56 4 -13 28 -66 $P(-2) = -66$

The Remainder Theorem: Test Prep	
Try Them	
Given $P(x) = x^4 - 25x^2 + 144$, use synthetic division and the Remainder Theorem to ascertain if $c = 3$ is a zero.	Yes. $P(3) = 0$.
Given $P(x) = x^4 - 2x^2 - 100x - 75$, find $P(-5)$.	$P(-5) = 1000$

Factor Theorem

Zeros and factors go together!

Factor Theorem

A polynomial $P(x)$ has a factor $(x - c)$ if and only if $P(c) = 0$. That is, $(x - c)$ is a factor of $P(x)$ if and only if c is a zero of P.
(page 266 of your textbook)

Remember that being a factor means that said factor divides evenly, without a remainder, into the dividend. There you go -- without a remainder. And by the Remainder Theorem, we know that the value of the polynomial for a specified x is given by its remainder. So the value of the polynomial must be zero if there is no remainder and no remainder means you've found a factor. At least it's all completely consistent!

★ **For Fun**

Consider $P(x) = x^4 - 8x^3 + 22x^2 - 29x + 24$. Determine whether *(x + 1)* and *(x - 3)* are factors. Now check to see if $x = -1$ and $x = 3$ are zeros. Use whatever method you prefer, but start to notice how the methods complement each other.

Take note: The Factor Theorem does not give you a method to *find* a factor, only to check and see if you have guessed one correctly.

The Factor Theorem: Test Prep	
Use synthetic division to determine if *(x + 2)* is a factor of $P(x) = x^3 + 8$.	$\begin{array}{r\|rrrr} -2 & 1 & 0 & 0 & 8 \\ & & -2 & 4 & -8 \\ \hline & 1 & -2 & 4 & 0 \end{array}$ *P(-2) = 0*; therefore, *x = -2* is a zero of *P(x)* and *(x + 2)* is a factor.
Try Them	
If the zeros of a polynomial *f(x)* are given to be *x = -1, 1, 3,* list three distinct factors of *f(x)*.	*(x + 1), (x - 1), (x - 3)*
Which is a factor of $2x^3 - 3x^2 - 11x + 6$ -- *(x + 3)* or *(x - 3)*?	*(x - 3)*

Reduced Polynomials

Just like we have to reduce fractions, it's possible to reduce polynomials. This is easy if you can completely factor the given polynomial. We've got the Remainder and Factor Theorems to help us check for factors now -- that will help.

❖ **reduced polynomial, depressed polynomial**
A reduced polynomial (or depressed polynomial, but that sounds, well, depressing) is simply the quotient resulting from a division by a factor of that polynomial. Does that sound confusing? Don't panic: reducing polynomials is the same as reducing fractions. Find common factors and cancel them!

For example:

$$\frac{x^4 - 8x^3 + 22x^2 - 29x + 24}{x - 3}$$
$$= \frac{(x-3)(x^3 - 5x^2 + 7x - 8)}{x - 3}$$
$$= x^3 - 5x^2 + 7x - 8$$

It will be very important in coming sections to recognize these reduced polynomials.

★ **For Fun**
Think about this. Say you have a rational function, $R(x) = \frac{x^4 - 8x^3 + 22x^2 - 29x + 24}{x - 3}$. Do you think that $R(x) = x^3 - 5x^2 + 7x - 8$? Hint: think about domains.

Reduce these:

$$f(x) = \frac{2x^2 + x - 1}{x + 1}$$

$$g(x) = \frac{x^3 - 4x^2 + x - 4}{x - 4}$$

$$h(x) = \frac{x^3 + x^2 - 3x}{x^2 + 4x}$$

3.2: Polynomial Functions of Higher Degree

Now we embark on the great undertaking of learning about polynomials of degree higher than those we have already studied. We have studied degree 1 -- lines -- and degree 2 -- parabolas.

- **smooth continuous curves**
  There are rigorous criteria for smooth and continuous curves in calculus, but luckily they wind up meaning just about the same thing the words mean in English. A curve that is the graph of a function and doesn't have any pointy, sharp turns or gaps or holes or jumps of any kind is a smooth continuous curve. Polynomials are smooth continuous curves.

Far-Left and Far-Right Behavior

Since polynomials behave so nicely (for instance, they are smooth and continuous), there are a lot of generalizations we can make about their behavior and thereby their appearance on a graph.

General Form for a Polynomial

In general a polynomial can be defined by this expression:

$$P(x) = a_n x^n + a_{n-1}x^{n-1} + a_{n-2}x^{n-2} + \cdots + a_2 x^2 + a_1 x + a_0$$

This is the most general form. The degree is only specified as *n*, a natural number. All the coefficients a_n through a_0 are real numbers (remember they may be positive or negative or even zero in some cases). Ponder the form of this expression for a while; see if you can recognize the pieces in other polynomials you encounter. After all, math is about pattern recognition.

✤ **leading term, dominant term**

$a_n x^n$ is the leading term, or dominant term, of any polynomial. Recognize that this is the term of highest degree in the polynomial.

✤ **leading coefficient**

In the leading term, $a_n x^n$, a_n is the leading coefficient.

★ **For Fun**

Find the degree of the polynomial, leading term, and leading coefficient in these polynomials:

$$-x+3x^2-2$$

$$x-2x^4+\frac{1}{3}x^3-12+x^2$$

$$\frac{x^4}{4}+2x^3-x^2+\frac{x}{2}$$

$$a_n x^n + a_{n-1}x^{n-1} + a_{n-2}x^{n-2} + \ldots + a_2 x^2 + a_1 x + a_0$$

While you're at it, make sure you can identify the *constant* term in each of them, too. Of course, you noticed that there is no reason (except politeness) for writing a polynomial in descending order, is there?

Leading Term Test: Test Prep	
Identify the leading term of *P(x)* and sketch its far-left and far-right behavior: $P(x)=1-2x-4x^3$	leading term: $-4x^3$ 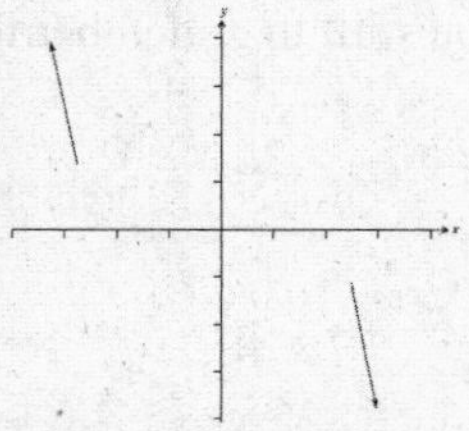

Leading Term Test: Test Prep	
Try Them	
Describe the far-left and far-right behavior of $f(x) = -\frac{x^4}{4} + x - 16$	*f(x)* goes down to both the far-left and the far-right.
Describe the far-left and far-right behavior of $T(x) = x^3 - x - 8$	*T(x)* goes down to the far-left and up to the far-right.

❖ **far-left behavior**

The far-left behavior of a polynomial alludes to the value of the polynomial as x gets extremely small (goes towards $-\infty$). The value on the y-axis of any of these smooth and continuous (and unbounded) polynomials will always approach either ∞ or $-\infty$ (hence unbounded).

❖ **far-right behavior**

Similarly, the far-right behavior refers to the value on the y-axis of the polynomial as x approaches ∞.

★ **For Fun**

Describe the far-left and far-right behaviors of these polynomials:

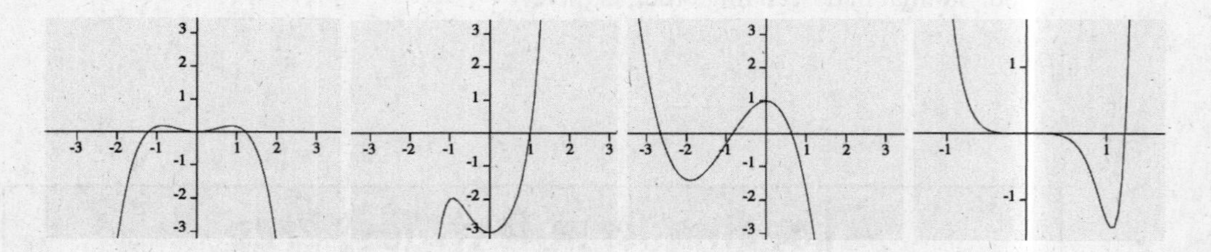

See the table on page 272 of your textbook to help you identify the sign of the leading coefficients and the even/odd nature of the degrees of these polynomial functions. By the way, this is all really good stuff to remember for calculus, if you're headed that way!

Maximum and Minimum Values

In between the far-left and far-right, a polynomial may "go up and down" quite a bit. One way to classify the behavior in these regions is to talk about where the value of the polynomial has a maximum or a minimum. We talked about the maximum and minimum of quadratic polynomials (parabolas) already.

✤ **turning points**

Turning points are points where the function changes from increasing to decreasing, where the graph of the function changes from going up to going down or vice versa. We talked about increasing and decreasing in Chapter 2.

A polynomial of degree n has at the very most n - 1 turning points. So what is the maximum number of turning points of $f(x) = -x^4 + 3x^2 - x + 11$? Does knowing the possible number of turning points help you picture the graph of a polynomial?

✤ **absolute minimum**

The absolute minimum value of the function is the smallest value in the function's range.

✤ **absolute maximum**

Similarly, the absolute maximum value of a function, as read on the y-axis, is the maximum value of the function's range.

★ **For Fun**

Reason this one out: What are the absolute maximum and minimum values of a polynomial of odd degree?

✤ **relative minimum, local minimum**

Relative minima and maxima of functions occur at turning points. A relative minimum of a function will occur where the function changes from decreasing to increasing as you read a graph from left to right. In other words, a relative minimum occurs at the bottom of a "trough" in the polynomial curve. It may not be the absolute minimum but locally it is a minimum.

✤ **relative maximum, local maximum**

A relative maximum occurs at a turning point where the function changes from increasing to decreasing as you read from left to right. It may or may not also be an absolute maximum, but it is a maximum in the local vicinity of values of the polynomial.

★ **For Fun**
Can you point out absolute and relative min's and max's on the graphs from the previous page?

Definition of Relative Minimum and Relative Maximum: Test Prep

Find the vertex of the parabola described by $f(x)=-(x-1)^2-3$. Is it a minimum or a maximum of $f(x)$?	By inspection (because the function is in the standard form for an equation of a parabola), the vertex is at *(1, -3)*. The vertex is a maximum because the leading coefficient is negative.
Try Them	
Mark the relative maximum values of *P(x)* from the graph: 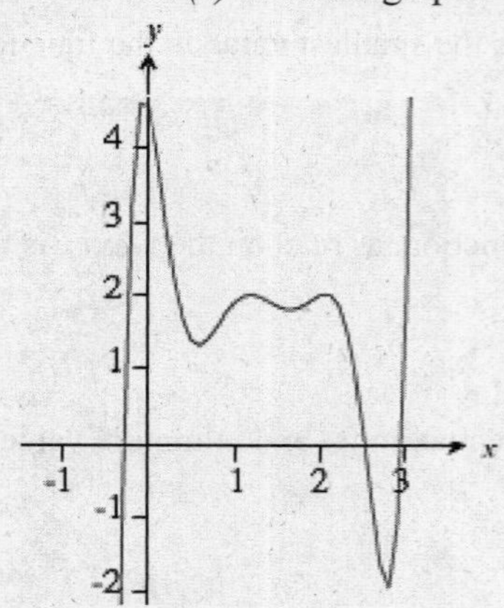	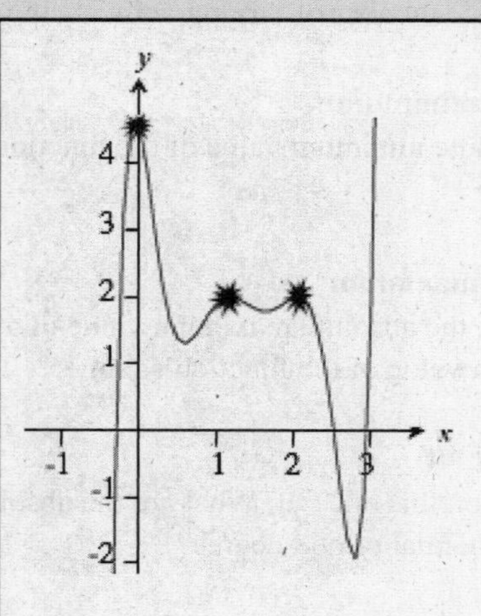
Use the minimum and maximum value functions on a calculator or computer to estimate relative minima and maxima of $H(x)=-2x^3-3x^2+12x+1$ Use what you know from the leading term test to name them relative or absolute extremes.	Relative maximum at $x\approx 1.0$ and $y\approx 8.0$; Relative minimum at $x\approx -2.0$ and $y\approx -19.0$

Real Zeros of a Polynomial Function

The *real* zeros (not imaginary zeros, those come later) of a polynomial are the self-same things as x's of the x-intercepts. Does that make sense? Isn't the value of the y-coordinate zero on the x-axis? Isn't where y equals zero exactly what we mean by the zero of the function?

At the very most, a polynomial of degree n can have n real zeros. Mind you, it may have fewer than n zeros, though.

Intermediate Value Theorem

Since a polynomial is continuous -- no skips or jumps or holes -- we can draw the very significant conclusion of the Intermediate Value Theorem, which basically says that in order to get from here to there without any jumping, you have to go through all the values in between the two places. Here it is more "officially":

The Intermediate Value Theorem

If P is a polynomial function and $P(a)$ is not equal to $P(b)$ for all $a < b$, then P takes on every value between $P(a)$ and $P(b)$ for all x's in the interval [a, b].

Try to visualize that every polynomial, whose domain is all real numbers, accepts each of the values in the interval [a, b] on the x-axis. Polynomials are continuous after all. This means that the polynomials also assign values on the y-axis between $P(a)$ and $P(b)$.

One of the most useful procedures that the Intermediate Value Theorem is used in is finding x-intercepts of polynomial functions. Doesn't it make sense that if a polynomial is negative in one region:

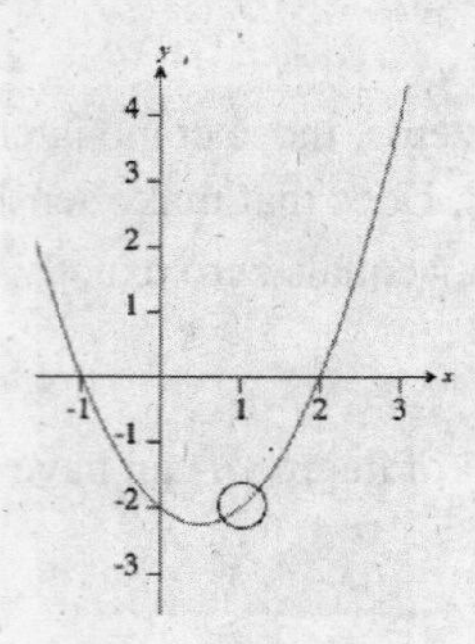

Then is positive in another region:

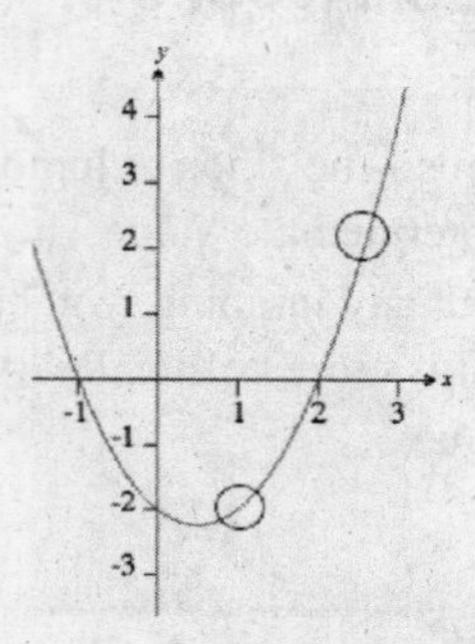

That there must be a point where it crosses the x-axis, and hence where there must be a zero of the function:

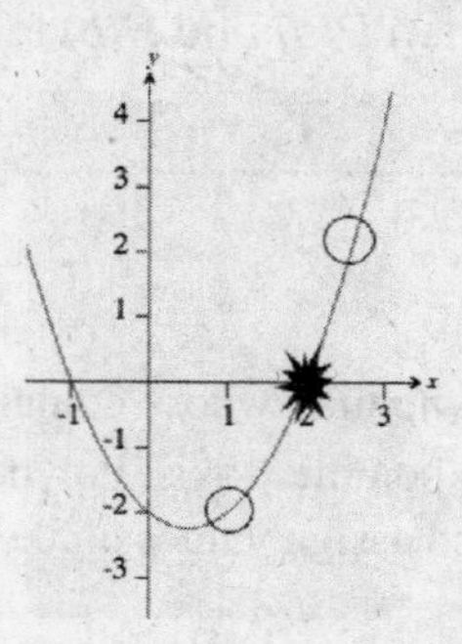

Helpful -- now we know where to start looking for an x-intercept!

Intermediate Value Theorem: Test Prep	
Use the Intermediate Value Theorem to show $P(x) = -5x^3 + 11x^2 - 3x - 1$ has a zero between *a = 0* and *b = 1.*	$P(0) = -1$ $P(1) = 2$ Since *P(0)* and *P(1)* have opposite signs, *P(x)* has at least one real zero between *x = 0* and *x = 1.*
Try Them	
Can you use the Intermediate Value Theorem to determine if $P(x) = -5x^3 + 11x^2 - 3x - 1$ has a zero between *a = -1* and *b = 1*?	Inconclusive; no.
Use the Intermediate Value Theorem to show $P(x) = x^4 + 3x^3 - 2x^2 + 3$ has a zero between *a = -4* and *b = -3*.	$P(-4) = 35$ $P(-3) = -15$

Real Zeros, x-Intercepts, and Factors of a Polynomial Function

To review and collate important results regarding the zeros, *x*-intercepts, and factors of a polynomial function so far:

If $(x - c)$ is a factor of *P(x)*, then $x = c$ is a zero of *P(x)*, $P(c) = 0$, and $(c, 0)$ is an *x*-intercept. In fact all four conditions here are equivalent; that is, if one is true they all are, and if one is not true, none are.

Even and Odd Powers of (x - c) Theorem

This topic deals with something that makes graphing a polynomial easier.

Say you have a polynomial of higher order that can be factored thus: $f(x)=(x-1)^2(x+2)^3$. Since $(x-1)$ is a factor, then *(1, 0)* is an x-intercept. Likewise for *(-2, 0)*. If you notice "how many times" $(x-1)$ is a factor of *f(x)*, you are identifying the power of the factor $(x-1)$. Here it is 2, an even number, and the power of $(x+2)$ is 3, an odd number.

Here's the thing, if that power is even then the graph of the polynomial only touches the x-axis and turns around; whereas, if that power is odd, then the graph crosses the x-axis. Like this:

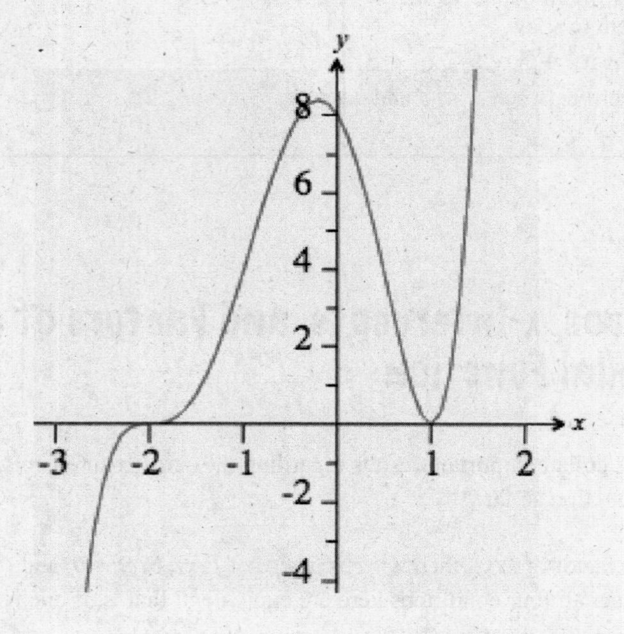

Even and Odd Powers of (x - c) Theorem: Test Prep

Given that $P(x) = x(x-6)^2(x+1)^2$, name its zeros and determine whether its graph crosses or only touches the x-axis at each zero.	• $x = 0$ -- multiplicity 1, odd; therefore, crosses • $x = 6$ -- multiplicity 2, even; therefore, only touches • $x = -1$ -- multiplicity 2, even; therefore, only touches
Try Them	
Given that $P(x) = -(x-1.6)^3(x+11)$, name its zeros and determine whether its graph crosses or only touches the x-axis at each zero.	• $x = 1.6$ -- multiplicity odd; therefore, crosses • $x = -11$ -- multiplicity odd; therefore, crosses
Given that $P(x) = (x^2-1)^2$, name its zeros and determine whether its graph crosses or only touches the x-axis at each zero.	• $x = 1$ -- multiplicity even; therefore, only touches • $x = -1$ -- multiplicity even; therefore, only touches

Procedure for Graphing Polynomial Functions

Already we know a great deal about the behavior of a polynomial graph. Piece by piece you can put together a reasonable sketch. Here are some ideas to keep in mind while you work on graphing a polynomial function:

1. far-left and far-right behavior
2. y-intercept
3. x-intercept(s) and their powers
4. maximum number of turning points
5. symmetries
6. plot some more points using a T-table

★ **For Fun**
Why can there never be more than one *y*-intercept on the graph of a polynomial function?

Matching:

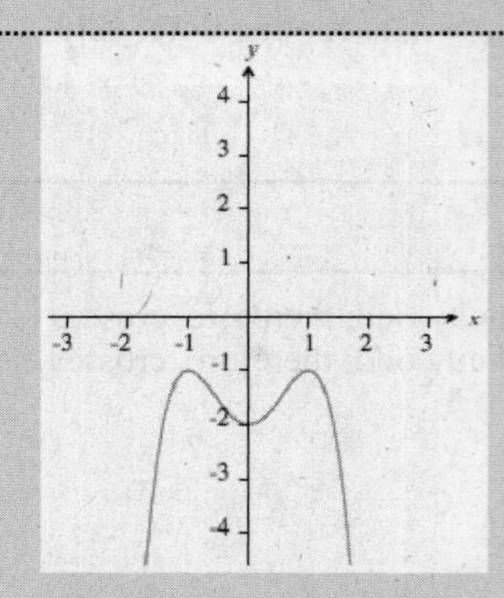

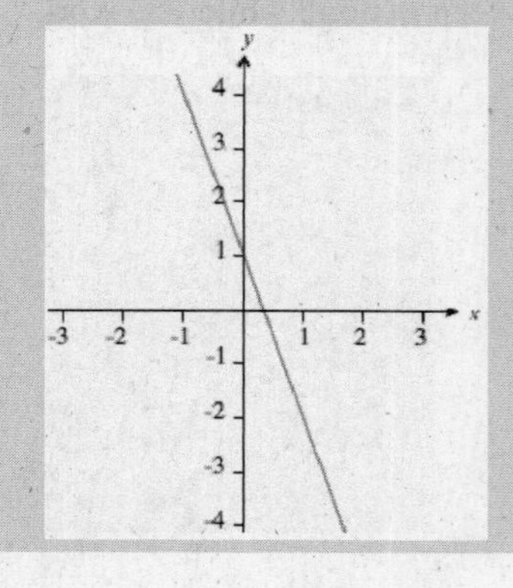

$f(x) = -x^4 + 2x^2 - 2$

$f(x) = x^5 - 2x^3 + x$

$f(x) = -3x + 1$

Procedure for Graphing Polynomial Functions: Test Prep

Identify which of the following is the graph of
$y = -x^4 - 4x^3 - 3x^2 + 4x + 4$:

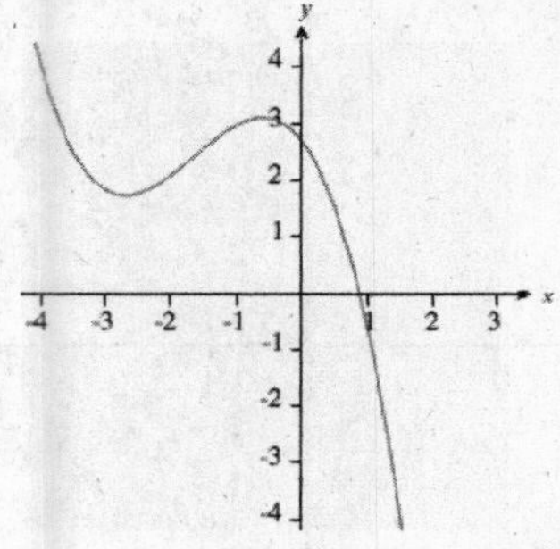

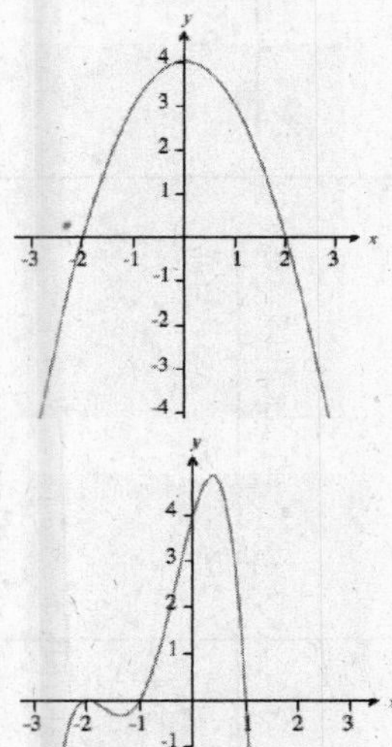

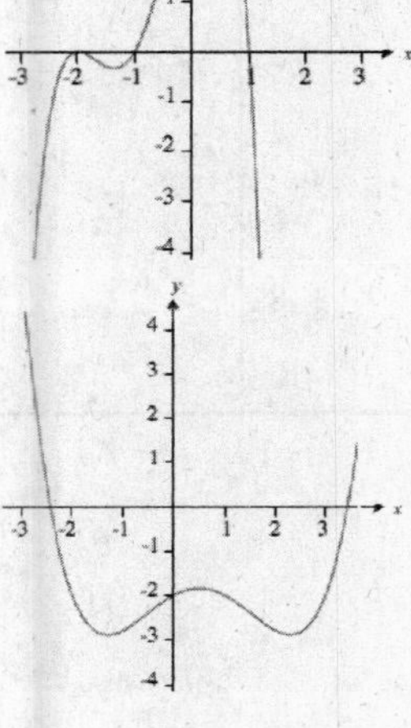

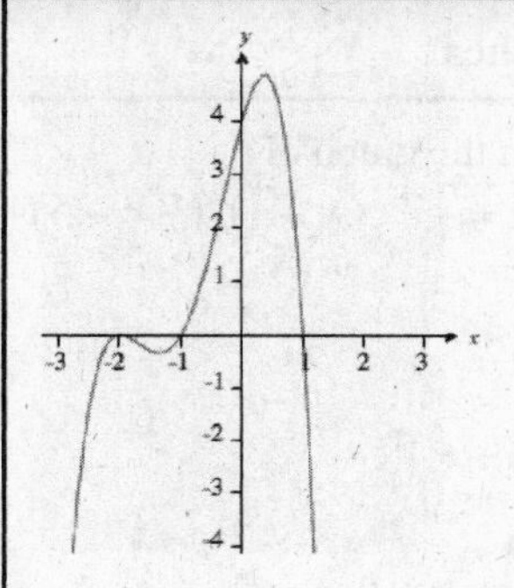

Procedure for Graphing Polynomial Functions: Test Prep

Try Them

Sketch the graph of $P(x) = -2x(x+1)^2(7x-5)$	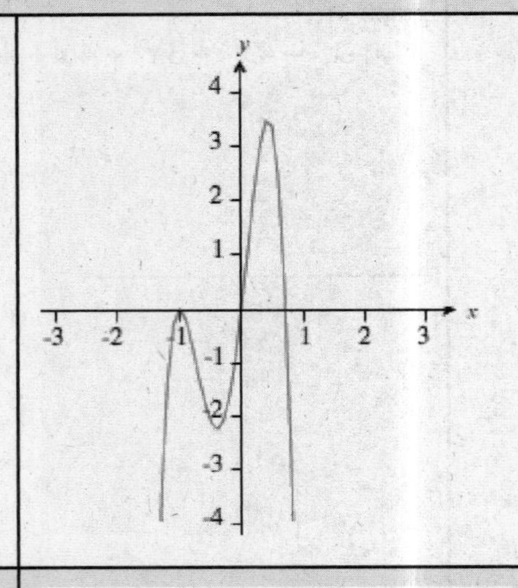
Sketch the graph of $y = x^4 - 6x^3 + 8x^2$	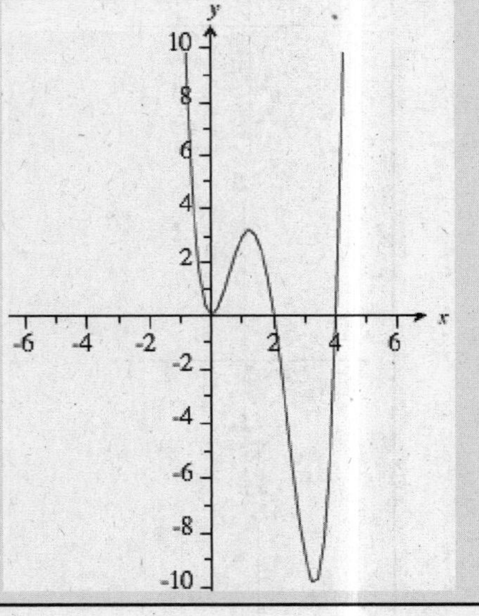

Cubic and Quartic Regression Models

Since we can now do something with polynomials of higher degree, we can also use regression of higher degree. Once again, research how your graphing calculator or spreadsheet program handles this.

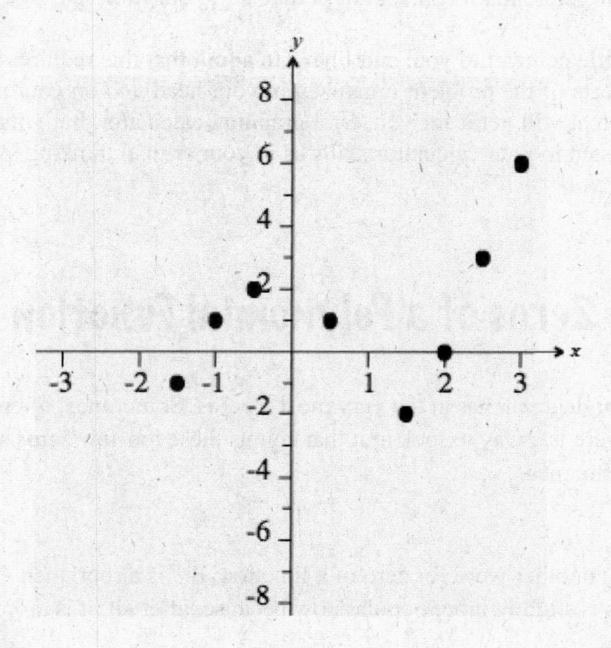

Have you developed the eye to recognize that perhaps a cubic regression might fit this data best? If you don't see that yet, keep looking and thinking -- you'll develop "the eye."

3.3: Zeros of Polynomial Functions

The skills of algebra and critical thinking (that you need to follow all the way through a problem of finding the zeros of a higher degree polynomial) are historical ... and they will stand you in good stead in future math courses and, I dare say, your life.

Of course, that's little comfort to you, and I have to admit that this requires a lot of mental work to keep all the aspects of the problem organized in your head *and* on your paper. Practice is the only thing I know that will get it for you. Or a graphing calculator, but your teacher is probably not so much interested in your calculator skills as in your critical thinking skills.

Best foot forward!

Multiple Zeros of a Polynomial Function

A polynomial of degree n has at the very most n zeros. Sometimes, often even, it has fewer. Well, in fact, there is a way to look at it that counts these too-few zeros several times to make up the difference.

- **roots**
  Root is just another word for zero of a function. If c is a root, then $P(c) = 0$. Calling roots zeros is slightly more popular now because, after all, it is more descriptive.

- **multiple zero**
  A polynomial like we had last section, $f(x) = (x-1)^2(x+2)^3$, has zeros at $x = 1$ and $x = -2$. Because the factors that go with these zeros occur more than once -- $(x-1)$ occurs twice, and $(x+2)$ occurs 3 times -- their zeros are multiple zeros. It's as if this polynomial had 5 zeros, but the zeros were of only two distinct values.

- **zero of multiplicity *k***
  Determining the multiplicity of a zero was what we did last section when we learned if the graph would show an x-intercept that crossed the x-axis or only touched it and turned around. The multiplicity of the zero $x = -2$ in the above polynomial is 3.

✤ **simple zero**
A simple zero has multiplicity of one.

★ **For Fun**
Would the graph cross or just touch the x-axis at a simple zero?

Rational Zero Theorem

Remember the general form for a polynomial function:

$$P(x) = a_n x^n + a_{n-1}x^{n-1} + a_{n-2}x^{n-2} + \cdots + a_2 x^2 + a_1 x + a_0$$

Have you learned to pick out the leading coefficient and the constant term? Good. Now we will use those numbers to list all the possible *rational zeros* of *P(x)*. Rational zeros will be only those zeros that can be written as a rational number -- that means that irrational numbers like $\sqrt{2}$ will never appear on this list. Also, the list can contain many, many more rational numbers than there could possibly be zeros. That's okay; just remember there is a firm maximum number of zeros possible for a polynomial function of degree *n*. (And that max is *n*.)

Let's list all possible rational zeros for $f(x) = 3x^3 - x^2 + 5x - 6$.

1. First list all possible factors of -6, including both positive and negative possibilities: $\pm1, \pm2, \pm3, \pm6$. Call those *p* values.

2. Then list all the possible factors of the leading coefficient: $\pm1, \pm3$. Call those *q* values.

3. Now construct all the possible ratios of *p*'s over *q*'s:

   $$\pm1, \pm2, \pm3, \pm6, \pm\frac{1}{3}, \pm\frac{2}{3}, \pm\frac{3}{3}, \pm\frac{6}{3}$$

   But some of those can be reduced: $\pm1, \pm2, \pm3, \pm6, \pm\frac{1}{3}, \pm\frac{2}{3}, \pm1, \pm2$. And we can remove the duplicates so that we have this list finally:

$\pm 1, \pm 2, \pm 3, \pm 6, \pm \frac{1}{3}, \pm \frac{2}{3}$. Now that's a list with 12 elements. Our polynomial is only degree 3, and so it's impossible for all of these to be zeros. In fact, it's possible for none of them to be zeros.

★ **For Fun**
Do you have an idea of how you might check to see if any of these values actually *are* zeros? Are you thinking "Remainder Theorem and synthetic division" like I am?

★ **For Fun**
How would this list be simpler to construct if the leading coefficient were 1?

Rational Zero Theorem: Test Prep	
List all possible rational zeros of $P(x) = -3x^3 - x + 1$	$p: \pm 1$ $q: \pm 1, \pm 3$ $\frac{p}{q}: \pm 1, \pm \frac{1}{3}$
Try Them	
List all possible rational zeros of $T(x) = -x^4 - 4x^3 - 3x^2 + 4x + 4$	$\frac{p}{q}: \pm 1, \pm 2, \pm 4$
List all possible rational zeros of $f(x) = -2x^4 + 3x - 4$	$\frac{p}{q}: \pm 1, \pm \frac{1}{2}, \pm 2, \pm 4$

Upper and Lower Bounds for Real Zeros

There are some swell little procedures that help rule out some of the surplus possible rational zeros in your list. One of them is to discover when your x-value is either larger than the largest possible zero or when it's smaller than the smallest possible zero.

By the way, you need to be pretty good at synthetic division by now in order to proceed with these tests.

✤ **upper bound**

In the context of finding the zeros of polynomials, "upper bound" means a number past which there are no more zeros. Say if you know the polynomial value goes towards infinity by its far-right behavior. Isn't there a point on the x-axis past which (in the positive direction) there are not more x-intercepts (and hence no more zeros)?

How do you recognize one? Carry out synthetic division using your possible *positive* zero. If the bottom, result row has only positive values listed, then there cannot be any zeros larger than the one you are testing -- whether or not it turned out to be a zero.

By the way, a zero on the bottom row can be counted as positive for this test. And remember this test only works for *positive* test values.

This synthetic division shows that 5 is an upper bound. All possible zeros larger than 5 can now be disregarded:

$$\begin{array}{r|rrrr} 5 & 1 & 0 & -19 & -28 \\ & & 5 & 25 & 30 \\ \hline & 1 & 5 & 6 & 2 \end{array}$$

Meaning and the number line:

If you discover that $x = 2$ is an upper bound for the zeros of this polynomial, is it still possible that $x = 1.5$ is also an upper bound?

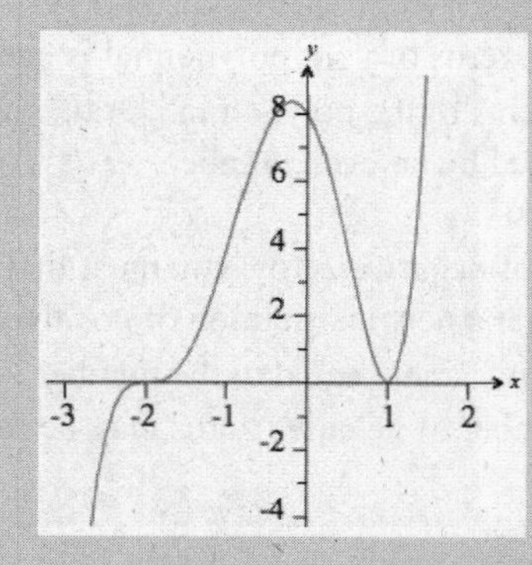

✤ **lower bound**
And a lower bound is an x-value past which (in the negative direction) there can be no more zeros.

This time the test is a tiny bit harder. The values in the bottom row of your synthetic division must alternate in sign in order for the test value to be a lower bound. If the values do alternate, you need not look for a zero smaller than your test.

Oh, yeah, this test only works for *negative* test values. Also, zero can be either positive or negative, but it cannot be both -- that is, you cannot change its sign once you've fixed it.

Here's a result that shows -5 is a lower bound:

-5	1	0	-19	-28
		-5	25	-30
	1	-5	6	-58

Descartes' Rule of Signs

Descartes' Rule of Signs is another way to discard possibilities from your list of potential rational zeros. It does require more algebra -- nothing challenging, but a little bit time-consuming.

✤ **Descartes' Rule of Signs**
This rule has two parts, and it applies to polynomials with their terms listed in strictly descending order:

1. The number of *positive* zeros of your polynomial is equal to the number of sign changes of the coefficients of the polynomial. Or the number of positive zeros may be that number decreased by an even number.

2. To test for the number of *negative* zeros, you must first replace x with $-x$ in your polynomial, simplify, get a new distribution of positive/negative coefficients. The number of negative zeros is now equal to the number of sign changes in the new polynomial. Or the number of negative zeros may be decreased by an even number.

✤ **variations in sign**
Variations in sign refers to the number of sign changes in the coefficients of a polynomial when the polynomial is written in descending order of terms. We also assume all the coefficients are real.

★ **For Fun**
Count the number of sign changes in these polynomials:

$3x^3 - x^2 + 5x - 6$

$-x^4 + 3x^2 - x + 11$

$-x + 3x^2 - 2$

$\frac{x^4}{4} + 2x^3 - x^2 + \frac{x}{2}$

$x^3 - 5x^2 + 7x - 8$

Now change all the x's to $-x$'s, simplify, and count the sign changes again.

Descartes' Rule of Signs: Test Prep

Use Descartes' Rule of Signs to state the number of possible positive and negative real zeros of $f(x) = 2x^4 - 19x^3 + 51x^2 - 31x + 5$	• There are four sign changes in $f(x)$, so 4, 2, or no positive real zeros. • $f(-x) = 2x^4 + 19x^3 + 51x^2 + 31x + 5$ has no sign changes, so there are no negative real zeros.
Try Them	
Use Descartes' Rule of Signs to state the number of possible positive and negative real zeros of $P(x) = 3x^3 + 11x^2 - 6x - 8$	• one positive real zero • two or no negative real zeros
Use Descartes' Rule of Signs to state the number of possible positive and negative real zeros of $P(x) = x^3 - 4x^2 - 3x$	• one positive real zero • one negative real zero

Zeros of a Polynomial Function

Here is a list of suggestions to consider when you begin to find all the zeros of a polynomial of higher degree:

- the degree of the polynomial
- the list of potential rational zeros: $\frac{p}{q}$
- Descartes' Rule of Signs and the possible number of positive or negative zeros
- upper and lower bounds on zeros
- reducing polynomials -- the coefficients are in the bottom, result row once you have found a zero
- all your quadratic skills once the polynomial is reduced to second degree

This can be a slightly long procedure. Have a care and be patient. Finding the zeros of a higher order polynomial is really a reward in itself. Think of it like this: since time immemorial man has been working on this problem and you've joined those ranks!

Besides the quadratic formula, which is used to find the zeros of a second-degree polynomial, there are also general solutions up to quartic polynomials. However, it has been proven that polynomials of degree 5 or more do not produce zeros according to a formula in terms of their coefficients.

You know the quadratic formula. Can you find the "cubic formula" somewhere? One look and you'll understand why we don't memorize this one. It's easier and quicker to try the methods of this chapter!

Guidelines for Finding the Zeros of a Polynomial Function with Integer Coefficients: Test Prep

Find the zeros of $P(x) = 2x^3 + 9x^2 - 2x - 9$ and state their multiplicities.	$p{:}\pm1,\pm3,\pm9$ $q{:}\pm1,\pm2$ $\frac{p}{q}{:}\pm\frac{1}{2},\pm1,\pm\frac{3}{2},\pm3,\pm\frac{9}{2},\pm9$ $\quad[ordered]$ $\begin{array}{r\|rrrr} 3 & 2 & 9 & -2 & -9 \\ & & 6 & 45 & 129 \\ \hline & 2 & 15 & 43 & 120 \end{array}$ Three is not a zero, but it is an upper bound. So $\frac{p}{q}{:}\pm\frac{1}{2},\pm1,\pm\frac{3}{2},-3,-\frac{9}{2},-9$ has fewer elements. $\begin{array}{r\|rrrr} 1 & 2 & 9 & -2 & 9 \\ & & 2 & 11 & 9 \\ \hline & 2 & 11 & 9 & 0 \end{array}$ One is a zero. (Not only is it, too, an upper bound, but it is the only positive zero by Descartes' Rule.) Now find the zeros of the reduced polynomial, which can be factored: $2x^2 + 11x + 9 = (2x+9)(x+1)$ The zeros are $x = 1, -1, -\frac{9}{2}$
Try Them	
Determine the zeros, and their multiplicity, of $f(x) = (x-1)^2(x+2)(x+4)$	• $x = 1$, multiplicity 2 • $x = -2$, multiplicity 1 • $x = -4$, multiplicity 1
Find all the real zeros of $P(x) = 3x^5 + 10x^4 - 2x^3 - 28x^2 - 8x + 16$	$x = \frac{2}{3}, -2, \sqrt{2}, -\sqrt{2}$

Applications of Polynomial Functions

Now translate word problems into polynomial functions and solve. You're not restricted to degree one or two polynomials for these purposes any more!

MID-CHAPTER QUIZ: TEXTBOOK PAGE 299.

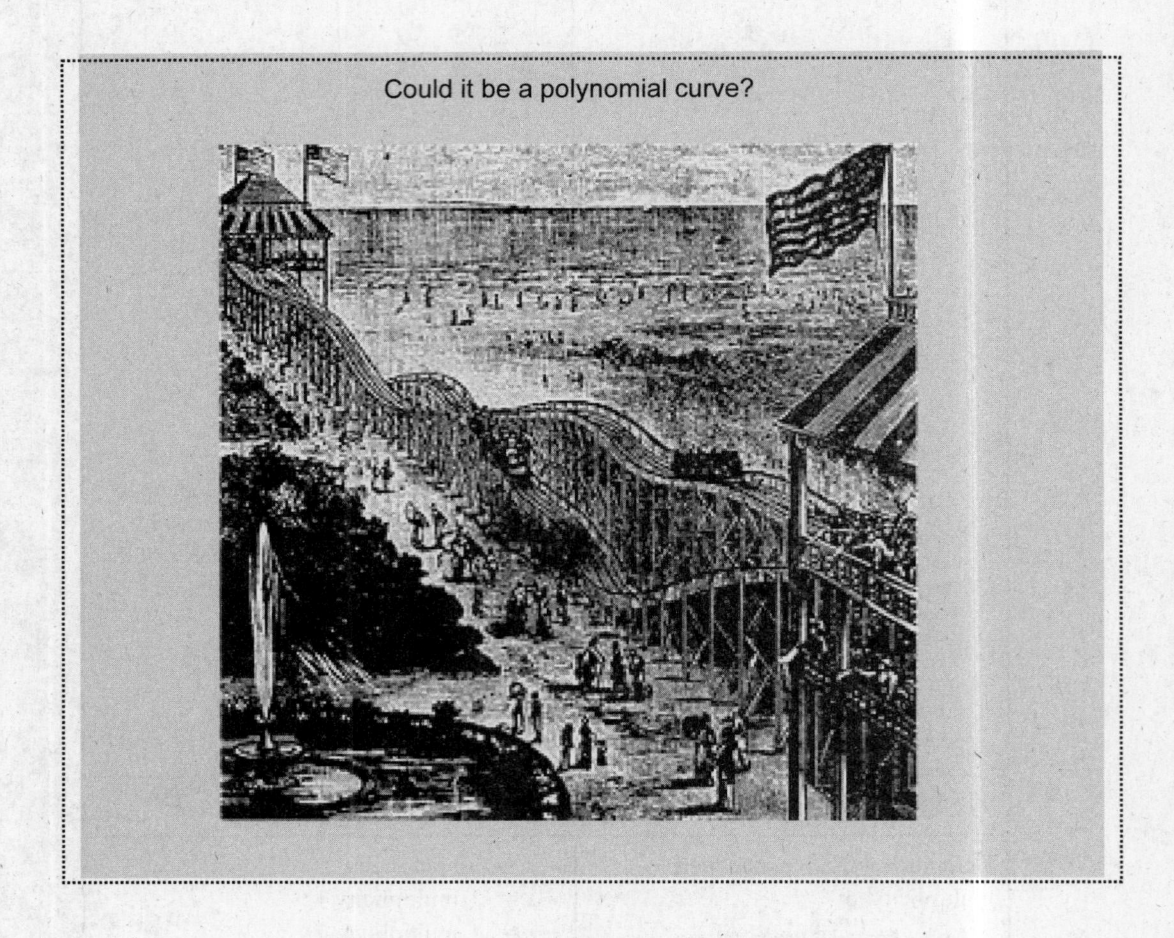

Could it be a polynomial curve?

3.4: Fundamental Theorem of Algebra

Up until now all we've said is about the number of zeros that a polynomial can have is the *maximum* number of zeros. Now we're going to say something more definitive about the number of zeros of a polynomial (and, by the way, it will be consistent with what has gone before). This ought to help our comprehension and our graphs.

Fundamental Theorem of Algebra

What this subsection really does is introduce us back to *complex* numbers (remember them from Chapter P?) and to consideration of complex numbers as possible zeros of higher order polynomials. We're always adding more layers of subtlety and complexity to our algebraic machinations!

✤ **Fundamental Theorem of Algebra**
The Fundamental Theorem of Algebra states that if your polynomial is of degree greater than one (greater than linear, that is) and if it has any complex coefficients, then it will have at least one complex zero.

Number of Zeros of a Polynomial Function

If you count each zero of a polynomial according to its multiplicity (section 3.3), then every polynomial of degree n has n zeros. By extension then, the polynomial must have n linear factors.

I make no promises that these zeros and factors will be easy to find, but they do exist.

✤ **existence theorems**
Existence theorems are the sort of theorems that assert the existence of *something* mathematical, like the existence of a certain number of zeros for one example.

How's this for an existence theorem?
"I think, therefore I am."
Do you know who said that? Yep, it's our pal René Descartes *again*.

The Fundamental Theorem of Algebra: Test Prep	
How many complex zeros, counted multiplicity-wise, does $P(x) = -2x^5 + 3x^4 + x^2 - x + 12$ have?	Since the degree of *P(x)* is 5, *P(x)* has 5 complex zeros, counted multiplicity-wise.
Try Them	
Name all zeros, and their multiplicities, of $S(x) = x^2(x-1)^3(x+\sqrt{2})(x-\sqrt{2})(x-3-i)(x-3+i)$	• $x = 0$, multiplicity 2 • $x = 1$, multiplicity 3 • $x = -\sqrt{2}$, multiplicity 1 • $x = \sqrt{2}$, multiplicity 1 • $x = 3+i$, multiplicity 1 • $x = 3-i$, multiplicity 1
Solve $2x^4 - 3x^3 - 7x^2 - 8x + 6 = 0$	$x = 3, \frac{1}{2}, -1+i, -1-i$

Conjugate Pair Theorem

Remember the complex conjugate (section P.6)? The Conjugate Pair Theorem gives up two zeros for the price of one:

Conjugate Pair Theorem

If $a + bi$ (where b is not equal to zero) is a complex zero of a polynomial with real coefficients, then the conjugate, $a - bi$, is automatically a zero too.

★ **For Fun**

Consider that if you already know that, say, $3-2i$ is a zero of your polynomial. Then you know that its conjugate $3+2i$ is also a zero. You also know that $\left(x-(3-2i)\right)$ *and* $\left(x-(3+2i)\right)$ are also factors of the polynomial. If you multiply those two linear factors together, will the result be a factor of your polynomial, too? What degree will that resulting factor be? If it is a factor, can you find the reduced polynomial given this new factor? At this point, at least try simplifying and multiplying out the two factors above.

Conjugate Pair Theorem: Test Prep

Given that $x = -i$ is a zero of $P(x)=3x^4+2x^3+2x^2+2x-1$, find all the zeros.	$x=i,-i$ are zeros; therefore, $(x-i),(x+i)=\left(x^2+1\right)$ is a factor. By long division $\dfrac{3x^4+2x^3+2x^2+2x-1}{x^2+1}=3x^2+2x-1$ $=(3x-1)(x+1)$ So all four zeros are $x=i,-i,-1,\dfrac{1}{3}$
Try Them	
If $x=2-2i$ is one zero of a polynomial with real coefficients, list another zero and the factors that correspond to these two zeros.	$x=2+2i$ $(x-2+2i),(x-2-2i)$
Solve $x^4-4x^3+6x^2-16x+8=0$ given that $x = 2i$ is one zero.	$x=2i,-2i,2+\sqrt{2},2-\sqrt{2}$

Finding a Polynomial Functions with Given Zeros

By the Factor Theorem, you know that zeros and factors go together. Also now that we know the Conjugate Pair Theorem, we know about conjugates that are automatically included, along with their factors, in our knowledge of the zeros of a given polynomial. Well, if you know the zeros, then you know the factors. If you know the factors, couldn't you multiply them and "reconstitute" a polynomial from the zeros? In order to get a more precise polynomial, you may also need information about its degree and maybe even its leading coefficient. (In other words, specifying the zeros is not enough information to make a polynomial that is unique to those zeros.)

Find a Polynomial Function With Given Zeros: Test Prep	
Find a polynomial function P, with real coefficients, that has zeros $x = 4, 1+2i$ and degree 3.	$P(x) = (x-4)(x-1-2i)(x-1+2i)$ $= (x-4)(x^2-2x+5)$ $= x^3-6x^2+13x-20$ (for example)
Try Them	
Write a polynomial P, with integer coefficients having the following zeros: $x = 2, -3, -\frac{1}{2}$.	$P(x) = 2x^3+3x^2-11x-6$ (for example)
Find a polynomial function P, of lowest degree with real coefficients, lead coefficient of 1, and the following zeros: $x = 3i$, multiplicity 1, and $x = 2$, multiplicity 2.	$P(x) = x^4-4x^3+13x^2-36x+36$

3.5: Graphs of Rational Functions and Their Applications

Now that we have pretty thoroughly investigated finding the zeros of polynomials, we'll consider rational functions again. After all, they are nothing more than a ratio of polynomials, and knowing how to find the zeros of the numerator and denominator might come in handy!

Vertical and Horizontal Asymptotes

Finding the asymptotes in order to graph a rational function has everything to do with being able to find zeros of polynomials.

★ **For Fun**
Look up the word *asymptote*. What other English words is it related to? Make sure you understand what it means to have an asymptote on your graph. Can you make that meaning hang together with the word's meaning in English?

✤ **rational function**

A rational function is a ratio of polynomials, like so: $R(x) = \dfrac{P(x)}{Q(x)}$ where P and Q are polynomials. We studied rational functions in a more general way back in Chapter P.

Partial fractions decomposition is a method that depends on rational functions. We'll meet them in Section 6.4.

✤ **vertical asymptote**
A vertical asymptote is a vertical line (of the form $x = a$) on the graph of a function that is not actually a part of the graph of the function (hence it is a dashed line) but is a boundary that the function never crosses. However, the function will get extremely close, horizontally, to a vertical asymptote.

Here is a graph with a vertical asymptote:

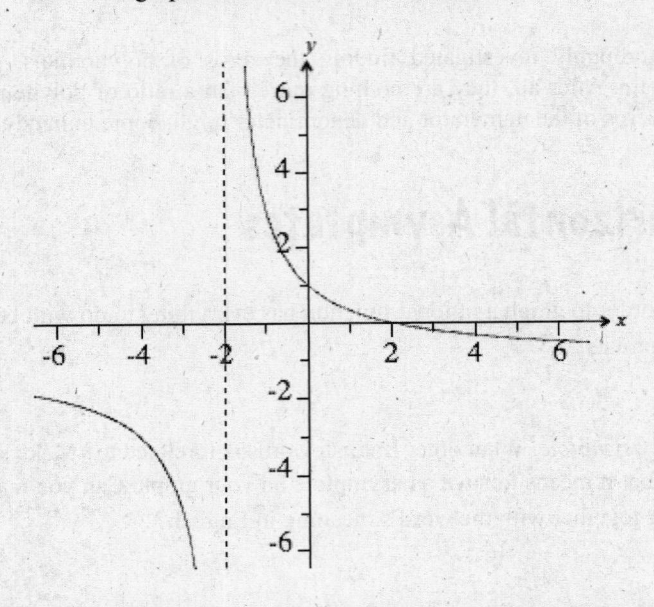

You'll notice that the vertical asymptote has to do with x values that are excluded from the domain. What do you remember about the domain of rational functions? Remember you cannot divide by zero, and so if you find where the denominator of a rational function goes to zero (but where the numerator does not also go to zero), then you've found that function's vertical asymptote(s).

✤ **horizontal asymptote**
Horizontal asymptotes are horizontal lines ($y = b$) also included on the graph of a function to show horizontal lines that the actual graph of the function cannot cross. (Whereas a function never crosses its vertical asymptote, there are some cases where a graph can cross a horizontal asymptote but only near the origin. More about that very soon!) Again, horizontal asymptotes are also shown as dashed lines because they are not actually part of the function.

In order to find your horizontal asymptotes, you have to be able to ascertain the degrees and sometimes the leading coefficients of the polynomials in the numerator and denominator. There are three cases:

1. If the degree of the numerator is larger than the degree of the denominator, then there are *no* horizontal asymptotes.

2. If the degree of the denominator is larger than the degree of the numerator, you automatically get exactly one horizontal asymptote: $y = 0$ (that's the x-axis).

3. If the degree of the numerator and the denominator are equal then you have a horizontal asymptote at the ratio of the leading coefficients, like so:

   $R(x) = \dfrac{2x-1}{x+5}$ yields the horizontal asymptote: $y = \dfrac{2}{1} = 2$. (By the way, do you have a clear idea in your mind of what the line $y = 2$ looks like on a graph?)

   Here's an example of a graph with a horizontal asymptote at $y = 2$.

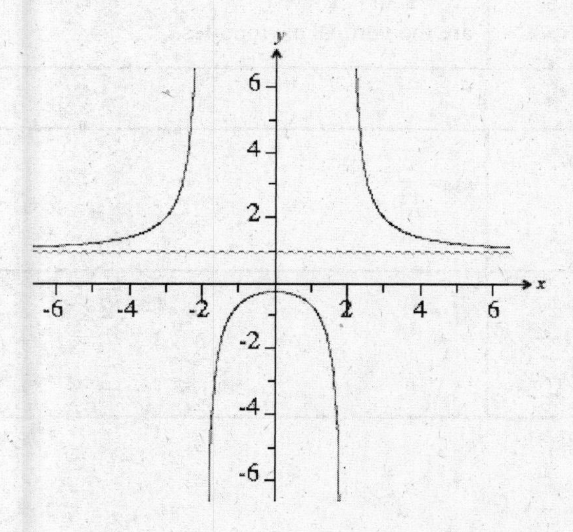

★ **For Fun**
Do you detect the presence of any vertical asymptotes in the above graph? Could you guess at what the denominator might be like?

✓ **Questions to Ask Your Teacher**
Be sure you understand in what form you should answer questions about asymptotes. After all asymptotes are actually lines and they ought to be expressed as equations for lines -- it's the least they deserve.

✓ **Questions to Ask Your Teacher**
Sometimes there is an "escape hatch" near the origin where a function may actually cross its horizontal asymptote. See if you can find out why, but be sure you know how to graph these instances!

Vertical Asymptotes: Test Prep	
Identify the vertical asymptote(s) of $\frac{x}{x^2-1}$	$x^2-1=(x-1)(x+1)=0$ $x=1$ and $x=-1$ are the vertical asymptotes.
Try Them	
Determine the domain of $f(x)=\frac{x+1}{3x-5}$	$x \neq \frac{5}{3}$
Identify the vertical asymptote(s) of $\frac{3x-1}{x^3-2x}$	$x=0, \quad x=\sqrt{2}, \quad x=-\sqrt{2}$

Horizontal Asymptotes: Test Prep	
Find the horizontal asymptote of $R(x)=\frac{6x^2-5}{2x^2+6}$ if any.	The degree of the numerator and the denominator are the same; therefore, there is a horizontal asymptote at the ratio of the leading coefficients: $y=\frac{6}{2}$ $y=3$ is the horizontal asymptote.

Horizontal Asymptotes: Test Prep

Try Them	
Find all vertical and horizontal asymptotes of $R(x) = -\dfrac{2}{x^2 - 7x + 10}$	$v.a.: x = 5, \quad x = 2$ $h.a.: y = 0$
Find all asymptotes of $R(x) = \dfrac{2x^2}{x^2 + 4}$	no vertical asymptotes $h.a.: y = 2$

Sign Property of Rational Functions

If you find all the zeros of the numerator and the denominator, you can set up a test number line and discover the *sign* of the rational function in each region:

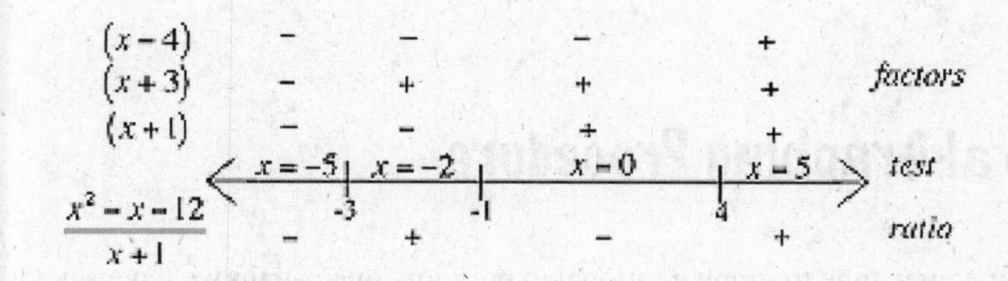

(We investigated this in Section 1.4.)

For a graph of the rational function $\dfrac{x^2 - x - 12}{x + 1}$, this number line lets me know that the graph will be *below* the *x*-axis from $-\infty$ to -3, then above the *x*-axis from -3 to -1, and so on.

Match regions of *positive* and *negative* function value according to our test number line from the previous page and this graph:

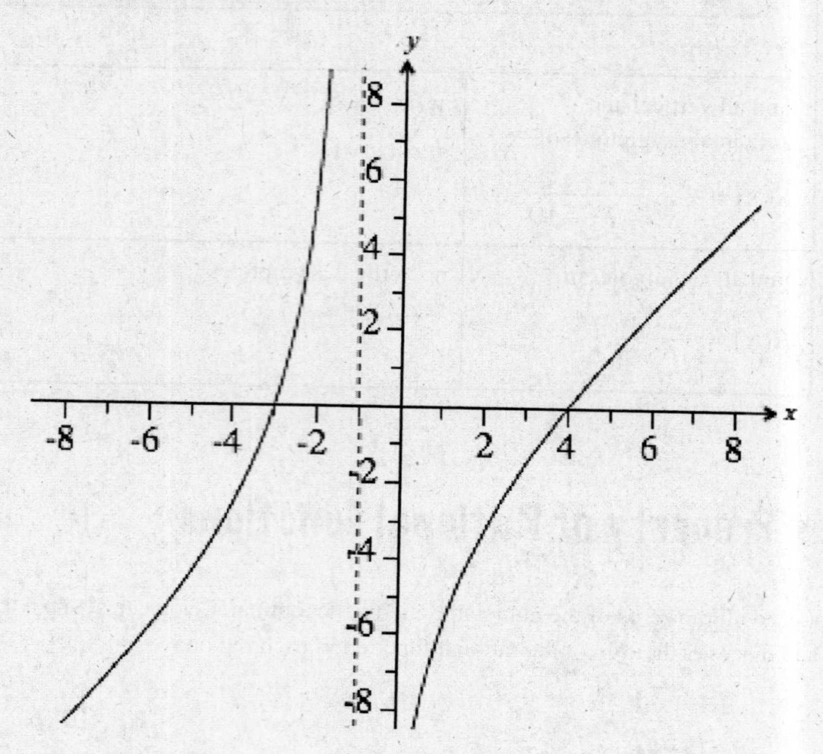

General Graphing Procedure

So when it comes time to graph a rational expression, once again we have several bits of information available to us to help construct a more complete picture. Consider many, if not all, of these aspects while you work:

1. asymptotes and behavior near asymptotes
2. intercepts
   (The y-intercept arises from an x-value of zero. x-intercepts arise from setting the numerator equal to zero.)
3. symmetry
4. sign behavior
5. plotting some extra points
   (Particularly consider each region of the domain as it might be broken up by vertical asymptotes. Also pay special attention near the origin where a horizontal asymptote "escape hatch" might exist.)

Slant Asymptotes

If you consider the graph of $\dfrac{x^2 - x - 12}{x+1}$ from a page ago, do you see possible evidence of an asymptote that is neither horizontal or vertical, but instead *slants* from the third quadrant into the first quadrant? There is indeed a slant asymptote on that graph.

✣ **slant asymptote**
A slant asymptote is an asymptote that adheres to the general form for a line, $y = mx + b$, instead of being strictly vertical or horizontal. Once again a function will approach but not cross a slant asymptote as the function grows towards positive or negative infinity in either direction.

A slant asymptote arises when the degree of the numerator is exactly one larger than the degree of the denominator, like $\dfrac{x^2 - x - 12}{x+1}$. The equation for said asymptote is found by carrying out the division implied by the rational function itself, which I can do by synthetic division in this case:

$$\begin{array}{r|rrr} -1 & 1 & -1 & -12 \\ & & -1 & 2 \\ \hline & 1 & -2 & 10 \end{array}$$

Disregard the remainder altogether and write the *linear* result. In the case it is, from the bottom line of my synthetic division above, $(y =)x - 2$. (I include the $y =$ to point out that this is an equation for a line in slope-intercept form. You can graph that (dashed line again since it's an asymptote and not actually part of the graph of the function).

Look back at our graph of the rational function $\dfrac{x^2 - x - 12}{x+1}$ and see if this equation doesn't match the asymptote behavior you can observe right there.

Slant Asymptotes: Test Prep	
Find the slant asymptote of $R(x) = \dfrac{x^3 - 4x}{x^2 - 1}$	By long division, $\dfrac{x^3 - 4x}{x^2 - 1} = x + \dfrac{-3x}{x^2 - 1}$ So $y = x$ is the slant asymptote (disregarding the remainder).
Try Them	
Pick the functions with slant asymptotes: • $\dfrac{x+1}{2x-1}$ • $\dfrac{3x^2 - x - 1}{x}$ • $-4x^3 - x + 8$ • $\dfrac{11 - x - 3x^3}{4 - x^2}$	• $\dfrac{3x^2 - x - 1}{x}$ • $\dfrac{11 - x - 3x^3}{4 - x^2}$
Determine the slant asymptote of $R(x) = \dfrac{-x^3 - 4x^2 - x + 4}{x^2 - x - 6}$	$y = -x - 5$

Can you tell when your graphing utility is/is not literally displaying a slant asymptote for a function that does, indeed, have one?
This graph has one. Can you picture it?

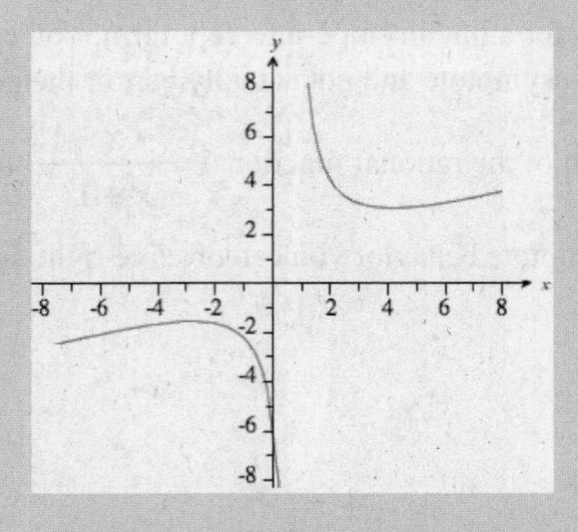

A General Procedure for Graphing Rational Functions That Have No Common Factors: Test Prep

Choose the graph of $R(x) = \dfrac{x^3 - 4x}{x^2 - 1}$:

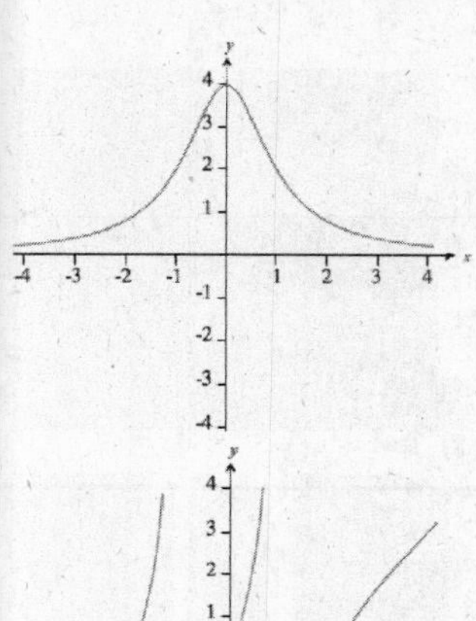

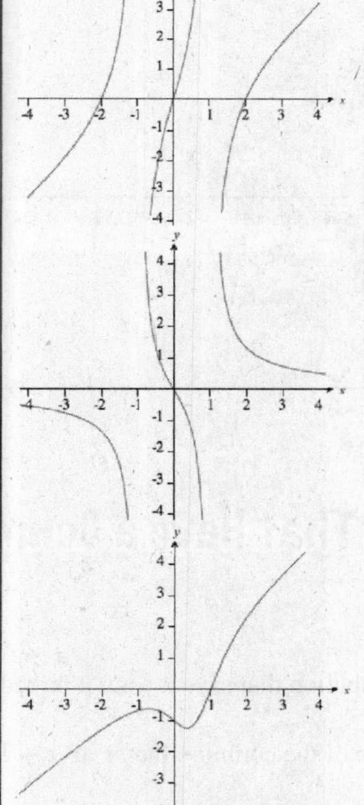

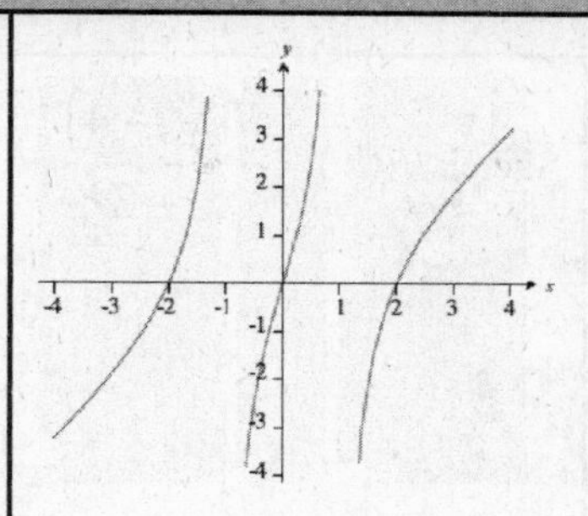

A General Procedure for Graphing Rational Functions That Have No Common Factors: Test Prep

Try Them

Sketch $R(x) = \dfrac{7(x^2 - 4)}{x^3 + 3x}$

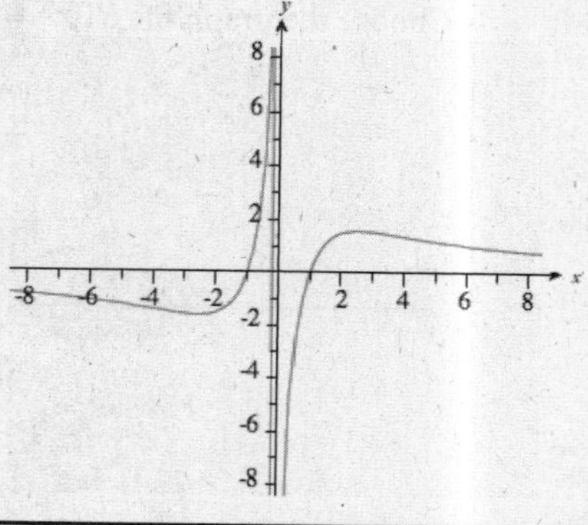

Sketch $R(x) = \dfrac{8 + 3x + 2x^2}{9 - x^2}$

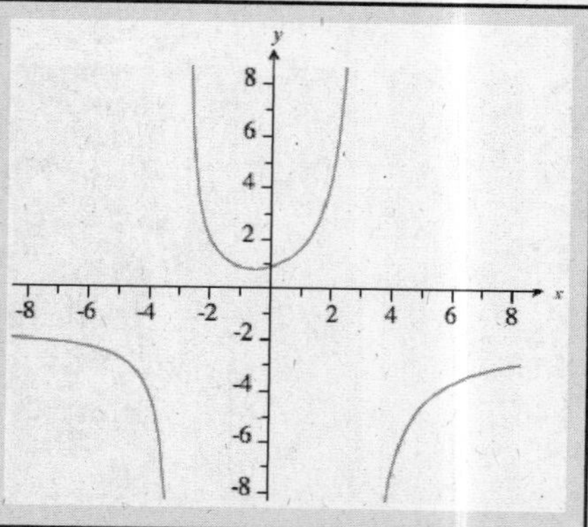

Graphing Rational Functions That Have a Common Factor

There is nothing about the definition of a rational function that says it cannot be reduced like $\dfrac{x^2 - 1}{x + 1} = \dfrac{(x - 1)(x + 1)}{x + 1}$ can be "reduced" because of the common factor of $x + 1$, right?

If you notice, both the denominator and the numerator will be equal to zero if you plug in $x = -1$. This means there is a common factor.

This also means that although the x-value in question is still *not* in the domain of the rational function, a vertical asymptote does *not* arise on the graph. Have you ever seen a graph with a hole in it?

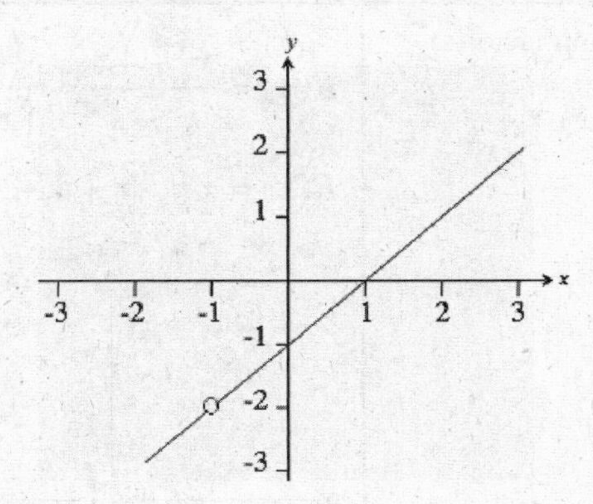

That's what happens when a rational function can be "reduced." That zero shared by the numerator and the denominator becomes a hole! And the rest of the graph looks like the reduced fraction. Simple.

Regarding Calculators:

Graphing your rational functions by using a graphing calculator is one of the modern marvels that we can be very thankful for.

However, be sure you thoroughly understand how your calculator displays asymptotes and the behavior of the graph of the function near asymptotes.

1. Asymptotes are not actually part of the graph of the function itself -- asymptotes are more of an aid to graphing. They should be dashed lines if they appear, but your calculator may show them as solid. Do you understand rational functions enough to tell that?

2. Except for horizontal asymptotes near the origin, graphs of rational functions do not cross their asymptotes. Nevertheless, your viewing window may not be able to clearly abide by that constraint. Just because your calculator makes it look as if a graph is touching its asymptote, will you be able to decipher when that is an artifact of the resolution of your graphing calculator and not actually taking place?

A General Procedure for Graphing Rational Functions That Have a Common Linear Factor: Test Prep

Find the common factor(s) and sketch $R(x) = \dfrac{x^3 - 4x}{x^2 - 4}$	$R(x) = \dfrac{x^3 - 4x}{x^2 - 4} = \dfrac{x(x^2 - 4)}{(x^2 - 4)}$ $R(x) = x \qquad x \neq 2, -2$ 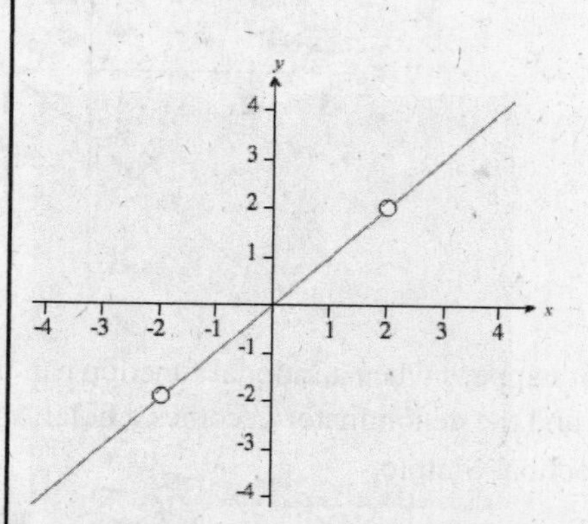

Try Them

Sketch $y = \dfrac{x^2 - 3x}{x}$	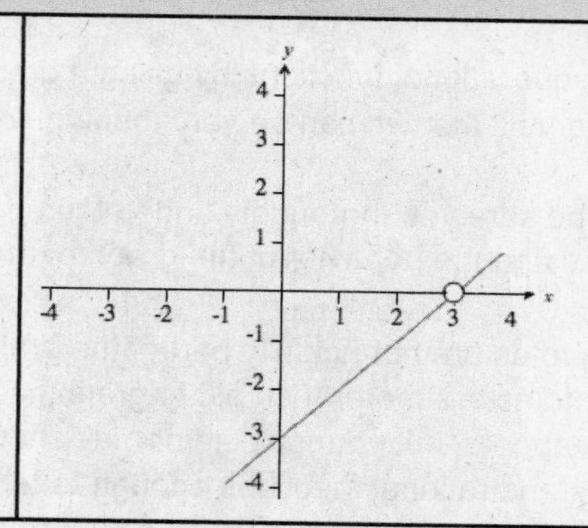

A General Procedure for Graphing Rational Functions That Have a Common Linear Factor: Test Prep

Sketch

$$f(x) = \frac{x^3 + x^2 + 11x}{3x^2 - 2x}$$

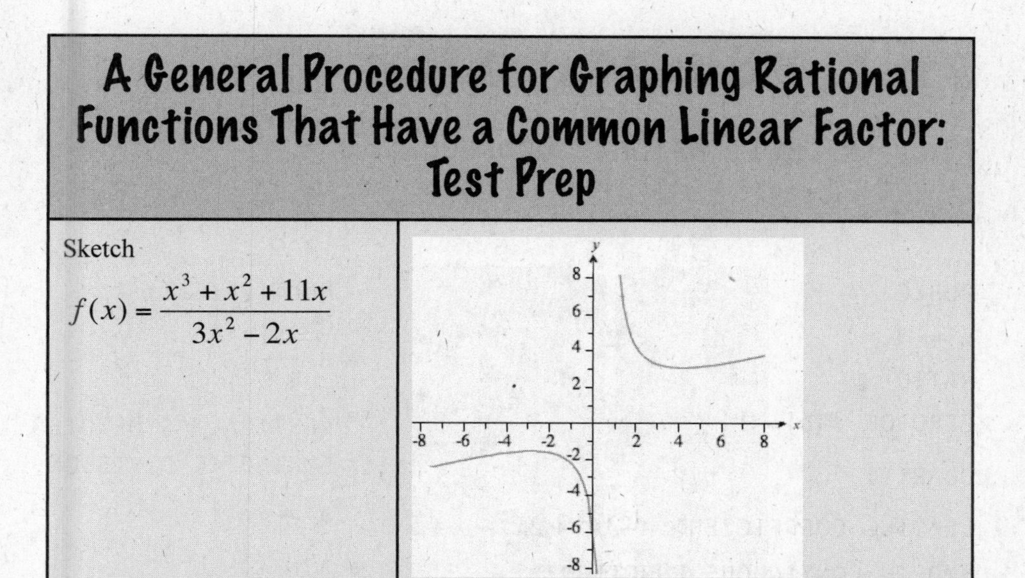

Applications of Rational Functions

Now your task will be to translate word problems into rational functions and solve them using your skills with rational functions. (This is, indeed, getting challenging!)

Final Fun: Matching

FIND THE BEST MATCH. GOOD LUCK.

CUBIC	MULTIPLICITY k
$x = -1$	$\frac{x^2 - 1}{x + 1}$
RATIO	VERTICAL ASYMPTOTE
ZERO OF $P(c) = 0$	SYNTHETIC DIVISION
QUARTIC	$x^3 - x$
LEADING COEFFICIENT: $-3x^3 + 2x^2 - x + 5$	$x^2 - x^4 + 1$
SMOOTH CONTINUOUS FUNCTION	c
$(x-2)^k$	$\frac{p}{q}$
$\begin{array}{r\|rrrr} 3 & 1 & 0 & -19 & -28 \\ & & 3 & 9 & -30 \\ \hline & 1 & 3 & -10 & -58 \end{array}$	-3
	MULTIPLICITY k
RATIONAL ZEROS	$f(x) = x^3 - x + 4$

NOW TRY THESE:

REVIEW EXERCISES: TEXTBOOK PAGE 328,

PRACTICE TEST: TEXTBOOK PAGE 331.

STAY UP ON YOUR GAME -- DO THE CUMULATIVE REVIEW EXERCISES: TEXTBOOK PAGE 332

Chapter 4: Exponential and Logarithmic Functions

Word Search

```
N I M A N C S T Q C K C C P E
D O T A O O S W O L I D L R F
N Z I M G E I N A M S R A I I
B A M T R N C T H D F I I N L
Q O T E I A I T A D N C T C F
N E T U V S I T K U D H N I L
D N A I R R O Y U J Q T E P A
I B T T A A Z P A D E E N A H
P Y R G A T L D M C E R O L O
B W O H T W O R G O E S P E L
U L Y T R E P O R P C D X S U
U A W H I J I N V E R S E A Z
U J Q L Y L D Z D P W T P B M
S E I S M O G R A P H T M L W
T E L C W S I B K B Z L O G Q
```

BASE	EXPONENTIAL	MAGNITUDE
COMMON	GROWTH	NATURAL
COMPOSITION	HALFLIFE	PRINCIPAL
CONCAVITY	INTEREST	PROPERTY
DECAY	INVERSE	RICHTER
EQUATION	LOGARITHMIC	SEISMOGRAPH

4.1: Inverse Functions

Every time you solve even the simplest equations, you are using inverse functions to help you. The idea is that multiplication by 3 is the inverse of division by 3 and vice versa. Keep in mind that when you're solving equations, you're working towards a result that has the variable all by itself, like: x = *something*. Inverse functions *do* exactly this. We'll see how a little bit more formally.

Introduction to Inverse Functions

In this section, we check into some of the implications of how inverse functions work with/against each other -- in equations and in graphs. This is good stuff by itself, but it's also preparation for two new functions that we are due to meet later in the chapter -- two new functions that are inverses of each other. Understanding now the implications (not just mechanics) of inverse functions will help you more fully understand what you're doing later in this very chapter.

✤ **inverse function**

Simply put, one function is the inverse of another if every ordered pair in one is exactly reversed in the other. The domain of one becomes the range of the other and vice versa.

If you think about it, this means that what one function does is undone by its inverse. If you put a 3 in and get a 2 out of a function -- *(3, 2)* -- its inverse would undo that -- *(2, 3)* -- and you get your original 3 back. See?

★ **For Fun**

Think about that in terms of the generic ordered pair, $(x, f(x))$. Can you see how getting an x back is like solving an equation.

In order for a function to have an inverse, said function must be one-to-one (see Chapter Two).

Graphs of Inverse Functions

Since an inverse function swaps domain and range with its counterpart, its graph has a very distinct relationship with the original function.

That relationship is one of a reflection about the line $y = x$:

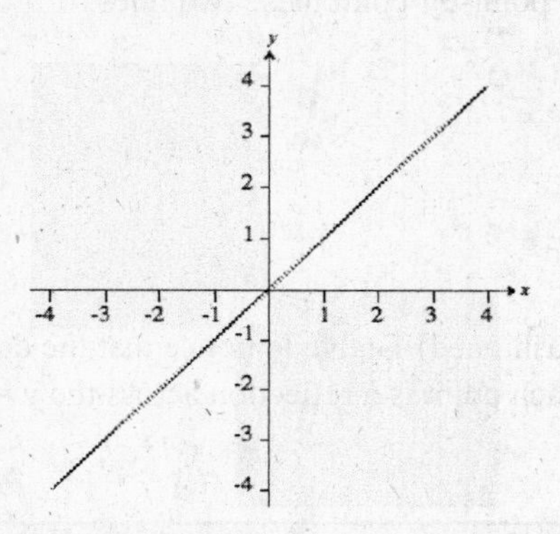

So when you graph functions that are inverses of each other on the same coordinate plane, they look something like this:

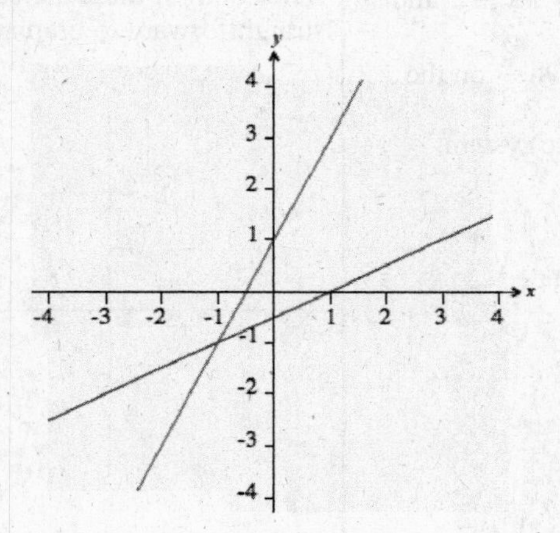

Can you see that every point on each line is a reflection of a point on the other line directly across the line $y = x$? (I've left $y = x$ off this graph to make you work a little bit harder at it.)

★ **For Fun**

If f is the inverse of g, is g necessarily the inverse of f? This is really an easy question; don't let it bog you down.

★ **For Fun**

Try out plotting point-by-point these two lines:

$$y = \frac{1}{3}x - 2$$

$$y = 3x + 6$$

Use your (old-fashioned) T-table to notice that the domain of one is the range of the other and that each point is a reflection across the $y = x$ line.

Graph the Inverse of a Function: Test Prep

Graph $f(x) = 3x - 2$ and $f^{-1}(x) = \frac{1}{3}x + \frac{2}{3}$ on the same coordinate system.

Since both of these are equations for lines, it's straightforward to graph them:

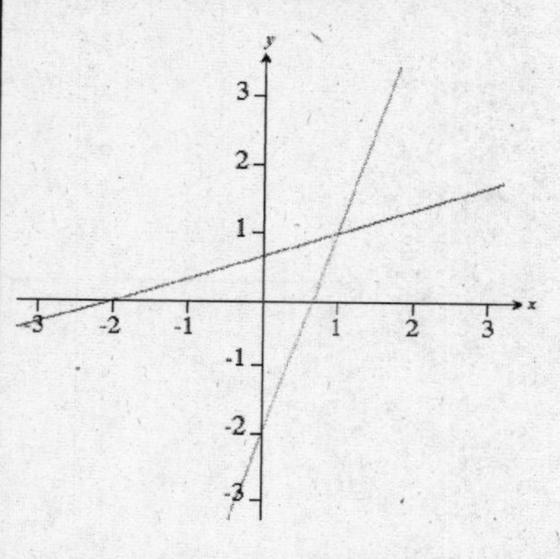

Try Them

Given the graph of $f(x)$, plot its inverse function on the same coordinate system:

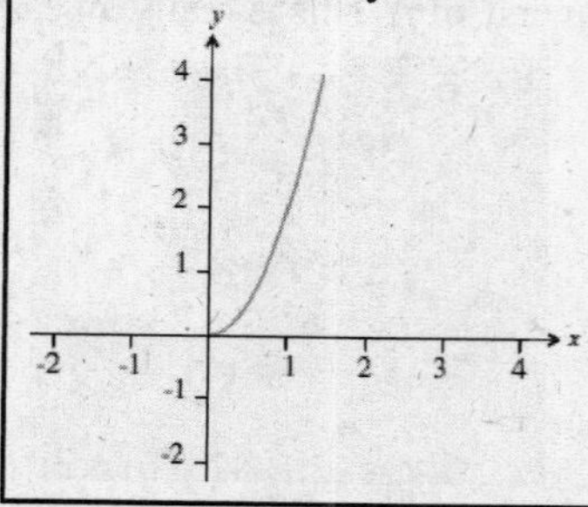

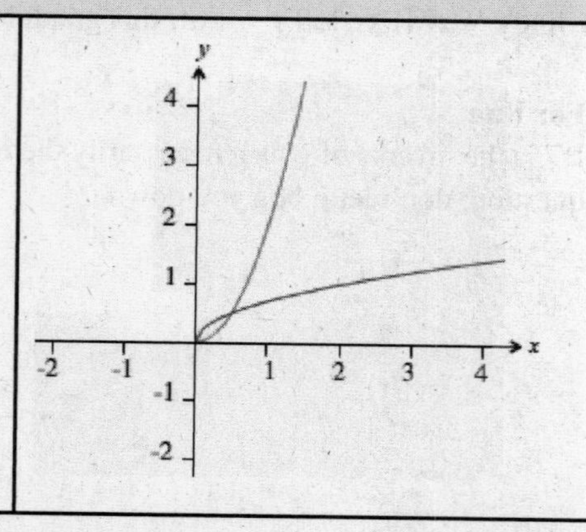

Graph the Inverse of a Function: Test Prep

Sketch
$f(x) = -x^4 - x^3 + x^2$.
Does this function have an inverse?

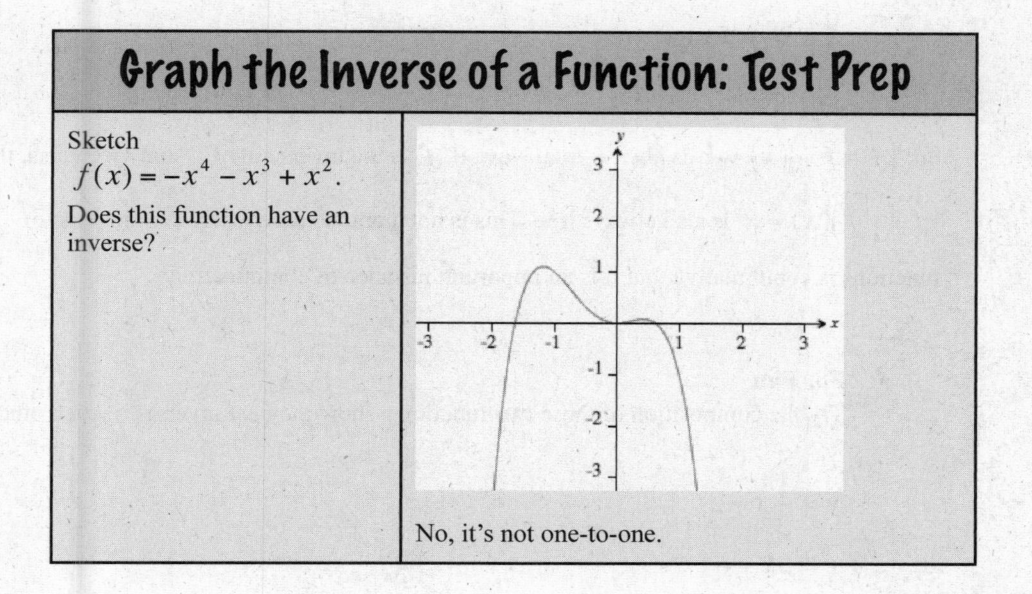

No, it's not one-to-one.

Composition of a Function and Its Inverse

Not only do functions and their inverses have special graphical relations, their compositions have a very special characteristic!

> Composition of Functions was covered in Section 2.6.

The composition of two functions that are inverses of each other behaves like so:

$$(f \circ g)(x) = x$$

And now perhaps you can begin to believe that solving for x is intimately related to inverse functions and their compositions.

Before we go any further, we need to cover the special notation that lets you know that one function is the inverse of another.

> The inverse of $f(x)$ is signified by $f^{-1}(x)$.

So, $\left(f \circ f^{-1}\right)(x) = x$ is always true. Now, if f is the inverse of f^{-1} and vice versa, then $\left(f^{-1} \circ f\right)(x) = x$ is also always true. This is not a unique case where composition of functions is commutative, but it is an important instance of commutativity.

★ **For Fun**

Try the composition of these two functions; show they are inverses of each other:

$$y = \frac{1}{3}x - 2$$
$$y = 3x + 6$$

If so, then we could name these functions so:

$$f(x) = \frac{1}{3}x - 2$$
$$f^{-1}(x) = 3x + 6$$

Do not confuse $f^{-1}(x)$ with the negative exponent. If you actually want the reciprocal of *f(x)* then the notation actually goes like this: $\left[f(x)\right]^{-1}$ (How's that for convoluted ... but a completely necessary distinction?)

Composition of Inverse Functions Property: Test Prep

Use composition of functions to verify that $g(x)=\dfrac{x+1}{x}$ is the inverse function of $f(x)=\dfrac{1}{x-1}$.	$(g\circ f)(x)=g(f(x))$ $=g\left(\dfrac{1}{x-1}\right)$ $=\dfrac{\dfrac{1}{x-1}+1}{\dfrac{1}{x-1}}\cdot\dfrac{x-1}{x-1}$ $=\dfrac{1+x-1}{1}$ $=x$ $(g\circ f)(x)=x$ so they are inverses of each other.
Try Them	
Use composition of functions to ascertain if the following are inverse functions of each other: $f(x)=(2x+8)^3$ $g(x)=\dfrac{1}{2}\sqrt[3]{x}-8$	No.
Use composition of functions to ascertain if converting Fahrenheit temperature T to Celsius, $f(T)=\dfrac{5}{9}(T-32)$, and back, $F(T)=\dfrac{9}{5}T+32$, will give you the Fahrenheit temperature you started with.	Yes, $f(F(T))=T$

Finding an Inverse Function

If you have ascertained that your function is one-to-one and that you want to know its inverse, then there is a straightforward procedure to follow (page 338 in your textbook):

1. Call *f(x)* by *y*.

2. Exchange *y* for *x* everywhere in the equation.

3. Solve the resulting equation for *y*. That is, isolate *y* on one side of the equation.

4. Now name that *y* by $f^{-1}(x)$.

5. Check your result by performing the composition $(f \circ f^{-1})(x)$ (or $(f^{-1} \circ f)(x)$).

   Either way, you must find the result to be *x* by itself, à la $(f^{-1} \circ f)(x) = x$ for instance.

Here's an example:

$$f(x) = \frac{1}{3}x - 2$$
$$y = \frac{1}{3}x - 2$$
$$x = \frac{1}{3}y - 2$$
$$x + 2 = \frac{1}{3}y$$
$$3(x + 2) = y$$
$$f^{-1}(x) = 3(x + 2)$$

And if you check it:

$$(f \circ f^{-1})(x) = f(f^{-1}(x))$$
$$= f(3(x+2))$$
$$= \frac{1}{3}(3(x+2)) - 2$$
$$= x + 2 - 2$$
$$= x$$

Yes -- x -- perfect!

Find the Inverse of a Function: Test Prep

Find the inverse of $f(x) = \frac{1}{x}$	$f(x) = \frac{1}{x}$ $y = \frac{1}{x}$ $x = \frac{1}{y}$ $xy = 1$ $y = \frac{1}{x}$ $f^{-1}(x) = \frac{1}{x}$
Try Them	
Find the inverse of $\{(0,0),(1,2),(-3,-1),(5,9),(-4,1),(3,4)\}$	$f^{-1}(x) = \{(0,0),(2,1),(-1,-3),(9,5),(1.-4),(4,3)\}$
a. Does $f(x) = -3(x+1)^2 + 2$ have an inverse? b. Does $f(x) = -3(x+1)^2 + 2$ have an inverse if the domain is restricted to $x \le -1$?	a. No b. Yes

✓ **Questions to Ask Your Teacher**

Sometimes functions that are not one-to-one can be made one-to-one by limiting their domain. Limiting the domain of a function limits the range of its inverse, which will likely limit in turn the domain of the inverse function. This is covered in example 5 on page 340 in your textbook, and you may want to ask your teacher to do an example for you too.

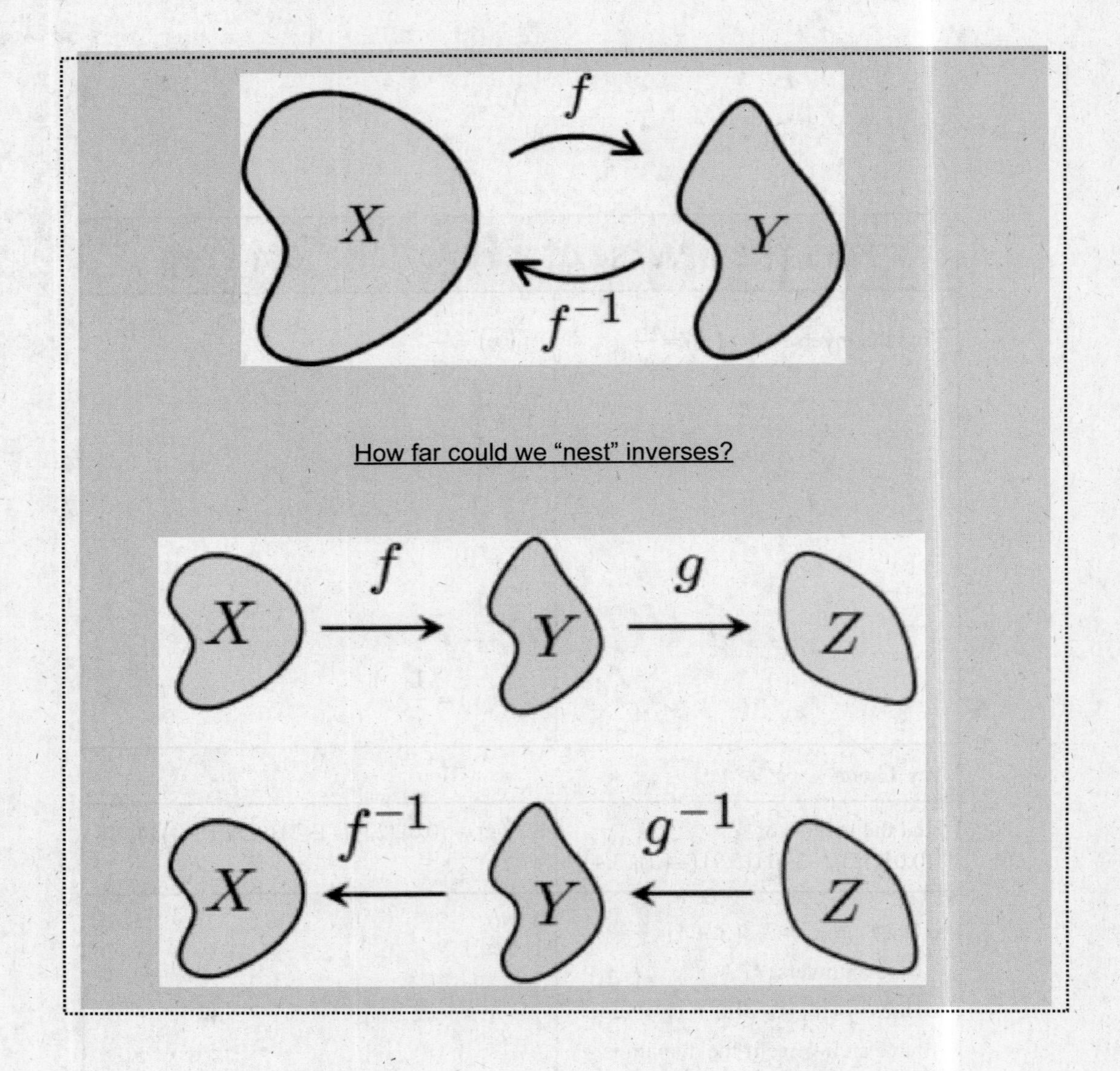

4.2: Exponential Functions and Their Applications

Believe it or not, our discussion of inverse functions has brought us to a new function. We've talked about polynomial, radicals, and rational functions. Now we will examine something about the exponential function.

Exponential Functions

You already know a lot about exponents, for example:

$$a^m a^n = a^{m+n}$$

$$\left(a^m\right)^n = a^{mn}$$

and so forth. All that will still apply, but now we will look at functions of this type:

$$f(x) = a^x$$

Do you notice that this is a function of x and x, the variable, is in the exponent. This is an exponential function!

✤ **exponential function with base *b***

$f(x) = b^x$ is an exponential function where b, the base, is a positive number and not equal to zero and x, the exponent, is a real number.

★ **For Fun**

What would $f(x) = a^x$ look like if a were allowed to be equal to one? Also sometime when you get a chance you might try plotting points and graphing $f(x) = a^x$ when a is a negative number -- good luck with that one!

Calculator note:

You must be able to evaluate exponents on your scientific calculator. Locate the appropriate buttons now. Try these:

3^4

2^{-3}

7^{π}

π^{-2}

(It used to be that textbooks had tables of exponential function values in them. They no longer do. More room for more problems. And more demand on and for calculators.)

By the way, do you know what 3^0 is? What is a^0 (as long as *a* is not zero)?

Properties of f(x) = bˣ: Test Prep

Given $f(x) = 3^x$ evaluate: a. $f(3)$ b. $f(-4)$ c. $f(x+2)$	a. $f(3) = 3^3 = 27$ b. $f(-4) = 3^{-4} = \frac{1}{3^4} = \frac{1}{81}$ c. $f(x+2) = 3^{x+2} = 3^x \cdot 3^2 = 9 \cdot 3^x$
Try Them	
Use your calculator to estimate these values to the nearest thousandth: a. $2^{1.2}$ b. $5^{-0.2}$ c. $\pi^{3.7}$	a. 2.297 b. 0.725 c. 69.096
When $f(x) = \left(\frac{2}{5}\right)^x$, find $f(3)$.	$\frac{8}{125}$

Graphs of Exponential Functions

Now that you know how to evaluate exponential functions on your calculator, it's a small matter to graph. At the very least you can plot points. For example:

$y = 3^x$

x	y
-2	$3^{-2} = \frac{1}{3^2} = \frac{1}{9}$
-1	$3^{-1} = \frac{1}{3}$
0	$3^0 = 1$
1	$3^1 = 3$
2	$3^2 = 9$

Plotting these gives us this picture:

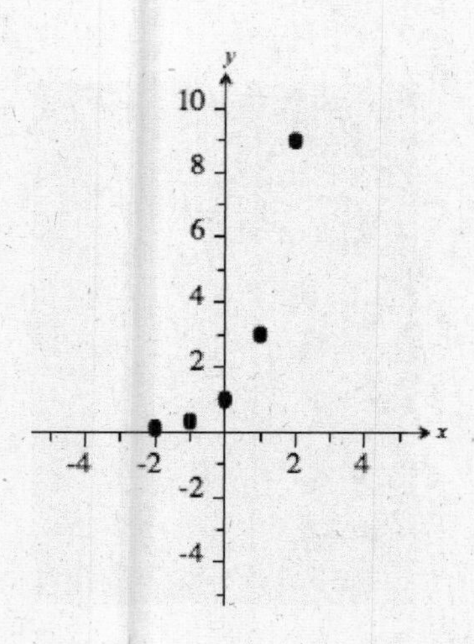

Connecting the dots, we see something like this:

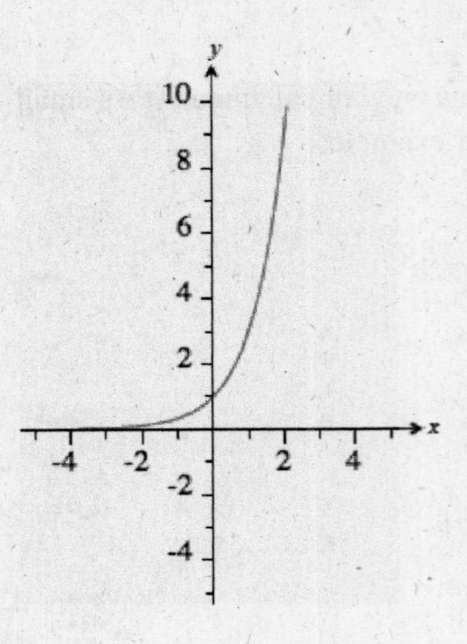

Notice the horizontal asymptote on the *x*-axis? Think about it: is there any *x* that will make 3^x become zero or negative? One prominent feature of the graph of an exponential function is that it has this asymptote.

And now you can apply all the techniques of translation, reflection, stretching, and compressing that we studied in Chapter 2. So the graph of, say, $y = 3^{-x}$ would look like this:

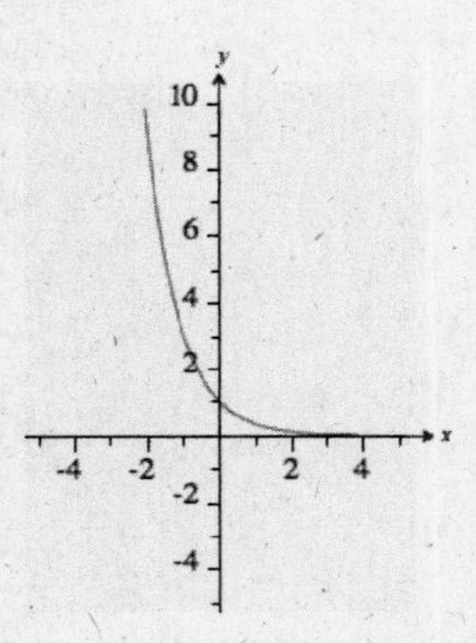

That's a reflection about the *y*-axis, caused by changing *x* into *-x* in the function, exactly as expected.

Match these:
(being sure to notice the horizontal asymptote in each)

$y = 2 + 3^x$

$y = -2^x$

$y = 2^{-x} - 1$

$y = -3^{-x}$

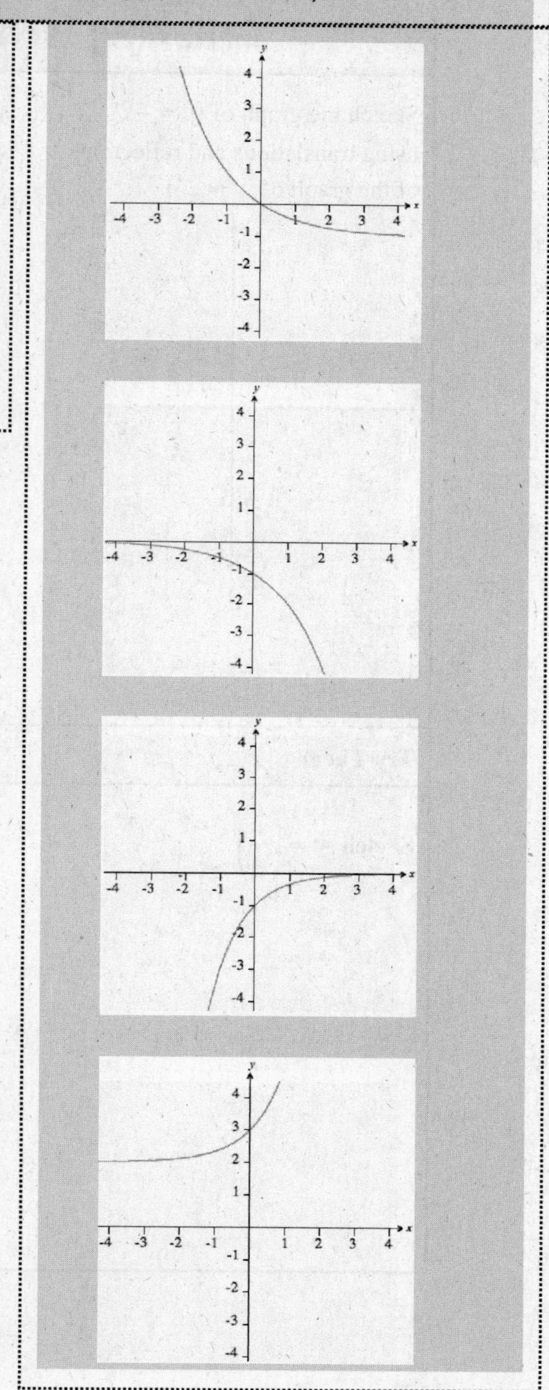

Graphing Techniques: Test Prep

Sketch the graph of $y = -2^{(x+1)}$ using translations and reflections of the graph of $y = 2^x$:

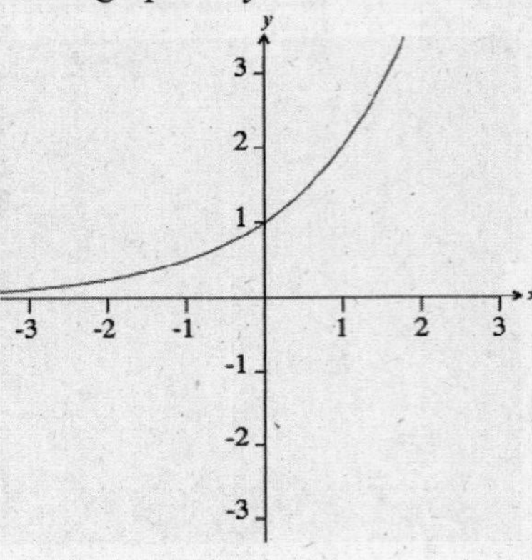

$y = -2^{(x+1)}$ is a horizontal translation of $y = 2^x$ to the left by one unit and a reflection about the x-axis:

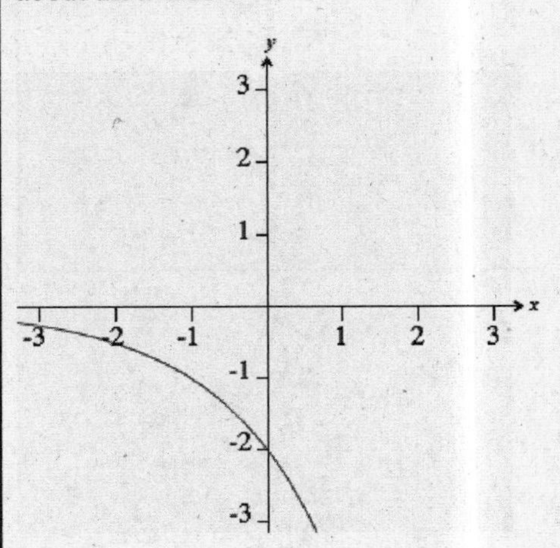

Try Them

Sketch $y = \left(\frac{2}{3}\right)^x$

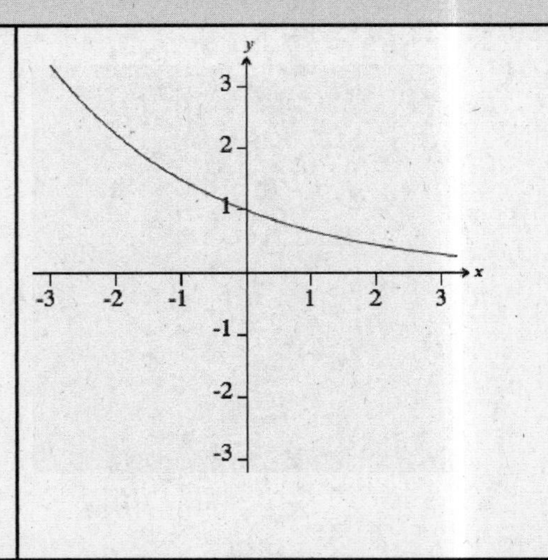

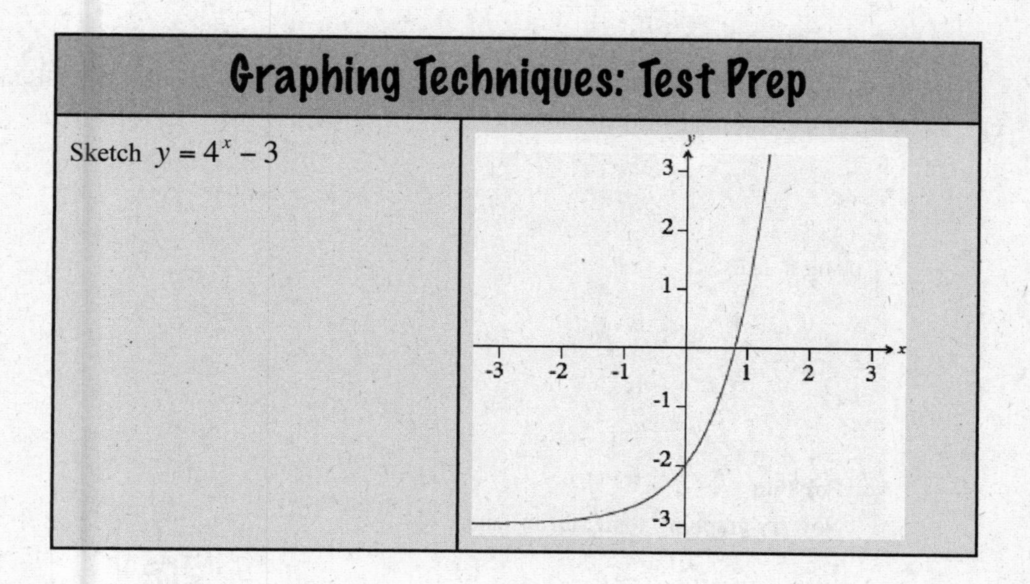

Natural Exponential Function

Every exponential function has a base that is strictly greater than zero and not equal to one. One such number is the irrational number *e*. *e* is called the natural base. Since we use a base-ten counting system, 10 is also a nice, clean number to use for an exponential base, but *e* has more "natural" applications.

- ***e***

  e is a real number between 2 and 3. It is approximately equal to 2.718281828459045. Since it is a non-repeating, non-terminating decimal, it is an irrational number (that is, you cannot express it as a ratio of integers).

  - ★ **For Fun**

    Who named it *e*? Any ideas why?

- **natural exponential function**

  $f(x) = e^x$ is the natural exponential function.

★ **For Fun**

Try these on your calculator; notice if the results are always positive. What is the range of the exponential function again?

e^2

$e^{4.289}$

e^{-5}

e^{π}

e^1

★ **For Fun**

Now try graphing some, for instance:

$y = -e^x$

$y = e^{2x}$

$y = 3 - e^x$

A double exponential function is a constant raised to the power of an exponential function. The general formula is $f(x) = a^{b^x}$, which gets big faster than our "ordinary" exponential function.

For example, if we let both *a* and *b equal 10*, then:

$f(-1) \approx 1.26$

$f(0) = 10$

$f(1) = 10^{10}$

$f(2) = 10^{100}$ (a googol)

$f(3) = 10^{1000}$

$f(100) = 10^{10^{100}}$ (a googolplex.)

Natural Exponential Function: Test Prep	
Use your calculator to estimate these to four decimal places: a. $e^{1,429}$ b. e^{π} c. $e^{-2.1}$	a. $e^{1,429} \approx 4.1745$ b. $e^{\pi} \approx 23.1407$ c. $e^{-2.1} \approx 0.1225$
Try Them	
Sketch $y = 2e^{-x}$	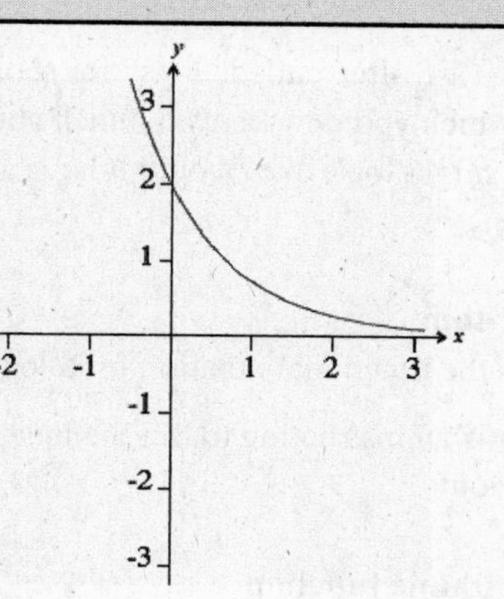
Write an equation for the horizontal asymptote of the graph of $y = 1 - e^{(x-3)}$	$y = 1$

4.3: Logarithmic Functions and Their Applications

The logarithmic function is the inverse of the exponential function, so like the exponential function, the logarithmic function has a *base*. Actual inverse exponential/logarithmic functions must have the same bases -- otherwise how could they swap domains and ranges with each other? At any rate, do you already have any ideas about logarithmic functions just because you know they are the inverse of exponentials?

Logarithmic Functions

Logarithms are often called "logs" for short. Since the log is the inverse of the exponential function which you now know a bunch about, perhaps you can imagine what the graph of a logarithm might look like. Would it have an asymptote? What would be the domain and range?

✤ **logarithm**
log is the functional notation for a logarithm. $\log_b x$ would be a logarithm, base *b*, of *x*. *log* is an instruction to run the independent value through the *log* function and get a value out.

✤ **logarithmic function**
$f(x) = \log_b x$ is the logarithmic function, base *b*. Like the base in an exponential function, the base must be strictly positive and not equal to one. $f(x) = \log_b x$ is the inverse function of $g(x) = b^x$.

There is a relationship between exponentials and logarithms because they are inverses of each other. This relationship can be formalized thus:

If $y = \log_b x$, then $b^y = x$

And vice versa:

If $b^x = y$, then $x = \log_b y$.

- **exponential form**

  A logarithmic function may be written in its equivalent exponential form, like so:

  $\log_3 81 = 4$. This can be written in logarithmic form: $3^4 = 81$. (By the way, is $3^4 = 81$ true?)

- **logarithmic form**

  Likewise an exponential might be rephrased in logarithmic form:

  $2^{-3} = \frac{1}{8}$ becomes $\log_2 \frac{1}{8} = -3$

There is a distinct pattern to these transformations between logarithmic and exponential forms.

One point to notice would be that the base of the exponential stays the base of a logarithmic equivalence.

What pattern can you see involving the other 2 parameters? (I see a sort of triangle, but maybe that's just me and my visual tendencies.) Figure out for yourself a method that helps you put all the pieces in the right places.

★ **For Fun**

Try to rewrite this one in logarithmic form: $2^{\log_2 3} = 3$. If you can figure that one out, you might start to see the *inverse* nature of the exponential and logarithmic functions.

Exponential and Logarithmic Form: Test Prep

Convert to logarithmic form: a. $3^{-2} = \frac{1}{9}$ b. $4^x = 42.5$ c. $a^m = n$	The base becomes the base of the logarithm; the right-hand-side number becomes the argument of the logarithm; the exponent becomes the answer: a. $3^{-2} = \frac{1}{9}$ → $\log_3 \frac{1}{9} = -2$ b. $4^x = 42.5$ → $\log_4 42.5 = x$ c. $a^m = n$ → $\log_a n = m$
Try Them	
Convert to exponential form: a. $\log_2 16 = 4$ b. $\log 142 \approx 2.15$ c. $\log_x 3 = .004$	a. $2^4 = 16$ b. $10^{2.15} \approx 142$ c. $x^{.004} = 3$
Solve $\log_2(x-3) = 2$ by converting it into its exponential form.	$x = 7$

★ **For Fun**

Try these, too:

$\log_2 2 =$ _______

$\log_2 1 =$ _______

You know enough about converting between exponentials and logarithms to do these, and you'll learn some basic properties of logarithms along the way. (Or see below for help.)

Basic Properties of Logarithms:

1. $\log_b b = 1$ 2. $\log_b 1 = 0$

3. $\log_b\left(b^x\right) = x$ 4. $b^{\log_b x} = x$

Basic Logarithmic Properties: Test Prep

Evaluate: a. $\log_2 2^{-3}$ b. $2\log_5 1$ c. $10^{\log 10}$	a. Since $\log_b\left(b^x\right) = x$, $\log_2 2^{-3} = -3$ b. Since $\log_b 1 = 0$, $2\log_5 1 = 2 \cdot 0 = 0$ c. Since $b^{\log_b x} = x$, $10^{\log 10} = 10$
Try Them	
Solve $\log_7 x = 1$	*x = 7*
Solve $\log_4 4^x = 12$	*x = 12*

LOGARITHMS, (from λόγος *ratio*, and ἀριθμὸς *number*), the indices of the ratios of numbers to one another; being a ſeries of numbers in arithmetical progreſſion, correſponding to others in geometrical progreſſion; by means of which, arithmetical calculations can be made with much more eaſe and expedition than otherwiſe.

From a 18th century Encyclopedia Britannica article.

Graphs of Logarithmic Functions

You know that $y = a^x$ has a graph that looks something like the following when $a > 1$.

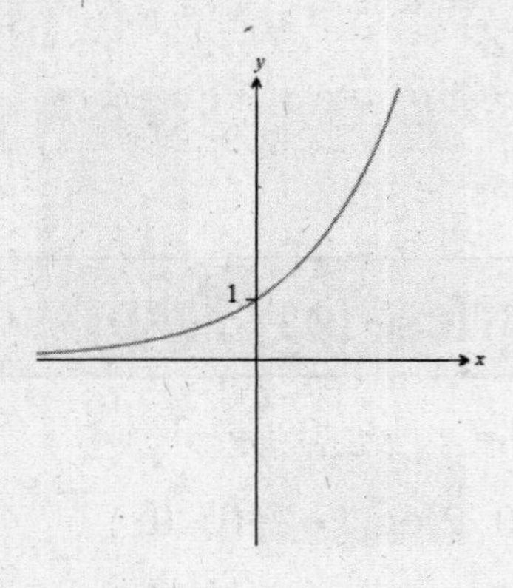

And you know that the appropriate inverse function is expressed as $\log_a x = y$, so its inverse graph would be a reflection about $y = x$:

But if you want to plot points on the graph of a function like $y = \log_2 x$, you have to use your knowledge of its exponential form (namely $2^y = x$) in order to actually construct a T-table. This time chose your y-values and find the values that go in the x-column:

x	y
___	–2
___	–1
___	0
___	1
___	2

Do you get points like this?

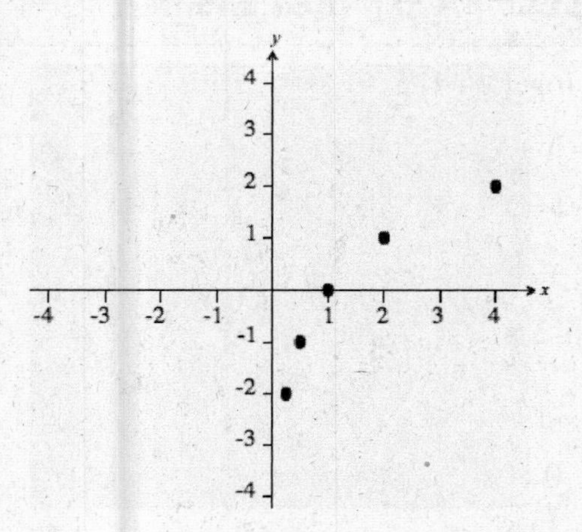

And you can see the logarithm curve now, right? By the way, can you see a *vertical* asymptote on this graph, too?

★ **For Fun**
Why do you have to do this sort of turn-inside-out in order to graph a logarithm? Think about the domain of an exponential, and realize that the domain becomes the range of the logarithm and vice versa -- can you see that switch in the graphs? Also, look on your calculator: can you find a key that returns a log *base 2*? (This is a shortcoming that we will overcome a little later in this chapter.)

Now that you know about the basic logarithmic graph, try some of these by translations, reflections, stretches, and compressions:

$$y = -\log_2 x$$
$$y = 1 + 3\log_2 x$$
$$y = \log_3(-x)$$
$$y = \log_3(x-2)$$

Pay attention to the asymptote!

Properties of $f(x)=\log_b x$: Test Prep

Sketch $y=\log_2(x+1)$ by changing the equation into its exponential form, solving for x, and plotting points.	$y=\log_2(x+1)$ $2^y=x+1$ $2^y-1=x$ x | y: $-\frac{3}{4}$ | -2; $-\frac{1}{2}$ | -1; 0 | 0; 1 | 1; 3 | 2 ... and so ...

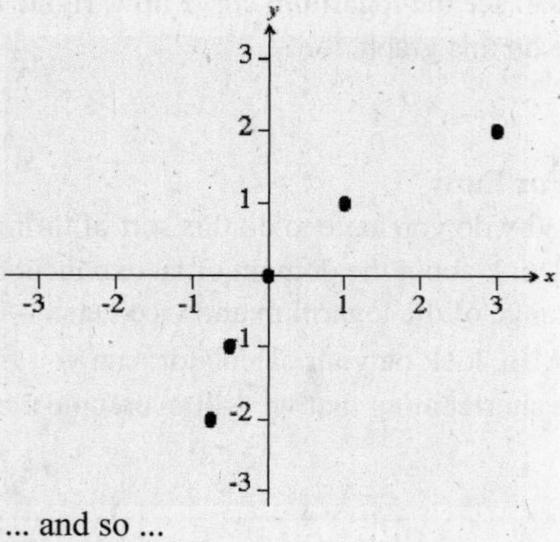

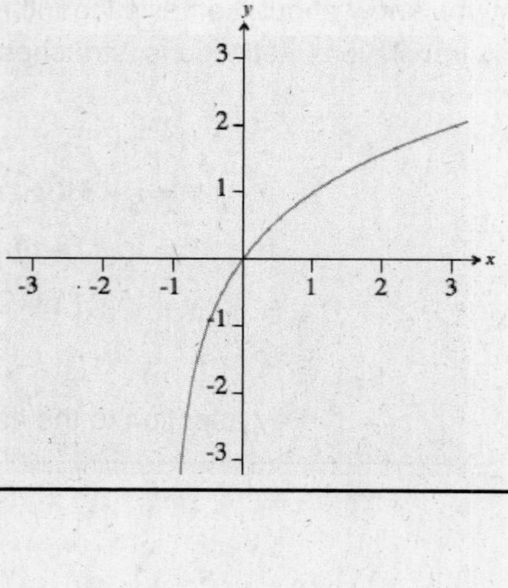

x	y
$-\frac{3}{4}$	-2
$-\frac{1}{2}$	-1
0	0
1	1
3	2

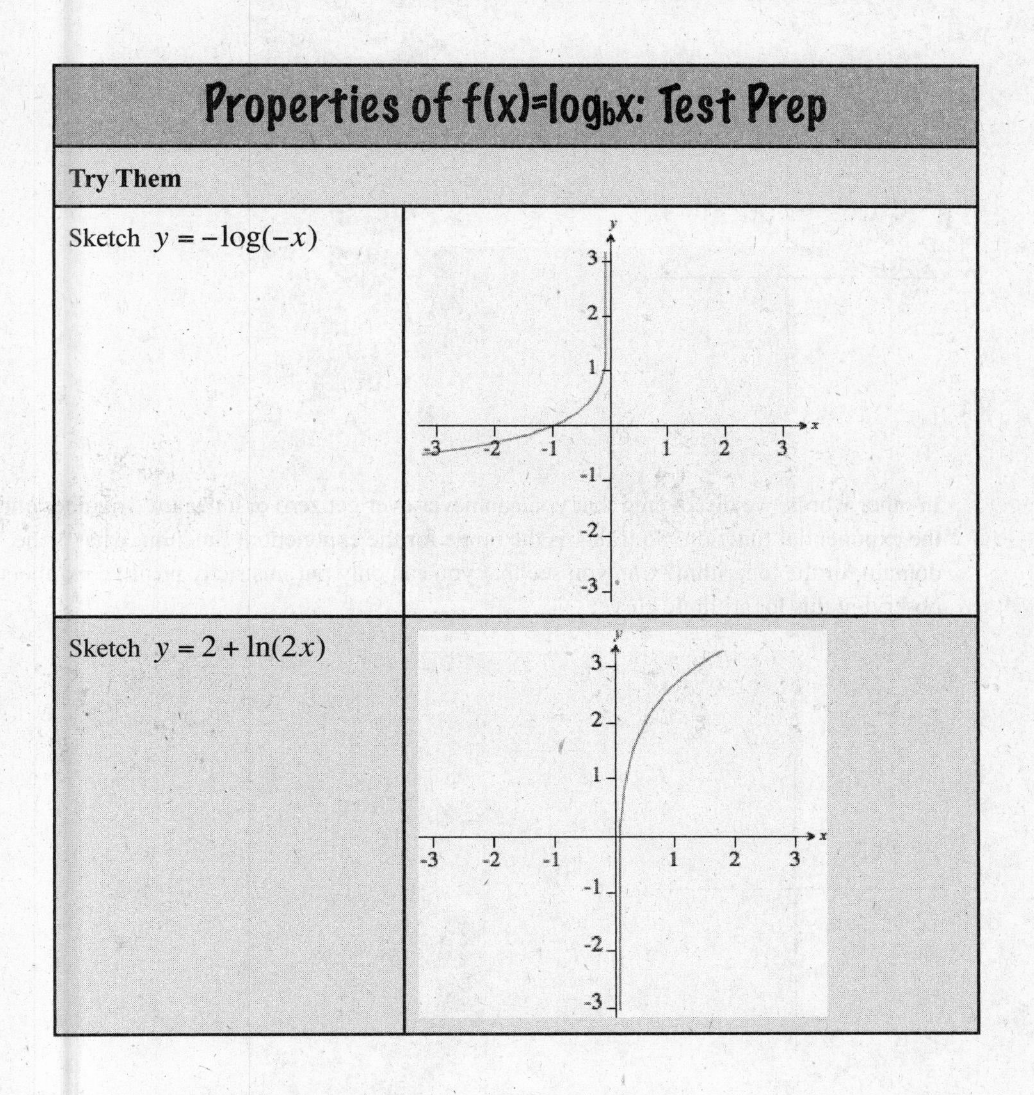

Domains of Logarithmic Functions

What is the domain of the exponential function? We have no concern when we pick out x-values for a T-table to graph an exponential. And when we picked values for the y variable in our last T-table for graphing the logarithm, we had the same lack of concern. (After all the domain of the exponential becomes the *range* of the logarithm.) But what is the range of the exponential? Didn't we investigate that horizontal asymptote?

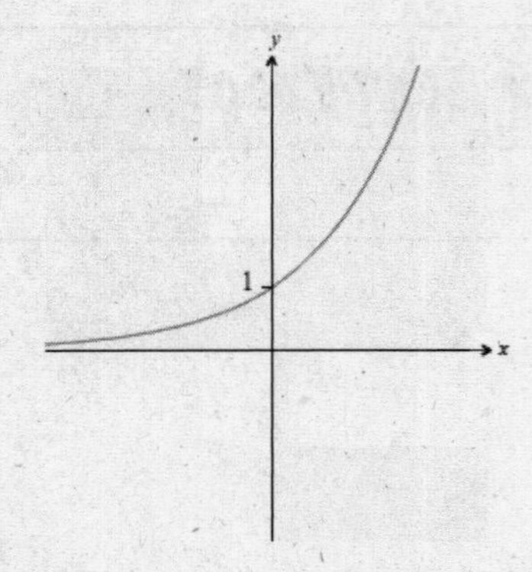

In other words, we discovered that you can never ever get zero or a negative number out of the exponential function. So if that is the range for the exponential function, what is the domain for the logarithm? Can you see that you can only put in strictly positive numbers by observing this logarithmic curve:

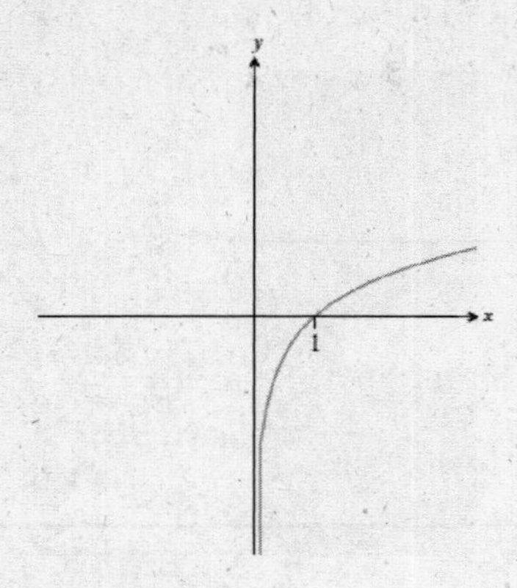

★ **For Fun**
Try asking for *log(-5)* in your calculator. Or even *log(0)*. What result do you see? By the way, you *have* found the *log* and *ln* keys on your calculator, right? You know you'll need these very, very soon.

Common and Natural Logarithms

These are the logarithmic functions that actually do have keys on your scientific calculator.

✤ **common logarithmic function**

The common logarithm is base 10. It is usually referred to simply as "log" without denoting the base explicitly. So a missing base means "base 10."

★ **For Fun**

Evaluate some of these in your calculator:

$$\log 3$$
$$\log 10$$
$$\log 0$$
$$\log \pi$$

✤ **natural logarithmic function**

The natural log, symbolized by *ln*, is the logarithm base *e*. The symbol *ln* means exactly the same as $\log_e$.

★ **For Fun**

Try these on your calculator:

$$\ln e$$
$$\ln 2$$
$$\ln(-1)$$
$$\ln 1$$

U.S. National Institute of Standards and Technology

NIST has proposed even newer notation for commonly used logarithms of different bases:

"ln *x*" means $\log_e x$

"lg *x*" means $\log_{10} x$

"lb *x*" means $\log_2 x$

However, we will stick with the much-better-known conventions of our textbook on these matters!

4.4: Properties of Logarithms and Logarithmic Scales

Since you already spent quite a bit of time in algebra classes learning all the rules of exponentials -- like $\left(\frac{a}{b}\right)^n = \frac{a^n}{b^n}$ -- you might not realize how much more easily you can understand exponentials than logarithms, or why. There are comparable rules for logarithms that you have to learn to use now.

Slide Rule, or how they used to look up logs

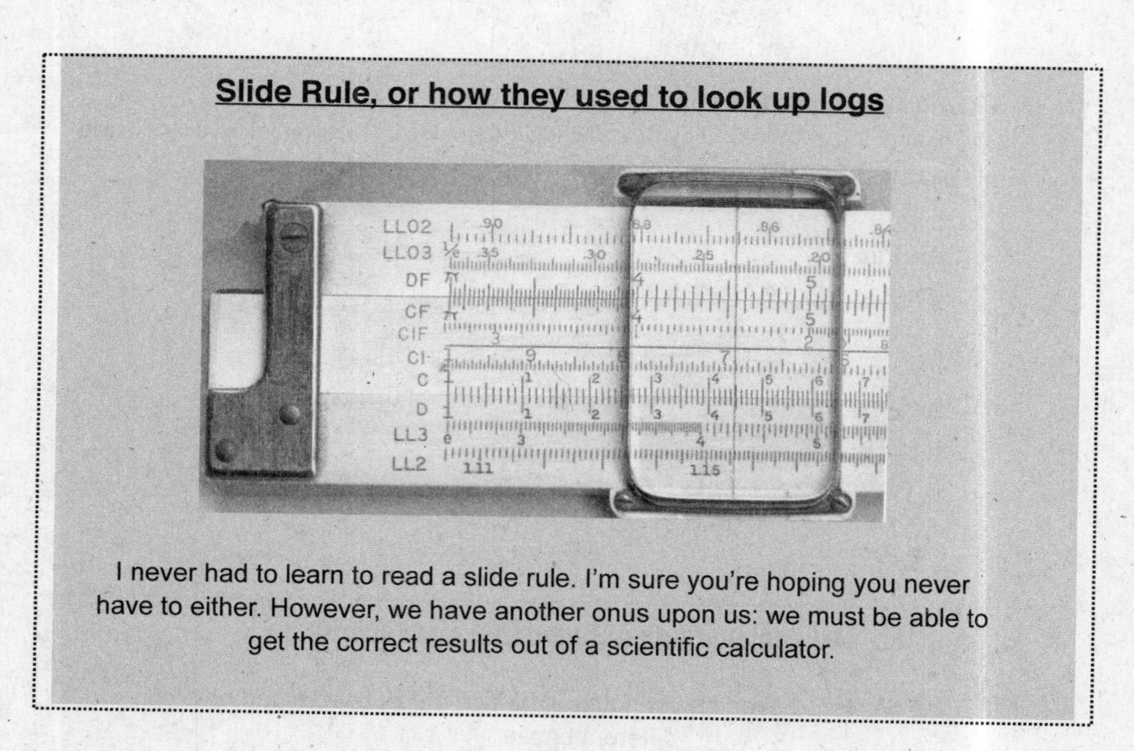

I never had to learn to read a slide rule. I'm sure you're hoping you never have to either. However, we have another onus upon us: we must be able to get the correct results out of a scientific calculator.

Properties of Logarithms

Many of these properties correspond to exponential properties, but whether you can see the parallels or not, you have to become as familiar and handy with these properties as quickly as you can. Ideally you would know them just as well as you know that $\left(a^m\right)^n = a^{mn}$, for instance.

✤ **product property of logarithms**
This property says that the log of a product is equal to the sum of logs:

$$\log_b(MN) = \log_b M + \log_b N$$

For example:

$$\log_2 2x = \log_2 2 + \log_2 x = 1 + \log_2 x$$

✤ **quotient property of logarithms**
The quotient property says that the log of a quotient is the difference of logs:

$$\log_b\left(\frac{M}{N}\right) = \log_b M - \log_b N$$

This one you must be extra careful with because, of course, subtraction is not commutative. The log of the denominator must be subtracted from the log of the numerator (and certainly not the other way around).

For example:

$$\log_2\left(\frac{4}{3}\right) = \log_2 4 - \log_2 3 = 2 - \log_2 3$$

With these first two properties, please be sure to notice exactly what sort of logarithmic expression they work on ... and which they do not.

For instance, none of these rules will allow you to simplify $\ln(x+2)$. You see there is a rule for the log of a product but not for the log of a sum.

✤ **power property of logarithms**
The power property runs thus:

$$\log_b M^p = p\log_b M$$

In other words, this rule makes a log of a power into the product of logs. For example:

$$\log_3 3^4 = 4\log_3 3 = 4 \cdot 1 = 4$$

✤ **logarithm-of-each-side property**
This property and the next basically communicate that just like adding or multiplying both sides of an equation does not change it, taking the logarithm, same base, of both sides of an equation does not change the truth value of the equation either. Formally:

If $M = N$, then $\log_b M = \log_b N$

✤ **one-to-one property of logarithms**
Conversely, if $\log_b M = \log_b N$, then $M = N$.

✤ **expanding the logarithmic expression**
Now we can use these properties to manipulate logarithmic expressions (in preparation for solving equations). One of the ways we can rewrite a logarithmic expression might be to expand it, for example:

$$\ln\left(\frac{x^2}{3}\right) = \ln x^2 - \ln 3$$
$$= 2\ln x - \ln 3$$

Each remaining logarithm is the logarithm of a single variable or expression.

Properties of Logarithms: Test Prep	
Write $\frac{1}{2}\log x + 3\log(x+1)$ as a single logarithm.	$\frac{1}{2}\log x + 3\log(x+1)$ $= \log x^{\frac{1}{2}} + \log(x+1)^3$ $= \log\left(x^{\frac{1}{2}}(x+1)^3\right)$ $= \log\left(\sqrt{x}(x+1)^3\right)$
Try Them	
Expand the expression $\ln(5x^3y)$	$\ln 5 + 3\ln x + \ln y$
Condense the expression $3\log_5 x + \log_5(y-2) - 2\log_5 a$	$\log_5\left(\frac{x^3(y-2)}{a^2}\right)$

Change-of-Base Formula

I've already mentioned that you cannot find a key for $\log_2$, for example, on your calculator. Only the common (base 10) and natural (base *e*) logarithms rate a button on your calculator. Never fear -- there is a way to evaluate an expression like $\log_2 3$ using your calculator. But you must learn the change-of-base formula:

$$\log_b x = \frac{\log x}{\log b}$$

Or, if you prefer the natural log like I do:

$$\log_b x = \frac{\ln x}{\ln b}$$

This formula is not hard, but please practice. It's easy to get the formula turned upside down at which point it represents a very wrong number!

Change-of-Base Formula: Test Prep

Use the Change-of-Base Formula and a scientific calculator to estimate the value of $\log_{50} 5120$ to three decimal places.	$\log_{50} 5120 = \dfrac{\log 5120}{\log 50} \approx 2.183$
Try Them	
Use the Change-of-Base Formula to show that the following are equivalent: $\frac{1}{2}\log_6 44$ $\log_6 \sqrt{44}$	Both are approximately 1.056
Use the Change-of-Base Formula to sketch $y = 3 + \log_5(x-2)$	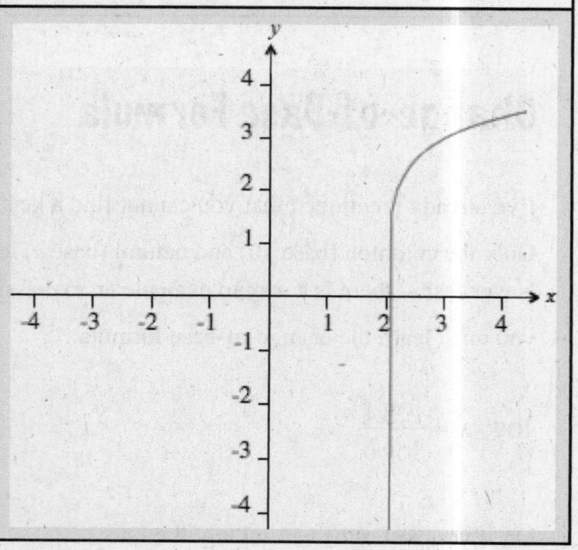

★ **For Fun**

Can you envision a way to use the change-of-base formula to help with plotting points in order to graph? (So that you wouldn't have to rewrite the equation to be graphed in exponential form and chose y's instead of x's?)

Logarithmic Scales

You might notice that $\log_3 81 = 4$ returns a much smaller number than its exponential counterpart, $3^4 = 81$. Sometimes displaying data with the considerably smaller logarithm value is favored. Some common uses for these logarithmic "scales" are the Richter scale, which measures earthquake intensity, and the pH scale, which measures the acidity of a substance.

- **zero-level earthquake**
  The smallest intensity, called I_0, that can be measured on a seismograph at the earthquake's epicenter. This will be a constant.

- **Richter scale magnitude**
  The Richter scale magnitude, *M*, of an earthquake is given by this formula:

  $M = \log\left(\frac{I}{I_0}\right)$ where *I* is the measured intensity of the quake and I_0 is the known zero-level earthquake intensity.

  Because logarithms return small numbers, realize that a change of just one order on the Richter scale is a tremendous change in the magnitude of a quake.

- **seismogram**
  A seismograph is an instrument that measures the details of an earthquake, such as amplitude and duration. A seismogram is the record that the seismograph produces. A seismogram might look something like

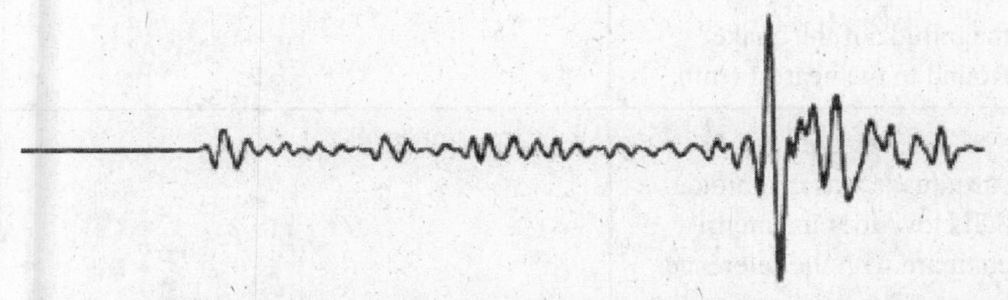

This is a record over time where earlier is to the left and later is on the right end. The individual seismograph would have its usual scales for time (the *x*-axis) and amplitude (the *y*-axis).

- **p-waves**
  The p-waves are the smaller precursor waves that precede the first large s-wave. (Can you spot them in the seismogram above?)

- **s-waves**
  The s-waves are the bigger waves that start after the p-waves.

Richter Scale Magnitude: Test Prep	
The 1989 San Francisco earthquake was a magnitude 7.1, but the 1906 earthquake has been estimated to have been an 8.3-magnitude earthquake based on damage done. What is the ratio of the 1906 earthquake to the 1989?	$M = \log\left(\frac{I}{I_0}\right) = \log I - \log I_0$ $M + \log I_0 = \log I$ $10^{M + \log I_0} = 10^{\log I}$ $10^M 10^{\log I_0} = I$ $I = I_0 \cdot 10^M$ *so* $\frac{I_{1906}}{I_{1989}} = \frac{I_0 \cdot 10^{8.3}}{I_0 \cdot 10^{7.1}}$ $= 10^{1.2} \approx 15.85$
Try Them	
If the intensity of an earthquake is determined to be 40,000 times the reference intensity I_0, what is the magnitude of the quake? Round to the nearest tenth.	4.6
The 1933 Japanese earthquake was magnitude 8.3. How does its intensity compare with the reference intensity?	$I \approx 200{,}000{,}000 I_0$

✤ **acidity**
Acidity is a property of a substance (usually a liquid) with more available hydrogen ions than pure water (hence "potential hydrogen" of the pH scale. "H" because that's the abbreviation for hydrogen on the periodic table.). Acids rank between 0 (very acid) up to 7 (neutral) on the pH scale.

✤ **alkalinity**
Alkalinity is a measure of a substances ability to neutralize an acidic solution. Alkalis measure between 14 (very alkali) to 7 (neutral) on the pH scale.

✤ **pH of a solution**
The pH of a solution is measured according to this logarithmic formula:

$pH = -\log\left[H^+\right]$ where H^+ is the measured hydrogen ion concentration in a mole of solution.

A mole is a measurement of the *amount* of stuff you have.

pH: Test Prep	
The hydrogen ion concentration, in moles per liter, of a certain hair gel is 8.9×10^{-8}. Find its pH to the nearest hundredth.	$pH = -\log(8.9\times10^{-8})$ $= -\left(\log 8.9 + \log 10^{-8}\right)$ $\approx -(0.949 - 8)$ $= 7.05$
Try Them	
Find the pH of blood if its concentration of hydrogen ions is 4.02×10^{-8}. Is this blood acidic or basic?	~7.40, basic
Determine the hydrogen ion concentration of a solution with a pH of 2.2.	$\sim 6.31\times10^{-3}$ moles per liter

MID-CHAPTER QUIZ: TEXTBOOK PAGE 380

4.5: Exponential and Logarithmic Equations

We made a bit of a deal out of inverse functions at the beginning of this chapter and how they are used to solve equations of all sorts. Now we're going to use these particular inverse functions, the exponential and logarithmic, to solve exponential and logarithmic equations. These procedures are just as natural as the skills you already use to solve simpler equations, but, unfortunately it's likely to take quite a bit of practice in order for these exercises to feel as natural.

For my money, I believe this section is the crux of this chapter!

✓ **Questions to Ask Your Teacher**
Ask your teacher what section of this chapter he or she believes is the most important.

Solving Exponential Equations

We'll start off with some of the easier types of exponential and logarithmic equations to solve.

✤ **exponential equation**
An exponential equation is an equation where a variable appears in the exponent, for instance: $3^x = \frac{1}{27}$. To solve such an equation, the idea is the same as it's always been -- get x by itself. Now the question is: How does one get the variable out of the exponent? That's new.

One way to solve exponential equations is to equate exponents. Consider our example: $3^x = \frac{1}{27}$. Can you see that *both* sides of the equation can be written as a power of 3?

$$3^x = \frac{1}{27}$$
$$= \frac{1}{3^3}$$
$$= 3^{-3}$$

So $3^x = 3^{-3}$. Now we can use the Equality of Exponents Theorem to equate x and -3. That is, to express the solution: $x = -3$. (And you can check this solution in the original equation as usual.)

Equality of Exponents

If $a^x = a^y$, then $x = y$.

★ **For Fun**

Check out an example, see what you think: $3^5 = 3^{3+2}$. Does $5 = 3 + 2$?

Equality of Exponents Theorem: Test Prep	
Solve: $2^{3x} = 64$	$2^{3x} = 64$ $2^{3x} = 2^6$ $3x = 6$ $x = 2$
Try Them	
Solve: $\left(\frac{1}{3}\right)^{x+4} = 81$	$x = -8$
Solve: $2^{-x} = -4$	no solution

★ **For Fun**
Check the answers in the first two examples in the preceding Test Prep block.

Now, I promise you will not always be able to write both sides of the equation in terms of the same base. In those cases, you will need to use the inverse function, same base *logarithm* of both sides, in order to "rescue" the variable out of the exponent. For example:

$$2^{x+1} = 44$$
$$\ln 2^{x+1} = \ln 44$$
$$(x+1)\ln 2 = \ln 44$$
$$x+1 = \frac{\ln 44}{\ln 2}$$
$$x = -1 + \frac{\ln 44}{\ln 2}$$
$$x \approx 4.4594$$

How do we "liberate" the variable from its exponentiation onto that base 2? The inverse function (applied on both sides of the equation) is the logarithm base 2. Remember: $\log_b b^x = x$. See how the logarithm can rescue that x from the exponent? You can also use the power law of logarithms to rescue the x: $\ln b^x = x\ln b$, for instance.

"log" is a function

Please do not think for one instant that you multiply both sides of your equation by some mysterious number named "log". You are performing the same function on both sides of the equation -- same as if you were taking the square root of both sides. Again the "log" in log*x* is much like the *f* in *f(x)*. It's a function; this is functional notation.

Exponential Equations: Test Prep

Solve: $4e^{3x+1} = 10$. Round your answer to 3 decimal places.	$4e^{3x+1} = 10$ $e^{3x+1} = \frac{10}{4} = \frac{5}{2}$ $\ln e^{3x+1} = \ln\left(\frac{5}{2}\right)$ $3x + 1 = \ln\left(\frac{5}{2}\right)$ $x = \frac{\ln\left(\frac{5}{2}\right) - 1}{3}$ $x \approx -0.028$
Try Them	
Solve $e^{2x} - 4e^{x} + 3 = 0$ by factoring.	$x = 0, \ln 3$
Solve: $e^{x} - e^{-x} = 30$	$x = \ln\left(15 + \sqrt{226}\right)$

Solving Logarithmic Equations

Let's try some of the same or corresponding techniques to solve logarithmic equations.

✤ **logarithmic equation**
Similar to an exponential equation, a logarithmic equation is an equation where a variable is inside a logarithmic function: $\log(2x - 1) = 3$.

Now, not only do you have to know how to use *exponentiation* as the inverse function to solve logarithmic equations, you have to sometimes know how best to use all the properties of logarithms in order to *condense* logarithms so that you can exponentiate meaningfully.

For example:

$$1+\log(3x-1)=\log(2x+1)$$
$$1=\log(2x+1)-\log(3x-1)$$
$$1=\log\left(\frac{2x+1}{3x-1}\right)$$
$$10^1=10^{\log\left(\frac{2x+1}{3x-1}\right)}$$
$$10=\frac{2x+1}{3x-1}$$
$$10(3x-1)=2x+1$$
$$30x-10=2x+1$$
$$28x=11$$
$$x=\frac{11}{28}$$

I had to use lots of steps to solve this problem. Can you follow carefully and describe what I did at each step? Do you recognize when I used exponentiation? properties of logs? plain ol' algebra?

Logarithmic Equations: Test Prep	
Solve for a: $\log a = 2\log b$	$\log a = 2\log b$ $10^{\log a} = 10^{\log b^2}$ $a = b^2$
Try Them	
Solve: $\log_3 124 = \log_3 4 - 2\log_3 x$	$x=\frac{\sqrt{31}}{31}$
Solve: $5^{2x+2}=3^{5x-1}$. Round your answer to 3 decimal places.	$x \approx 1.898$

4.6: Exponential Growth and Decay

The exponential function has some very important features. One is that it handily describes lots of systems -- systems that grow and decay -- from populations of bacteria to money in the bank.

Exponential Growth and Decay

The form of the exponential growth/decay functions is well known. You can learn how to put one together for yourself!

First of all, instead of *x* and *y* variables, the independent variable in an exponential growth or decay function is usually *t* for time. This function will model how systems change over time. Secondly, we name the function *N* for the *number* left in the population. Lastly, N_0 is going to be the *initial* (when $t = 0$) population, or simply "number of things."

- **growth rate constant**

  In the exponential function, $N(t) = N_0 e^{kt}$, the *k* is the growth constant. Can you imagine that if it's bigger, then the effect of time, *t*, is larger and vice versa if the growth constant is smaller?

- **exponential growth function**

  If *k* is positive, then *N* is increasing, and that means $N(t) = N_0 e^{kt}$ describes a system that is *growing* exponentially.

- **exponential decay function**

  Contrarily, if *k* is negative, then *N(t)* is an exponential decay function because it describes a shrinking population.

- **half-life**

  The half-life is the value of *t* (time, remember) when the population *N(t)* is exactly half of what it started out being. This applies, obviously?, to *decay* functions. "Half-life" is usually reserved for discussions of radioactive decay in particular.

Carbon Dating

In living tissue, the ratio of the carbon isotopes -14 and -12 stay very constant. Once the tissue dies, the carbon-14, which is slightly radioactive with a half-life of 5730 years, begins to decay. That decay is modeled mathematically by the function $P(t) = 0.5^{\frac{t}{5730}}$.

★ **For Fun**

First of all, can you see how the half-life and the growth constant are related?

Second of all, isn't this supposed to be a *decay* function?! How come the exponent isn't negative? Well, isn't 0.5 equal to 2^{-1}? Could you rewrite *P(t)* so that it had a negative exponent now?

Thirdly, how come this exponential decay function isn't in terms of *e*?

✤ **carbon dating**

Carbon dating is the Nobel prize-winning process that used the amount of the carbon-14 isotope present in an old bone to estimate its age.

Exponential Growth and Decay Functions: Test Prep

At the start of an experiment, there are 100 bacteria. If the bacteria follow an exponential growth pattern with *t* measured in hours and rate constant $k = 0.02$, how long will it take for the population to double?	The growth equation is $N = 100e^{0.02t}$ because *k* is 0.02 and the initial population of bacteria is 100. When the population is doubled, we have $200 = 100e^{0.02t}$ $2 = e^{0.02t}$ $\ln 2 = 0.02t$ $t = \frac{\ln 2}{0.02} \approx 35hr.$

Exponential Growth and Decay Functions: Test Prep

Try Them	
Suppose that at the start of an experiment there are 8,000 bacteria. A growth inhibitor and a lethal pathogen are introduced into the colony. After two hours 1,000 bacteria are dead. If the death rates are exponential, how long will it take for the population to drop below 5,000? Round your answers to the nearest tenth.	$t \approx 7.1hr.$
Say a fossil is found that has 35% carbon 14 compared to the living sample. How old is the fossil?	~8,680 years old

Compound Interest Formulas

How money makes money is a mathematical proposition. It's also exponential.

- **interest**
  Interest is paid for using money. If you loan money, you earn interest. If you borrow it, unfortunately you pay interest.

- **simple interest**
  Simple interest is easy to calculate. It is the amount of money earned when the percentage earned and the compounding term are constant. It's like adding 1% every 3 months (at which point the process can start all over again on the new amount). The equation for simple interest is $I = Prt$ where I is the interested earned, P is the principal invested, r is the rate of return per investment term (usually expressed as a percentage), and t is the number of terms.

- **principal**
  Principal is the amount of money invested, or the amount of money to be used (or loaned) in order for it to earn interest.

- **compound interest**
  Compounding interest is exactly the act of starting over again after the period of simple interest is over. The equation for compound interest looks like this: $A = P\left(1+\frac{r}{n}\right)^{nt}$.

  A is the total amount in the account; P is the principal invested; r is the annual rate; n is times per year that the interest will be compounded; t is the amount of time, in years, that the principal has been invested. (Do you see that this is an exponential function? The variable t is in the exponent, and all the other parameters are fixed in a given investment scenario.)

Compound Interest Formula: Test Prep	
An amount of $1,500.00 is deposited in a bank paying an annual interest rate of 4.3%, compounded quarterly. What is the balance after 6 years?	$P = 1500$ $r = 0.043$ $n = 4$ $t = 6$ $A = 1500\left(1+\frac{0.043}{4}\right)^{4\cdot 6} \approx \$1938.84$
Try Them	
The first credit card that you got charges 12.49 % interest to its customers and compounds that interest monthly. Within one day of getting your first credit card, you max out the credit limit by spending $1,200.00 . If you do not buy anything else on the card and you do not make any payments, how much money would you owe the company after 6 months?	$1276.92

Compound Interest Formula: Test Prep	
If $12,500 is invested at an annual interest rate of 8% for 10 years, find the balance if the interest is compounded hourly.	$27,819.16

- **compound continuously**
For interest to compound continuously -- that is, not stop to count up every year or month or other period then start over again -- some mathematical manipulation is done to produce the Continuous Compounding Interest Formula: $A = Pe^{rt}$. Now the function is clearly the exponential growth function.

Continuous Compounding Interest Formula: Test Prep	
If I invest n dollars in a bank account that compounds 5% interest continuously, how long will it take that account to double in value?	$2n = ne^{0.05t}$ $t = \dfrac{\ln 2}{0.05} \approx 14\,yr.$
Try Them	
$10,000 earning 5% interest, compounded continuously, will be worth how much after one year?	$10.512.71
Find the balance if $12,500 is invested at an annual interest rate of 8% for 10 year and the interest is compounded continuously.	$27,819.26

Restricted Growth Models

The logistic model is a growth model, like the exponential growth model, but it takes into account limited resources.

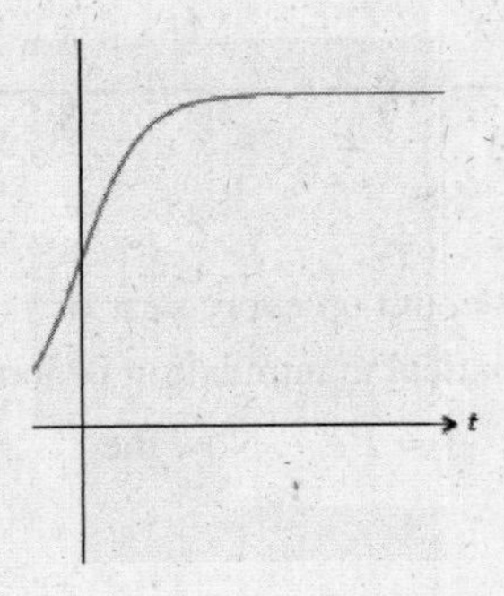

There is a portion of this curve which proceeds like an exponential curve, but then the curve flattens out when the limits of the resources supporting the population are approached.

The equation representing a logistic system looks like this: $P(t) = \dfrac{c}{1+ae^{-bt}}$. This is still an exponential function -- the variable is still in the exponent some place.

- **logistic model**
  The logistic model is also called a restricted growth model because is takes into account the limited resources available to a growing system.

- **carrying capacity**
  The parameter c above is the carrying capacity. c is the maximum population that a system can carry.

- **growth rate constant**
  b, a positive real number, is the growth rate constant in a logistic model. (You might notice that if b is always positive, then $-b$ is always negative in the logistic model.)

- **initial population**
  P_0 is the initial population. It's related to the parameter a in the logistic model by the equation: $a = \dfrac{c - P_0}{P_0}$.

Logistic Model: Test Prep

Given the logistic population model $N = \dfrac{300}{1+186e^{-0.52t}}$, estimate, to one decimal place, the t when N will reach 170.	$140 = \dfrac{300}{1+186e^{-0.52t}}$ $1+186e^{-0.52t} = \dfrac{300}{140}$ $e^{-0.52t} = \dfrac{\frac{300}{140}-1}{186}$ $t = \dfrac{1}{-0.52}\ln\dfrac{\frac{300}{140}-1}{186} \approx 9.8$
Try Them	
Sketch the logistic function $P(t) = \dfrac{100}{1+25e^{-0.09t}}$	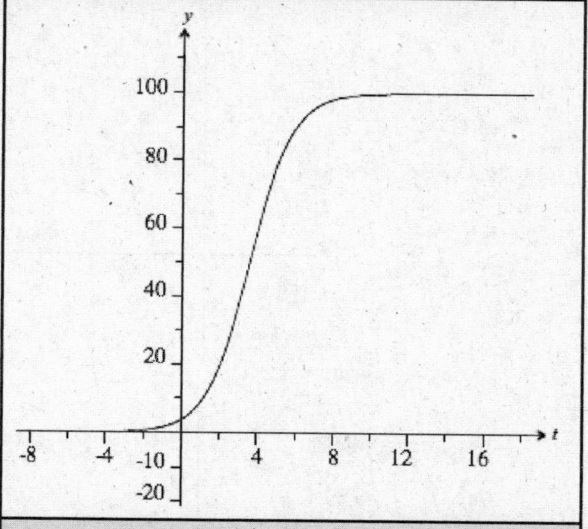
Find the initial population for the following logistic model: $P(t) = \dfrac{320}{1+15e^{-0.12t}}$	20

4.7: Modeling Data With Exponential and Logarithmic Functions

Now, we can consider models both exponential and logistic!

Analyzing Scatter Plots

Can you imagine a scatter plot of points that might fit an exponential (or logistic) function?

Which one of these do you think would be a better candidate for an exponential model?

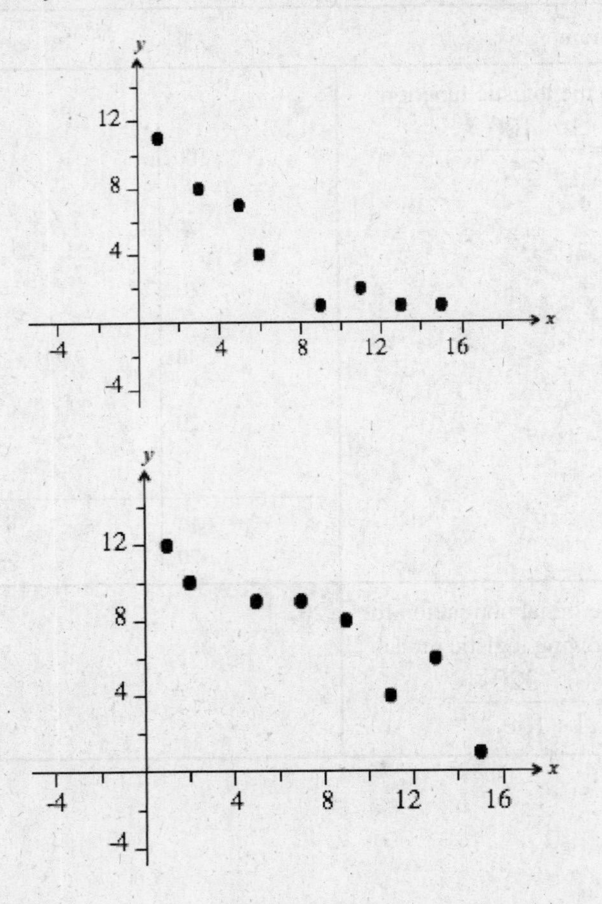

- **concave upward**
  Technically, a graph all of whose tangents lie below said graph is concave upward. You can just remember "u" for "upward" and remember that *curved* curves that open up are concave upward.

- **concave downwards**
  Likewise, although technically concave downward means that all the tangents of a curve are above said curve, a curve that *curves* downward is concave down.

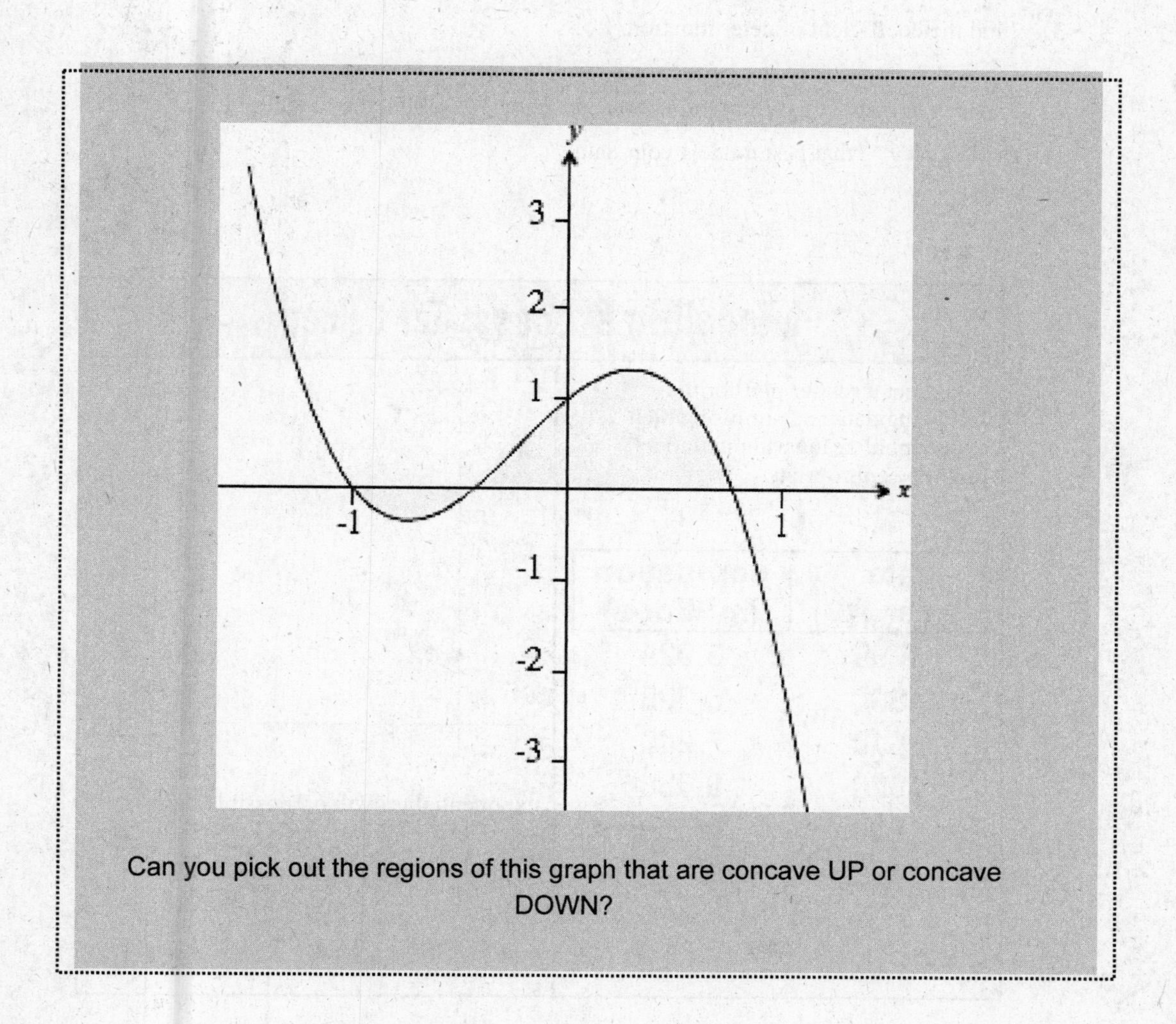

Can you pick out the regions of this graph that are concave UP or concave DOWN?

Modeling Data

Check out your spreadsheet or calculator regression options. Besides the linear, quadratic, cubic, and quartic models we've looked at, do you have exponential and logarithmic options? Now you can take those into consideration and continue to use your experience with regression parameters to make good choices of models:

1. Construct your scatter plot.

2. Chose a function to model your data.

3. Find the coefficient of determination, r^2.

4. If necessary, chose a different function and find its r^2. Find the function (the one with the largest r^2) that best models your data.

Modeling Process: Test Prep

Construct a scatter plot for the following data and surmise which exponential or logarithmic model might be appropriate.

date (year AD)	population (millions)
1790	3.929
1800	5.308
1810	7.240
1820	9.638
1830	12.866
1840	17.069

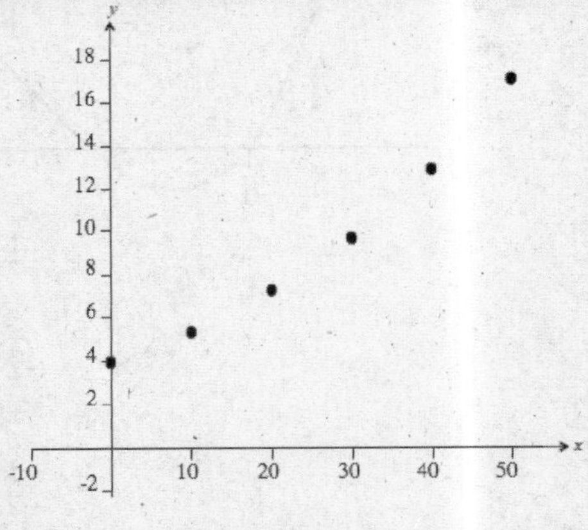

exponential growth or, possibly, logistic

Modeling Process: Test Prep

Try Them

Find an exponential model for the following data:

x	y
10	6.8
12	6.9
14	15
16	16.1
18	50
19	20

$f(x) = 1.2^x$

Find a logarithmic model for the following data:

x	y
8	67.1
10	67.8
12	68.4
14	69
16	69.4

$f(x) = 60.09 + 3.36 \ln x$

Finding a Logistic Growth Model

If you're lucky and your calculator (or spreadsheet application) includes functions for modeling data logistically, you can include that in your repertoire also ... if, of course, your data warrants a logistic model:

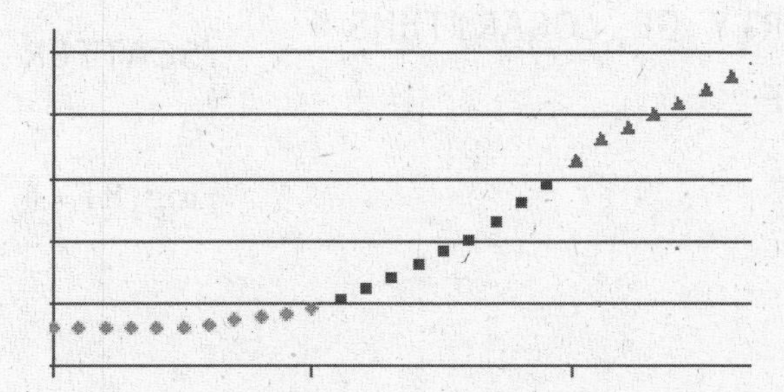

Final Fun: Matching

FIND THE BEST MATCH. GOOD LUCK.

LOGISTIC FUNCTION

$P(t) = 0.5^{\frac{t}{5730}}$

SEISMOGRAM

CHANGE OF BASE FORMULA

COMPOUNDING INTEREST FORMULA

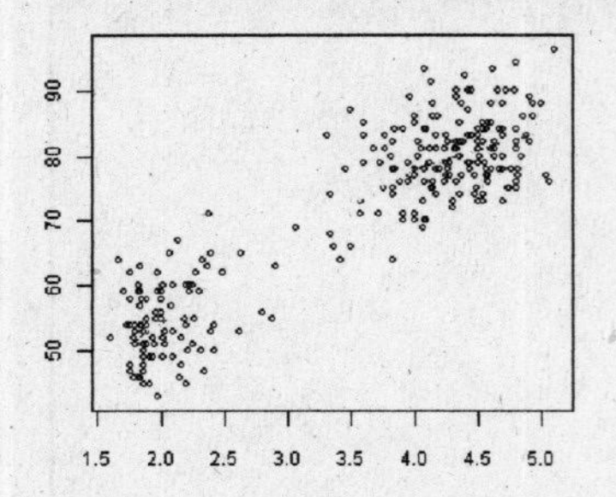

$f^{-1}(x)$

$f(x) = a^x$

LOGARITHMIC FUNCTION

$\log_4 \frac{1}{16}$

COMPOSITION OF INVERSES

POWER PROPERTY OF LOGARITHMS

NATURAL BASE

$-\log[H^+]$

$3^4 = 81$

$\log_b M^p = p \log_b M$

$P(t) = \frac{c}{1 + ae^{-bt}}$

EXPONENTIAL DECAY

$A = P\left(1 + \frac{r}{n}\right)^{nt}$

EXPONENTIAL FUNCTION

pH

$f(x) = \log_b x$

-2

INVERSE FUNCTION

$\log_b x = \frac{\ln x}{\ln b}$

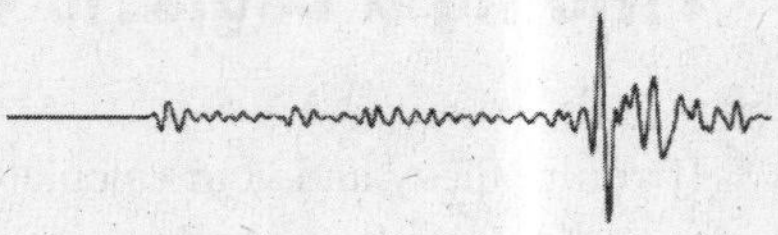

$(f^{-1} \circ f)(x)$

SCATTER PLOT

e

$\log_3 81 = 4$

NOW TRY THESE:
REVIEW EXERCISES: TEXTBOOK PAGE 421,
PRACTICE TEST: TEXTBOOK PAGE 424.

STAY UP ON YOUR GAME -- DO THE CUMULATIVE REVIEW EXERCISES: TEXTBOOK PAGE 425

Chapter 5: Topics in Analytic Geometry

Word Search

```
D E A S T C R E F O C I V C D
R I M X I X M C A O T M K O G
X W R N I V B C R R U R R N X
J E O E K S E E A X B O L J V
J C T Y C L I N T W J U U U M
Q Y O R L T S T A A A A S G U
T K A I E V R R M Q L L S A D
M F P G E V C I P D O O T T W
I S Y R J P M C X D B B A E V
E B S U Y E I I Z Y A R N S Y
F E X P S A C T C R R E D U A
U E T O T P M Y S A A P A C H
X D I O L O B A R A P Y R O E
S E M I M I N O R W O H D F T
R E T N E C V S Y M M E T R Y
```

ASYMPTOTE
AXIS
CENTER
CONIC
CONJUGATE
DIRECTRIX
ECCENTRICITY
ELLIPSE
FOCI
FOCUS
HYPERBOLA
PARABOLA
PARABOLOID
SEMIMAJOR
SEMIMINOR
STANDARD
SYMMETRY
TRANSVERSE
VERTEX

5.1: Parabolas

We already studied parabolas -- quadratic functions -- in chapter 2. We are going to add a little more finesse and general applicability to that knowledge now. But first, since parabolas are one of the "conic sections," we consider cones and how they can be sectioned by passing a plane through them:

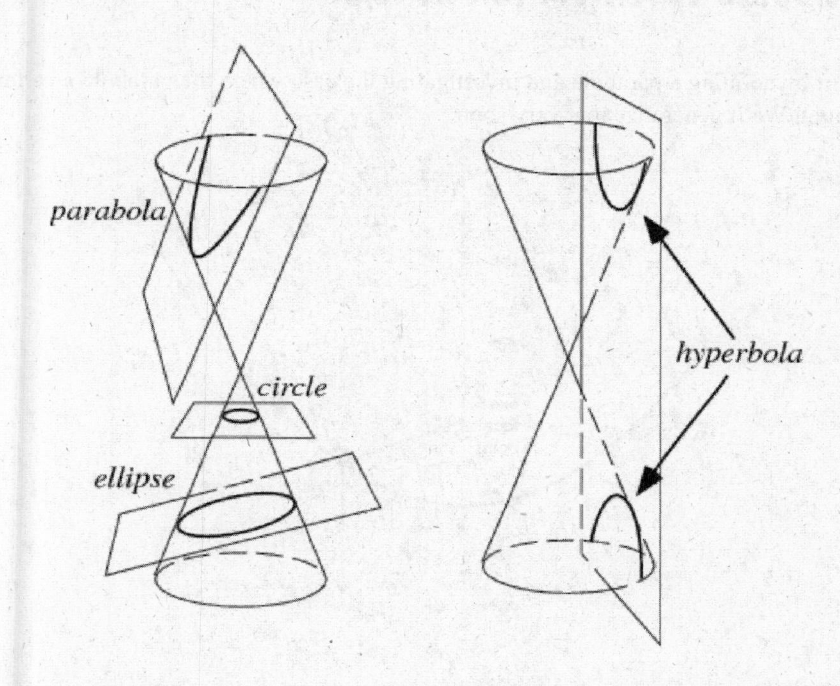

Do you notice how a mathematician envisions the "complete" cone? It's not just a dunce hat or an ice cream cone -- the complete cone continues on in *both* directions: up and down.

Be forewarned: just as we had to complete the square in chapter 2, we will have to revisit those skills throughout this chapter.

Appolonious of Perga, a 3rd century B.C. Greek geometer, wrote the "Conics": the first place to show how all three curves, along with the circle, could be obtained by slicing the same cone at various angles.

✣ **degenerate conic section**
There are a few special cases of passing a plane through a cone that result in something very simple: a line, a point, or two intersecting lines. These cases are called degenerate conic sections. You already know how to plot points and graph lines, so these are nothing new.

Parabolas With Vertex at (0,0)

We start by defining a parabola and investigating the case where the parabola's vertex is at the origin. We'll generalize that very soon.

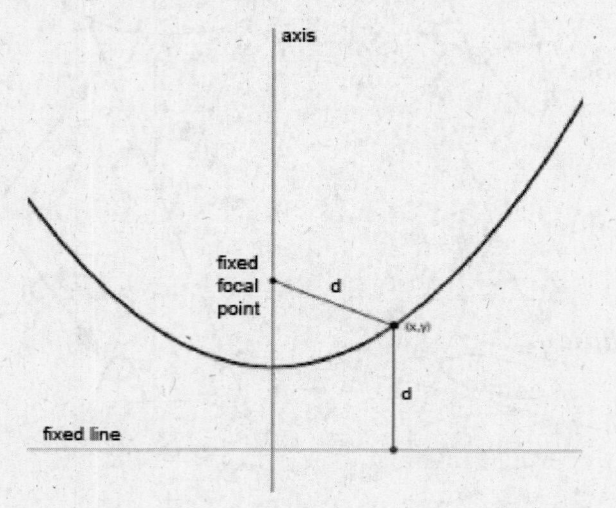

✣ **parabola**
You already know a parabola as a U-shaped curve. The technical definition says that a parabola is the set of all points that are the *same* distance from a fixed focal point and a fixed line.

✣ **directrix**
The fixed line that all points in a parabola must be a certain distance from is called the directrix.

✣ **focus**
And the focus is that fixed point that, along with the directrix, defines the positions of all the points that actually are on the parabola. (Like the center of the circle, the directrix and focus help define the parabola, but they are not actually a part of it. Can you see that on the pictures of the cones on the previous page?)

✤ **axis of symmetry**
The "axis" in the previous illustration is the axis of the parabola, its "axis of symmetry." We've investigated symmetries before, so I'm sure you can *see* the symmetry of a parabola. The challenge here might be actually writing the equation for the *line* that is the axis of symmetry. Do you remember the forms of the equations for vertical and horizontal lines? Linear functions were in section 2.3.

✤ **vertex**
The vertex of the parabola is the (x, y) point where the axis of symmetry hits the parabola itself. If you think about symmetry, you'll be able to grasp that there is some sort of "turning around" going on at the vertex of a parabola.

We used this equation for the parabolas we looked at in Chapter 2: $f(x) = a(x-h)^2 + k$.

We're about to get somewhat more sophisticated. First call *f(x)* by *y*, then also let the vertex be at the origin, so both *h* and *k* are zero: $y = ax^2$. Solve this for that x^2 and you have:

$x^2 = \frac{1}{a}y$. Now call that coefficient of *y* by *4p*.

Standard Form of the Equation of a Parabola With Vertex at the Origin

1. axis of symmetry is coincident with the *y*-axis: $x^2 = 4py$

   For example:

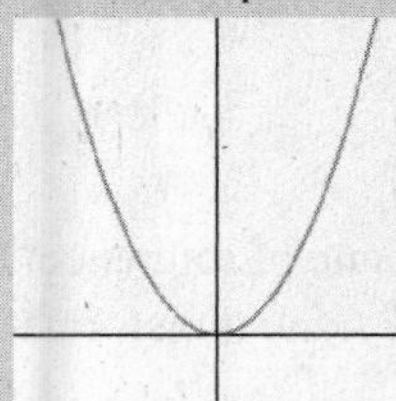

2. axis of symmetry is coincident with the *x*-axis: $y^2 = 4px$

   For example:

Since you know that the sign of our "old" parameter a told us whether the parabola opened up or down, you know the sign of p does the same thing. In the second case, the sign of p will tell you if the parabola opens left or right -- right being reserved for positive values of p.

The distance from the vertex to the focus and the vertex to the directrix has to be the same (from the definition of a parabola, right?). This particular distance -- the one from focus to vertex and vertex to directrix -- is the parameter p. The size of p ultimately tells you about the shape of the parabola.

★ **For Fun**
Can you identify the distance p on the parabola pictured at the beginning of this subsection?

Parabolas With Vertex at (h,k)

The equations for parabolas with vertex at any point (h, k) are as follows:

1. $(x-h)^2 = 4p(y-k)$ -- for a parabola with a vertical axis of symmetry (it opens up or down)

2. $(y-k)^2 = 4p(x-h)$ -- for a parabola with a horizontal axis of symmetry (it opens to the left or right)

✤ **transformation equations**
Transformation equations are expressions of a change in the frame of reference. In the case of changing from a vertex at the origin to a vertex at (h, k), the transformation equations are

$$x' = x - h$$
$$y' = y - k$$

It's a pretty straightforward operation to plug those transformation equations into the centered-at-origin parabola equations. That's substitution!

Standard Forms of the Equation of a Parabola With Vertex (h,k): Test Prep

Find the standard form of the equation for the parabola with vertex at (2, -3) and focus at (0, -3).	From the configuration of these two points, we have a parabola with horizontal axis and opening to the left. So the difference in the x-coordinates is $2-0=2=\lvert p \rvert$, and $p=-2$ (opening left). Vertex at (2, -3) tells us $h=2$ and $k=-3$. Therefore, we have $(y+3)^2=4(-2)(x-2)$ $(y+3)^2=-8(x-2)$
Try Them	
Sketch $(x-1)^2=-2(y+2)$. Include the focus and the directrix.	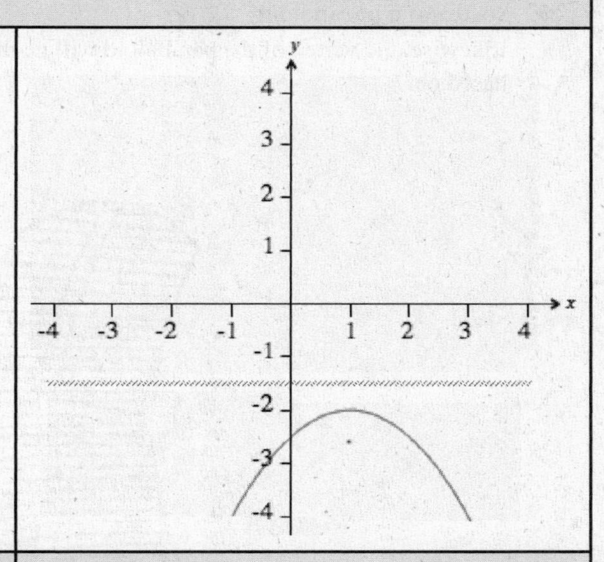
Find the standard form of the parabola: $x-y^2-4y-1=0$. Also find the x- and y-intercepts, if they exist.	$(y+2)^2=x+3$ y-intercepts: $\left(0,-2+\sqrt{3}\right)$, $\left(0,-2-\sqrt{3}\right)$ x-intercept: (1, 0)

Applications of Parabolas

Parabolic mirrors are one of the most well-known applications of the parabola. There might be, after all, a reason the "focus" is called the focus -- light rays come together there. Or maybe radio waves -- something a satellite dish would pick up.

✤ **paraboloid**
If you can imagine revolving your 2-dimensional parabola around its axis of symmetry, the 3-dimensional shape you would make is called a paraboloid.

✤ **focus of a paraboloid**
The focus of such a paraboloid remains the same as the focus of the parabola that you created it from.

✤ **vertex of a paraboloid**
Likewise, the vertex of the paraboloid will be the same as the vertex of the parabola it is based on.

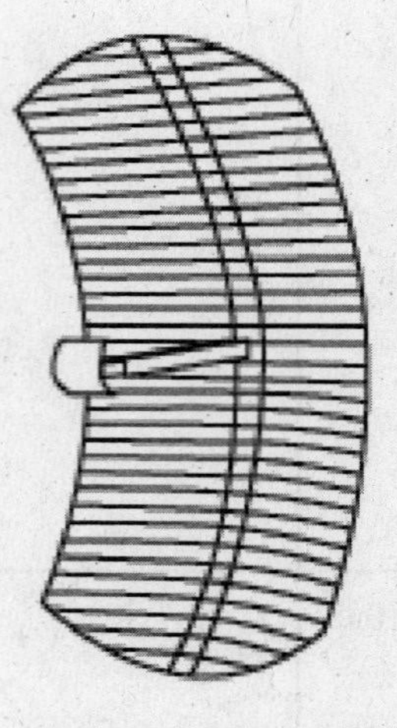

A typical parabolic antenna.
Do you know where the focus of this paraboloid is?

5.2: Ellipses

Next come ellipses. Keep in mind that a circle is a sort of special case of the ellipse, and remember the standard form of the equation for a circle (section 2.1). You'll see some similar features in the equation for an ellipse, so it might help.

✤ **ellipse**
The formal definition of an ellipse goes something like this: An ellipse is the set of points the sum of whose distances from two fixed foci is a constant.

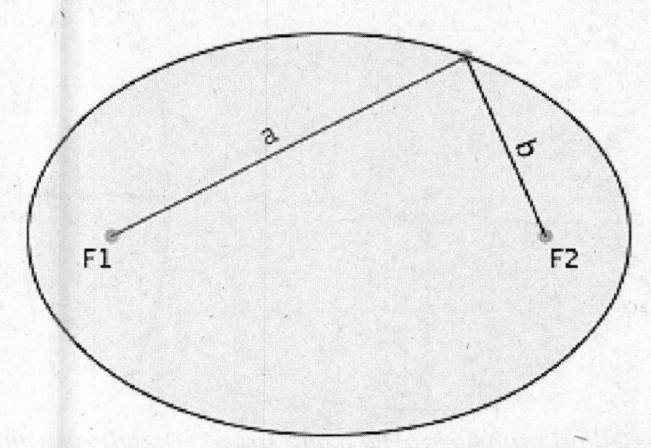

(In this depiction, $a + b$ is always the same, no matter what point you visit on the ellipse.)

✤ **foci**
The foci are the two fixed points required to describe the ellipse.

★ **For Fun**
If the two foci were coincident, what shape would you have?

Ellipses With Center at (0,0)

Like the circle, which was based on all points being the same distance from the center, deriving the equation for an ellipse depends on the distance formula. This time it's the sum of two distances that is constant. I'm sure you can already imagine that is more complicated, and if you want to investigate further, the derivation is in your textbook on page 440. For us, let's look at the pieces that make up the ellipse.

- **major axis**
  An ellipse is like an elongated circle. Picture that flattened circle with a longest diameter and a shortest diameter. That longest diameter would be the major axis. (And it's not called a diameter at all!)

- **minor axis**
  The "shortest diameter" of a flattened circle is what we call the minor axis of an ellipse.

For ellipses centered at the origin, the axes will be either the x-axis or y-axis. Which way is the circle squashed out?

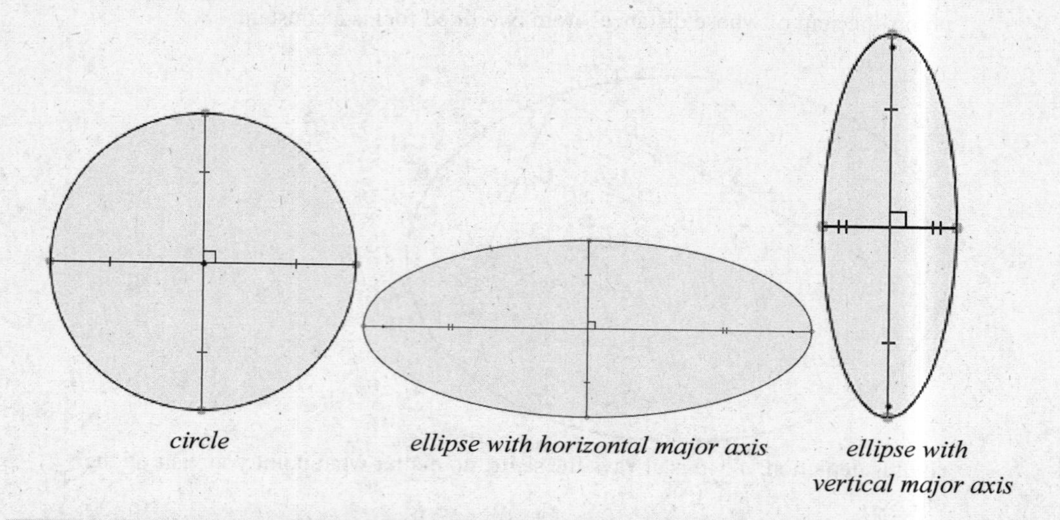

circle *ellipse with horizontal major axis* *ellipse with vertical major axis*

Standard Forms for an Ellipse Centered at the Origin

1. An ellipse whose major axis falls on the x-axis has an equation of the form:
   $\frac{x^2}{a^2} + \frac{y^2}{b^2} = 1$ where $a > b$.

2. An ellipse whose major axis falls on the y-axis has an equation of the form:
   $\frac{x^2}{b^2} + \frac{y^2}{a^2} = 1$ where $a > b$.

✤ **center**
The center of an ellipse is the point where the major and minor axes intersect.

✤ **vertices**
The two vertices of an ellipse are the points where the ellipse and the major axis intersect.

✤ **semimajor axis**
The semimajor axis is one half of the major axis that connects the center with a vertex. The semimajor axis is a long. (How long is the major axis then?)

✤ **semiminor axis**
The semiminor axis is, likewise, one half of the minor axis; it connects the center of the ellipse with the point on the ellipse that is closest to the center. The length of the semiminor axis is b.

★ **For Fun**
Can you label the lengths of the axes, semi and otherwise, on the ellipse on the previous page?

Ellipses With Center at (h,k)

Use the same transformation equations that we used to generalize the position of the parabola in the last section, and you will get the generalized ellipse equations:

$$\frac{(x-h)^2}{a^2}+\frac{(y-k)^2}{b^2}=1$$

$$\frac{(x-h)^2}{b^2}+\frac{(y-k)^2}{a^2}=1$$

$a > b$ still, so can you pick out the ellipse with a vertical major axis? a horizontal major axis?

★ **For Fun**
A circle has one and only one diameter and radius for all points on its circumference. So it's like an ellipse where the axes are the same length. If you let $a = b$ (which is *not* an ellipse because for an ellipse $a > b$), could you recover the standard form of the equation for a circle? I suggest you let $a^2 = b^2 = r^2$.

Standard Forms of the Equation of an Ellipse With Center at (h,k): Test Prep

Sketch $\frac{x^2}{4} + \frac{y^2}{9} = 1$

Because $a=3$ is greater than $b=2$, this ellipse has a vertical major axis. The center is at the origin. From the origin, I plot points 3 up and 3 down, 2 to the left and 2 to the right:

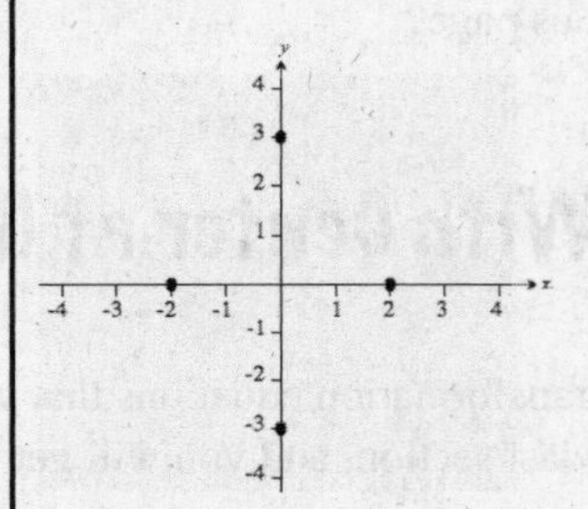

Then connect those to form an ellipse shape:

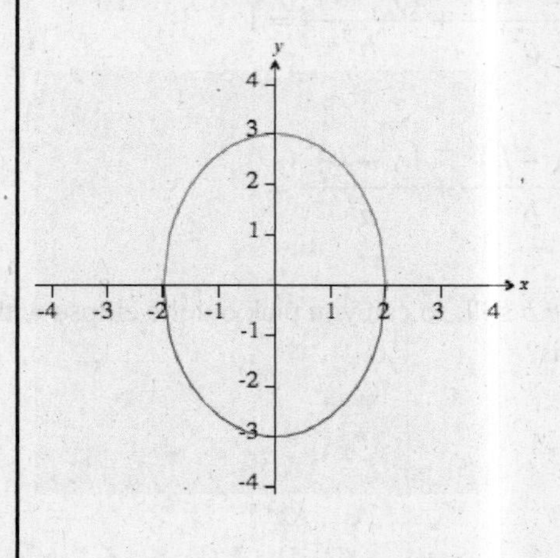

Standard Forms of the Equation of an Ellipse With Center at (h,k): Test Prep

Try Them	
Find the standard form of the equation for the ellipse $18x^2 + 12y^2 - 144x + 48y + 120 = 0$	$\dfrac{(x-4)^2}{12} + \dfrac{(y+2)^2}{18} = 1$
Sketch $\dfrac{(x+1)^2}{100} + \dfrac{(y-2)^2}{64} = 1$	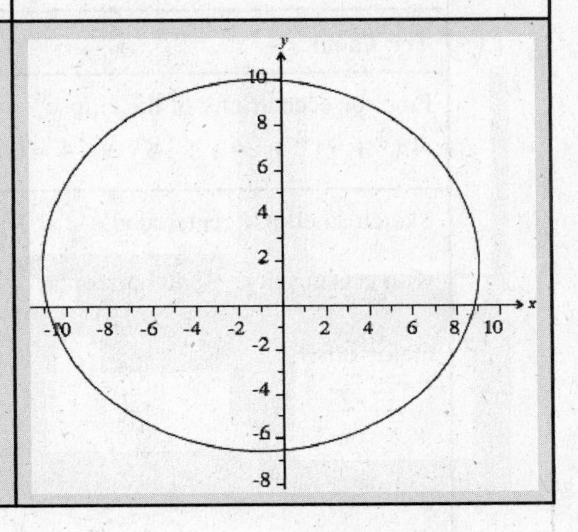

Eccentricity of an Ellipse

Some ellipses are more squashed than others. The measure of their "squashedness" is called the eccentricity.

✤ **eccentricity**
The eccentricity of an ellipse is given by

$$e = \frac{c}{a}$$

where $c^2 = a^2 - b^2$.

The measurement e will always be between 0 and 1. (e close to 0 means an ellipse close to circular; does that make sense knowing what "eccentricity" means in English? e close to 1 means an ellipse so elongated it's almost a line -- very eccentric!)

Eccentricity of an Ellipse: Test Prep	
Find the eccentricity of the ellipse: $\frac{(y-4)^2}{289}+\frac{(x-3)^2}{225}=1$	$a^2=289$ $b^2=225$ $e=\frac{\sqrt{289-225}}{\sqrt{289}}=\frac{8}{17}$
Try Them	
Find the eccentricity of the ellipse: $4x^2+9y^2+24x+18y+44=0$	$e=\frac{\sqrt{5}}{3}$
Sketch an ellipse centered at (-2, 1) with eccentricity $\frac{2}{9}$ and horizontal major axis.	for example,

Applications of Ellipses

Planetary orbits and elliptical ceilings are a couple examples of the application of the ellipse. There are also elliptical gears in machines and on bicycles, lithotripsy in medicine, oval tankers, and elliptical pool tables. By the way, the American football is an oval!

The Oval Office in Washington, D.C.: Do you imagine the desk is in the center or at one of the focuses?

MID-CHAPTER QUIZ: TEXTBOOK PAGE 452

5.3: Hyperbolas

Check back to the beginning picture about these conic sections. The hyperbola is the one with two different curves. This makes it somewhat more difficult than the other two, mostly with respect to graphing.

✤ **hyperbola**
This two-part curve is made of flattened-out U-shapes is the conic section called the hyperbola. Defined in terms of the points that make it up, a hyperbola is all the points the *difference* of whose distances from two foci are a constant -- in the following illustration a - b is constant. (Whereas, it was the *sum* of these parameters that remained constant for the ellipse.)

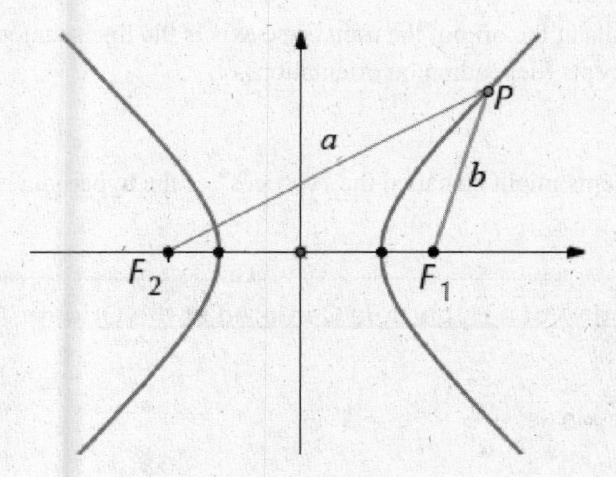

★ **For Fun**
The words "hyperbola" and "hyperbole" are related. Do you know what a hyperbole is? Can you see why they are related?

✤ **foci**
Like the focus of the parabola, the foci of a hyperbola are "inside" the U-shaped curves.

Hyperbolas With Center at (0,0)

We'll start again with hyperbolas whose center is at the origin. But where is the center of a hyperbola? The center is the point midway between the two vertices, which are the turning-around points of the two curves. So now we'll have hyperbolas that either open up-and-down or left-and-right:

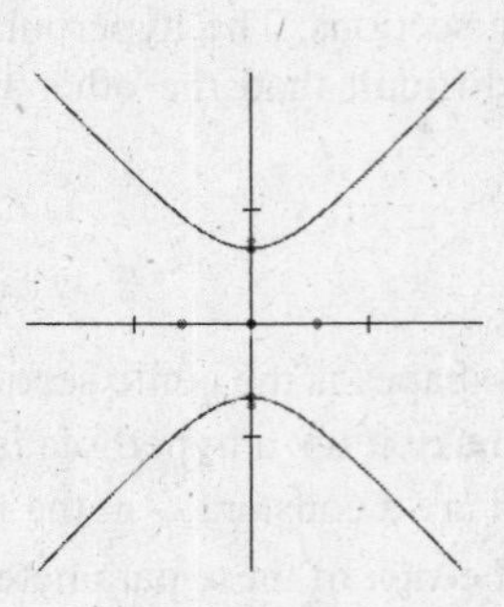

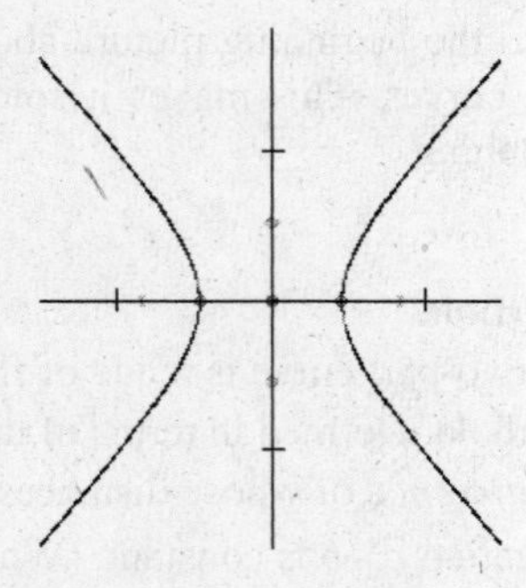

✤ **transverse axis**
With the center of the hyperbola at the origin, the transverse axis is the line segment that connects the x- or y-intercepts (depending on orientation).

★ **For Fun**
Do you think these intercepts might be called the "vertices" of the hyperbola?

Standard Forms of the Equation of a Hyperbola Centered at the Origin

1. With transverse axis on the x-axis: $\dfrac{x^2}{a^2} - \dfrac{y^2}{b^2} = 1$

2. With transverse axis on the y-axis: $\dfrac{y^2}{a^2} - \dfrac{x^2}{b^2} = 1$

Notice that there is not a requirement that *a* be greater than *b* -- instead it is the variable with or without the subtraction symbol that tells you the orientation of the hyperbola.

- **center**
  The center of the hyperbola is the midpoint of the transverse axis. The center is exactly halfway between the points where the two curves of the hyperbola are closest.

- **conjugate axis**
  The conjugate axis is a line segment of length $2b$ perpendicular to the transverse axis and centered at the center of the hyperbola.

- **vertices**
  The vertices of the hyperbola are the turning-around points where the two curves are nearest to each other. The vertices are each exactly a away from the center in the direction of the transverse axis.

- **asymptotes**
  The asymptotes of a hyperbola are important graphing aids. They are two lines that pass through the center of the hyperbola and show the shape of the curves because each half of each curve of the hyperbola will approach these asymptotes

Equations for the Asymptotes of a Hyperbola Centered at the Origin

1. Horizontal transverse axis: $y = \frac{b}{a}x$ and $y = -\frac{b}{a}x$

2. Vertical transverse axis: $y = \frac{a}{b}x$ and $y = -\frac{a}{b}x$

   (The parameters *a* and *b* come from the equation of the hyperbola!)

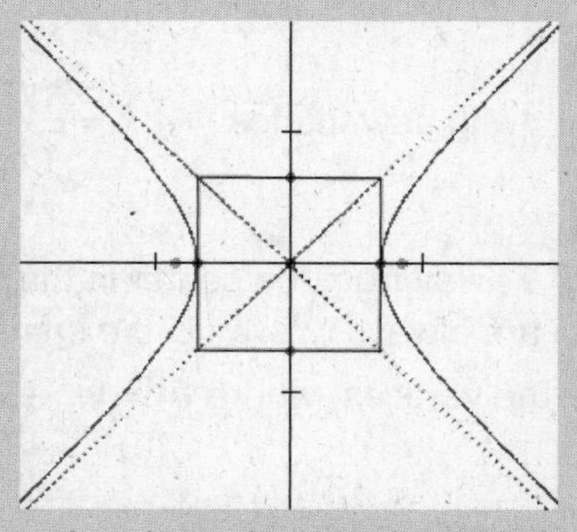

Notice that if you make a box based on the length of the transverse axis and the conjugate axis, then your asymptotes coincide with the diagonals of that box.

✓ **Questions to Ask Your Teacher**
There is a nifty method to remember the equations for the asymptotes of a hyperbola centered at the origin at the bottom of page 454 of your textbook! Ask your teacher to go over it if you're interested.

Hyperbolas With Center at (h,k)

Again, we have to generalize to a hyperbola that is centered anywhere in the 2-dimensional plane. We use the same transformation equations as we used in the last two sections.

Standard Forms of the Equations of a Hyperbola With Center at *(h, k)*

1. Horizontal transverse axis:

$$\frac{(x-h)^2}{a^2}-\frac{(y-k)^2}{b^2}=1 \text{ (with asymptotes } y-k=\pm\frac{b}{a}(x-h))$$

2. Vertical transverse axis:

$$\frac{(y-k)^2}{a^2}-\frac{(x-h)^2}{b^2}=1 \text{ (with asymptotes } y-k=\pm\frac{a}{b}(x-h))$$

The vertices are always *a* away from the center in the direction of the transverse axis. The foci are always *c* away in the same direction (and farther than the vertices) where $c^2=a^2+b^2$.

Standard Forms of the Equation of a Hyperbola With Center at (h,k): Test Prep

Sketch $\dfrac{(x-4)^2}{9}-\dfrac{(y-5)^2}{16}=1$ Include the asymptotes.	This form is for a hyperbola with a horizontal transverse axis. The center is at (4, 5). The asymptotes are $y=\dfrac{3}{4}x-\dfrac{1}{3}$ and $y=-\dfrac{3}{4}x+\dfrac{31}{3}$. 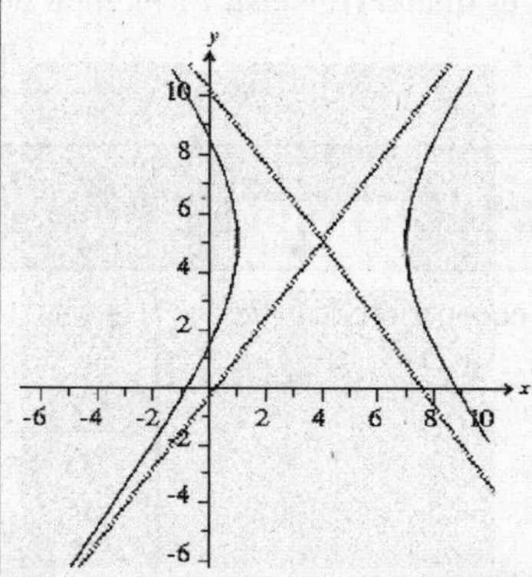
Try Them	
Find the standard form of the equation for the hyperbola: $4x^2-9y^2-24x-90y-153=0$	$\dfrac{(y+5)^2}{4}-\dfrac{(x-3)^2}{9}=1$
Sketch $25x^2-9y^2-100x-72y-269=0$	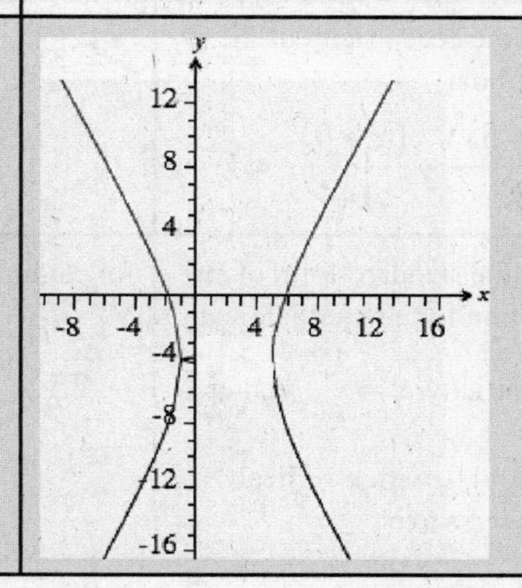

Eccentricity of a Hyperbola

Like the ellipse, some hyperbolas are more eccentric than others!

✤ **eccentricity**

The eccentricity of a hyperbola is a measure of its "wideness." The eccentricity of a hyperbola is calculated using $e = \frac{c}{a}$ while remembering that $c^2 = a^2 + b^2$. Since c is always bigger than a (it is, isn't it?), then e for hyperbola is always greater than 1.

Eccentricity of a Hyperbola: Test Prep

Find the eccentricity of the hyperbola $4x^2 - y^2 = 16$	$4x^2 - y^2 = 16$ $\frac{4x^2}{16} - \frac{y^2}{16} = 1$ $\frac{x^2}{4} - \frac{y^2}{16} = 1$ So $a^2 = 4$ and $b^2 = 16$. And $e = \frac{\sqrt{4+16}}{\sqrt{4}} = \frac{\sqrt{20}}{2} = \frac{2\sqrt{5}}{2} = \sqrt{5}$
Try Them	
Find the eccentricity of the hyperbola $\frac{(y+5)^2}{25} - \frac{(x+9)^2}{144} = 1$	$e = \frac{13}{5}$
Find the standard form of an equation for a hyperbola with eccentricity $e = \frac{12}{5}$, centered at (0, -1), with a vertical transverse axis.	for example, $\frac{(y+1)^2}{25} - \frac{x^2}{119} = 1$

Applications of Hyperbolas

Like the other conic sections, hyperbolas are everywhere. Some comets (those that visit our sun only once) travel in hyperbolic orbits. Much architecture uses hyperbolas, and industry uses gears that are hyperboloids. The sonic boom is hyperbolic in shape. Radio transmission depends on the foci of hyperbolas, and mirrors and lenses that are hyperbolic in shape are widespread.

The pattern cast on a wall by a lamp with a cylindrical shade is a hyperbola.

The hyperboloid is the design standard for nuclear cooling towers.

When two stones are thrown simultaneously into a pool of still water, ripples move outward in concentric circles, which intersect at points which form a hyperbola.

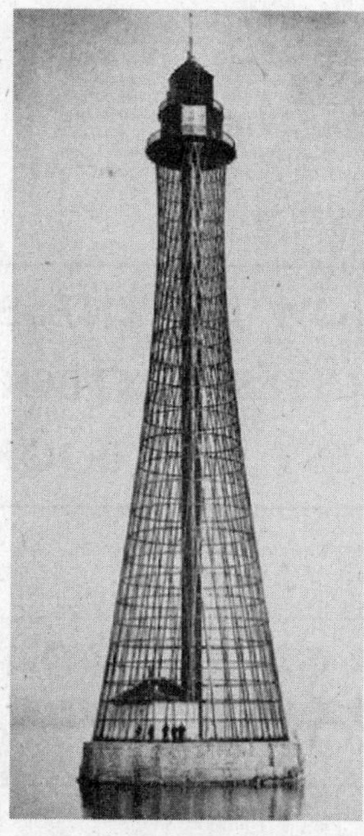

A hyperboloid lighthouse from 1911.

Final Fun: Matching

FIND THE BEST MATCH. GOOD LUCK.

ECCENTRICITY OF ELLIPSE	$y = -\frac{b}{a}x$
PARABOLA WITH VERTICAL AXIS	$e = \frac{\sqrt{a^2 + b^2}}{a}$
ECCENTRICITY OF HYPERBOLA	$(x-h)^2 = 4p(y-k)$
ELLIPSE WITH VERTICAL MAJOR AXIS	$e = \frac{\sqrt{a^2 - b^2}}{a}$
$x^2 + y^2 = r^2$	CIRCLE
HYPERBOLIC ASYMPTOTE	$\frac{(x-h)^2}{b^2} + \frac{(y-k)^2}{a^2} = 1$
$\frac{y^2}{a^2} - \frac{x^2}{b^2} = 1$	HYPERBOLA

NOW TRY THESE:
REVIEW EXERCISES: TEXTBOOK PAGE 467,
PRACTICE TEST: TEXTBOOK PAGE 468.

STAY UP ON YOUR GAME -- DO THE CUMULATIVE
REVIEW EXERCISES: TEXTBOOK PAGE 469

Chapter 6: Systems of Equations and Inequalities

Word Search

N V E H R R O P K I R T D N I
E O I L A A A R N D N J S O N
T G I E I R E E Q E F T O I D
Q R N T T M Q N T M N J L T E
D I I I I U I S I I F E U U P
L E A A A S I N A L Q G T T E
M L P L N S O R A U N R I I N
V E I E N G T P I T I O O T D
V T T O N S U L M V I O N S E
Y F C S N D I L I O H O R B N
R N E O Y B E A A S C V N U T
I B C T R S L N X R N E I S H
O P T I M I Z A T I O N D V N
N S U O E N E G O M O H M B M
Y M E Q U I V A L E N T Q S S

CONSTRAINTS
DECOMPOSITION
DEPENDENT
ELIMINATION
EQUILIBRIUM
EQUIVALENT
HOMOGENEOUS
INCONSISTENT
INDEPENDENT
INEQUALITY
LINEAR
NONLINEAR
OPTIMIZATION
PARTIAL
SOLUTION
SUBSTITUTION
SYSTEM
TRIANGULAR
TRIVIAL

6.1: Systems of Linear Equations in Two Variables

We've discussed linear equations pretty thoroughly. If you can graph one line, you can graph more than one line on the same 2-dimensional plane. Now that you have a system of linear equations in 2-dimensions, what can we do with this concept?

✤ **system of equations**
A system of equations is a group of two or more equations that are all intended to be considered together -- as if they were all part of the description of the same "universe."

✤ **linear system of equations**
A *linear* system of equations is a system of equations whose members are all linear equations. You remember that a linear equation has the general form: $Ax + By = C$
from Section 2.3, right?

✤ **solution of a system of equations**
Remembering that a solution to a linear equation is an ordered pair whose x- and y-values make the equation *true*, it makes sense to say that the solution to a system of such equations is an ordered pair where *all* the equations are true.

✓ **Questions to Ask Your Teacher**
Be sure you know the format your teacher wants solutions to systems in. By what's right, answers should appear as an ordered pair or some other sensible representation of two (or more -- that's for later) variables.

✤ **consistent system**
A consistent system of equations has at least one solution.

✤ **inconsistent system**
An inconsistent system has no solutions.

★ **For Fun**
Can you picture the case where, say, two linear equations do not share an ordered pair in their solution sets? There's a lot of language in that question to comprehend in order to get at the very simple answer.

✤ **independent system**
An independent system is a consistent system that has exactly *one* solution.

✤ **dependent system**
A dependent system of equations is a consistent system that has an infinite number of solutions.

★ **For Fun**
Now can you picture a system of two linear equations which share an infinite number of points, share their *entire* solutions set? Again, lots of math language to get at what is really a very simple configuration.

Substitution Method for Solving a System of Equations

One method for finding the solution(s), or lack thereof, to a system of equations is this *substitution* method. It's tried and true, a method you can always call on, and one you'll need to call on at some time. So practice a few of these!

✤ **substitution method**
The method of substitution involves the following steps:

1. If necessary, solve one of the equations for one of the variables. (It neither matters which equation nor which variable, so pick the easiest.)

   For example:

   $$-3x + 7y = 14$$
   $$2x - y = -13$$

   Pick the second equation and solve for y:

   $$y = 2x + 13$$

2. Rewrite the other equation by substituting the expression that you just derived for the variable you just solved for.

   For example:

   $$-3x + 7(2x + 13) = 14$$

3. Now solve that equation for the *one* variable it is now in terms of.

   For example:

   $$-3x + 7(2x + 13) = 14$$
   $$-3x + 14x + 91 = 14$$
   $$11x = -77$$
   $$x = -7$$

4. Substitute that *value* back into the other equation and find the corresponding value of the other variable.

   For example:

   $$2(-7) - y = -13$$
   $$-14 - y = -13$$
   $$y = -1$$

5. Write your answer in the form of an ordered pair.

   For example:

   (-7, -1)

Elimination Method for Solving Systems of Equations

Here's a different method that basically relies on the idea that the equation has the same truth value if you do the same thing to both sides! That's not a new idea.

✤ **equivalent**
Two systems of equations are equivalent as long as they have the same solution. (So if we can rewrite a system into an equivalent system that for some reason might be easier to solve, we're still finding the same solution that we would for the original system. And if you simply are using the rules of algebra to modify any given equation, it'll still have the same solution set, right?)

Allowed Actions for Producing Equivalent Systems:

1. Interchanging any two equations.
2. Multiplying both sides of any equation by the same non-zero value.
3. Adding (or subtracting) the same thing to both sides of an equation, usually in the form of a multiple of one of the other equations, whose two sides are, after all, *equal*.

✤ **elimination method**
The elimination method uses the properties of equivalent systems to *eliminate* one of the variables (you did that with substitution, too) and then proceed to solve the system of equations.

For example, consider this system:

$$3x - 8y = -6$$
$$-5x + 4y = 10$$

Multiplying the second equation by *2* re-renders the system like this:

$$3x - 8y = -6$$
$$-10x + 8y = 20$$

Adding the new second equation to both sides of the first, the left sides together and rights, respectively, eliminates the y-variable and results in a new equation:

$$-7x = 14$$

The solution of this equation is easily arrived at:

$$x = -2$$

Plugging this value back into one of our original (equivalent-system) equations gives us the value for the second variable:

$$3(-2) - 8y = -6$$
$$-6 - 8y = -6$$
$$-8y = 0$$
$$y = 0$$

And so the solution is *(-2, 0)*.

You can now *check* your solution in the other equation by plugging in this x- and y-value:

$$-5(-2) + 4(0) \stackrel{?}{=} 10$$

(I think our solution is correct.)

✓ **Questions to Ask Your Teacher**
How might you recognize an inconsistent or dependent system of equations? Both of my examples have been for independent systems -- they had exactly one ordered-pair solution. Be sure you know how to format an answer for inconsistent and dependent systems, too.

★ **For Fun**
Besides the methods of substitution and elimination, do you think you could find solutions by *graphing*?

Systems of Linear Equations in Two Variables: Test Prep

Solve the system of linear equations: $\begin{array}{l} 3x - 4y = 8 \\ 6x - 8y = 9 \end{array}$	$3x - 4y = 8$ (1) $6x - 8y = 9$ (2) Multiply (1) by 2: $6x - 8y = 16$ (1) $6x - 8y = 9$ (2) Subtract (2) from (1) with the result: $0 = 7$ which is a contradiction; therefore, there is no solution.
Try Them	
Solve the system of equations: $\begin{cases} \frac{3}{4}x + \frac{2}{5}y = 1 \\ \frac{1}{2}x - \frac{3}{5}y = -1 \end{cases}$	$\left(\frac{4}{13}, \frac{25}{13}\right)$
Solve the system: $\begin{cases} y = 3x + 4 \\ x = 4y - 5 \end{cases}$	$(-1,1)$

A system of linear equations can be solved four different ways:

- graphing
- substitution
- elimination
- matrices (next chapter)

Classification of Systems of Linear Equations: Test Prep

Are the linear equations $\begin{cases} 2x + y = 15 \\ x - 4y = -1 \end{cases}$ independent, dependent, or inconsistent?	$2x + y = 15$ $x - 4y = -1$ becomes $2x + y = 15$ $2x - 8y = -2$ Subtracting the equations results in $9y = 17$ $y = \frac{17}{9}$ Substituting back in for y $x - 4\left(\frac{17}{9}\right) = -1$ $x = \frac{-9 + 68}{9}$ $x = \frac{59}{9}$ The lines do intersect at one point, $\left(\frac{59}{9}, \frac{17}{9}\right)$, and so they are independent.
Try Them	
Are $\begin{cases} x + 3y = 4 \\ 2x - y = 1 \end{cases}$ inconsistent, dependent, or independent?	independent
Are $\begin{cases} \frac{x+3}{4} = \frac{2y-1}{6} \\ 3x - 4y = 2 \end{cases}$ independent, dependent, or inconsistent?	inconsistent

Applications of Systems of Equations

Economics, biology, chemistry, and physics are some of the places you may need to solve systems of linear equations. Any "universe" where two equations are operating simultaneously is a possible candidate. (And most "universes" have many things going on at once.)

✤ **equilibrium price**
The price that balances income and expenses is the equilibrium price.

★ **For Fun**
Equilibrium is an interesting word. You might look it up!

✤ **supply-demand problems**
A supply-demand problem entails two equations: a supply equation and a demand equation. Solving such a system solves the supply-demand problem by returning the correct amount to be manufactured and the best price to charge for it!

Solve a Supply-Demand Problem:

If the number of computers to be manufactured for a price p is x, and the number that will be purchased by a retailer depends on p according to the following two equations, can you find the (x, p) data point that is the best solution to this system? What does this x mean, and what is this p?

$$x = \frac{9}{5}p - 1530$$

$$x = -\frac{3}{10}p + 780$$

6.2: Systems of Linear Equations in Three Variables

In the last section, we worked in two dimensions. Now we extend that practice into three dimensions:

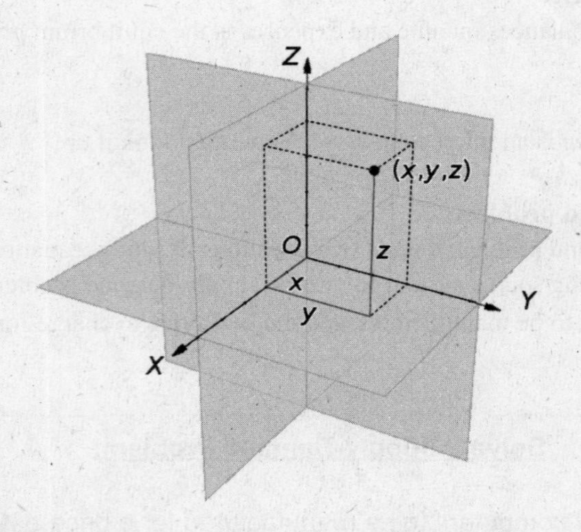

Three dimensional space is difficult to portray in two dimensions. Can you tell which way these axes are pointing? Can you pick out the three 2-dimensional planes that partition this 3-d space?

Systems of Linear Equations in Three Variables

The methods for solving linear equations in three dimensions "by hand" are exactly the same as those we used for 2-dimensional systems ... only much more complicated. You'll need to bring all your organizational and focusing skills to bear now!

✤ **ordered triple**

Like an ordered (x, y) pair, an ordered triple can be the solution to a system of linear equations ... in three dimensions: (x, y, z).

✤ **z-axis**
Three dimensions generally require another axis, or direction, to be specified and added to the 2-dimensional cartesian plane. Most often that axis is called the z-axis, and coordinate distances measured along the scale on the z-axis come third in an ordered triple.

★ **For Fun**
Look carefully at the previous graphic for the 3-dimensional system. This is the usual orientation of the three axes. Can you picture these axes in three dimensions? The x-axis would actually be poking out of the page -- straight out, orthogonal to the plane of the page. The y- and z-axes are orthogonal to each other and in the plane of the page. Can you imagine where in space the ordered triple in the graphic is situated? This is not easy or trivial stuff.

Triangular Form

Triangular form is a particularly easy form for three or more linear equations to be solved from. So the idea is to find a triangular system of equations *equivalent* to the system you have to solve. This is accomplished by just the same manipulations we did earlier to find equivalent systems.

✤ **triangular form**
The triangular form of a system of equations looks like this, for example:

$$\begin{array}{rrrcr} x & +2y & -z & = & 1 \\ & -5y & +3z & = & 4 \\ & & -2z & = & -6 \end{array}$$

This form makes it possible to solve the bottom equation for one variable, use that result to solve the next equation up, and use those results to solve the top equation.

★ **For Fun**
Try it for the above system. See if you get the solution, an ordered triple, *(2, 1, 3)*.

The careful manipulations needed to get each row of a system of three equations so that they become their equivalent triangular form is exacting but necessary, and learning this now will make the material in the next chapter come so much more easily. Here comes an example:

$$\begin{aligned} x \quad -2y \quad +3z &= 5 \\ 3x \quad -3y \quad +z &= 9 \\ 5x \quad +y \quad -3z &= 3 \end{aligned}$$

And multiply the top row by 5, subtract the top row from the bottom row in the next step:

$$\begin{aligned} 5x \quad -10y \quad +15z &= 25 \\ 3x \quad -3y \quad +z &= 9 \\ 11y \quad -18z &= -22 \end{aligned}$$

Then divide the top equation by 5 and multiply it by 3 and subtract the top row from the second row:

$$\begin{aligned} 3x \quad -6y \quad +9z &= 15 \\ 3y \quad -8z &= -6 \\ 11y \quad -18z &= -22 \end{aligned}$$

Next, divide the top row by 3 (gets it back to its original form), multiply the bottom row by 3 and the second row by 11, subtract the second row from the bottom row:

$$\begin{aligned} x \quad -2y \quad +3z &= 5 \\ 33y \quad -88z &= -66 \\ 34z &= 0 \end{aligned}$$

Divide the middle row by 11 to get it back to smaller numbers and begin your back-solving:

$$\begin{aligned} x \quad -2y \quad +3z &= 5 \\ 3y \quad -8z &= -6 \\ 34z &= 0 \end{aligned}$$

From the third equation we see that $z = 0$. The second equation becomes $3y - 8(0) = -6$. So $y = -2$. And the top equation becomes $x - 2(-2) + 3(0) = 5$, which means $x = 1$.

The solution is *(1, -2, 0)*. Lots of work? You betcha! But it's kinda fun once you get the rhythm of it.

✓ **Questions to Ask Your Teacher**
Once again, be sure you know how to recognize and specify when a 3-dimensional system of equations is inconsistent or dependent.

➡ Now shall we do a 4-dimensional system? ... Yeah, I thought not! Next chapter covers handier ways of handling bigger systems.

Systems of Linear Equations in Three Variables: Test Prep

Solve, if possible, $\begin{cases} 2x - y + z = 3 \\ x + 3y - 2z = 11 \\ 3x - 2y + 4z = 1 \end{cases}$	$\begin{cases} 2x - y + z = 3 \\ 3x + 9y - 6z = 33 \\ 3x - 2y + 4z = 1 \end{cases}$ $\begin{cases} 2x - y + z = 3 \\ x + 3y - 2z = 11 \\ -11y + 10z = -32 \end{cases}$ $\begin{cases} 2x - y + z = 3 \\ 2x + 6y - 4z = 22 \\ -11y + 10z = -32 \end{cases}$ $\begin{cases} 2x - y + z = 3 \\ 7y - 5z = 19 \\ -11y + 10z = -32 \end{cases}$ $\begin{cases} 2x - y + z = 3 \\ 77y - 55z = 209 \\ -77y + 70z = -224 \end{cases}$ $\begin{cases} 2x - y + z = 3 \\ 7y - 5z = 19 \\ 15z = -15 \end{cases}$ $z = -1$ $7y + 5 = 19; \quad y = 2$ $2x - 2 - 1 = 3; \quad x = 3$ $(3,2,-1)$

Systems of Linear Equations in Three Variables: Test Prep

Try Them	
Solve: $\begin{cases} 4x + 6y - 3z = 24 \\ 3x + 4y - 6z = 2 \\ 6x - 3y + 4z = 46 \end{cases}$	$(6, 2, 4)$
Solve: $\begin{cases} 3x + y - z = 4 \\ x + y + 4z = 3 \\ 9x + 5y + 10z = 8 \end{cases}$	inconsistent

Nonsquare Systems of Equations

A system of equations with fewer equations than there are variables is nonsquare. Nonsquare systems either have no solutions or infinitely many solutions.

Nonsquare Systems of Equations: Test Prep

Express the infinite solutions of this system as an ordered triple. $\begin{cases} x + 2y - 3z = 5 \\ 3x + 7y - 10z = 13 \end{cases}$	$\begin{cases} x + 2y - 3z = 5 \\ 3x + 7y - 10z = 13 \end{cases}$ $\begin{cases} 3x + 6y - 9z = 15 \\ 0 \quad y - 1z = -2 \end{cases}$ $\begin{cases} x + 2y - 3z = 5 \\ 0 \quad y - 1z = -2 \end{cases}$ $y = z - 2$ $x + 2z - 4 - 3z = 5; \quad x = z + 9$ Let z be any c, and the solution is $(c+9, c-2, c)$.

Nonsquare Systems of Equations: Test Prep

Try Them	
Does this nonsquare system have infinite or no solutions? $\begin{cases} 6x & -9y & +6z & = & 7 \\ 4x & -6y & +4z & = & 9 \end{cases}$	no solutions
Does this nonsquare system have infinite or no solutions? $\begin{cases} x & -3y & +4z & = & 9 \\ 3x & -8y & -2z & = & 4 \end{cases}$	infinite solutions

Homogeneous Systems of Equations

A homogeneous system of equations simply have zero as the "answer" to each equation. In other words, every equation is equal to zero, or the right-hand side of every equation is zero.

✤ **homogeneous systems of equations**
An example of a homogeneous system of equations might look like

$$\begin{array}{cccc} x & -4y & +2z & = 0 \\ -x & +7y & -4z & = 0 \\ 2x & -17y & +10z & = 0 \end{array}$$

★ **For Fun**
This system is actually dependent. You want to solve it?

✤ **trivial solution**
Every system of homogeneous equations admits the solution *(0, 0, 0)*. (Think about it.) This is the trivial solution. Usually we try to discover if there are other solutions (as with the example above).

Some Other Uses of "Homogeneous" in Mathematics:

Homogeneous function
Homogeneous polynomial
Homogeneous differential equation
Homogeneous linear transformation
Homogeneous coordinates
Homogeneous space
Homogeneous ideal
Homogeneous (large cardinal property)
Homogeneous model

✓ **For Fun**
Maybe you'd like to look up homogeneous.

Homogeneous Systems of Equations: Test Prep

Find the ordered triple that expresses the infinite solutions of:

$$\begin{cases} 2x - 3y + 5z = 0 \\ 3x + 2y + z = 0 \\ x - 4y + 5z = 0 \end{cases}$$

$$\begin{cases} 2x - 3y + 5z = 0 \\ 3x + 2y + z = 0 \\ x - 4y + 5z = 0 \end{cases}$$

$$\begin{cases} x - 4y + 5z = 0 \\ 6x - 9y + 15z = 0 \\ 6x + 4y + 2z = 0 \end{cases}$$

$$\begin{cases} x - 4y + 5z = 0 \\ -13y + 13z = 0 \\ 13y - 13z = 0 \end{cases}$$

$$\begin{cases} x - 4y + 5z = 0 \\ -y + z = 0 \\ 0 = 0 \end{cases}$$

$y = z$

$x - 4z + 5z = 0; \quad x = -z$

Let z be any c, and the solution is $(-c, c, c)$

Homogeneous Systems of Equations: Test Prep

Try Them	
Show the trivial solution is the only solution of: $$\begin{cases} 5x + 2y + 3z = 0 \\ 3x + y - 2z = 0 \\ 4x - 7y + 5z = 0 \end{cases}$$	$$\begin{cases} 5x + 2y + 3z = 0 \\ 3x + y - 2z = 0 \\ 4x - 7y + 5z = 0 \end{cases}$$ $$\begin{cases} 20x + 8y + 12z = 0 \\ 3x + y - 2z = 0 \\ 20x - 35y + 25z = 0 \end{cases}$$ $$\begin{cases} 15x + 6y + 9z = 0 \\ 15x + 5y - 10z = 0 \\ 43y - 13z = 0 \end{cases}$$ $$\begin{cases} 5x + 2y + 3z = 0 \\ -43y - 817z = 0 \\ 43y + 13z = 0 \end{cases}$$ $$\begin{cases} 5x + 2y + 3z = 0 \\ -43y - 817z = 0 \\ -804z = 0 \end{cases}$$ $z = 0$ $-43y = 0;\quad y = 0$ $5x = 0;\quad x = 0$ $(0,0,0)$
Solve: $$\begin{cases} 5x - 2y - 3z = 0 \\ 3x - y - 4z = 0 \\ 4x - y - 9z = 0 \end{cases}$$	*(5c, 11c, c)*

Applications of Systems of Equations

Systems of linear equations in many dimensions are used frequently in mathematics (curve fitting and differential equations), chemistry (simultaneous chemical processes), business (time/cost/demand and other simultaneous parameters), etc.

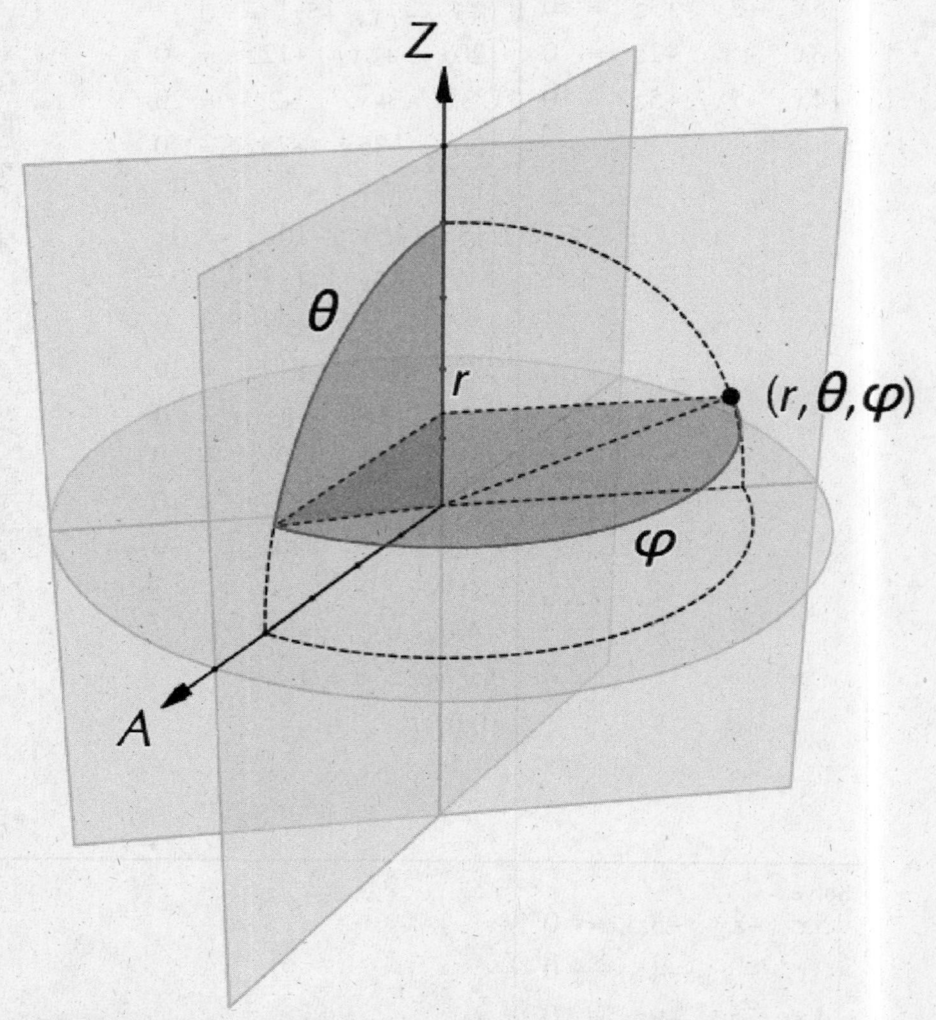

The 3-dimensional space in spherical coordinates, for those who may be tired of cartesian coordinates.

6.3: Nonlinear Systems of Equations

Linear systems are by no means the only sorts of systems available -- no more-so than linear functions are the only kinds of functions. We've seen linear, rational, radical, quadratic, conic, exponential, and logarithmic functions already in this course.

Solving Nonlinear Systems of Equations

You may use the method of substitution or elimination to solve nonlinear systems. However, I personally prefer substitution method for nonlinear systems ... at least until a bit of experience is built up. (Elimination requires a careful matching and accounting for like terms, and now that we can have more than just linear terms, it can get ever more complicated, but it certainly does work if applied correctly!)

✤ **nonlinear system of equations**
A nonlinear system of equations is a system of equations, just like we've been dealing with, but where one or more of the equations is *not* linear. An example might look like this:

$$5x + y = 3$$
$$y = x^2 - 3x - 5$$

The graphical solution to the system looks like this:

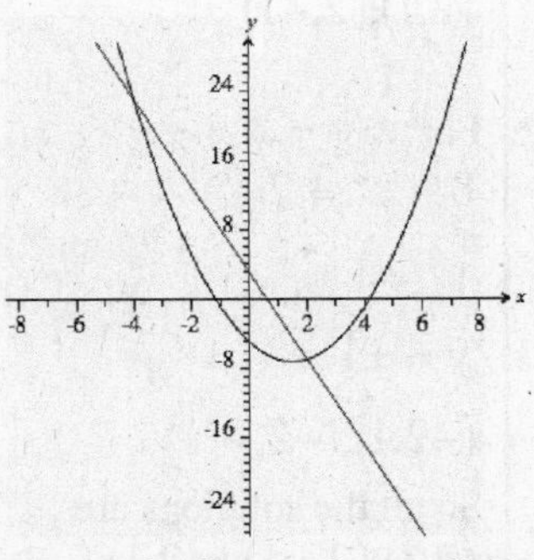

(Did you catch that this was a system of a line and a parabola? Do you see that now we can have more than just *one* point where these two equations "agree"?)

★ **For Fun**

Is this system nonlinear? Could you graph it?

$x + y = 1$

$xy = 1$

★ **For Fun**

Can you visualize a nonlinear system with only one solution? ... no solutions? ... infinitely many solutions?

Solutions of Nonlinear Systems of Equations: Test Prep

Solve the system of quadratic equations by elimination: $\begin{cases} 2x^2 & -y^2 & = & 7 \\ 3x^2 & +2y^2 & = & 14 \end{cases}$	$\begin{cases} 4x^2 & -2y^2 & = & 14 \\ 3x^2 & +2y^2 & = & 14 \end{cases}$ Adding the equations yields $7x^2 = 28$ $x^2 = 4$ $x = \pm 2$ For $x = 2$: $8 - y^2 = 7$ $y^2 = 1$ $y = \pm 1$ $(2,1),(2,-1)$ For $x = -2$: $8 - y^2 = 7$ $y^2 = 1$ $y = \pm 1$ $(-2,1),(-2,-1)$ So all the solutions are $(2,1),(2,-1),(-2,1),(-2,-1)$

Solutions of Nonlinear Systems of Equations: Test Prep

Try Them

Graph and solve: $\begin{cases} x^2+y^2=25 \\ x+2y=10 \end{cases}$	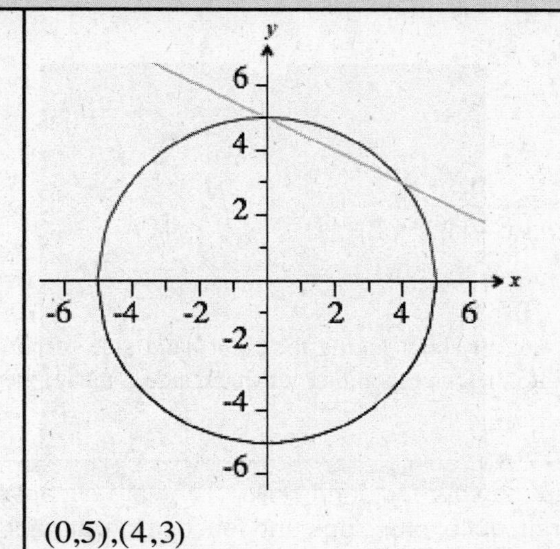 (0,5),(4,3)
Solve: $\begin{cases} x^2+y^2=40 \\ xy=12 \end{cases}$	(6,2),(−6,−2),(2,6),(−2,−6)

MID-CHAPTER QUIZ: TEXTBOOK PAGE 501

6.4: Partial Fractions

By now the idea of adding or subtracting fractions is hopefully not too scary. Are these do-able?

$$\frac{3}{5}-\frac{2}{7}=\frac{11}{35}$$

$$\frac{5}{x-1}+\frac{1}{x+2}=\frac{6x+9}{(x-1)(x+2)}$$

Partial fractions are all about taking the right-hand side of an equation, like those above and *decomposing* it into the corresponding left-hand side. I mean, they're equal after all, right?

> Are you "up" on your factoring skills? They are essential here, just like they were with rational expressions and functions. You need to start by factoring the denominators of every ratio you need to "decompose." Can you see how the factors on the right-hand side of the example above show up on the left-hand side?

Partial Fraction Decomposition

In order to do partial fraction decomposition, there are quite a few cases and methods and procedures to explore and learn. Practicing these can be fun, though, especially if you like puzzles! Sorting out the options is like flow-charting in computer science.

If you need more motivation than "fun" and "flow-charting," then I can assure you that this is an essential skill for calculus and, by extension, everywhere calculus is used.

✤ **partial fraction decomposition**
The process of writing a more complicated rational expression in terms of sums of simpler ratios is called partial fraction decomposition.

> The following cases sorta go from easiest to harder and harder. Hang on!

Case 1: Nonrepeated Linear Factors

Each non-repeated linear factor, $ax+b$, in the denominator will be included in the decomposition in the form $\frac{A}{ax+b}$.

For example

$$\frac{x-5}{x^2-1}=\frac{x-5}{(x+1)(x-1)}=\frac{A}{x+1}+\frac{B}{x-1}$$

$$(x+1)(x-1)\left(\frac{x-5}{(x+1)(x-1)}\right)=(x+1)(x-1)\left(\frac{A}{x+1}+\frac{B}{x-1}\right)$$

$$x-5=A(x-1)+B(x+1)$$

Let $x=1$ in this last equation:

$$-4=A\cdot 0+B\cdot 2$$

$$B=-2$$

Now let $x=-1$ in the same equation:

$$-6=A\cdot(-2)+B\cdot 0$$

$$A=3$$

And finally: $\frac{x-5}{x^2-1}=\frac{3}{x+1}+\frac{-2}{x-1}$

✓ **Questions to Ask Your Teacher**

I've used an alternate method to find my unknown parameters in this example. In all other examples, I'll use the equality of polynomials mentioned on the next page, consistent with your textbook. You can ask your teacher to explore both methods if you want to expand those horizons into something more robust.

Case 2: Repeated Linear Factors

Each repeated linear factor in the denominator will be included in the decomposition in the form $\frac{A_1}{ax+b}+\frac{A_2}{(ax+b)^2}+\ldots+\frac{A_m}{(ax+b)^m}$ where m is the multiplicity of the factor in the denominator.

For example

$$\frac{-7x+27}{x(x-3)^2}=\frac{A}{x}+\frac{B}{x-3}+\frac{C}{(x-3)^2}$$

$$x(x-3)^2\left(\frac{-7x+27}{x(x-3)^2}\right)=x(x-3)^2\left(\frac{A}{x}+\frac{B}{x-3}+\frac{C}{(x-3)^2}\right)$$

$$-7x+27=A(x-3)^2+Bx(x-3)+Cx$$

Now we'll use the definition of equality of polynomials from page 504 in your textbook: If two polynomials are equal then the coefficients of equal-degree terms are the same.

$$-7x+27=A(x-3)^2+Bx(x-3)+Cx$$

$$-7x+27=A(x^2-6x+9)+B(x^2-3x)+Cx$$

$$-7x+27=Ax^2-6Ax+9A+Bx^2-3Bx+Cx$$

$$-7x+27=(A+B)x^2+(-6A-3B+C)x+9A$$

★ **For Fun**
What is the coefficient of the quadratic term on the left-hand side?

And now we have a system of equations that arises out of equating coefficients. The first equation here comes from equating the quadratic coefficients on each side of the last equation; the next equation from equating linear coefficients; the last from equating constant terms:

$$\begin{array}{cccccc} A & +B & & = & 0 \\ -6A & -3B & +C & = & -7 \\ 9A & & & = & 27 \end{array}$$

From the last equation: $A = 3$. Substitute that result into the top equation and you have:

$$3 + B = 0$$

$$B = -3$$

Substitute both those results into the middle equation and

$$-6(3) - 3(-3) + C = -7$$

$$C = 2$$

And here's the final decomposition:

$$\frac{-7x + 27}{x(x-3)^2} = \frac{3}{x} + \frac{-3}{x-3} + \frac{2}{(x-3)^2}$$

★ **For Fun**

Could you actually add the rational expressions on the right-hand side and reproduce the left-hand side? At least, do you see that if you added the rational expressions on the right-hand side, what would the least common denominator be?

Case 3: Nonrepeated quadratic factors

Each nonrepeated quadratic factor in the denominator will be included in the decomposition in the form $\dfrac{Ax + B}{ax^2 + bx + c}$. Notice the numerator!

<u>*For example*</u>

$$\frac{3x^2+6x-21}{(x-5)(x^2+3)}=\frac{A}{x-5}+\frac{Bx+C}{x^2+3}$$

$$(x-5)(x^2+3)\left(\frac{3x^2+6x-21}{(x-5)(x^2+3)}\right)=(x-5)(x^2+3)\left(\frac{A}{x-5}+\frac{Bx+C}{x^2+3}\right)$$

$$3x^2+6x-21=A(x^2+3)+(Bx+C)(x-5)$$

$$3x^2+6x-21=Ax^2+3A+Bx^2-5Bx+Cx-5C$$

$$3x^2+6x-21=(A+B)x^2+(-5B+C)x+(3A-5C)$$

Which gives us the "coefficient" system of equations:

$$\begin{array}{llllll} A & +B & & = & 3 \\ & -5B & +C & = & 6 \\ 3A & & -5C & = & -21 \end{array}$$

And there's our system of equations to find *A*, and *B*, and *C*. Now you might want to really try to get this in triangular form to solve these! Here's my triangular system:

$$\begin{array}{lllll} A & +B & & = & 3 \\ & -5B & +C & = & 6 \\ & & C & = & 6 \end{array}$$

So:

$C=6$

$-5B+6=6$

$B=0$

$A+0=3$

$A=3$

Finally, the decomposition we've been working towards:

$$\frac{3x^2+6x-21}{(x-5)(x^2+3)}=\frac{3}{x-5}+\frac{6}{x^2+3}$$

Case 4: Repeated Quadratic Factors

Each repeated quadratic factor in the denominator will be included in the decomposition in the form

$$\frac{A_1x+B_1}{ax^2+bx+c}+\frac{A_2x+B_2}{(ax^2+bx+c)^2}+\ldots+\frac{A_mx+B_m}{(ax^2+bx+c)^m}$$ where, again, m is the multiplicity of the quadratic factor in the denominator.

For example

$$\frac{2x^3-x^2-6x-7}{(x^2-x-3)^2}=\frac{Ax+B}{x^2-x-3}+\frac{Cx+D}{(x^2-x-3)^2}$$

$$(x^2-x-3)^2\left(\frac{2x^3-x^2-6x-7}{(x^2-x-3)^2}\right)=(x^2-x-3)^2\left(\frac{Ax+B}{x^2-x-3}+\frac{Cx+D}{(x^2-x-3)^2}\right)$$

$$2x^3-x^2-6x-7=(Ax+B)(x^2-x-3)+Cx+D$$

$$2x^3-x^2-6x-7=Ax^3-Ax^2-3Ax+Bx^2-Bx-3B+Cx+D$$

$$2x^3-x^2-6x-7=Ax^3+(-A+B)x^2+(-3A-B+C)x+(-3B+D)$$

Equating coefficients again:

$$\begin{array}{lllll} A & & & & = 2 \\ -A & +B & & & = -1 \\ -3A & -B & +C & & = -6 \\ & -3B & & D & = -7 \end{array}$$

And here's my triangular version:

$$\begin{array}{lllll} A & & & & = 2 \\ & B & & & = 1 \\ & & C & & = 1 \\ & & & D & = -4 \end{array}$$

And you might notice that you can read the solution directly from this system (because not only is it triangular, it turned out to be "diagonal." More about that in the next chapter). Here's the final decomposition:

$$\frac{2x^3 - x^2 - 6x - 7}{\left(x^2 - x - 3\right)^2} = \frac{2x + 1}{x^2 - x - 3} + \frac{x - 4}{\left(x^2 - x - 3\right)^2}$$

Partial Fraction Decomposition: Test Prep

Give the *form* of the partial fraction decomposition of: $\frac{x-1}{x^3 - 3x^2 + 7x - 21}$. Do not solve for the unknown constants.	Factoring the denominator by grouping and choosing the appropriate numerators for the factors gives us: $\frac{x-1}{x^3 - 3x^2 + 7x - 21} = \frac{Ax + B}{x^2 + 7} + \frac{C}{x - 3}$
Try Them	
Find the partial fraction decomposition of: $\frac{-2x + 15}{(x-4)(x+3)}$	$\frac{-2x + 15}{(x-4)(x+3)} = \frac{1}{x-4} + \frac{-3}{x+3}$

Partial Fraction Decomposition: Test Prep

Find the partial fraction decomposition of: $\dfrac{3x^3+13x^2+18x+14}{(x^2+3x+4)(x-1)(x+2)}$	$\dfrac{3x^3+13x^2+18x+14}{(x^2+3x+4)(x-1)(x+2)}$ $=\dfrac{2x+3}{x^2+3x+4}+\dfrac{2}{x-1}+\dfrac{-1}{x+2}$

Dealing with "Improper" Rational Expressions

If the degree of the numerator is *larger* than the degree of the denominator, then we have a rational expression that is analogous to an improper fraction. This will cause a mess in partial fraction decomposition!

If you have an "improper" rational expression to decompose, then first thing to do is actually *divide* the polynomials. The remainder (if there is one) will then be the rational expression you need to find the partial fraction decomposition for.

Try it:

$$\frac{2x^3-7x^2+6x-21}{(x+1)(x-2)(x-3)}\left[=2+\frac{x^2+4x-33}{(x+1)(x-2)(x-3)}\right]$$

6.5: Inequalities in Two Variables and Systems of Inequalities

Back in Chapter One, we considered solving inequalities. In this section we'll work on the graphs of inequalities and then extend everything to *systems* of inequalities.

Graphing an Inequality

The graph of an inequality is a picture of the solution set of the inequality.

✤ **solution set of an inequality**
The solution set of an inequality is all the ordered pairs that make the inequality true. Please note that $x > 3$ includes a lot more "true" values than $x = 3$. Expect a bigger and qualitatively different solution set for an inequality.

✤ **graph of an inequality**
Exactly like a curve on a graph is an illustration of the *solution set* of an equation, the graph of an inequality is a depiction of all the points that are in the solution set of the inequality. Normally this will now be *more* than a curve, usually a whole shaded-in *region* of the 2-dimensional plane.

To Graph an Inequality

1. Graph the corresponding equality.

    a. If the inequality *includes* the equality (for example, ≤ or ≥), then graph the corresponding equation with a solid line.

    b. If the inequality does not include the equality (for example, < or >), then graph the corresponding equality using a dashed line.

2. Pick a test point (ordered pair) that does not fall on the curve you just graphed. (The origin is usually the easiest if it's available.)

3. Test the truth value of the inequality using the ordered pair you just chose.

   a. If the inequality is true, shade in the region on that side of the curve.

   b. If the inequality is false at the test point, shade in all regions on the "other:" side of the curve.

For example, the graph of $y < |x - 1|$ looks like this:

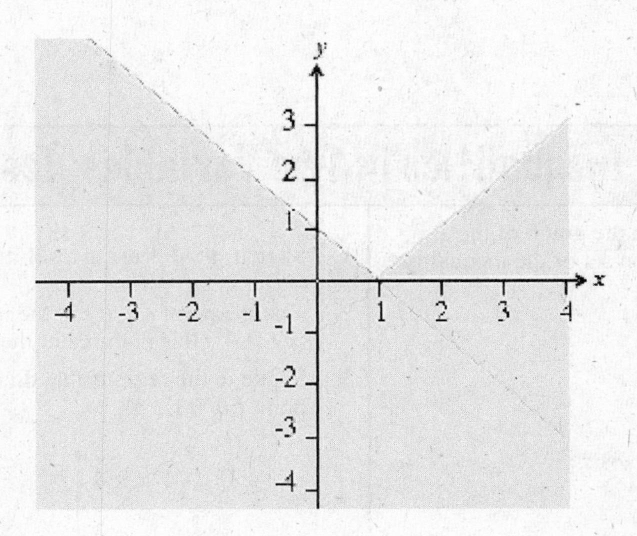

★ **For Fun**
Check the test point *(0, 0)* in the above inequality. Does it render the inequality true or not? Is the region which includes the origin in the shaded region of the graph?

Systems of Inequalities in Two Variables

You can illustrate and solve a system of inequalities graphically. A region on the graph that is shaded in for *every* inequality in the system will be the region of interest -- the solution set. If you recall the meaning of an intersection of sets, then you'll realize that the region of overlapping shading is the *intersection* of the solution sets of all the inequalities in the system. (So beware that if there is no overlapping region, there is no solution.)

✓ **Questions to Ask Your Teacher**
By the way, there are algebraic ways of solving systems of inequalities that compare to our methods of substitution and elimination, but we'll stick with the method of graphing for now. You could ask your teacher about these methods if you want to push the point!

✤ **solution set of a system of inequalities**
The solution set for a system of inequalities is the *region* that makes all the inequalities in the system true.

Inequalities in Two Variables: Test Prep

<table>
<tr><td>Sketch the graph of the solution set of the inequality:

$xy \geq 4$</td><td>1. Sketch $y = \frac{4}{x}$ using a solid line.
2. Test the point $(0, 0)$ in the inequality $xy > 4$: 0 is not greater than 4.
3. Shade in the regions that do not include the point $(0, 0)$.
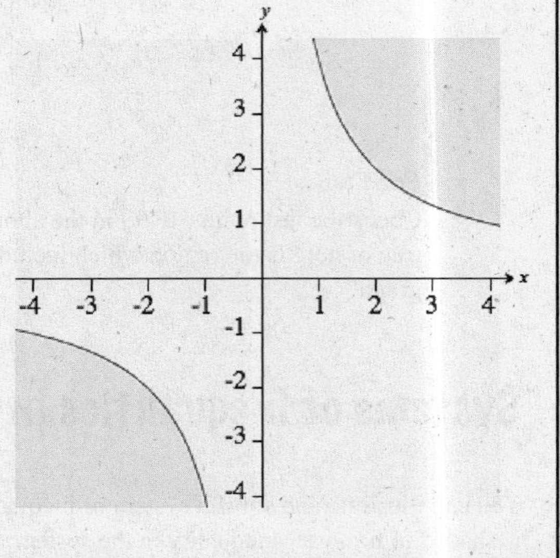
</td></tr>
</table>

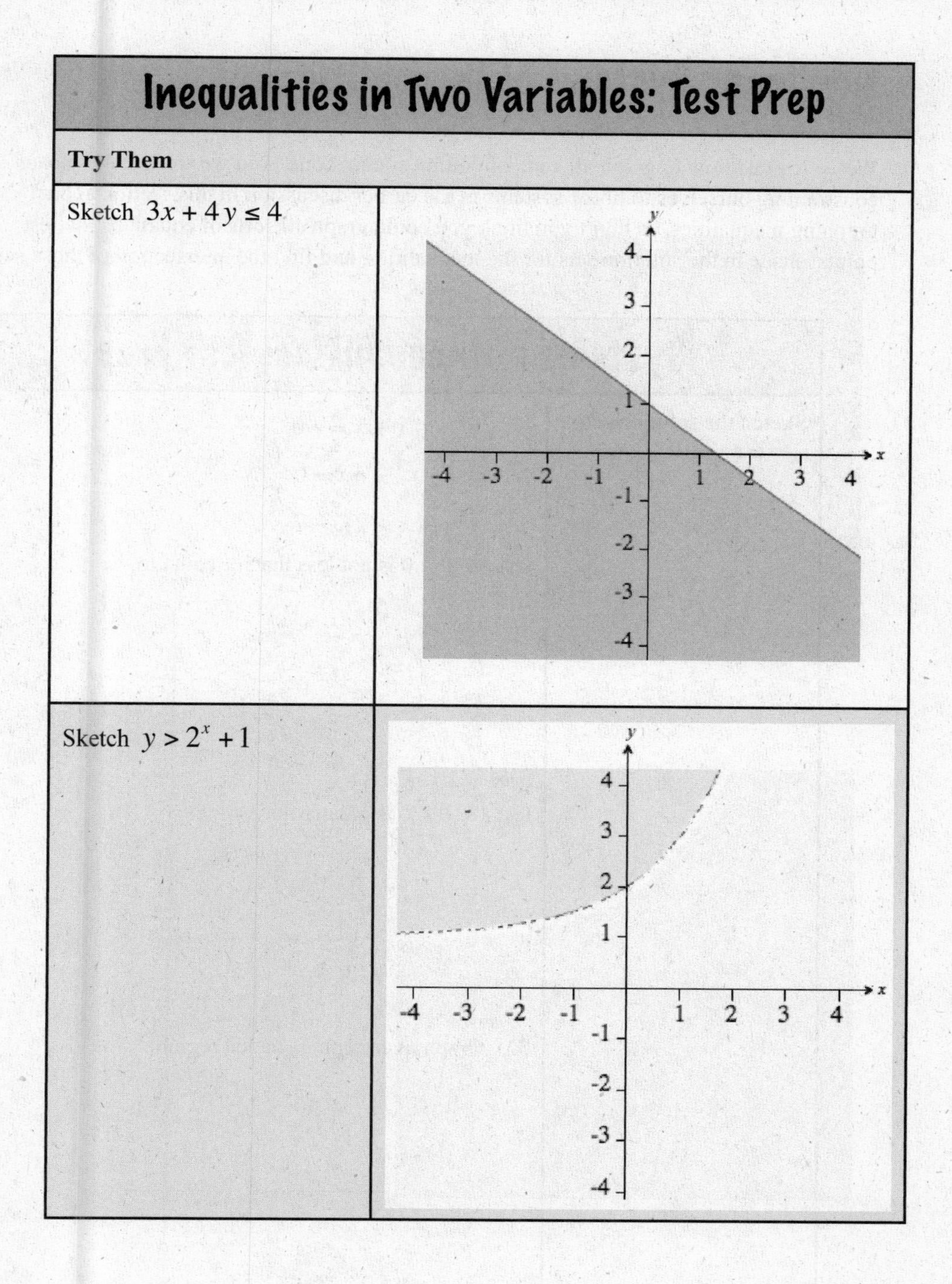
Inequalities in Two Variables: Test Prep
Try Them
Sketch $3x + 4y \leq 4$
Sketch $y > 2^x + 1$
y
x
4
3
2
1
-1
-2
-3
-4

Nonlinear Systems of Inequalities

We've learned how to graph all sorts of nonlinear equations. And we were by no means constraining ourselves to linear systems in the earlier discussion in this section about graphing inequalities, so don't you think you could graph all sorts of equations, do test points, shade in the solution sets for the inequalities, and find the intersection of those sets?

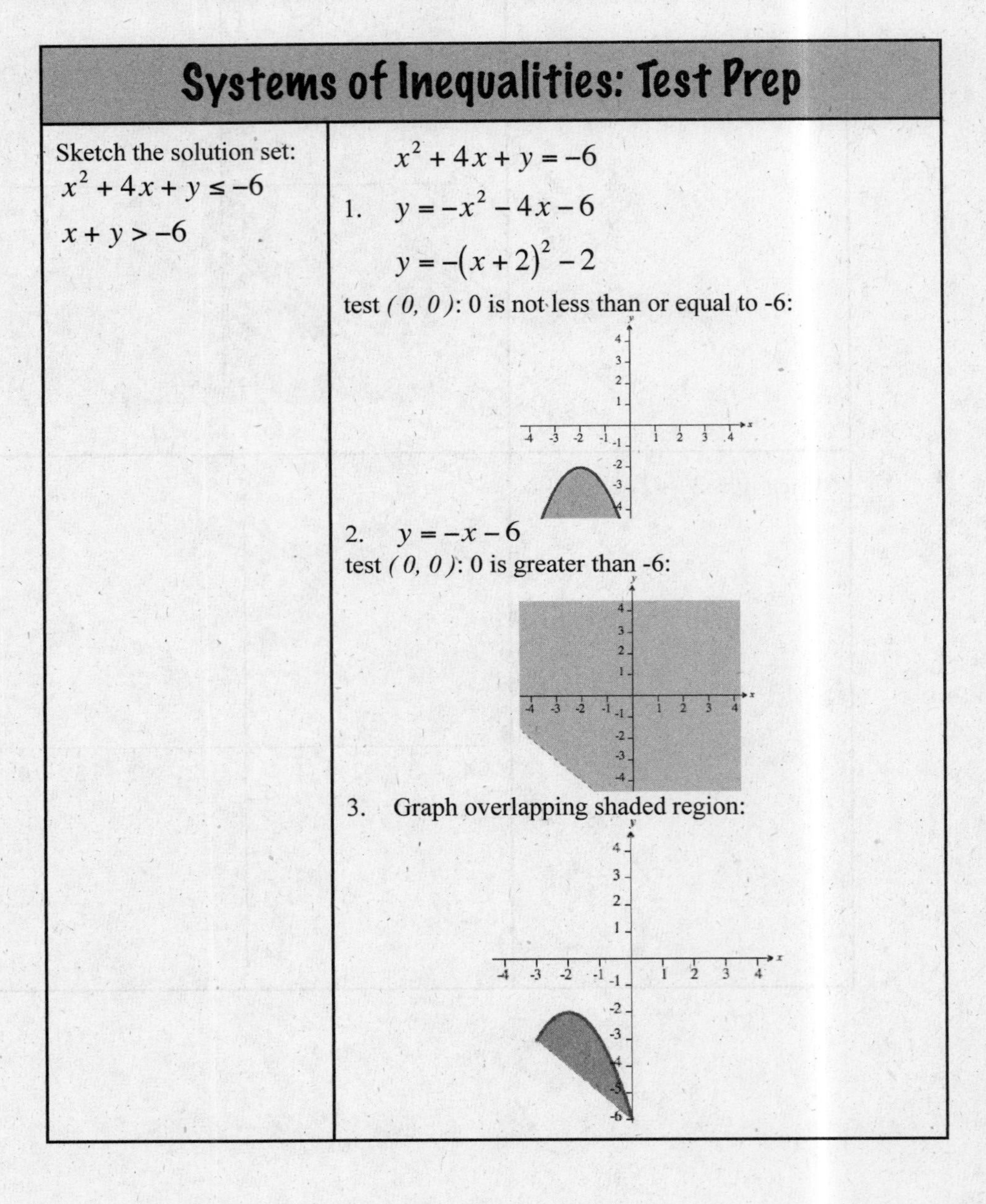

Systems of Inequalities: Test Prep

Sketch the solution set:

$x^2 + 4x + y \le -6$

$x + y > -6$

$x^2 + 4x + y = -6$

1. $y = -x^2 - 4x - 6$

$y = -(x + 2)^2 - 2$

test *(0, 0)*: 0 is not less than or equal to -6:

2. $y = -x - 6$

test *(0, 0)*: 0 is greater than -6:

3. Graph overlapping shaded region:

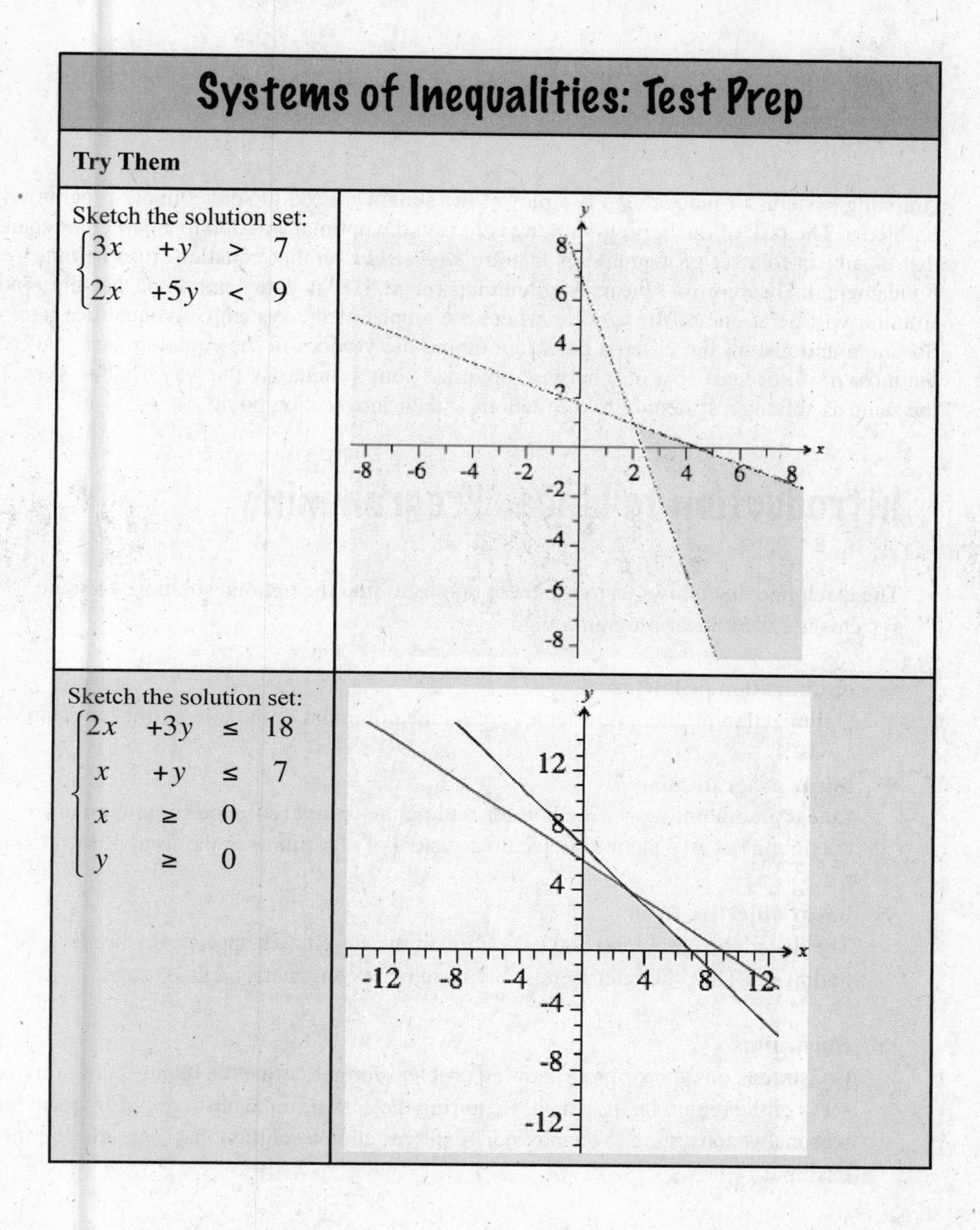

Systems of Inequalities: Test Prep

Try Them

Sketch the solution set:

$$\begin{cases} 3x + y > 7 \\ 2x + 5y < 9 \end{cases}$$

Sketch the solution set:

$$\begin{cases} 2x + 3y \le 18 \\ x + y \le 7 \\ x \ge 0 \\ y \ge 0 \end{cases}$$

6.6: Linear Programming

Graphing systems of inequalities is a part of the skills we need to solve "linear programming" problems. The task of linear programming is to find the optimal solution to a particular equation that is subject to a set of inequalities that are *constraints* on that equation. It turns out, by the Fundamental Theorem of Linear Programming (page 518 in your textbook), that the optimal solution will be at one of the *vertices* where the graphs of the system of inequalities intersect. So graph and test all the ordered pairs that define the vertices in the equation, and you'll find the most profit or least cost or otherwise optimize your system. By the way, finding vertices is the same as solving a system of two equations -- their intersection point!

Introduction to Linear Programming

The discipline that allows us to use these graphs to find the optimal solutions to some systems is called linear programming.

- **optimization problems**
  Optimization problems are set up to find the minimum or maximum of a situation.

- **linear programming**
  Linear programming is a method for finding the optimal solution (minimum or maximum) of an equation subject to a system of constraints in the form of inequalities.

- **linear objective form**
  The linear objective form is the equation in the system that specifically needs to be optimized. The other members of the system are constraints on the system.

- **constraints**
  Constraints on a linear programming problem further define the realm of the solution set -- or the region on the graph. Requiring the answer to be positive is a frequent and reasonable constraint. (You may not be interested in a solution that loses money, for example.)

  - **For Fun**
    If the answer is constrained to be positive, which quadrant of the cartesian coordinate system must that solution lie in?

- **set of feasible solutions**
  The set of feasible solutions is the region that ends up in the intersection of all the shaded regions on the graph of the constraints.

Solving Optimization Problems

Here is an example of finding the solution to a linear programming problem. The objective function is $p = 3x + 5y$, and we want to maximize that. The system of constraints is

$$x + y \le 5$$
$$2x + y \le 6$$
$$x \ge 0$$
$$y \ge 0$$

The graph of the feasible solutions looks like this:

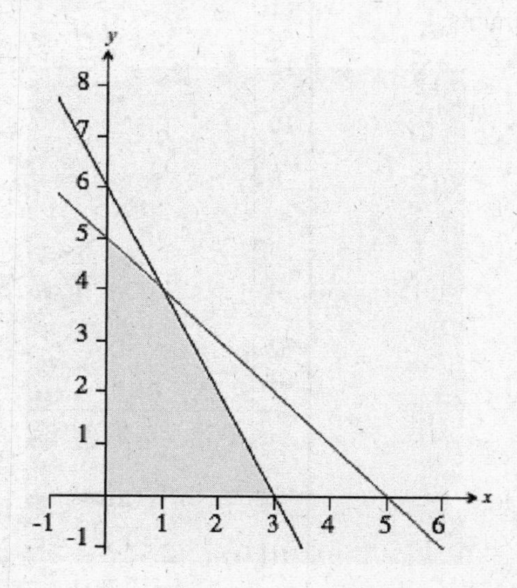

The vertices of the region are *(0, 0), (3, 0), (0, 5),* and *(1, 4).* The last point can be found by solving the system: $\begin{cases} x + y = 5 \\ 2x + y = 6 \end{cases}$.

Now test these points in the object function:

$(0, 0)$: $p = 0$

$(3, 0)$: $p = 9$

$(0, 5)$: $p = 25$

$(1, 4)$: $p = 23$

The maximum value of the objective function occurs at the point $(0, 5)$ and results in a value of 25 for p.

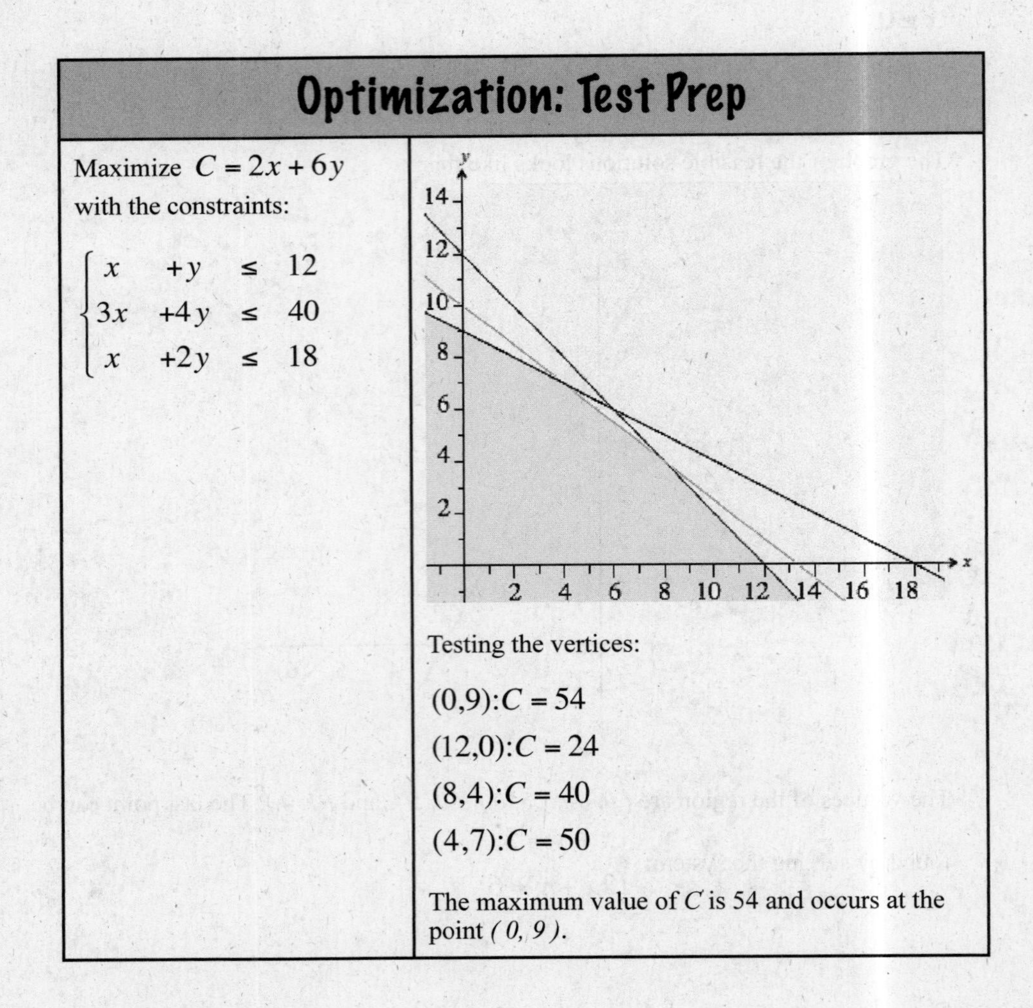

Optimization: Test Prep

Maximize $C = 2x + 6y$
with the constraints:

$$\begin{cases} x + y \le 12 \\ 3x + 4y \le 40 \\ x + 2y \le 18 \end{cases}$$

Testing the vertices:

$(0,9)$: $C = 54$

$(12,0)$: $C = 24$

$(8,4)$: $C = 40$

$(4,7)$: $C = 50$

The maximum value of C is 54 and occurs at the point $(0, 9)$.

Optimization: Test Prep

Try Them

Maximize $R = 40x + 30y$ subject to the constraints: $\begin{cases} 4x + 2y \le 800 \\ x + y \le 300 \\ x \ge 0 \\ y \ge 0 \end{cases}$	The maximum value of R is 10,000 and occurs at the point *(100, 200)*.
Minimize $C = 5x + 4y$ within the constraints: $\begin{cases} 3x + 4y \ge 32 \\ x + 4y \ge 24 \\ 0 \le x \le 12 \\ y \le 15 \end{cases}$	The minimum value of C is 32 and occurs at the point *(0, 8)*.

Some of the computer applications available to solve linear programming problems:

BPMPD
CPLEX
C-WHIZ
Excel and Quattro Pro Solvers
FortMP
GAUSS
GIPALS32
HS/LP Linear Optimizer
KORBX
LAMPS
LINDO Callable Library
LINGO
LOQO
LP88 and BLP88
MINOS
MOSEK
MPSIII
OML
OSL
PCx
PORT 3
PROC LP.
QPOPT
SQOPT
What'sBest
WHIZARD
XPRESS-MP

(In other words this is a kind of problem that is in big demand to be solved.)

Final Fun: Matching

FIND THE BEST MATCH. GOOD LUCK.

$$\begin{cases} 4x - 2y = 9 \\ 2x - y = 3 \end{cases}$$

QUADRATIC FACTOR MULTIPLICITY 2

$$\begin{cases} 2x - 4y + z = -3 \\ \phantom{2x - {}} 3y - 2z = 9 \\ \phantom{2x - 4y + {}} 3z = -9 \end{cases}$$

NONSQUARE SYSTEM

FEASIBLE SOLUTIONS

(0,0,0)

HOMOGENEOUS SYSTEM

$Ax + B$

NONLINEAR SYSTEM

$$\frac{x^3 + 2x}{(x^2 + 1)^2} = \frac{Ax + B}{x^2 + 1} + \frac{Cx + D}{(x^2 + 1)^2}$$

SYSTEM OF INEQUALITIES

PARTIAL FRACTIONS

$$\begin{cases} x - 4y = 0 \\ -x + 7y = 0 \end{cases}$$

TRIVIAL SOLUTION

$$\begin{cases} 3x - 2y > 6 \\ 2x - 5y \le 10 \end{cases}$$

INCONSISTENT SYSTEM

TRIANGULAR FORM

$$\begin{cases} x^2 + y^2 - 4y = 4 \\ \phantom{x^2 + {}} 5x - 2y = 2 \end{cases}$$

LINEAR FACTOR

$(x^2 + 3)^2$

$$\begin{cases} x + 2y - 3z = 5 \\ 3x + 7y - 10z = 13 \end{cases}$$

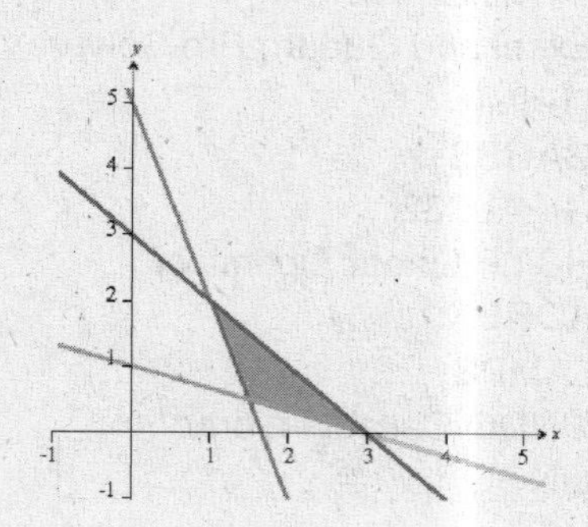

NOW TRY THESE:
REVIEW EXERCISES: TEXTBOOK PAGE 529,
PRACTICE TEST: TEXTBOOK PAGE 531.

STAY UP ON YOUR GAME -- DO THE CUMULATIVE
REVIEW EXERCISES: TEXTBOOK PAGE 531

Chapter 7: Matrices

Word Search

```
W S D E S R E V N I N M G M Y
N K I H Z C O X Z O O N C J C
E O M N H M E X N I I H O T N
D E I E G T I S W T T N F R E
G X L T R U I N A V A Y A A C
E O I E C N L L O I M R C N A
N D V R G E O A S R R A T S J
M R S U T P L S R O O L O L D
Z Q L H R A U F W L F A R A A
B A U E A A M K E L S C H T T
R O T V G Q L K V R N S O I G
L N L M E L I M I N A T I O N
I C R A M E R G P A R E H N G
Y W T N A N I M R E T E D E C
E F R O T A T I O N C R R Z D
```

ADJACENCY
COFACTOR
CRAMER
DETERMINANT
ECHELON
EDGE
ELIMINATION
GAUSSIAN
INTERPOLATING
INVERSE
MATRIX
MINOR
NONSINGULAR
REFLECTION
ROTATION
ROW
SCALAR
SINGULAR
TRANSFORMATION
TRANSLATION
VERTEX

7.1: Gaussian Elimination Method

The Gaussian Elimination Method for matrices is a convenient and pretty powerful tool for solving systems of equations. We're going to build up to it slowly in this section.

✓ **Questions to Ask Your Teacher**
Have a look-see in your calculator manual and find out what it can do with matrices. Then check with your teacher to make certain sure you know how much and when you have to show your own work ... and when it might be okay to get your calculator to manipulate a matrix for you . That is, of course, once you figure out how to get your calculator to manipulate matrices for you if you want to!

Introduction to Matrices

Here we will study some of the basic configurations of matrices and their elements and learn the vocabulary that goes with these ideas.

✤ **matrix**
A matrix is an array, a table, or a rectangular "holder" of numbers made up of rows (they go horizontally) and columns (they go vertically). By the way, working with matrices rather naturally suggests the use of graph paper since you probably need to keep your "holder" well organized.

✤ **element**
What's in each "compartment" of a matrix is called an element. You can specify an element by which row and column it's in -- where the row/column intersects -- it takes two pieces of information to locate an element. For example, the "first" element in a matrix, the element in the top left-hand position, is indicated by specifying its column and row in a subscript like so: a_{11}. Moving one element to the right of a_{11} means we have remained on the same row, but added one to the column tally, and we specify this element like so: a_{12}. The element below a_{11} is element a_{21}. Do you see that the first subscript denotes the row and the second denotes the column?

✣ **order *m x n*, dimension *m x n***

The "size" of a matrix is called its order or dimension. Like the area of a rectangle, the dimension of a matrix is like multiplying length times width, only it's $m \times n$ ("m by n") where m is the number of rows and n is the number of columns ... and you don't actually carry out the multiplication either. This is, for instance, a 2 by 4 matrix:

$$\begin{bmatrix} -1 & 0 & 3 & 5 \\ -7 & 2 & -2 & 0 \end{bmatrix}$$

✣ **square matrix of order *n***

If a matrix is square, then the number of rows is the same as the number of columns, and one need only specify one dimension n and that the matrix is square. Here is a square matrix of dimension 2:

$$\begin{bmatrix} 2 & -1 \\ 3 & 0 \end{bmatrix}$$

✣ **main diagonal**

The main diagonal of a matrix is the diagonal that runs from the upper left diagonally towards the lower right. (It will not end in the lower right unless the matrix is square.)

✣ **augmented matrix**

You can translate a system of equations (Chapter 6) into an array using the coefficients and constants. This is how it's done:

$$\begin{aligned} 2x - y &= 9 \\ x + 3y &= -7 \end{aligned} \quad \text{becomes} \quad \left[\begin{array}{cc|c} 2 & -1 & 9 \\ 1 & 3 & -7 \end{array}\right]$$

This matrix is an augmented matrix because it includes the column of "answers" from the system -- the constants on the left-hand side of the equations.

★ **For Fun**
Make an augmented matrix of this system. Be careful! All the x's and y's and constants have to line up correctly -- a sort of "place value" for matrices.

$$x = -2y + 1$$
$$y - x = 4$$

✤ **coefficient matrix**
If you leave off the answer/constant column from your augmented matrix, you'll have a matrix made up of only the coefficients of the equations in your system:

$$\begin{bmatrix} 2 & -1 \\ 1 & 3 \end{bmatrix}$$

★ **For Fun**
Can you name the a_{12} element of this coefficient matrix? ... the a_{21}? ... the elements of the main diagonal?

✤ **constant matrix**
The (left-over) constant column that you took off your augmented matrix makes up a *column* matrix (of one column only) that looks like this:

$$\begin{bmatrix} 9 \\ -7 \end{bmatrix}$$

★ **For Fun**
What are the dimensions of this constant matrix? Get the order of m and n correct.

Matrix: Test Prep	
What order is this matrix? $\begin{bmatrix} 0 & 1 & 3 & 9 & 0 \\ 2 & -2 & 5 & 0 & -1 \\ 3 & 11 & -7 & 2 & -3 \end{bmatrix}$	Since the matrix has 3 rows and 5 columns it is an order *3 x 5* matrix

Matrix: Test Prep

<table>
<tr><th colspan="2">Try Them</th></tr>
<tr><td>Matching, matrix to its dimension:
1. $\begin{bmatrix} 4 & 0 & -1 \\ 2 & 3 & -7 \\ 5 & 0 & 9 \end{bmatrix}$
2. $\begin{bmatrix} 4 \\ 1 \\ -1 \\ 0 \\ -2 \end{bmatrix}$
3. $\begin{bmatrix} 0 & 9 \\ -2 & 3 \\ 5 & -7 \\ 10 & 1 \end{bmatrix}$
4. $\begin{bmatrix} \frac{1}{2} & 0 & -1 & \frac{2}{3} \\ -4 & 3 & \frac{2}{3} & 0 \\ 9 & -5 & 1 & \frac{1}{6} \end{bmatrix}$

A. 4 x 2
B. 5 x 1
C. 3 x 3
D. 3 x 4</td><td>1. C
2. B
3. A
4. D</td></tr>
<tr><td>Rewrite this system of equations as an augmented matrix:
$2x - 6y + 5z = 1$
$8x + 4y + z = 13$
$-2x + 3y + 4z = 5$</td><td>$\left[\begin{array}{ccc|c} 2 & -6 & 5 & 1 \\ 8 & 4 & 1 & 13 \\ -2 & 3 & 4 & 5 \end{array}\right]$</td></tr>
</table>

✤ **row echelon form**
A matrix in a form that corresponds to the *triangular* form for a set of equations is in row echelon form. Row echelon form is desirable for the same reason triangular form was -- you can "back substitute" to solve the system of equations (that corresponds to your matrix). A matrix is in row echelon form if all the elements "under" the main diagonal are zero.

Elementary Row Operations

Just like we learned how to use equivalence of equations to manipulate systems of equations, there is an analogous set of operations that will help us to get a matrix into, say, row echelon form.

✤ **elementary row operations**
Elementary row operations are those analogous manipulations allowed to be performed on a matrix without changing the matrix's underlying validity.

Elementary Row Operations

Here are the operations you can perform on a matrix that correspond to a (still) equivalent system of equations:

1. You may interchange any two rows.
2. You may multiply all the elements in a row by the same (non-zero) constant.
3. You can replace any row with the sum of itself and a non-zero multiple of any other row.

★ **For Fun**
Can you see the correspondences between these three allowed row operations and the operations that were allowed for equivalent systems (page 475 of your textbook)?

★ **For Fun**

See if you can name the operation I perform on each step of reducing this matrix to row echelon form.

$$\begin{bmatrix} 5 & 1 & -2 & 4 \\ -2 & 4 & 4 & 6 \\ 3 & 5 & 2 & 10 \end{bmatrix} \to \begin{bmatrix} 5 & 1 & -2 & 4 \\ -6 & 12 & 12 & 18 \\ 6 & 10 & 4 & 20 \end{bmatrix} \to \begin{bmatrix} 5 & 1 & -2 & 4 \\ -2 & 4 & 4 & 6 \\ 0 & 22 & 16 & 38 \end{bmatrix} \to$$

$$\begin{bmatrix} 10 & 2 & -4 & 8 \\ -10 & 20 & 20 & 30 \\ 0 & 11 & 8 & 19 \end{bmatrix} \to \begin{bmatrix} 5 & 1 & -2 & 4 \\ 0 & 22 & 16 & 38 \\ 0 & 11 & 8 & 19 \end{bmatrix} \to \begin{bmatrix} 5 & 1 & -2 & 4 \\ 0 & 11 & 8 & 19 \\ 0 & 0 & 0 & 0 \end{bmatrix}$$

★ **For Fun**

What does that last row of all-zeros in the example above tell you about the system of equations that might correspond to this matrix? (consistent, dependent, independent?)

Row Echelon Form: Test Prep

Use elementary row operations to rewrite this matrix in row echelon form: $\begin{bmatrix} 2 & 6 & -1 & 1 \\ 1 & 3 & -1 & 1 \\ 3 & 10 & -2 & 1 \end{bmatrix}$	$\begin{bmatrix} 2 & 6 & -1 & 1 \\ 1 & 3 & -1 & 1 \\ 3 & 10 & -2 & 1 \end{bmatrix} \to \begin{bmatrix} 2 & 6 & -2 & 2 \\ 3 & 10 & -2 & 1 \\ 2 & 6 & -1 & 1 \end{bmatrix} \to \begin{bmatrix} 3 & 9 & -3 & 3 \\ 3 & 10 & -2 & 1 \\ 0 & 0 & 1 & -1 \end{bmatrix} \to \begin{bmatrix} 1 & 3 & -1 & 1 \\ 0 & 1 & 1 & -2 \\ 0 & 0 & 1 & -1 \end{bmatrix}$
Try Them	
Write in row echelon form: $\begin{bmatrix} -3 & 6 & -2 \\ 6 & 6 & -8 \end{bmatrix}$	possibly: $\begin{bmatrix} 1 & -2 & \frac{2}{3} \\ 0 & 1 & -\frac{2}{3} \end{bmatrix}$

Row Echelon Form: Test Prep	
Write in row echelon form: $\begin{bmatrix} 2 & 5 & 2 & -1 \\ 1 & 2 & -3 & 5 \\ 5 & 12 & 1 & 10 \end{bmatrix}$	possibly: $\begin{bmatrix} 1 & 2 & -3 & 5 \\ 0 & 1 & 8 & -11 \\ 0 & 0 & 0 & 1 \end{bmatrix}$

Gaussian Elimination Method

The goal of Gaussian Elimination Method is to rewrite an augmented matrix in row echelon form so as to solve the corresponding system of equations.

★ **For Fun**
You might pay careful attention to just how this "elimination" method is an awful lot like the "Method of Elimination" we used to solve systems of equations in Chapter 6.

✤ **Gaussian elimination method**
The Gaussian elimination method is (simply) an algorithm using matrices to solve a system of equations by getting the corresponding matrix into a row echelon form.

★ **For Fun**
What is the difference between the words "algebra" and "algorithm?"

Gaussian Elimination Method: Test Prep	
Solve using Gaussian Elimination method: $x-3y+2z=6$ $4x-y+3z=10$ $7x+y+4z=-2$	$\begin{bmatrix} 1 & -3 & 2 & 6 \\ 4 & -1 & 3 & 10 \\ 7 & 1 & 4 & -2 \end{bmatrix} \to \begin{bmatrix} 7 & -21 & 14 & 42 \\ 4 & -1 & 3 & 10 \\ 0 & 22 & -10 & -44 \end{bmatrix}$ $\to \begin{bmatrix} 4 & -12 & 8 & 24 \\ 0 & 11 & -5 & -14 \\ 0 & 11 & -5 & -22 \end{bmatrix} \to \begin{bmatrix} 1 & -3 & 2 & 6 \\ 0 & 11 & -5 & -14 \\ 0 & 0 & 0 & -8 \end{bmatrix}$ $0 \neq -8$; therefore, no solution.

Gaussian Elimination Method: Test Prep

Try Them	
Solve using Gaussian Elimination method: $2x - 5y = 10$ $5x + 2y = 4$	$\left(\frac{40}{29}, -\frac{42}{29}\right)$
Solve using Gaussian Elimination method: $w + 2x - 3y + 2z = 11$ $2w + 5x - 8y + 5z = 28$ $-2w - 4x + 7y - z = -18$	$(3c - 5, -7c + 14, 4 - 3c, c)$

Application: Interpolating Polynomials

Here's a neat little application of matrices that help us find a polynomial model ... without a calculator!

✤ **interpolating polynomial**

An interpolating polynomial is a sort of best-guess polynomial that models an application based on knowledge of a finite number of data points that have been observed to hold for the situation. After all, each ordered pair assumes a function. We then assume the function is a polynomial of an arbitrary degree (depending on the number of ordered pairs we have). Then we use a matrix to find the unknown coefficients of said polynomial like we used systems of equations to find the unknown constants in partial fraction decomposition (section 6.4).

Interpolating Polynomial: Test Prep

Write a system of equations to describe a second degree polynomial passing through the points *(-3, 28)*, *(-1, 6)*, and *(2, 3)*.	$a_2(-3)^2 + a_1(-3) + a_0 = 28$ $a_2(-1)^2 + a_1(-1) + a_0 = 6$ $a_2(2)^2 + a_1(2) + a_0 = 3$ becomes $9a_2 - 3a_1 + a_0 = 28$ $a_2 - a_1 + a_0 = 6$ $4a_2 + 2a_1 + a_0 = 3$
Try Them	
Write the above polynomial by solving the system.	$p(x) = 2x^2 - 3x + 1$
Find a polynomial whose graph passes through the points *(-2, -3)*, *(0, -1)*, and *(3, 17)*.	$p(x) = x^2 + 3x - 1$

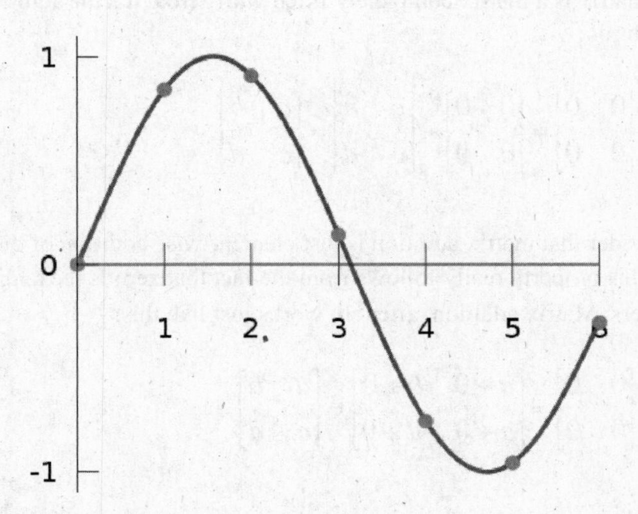

The dots denote data points while the curve shows the interpolation polynomial.

7.2: Algebra of Matrices

Matrices are, after all, just arrays of numbers. In the last section we considered them mostly as stand-ins for systems of equations. However, in many cases, they are nothing more or less than a table of values, and as such we may wish to do any number of arithmetic or algebraic operations with tables-cum-matrices.

Addition and Subtraction of Matrices

If you use a table to represent data, such as sales figures or rainfall, you might want to add two matrices together to get a grand total. You might want to subtract matrices of data if you wanted to know how much you saved from one month to the next or something similar. Adding and subtracting matrices is easy ... just be careful to apply each operation to each element correctly.

First we have to set up the whole arena of matrices for arithmetic operations by talking about the same sorts of underlying ideas we had to address for arithmetic with real numbers -- additive identities and so forth.

- **zero matrix, *0***
  The zero matrix is a matrix completely filled with zeros. It's the additive identity for matrix addition:

$$\begin{bmatrix} a & b \\ c & d \end{bmatrix} + \begin{bmatrix} 0 & 0 \\ 0 & 0 \end{bmatrix} = \begin{bmatrix} 0 & 0 \\ 0 & 0 \end{bmatrix} + \begin{bmatrix} a & b \\ c & d \end{bmatrix} = \begin{bmatrix} a & b \\ c & d \end{bmatrix}$$

  If you consider that matrix addition is just element-wise addition of the numbers in the matrices, this property really follows from the fact that zero is the additive identity for real numbers. Matrix addition, after all, works just like this:

$$\begin{bmatrix} a & b \\ c & d \end{bmatrix} + \begin{bmatrix} 0 & 0 \\ 0 & 0 \end{bmatrix} = \begin{bmatrix} a+0 & b+0 \\ c+0 & d+0 \end{bmatrix} = \begin{bmatrix} a & b \\ c & d \end{bmatrix}$$

- **commutative**
  Likewise, since addition of real numbers is commutative, matrix addition would be commutative, too right? So if $\boldsymbol{A}$ is an m x n matrix as is $\boldsymbol{B}$, then $\boldsymbol{A} + \boldsymbol{B} = \boldsymbol{B} + \boldsymbol{A}$.

✤ **associative**

Again, since adding matrices involves adding the real numbers element-wise in the matrices, then matrix addition is associative: $\boldsymbol{A} + (\boldsymbol{B} + \boldsymbol{C}) = (\boldsymbol{A} + \boldsymbol{B}) + \boldsymbol{C}$.

★ **For Fun**

Do you understand why the matrices being added must be the same order/ dimensions?

✤ **additive inverse**

In order to talk about subtraction of matrices, we can first define the additive inverse (which is formally what we have to do to define subtraction of real numbers, isn't it?). So in order to construct the additive inverse of matrix ***A***, we'll simply take the additive inverse of every element in ***A***, and we can call that ***-A***:

$$A = \begin{bmatrix} a & b \\ c & d \end{bmatrix} \text{ and } -A = \begin{bmatrix} -a & -b \\ -c & -d \end{bmatrix}$$

So, to subtract matrices, you wind up simply adding the additive inverse of each element:

$$\begin{bmatrix} 3 & -1 \\ 0 & 2 \end{bmatrix} - \begin{bmatrix} 0 & 3 \\ 9 & -3 \end{bmatrix} = \begin{bmatrix} 3 & -1 \\ 0 & 2 \end{bmatrix} + \begin{bmatrix} 0 & -3 \\ -9 & 3 \end{bmatrix} = \begin{bmatrix} 3+0 & -1+(-3) \\ 0+(-9) & 2+3 \end{bmatrix} = \begin{bmatrix} 3 & -4 \\ -9 & 5 \end{bmatrix}$$

(This is not mysterious stuff!)

✤ **additive identity**

The additive identity matrix is the zero matrix. The dimensions of the additive identity matrix will match exactly the dimensions of the matrix you are adding it to -- that is, add *nothing* to.

Scalar Multiplication

If we need to be able to change every element in a matrix (or table) by some constant factor, we need to have a way of wholesale multiplying that number by every element in that matrix. This is scalar multiplication.

★ **For Fun**

What does the word "scalar" mean in this context?

✤ **scalar multiplication**

Scalar multiplication is basically the operation of multiplying each element in a given matrix by a constant (or scalar). For example:

$$-2\begin{bmatrix}3 & -1\\0 & 2\end{bmatrix}=\begin{bmatrix}-6 & 2\\0 & -4\end{bmatrix}$$

★ **For Fun**

Could we describe the additive inverse matrix as a product of scalar multiplication?

Operations on Matrices: Test Prep	
Given $\mathbf{A}=\begin{bmatrix}0 & -1 & 2\\5 & 7 & 9\\1 & 3 & 1\end{bmatrix}$ and $\mathbf{B}=\begin{bmatrix}4 & 1 & 1\\2 & 3 & 1\\0 & 1 & 0\end{bmatrix}$, find $-3A+2B$.	$-3\mathbf{A}+2\mathbf{B}=\begin{bmatrix}0 & 3 & -6\\-15 & -21 & -27\\-3 & -9 & -3\end{bmatrix}+\begin{bmatrix}8 & 2 & 2\\4 & 6 & 2\\0 & 2 & 0\end{bmatrix}$ $=\begin{bmatrix}8 & 5 & -4\\-11 & -15 & -25\\-3 & -7 & -3\end{bmatrix}$
Try Them	
Given $\mathbf{A}=\begin{bmatrix}2 & 0\\-1 & 3\end{bmatrix}$ and $\mathbf{B}=\begin{bmatrix}3 & 4 & 0\end{bmatrix}$, find $A+B$.	Cannot add matrices of different order.
Carry out $2\begin{bmatrix}4 & 0\\-1 & 1\end{bmatrix}-\begin{bmatrix}5 & 7\\1 & 3\end{bmatrix}+5\begin{bmatrix}1 & -2\\0 & 4\end{bmatrix}$	$\begin{bmatrix}8 & -17\\-3 & 19\end{bmatrix}$

Matrix Multiplication

Scalar multiplication is *one* kind of multiplication that is carried out on matrices. In real numbers (scalars), we really have only one kind of multiplication. Matrices have another kind: matrix multiplication where two or more matrices are multiplied by each other. This is the beginning of a strangeness in matrices that is unlike anything you've done with real numbers. This process will take some careful place-keeping and practice.

✤ **column matrix**
A column matrix is a matrix with only one column (but any number of rows), for instance:

$$\begin{bmatrix} 2 \\ -5 \\ 0 \end{bmatrix}$$

✤ **row matrix**
Similarly, a row matrix is a matrix with only one row:

$$\begin{bmatrix} 1 & -1 & 0 & 2.5 \end{bmatrix}$$

★ **For Fun**
Can you give the dimensions of these last two matrices?

✤ **associative property**
Without defining matrix multiplication (yet), we can go ahead and affirm that if the matrices can be multiplied together in this order, then the multiplication follows the associative rule: $A(BC)=(AB)C$.

Matrix multiplication is not commutative in general.

✤ **distributive property**
Given that the matrices can be multiplied together in the order given, then matrix multiplication does follow a distributive property of matrix multiplication over matrix addition: $A(B+C)=AB+AC$ or $(A+B)C=AC+BC$.

✤ **identity matrix**
The identity matrix is always a square matrix. The identity matrix of order or dimension n is the $n \times n$ square matrix with all 1's on the main diagonal and all 0's everywhere else.

Order/Dimension Considerations in Matrix Multiplication

- In order for two matrices to be multiplied together, the number of columns in the first matrix must match the number of rows in the second matrix. For example, a 3×2 matrix can be multiplied by a 2×4 matrix, but a 3×2 matrix cannot be multiplied by a 3×4 matrix.

- The matrix that results from matrix multiplication will have the number of rows of the first matrix in the multiplication and the number of columns in the second matrix in the multiplication. So multiplying a 3×2 matrix by a 2×4 matrix will result in a new matrix of order 3×4.

Method of Matrix Multiplication

- The elements of the first row of the first matrix get multiplied in a one-by-one matching with the elements in the first column of the second matrix. Those products are added together, and that sum becomes the new a_{11} of the product matrix.

- The elements of the first row of the first matrix are multiplied one-by-one by the elements of the second column of the second matrix, and the sum of those products becomes the a_{12} element of the product matrix.

- ... and so on.

<u>*For example:*</u>

$$\begin{bmatrix} 1 & -2 \\ 0 & 3 \\ -1 & 5 \end{bmatrix}\begin{bmatrix} -3 & 2 \\ -2 & 0 \end{bmatrix}$$

$$= \begin{bmatrix} (1)(-3)+(-2)(-2) & (1)(2)+(-2)(0) \\ (0)(-3)+(3)(-2) & (0)(2)+(3)(0) \\ (-1)(-3)+(5)(-2) & (-1)(2)+(5)(0) \end{bmatrix}$$

$$= \begin{bmatrix} -3+4 & 2+0 \\ 0-6 & 0+0 \\ 3-10 & -2+0 \end{bmatrix}$$

$$= \begin{bmatrix} 1 & 2 \\ -6 & 0 \\ -7 & -2 \end{bmatrix}$$

★ **For Fun**
Can you do the dimensional analysis of these matrices to verify that they are multipliable and that my result matrix has the correct order?

Matrix multiplication is a pattern-heavy, almost kinesthetic, surely visual operation. Matrix multiplication involves lots of small steps, so it is error-prone. Practice is my advice.

<u>Think About It:</u>
Do you think that matrix multiplication could be used to *encrypt* messages? How about a series of matrix multiplications?

Multiplication of Two Matrices: Test Prep

Multiply, if possible: $\begin{bmatrix} 3 & 6 & 1 \\ 0 & -1 & -2 \end{bmatrix}\begin{bmatrix} 5 \\ -4 \\ 0 \end{bmatrix}$	$\begin{bmatrix} 3 & 6 & 1 \\ 0 & -1 & -2 \end{bmatrix}\begin{bmatrix} 5 \\ -4 \\ 0 \end{bmatrix}$ $= \begin{bmatrix} 15-24+0 \\ 0+4+0 \end{bmatrix}$ $= \begin{bmatrix} -9 \\ 4 \end{bmatrix}$
Try Them	
Name the dimension of these two matrices and the dimension of their product, assuming they are multiplied in the order given: $\begin{bmatrix} 3 & 0 & 1 \\ 5 & -1 & 4 \end{bmatrix}$, $\begin{bmatrix} 4 & 5 & 0 & -1 \\ 5 & 1 & -1 & 0 \\ 2 & 3 & 7 & -4 \end{bmatrix}$	*2 x 3, 3 x 4, 2 x 4*
Carry out the matrix multiplication, if possible: $\begin{bmatrix} 1 & -2 & 3 \\ 2 & -1 & 8 \\ -1 & 3 & -2 \end{bmatrix}\begin{bmatrix} -1 & 3 & 2 \\ 1 & 4 & -1 \end{bmatrix}$	not possible

Matrix Products and Systems of Equations

Systems of equations can be expressed as multiplication of matrices in the following manner, for example:

$$\begin{cases} 3x & -y & +2z & = & 5 \\ -x & +y & & = & -2 \\ 2x & -3y & +z & = & 1 \end{cases} \rightarrow \begin{bmatrix} 3 & -1 & 2 \\ -1 & 1 & 0 \\ 2 & -3 & 1 \end{bmatrix} \begin{bmatrix} x \\ y \\ z \end{bmatrix} = \begin{bmatrix} 5 \\ -2 \\ 1 \end{bmatrix}$$

Check the orders of the matrices. Can they be multiplied together? Can you go ahead and multiply the matrices together? Can you rewrite the system as an augmented matrix? Can you recover the system of equations from the augmented matrix?

Matrix Form of a System of Equations: Test Prep

<table>
<tr>
<td>Rewrite this matrix equation as a system of equations. Use matrix multiplication.
$$\begin{bmatrix} 2 & 0 & -3 & 5 \\ 0 & 1 & -1 & 3 \\ -4 & 3 & 2 & -1 \\ 0 & 5 & 4 & 0 \end{bmatrix} \begin{bmatrix} w \\ x \\ y \\ z \end{bmatrix} = \begin{bmatrix} 15 \\ -3 \\ 5 \\ 11 \end{bmatrix}$$</td>
<td>$$\begin{bmatrix} 2 & 0 & -3 & 5 \\ 0 & 1 & -1 & 3 \\ -4 & 3 & 2 & -1 \\ 0 & 5 & 4 & 0 \end{bmatrix} \begin{bmatrix} w \\ x \\ y \\ z \end{bmatrix} = \begin{bmatrix} 15 \\ -3 \\ 5 \\ 11 \end{bmatrix}$$
$$\begin{bmatrix} 2w - 3y + 5z \\ x - y + 3z \\ -4w + 3x + 2y - z \\ 5x + 4y \end{bmatrix} = \begin{bmatrix} 15 \\ -3 \\ 5 \\ 11 \end{bmatrix}$$
$2w - 3y + 5z = 15$
$x - y + 3z = -3$
$-4w + 3x + 2y - z = 5$
$5x + 4y = 11$
by equality of matrices</td>
</tr>
</table>

Matrix Form of a System of Equations: Test Prep

Try Them	
Write a system of equations from the matrix equation: $\begin{bmatrix} 2 & -1 \\ 0 & 5 \end{bmatrix}\begin{bmatrix} x \\ y \end{bmatrix} = \begin{bmatrix} 3 \\ -2 \end{bmatrix}$	$2x - y = 3$ $5y = -2$
Write a system of equations from the matrix equation: $\begin{bmatrix} 3 & 0 & \frac{1}{2} \\ -1 & 5 & -6 \\ \frac{2}{3} & 4 & 0 \end{bmatrix}\begin{bmatrix} x \\ y \\ z \end{bmatrix} = \begin{bmatrix} \frac{1}{3} \\ -2 \\ -7 \end{bmatrix}$	$\begin{bmatrix} 3 & 0 & \frac{1}{2} \\ -1 & 5 & -6 \\ \frac{2}{3} & 4 & 0 \end{bmatrix}\begin{bmatrix} x \\ y \\ z \end{bmatrix} = \begin{bmatrix} \frac{1}{3} \\ -2 \\ -7 \end{bmatrix}$ $3x + \frac{1}{2}z = \frac{1}{3}$ $-x + 5y - 6z = -2$ $\frac{2}{3}x + 4y = -7$

Transformation Matrices

Using matrix multiplication and some special matrices you can transform points (and thereby objects that are defined by their vertices or similarly) by moving (translating) them through space, reflecting them, and rotating them. There are a set of well defined, well used matrices available to accomplish these feats!

✣ **transformation matrices**

A matrix that, when it is multiplied by another (frequently column) matrix, transforms that (column) matrix into a different matrix in a regular way.

❖ **translation matrix $\mathbf{T}_{a,b}$**

The translation matrix in three dimensions usually looks like this:

$$\begin{bmatrix} 1 & 0 & a \\ 0 & 1 & b \\ 0 & 0 & 1 \end{bmatrix}$$

In this form, the translation matrix will move a *3 x 1* three dimensional point a units in the x-direction and b units in the y-direction. For example:

$$\begin{bmatrix} 1 & 0 & 3 \\ 0 & 1 & -2 \\ 0 & 0 & 1 \end{bmatrix}\begin{bmatrix} 1 \\ 1 \\ 1 \end{bmatrix} = \begin{bmatrix} 1+3 \\ 1-2 \\ 1 \end{bmatrix} = \begin{bmatrix} 4 \\ -1 \\ 1 \end{bmatrix}$$

... and we've moved the ordered triple *(1, 1, 1)* +3 in the x-direction and -2 in the y-direction so that the triple is now *(4, -1, 1)*.

❖ **reflection matrix R_x**

The reflection matrix that reflects a point across the x-axis looks like this:

$$\begin{bmatrix} 1 & 0 & 0 \\ 0 & -1 & 0 \\ 0 & 0 & 1 \end{bmatrix}$$

❖ **reflection matrix R_y**

A matrix the accomplishes the reflection of a point about the y-axis is

$$\begin{bmatrix} -1 & 0 & 0 \\ 0 & 1 & 0 \\ 0 & 0 & 1 \end{bmatrix}$$

✤ **reflection matrix** R_{xy}

And the matrix that will reflect a point about the line $y = x$ look like this:

$$\begin{bmatrix} 0 & 1 & 0 \\ 1 & 0 & 0 \\ 0 & 0 & 1 \end{bmatrix}$$

✤ **rotation matrices**

There are several versions of rotational matrices that depend on which way and how much you wish to rotate your point. For example, this matrix rotates a point ninety degrees about the origin:

$$\mathbf{R}_{90} = \begin{bmatrix} 0 & -1 & 0 \\ 1 & 0 & 0 \\ 0 & 0 & 1 \end{bmatrix}$$

Some other popular rotation matrices are given on page 558 in your textbook.

★ **For Fun**

Try some of these matrices out on a point in 3 dimensions written in the form of a *3 x 1* column matrix.

Combinations of Translations, Reflections, and Rotations

A new matrix that does many transformations can be built by matrix multiplication of the individual transformation matrices in a sort of reverse order (remember that matrix multiplication is not commutative so the order can matter). The first transformation you wish to use will be the right-most, and each successive transformation desired will be one more place to the left. For example, translating in *x* and *y* and then rotating by 90 degrees might look like this:

$$\begin{bmatrix} 0 & -1 & 0 \\ 1 & 0 & 0 \\ 0 & 0 & 1 \end{bmatrix}\begin{bmatrix} 1 & 0 & 3 \\ 0 & 1 & -2 \\ 0 & 0 & 1 \end{bmatrix} = \begin{bmatrix} 0 & -1 & 2 \\ 1 & 0 & 3 \\ 0 & 0 & 1 \end{bmatrix}$$

★ **For Fun**

Try the above transformation matrix on a column matrix. See if it does the expected translation and rotation. By the way, you might also make sure your matrix multiplication skills are getting tuned up by trying to build a few transformation matrices like this for yourself.

✓ **Questions to Ask Your Teacher**

These transformations can be used on whole shapes, too, not just points in the form of a *3 x 1* column matrix. This is discussed further in section 7.2 in your textbook. Ask you teacher to go over these expanded examples. It may be required that you know how to move an entire shape in one transformation matrix.

Transformation Matrices: Test Prep

Rotate the point $\begin{bmatrix} 5 \\ -3 \\ 2 \end{bmatrix}$ about the origin of the 2-dimensional cartesian plane 180 degrees counterclockwise.	$\begin{bmatrix} -1 & 0 & 0 \\ 0 & -1 & 0 \\ 0 & 0 & 1 \end{bmatrix}\begin{bmatrix} 5 \\ -3 \\ 2 \end{bmatrix} = \begin{bmatrix} -5+0+0 \\ 0+3+0 \\ 0+0+2 \end{bmatrix} = \begin{bmatrix} -5 \\ 3 \\ 2 \end{bmatrix}$
Try Them	
Define a matrix that will rotate a 3-dimensional point 90 degrees counterclockwise about the point *(2, -3, 0)*.	$\begin{bmatrix} 0 & -1 & 3 \\ 1 & 0 & 2 \\ 0 & 0 & 1 \end{bmatrix}$
Define a matrix that will reflect a point about the *x*-axis and about the *y*-axis.	$\begin{bmatrix} -1 & 0 & 0 \\ 0 & -1 & 0 \\ 0 & 0 & 1 \end{bmatrix}$

Adjacency Matrices

Adjacency matrices refer to ways that points can be connected to each other. This is part of what is called "Graph Theory."

✤ **graph**
A graph in this context is a set of points, vertices, and the line segments, edges, that connect those points. Here is an example of a graph with vertices labeled:

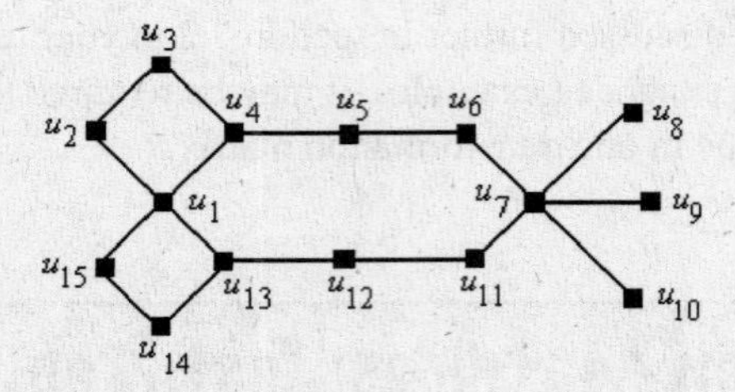

✤ **vertices**
The vertices of a graph are the points.

✤ **edges**
The edges of the graph are the line segments that connect the vertices.

✤ **step**
A step on a graph is a movement from one vertex to another along the edge that connects them. Only two vertices are involved in a step.

✤ **walk**
A walk is a string of several steps.

✤ **length**
The number of steps in a walk is the length of the walk. Frequently, graphs are used as a way to explore finding walks that are of the shortest possible length -- naturally!

✤ **adjacency matrix**
The elements of an adjacency matrix signify how many vertices connect one vertex to another. In other words, the element $a_{13} = 2$ means that there are two vertices between vertex #1 and vertex #3. $a_{13} = 0$ would mean that there are no edges between these two vertices (though one may still find a *walk* from vertex #1 and #3).

★ **For Fun**

What does the word "adjacent" mean?

★ **For Fun**

Think about it: would an adjacency matrix have to have some *symmetries*?

★ **For Fun**

Here is an example of an adjacency matrix for a graph with 4 vertices:

$$\begin{bmatrix} 0 & 1 & 1 & 0 \\ 1 & 1 & 2 & 0 \\ 1 & 2 & 0 & 1 \\ 0 & 0 & 1 & 0 \end{bmatrix}$$

Could you sketch a graph that would correspond to this adjacency matrix?

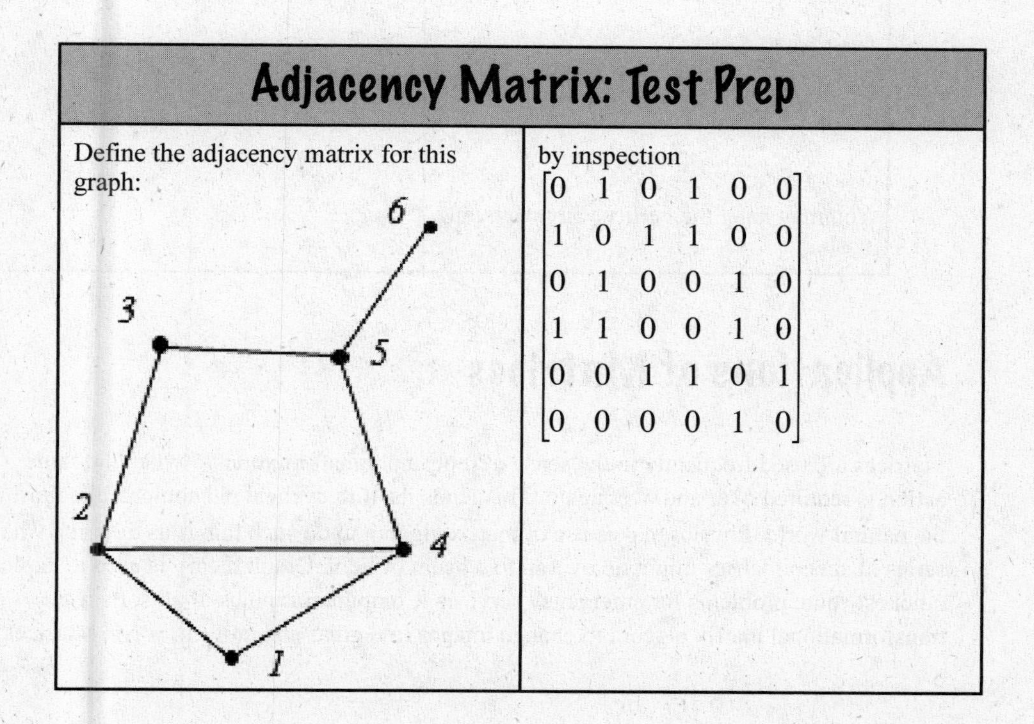

Adjacency Matrix: Test Prep

Define the adjacency matrix for this graph:

by inspection

$$\begin{bmatrix} 0 & 1 & 0 & 1 & 0 & 0 \\ 1 & 0 & 1 & 1 & 0 & 0 \\ 0 & 1 & 0 & 0 & 1 & 0 \\ 1 & 1 & 0 & 0 & 1 & 0 \\ 0 & 0 & 1 & 1 & 0 & 1 \\ 0 & 0 & 0 & 0 & 1 & 0 \end{bmatrix}$$

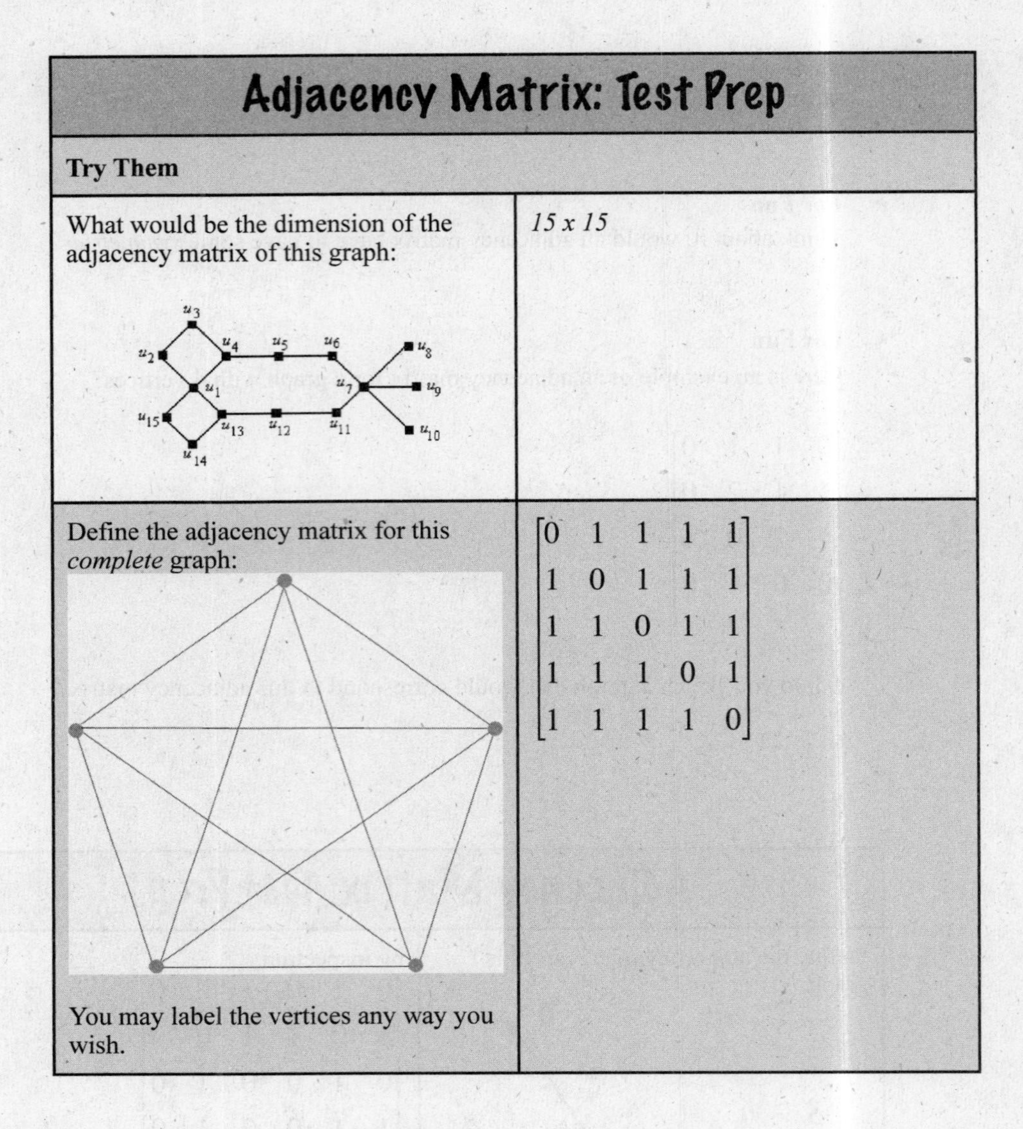

Adjacency Matrix: Test Prep

Try Them	
What would be the dimension of the adjacency matrix of this graph: u_1, u_2, u_3, u_4, u_5, u_6, u_7, u_8, u_9, u_{10}, u_{11}, u_{12}, u_{13}, u_{14}, u_{15}	*15 x 15*
Define the adjacency matrix for this *complete* graph: You may label the vertices any way you wish.	$\begin{bmatrix} 0 & 1 & 1 & 1 & 1 \\ 1 & 0 & 1 & 1 & 1 \\ 1 & 1 & 0 & 1 & 1 \\ 1 & 1 & 1 & 0 & 1 \\ 1 & 1 & 1 & 1 & 0 \end{bmatrix}$

Applications of Matrices

Matrices are used frequently in the sense of repeated transformations -- when the same action is required over and over again. This lends itself to cyclical phenomena like those in the natural world. Physics makes use of matrix algebra to do such things as describe what a series of several lenses might do overall to a beam of light. Graph theory is used to explore quickest-route problems for emergency services. Computer graphics of all sorts use transformational matrix algebra to change images in regular and uniform ways. Matrices have even been used in the arts as generators of patterns to be exploited in musical or visual pieces.

Applications: Test Prep

If each of these matrices give sales figures for the summer months at a retail store, what would the result of adding these two matrices together and what would the meaning of that result be? $\begin{bmatrix}12\\4\\7\\16\end{bmatrix}\begin{bmatrix}2\\10\\8\\20\end{bmatrix}$	$\begin{bmatrix}14\\14\\15\\36\end{bmatrix}$ The resulting matrix shows total sales for two years.
Try Them	
A town has two grocery stores, *A* and *B*. Each month store *A* retains 98% of is customers and loses 2% to store *B*, and store *B* retains 95% of its customers and loses 5% to store *A*. To start with store *B* has 75% of the town's customers and store *A* has the other 25%. After *n* months, the percent of the customers who shop at store *A*, denoted by *a*, and the percent of customers who shop at store *B*, denoted by *b*, are given by $\begin{bmatrix}0.25 & 0.75\end{bmatrix}\begin{bmatrix}0.98 & 0.02\\0.05 & 0.95\end{bmatrix}^n=\begin{bmatrix}a & b\end{bmatrix}.$ Find the percent, to the nearest 0.1% of the customers who shop at store *A* after 5 months.	39.10%
Experiment with different values of *n* in the above problem to find out how long, to the nearest month, it will be before store *A* has 50% of the customers.	11 months

7.3: Inverse of a Matrix

Much like inverse functions are used to solve equations, inverse matrices have a major use in the solutions of systems of equations. Be sure your elementary row operation skills are in good working order as we embark on this exercise.

Finding the Inverse of a Matrix

One of the main reasons for finding the inverse of a matrix is to use it to solve systems of equations. This is a really neat trick, but first you have to find the inverse ... which is labor-intensive (when done by hand).

- **inverse**
  Some square matrices have inverse matrices. The inverse of the matrix $\mathbf{A}$ is signified by $\mathbf{A}^{-1}$. Much as with multiplicative inverses of real numbers result in 1 when multiplied together, matrix multiplication of a matrix with its inverse results in the identity matrix.

The method for finding the inverse of a matrix (if it has one) involves identity matrices and elementary row operations. Here is an example of finding the inverse of a *3 x 3* square matrix:

First augment the matrix with the identity matrix of the same order:

$$\begin{bmatrix} 1 & 1 & 4 \\ 2 & 3 & 6 \\ -1 & -1 & 2 \end{bmatrix}\begin{bmatrix} 1 & 0 & 0 \\ 0 & 1 & 0 \\ 0 & 0 & 1 \end{bmatrix}$$

Now, go about row operations with an eye to turning the left-hand matrix into the *3 x 3* identity matrix. Meanwhile, let the right-hand matrix evolve thereby into whatever it must become according to those row operations.

$$\left[\begin{array}{ccc|ccc} 1 & 1 & 4 & 1 & 0 & 0 \\ 2 & 3 & 6 & 0 & 1 & 0 \\ -1 & -1 & 2 & 0 & 0 & 1 \end{array}\right] \rightarrow \left[\begin{array}{ccc|ccc} 2 & 2 & 8 & 2 & 0 & 0 \\ 2 & 3 & 6 & 0 & 1 & 0 \\ 0 & 0 & 6 & 1 & 0 & 1 \end{array}\right] \rightarrow$$

$$\left[\begin{array}{ccc|ccc} 1 & 1 & 4 & 1 & 0 & 0 \\ 0 & 1 & -2 & -2 & 1 & 0 \\ 0 & 0 & 2 & \frac{1}{3} & 0 & \frac{1}{3} \end{array}\right] \rightarrow \left[\begin{array}{ccc|ccc} 1 & 0 & 6 & 3 & -1 & 0 \\ 0 & 1 & 0 & -\frac{5}{3} & 1 & \frac{1}{3} \\ 0 & 0 & 6 & 1 & 0 & 1 \end{array}\right] \rightarrow$$

$$\left[\begin{array}{ccc|ccc} 1 & 0 & 0 & 2 & -1 & -1 \\ 0 & 1 & 0 & -\frac{5}{3} & 1 & \frac{1}{3} \\ 0 & 0 & 1 & \frac{1}{6} & 0 & \frac{1}{6} \end{array}\right]$$

The resulting right-hand matrix *is* the inverse matrix. You can check it by performing the matrix multiplication $\mathbf{A} \cdot \mathbf{A}^{-1}$ or $\mathbf{A}^{-1} \cdot \mathbf{A}$. They're inverses if the result is the *3 x 3* identity matrix.

✓ **Questions to Ask Your Teacher**

If you cannot decipher all the steps I took in the above row operations, you might want to step through it in class with your teacher and for the benefit of all your classmates. Probably several examples will help everybody in class!

Finding the Inverse of a Matrix

We are using Gaussian elimination to find the inverse of a matrix. There are other methods that you are welcome to research!

1. Analytically, using Cramer's Rule (section 7.5)
2. Neumann Series
3. Blockwise Inversion

Inverse of a Matrix: Test Prep

Verify that these matrices are inverses of each other by using matrix multiplication: $\begin{bmatrix} 1 & 1 & 4 \\ 2 & 3 & 6 \\ -1 & -1 & 2 \end{bmatrix}$ $\begin{bmatrix} 2 & -1 & -1 \\ -\frac{5}{3} & 1 & \frac{1}{3} \\ \frac{1}{6} & 0 & \frac{1}{6} \end{bmatrix}$	$\begin{bmatrix} 1 & 1 & 4 \\ 2 & 3 & 6 \\ -1 & -1 & 2 \end{bmatrix}\begin{bmatrix} 2 & -1 & -1 \\ -\frac{5}{3} & 1 & \frac{1}{3} \\ \frac{1}{6} & 0 & \frac{1}{6} \end{bmatrix}$ $= \begin{bmatrix} 2-\frac{5}{3}+\frac{4}{6} & -1+1+0 & -1+\frac{1}{3}+\frac{4}{6} \\ 4-\frac{15}{3}+\frac{6}{6} & -2+3+0 & -2+\frac{3}{3}+\frac{6}{6} \\ -2+\frac{5}{3}+\frac{2}{6} & 1-1+0 & 1-\frac{1}{3}+\frac{2}{6} \end{bmatrix}$ $= \begin{bmatrix} 1 & 0 & 0 \\ 0 & 1 & 0 \\ 0 & 0 & 1 \end{bmatrix} = \mathbf{I}$
Try Them	
Find the inverse of the matrix, if it exists: $\begin{bmatrix} 3 & 4 \\ 2 & 3 \end{bmatrix}$ Verify the result using matrix multiplication.	$\begin{bmatrix} 3 & -4 \\ -2 & 3 \end{bmatrix}$
Find the inverse of the matrix, if it exists: $\begin{bmatrix} -3 & -9 & 0 \\ -1 & -5 & -2 \\ 5 & 9 & -6 \end{bmatrix}$ Verify the result using matrix multiplication.	The matrix has no inverse.

✤ **singular matrix**

A singular matrix does not have an inverse. You'll discover that your matrix is singular if, during the row manipulations to find the inverse, one of the rows of the original left-hand matrix becomes all zeros.

★ **For Fun**

If one row becomes all zeros, wouldn't the system be dependent? Would the system have a unique solution?

✤ **nonsingular matrix**

A nonsingular matrix is a square matrix that does have an inverse.

★ **For Fun**

Do you imagine that nonsquare matrices can have inverses?

Singular Matrix: Test Prep

Show that this matrix is singular: $\begin{bmatrix} 1 & 3 & 4 \\ 2 & 5 & 3 \\ 1 & 4 & 9 \end{bmatrix}$	$\left[\begin{array}{ccc\|ccc} 1 & 3 & 4 & 1 & 0 & 0 \\ 2 & 5 & 3 & 0 & 1 & 0 \\ 1 & 4 & 9 & 0 & 0 & 1 \end{array}\right] \rightarrow$ $\left[\begin{array}{ccc\|ccc} 1 & 3 & 4 & 1 & 0 & 0 \\ 0 & 1 & 5 & 2 & -1 & 0 \\ 0 & 1 & 5 & -1 & 0 & 1 \end{array}\right] \rightarrow$ $\left[\begin{array}{ccc\|ccc} 1 & 0 & -11 & -5 & -3 & 0 \\ 0 & 1 & 5 & 2 & 1 & 0 \\ 0 & 0 & 0 & -3 & 1 & 1 \end{array}\right]$ Zero row on left-hand matrix means a singular matrix.

Singular Matrix: Test Prep	
Try Them	
Find the inverse of the matrix, if it exists: $\begin{bmatrix} 1 & -6 & 4 \\ 3 & 4 & 2 \\ 5 & 3 & 5 \end{bmatrix}$	The matrix is singular.
Given $\mathbf{A} = \begin{bmatrix} 2 & 5 & 4 \\ 1 & 4 & 3 \\ 1 & -3 & -2 \end{bmatrix}$ find $\mathbf{A}^{-1}$. Is A singular or nonsingular?	$\mathbf{A}^{-1} = \begin{bmatrix} -1 & 2 & 1 \\ -5 & 8 & 2 \\ 7 & -11 & -3 \end{bmatrix}$ A is nonsingular.

Solving Systems of Equations Using Inverse Matrices

If you rewrite a 2-dimensional system of equations symbolically like this $\mathbf{A} \cdot \begin{bmatrix} x \\ y \end{bmatrix} = \begin{bmatrix} a \\ b \end{bmatrix}$,

notice what can be done with the inverse of the *coefficient* matrix $\mathbf{A}$.

$$\mathbf{A}^{-1} \cdot \mathbf{A} \cdot \begin{bmatrix} x \\ y \end{bmatrix} = \mathbf{A}^{-1} \cdot \begin{bmatrix} a \\ b \end{bmatrix}$$

$$\begin{bmatrix} 1 & 0 \\ 0 & 1 \end{bmatrix} \cdot \begin{bmatrix} x \\ y \end{bmatrix} = \mathbf{A}^{-1} \cdot \begin{bmatrix} a \\ b \end{bmatrix}$$

$$\begin{bmatrix} x \\ y \end{bmatrix} = \mathbf{A}^{-1} \cdot \begin{bmatrix} a \\ b \end{bmatrix}$$

And you might notice that you have solved the system for the "coordinate" column matrix. That boils down to having solved for the variables x and y. (Of course, finding the inverse of the coefficient matrix can still be a big job.)

This can clearly be extended to arbitrarily large systems of equations with many variables to solve for.

Solving Systems of Equations Using Inverse Matrices: Test Prep

Solve the matrix equation: $\begin{bmatrix} 7 & -5 \\ 2 & -3 \end{bmatrix}\begin{bmatrix} x \\ y \end{bmatrix} = \begin{bmatrix} 12 \\ 6 \end{bmatrix}$	If $\mathbf{A} = \begin{bmatrix} 7 & -5 \\ 2 & -3 \end{bmatrix}$, then $\mathbf{A}^{-1} = \begin{bmatrix} \frac{3}{11} & \frac{-5}{11} \\ \frac{2}{11} & \frac{-7}{11} \end{bmatrix} = -\frac{1}{11}\begin{bmatrix} -3 & 5 \\ -2 & 7 \end{bmatrix}$ and $\begin{bmatrix} x \\ y \end{bmatrix} = -\frac{1}{11}\begin{bmatrix} -3 & 5 \\ -2 & 7 \end{bmatrix}\begin{bmatrix} 12 \\ 6 \end{bmatrix}$ $= -\frac{1}{11}\begin{bmatrix} -36+30 \\ -24+42 \end{bmatrix} = \begin{bmatrix} \frac{6}{11} \\ -\frac{18}{11} \end{bmatrix}$
Try Them	
Use inverse matrices to solve the system of equations: $\begin{cases} 4x + 7y = 2 \\ 3x + 5y = 1 \end{cases}$	*(-3, 2)*
Use inverse matrices to solve the system of equations: $\begin{cases} x + y - z = -3 \\ x + 2y - z = 6 \\ -x - y + 2z = 4 \end{cases}$	*(-11, 9, 1)*

Input-Output Analysis

The labor required to find the inverse of a coefficient matrix is similar to the effort needed to simply put the matrix form of a system of equations into row-echelon form and back-substitute in order to solve for the variables under consideration. But if the result matrix changes while the coefficient matrix doesn't, then it's well worth your time to find the one key that will solve every variable with just some matrix multiplication and algebraic interpretation. Input-output analysis makes use of this convenience.

✤ **input-output analysis**
Input-output analysis is a discipline that seeks to understand quantitatively the relationships between the demands of industries and customers and their outputs.

✤ **input-output matrix**
A particular matrix that gives the monetary relationship between input and output costs. Each column represents the cost of an input that will generate $1 of output.

✤ **final demand**
Final demand is represented by a column matrix and signifies the amount of output from the industries that consumers (and other industries) will purchase.

Paddy
Other agricultural products
Livestock and poultry
Forestry
Fishery
Crude petroleum and natural gas
Other mining
Food, beverage and tobacco
Textile, leather, and the products thereof
Timber and wooden products
Pulp, paper and printing
Chemical products
Petroleum and petro products
Rubber products
Non-metallic mineral products
Metal products
Machinery
Transport equipment
Other manufacturing products
Electricity, gas, and water supply
Construction
Trade and transport
Services
Public administration

Some possible economic sectors

Applications: Test Prep

Each edge of a metal plate is kept at a constant temperature as shown below. Find the temperature at x_1 and x_2.

Round to the nearest tenth of a degree.

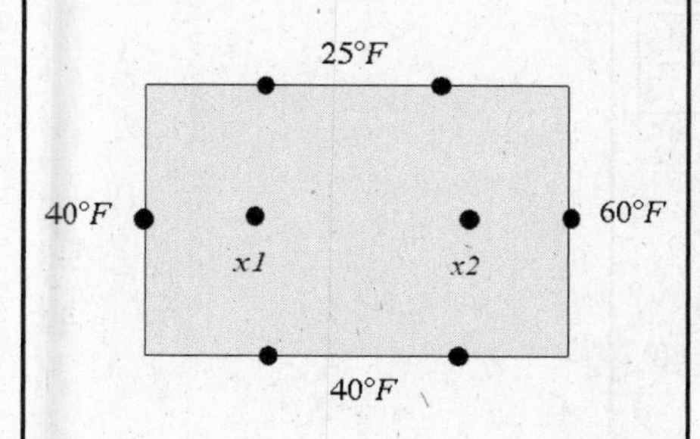

Write a system of equations:

$$\begin{cases} x_1 = \dfrac{40 + 40 + 25 + x_2}{4} = \dfrac{105 + x_2}{4} \\ x_2 = \dfrac{25 + 60 + 40 + x_1}{4} = \dfrac{125 + x_1}{4} \end{cases}$$

Rewrite the system of equations:

$$\begin{cases} 4x_1 - x_2 = 105 \\ -x_1 + 4x_2 = 125 \end{cases}$$

Solve the equations using the inverse matrix:

$$\begin{bmatrix} 4 & -1 \\ -1 & 4 \end{bmatrix}\begin{bmatrix} x_1 \\ x_2 \end{bmatrix} = \begin{bmatrix} 105 \\ 125 \end{bmatrix}$$

$$\begin{bmatrix} x_1 \\ x_2 \end{bmatrix} = \frac{1}{15}\begin{bmatrix} 4 & 1 \\ 1 & 4 \end{bmatrix}\begin{bmatrix} 105 \\ 125 \end{bmatrix}$$

$$\begin{bmatrix} x_1 \\ x_2 \end{bmatrix} = \begin{bmatrix} \dfrac{109}{3} \\ \dfrac{121}{3} \end{bmatrix}$$

And the temperatures, to the nearest tenth of a degree, are $36.3°$ and $40.3°$.

Applications: Test Prep

Try Them

<table>
<tr><td>

The following table shows the carbohydrate, fat, and protein content of three food types in grams:

	carbo-hydrate	fat	protein
type I	13	10	13
type II	4	4	3
type III	1	0	10

A nutritionist must prepare two diets from these three food groups. The first diet must contain 23 grams of carbohydrate, 18 grams of fat, and 39 grams of protein. The second diet must contain 35 grams of carbohydrate, 28 grams of fat, and 42 grams of protein. How many grams of each food type are required for the first diet, and how many grams of each food type are required for the second diet?

</td><td>

For the first diet, 100g of type I, 200g of type II, and 200g of type III; for the second diet, 200g of type I, 200g of type II, and 100g of type III.

</td></tr>
<tr><td>

A four-sector economy consists of manufacturing, agriculture, service, and transportation. The input-output matrix for this economy is

$$\begin{bmatrix} 0.10 & 0.05 & 0.200 & 0.15 \\ 0.20 & 0.10 & 0.30 & 0.10 \\ 0.05 & 0.30 & 0.20 & 0.40 \\ 0.10 & 0.20 & 0.15 & 0.20 \end{bmatrix}$$

Find the gross output needed to satisfy the consumer demand of $80 million worth of transportation.

</td><td>

$219.0 million of manufacturing, $294.3 million agriculture, $316.7 million service, and $260.3 million transportation

</td></tr>
</table>

MID-CHAPTER QUIZ: TEXTBOOK PAGE 577.

7.4: Determinants

A determinant is some sort of "characteristic value" associated with a matrix. First we learn how to calculate them, then what they are useful for.

Determinant of a 2 X 2 Matrix

At the root of most methods of calculating the value of a determinant is the value of a *2 x 2* matrix's determinant. The determinant for a *2 x 2* is easy, so it's a good place to start. Ultimately it will generalize to larger matrices, too.

✤ **determinant**

The determinant of a matrix A is denoted by $|\mathbf{A}|$ or sometimes by det(A). The absolute value-like symbol can help to remind you that the determinant of a matrix is a scalar -- that is, a number. What this number tells you remains to be seen; however, in the meantime, here is how one calculates the determinant of a *2 x 2* matrix:

$$\begin{vmatrix} a_{11} & a_{12} \\ a_{21} & a_{22} \end{vmatrix} = a_{11} \cdot a_{22} - a_{12} \cdot a_{21}$$

For example,

$$\begin{vmatrix} 1 & -2 \\ 3 & 0 \end{vmatrix} = 1 \cdot 0 - (-2) \cdot 3 = 6$$

Maybe you can see from this example that this is a place to be very mindful of the positive/negative natures of the numbers involved. Be careful with the signs of the numbers and subtracting!

★ **For Fun**

Can you figure out how to get your calculator to calculate a *2 x 2* determinant?

Determinant of a 2 X 2 Matrix: Test Prep	
Find the determinant of $\begin{bmatrix} 9 & 1 \\ 8 & 2 \end{bmatrix}$	$\begin{vmatrix} 9 & 1 \\ 8 & 2 \end{vmatrix} = 9 \cdot 2 - 1 \cdot 8 = 18 - 8 = 10$
Try Them	
Evaluate $\begin{vmatrix} -1 & 5 \\ 3 & -7 \end{vmatrix}$	-8
Evaluate $\begin{vmatrix} 3 & 7 \\ 0 & 0 \end{vmatrix}$	0

Minors and Cofactors

Minors and cofactors are going to help us work up to figuring the determinant of matrices of order larger than *2 x 2*.

✤ **minor**

The minor of a matrix is the determinant of a smaller portion of the matrix. Consider this *3 x 3* matrix:

$$\begin{bmatrix} 3 & 0 & -1 \\ 1 & 2 & -3 \\ 5 & 7 & -4 \end{bmatrix}$$

The M_{13} minor is, as follows:

$$\begin{vmatrix} 1 & 2 \\ 5 & 7 \end{vmatrix}$$

Do you see those elements in the above matrix? It is the determinant of what's leftover when you remove the 1st row and 1st column of the matrix, and the subscripts of M_{11} tell you which row/column to "remove".

★ **For Fun**

Could you calculate M_{11} for the above matrix? How about a different minor like, maybe, M_{32}?

✤ **cofactor**

The cofactor C_{ij} of a matrix is the minor determinant with a ± 1 factor in front of it. The ± 1 depends on which minor like this:

$$C_{ij} = (-1)^{i+j} M_{ij}$$

So the cofactor C_{22}, for instance, would have entail a factor of positive 1; whereas, the C_{32} cofactor would get a negative one.

★ **For Fun**

Figure that ± 1 factor for every element in a *3 x 3* matrix, for instance. You could use my *3 x 3* matrix above. Look for a pattern.

★ **For Fun**

Now look at the formula for calculating the determinant of a *2 x 2*. Do you see an alternating of signs even in it?

Minor and Cofactor: Test Prep

Given the matrix $\begin{bmatrix} 5 & -2 & 1 \\ 0 & -1 & 3 \\ -4 & 2 & 0 \end{bmatrix}$ find M_{23} and C_{23}	$M_{23} = \begin{vmatrix} 5 & -2 \\ -4 & 2 \end{vmatrix} = 10 - 8 = 2$ $C_{23} = (-1)^5 \begin{vmatrix} 5 & -2 \\ -4 & 2 \end{vmatrix} = -2$

Minor and Cofactor: Test Prep	
Try Them	
Find M_{11} and C_{11} for the matrix above.	$M_{11} = -6$ $C_{11} = -6$
Find M_{21} and C_{21} for the matrix above.	$M_{21} = -2$ $C_{21} = 2$

Evaluating a Determinant Using Expanding by Cofactors

Now we'll see how we can carefully combine cofactors, which include their corresponding minors, and the elements of a *3 x 3* or larger square matrix to calculate a determinant. Again, to me this is a very visual exercise, so watch plenty of examples and then do plenty of examples of your own.

First of all, you have to pick one particular row *or* column to expand your determinant. Personally, I like to pick a row with as many small and positive numbers as I can with special preference given to zeros. For instance, considering the matrix

$$\begin{bmatrix} 2 & 0 & -1 \\ -2 & 0 & -3 \\ 5 & 7 & -4 \end{bmatrix}$$

I would definitely choose to expand using the 2nd column. Try a few, you'll develop a sense for this quickly.

Having chosen the 2nd column above, this is how the determinant would be calculated:

$$\begin{vmatrix} 2 & 0 & -1 \\ -2 & 0 & -3 \\ 5 & 7 & -4 \end{vmatrix} = (0)\cdot(-1)^{1+2}\cdot M_{12} + (0)\cdot(-1)^{2+2}\cdot M_{22} + (7)\cdot(-1)^{3+2}\cdot M_{32}$$

Do you see the elements of the column multiplied by their corresponding cofactors? Here's the next step:

$$(0)\cdot(-1)^{1+2}\cdot\begin{vmatrix} -2 & -3 \\ 5 & -4 \end{vmatrix} + (0)\cdot(-1)^{2+2}\cdot\begin{vmatrix} 2 & -1 \\ 5 & -4 \end{vmatrix} + (7)\cdot(-1)^{3+2}\cdot\begin{vmatrix} 2 & -1 \\ -2 & -3 \end{vmatrix}$$

See the minors, which are now determinants of *2 x 2*'s? Do you see that the elements of those minors are the elements of the matrix once you've "removed" the corresponding row/column specified for the minor? Next step:

$$-7\cdot\begin{vmatrix} 2 & -1 \\ -2 & -3 \end{vmatrix} = -7[(2)\cdot(-3)-(-1)\cdot(-2)] = -7(-6-2) = 56$$

So $\begin{vmatrix} 2 & 0 & -1 \\ -2 & 0 & -3 \\ 5 & 7 & -4 \end{vmatrix} = 56.$

This will work for any square matrix. I'm sure you can imagine the care and attention it would take to calculate the determinant of matrices of still larger dimension than this one. More help coming for that.

(Meanwhile, let me urge you to hang on. Determinants are very helpful for solving systems of equations. This is more than an exercise in curiosity, okay?)

★ **For Fun**

Use cofactors to derive the general rule for a *3 x 3* determinant of $\begin{bmatrix} a & b & c \\ d & e & f \\ g & h & i \end{bmatrix}$. Is it $aei - afh + bfg - bdi + cdh - ceg$?

Evaluate a Determinant by Expanding by Cofactors: Test Prep

Evaluate the determinant by expanding the cofactors: $\begin{bmatrix} 0 & 1 & -2 \\ -1 & 2 & 1 \\ 4 & 0 & 3 \end{bmatrix}$	$\begin{vmatrix} 0 & 1 & -2 \\ -1 & 2 & 1 \\ 4 & 0 & 3 \end{vmatrix} = 1\cdot 0\cdot \begin{vmatrix} 2 & 1 \\ 0 & 3 \end{vmatrix} - 1\cdot 1\cdot \begin{vmatrix} -1 & 1 \\ 4 & 3 \end{vmatrix} + 1\cdot(-2)\cdot \begin{vmatrix} -1 & 2 \\ 4 & 0 \end{vmatrix}$ $= -1(-3-4) - 2(0-8) = 7 + 16 = 23$
Try Them	
Find the determinant of this matrix by expanding the cofactors: $\begin{bmatrix} 6 & 0 & 0 \\ 2 & -3 & 0 \\ 7 & -8 & 2 \end{bmatrix}$	-36
Evaluate: $\begin{vmatrix} -2 & 3 & 9 \\ 4 & -2 & -6 \\ 0 & -8 & -24 \end{vmatrix}$	0

Evaluating a Determinant Using Elementary Row Operations

We can use elementary row operations to help us simplify finding the determinant of larger matrices. First some rules about row operations and how they effect the determinant itself.

Elementary Row Operations on Determinants:

1. Interchanging any two rows of the matrix ***A*** changes the sign of the determinant of ***A***.

$$|\mathbf{A}| \quad \rightarrow \quad -|\mathbf{A}|$$

2. Multiplying a row of ***A*** by a constant *k* changes the determinant of ***A*** by the same factor.

$$|\mathbf{A}| \quad \rightarrow \quad k|\mathbf{A}|$$

3. Adding a multiple of a row of ***A*** to another row of ***A*** doesn't change the value of the determinant of ***A.***

$$|\mathbf{A}| \quad \rightarrow \quad |\mathbf{A}|$$

✤ **triangular form**

We've met the triangular form of a matrix before. For instance:

$$\begin{bmatrix} 2 & -3 & 0 & 1 \\ 0 & 1 & -1 & 5 \\ 0 & 0 & 3 & -2 \\ 0 & 0 & 0 & 4 \end{bmatrix}$$

We'll expand what we mean as a matrix in triangular form to include matrices like this now:

$$\begin{bmatrix} -1 & 0 & 0 & 0 \\ 5 & 2 & 0 & 0 \\ 1 & -3 & 4 & 0 \\ 9 & 7 & 8 & 1 \end{bmatrix}$$

In other words, any square matrix whose elements either below *or* above the main diagonal are all zeros qualify for the triangular form label.

If your matrix is in triangular form, then the determinant of that matrix is simply the product of the elements on the main diagonal.
(Think about it.)

Evaluate a Determinant by Using Elementary Row Operations: Test Prep

Find the determinant of $\begin{bmatrix} 2 & 1 & 3 \\ 1 & 0 & -4 \\ -2 & 0 & 10 \end{bmatrix}$ by using row operations and their rules for determinants to turn this into a determinant of a triangular form.	$\begin{vmatrix} 2 & 1 & 3 \\ 1 & 0 & -4 \\ -2 & 0 & 10 \end{vmatrix}$ $= \begin{vmatrix} 2 & 1 & 3 \\ 1 & 0 & -4 \\ 0 & 1 & 13 \end{vmatrix}$ $= \frac{1}{2}\begin{vmatrix} 2 & 1 & 3 \\ 2 & 0 & -8 \\ 0 & 1 & 13 \end{vmatrix}$ $= \frac{1}{2}\begin{vmatrix} 2 & 1 & 3 \\ 0 & -1 & -11 \\ 0 & 1 & 13 \end{vmatrix}$ $= -\frac{1}{2}\begin{vmatrix} 2 & 1 & 3 \\ 0 & 1 & 11 \\ 0 & 1 & 13 \end{vmatrix}$ $= -\frac{1}{2}\begin{vmatrix} 2 & 1 & 3 \\ 0 & 1 & 11 \\ 0 & 0 & 2 \end{vmatrix}$ $= \left(-\frac{1}{2}\right)(2)(1)(2) = -2$

Evaluate a Determinant by Using Elementary Row Operations: Test Prep

Try Them	
Evaluate: $\begin{vmatrix} 3 & 0 & 0 \\ 2 & -1 & 0 \\ 3 & 4 & 5 \end{vmatrix}$	-15
Evaluate the determinant by first rewriting it in triangular form: $\begin{vmatrix} 1 & 2 & 0 & -2 \\ -1 & 1 & 3 & 5 \\ 2 & 1 & 4 & 0 \\ -2 & 5 & 2 & 6 \end{vmatrix}$	0

Condition for a Square Matrix to Have a Multiplicative Inverse

Determinants are handy for discovering if a square matrix has an inverse. And we've seen that an inverse means there exists a unique solution to the system back in section 7.3. A singular matrix does not have an inverse. This is exactly analogous to a matrix whose determinant is zero. Here are the conditions when the determinant will be equal to zero:

Given that ***A*** is a square matrix, $|\mathbf{A}| = 0$ when ...

1. Any row or column is all-zeros.
2. Any two rows or columns are the same.
3. Any one row is a constant multiple of another row. Or any one column is a constant multiple of another column.

These conditions need not hold for the original matrix but do guarantee a zero determinant if they appear after row reductions!

Now that we know how to tell if a determinant is zero, we can say the following:

If $\boldsymbol{A}$ is a square matrix of order n, then $\boldsymbol{A}$ has a multiplicative inverse if and only if $|\mathbf{A}|$ is not equal to zero, which also implies $|\mathbf{A}^{-1}| = \dfrac{1}{|\mathbf{A}|}$.

★ **For Fun**
So do you imagine that a determinant that is equal to zero implies anything about the solution set of a system of equations?

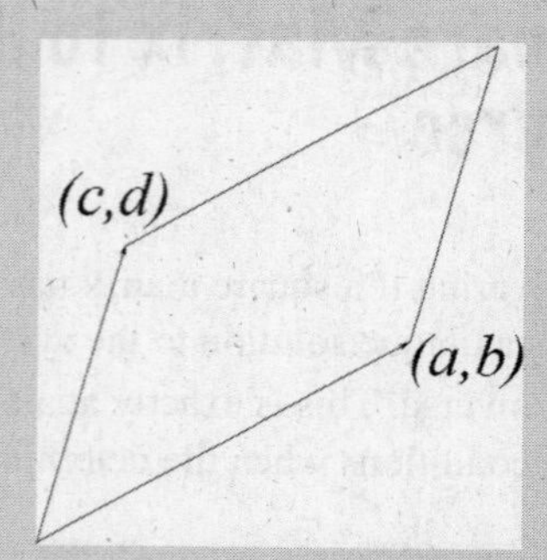

The area of a parallelogram is related to the determinant of the matrix formed from the points of the vertices of the shape: area = $|ad - bc|$

7.5: Cramer's Rule

Determinants can be used to solve a system of equations! (Does that start to sound more appealing than elementary row operations on an augmented matrix representation of a system?)

Cramer's Rule

This is Cramer's Rule, from page 589 of your textbook:

$$\text{Let}\begin{cases} a_{11}x_1 + a_{12}x_2 + a_{13}x_3 + \ldots + a_{1n}x_n = b_1 \\ a_{21}x_1 + a_{22}x_2 + a_{23}x_3 + \ldots + a_{2n}x_n = b_2 \\ a_{31}x_1 + a_{32}x_2 + a_{33}x_3 + \ldots + a_{3n}x_n = b_3 \\ \vdots \quad\quad \vdots \quad\quad \vdots \quad\quad \vdots \quad\quad \vdots \quad = \quad \vdots \\ a_{n1}x_1 + a_{n2}x_2 + a_{n3}x_3 + \ldots + a_{nn}x_n = b_n \end{cases}$$

be a system of n equations in n variables. The solution of the system is given by $(x_1, x_2, x_3, \ldots, x_n)$, where

$$x_1 = \frac{D_1}{D} \quad x_2 = \frac{D_2}{D} \quad \ldots \quad x_i = \frac{D_i}{D} \quad \ldots \quad x_n = \frac{D_n}{D}$$

and D is the determinant of the coefficient matrix, $D \neq 0$. D_i is the determinant formed by replacing the ith column of the coefficient matrix with the column of constants $b_1, b_2, b_3, \ldots, b_n$.

✓ **Questions to Ask Your Teacher**
Read carefully. Cramer's Rule tells you how to solve a system of equations using determinants, provided the system can be solved uniquely. This rule includes some pretty intense notation and lots of subscripts. Ask your teacher to help you understand every little bit if you are unclear about any of the pieces.

★ **For Fun**
Can you see why the determinant of the coefficient matrix D cannot be equal to zero?

This method of solving a system of equations is pretty intense on the determinant front. Have you practiced your determinants?

Cramer's Rule: Test Prep	
Solve the following system of equations using Cramer's Rule: $\begin{cases} 4x_1 - 5x_2 = 12 \\ 3x_1 - 4x_2 = 10 \end{cases}$	$x_1 = \dfrac{\begin{vmatrix} 12 & -5 \\ 10 & 4 \end{vmatrix}}{\begin{vmatrix} 4 & -5 \\ 3 & 4 \end{vmatrix}} = \dfrac{2}{-1} = -2$ $x_2 = \dfrac{\begin{vmatrix} 4 & 12 \\ 3 & 10 \end{vmatrix}}{\begin{vmatrix} 4 & -5 \\ 3 & -4 \end{vmatrix}} = \dfrac{4}{-1} = -4$ *(-2, -4)*
Try Them	
Consider the system $\begin{cases} 4x - 6y = 8 \\ 2x - 3y = 4 \end{cases}$ 1. Can this system be solved using Cramer's Rule? 2. Can this system be solved using Gaussian elimination method?	1. no 2. yes
Solve the following system of equations using Cramer's Rule: $\begin{cases} x + 3y = -2 \\ 2x - 3y + z = 1 \\ 4x + 5y - 2z = 0 \end{cases}$	$\left(\dfrac{4}{25}, -\dfrac{18}{25}, -\dfrac{37}{25}\right)$

Final Fun: Matching

FIND THE BEST MATCH. GOOD LUCK.

AUGMENTED MATRIX

$$\begin{vmatrix} 2 & -1 & 0.5 \\ 3 & 5 & -4 \\ 0 & 7 & -0 \end{vmatrix}$$

TRIANGULAR MATRIX

ROW MATRIX

DETERMINANT OF 2 X 2

$(-1)^{i+j}\mathbf{M}_{ij}$

IDENTITY

$$\begin{bmatrix} 2 & -3 & 1 & 0 & 5 \\ -5 & 2 & -2 & 1 & 3 \end{bmatrix}$$

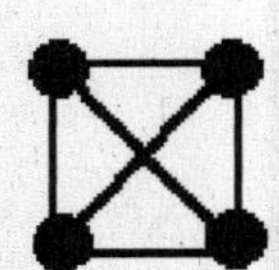

$\mathbf{B}^{-1}$

$$\begin{bmatrix} 0 & 0 \\ 0 & 0 \end{bmatrix}$$

COFACTOR

$$\begin{bmatrix} 1 & 0 \\ -2 & 3 \end{bmatrix}$$

DETERMINANT

GRAPH

$$\begin{bmatrix} -3 & 0 & 1 & -2 \end{bmatrix}$$

ADDITIVE IDENTITY

INVERSE MATRIX

2 X 5 MATRIX

$$\left[\begin{array}{cc|c} 2 & -1 & 9 \\ 1 & 3 & -7 \end{array}\right]$$

$$\begin{bmatrix} 1 & 0 & 0 \\ 0 & 1 & 0 \\ 0 & 0 & 1 \end{bmatrix}$$

$a_{11} \cdot a_{22} - a_{12} \cdot a_{21}$

NOW TRY THESE:
REVIEW EXERCISES: TEXTBOOK PAGE 594,
PRACTICE TEST: TEXTBOOK PAGE 598.

STAY UP ON YOUR GAME -- DO THE CUMULATIVE
REVIEW EXERCISES: TEXTBOOK PAGE 599

Chapter 8: Sequences, Series, and Probability

Word Search

```
L A I M O N I B H Z D J X F I
S C I N D E P E N D E N T O N
S X S W B K V C G N S A E N F
N E I P H G O E U W T H X O I
A O Q V N U O Q N C E Y P I N
S R I U N M U N A T C T E T I
Z E I T E H S E V G I R C T
N O I T A N I B M O C L I U E
K N R R H T C M W I N I M D T
G I F N E M U E Z H A B E N I
C B X C B S E M X R W A N I N
P A S C A L U T R S K B T V I
Z K T J O E J K I E G O X C F
L A I R O T C A F C P R S C N
E X C L U S I V E B P P J F M
```

ARITHMETIC	EXPERIMENT	INFINITE
BINOMIAL	FACTORIAL	PASCAL
COMBINATION	FINITE	PERMUTATION
COUNTING	GEOMETRIC	PROBABILITY
EVENT	INDEPENDENT	SEQUENCE
EXCLUSIVE	INDUCTION	SERIES

8.1: Infinite Sequences and Summation Notation

This section is really just about learning a new notation: the summation notation. Just like x_2 means something entirely different from x^2, you need to be able to assign the correct *meaning* to a subscript, or an *index*, in an expression. By the way, you are clear about the difference between x_2 and x^2, right?

Infinite Sequences

A sequence is a set of elements that come in a particular order. Remember from the discussion of sets way back in Chapter P that there is more than one way to describe a set. Now we get the way one can describe an infinite *ordered* set.

- **infinite sequence**
  An infinite sequence can be described in terms of functions: An infinite sequence is a function whose domain is the positive integers and whose range is a set of real numbers (page 602 in your textbook). The thing with sequences is that we're usually very interested in the order of the elements of the range of the function that defines the sequence, and we are usually very particular about starting off our domain at one (or sometimes zero). This way, the correspondence between the input and the output can become meaningful.

- **terms**
  A term is an element of the range of the sequence. If the function defining the sequence is something like $f(n)=(n+1)^2$, then the first *term* of the sequence is $f(1)=4$.

- **first term**
  Unless otherwise noted, the "positive integers" of the domain of a sequence starts at one. So the first term of a sequence is $f(1)$.

- **second term**
  The second term will be $f(2)$ ... and so on.

✤ ***nth* term, general term**

The *nth* term definition of an element of a sequence is just exactly the definition of the sequence that depends on *n*. The *nth* term, or general term, of $f(n) = (n+1)^2$ is just $(n+1)^2$.

★ **For Fun**

So finding a particular term of a sequence is pretty much the same proposition as evaluating a function, right?

✤ **alternating sequence**

An alternating sequence is a sequence whose terms alternate between negative and positive. (Frequently such a sequence will include an expression like $(-1)^n$ or some other evident dependence on *-1*.

✤ **recursively-defined sequence**

A recursive definition is one that depends on itself. If you are programming and your function calls itself, that is a recursive relationship. A recursively-defined term in a sequence will depend on itself something like this:

$$a_n = \frac{a_{n-1} + a_{n+1}}{2}$$

✤ **Fibonacci sequence**

The famous Fibonacci sequence is 1, 1, 2, 3, 5, 8, 13, 21, ...

★ **For Fun**

Find out more about this Fibonacci character and his tremendous contributions to mathematics. You might found out how the sequence of Fibonacci numbers is defined also!

Infinite Sequence: Test Prep	
Find the eleventh term of this sequence: $a_n = \frac{(-1)^{n+1}}{n^2}$	$a_{11} = \frac{(-1)^{11+1}}{11^2} = \frac{1}{121}$

Infinite Sequence: Test Prep	
Try Them	
List the first five terms of this infinite sequence: $a_n = \left(\frac{3}{5}\right)^n$	$\frac{3}{5}, \frac{9}{25}, \frac{27}{125}, \frac{81}{625}, \frac{243}{3125}$
List the first three terms of the recursively defined sequence: $a_1 = 2, \quad a_n = (-3)na_{n-1}$	2, -12, 108

Factorials

Although you probably have a factorial function on your calculator (it looks like an exclamation mark), it's necessary to understand how factorials work ... because they get really big really quickly. Your calculator is limited in how well it can handle big numbers.

✤ ***n* factorial**
n-factorial is defined (as a function of *n*) like this:

$$n! = n \cdot (n-1) \cdot (n-2) \cdot \cdots \cdot 2 \cdot 1.$$

So, for instance, $5! = 5 \cdot 4 \cdot 3 \cdot 2 \cdot 1 = 120$.

★ **For Fun**
Does the definition of the factorial seem recursive to you?

n Factorial: Test Prep	
Evaluate the factorial expression without using a calculator: $\frac{26!}{24!}$	$\frac{26!}{24!} = \frac{26 \cdot 25 \cdot 24 \cdot \cdots \cdot 3 \cdot 2 \cdot 1}{24 \cdot \cdots \cdot 3 \cdot 2 \cdot 1} = 26 \cdot 25 = 650$

n Factorial: Test Prep	
Try Them	
Evaluate $5! + 4!$	144
Find the fifth term in the infinite sequence: $a_n = \dfrac{(-1)^n n!}{n-1}$	-30

Partial Sums and Summation Notation

So far every time we've mentioned sequence, we've been referring to an infinite sequence, like the Fibonacci sequence whose ellipses say "This goes on forever." Partial sums happen when you take a finite part of a sequence *and add the terms up.*

- ***n*th partial sum**
  S_n is the nth partial sum of the a sequence. For instance, if you need to find S_4, you would add the first four terms of your sequence together to find the fourth partial sum.

- **sequence of partial sums**
  Okay, now you can make a new infinite sequence out of the successive partial sums of a sequence. That is $S_1, S_2, S_3, \cdots, S_n, \cdots$

  ★ **For Fun**
  You wanna try to write the first few terms of the sequence of partial sums of $f(n) = (n+1)^2$?

- **summation notation**
  And now it's time to dissect and reconstitute summation notation which might look like this: $\sum_{i=1}^{\infty} a_i$.

★ **For Fun**

Do you know which Greek letter Σ is? Does it make sense that this stands for "summation"?

✤ **index of the summation**

In the example $\sum_{i=1}^{\infty} a_i$, i is the index of the summation. The index starts where the statement at the bottom of the summation notation informs you (1 in this case) and continues in integer steps to the value at the top of the summation symbol. This example is an *infinite* sum because the index goes until ∞ (not that ∞ is a number, mind you!).

✤ **upper limit**

The number at the top of the summation symbol is the upper limit. That's the value of the index that arrests the sum (if it is not an infinite sum).

✤ **lower limit**

The lower limit is the value where the index starts. This is given at the bottom of the summation symbol.

nth Partial Sum: Test Prep	
Find the 3rd partial sum of $a_n = (-1)^n \dfrac{n}{n!}$	$a_1 = -1;\quad a_2 = \dfrac{2}{2!} = 1;\quad a_3 = -\dfrac{3}{3!} = -\dfrac{1}{2}$ $a_1 + a_2 + a_3 = -1 + 1 - \dfrac{1}{2} = -\dfrac{1}{2}$
Try Them	
Evaluate the series: $\sum_{i=1}^{3} \dfrac{i-1}{i+1}$	$\dfrac{5}{6}$
Evaluate the series: $\sum_{k=3}^{6} (-1)^k k!$	618

8.2: Arithmetic Sequences and Series

The next couple of sections deal with some standard examples of sequences and series (sums). They're good practice ... and you'll see them in calculus again.

Arithmetic Sequences

A sequence whose terms differ by a constant quantity is an arithmetic sequence -- I guess because you have to do arithmetic to figure out the next term!

- **arithmetic sequence**
  If you calculate the first few terms of a sequence defined like, say, $5n - 2$, you'll find that each term differs from the next by five. That makes it an arithmetic sequence: a sequence where $a_{i+1} - a_i = d$, where *d* is a real number.

  ★ **For Fun**
  Try a few: $-2n + 3,\ n - 1,\ 3(2n - 1)$.

- **common difference**
  d in the definition above is the common difference between the terms of an arithmetic sequence.

  ★ **For Fun**
  What is the numeric value of the common differences in the examples given above?

Arithmetic Sequence: Test Prep	
Consider the arithmetic sequence *7, 9, 11, ... , 2n+5, ...* What is the common difference?	Since all the terms of the series increase by 2, it is 2; also the coefficient of the *nth* term is 2.

Arithmetic Sequence: Test Prep	
Try Them	
Find the 21st term of the above sequence.	47
Find the 13th element of the sequence: *-3, 4, 11, ... , 7n-10*	81

Arithmetic Series

Now we're adding terms together that have a common difference. Don't forget how to do partial sums.

✤ **arithmetic series**
The third partial sum of an arithmetic series might look something like this:

$$\sum_{n=1}^{3}(-4n+1)$$

★ **For Fun**
Compute the above partial arithmetic sum. What is the common difference of the terms?

Formulae for the *n*th Partial Sum of an Arithmetic Series:

1. $S_n = \frac{n}{2}(a_1 + a_n)$

2. $S_n = \frac{n}{2}[2a_1 + (n-1)d]$

[Practice these formulae.
They will help clarify variables and indices in your mind.]

Sum of an Arithmetic Series: Test Prep	
Find the 14th partial sum of the arithmetic sequence $a_n = 7n$	$S_{14} = \frac{14}{2}(7+98) = 7 \cdot 105 = 735$
Try Them	
Find the sum of the first 75 terms of the arithmetic sequence whose first 3 terms are $\frac{1}{2}, \frac{9}{4}, 4$.	$\frac{19979}{4}$
Find the 25th partial sum of the arithmetic sequence $a_n = n - 4$	225

Arithmetic Means

Can you build an arithmetic sequence from a series of halfway points?

✤ **arithmetic mean**
The arithmetic mean is what we usually refer to simply as the mean: the arithmetic mean of two numbers a and b is $\frac{a+b}{2}$. If you think about it, since the mean is situated exactly between a and b, there is a common difference between a and b and their mean. So those three numbers form an arithmetic sequence. Since you can splint any interval exactly in half, then one can create any arithmetic sequence using successive means. Do you see what I mean?

★ **For Fun**
This reminds me of Zeno's paradox, "Arguments Against Motion." Are you familiar with it?

Arithmetic Mean: Test Prep

Insert five arithmetic means between 4 and 20.	$a = 4, c_1, c_2, c_3, c_4, c_5, 20 = b$ $n = 7$ $20 = 4 + (7-1)d = 4 + 6d$ $d = \frac{16}{6} = \frac{8}{3}$ $c_1 = 4 + \frac{8}{3} = \frac{20}{3}$ $c_2 = 4 + 2\left(\frac{8}{3}\right) = \frac{12}{3} + \frac{16}{3} = \frac{28}{3}$ $c_3 = 4 + 3\left(\frac{8}{3}\right) = 12$ $c_4 = \frac{44}{3}$ $c_5 = \frac{52}{3}$ $4, \frac{20}{3}, \frac{28}{3}, 12, \frac{44}{3}, \frac{52}{3}, 20$
Try Them	
Insert 5 arithmetic means between 7 and 19.	*7, 9, 11, 13, 15, 17, 19*
Insert 4 arithmetic means between $\frac{11}{3}$ and 6.	$\frac{11}{3}, \frac{62}{15}, \frac{69}{15}, \frac{76}{15}, \frac{83}{15}, 6$

★ **For Fun**

There's a famous comparison of arithmetic means with geometric means. You could look it up!

8.3: Geometric Sequences and Series

More on common sequences and series: the geometric sequence and series shows up again in calculus. Meanwhile, it's great practice sorting out notation.

Geometric Sequences

The separation between the terms in a geometric series is not a common difference like what holds sway in the land of arithmetic sequences; the relation between geometric terms is more of a common *ratio.*

✤ **geometric sequence**
Here is the relation between terms in a geometric sequence:

$$\frac{a_{j+1}}{a_j} = r.$$

★ **For Fun**
Do you see a common ratio between terms?

✤ **common ratio**
The parameter r in the above definition is the common ratio of the geometric sequence (or series).

The *n*th Term of a Geometric Sequence:

$$a_n = a_1 r^{n-1}$$

Geometric Sequence: Test Prep	
Find the *nth* term of the geometric sequence whose first three terms are 36, 30, 25.	$r = \frac{30}{36} = \frac{5}{6}$ $a_1 = 36$ $a_n = 36\left(\frac{5}{6}\right)^{n-1}$
Try Them	
Determine if this sequence is geometric: *16, 9, 2, -5, ... , -7n+23, ...*	No (it is arithmetic, however)
Determine if the sequence is geometric, arithmetic, or neither: a. e^n b. $\frac{(-1)^n}{n}$ c. $3n + 3$	a. Geometric b. Neither c. Arithmetic

Finite Geometric Series

We can begin by practicing with the partial sums of geometric series, but next we will progress to using and understanding the infinite geometric series. It has some interesting usefulness.

✤ **finite geometric series**
The *n*th partial sum of a geometric series is a finite geometric series. An example might appear like this:

$$\sum_{n=1}^{7} 2^n$$

★ **For Fun**

Which partial sum is the above? Could you evaluate it term-by-term? Can you discover that the common ratio is 2? Could you compare the term-by-term result with the result obtained by trying the following formula from page 617 in your textbook?

Formula for the *n*th Partial Sum of a Geometric Series:

$$S_n = \frac{a_1(1-r^n)}{1-r} \quad \text{where } r \neq 1$$

Sum of a Finite Geometric Series: Test Prep

Find the sum of the first 5 terms of the geometric sequence $-4, 20, -100, \ldots, -4(-5)^{n-1}, \ldots$	$a_1 = -4$ $r = -5$ $n = 5$ $S_5 = \frac{-4\left[1-(-5)^5\right]}{1-(-5)} = \frac{-4(1+3125)}{6}$ $= \frac{-2(3126)}{3} = -2(1042)$ $= -2084$
Try Them	
Evaluate the finite geometric series: $\sum_{n=0}^{6} 5\left(\frac{2}{3}\right)^n$	$\frac{10295}{729}$
Find the sum of the finite geometric series: $\sum_{n=1}^{7} 2^n$	254

Infinite Geometric Series

Under certain circumstances the *infinite* sum of a geometric series is absolutely defined and easy to evaluate.

★ **For Fun**
Look up a proof for fun.

✤ **infinite series**
The upper limit of an infinite series is infinity: ∞.

✤ **infinite geometric series**
A geometric series that includes *all* its terms (whose upper limit is ∞) is an infinite geometric series.

The Sum of an Infinite Geometric Series

Given an infinite geometric series with terms a_n and where $|r| < 1$, then the sum S is $\dfrac{a_1}{1-r}$.

Sum of an Infinite Geometric Series: Test Prep

Evaluate $\sum_{n=1}^{\infty}\left(\frac{3}{8}\right)^n$	$a_1 = \frac{3}{8}$ $r = \frac{3}{8}$ $\sum_{n=1}^{\infty}\left(\frac{3}{8}\right)^n = \frac{\frac{3}{8}}{1-\frac{3}{8}} \cdot \frac{8}{8} = \frac{3}{8-3} = \frac{3}{5}$

Sum of an Infinite Geometric Series: Test Prep	
Try Them	
Evaluate $\sum_{n=1}^{\infty}(0.5)^n$	1
Write $0.\overline{63}$ as a ratio of two integers in simplest form.	$\frac{7}{11}$

Applications of Geometric Sequences and Series

There are some financial applications of these geometric sequences and series that we will explore briefly now.

- **future value**
  The future value of an investment, or the value after some *last* deposit, is the sum of the values of all the deposits.

- **annuities**
  Equal amounts deposited at equal intervals of time are called annuities.

- **ordinary annuity**
  Ordinary annuities entail deposits made at the end of investment periods -- for example, the 31st of December.

Future Value of an Ordinary Annuity

Let $r = \frac{i}{n}$ and $m = nt$, where i is the annual interest rate, n is the number of compounding periods per year, and t is the number of years. Then the future value A of an ordinary annuity after m compounding periods is given by

$$A = \frac{P\left[(1+r)^m - 1\right]}{r}$$ where P is the amount of each deposit

(page 621 of your textbook).

✤ **Gordon model of stock valuation**

This model of stock valuation is used to determine the value of a stock whose dividend is expected to increase by the same percentage each year.

✤ **multiplier effect**

The multiplier effect has to do with an idea that an initial amount of spending leads to an increase in consumer spending that ideally results in an increase in overall income that outweighs the initial expense. In other words, the increase is a multiple of the initial expense.

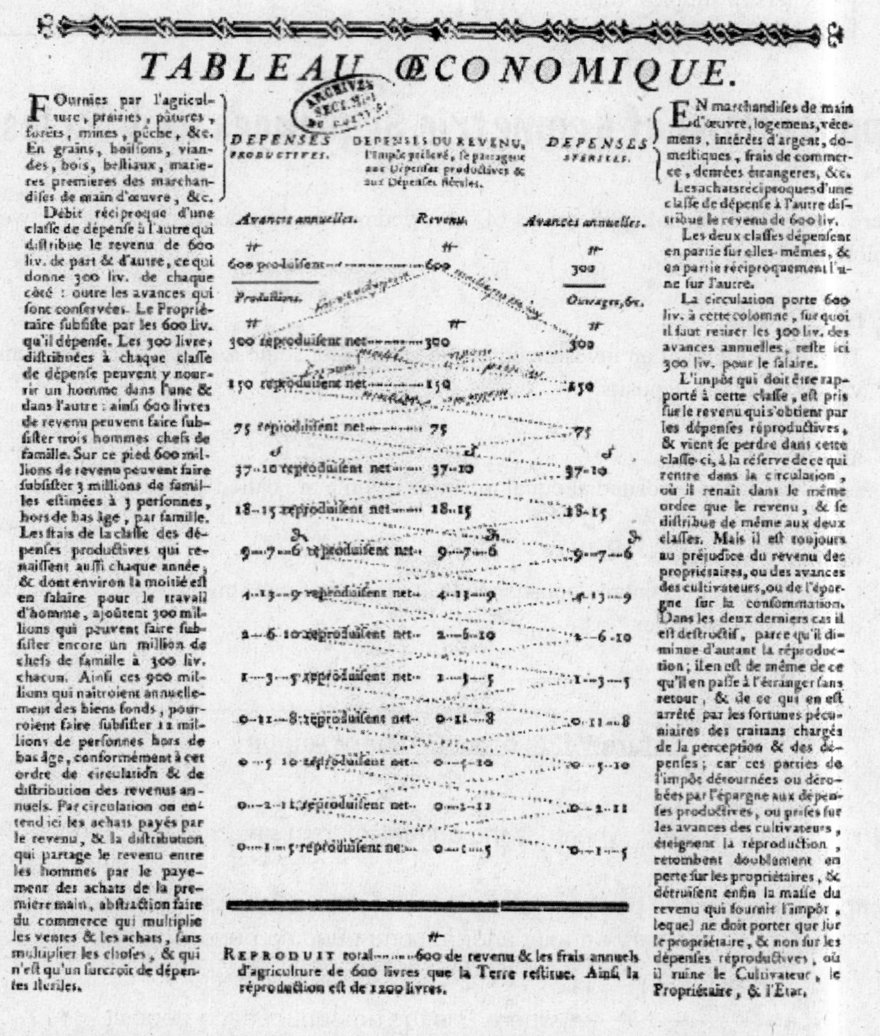

TABLEAU ŒCONOMIQUE.

Fournies par l'agriculture, prairies, pâtures, forêts, mines, pêche, &c. En grains, boissons, viandes, bois, bestiaux, matieres premieres des marchandises de main d'œuvre, &c.

Débit réciproque d'une classe de dépense à l'autre qui distribue le revenu de 600 liv. de part & d'autre, ce qui donne 300 liv. de chaque côté : outre les avances qui sont conservées. Le Propriétaire subsiste par les 600 liv. qu'il dépense. Les 300 livres distribuées à chaque classe de dépense peuvent y nourrir un homme dans l'une & dans l'autre : ainsi 600 livres de revenu peuvent faire subsister trois hommes chefs de famille. Sur ce pied 600 millions de revenu peuvent faire subsister 3 millions de familles estimées à 3 personnes, hors de bas âge, par famille. Les frais de la classe des dépenses productives qui renaissent aussi chaque année, & dont environ la moitié est en salaire pour le travail d'homme, ajoûtent 300 millions qui peuvent faire subsister encore un million de chefs de famille à 300 liv. chacun. Ainsi ces 900 millions qui naitroient annuellement des biens fonds, pourroient faire subsister 12 millions de personnes hors de bas âge, conformément à cet ordre de circulation & de distribution des revenus annuels. Par circulation on entend ici les achats payés par le revenu, & la distribution qui partage le revenu entre les hommes par le payement des achats de la premiere main, abstraction faite du commerce qui multiplie les ventes & les achats, sans multiplier les choses, & qui n'est qu'un surcroît de dépenses stériles.

DEPENSES PRODUCTIVES.	DEPENSES DU REVENU, l'Impôt prélevé, se partagent aux Dépenses productives & aux Dépenses stériles.	DEPENSES STERILES.
Avances annuelles.	Revenu.	Avances annuelles.
₶	₶	₶
600 produisent	600	300
Productions.		Ouvrages, &c.
₶	₶	₶
300 réproduisent net	300	300
150 réproduisent net	150	150
75 réproduisent net	75	75
37-10 réproduisent net	37-10	37-10
18-15 réproduisent net	18-15	18-15
9-7-6 réproduisent net	9-7-6	9-7-6
4-13-9 réproduisent net	4-13-9	4-13-9
2-6-10 réproduisent net	2-6-10	2-6-10
1-3-5 réproduisent net	1-3-5	1-3-5
0-11-8 réproduisent net	0-11-8	0-11-8
0-5-10 réproduisent net	0-5-10	0-5-10
0-2-11 réproduisent net	0-2-11	0-2-11
0-1-5 réproduisent net	0-1-5	0-1-5

REPRODUIT total ₶ 600 de revenu & les frais annuels d'agriculture de 600 livres que la Terre restitue. Ainsi la réproduction est de 1200 livres.

En marchandises de main d'œuvre, logemens, vêtemens, intérêts d'argent, domestiques, frais de commerce, denrées étrangeres, &c.

Les achats réciproques d'une classe de dépense à l'autre distribue le revenu de 600 liv.

Les deux classes dépensent en partie sur elles-mêmes, & en partie réciproquement l'une sur l'autre.

La circulation porte 600 liv. à cette colonne, sur quoi il faut retirer les 300 liv. des avances annuelles, reste ici 300 liv. pour le salaire.

L'impôt qui doit être rapporté à cette classe, est pris sur le revenu qui s'obtient par les dépenses réproductives, & vient se perdre dans cette classe-ci, à la réserve de ce qui rentre dans la circulation, où il renaît dans le même ordre que le revenu, & se distribue de même aux deux classes. Mais il est toujours au préjudice du revenu des propriétaires, ou des avances des cultivateurs, ou de l'épargne sur la consommation. Dans les deux derniers cas il est destructif, parce qu'il diminue d'autant la réproduction; il en est de même de ce qu'il en passe à l'étranger sans retour, & de ce qui en est arrêté par les fortunes pécuniaires des traitans chargés de la perception & des dépenses; car ces parties de l'impôt détournées ou dérobées par l'épargne aux dépenses productives, ou prises sur les avances des cultivateurs, éteignent la réproduction, retombent doublement en perte sur les propriétaires, & détruisent enfin la masse du revenu qui fournit l'impôt, lequel ne doit porter que sur le propriétaire, & non sur les dépenses réproductives, où il ruine le Cultivateur, le Propriétaire, & l'Etat.

This "Tableau", from about 1759, is considered one of the first examples of multiplier theory. [Francois Quesney, Tableau Économique *(2nd. Ed., 1759)]*

MID-CHAPTER QUIZ: TEXTBOOK PAGE 625.

8.4: Mathematical Induction

Mathematical induction -- not the same as deduction -- is based on this Induction Axiom from page 627 of your textbook. It is deceptively simple.

Induction Axiom

Suppose S is a set of positive integers with the following two properties:
1. 1 is an element of S.
2. If the positive integer k is in S, then $k + 1$ is in S.
Then S contains all positive integers.

Principle of Mathematical Induction

Expanding the Induction Axiom, so that it can be used to prove the general case from specific cases in mathematics, forms the Principle of Mathematical Induction.

Principle of Mathematical Induction

Let P_n be a statement about a positive integer n. If **1.** P_1 is true and **2.** the truth of P_k implies the truth of P_{k+1}, then P_n is true for all positive integers.

It has been suggested that a mathematical proof by induction is like knocking over dominoes -- if knocking over any domino will cause the next to fall and you knock over the first one, then all the dominoes will fall.

✣ **induction hypothesis**
So there are basically three steps to an induction proof. First show the premise holds for the number 1. Then assume it holds true for k -- this step is the induction hypothesis. Third, show that given truth for k that the premise is still true for $k + 1$.

Principle of Mathematical Induction: Test Prep

Prove that $1^2+3^2+5^2+\cdots+(2n-1)^2 = \dfrac{n(2n+1)(2n-1)}{3}$, by mathematical induction.	a. Let $n = 1$: $1^2 = 1$ $\dfrac{1(2+1)(2-1)}{3} = 1$ (satisfied) b. Assume $1^2+3^2+5^2+\cdots+(2k-1)^2 = \dfrac{k(2k+1)(2k-1)}{3}$ c. Show: $1^2+3^2+5^2+\cdots+(2k-1)^2+(2k+1)^2 = \dfrac{(k+1)(2(k+1)+1)(2(k+1)-1)}{3} = \dfrac{4k^3+12k^2+11k+3}{3}$ Adding the new term onto both sides of the assumption from step two: $1^2+3^2+5^2+\cdots+(2k-1)^2+(2k+1)^2 = \dfrac{k(2k+1)(2k-1)}{3}+(2k+1)^2 = \dfrac{4k^3-k}{3}+\dfrac{3(4k^2+4k+1)}{3} = \dfrac{4k^3+12k^2+11k+3}{3}$ (satisfied)

Principle of Mathematical Induction: Test Prep

Try Them	
Prove that $\left(1-\frac{1}{2}\right)\left(1-\frac{2}{3}\right)\left(1-\frac{3}{4}\right)\times\cdots\times\left(1-\frac{n}{n+1}\right)$ $=\frac{1}{(n+1)!}$ for all positive integers *n*.	a. Let $n = 1$: $\left(1-\frac{1}{2}\right)=\frac{1}{(1+1)!}=\frac{1}{2}$ (satisfied) b. Assume: $\left(1-\frac{1}{2}\right)\left(1-\frac{2}{3}\right)\left(1-\frac{3}{4}\right)\times\cdots\times\left(1-\frac{k}{k+1}\right)$ $=\frac{1}{(k+1)!}$ c. Show $\left(1-\frac{1}{2}\right)\left(1-\frac{2}{3}\right)\left(1-\frac{3}{4}\right)\times\cdots\times\left(1-\frac{k}{k+1}\right)\left(1-\frac{k+1}{k+2}\right)$ $=\frac{1}{(k+2)!}$ Substituting from step two: $\left(\frac{1}{(k+1)!}\right)\left(1-\frac{k+1}{k+2}\right)$ $=\left(\frac{1}{(k+1)!}\right)\left(\frac{k+2-k-1}{k+2}\right)$ $=\left(\frac{1}{(k+1)!}\right)\left(\frac{1}{k+2}\right)=\frac{1}{(k+2)!}$ (satisfied)
Prove that $2n-1<2^n$ for all positive integers *n*.	a. Let $n = 1$: $2-1<2$ (satisfied) b. Assume: $2k-1<2^k$ c. Show $2(k+1)-1<2^{k+1}$. But since $2(2k-1)=4k-2>2k+1$, then $2^{k+1}>2k+1$ (satisfied)

Extended Principle of Mathematical Induction

In the case where the beginning index for the desired induction is greater than 1, we have the Extended Principle of Mathematical Induction (page 630 in your textbook). Look for similarities between the structure of this argument and the structure of the previous induction principle.

The Extended Principle of Mathematical Induction

Let P_n be statement about a positive integer *n*. If **1.** P_j is true for some positive integer *j*, and **2.** for $k \ge j$ the truth of P_k implies the truth of P_{k+1}, then P_n is true for all positive integers $n \ge j$

Extended Principle of Mathematical Induction: Test Prep

Prove that for each natural number $n \ge 6$, $(n+1)^2 \le 2^n$	Let m be a natural number such that $(m+1)^2 \le 2^m$. The number 6 is a possible value of m because $(6+1)^2 = 49 \le 64 = 2^6$. Suppose that $n \ge 6$ and $(n+1)^2 \le 2^n$. Then $\frac{(n+1)^2}{2^n} \le 1$. Compute $\frac{(n+2)^2}{2^{n+1}} = \frac{1}{2}\left(\frac{n+2}{n+1}\right)^2 \frac{(n+1)^2}{2^n} \le \frac{1}{2}\left(1+\frac{1}{n+1}\right)^2$. Since $n \ge 6$, $1+\frac{1}{n+1} \le 1.4 \le \sqrt{2}$. Therefore, $\frac{(n+2)^2}{2^{n+1}} \le 1$, and we have proved $m = n+1$ is valid. By the Extended Principle of Mathematical Induction we have shown $(n+1)^2 \le 2^n$ for $n \ge 6$.

Extended Principle of Mathematical Induction: Test Prep

Try Them	
Prove that for natural numbers $n \geq 2$ we have $2^{n+1} \leq 3^n$.	For $n=2$ we have $2^{2+1} = 8 \leq 9 = 3^2$. Now assume for $k \geq 2$ that $2^{k+1} \leq 3^k$. For $k+1$, we have $2^{k+2} = 2 \cdot 2^{k+1} \leq 2 \cdot 3^k \leq 3 \cdot 3^k = 3^{k+1}$. So the assumption holds for $k+1$, and by the Extended Principle of Mathematical Induction we've shown that $2^{n+1} \leq 3^n$ for $n \geq 2$.
Prove that $n^2 \geq 3n$ for $n \geq 3$.	For $n = 3$ we have $3^2 = 9 = 3 \cdot 3$. Assume that for $n \geq 3$, $n^2 \geq 3n$. For $n+1$: $(n+1)^2 = n^2 + 2n + 1 \geq 3n + 2n + 1 \geq 3n + 3$ for $n \geq 3$. And so $(n+1)^2 \geq 3n + 3 = 3(n+1)$ or $(n+1)^2 \geq 3(n+1)$ for $n \geq 3$ by the Extended Principle of Mathematical Induction.

★ **For Fun**

The following is a statement of the Induction Axiom in logical symbols. Do you know any of these symbols? Look them up. At the very least, you have to admit that it's a compact way to state the Axiom (even if incomprehensible):

$$\forall X(1 \in X \wedge \forall x(x \in X \Rightarrow x + 1 \in X) \Rightarrow \forall x(x \in X))$$

8.5: Binomial Theorem

Not only will this section be about learning a pattern to simplify a big multiplication problem, but the notation carries over into calculus, statistics, and probability. I actually like to think of this section as "fun with numbers"!

Binomial Theorem

Actually carrying out the multiplication of a binomial like $(x+3)^7$ can be a pretty daunting proposition. Do you want to perform this operation:

$$(x+3)(x+3)(x+3)(x+3)(x+3)(x+3)(x+3)$$

And be assured that you haven't made an error?

There are some very regular patterns in this enterprise that we can learn and use.

- **expanding the binomial**
  Expanding the binomial refers to any process of carrying out the multiplication implied by the exponent applied to a binomial: $(a+b)^n$

- **binomial coefficient**
  The binomial coefficient is symbolized like this:

  $$\binom{n}{k}$$

  This is not a fraction -- although it looks somewhat like a fraction. What this symbol means is

  $$\frac{n!}{k!(n-k)!}$$

The Binomial Coefficient

$$\binom{n}{k} = \frac{n!}{k!(n-k)!}$$

★ **For Fun**
Are factorials defined for negative integers? Can you tell which *has* to be bigger, n or k, from the denominator of the definition of the binomial coefficient?

Binomial Coefficient: Test Prep

Evaluate $\binom{10}{3}$	$\binom{10}{3} = \frac{10!}{3!(10-3)!} = \frac{10!}{3!7!}$ $= \frac{10 \cdot 9 \cdot 8 \cdot 7 \cdot 6 \cdot 5 \cdot 4 \cdot 3 \cdot 2 \cdot 1}{3 \cdot 2 \cdot 1 \cdot 7 \cdot 6 \cdot 5 \cdot 4 \cdot 3 \cdot 2 \cdot 1}$ $= \frac{10 \cdot 9 \cdot 8}{3 \cdot 2 \cdot 1} = 10 \cdot 3 \cdot 4$ $= 120$
Try Them	
Evaluate $\binom{7}{0}$	1
Evaluate $\binom{15}{13}$	105

The following formula is one place the binomial coefficient is used. This is how you can expand a binomial.

The Binomial Theorem for Positive Integers:

If n is a positive integer, then

$$(a+b)^n = \sum_{i=0}^{n} \binom{n}{i} a^{n-i} b^i$$

$$= \binom{n}{0} a^n + \binom{n}{1} a^{n-1} b + \binom{n}{2} a^{n-2} b^2 + \cdots + \binom{n}{n} b^n$$

The above formula has it all for expanding binomials. Every element of the expansion is described in the formula, but the formula is a little complicated to most newcomers. I would suggest some serious practice. It gets easier pretty quickly. And then you can amaze your friends with just how fast you can multiply something like $\left(x-\sqrt{2}\right)^6$.

Binomial Theorem for Positive Integers: Test Prep	
Expand $\left(x-\sqrt{2}\right)^6$ using the binomial theorem.	$\left(x-\sqrt{2}\right)^6 = \sum_{i=0}^{6}\binom{6}{i}x^{6-i}\left(-\sqrt{2}\right)^i$ $= \binom{6}{0}x^6 + \binom{6}{1}x^5\left(-\sqrt{2}\right) + \binom{6}{2}x^4\left(-\sqrt{2}\right)^2$ $+ \binom{6}{3}x^3\left(-\sqrt{2}\right)^3 + \binom{6}{4}x^2\left(-\sqrt{2}\right)^4$ $+ \binom{6}{5}x\left(-\sqrt{2}\right)^5 + \binom{6}{6}\left(-\sqrt{2}\right)^6$ $= x^6 - 6x^5\sqrt{2} + 30x^4 - 40x^3\sqrt{2} + 60x^2 - 24x\sqrt{2} + 8$
Try Them	
Expand $(a-2)^4$ using the binomial theorem.	$a^4 - 8a^3 + 24a^2 - 32a + 16$
Expand $(2x-1)^5$ using the binomial theorem.	$32x^5 - 80x^4 + 80x^3 - 40x^2 + 10x - 1$

ith Term of a Binomial Expansion

You can find a *particular* term in a binomial expansion using the following formula. All this takes is careful attention to indices and exponential evaluation.

The *i*th term of the binomial expansion of $(a+b)^n$ is given by

$$\binom{n}{i-1}a^{n-i+1}b^{i-1}$$

ith Term of a Binomial Expansion: Test Prep

Find the 3rd term of the expansion of $(x+8)^7$.	$a=x$ $b=8$ $n=7$ $i=3$ $\binom{7}{2}x^5\cdot 8^2=\frac{7!}{2!5!}64x^5=21\cdot 64x^5=1344x^5$
Try Them	
Find the sixth term of $(5x^2-2y^3)^8$	$-224000x^6y^{15}$
Find the 8th term of $\left(\frac{3}{x}-\frac{x}{3}\right)^{13}$	$\frac{5148}{x}$

Pascal's Triangle

Pascal's Triangle was known well before Blaise Pascal documented it. This Triangle has a lot of neat features and patterns. We'll use Pascal's Triangle as it pertains to the binomial expansion, however.

✤ **Pascal's Triangle**

Named for Blaise Pascal, Pascal's Triangle is a way to arrange the binomial coefficients. The Triangle also makes it possible to derive the smaller binomial coefficients quickly because each number in the triangle is the sum of the two diagonally above it.

If you consider the coefficient $\binom{n}{k}$ and label the rows as n's and the diagonals as k's -- in both cases you must start labeling with zero -- then you can discover the correspondence between these numbers and the binomial coefficients.

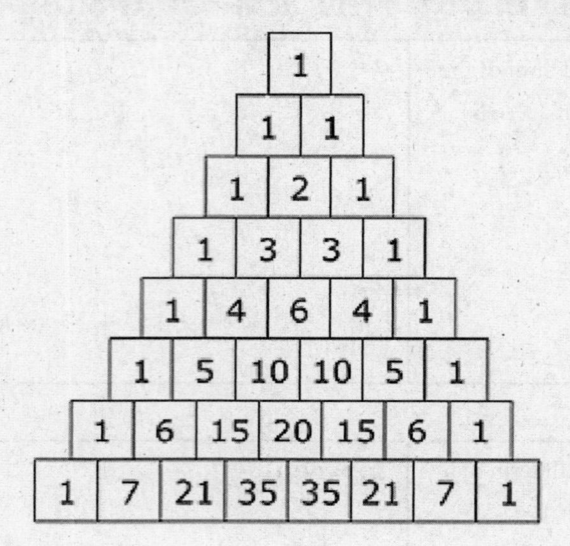

★ **For Fun**

Can you find all the values of the next row or two of the triangle above?

Here are a few of things you can find in Pascal's Triangle:

- Hockey stick pattern
- Prime numbers
- Fibonacci sequence
- Magic 11's
- Powers of 2
- Triangular numbers
- Square numbers

It's amazing!

8.6: Permutations and Combinations

Counting, permutations, and combinations are the beginning of probability theory. But to a large extent you may find that it is still "fun with numbers."

Fundamental Counting Principle

The Fundamental Counting Principle is a way to find out how many ways events -- each with many ways of happening -- can all combine to a larger list of choices. For instance, if you have 5 shirts, 3 pants, and 2 pair of shoes, you have 5 x 3 x 2 possible outfits:

Fundamental Counting Principle

Given a sequence of *n* events $E_1, E_2, E_3, E_4, E_5, \cdots, E_n$ where each event can occur in several ways w_n with *n* showing the correspondence to the event in the sequence, then the number of ways that all events can occur is given by $w_1 \cdot w_2 \cdot w_3 \cdot \cdots \cdot w_n$.

Fundamental Counting Principle: Test Prep	
If ID numbers are generated by choosing 4 capital letters and 2 digits, how many different ID numbers are possible?	$26 \cdot 26 \cdot 26 \cdot 26 \cdot 10 \cdot 10$ $= 45697600$

Fundamental Counting Principle: Test Prep

Try Them	
A home improvement store is offering a package on Christmas trees. The trees come in three colors, decoration packs in 5 colors, skirts in 2 patterns, and stands in 3 models. How many possible combinations are available in a package deal that includes one of each of the above?	90
A computer monitor produces color by blending colors on a palette. If a monitor has four palettes and each palette has four colors, how many blended colors can be formed? (Every palette has to be used in each blend.)	256

Permutations

The Fundamental Counting Principle can help us discover how many ways a distinct set of elements can be arranged in a definite order.

✤ **permutation**
A permutation is an arrangement of distinct objects in a definite order.

The number of permutations of n distinct objects taken r at a time is given by

$$P(n,r) = \frac{n!}{(n-r)!}$$

★ **For Fun**
If the objects were *not* distinct, do you think there would be more ways to combine them or fewer?

Permutations: Test Prep	
Evaluate $P(10,3)$.	$P(10,3)=\frac{10!}{(10-3)!}=\frac{10!}{7!}=10\cdot 9\cdot 8$ $=720$
Try Them	
How many ways can you choose 4 people to sit on a committee from a group of 12 people?	11880
How many ways can first, second, and third prizes be awarded to 100 entrants in a raffle?	970200

Combinations

Combinations have to do with selecting objects that are not distinct. There is a difference between choosing a president, a vice president, and a secretary and simply choosing three people.

✤ **combination**
A combination is an arrangement of objects where the order of selection is not meaningful.

The number of combinations of n objects taken r at a time is given by

$$C(n,r)=\frac{n!}{r!(n-r)!}$$

★ **For Fun**
Does the combinations formula remind you of another formula we've seen recently?

Combinations: Test Prep	
Evaluate $C(27{,}10)$.	$C(27{,}10) = \dfrac{27!}{10!(27-10)!} = \dfrac{27!}{10!17!}$ $= 27 \cdot 13 \cdot 5 \cdot 23 \cdot 11 \cdot 19$ $= 8{,}436{,}285$
Try Them	
Evaluate $C(20{,}17)$.	1140
How many 13-card hands can be chosen from a standard 52-card deck?	635,013,559,600

✓ **Questions to Ask Your Teacher**
It's difficult to get the hang of when to use the counting principle, permutations, or combinations. There are guidelines on page 641 of your textbook, and your instructor may have more examples to help.

Please make sure you never cancel factorials incorrectly:

$$\frac{7!}{5!} \neq 2!$$

Where in the order of operations does the factorial operation fall?

8.7: Introduction to Probability

The study of probabilities came into being the first time someone decided to *insure* something -- a ship, a life, a cargo. The pastime of gambling has also informed a lot of the interest and study of the science of predicting outcomes: Probability.

- **probability**
  Officially, probability is the mathematical study of random patterns. The word is derived from a Latin word that means "credible". "Probability" has the feel of "likelihood". I think we have a different sense of what we mean when we use "probability" in an English sentence. In math its meaning tends to surprise. Try to have an open mind about what probability means and signifies in math.

Sample Spaces and Events

The basic ideas of probability are a beginning to our study. What are we trying to predict with our *probabilities*? What are we looking at? What sort of systems does probability study pertain to? How is the subject organized?

- **experiment**
  An occurrence with an observable outcome is an experiment.

- **sample space**
  A sample space is the set of *all* possible outcomes of an experiment. For instance, you might write the sample space for flipping a coin like so: $\{T, H\}$.

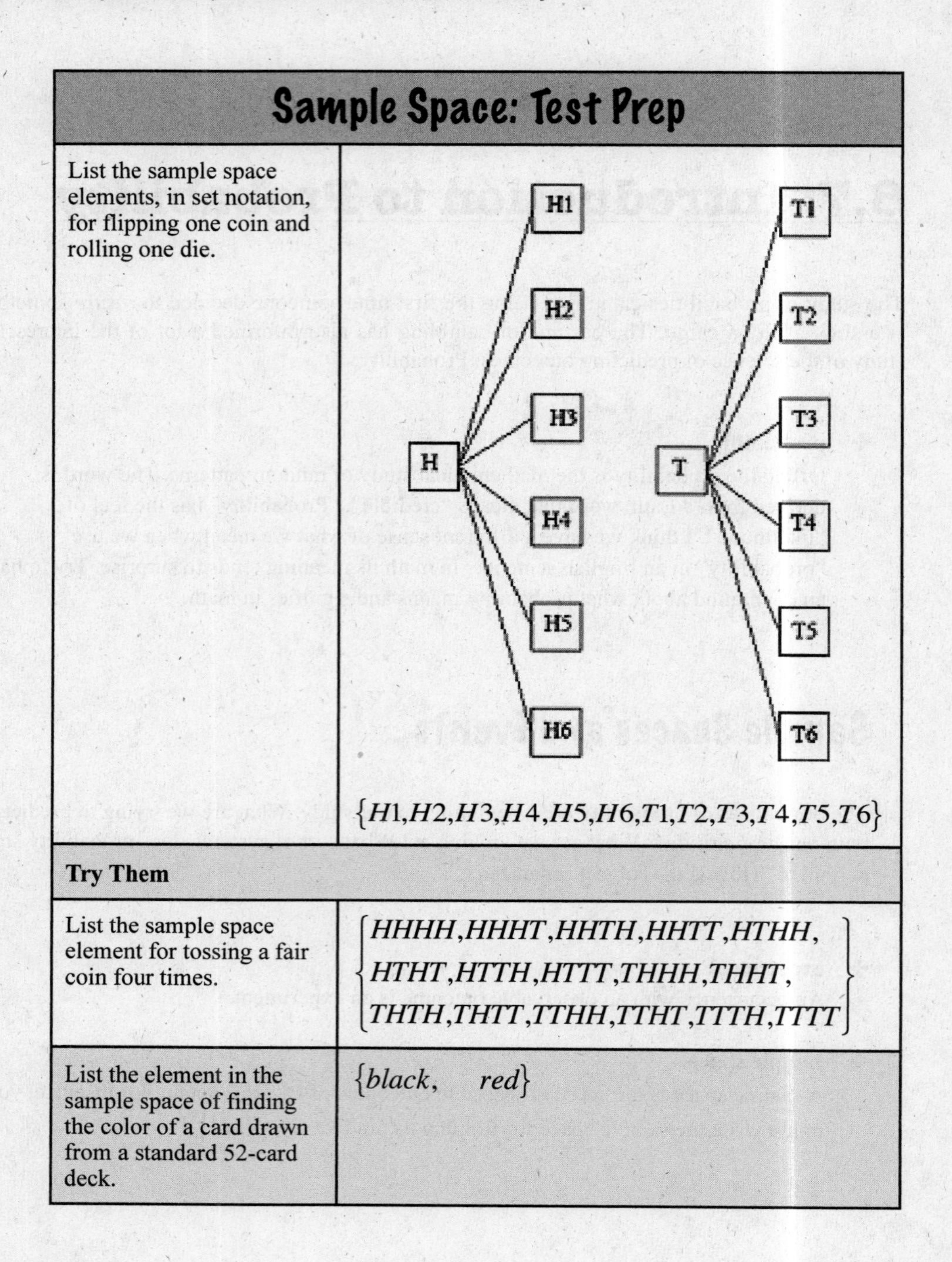

Sample Space: Test Prep	
List the sample space elements, in set notation, for flipping one coin and rolling one die.	$\{H1, H2, H3, H4, H5, H6, T1, T2, T3, T4, T5, T6\}$
Try Them	
List the sample space element for tossing a fair coin four times.	$\left\{\begin{matrix} HHHH, HHHT, HHTH, HHTT, HTHH, \\ HTHT, HTTH, HTTT, THHH, THHT, \\ THTH, THTT, TTHH, TTHT, TTTH, TTTT \end{matrix}\right\}$
List the element in the sample space of finding the color of a card drawn from a standard 52-card deck.	$\{black, \quad red\}$

✤ **event**

An event, frequently symbolized by E, is a specific subset of the sample space. One might ask what the probability of an event is.

The sample space for the experiment of tossing two coins might look like this: $\{TT, TH, HT, HH\}$. An event might be defined as "tossing two tails" which corresponds to the subset: $\{TT\}$. But an event might also be defined as "tossing at least one head, and that subset would be: $\{HT, TH, HH\}$.

Event: Test Prep	
You pull two marbles from an urn holding blue, green, and red marbles. Express the event where neither marbles are red.	These are the ways you can pull two marbles from the urn where neither marble is red: blue-green blue-blue green-blue green-green
Try Them	
A 2-digit number is formed by choosing from the digits *1, 2, 3, 4, 5,* and *6* without replacing the first choice and where both digits are odd. Describe the event space.	*13, 15, 31, 35, 51, 53*
Express the event space of three or more tails from flipping four fair coins.	$\{HTTT, TTTH, THTT, TTHT, TTTT\}$

Probability of an Event

Calculating the probability of an event can be just as simple as counting up the elements in an event subset and dividing it by the number of elements in the sample space.

Probability of an Event

Let $n(S)$ and $n(E)$ represent the number of elements in a sample space S and in an event E, respectively. The probability of an event E is given by

$$P(E) = \frac{n(E)}{n(S)}$$

Because of the relationship between the sizes of a set and its possible subsets, possible values for probabilities all fall between 0 and 1.

Choose the valid probabilities:

0.03	1.3%
21	$2/5$
21%	.123
$5/3$	1

★ **For Fun**
Can you name an event whose probability is 1, or 100%? Can you name an event whose probability is 0?

Probability of an Event: Test Prep	
What is the probability of getting exactly two heads when flipping four fair coins?	Sample space S is $\left\{\begin{matrix} HHHH, HHHT, HHTH, HHTT, HTHH, \\ HTHT, HTTH, HTTT, THHH, THHT, \\ THTH, THTT, TTHH, TTHT, TTTH, TTTT \end{matrix}\right\}$. Event space E is $\{HHTT, HTHT, HTTH, THHT, THTH, TTHH\}$. $P(2H) = \frac{n(E)}{n(S)} = \frac{6}{16} = \frac{3}{8}$

Probability of an Event: Test Prep	
Try Them	
What is the probability of tossing two fair dice whose sum is greater than three?	$\frac{11}{12}$
What is the probability of drawing any king from a standard 52-card deck?	$\frac{1}{13}$

✤ **mutually exclusive**

Two (or more) events that cannot occur at the same time are mutually exclusive. (Does that give you an idea of how to describe an event of zero probability?) For instance, if you have red marbles and yellow marbles, and you pull one out of a bag, it is impossible for the marble you have drawn to be both red *and* yellow -- "red" and "yellow" are mutually exclusive events -- there is no overlap in their Venn diagrams (remember sets?).

Addition Rules for Probabilities:

If E_1 and E_2 are two events, then the probability of both happening is given by

$$P(E_1 \cup E_2) = P(E_1) + P(E_2) - P(E_1 \cap E_2)$$

If E_1 and E_2 are mutually exclusive then their intersection ($E_1 \cap E_2$) is empty and the probability of two mutually exclusive events happening is given by

$$P(E_1 \cup E_2) = P(E_1) + P(E_2)$$

Addition Rules for Probabilities: Test Prep	
Toss a coin three times. Let E_1 be the event that the coin lands on heads two times and E_2 be the event that the third toss is a head. Find the probability of E_1 or E_2.	The events are not mutually exclusive. $S=\{HHH,HHT,HTH,THH,HTT,THT,TTH,TTT\}$ $P(E_1)=\frac{3}{8}$ $P(E_2)=\frac{4}{8}=\frac{1}{4}$ $P(E_1\cap E_2)=\frac{1}{8}$ $P(E_1\cup E_2)=\frac{3}{8}+\frac{1}{2}-\frac{1}{8}=\frac{3}{4}$
Try Them	
Toss a coin three times. Let E_1 be the event that the coin is a head all 3 times; let E_2 be the event that the coin is a tail all 3 times. Find the probability of E_1 or E_2.	$\frac{1}{4}$
Toss a pair of dice. Find the probability that one of the die is a two or the sum is six.	$\frac{7}{18}$

Independent Events

The question of whether events are independent or not is very important (and subtle) in the calculation of probabilities. If the outcome of one experiment effects the next experiment, wouldn't that possibly greatly influence the probability of an event in the second experiment?

✤ **independent**
Independent events are events whose outcomes do *not* have effects on each other. Rolling a fair dice twice is an example of independent events -- the result from the first roll has no effect on the second roll. Pulling a marble out of a box then pulling another marble out without replacing the first one is an example of events that are not independent -- the outcome of the first draw affects the probabilities of the second.

Probability Rule for Independent Events:

If E_1 and E_2 are independent events, then the probability that both E_1 and E_2 happen is give by the product of their individual probabilities:

$$P(E_1) \cdot P(E_2)$$

Independent Events: Test Prep

The probability Joe hits a target at the shooting range is $\frac{1}{4}$; the probability Ann hits is $\frac{2}{3}$. Find the probability that Joe and Ann hit the target.	Assuming the events are independent, the probability is $\frac{1}{4} \cdot \frac{2}{3} = \frac{1}{6}$
Try Them	
A card is drawn from a standard 52-card deck then a coin is tossed. What is the probability the card was an ace and the coin was a head?	$\frac{1}{26}$

Independent Events: Test Prep	
Let E_1 and E_2 be independent events. Let $P(E_1)=\frac{1}{4}$ and $P(E_2)=\frac{1}{3}$. Find the probability either one happens.	$\frac{1}{2}$

Binomial Probabilities

Binomial experiments are experiments with two and only two possible outcomes. (Most experiments can be *re*-phrased as binomial experiments, but then there is no partial credit!) This binomial scenario is well studied, modeled by a formula, and involves many calculations similar to those we learned in the previous section on The Binomial Theorem.

Binomial Probability Formula:

Given an experiment that consists of *n* independent trials where the probability of success on a single trial is *p* and the probability of failure is $q=1-p$, the probability of *k* successes out of *n* trials is

$$P(k)=\binom{n}{k}p^k q^{n-k}$$

There are quite a few "little pieces" in this formula, but once you get the hang of defining your experiment, it becomes pretty easy to use. Oh, yes, you need to be able to enter all these parameters into your calculator correctly and interpret the result the calculator gives you, too!

Binomial Probability Formula: Test Prep	
A true/false test consists of four questions. Given random guessing, what is the probability of getting exactly 2 questions correct?	$n = 4$ $p = q = \frac{1}{2}$ $k = 2$ $P(k = 2) = \binom{4}{2}\left(\frac{1}{2}\right)^2\left(\frac{1}{2}\right)^2 = 4 \cdot 3 \cdot \frac{1}{4} \cdot \frac{1}{4}$ $= \frac{3}{4}$
Try Them	
Calculate a binomial probability given $n = 20$ $p = 0.7.$ $k = 15$	approximately *0.1787*
The probability Duane scores a 3-point basketball shot is 0.45. He shoots from the 3-point zone 8 times. What is the probability he scores exactly one time?	approximately *0.0055*

★ **For Fun**

How do you think that last problem would change if you wanted to know the probability that Duane would score three or fewer 3-point goals?

Final Fun: Matching

FIND THE BEST MATCH. GOOD LUCK.

FIBONACCI SEQUENCE

$\sum_{i=1}^{\infty}(i-1)^2$

PASCAL'S TRIANGLE

$\frac{X+Y}{2}$

BINOMIAL COEFFICIENT

$\binom{6}{0}x^6+\binom{6}{1}x^5y+\binom{6}{2}x^4y^2+\cdots+\binom{6}{6}b^6$

PERMUTATION

PROBABILITY OF FAILURE

$\binom{7}{3}$

BINOMIAL EXPANSION

ARITHMETIC MEAN

q

$\frac{10!}{(10-7)!}$

1, 1, 2, 3, 5, 8, ...

INFINITE SERIES

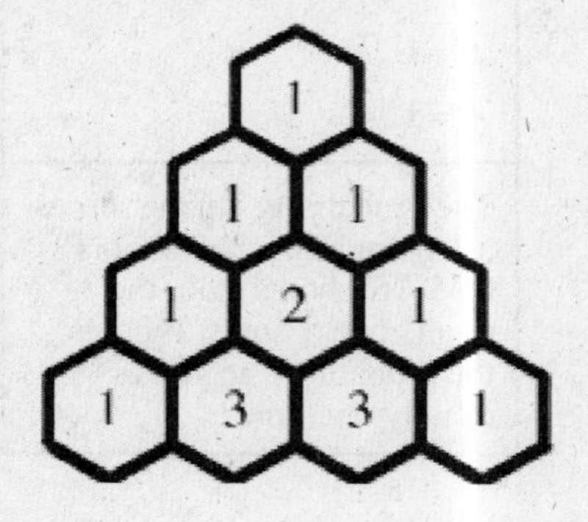

NOW TRY THESE:
REVIEW EXERCISES: TEXTBOOK PAGE 656,
PRACTICE TEST: TEXTBOOK PAGE 659.
STAY UP ON EVERYTHING IN THIS BOOK -- DO THE CUMULATIVE REVIEW EXERCISES: TEXTBOOK PAGE 660.

Student Solutions Manual

Chapter P: Preliminary Concepts

Section P.1 Exercises

1. Classify each number.

$-\frac{1}{5}$: rational, real; 0: integer, rational, real;

–44: integer, rational, real; π: irrational, real;

3.14: rational, real;

5.05005000500005…: irrational, real;

$\sqrt{81} = 9$: integer, rational, prime, real;

53: integer, rational, prime, real

3. List the four smallest elements of the set.

Let x = 1, 2, 3, 4.

Then $\{2x \mid x \text{ is a positive integer}\} = \{2, 4, 6, 8\}$

5. List the four smallest elements of the set.

Let x = 1, 2, 3, 4. (Recall 0 is not a natural number.)

Then $\{y \mid y = 2x + 1, x \text{ is a natural number}\} = \{3,5,7,9\}$

7. List the four smallest elements of the set.

Let x = 0, 1, 2, 3.

(We could have used x = –3, –2, –1, 0.)

Then $\{z \mid z = |x|, x \text{ is an integer}\} = \{0, 1, 2, 3\}$

9. Perform the operation.

$A \cup B = \{-3, -2, -1, 0, 1, 2, 3, 4, 6\}$

11. Perform the operation.

$A \cap C = \{0, 1, 2, 3\}$

13. Perform the operation.

$B \cap D = \varnothing$

15. Perform the operation.

$(B \cup C) = \{-2, 0, 1, 2, 3, 4, 5, 6\}$

$D \cap (B \cup C) = \{1, 3\}$

17. Perform the operation.

$(B \cup C) \cap (B \cup D)$

$= \{-2, 0, 1, 2, 3, 4, 5, 6\} \cap \{-3, -2, -1, 0, 1, 2, 3, 4, 6\}$

$= \{-2, 0, 1, 2, 3, 4, 6\}$

19. Perform the operation.

For any set A, $A \cup A = A$.

21. Perform the operation.

For any set A, $A \cap \varnothing = \varnothing$.

23. Perform the operation.

If A and B are two sets and a $A \cup B = A$, then all elements of B are contained in A. So B is a subset of A.

25. Graph the set and write in set builder notation.

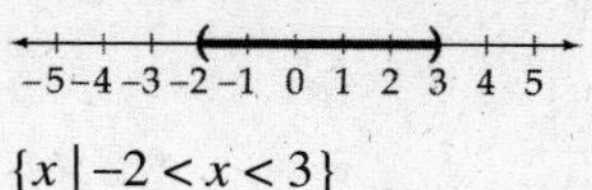

$\{x \mid -2 < x < 3\}$

27. Graph the set and write in set builder notation.

–5 –4 –3 –2 –1 0 1 2 3 4 5

$\{x \mid -5 \le x \le -1\}$

29. Graph the set and write in set builder notation.

–5 –4 –3 –2 –1 0 1 2 3 4 5

$\{x \mid x \ge 2\}$

31. Graph the set and write in interval notation.

–5 –4 –3 –2 –1 0 1 2 3 4 5

(3, 5)

33. Graph the set and write in interval notation.

–5 –4 –3 –2 –1 0 1 2 3 4 5

$[-2, \infty)$

35. Graph the set and write in interval notation.

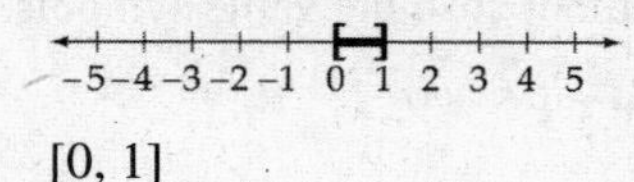

[0, 1]

37. Graph the set.

$(-\infty, 0) \cup [2, 4]$

–5 –4 –3 –2 –1 0 1 2 3 4 5

39. Graph the set.

$(-4, 0) \cap [-2, 5] = [-2, 0)$

–5 –4 –3 –2 –1 0 1 2 3 4 5

41. Graph the set.

$(1, \infty) \cup (-2, \infty) = (-2, \infty)$

–5 –4 –3 –2 –1 0 1 2 3 4 5

43. Graph the set.

$(1, \infty) \cap (-2, \infty) = (1, \infty)$

–5 –4 –3 –2 –1 0 1 2 3 4 5

45. Graph the set.

$[-2, 4] \cap [4, 5]$

–5 –4 –3 –2 –1 0 1 2 3 4 5

47. Graph the set.

$(-2, 4) \cap (4, 5) = \varnothing$

–5 –4 –3 –2 –1 0 1 2 3 4 5

49. Graph the set.

$\{x | x < -3\} \cup \{x | 1 < x < 2\}$

–5 –4 –3 –2 –1 0 1 2 3 4 5

51. Graph the set.

$\{x | x < -3\} \cup \{x | x < 2\} = \{x | x < 2\}$

–5 –4 –3 –2 –1 0 1 2 3 4 5

53. Write the expression without absolute value symbols.

$-|-5| = -5$

55. Write the expression without absolute value symbols.

$|3| \cdot |-4| = 3(4) = 12$

57. Write the expression without absolute value symbols.

$|\pi^2 + 10| = \pi^2 + 10$

59. Write the expression without absolute value symbols.

$|x-4| + |x+5| = 4 - x + x + 5 = 9$

61. Write the expression without absolute value symbols.

$$\begin{aligned} |2x| - |x-1| &= 2x - (1-x) \\ &= 2x - 1 + x \\ &= 3x - 1 \end{aligned}$$

63. Write in absolute value notation.

$|m-n|$

65. Write in absolute value notation.

$|x-3|$

67. Write in absolute value notation.

$$\begin{aligned} |x-(-2)| &= 4 \\ |x+2| &= 4 \end{aligned}$$

69. Write in absolute value notation.

$|a-4| < 5$

71. Write in absolute value notation.

$|x+2| > 4$

73. Write in absolute value notation.

$0 < |x-4| < 1$

75. Evaluate the expression.

$-5^3(-4)^2 = -5 \cdot 5 \cdot 5(-4)(-4) = -125(16) = -2000$

77. Evaluate the expression.

$$\begin{aligned} 4 + (3-8)^2 &= 4 + (-5)^2 \\ &= 4 + (-5)(-5) \\ &= 4 + 25 \\ &= 29 \end{aligned}$$

79. Evaluate the expression.

$$\begin{aligned} 28 \div (-7+5)^2 &= 28 \div (-2)^2 \\ &= 28 \div (-2)(-2) \\ &= 28 \div 4 \\ &= 7 \end{aligned}$$

81. Evaluate the expression.

$$\begin{aligned} &7 + 2[3(-2)^3 - 4^2 \div 8] \\ &= 7 + 2[3(-2)(-2)(-2) - (4 \cdot 4) \div 8] \\ &= 7 + 2[-24 - 16 \div 8] \\ &= 7 + 2[-24 - 2] \\ &= 7 + 2[-26] \\ &= 7 + -52 \\ &= -45 \end{aligned}$$

83. Evaluate.

$-(-2)^3 = -(-8) = 8$

85. Evaluate.

$2(3)(-2)(-1) = 12$

87. Evaluate.

$-2(3)^2(-2)^2 = -2(9)(4) = -72$

89. Evaluate.

$$\begin{aligned} 3(-2) - (-1)[3-(-2)]^2 &= 3(-2) - (-1)[3+2]^2 \\ &= (3)(-2) - (-1)[5]^2 \\ &= (3)(-2) - (-1)(25) \\ &= -6 + 25 = 19 \end{aligned}$$

91. Evaluate.

$$\frac{3^2+(-2)^2}{3+(-2)}=\frac{9+4}{1}=\frac{13}{1}=13$$

93. Evaluate.

$$\frac{3(-2)}{3}-\frac{2(-1)}{-2}=\frac{-6}{3}-\frac{-2}{-2}=-2-1=-3$$

95. State the property used.

$(ab^2)c=a(b^2c)$

Associative property of multiplication

97. State the property used.

$4(2a-b)=8a-4b$

Distributive property

99. State the property used.

$(3x)y=y(3x)$

Commutative property of multiplication

101. State the property used.

$1\cdot(4x)=4x$

Identity property of multiplication

103. State the property used.

$x^2+1=x^2+1$

Reflexive property of equality

105. State the property used.

If $2x+1=y$ and $3x-2=y$, then $2x+1=3x-2$

Transitive property of equality

107. State the property used.

$4\cdot\frac{1}{4}=1$

Inverse property of multiplication

109. No.

$(8\div4)\div2=2\div2=1$

$8\div(4\div2)=8\div2=4$

111. All but the multiplicative inverse property

113. Simplify the variable expression.

$$\begin{aligned}&2+3(2x-5)\\&=2+6x-15\\&=6x-13\end{aligned}$$

115. Simplify the variable expression.

$$\begin{aligned}&5-3(4x-2y)\\&=5-12x+6y\\&=-12x+6y+5\end{aligned}$$

117. Simplify the variable expression.

$$\begin{aligned}&3(2a-4b)-4(a-3b)\\&=6a-12b-4a+12b\\&=6a-4a-12b+12b\\&=2a\end{aligned}$$

119. Simplify the variable expression.

$$\begin{aligned}&5a-2[3-2(4a+3)]\\&=5a-2(3-8a-6)\\&=5a-2(-8a-3)\\&=5a+16a+6\\&=21a+6\end{aligned}$$

121. Simplify the variable expression.

$$\begin{aligned}&\frac{3}{4}(5a+2)-\frac{1}{2}(3a-5)\\&=\frac{15a}{4}+\frac{6}{4}-\frac{3a}{2}+\frac{5}{2}\\&=\frac{15a}{4}-\frac{6a}{4}+\frac{3}{2}+\frac{5}{2}\\&=\frac{9a}{4}+4\end{aligned}$$

123. Find the area.

$$\text{Area}=\frac{1}{2}bh=\frac{1}{2}(3\text{ in})(4\text{ in})=6\text{ in}^2$$

125. Find the heart rate.

$$\begin{aligned}\text{Heart rate}&=65+\frac{53}{4t+1}\\&=65+\frac{53}{4(10)+1}\\&=65+\frac{53}{41}\\&\approx66\end{aligned}$$

Heart rate is about 66 beats per minute.

127. Find the height.

$$\begin{aligned}\text{Height}&=-16t^2+80t+4\\&=-16(2)^2+80(2)+4\\&=-16(4)+80(2)+4\\&=-64+160+4\\&=100\end{aligned}$$

After 2 seconds, the ball will have a height of 100 feet.

Prepare for Section P.2

P1. Simplify.

$2^2 \cdot 2^3 = 4 \cdot 8 = 32$

Alternate method: $2^2 \cdot 2^3 = 2^{2+3} = 2^5 = 32$

P3. Simplify.

$(2^3)^2 = 8^2 = 64$

Alternate method: $(2^3)^2 = 2^{3(2)} = 2^6 = 64$

P5. False

$3^4 \cdot 3^2 = 3^6$, not 9^6.

Section P.2 Exercises

1. Evaluate the expression.

$-5^3 = -(5^3) = -125$

3. Evaluate the expression.

$\left(\frac{2}{3}\right)^0 = 1$

5. Evaluate the expression.

$4^{-2} = \frac{1}{4^2} = \frac{1}{16}$

7. Evaluate the expression.

$\frac{1}{2^{-5}} = 2^5 = 32$

9. Evaluate the expression.

$\frac{2^{-3}}{6^{-3}} = \left(\frac{2}{6}\right)^{-3} = \left(\frac{1}{3}\right)^{-3} = \left(\frac{3}{1}\right)^3 = 3^3 = 27$

11. Evaluate the expression.

$-2x^0 = -2$

13. Write in simplest form.

$2x^{-4} = 2\left(x^{-4}\right) = \frac{2}{x^4}$

15. Write in simplest form.

$\frac{5}{z^{-6}} = 5z^6$

17. Write in simplest form.

$(x^3y^2)(xy^5) = x^{3+1}y^{2+5} = x^4y^7$

19. Write in simplest form.

$(-2ab^4)(-3a^2b^5) = (-2)(-3)a^{1+2}b^{4+5} = 6a^3b^9$

21. Write in simplest form.

$$\begin{aligned}\left(-4x^{-3}y\right)\left(7x^5y^{-2}\right) &= (-4)(7)x^{-3+5}y^{1-2}\\ &= -28x^2y^{-1}\\ &= -\frac{28x^2}{y}\end{aligned}$$

23. Write in simplest form.

$\frac{6a^4}{8a^8} = \frac{6}{8}a^{4-8} = \frac{3}{4}a^{-4} = \frac{3}{4a^4}$

25. Write in simplest form.

$\frac{12x^3y^4}{18x^5y^2} = \frac{12}{18}x^{3-5}y^{4-2} = \frac{2}{3}x^{-2}y^2 = \frac{2y^2}{3x^2}$

27. Write in simplest form.

$\frac{36a^{-2}b^3}{3ab^4} = \frac{36}{3}a^{-2-1}b^{3-4} = 12a^{-3}b^{-1} = \frac{12}{a^3b}$

29. Write in simplest form.

$$\begin{aligned}(-2m^3n^2)(-3mn^2)^2 &= (-2m^3n^2)(9m^2n^4)\\ &= (-2)(9)m^{3+2}n^{2+4}\\ &= -18m^5n^6\end{aligned}$$

31. Write in simplest form.

$$\begin{aligned}(x^{-2}y)^2(xy)^{-2} &= (x^{-4}y^2)(x^{-2}y^{-2})\\ &= x^{-4-2}y^{2-2}\\ &= x^{-6}y^0\\ &= \frac{1}{x^6}\end{aligned}$$

33. Write in simplest form.

$$\begin{aligned}\left(\frac{3a^2b^3}{6a^4b^4}\right)^2 &= \frac{9a^4b^6}{36a^8b^8}\\ &= \frac{9}{36}a^{4-8}b^{6-8}\\ &= \frac{1}{4}a^{-4}b^{-2}\\ &= \frac{1}{4a^4b^2}\end{aligned}$$

35. Write in simplest form.

$\frac{(-4x^2y^3)^2}{(2xy^2)^3} = \frac{16x^4y^6}{8x^3y^6} = 2x^{4-3}y^{6-6} = 2x$

37. Write in simplest form.

$$\left(\frac{a^{-2}b}{a^3b^{-4}}\right)^2 = \frac{a^{-4}b^2}{a^6b^{-8}}$$
$$= a^{-4-6}b^{2-(-8)}$$
$$= a^{-4-6}b^{2+8}$$
$$= a^{-10}b^{10}$$
$$= \frac{b^{10}}{a^{10}}$$

39. Write in scientific notation.

$$2{,}011{,}000{,}000{,}000 = 2.011\times10^{12}$$

41. Write in scientific notation.

$$0.000000000562 = 5.62\times10^{-10}$$

43. Write in decimal notation.

$$3.14\times10^7 = 31{,}400{,}000$$

45. Write in decimal notation.

$$-2.3\times10^{-6} = -0.0000023$$

47. Perform the operation and write in scientific notation.

$$(3\times10^{12})(9\times10^{-5}) = (3)(9)\times10^{12-5}$$
$$= 27\times10^7$$
$$= 2.7\times10^8$$

49. Perform the operation and write in scientific notation.

$$\frac{9\times10^{-3}}{6\times10^8} = \frac{9}{6}\times10^{-3-8}$$
$$= 1.5\times10^{-11}$$

51. Perform the operation and write in scientific notation.

$$\frac{(3.2\times10^{-11})(2.7\times10^{18})}{1.2\times10^{-5}} = \frac{(3.2)(2.7)}{1.2}\times10^{-11+18-(-5)}$$
$$= 7.2\times10^{-11+18+5}$$
$$= 7.2\times10^{12}$$

53. Perform the operation and write in scientific notation.

$$\frac{(4.0\times10^{-9})(8.4\times10^5)}{(3.0\times10^{-6})(1.4\times10^{18})} = \frac{(4.0)(8.4)}{(3.0)(1.4)}\times10^{-9+5+6-18}$$
$$= 8\times10^{-16}$$

55. Evaluate the expression.

$$4^{3/2} = \sqrt{4}^3 = 2^3 = 8$$

57. Evaluate the expression.

$$-64^{2/3} = -\sqrt[3]{64}^2 = -4^2 = -16$$

59. Evaluate the expression.

$$9^{-3/2} = \frac{1}{9^{3/2}} = \frac{1}{\left(\sqrt{9}\right)^3} = \frac{1}{3^3} = \frac{1}{27}$$

61. Evaluate the expression.

$$\left(\frac{4}{9}\right)^{1/2} = \sqrt{\frac{4}{9}} = \frac{\sqrt{4}}{\sqrt{9}} = \frac{2}{3}$$

63. Evaluate the expression.

$$\left(\frac{1}{8}\right)^{-4/3} = 8^{4/3} = \sqrt[3]{8}^4 = 2^4 = 16$$

65. Evaluate the expression.

$$(4a^{2/3}b^{1/2})(2a^{1/3}b^{3/2}) = (4)(2)a^{2/3+1/3}b^{1/2+3/2}$$
$$= 8a^{3/3}b^{4/2} = 8ab^2$$

67. Evaluate the expression.

$$(-3x^{2/3})(4x^{1/4}) = (-3)(4)x^{2/3+1/4}$$
$$= -12x^{8/12+3/12}$$
$$= -12x^{11/12}$$

69. Evaluate the expression.

$$(81x^8y^{12})^{1/4} = 81^{1/4}x^{8/4}y^{12/4} = \sqrt[4]{81}x^2y^3 = 3x^2y^3$$

71. Evaluate the expression.

$$\frac{16z^{3/5}}{12z^{1/5}} = \frac{16z^{3/5-1/5}}{12} = \frac{4z^{2/5}}{3}$$

73. Evaluate the expression.

$$(2x^{2/3}y^{1/2})(3x^{1/6}y^{1/3}) = (2)(3)x^{2/3+1/6}y^{1/2+1/3}$$
$$= 6x^{5/6}y^{5/6}$$

75. Evaluate the expression.

$$\frac{9a^{3/4}b}{3a^{2/3}b^2} = \frac{9a^{3/4-2/3}b^{1-2}}{3} = 3a^{9/12-8/12}b^{-1} = \frac{3a^{1/12}}{b}$$

77. Simplify the radical expression.

$$\sqrt{45} = \sqrt{3^2\cdot5} = 3\sqrt{5}$$

79. Simplify the radical expression.

$$\sqrt[3]{24} = \sqrt[3]{2^3\cdot3} = 2\sqrt[3]{3}$$

81. Simplify the radical expression.

$$\sqrt[3]{-135} = \sqrt[3]{(-3)^3\cdot5} = -3\sqrt[3]{5}$$

83. Simplify the radical expression.

$$\sqrt{24x^2y^3} = \sqrt{2^2x^2y^2} \cdot \sqrt{6y} = 2|xy|\sqrt{6y}$$

85. Simplify the radical expression.

$$\sqrt[3]{16a^3y^7} = \sqrt[3]{2^3a^3y^6} \cdot \sqrt[3]{2y} = 2ay^2\sqrt[3]{2y}$$

87. Simplify and combine like radicals.

$$\begin{aligned} 2\sqrt{32} - 3\sqrt{98} &= 2\sqrt{16} \cdot \sqrt{2} - 3\sqrt{49} \cdot \sqrt{2} \\ &= 2(4)\sqrt{2} - 3(7)\sqrt{2} \\ &= 8\sqrt{2} - 21\sqrt{2} \\ &= -13\sqrt{2} \end{aligned}$$

89. Simplify and combine like radicals.

$$\begin{aligned} -8\sqrt[4]{48} + 2\sqrt[4]{243} &= -8\sqrt[4]{16 \cdot 3} + 2\sqrt[4]{81 \cdot 3} \\ &= -8\sqrt[4]{16} \cdot \sqrt[4]{3} + 2\sqrt[4]{81} \cdot \sqrt[4]{3} \\ &= -8\sqrt[4]{2^4} \cdot \sqrt[4]{3} + 2\sqrt[4]{3^4} \cdot \sqrt[4]{3} \\ &= -8(2)\sqrt[4]{3} + 2(3)\sqrt[4]{3} \\ &= -16\sqrt[4]{3} + 6\sqrt[4]{3} \\ &= -10\sqrt[4]{3} \end{aligned}$$

91. Simplify and combine like radicals.

$$\begin{aligned} 4\sqrt[3]{32y^4} + 3y\sqrt[3]{108y} &= 4\sqrt[3]{8y^3 \cdot 4y} + 3y\sqrt[3]{27 \cdot 4y} \\ &= 4\sqrt[3]{8y^3} \cdot \sqrt[3]{4y} + 3y\sqrt[3]{27} \cdot \sqrt[3]{4y} \\ &= 4\sqrt[3]{2^3y^3} \cdot \sqrt[3]{4y} + 3y\sqrt[3]{3^3} \cdot \sqrt[3]{4y} \\ &= 4(2y)\sqrt[3]{4y} + 3y(3)\sqrt[3]{4y} \\ &= 8y\sqrt[3]{4y} + 9y\sqrt[3]{4y} \\ &= 17y\sqrt[3]{4y} \end{aligned}$$

93. Simplify and combine like radicals.

$$\begin{aligned} &x\sqrt[3]{8x^3y^4} - 4y\sqrt[3]{64x^6y} \\ &= x\sqrt[3]{8x^3y^3 \cdot y} - 4y\sqrt[3]{64x^6 \cdot y} \\ &= x\sqrt[3]{8x^3y^3} \cdot \sqrt[3]{y} - 4y\sqrt[3]{64x^6} \cdot \sqrt[3]{y} \\ &= x\sqrt[3]{2^3x^3y^3} \cdot \sqrt[3]{y} - 4y\sqrt[3]{4^3x^6} \cdot \sqrt[3]{y} \\ &= x(2xy)\sqrt[3]{y} - 4y(4x^2)\sqrt[3]{y} \\ &= 2x^2y\sqrt[3]{y} - 16x^2y\sqrt[3]{y} \\ &= -14x^2y\sqrt[3]{y} \end{aligned}$$

95. Find the product and write in simplest form.

$$\begin{aligned} (\sqrt{5}+3)(\sqrt{5}+4) &= \sqrt{5}^2 + 4\sqrt{5} + 3\sqrt{5} + (3)(4) \\ &= 5 + 7\sqrt{5} + 12 \\ &= 17 + 7\sqrt{5} \end{aligned}$$

97. Find the product and write in simplest form.

$$\begin{aligned} (\sqrt{2}-3)(\sqrt{2}+3) &= \sqrt{2}^2 + 3\sqrt{2} - 3\sqrt{2} + (-3)(3) \\ &= 2 - 9 \\ &= -7 \end{aligned}$$

99. Find the product and write in simplest form.

$$\begin{aligned} &(3\sqrt{z}-2)(4\sqrt{z}+3) \\ &= (3)(4)\sqrt{z}^2 + 3(3\sqrt{z}) - 2(4\sqrt{z}) + (-2)(3) \\ &= 12z + 9\sqrt{z} - 8\sqrt{z} - 6 \\ &= 12z + \sqrt{z} - 6 \end{aligned}$$

101. Find the product and write in simplest form.

$$\begin{aligned} (\sqrt{x}+2)^2 &= \sqrt{x}^2 + 2(\sqrt{x})(2) + 2^2 \\ &= x + 4\sqrt{x} + 4 \end{aligned}$$

103. Find the product and write in simplest form.

$$\begin{aligned} (\sqrt{x-3}+2)^2 &= \sqrt{x-3}^2 + 2(\sqrt{x-3})(2) + 2^2 \\ &= x - 3 + 4\sqrt{x-3} + 4 \\ &= x + 4\sqrt{x-3} + 1 \end{aligned}$$

105. Rationalize the denominator.

$$\frac{2}{\sqrt{2}} = \frac{2}{\sqrt{2}} \cdot \frac{\sqrt{2}}{\sqrt{2}} = \frac{2\sqrt{2}}{\sqrt{2}^2} = \frac{2\sqrt{2}}{2} = \frac{\cancel{2}\sqrt{2}}{\cancel{2}} = \sqrt{2}$$

107. Rationalize the denominator.

$$\begin{aligned} \sqrt{\frac{5}{18}} &= \sqrt{\frac{5}{2 \cdot 3^2}} = \sqrt{\frac{5}{2 \cdot 3^2}} \cdot \sqrt{\frac{2}{2}} \\ &= \sqrt{\frac{5 \cdot 2}{2^2 \cdot 3^2}} = \frac{\sqrt{5 \cdot 2}}{\sqrt{2^2 \cdot 3^2}} = \frac{\sqrt{10}}{2 \cdot 3} \\ &= \frac{\sqrt{10}}{6} \end{aligned}$$

109. Rationalize the denominator.

$$\frac{3}{\sqrt[3]{2}} = \frac{3}{\sqrt[3]{2}} \cdot \frac{\sqrt[3]{2^2}}{\sqrt[3]{2^2}} = \frac{3\sqrt[3]{2^2}}{\sqrt[3]{2^3}} = \frac{3\sqrt[3]{4}}{2}$$

111. Rationalize the denominator.

$$\frac{4}{\sqrt[3]{8x^2}} = \frac{4}{\sqrt[3]{2^3x^2}} = \frac{4}{2\sqrt[3]{x^2}} = \frac{\overset{2}{\cancel{4}}}{\cancel{2}\sqrt[3]{x^2}} = \frac{2}{\sqrt[3]{x^2}}$$

$$= \frac{2}{\sqrt[3]{x^2}} \cdot \frac{\sqrt[3]{x}}{\sqrt[3]{x}} = \frac{2\sqrt[3]{x}}{\sqrt[3]{x^3}}$$

$$= \frac{2\sqrt[3]{x}}{x}$$

113. Rationalize the denominator.

$$\frac{3}{\sqrt{3}+4} = \frac{3}{\sqrt{3}+4} \cdot \frac{\sqrt{3}-4}{\sqrt{3}-4} = \frac{3(\sqrt{3}-4)}{(\sqrt{3}+4)(\sqrt{3}-4)}$$

$$= \frac{3(\sqrt{3}-4)}{\sqrt{3}^2-4^2} = \frac{3(\sqrt{3}-4)}{3-16}$$

$$= \frac{3\sqrt{3}-12}{-13}$$

$$= -\frac{3\sqrt{3}-12}{13}$$

115. Rationalize the denominator.

$$\frac{6}{2\sqrt{5}+2} = \frac{6}{2(\sqrt{5}+1)} = \frac{\overset{3}{\cancel{6}}}{\cancel{2}(\sqrt{5}+1)} = \frac{3}{\sqrt{5}+1}$$

$$= \frac{3}{\sqrt{5}+1} \cdot \frac{\sqrt{5}-1}{\sqrt{5}-1}$$

$$= \frac{3(\sqrt{5}-1)}{(\sqrt{5}+1)(\sqrt{5}-1)} = \frac{3(\sqrt{5}-1)}{\sqrt{5}^2-1}$$

$$= \frac{3(\sqrt{5}-1)}{5-1}$$

$$= \frac{3\sqrt{5}-3}{4}$$

117. Rationalize the denominator.

$$\frac{3+2\sqrt{5}}{5-3\sqrt{5}} = \frac{3+2\sqrt{5}}{5-3\sqrt{5}} \cdot \frac{5+3\sqrt{5}}{5+3\sqrt{5}}$$

$$= \frac{(3+2\sqrt{5})(5+3\sqrt{5})}{(5-3\sqrt{5})(5+3\sqrt{5})}$$

$$= \frac{3(5+3\sqrt{5})+2\sqrt{5}(5+3\sqrt{5})}{(5-3\sqrt{5})(5+3\sqrt{5})}$$

$$= \frac{15+9\sqrt{5}+10\sqrt{5}+30}{5^2-(3\sqrt{5})^2}$$

$$= \frac{45+19\sqrt{5}}{25-45}$$

$$= \frac{45+19\sqrt{5}}{-20} \text{ or } -\frac{45+19\sqrt{5}}{20}$$

119. Rationalize the denominator.

$$\frac{6\sqrt{3}-11}{4\sqrt{3}-7} = \frac{6\sqrt{3}-11}{4\sqrt{3}-7} \cdot \frac{4\sqrt{3}+7}{4\sqrt{3}+7}$$

$$= \frac{(6\sqrt{3}-11)(4\sqrt{3}+7)}{(4\sqrt{3}-7)(4\sqrt{3}+7)}$$

$$= \frac{6\sqrt{3}(4\sqrt{3}+7)-11(4\sqrt{3}+7)}{(4\sqrt{3}-7)(4\sqrt{3}+7)}$$

$$= \frac{72+42\sqrt{3}-44\sqrt{3}-77}{(4\sqrt{3})^2-7^2}$$

$$= \frac{-5-2\sqrt{3}}{48-49} = \frac{-5-2\sqrt{3}}{-1}$$

$$= 5+2\sqrt{3}$$

121. Rationalize the denominator.

$$\frac{2+\sqrt{x}}{3-2\sqrt{x}} = \frac{2+\sqrt{x}}{3-2\sqrt{x}} \cdot \frac{3+2\sqrt{x}}{3+2\sqrt{x}}$$

$$= \frac{(2+\sqrt{x})(3+2\sqrt{x})}{(3-2\sqrt{x})(3+2\sqrt{x})}$$

$$= \frac{2(3+2\sqrt{x})+\sqrt{x}(3+2\sqrt{x})}{(3-2\sqrt{x})(3+2\sqrt{x})}$$

$$= \frac{6+4\sqrt{x}+3\sqrt{x}+2x}{3^2-(2\sqrt{x})^2}$$

$$= \frac{6+7\sqrt{x}+2x}{9-4x}$$

123. Rationalize the denominator.

$$\frac{x-\sqrt{5}}{x+2\sqrt{5}} = \frac{x-\sqrt{5}}{x+2\sqrt{5}} \cdot \frac{x-2\sqrt{5}}{x-2\sqrt{5}} = \frac{(x-\sqrt{5})(x-2\sqrt{5})}{(x+2\sqrt{5})(x-2\sqrt{5})}$$

$$= \frac{x(x-2\sqrt{5})-\sqrt{5}(x-2\sqrt{5})}{(x+2\sqrt{5})(x-2\sqrt{5})}$$

$$= \frac{x^2-2x\sqrt{5}-x\sqrt{5}+10}{x^2-(2\sqrt{5})^2}$$

$$= \frac{x^2-3x\sqrt{5}+10}{x^2-20}$$

125. Rationalize the denominator.

$$\frac{3}{\sqrt{5}+\sqrt{x}} = \frac{3}{\sqrt{5}+\sqrt{x}} \cdot \frac{\sqrt{5}-\sqrt{x}}{\sqrt{5}-\sqrt{x}}$$

$$= \frac{3(\sqrt{5}-\sqrt{x})}{(\sqrt{5}+\sqrt{x})(\sqrt{5}-\sqrt{x})}$$

$$= \frac{3\sqrt{5}-3\sqrt{x}}{(\sqrt{5})^2-(\sqrt{x})^2}$$

$$= \frac{3\sqrt{5}-3\sqrt{x}}{5-x}$$

127. Rationalize the numerator.

$$\frac{\sqrt{4+h}-2}{h}=\frac{\sqrt{4+h}-2}{h}\cdot\frac{\sqrt{4+h}+2}{\sqrt{4+h}+2}$$
$$=\frac{(\sqrt{4+h})^2-2^2}{h(\sqrt{4+h}+2)}$$
$$=\frac{4+h-4}{h(\sqrt{4+h}+2)}=\frac{h}{h(\sqrt{4+h}+2)}$$
$$=\frac{1}{\sqrt{4+h}+2}$$

129. Find the number of seeds.

$$\frac{1\text{ seed}}{3.2\times10^{-8}\text{ ounce}}\cdot\frac{1\text{ ounce}}{\text{package}}$$
$$\frac{1}{3.2\times10^{-8}}\cdot1=\frac{1}{3.2}\times10^{8}$$
$$=0.3125\times10^{8}$$
$$=3.13\times10^{7}\text{ seeds per package}$$

131. Find the red shift.

$$\text{Red shift}=\frac{\lambda_r-\lambda_s}{\lambda_s}$$
$$=\frac{5.13\times10^{-7}-5.06\times10^{-7}}{5.06\times10^{-7}}$$
$$=\frac{(5.13-5.06)\times10^{-7}}{5.06\times10^{-7}}$$
$$=\frac{0.07}{5.06}\cdot\frac{10^{-7}}{10^{-7}}$$
$$\approx1.38\times10^{-2}$$

133. Find the number of minutes.

$$\frac{1\text{ sec}}{3\times10^{8}\text{ m}}\cdot1.44\times10^{11}\text{ m}\cdot\frac{1\text{ min}}{60\text{ sec}}$$
$$\frac{1}{3\times10^{8}}\cdot1.44\times10^{11}\cdot\frac{1}{60}=\frac{(1)(1.44)(1)\times10^{11}}{3(60)\times10^{8}}$$
$$=\frac{1.44}{180}\times10^{11-8}$$
$$=0.008\times10^{3}$$
$$=8\text{ minutes}$$

135. Find the mass.

$$\frac{1\text{ gram}}{6.023\times10^{23}\text{ atoms}}\cdot1\text{ atom}$$
$$\frac{1}{6.023\times10^{23}}\cdot1=\frac{1}{6.023}\times10^{-23}$$
$$\approx0.1660302175\times10^{-23}$$
$$\approx1.66\times10^{-24}$$

One hydrogen atom weighs approximately 1.66×10^{-24} gram.

137.a. Evaluate P when $d = 10$.

$$P=10^{2-d/40}=10^{2-10/40}=10^{2-0.25}=10^{1.75}\approx56$$

The amount of light that will pass to a depth of 10 feet below the ocean's surface is about 56%.

b. Evaluate P when $d = 25$.

$$P=10^{2-d/40}=10^{2-25/40}=10^{2-0.625}=10^{1.375}\approx24$$

The amount of light that will pass to a depth of 25 feet below the ocean's surface is about 24%.

139. First, we substitute $K.E_r$ and $K.E_n$ in the formula to find % error.

$$\%\text{ error}=\frac{|K.E_r-K.E_n|}{K.E_r}\times100$$
$$=\frac{\left|mc^2\left(\frac{1}{\sqrt{1-\frac{v^2}{c^2}}}-1\right)-\frac{1}{2}mv^2\right|}{mc^2\left(\frac{1}{\sqrt{1-\frac{v^2}{c^2}}}-1\right)}\times100$$
$$=\frac{\left|c^2\left(\frac{1}{\sqrt{1-\frac{v^2}{c^2}}}-1\right)-\frac{1}{2}v^2\right|}{c^2\left(\frac{1}{\sqrt{1-\frac{v^2}{c^2}}}-1\right)}\times100$$

Note that m, the mass is eliminated from the formula.

a. Substituting $v = 30$ and $c=3.0\times10^{8}$ in the above formula, we find % error = 7.5×10^{-13}

Due to the way some calculators make calculations, you may get an incorrect answer here of 2.354536986.

b. Substituting $v = 240$ and $c=3.0\times10^{8}$ in the above formula, we find % error = 4.8×10^{-11}

Due to the way some calculators make calculations,

you may get an incorrect answer here of 0.010539955.

c. Substituting $v = 3\times10^7$ and $c = 3.0\times10^8$ in the above formula, we find % error = 0.750628145

d. Substituting $v = 1.5\times10^8$ and $c = 3.0\times10^8$ in the above formula, we find % error = 19.1987

e. Substituting $v = 2.7\times10^8$ and $c = 3.0\times10^8$ in the above formula, we find % error = 68.7055

f. The percent error is very small for everyday speeds.

g. As the speed of the object approaches the speed of light, the denominator of the kinetic energy equation approaches 0, which implies that the kinetic energy is approaching infinity. Thus, it would require an infinite amount of energy to move a particle at the speed of light.

Prepare for Section P.3

P1. Simplify.

$-3(2a-4b)$

$=-6a+12b$

P3. Simplify.

$2x^2+3x-5+x^2-6x-1$

$=2x^2+x^2+3x-6x-5-1$

$=(2+1)x^2+(3-6)x-(5+1)$

$=3x^2-3x-6$

P5. $4-3x-2x^2 \overset{?}{=} -2x^2-4x+4$

$-2x^2-3x+4 \overset{?}{=} -2x^2-4x+4$

False.

Section P.3 Exercises

1. D

3. H

5. G

7. B

9. J

11. a. x^2+2x-7

b. 2

c. 1, 2, −7

d. 1

e. x^2, $2x$, -7

13. a. x^3-1

b. 3

c. 1, −1

d. 1

e. x^3, -1

15. a. $2x^4+3x^3+4x^2+5$

b. 4

c. 2, 3, 4, 5

d. 2

e. $2x^4$, $3x^3$, $4x^2$, 5

17. Determine the degree of the polynomial.

3

19. Determine the degree of the polynomial.

5

21. Determine the degree of the polynomial.

2

23. Perform operations, simplify, write in standard form.

$(3x^2+4x+5)+(2x^2+7x-2)=5x^2+11x+3$

25. Perform operations, simplify, write in standard form.

$(4w^3-2w+7)+(5w^3+8w^2-1)=9w^3+8w^2-2w+6$

27. Perform operations, simplify, write in standard form.

$(r^2-2r-5)-(3r^2-5r+7)$

$=r^2-2r-5-3r^2+5r-7$

$=-2r^2+3r-12$

29. Perform operations, simplify, write in standard form.

$(u^3-3u^2-4u+8)-(u^3-2u+4)$

$=u^3-3u^2-4u+8-u^3+2u-4$

$=-3u^2-2u+4$

31. Perform operations, simplify, write in standard form.

$(2x^2+7x-8)(4x-5)$

$=8x^3-10x^2+28x^2-35x-32x+40$

$=8x^3+18x^2-67x+40$

33. Perform operations, simplify, write in standard form.

$$\begin{aligned}&(3x^2-5x+6)(3x-1)\\&=(3x^2-5x+6)(3x)+(3x^2-5x+6)(-1)\\&=9x^3-15x^2+18x-3x^2+5x-6\\&=9x^3-18x^2+23x-6\end{aligned}$$

35. Perform operations, simplify, write in standard form.

$$\begin{aligned}&(2x+6)(5x^3-6x^2+4)\\&=2x(5x^3-6x^2+4)+6(5x^3-6x^2+4)\\&=10x^4-12x^3+8x+30x^3-36x^2+24\\&=10x^4+18x^3-36x^2+8x+24\end{aligned}$$

37. Perform operations, simplify, write in standard form.

$$\begin{aligned}&(x^3-4x^2+9x-6)(2x+5)\\&=(x^3-4x^2+9x-6)(2x)+(x^3-4x^2+9x-6)(5)\\&=2x^4-8x^3+18x^2-12x+5x^3-20x^2+45x-30\\&=2x^4-3x^3-2x^2+33x-30\end{aligned}$$

39. Perform operations, simplify, write in standard form.

$$\begin{array}{r}3x^2-2x+5\\ 2x^2-5x+2\\ \hline +\ 6x^2-\ 4x+10\\ -15x^3+10x^2-25x\quad\\ +\ 6x^4-\ 4x^3+10x^2\qquad\quad\\ \hline 6x^4-19x^3+26x^2-29x+10\end{array}$$

41. Find the product using the FOIL method.

$$\begin{aligned}(2x+4)(5x+1)&=10x^2+2x+20x+4\\&=10x^2+22x+4\end{aligned}$$

43. Find the product using the FOIL method.

$$\begin{aligned}(y+2)(y+1)&=y^2+y+2y+2\\&=y^2+3y+2\end{aligned}$$

45. Find the product using the FOIL method.

$$\begin{aligned}(4z-3)(z-4)&=4z^2-16z-3z+12\\&=4z^2-19z+12\end{aligned}$$

47. Find the product using the FOIL method.

$$\begin{aligned}(a+6)(a-3)&=a^2-3a+6a-18\\&=a^2+3a-18\end{aligned}$$

49. Find the product using the FOIL method.

$$\begin{aligned}(5x-11y)(2x-7y)&=10x^2-35xy-22xy+77y^2\\&=10x^2-57xy+77y^2\end{aligned}$$

51. Find the product using the FOIL method.

$$\begin{aligned}(9x+5y)(2x+5y)&=18x^2+45xy+10xy+25y^2\\&=18x^2+55xy+25y^2\end{aligned}$$

53. Find the product using the FOIL method.

$$\begin{aligned}(3p+5q)(2p-7q)&=6p^2-21pq+10pq-35q^2\\&=6p^2-11pq-35q^2\end{aligned}$$

55. Use the special product formulas.

$$(3x+5)(3x-5)=9x^2-25$$

57. Use the special product formulas.

$$(3x^2-y)^2=9x^4-6x^2y+y^2$$

59. Use the special product formulas.

$$(4w+z)^2=16w^2+8wz+z^2$$

61. Use the special product formulas.

$$\begin{aligned}[(x+5)+y][(x+5)-y]&=(x+5)^2-y^2\\&=x^2+10x+25-y^2\end{aligned}$$

63. Perform the operations and simplify.

$$\begin{aligned}&(4d-1)^2-(2d-3)^2\\&=(16d^2-8d+1)-(4d^2-12d+9)\\&=16d^2-8d+1-4d^2+12d-9\\&=12d^2+4d-8\end{aligned}$$

65. Perform the operations and simplify.

$$\begin{array}{r}r^2-rs+s^2\\ r+s\\ \hline +r^2s-rs^2+s^3\\ r^3-r^2s+rs^2\qquad\\ \hline r^3\qquad\qquad\quad+s^3\end{array}$$

67. Perform the operations and simplify.

$$(3c-2)(4c+1)(5c-2)=(12c^2-5c-2)(5c-2)$$

$$\begin{array}{r}12c^2-5c-2\\ 5c-2\\ \hline -24c^2+10c+4\\ 60c^3-25c^2-10c\qquad\\ \hline 60c^3-49c^2\qquad\quad+4\end{array}$$

69. Evaluate x^2+7x-1 for $x=3$.

$(3)^2+7(3)-1=9+21-1=29$

71. Evaluate $-x^2+5x-3$ for $x=-2$.

$-(-2)^2+5(-2)-3=-4-10-3=-17$

73. Evaluate $3x^3-2x^2-x+3$ for $x=-1$.

$3(-1)^3-2(-1)^2-(-1)+3$
$=3(-1)-2(1)+1+3$
$=-3-2+1+3$
$=-1$

75. Evaluate $1-x^5$ for $x=-2$.

$1-(-2)^5=1-(-32)=1+32=33$

77. Substitute the given value of v into $0.016v^2$. Then simplify.

a. $0.016v^2$

$0.016(10)^2=1.6$

The air resistance is 1.6 pounds.

b. $0.016v^2$

$0.016(15)^2=3.6$

The air resistance is 3.6 pounds.

79. Substitute the given value of h and r into πr^2h. Then simplify

a. πr^2h

$\pi(3)^2(8)=72\pi$

The volume is 72π in^3.

b. πr^2h

$\pi(5)^2(12)=300\pi$

The volume 300π cm^3.

81. Substitute the given value of v into $0.005x^2-0.32x+12$.

a. $0.005x^2-0.32x+12$

$0.005(20)^2-0.32(20)+12=7.6$

The reaction time is 7.6 hundredths of a second or 0.076 second.

b. $0.005x^2-0.32x+12$

$0.005(50)^2-0.32(50)+12=8.5$

The reaction time is 8.5 hundredths of a second or 0.085 second.

83. Find the number of chess matches.

$\frac{1}{2}n^2-\frac{1}{2}n=\frac{1}{2}(150)^2-\frac{1}{2}(150)=11{,}175$ matches

85. Estimate the times.

a. $1.9\times10^{-6}(4000)^2-3.9\times10^{-3}(4000)=14.8$ s

b. $1.9\times10^{-6}(8000)^2-3.9\times10^{-3}(8000)=90.4$ s

87. Evaluate $-16t^2+4.7881t+6$ when $t=0.5$

height $=-16t^2+4.7881t+6$
$=-16(0.5)^2+4.7881(0.5)t+6$
$=4.39$

Yes. The ball is approximately 4.4 feet high when it crosses home plate.

Mid-Chapter P Quiz

1. Evaluate $2x^3-4(3xy-z^2)$ for the given values.

$2(-2)^3-4(3(-2)(3)-(-4)^2)$
$=2(-8)-4(-18-16)$
$=-16-4(-34)$
$=-16+136$
$=120$

3. Simplify.

$$\frac{24x^{-3}y^4}{6x^2y^{-3}}=4x^{-3-2}y^{4+3}=4x^{-5}y^7=\frac{4y^7}{x^5}$$

5. Simplify.

$$\sqrt[3]{16a^4b^9c^8}=\sqrt[3]{8a^3b^9c^6}\times\sqrt[3]{2ac^2}=2ab^3c^2\sqrt[3]{2ac^2}$$

7. Multiply.

$$(3x-4y)(2x+5y)=6x^2+15xy-8xy-20y^2$$
$$=6x^2+7xy-20y^2$$

9. Multiply.

$$(2x-3)(4x^2+5x-7)$$
$$=2x(4x^2+5x-7)-3(4x^2+5x-7)$$
$$=8x^3+10x^2-14x-12x^2-15x+21$$
$$=8x^3-2x^2-29x+21$$

Prepare for Section P.4

P1. Simplify.

$$\frac{6x^3}{2x}=3x^{3-1}=3x^2$$

P3. a. $x^6=(x^2)^3$

b. $x^6=(x^3)^2$

P5. Replace "?" to make a true statement.

$$-3(5a-?)=-15a+21=-3(5a-7)$$

Thus, ? = 7

Section P.4 Exercises

1. Factor out the GCF.

$$5x+20=5(x+4)$$

3. Factor out the GCF.

$$-15x^2-12x=-3x(5x+4)$$

5. Factor out the GCF.

$$10x^2y+6xy-14xy^2=2xy(5x+3-7y)$$

7. Factor out the GCF.

$$(x-3)(a+b)+(x-3)(a+2b)$$
$$=(x-3)(a+b+a+2b)$$
$$=(x-3)(2a+3b)$$

9. Factor.

$$x^2+7x+12=(x+3)(x+4)$$

11. Factor.

$$a^2-10a-24=(a-12)(a+2)$$

13. Factor.

$$x^2+6x+5=(x+5)(x+1)$$

15. Factor.

$$6x^2+25x+4=(6x+1)(x+4)$$

17. Factor.

$$51x^2-5x-4=(17x+4)(3x-1)$$

19. Factor.

$$6x^2+xy-40y^2=(3x+8y)(2x-5y)$$

21. Factor.

$$6x^2+23x+15=(6x+5)(x+3)$$

23. Determine whether the trinomial is factorable.

$$b^2-4ac=26^2-4(8)(15)=196=14^2$$

The trinomial is factorable over the integers.

25. Determine whether the trinomial is factorable.

$$b^2-4ac=(-5)^2-4(4)(6)=-71$$

The trinomial is not factorable over the integers.

27. Determine whether the trinomial is factorable.

$$b^2-4ac=(-14)^2-4(6)(5)=76$$

The trinomial is not factorable over the integers.

29. Factor the difference of squares.

$$x^2-9=(x+3)(x-3)$$

31. Factor the difference of squares.

$$4a^2-49=(2a+7)(2a-7)$$

33. Factor the difference of squares.

$$1-100x^2=(1+10x)(1-10x)$$

35. Factor the difference of squares.

$$(x+1)^2-4=[(x+1)+2][(x+1)-2]$$
$$=(x+3)(x-1)$$

37. Factor the difference of squares.

$$6x^2-216=6(x^2-36)$$
$$=6(x+6)(x-6)$$

39. Factor the difference of squares.

$$x^4-625=(x^2+25)(x^2-25)$$
$$=(x^2+25)(x+5)(x-5)$$

41. Factor the difference of squares.

$$x^5-81x=x(x^4-81)$$
$$=x(x^2+9)(x^2-9)$$
$$=x(x^2+9)(x+3)(x-3)$$

43. Factor the perfect-square trinomial.

$$x^2+10x+25=(x+5)^2$$

45. Factor the perfect-square trinomial.

$$a^2-14a+49=(a-7)^2$$

47. Factor the perfect-square trinomial.

$$4x^2+12x+9=(2x+3)^2$$

49. Factor the perfect-square trinomial.

$$z^4+4z^2w^2+4w^4=(z^2)^2+2(2z^2w^2)+(2w^2)^2$$
$$=(z^2+2w^2)^2$$

51. Factor the sum or difference of cubes.

$$x^3-8=(x-2)(x^2+2x+4)$$

53. Factor the sum or difference of cubes.

$$8x^3-27y^3=(2x)^3-(3y)^3$$
$$=(2x-3y)(4x^2+6xy+9y^2)$$

55. Factor the sum or difference of cubes.

$$8-x^6=2^3-(x^2)^3$$
$$=(2-x^2)(4+2x+x^4)$$

57. Factor the sum or difference of cubes.

$$(x-2)^3-1=[(x-2)-1][(x-2)^2+(x-2)+1]$$
$$=(x-3)[x^2-4x+4+x-2+1]$$
$$=(x-3)(x^2-3x+3)$$

59. Factor polynomials that are quadratic in form.

$$x^4-x^2-6=(x^2)^2-x^2-6 \quad \text{Let } u=x^2.$$
$$=u^2-u-6$$
$$=(u-3)(u+2) \quad \text{Replace } u \text{ with } x^2.$$
$$=(x^2-3)(x^2+2)$$

61. Factor polynomials that are quadratic in form.

$$x^2y^2+4xy-5=(xy)^2+4xy-5 \quad \text{Let } u=xy.$$
$$=u^2+4u-5$$
$$=(u+5)(u-1) \quad \text{Replace } u \text{ with } xy.$$
$$=(xy+5)(xy-1)$$

63. Factor polynomials that are quadratic in form.

$$4x^5-4x^3-8x$$
$$=4x(x^4-x^2-2)$$
$$=4x\left[(x^2)^2-x^2-2\right] \quad \text{Let } u=x^2.$$
$$=4x[u^2-u-2]$$
$$=4x(u-2)(u+1) \quad \text{Replace } u \text{ with } x^2.$$
$$=4x(x^2-2)(x^2+1)$$

65. Factor polynomials that are quadratic in form.

$$z^4+z^2-20=(z^2)^2+z^2-20 \quad \text{Let } u=z^2.$$
$$=u^2+u-20$$
$$=(u-4)(u+5) \quad \text{Replace } u \text{ with } z^2.$$
$$=(z^2-4)(z^2+5)$$
$$=(z+2)(z-2)(z^2+5)$$

67. Factor by grouping.

$$3x^3+x^2+6x+2=x^2(3x+1)+2(3x+1)$$
$$=(3x+1)(x^2+2)$$

69. Factor by grouping.

$$ax^2-ax+bx-b=ax(x-1)+b(x-1)$$
$$=(x-1)(ax+b)$$

71. Factor by grouping.

$$6w^3+4w^2-15w-10=2w^2(3w+2)-5(3w+2)$$
$$=(3w+2)(2w^2-5)$$

73. Factor the polynomial, if not state nonfactorable.

$$18x^2-2=2(9x^2-1) \quad \text{difference of squares}$$
$$=2(3x+1)(3x-1)$$

75. Factor the polynomial, if not state nonfactorable.

$$16x^4-1=(4x^2)^2-1 \quad \text{difference of squares}$$
$$=(4x^2+1)(4x^2-1) \quad \text{difference of squares}$$
$$=(4x^2+1)(2x+1)(2x-1)$$

77. Factor the polynomial, if not state nonfactorable.

$$12ax^2-23axy+10ay^2=a(12x^2-23xy+10y^2)$$
$$=a(3x-2y)(4x-5y)$$

79. Factor the polynomial, if not state nonfactorable.

$$3bx^3+4bx^2-3bx-4b$$
$$=b(3x^3+4x^2-3x-4) \quad \text{factor by grouping}$$
$$=b[x^2(3x+4)-(3x+4)]$$
$$=b(3x+4)(x^2-1)$$
$$=b(3x+4)(x+1)(x-1)$$

81. Factor the polynomial, if not state nonfactorable.

$$72bx^2+24bxy+2by^2$$
$$=2b(36x^2+12xy+y^2) \quad \text{perfect-square trinomial}$$
$$=2b(6x+y)^2$$

83. Factor the polynomial, if not state nonfactorable.

$$(w-5)^3+8 \quad \text{sum of cubes}$$
$$=[(w-5)+2][(w-5)^2-2(w-5)+4]$$
$$=(w-3)(w^2-10w+25-2w+10+4)$$
$$=(w-3)(w^2-12w+39)$$

85. Factor the polynomial, if not state nonfactorable.

$$x^2+6xy+9y^2-1 \quad \text{perfect-square trinomial}$$
$$=(x+3y)^2-1 \quad \text{difference of squares}$$
$$=(x+3y+1)(x+3y-1)$$

87. Factor the polynomial, if not state nonfactorable.

$8x^2+3x-4$ is nonfactorable over the integers.

89. Factor the polynomial, if not state nonfactorable.

$$5x(2x-5)^2-(2x-5)^3 \quad \text{factor by grouping}$$
$$=(2x-5)^2[5x-(2x-5)]$$
$$=(2x-5)^2(3x+5)$$

91. Factor the polynomial, if not state nonfactorable.

$$4x^2+2x-y-y^2$$
$$=4x^2-y^2+2x-y \quad \text{factor by grouping}$$
$$=(2x+y)(2x-y)+(2x-y)$$
$$=(2x-y)(2x+y+1)$$

93. Find k.

$$x^2+kx+16$$
$$(x+4)^2=x^2+8x+16$$

Thus, $k=8$.

95. Find k.

$$x^2+16x+k$$
$$(x+8)^2=x^2+16x+64$$

Thus, $k=64$.

Prepare for Section P.5

P1. Simplify.

$$1+\frac{1}{2-\frac{1}{3}}=1+\frac{1}{2-\frac{1}{3}}\cdot\left(\frac{3}{3}\right)$$
$$=1+\frac{1\cdot 3}{2\cdot 3-\frac{1}{3}\cdot 3}$$
$$=1+\frac{3}{6-1}=1+\frac{3}{5}$$
$$=1\tfrac{3}{5} \text{ or } \frac{8}{5}$$

P3. Find the common binomial factor.

$$x^2+2x-3=(x+3)(x-1)$$
$$x^2+7x+12=(x+4)(x+3)$$

The common factor is $x+3$.

P5. Factor completely.

$$x^2-5x-6=(x-6)(x+1)$$

Section P.5 Exercises

1. Simplify.

$$\frac{x^2-x-20}{3x-15}=\frac{(x+4)(x-5)}{3(x-5)}=\frac{x+4}{3}$$

3. Simplify.

$$\frac{x^3-9x}{x^3+x^2-6x}=\frac{x(x^2-9)}{x(x^2+x-6)}=\frac{x(x-3)(x+3)}{x(x+3)(x-2)}=\frac{x-3}{x-2}$$

5. Simplify.

$$\frac{a^3+8}{a^2-4}=\frac{(a+2)(a^2-2a+4)}{(a-2)(a+2)}=\frac{a^2-2a+4}{a-2}$$

7. Simplify.

$$\frac{x^2+3x-40}{-(x^2-3x-10)}=\frac{(x-5)(x+8)}{-(x-5)(x+2)}=-\frac{x+8}{x+2}$$

9. Simplify.

$$\frac{4y^3-8y^2+7y-14}{-y^2-5y+14}=\frac{4y^2(y-2)+7(y-2)}{-(y^2+5y-14)}$$
$$=\frac{(y-2)(4y^2+7)}{-(y+7)(y-2)}$$
$$=-\frac{4y^2+7}{y+7}$$

11. Simplify.

$$\left(-\frac{4a}{3b^2}\right)\left(\frac{6b}{a^4}\right)=-\frac{24ab}{3a^4b^2}=-\frac{8}{a^3b}$$

13. Simplify.

$$\left(\frac{6p^2}{5q^2}\right)^{-1}\left(\frac{2p}{3q^2}\right)^2=\frac{5q^2}{6p^2}\cdot\frac{4p^2}{9q^4}=\frac{10}{27q^2}$$

15. Simplify.

$$\frac{x^2+x}{2x+3}\cdot\frac{3x^2+19x+28}{x^2+5x+4}=\frac{x(x+1)}{2x+3}\cdot\frac{(3x+7)(x+4)}{(x+4)(x+1)}$$
$$=\frac{x(3x+7)}{2x+3}$$

17. Simplify.

$$\frac{3x-15}{2x^2-50}\cdot\frac{2x^2+16x+30}{6x+9}$$
$$=\frac{3(x-5)}{2(x^2-25)}\cdot\frac{2(x^2+8x+15)}{3(2x+3)}$$
$$=\frac{3(x-5)}{2(x-5)(x+5)}\cdot\frac{2(x+3)(x+5)}{3(2x+3)}=\frac{x+3}{2x+3}$$

19. Simplify.

$$\frac{12y^2+28y+15}{6y^2+35y+25}\div\frac{2y^2-y-3}{3y^2+11y-20}$$
$$=\frac{(6y+5)(2y+3)}{(6y+5)(y+5)}\cdot\frac{(3y-4)(y+5)}{(2y-3)(y+1)}$$
$$=\frac{(2y+3)(3y-4)}{(2y-3)(y+1)}$$

21. Simplify.

$$\frac{a^2+9}{a^2-64}\div\frac{a^3-3a^2+9a-27}{a^2+5a-24}$$
$$=\frac{a^2+9}{(a-8)(a+8)}\cdot\frac{(a-3)(a+8)}{a^2(a-3)+9(a-3)}$$
$$=\frac{a^2+9}{(a-8)(a+8)}\cdot\frac{(a-3)(a+8)}{(a-3)(a^2+9)}=\frac{1}{a-8}$$

23. Simplify.

$$\frac{p+5}{r}+\frac{2p-7}{r}=\frac{p+5+2p-7}{r}=\frac{3p-2}{r}$$

25. Simplify.

$$\frac{5y-7}{y+4}-\frac{2y-3}{y+4}=\frac{(5y-7)-(2y-3)}{y+4}$$
$$=\frac{5y-7-2y+3}{y+4}$$
$$=\frac{3y-4}{y+4}$$

27. Simplify.

$$\frac{x}{x-5}+\frac{7x}{x+3}=\frac{x(x+3)+7x(x-5)}{(x-5)(x+3)}$$
$$=\frac{x^2+3x+7x^2-35x}{(x-5)(x+3)}$$
$$=\frac{8x^2-32x}{(x-5)(x+3)}$$
$$=\frac{8x(x-4)}{(x-5)(x+3)}$$

29. Simplify.

$$\frac{4z}{2z-3}+\frac{5z}{z-5}=\frac{4z(z-5)+5z(2z-3)}{(2z-3)(z-5)}$$
$$=\frac{4z^2-20z+10z^2-15z}{(2z-3)(z-5)}$$
$$=\frac{14z^2-35z}{(2z-3)(z-5)}$$
$$=\frac{7z(2z-5)}{(2z-3)(z-5)}$$

31. Simplify.

$$\frac{x}{x^2-9}-\frac{3x-1}{x^2+7x+12}=\frac{x}{(x-3)(x+3)}-\frac{3x-1}{(x+3)(x+4)}$$
$$=\frac{x(x+4)-(3x-1)(x-3)}{(x-3)(x+3)(x+4)}$$
$$=\frac{(x^2+4x)-(3x^2-10x+3)}{(x-3)(x+3)(x+4)}$$
$$=\frac{x^2+4x-3x^2+10x-3}{(x-3)(x+3)(x+4)}$$
$$=\frac{-2x^2+14x-3}{(x-3)(x+3)(x+4)}$$

33. Simplify.

$$\frac{1}{x}+\frac{2}{3x-1}\cdot\frac{3x^2+11x-4}{x-5}=\frac{1}{x}+\frac{2}{3x-1}\cdot\frac{(3x-1)(x+4)}{(x-5)}$$
$$=\frac{1}{x}+\frac{2(x+4)}{x-5}$$
$$=\frac{1(x-5)+x[2(x+4)]}{x(x-5)}$$
$$=\frac{x-5+2x^2+8x}{x(x-5)}$$
$$=\frac{2x^2+9x-5}{x(x-5)}$$
$$=\frac{(2x-1)(x+5)}{x(x-5)}$$

35. Simplify.

$$\frac{q+1}{q-3}-\frac{2q}{q-3}\div\frac{q+5}{q-3}=\frac{q+1}{q-3}-\frac{2q}{q-3}\cdot\frac{q-3}{q+5}$$
$$=\frac{q+1}{q-3}-\frac{2q}{q+5}$$
$$=\frac{(q+1)(q+5)-2q(q-3)}{(q-3)(q+5)}$$
$$=\frac{q^2+6q+5-2q^2+6q}{(q-3)(q+5)}$$
$$=\frac{-q^2+12q+5}{(q-3)(q+5)}$$

37. Simplify.

$$\frac{1}{x^2+7x+12}+\frac{1}{x^2-9}+\frac{1}{x^2-16}$$
$$=\frac{1}{(x+3)(x+4)}+\frac{1}{(x-3)(x+3)}+\frac{1}{(x-4)(x+4)}$$
$$=\frac{1(x-3)(x-4)+1(x-4)(x+4)+1(x-3)(x+3)}{(x+3)(x+4)(x-3)(x-4)}$$
$$=\frac{x^2-7x+12+x^2-16+x^2-9}{(x+3)(x+4)(x-3)(x-4)}$$
$$=\frac{3x^2-7x-13}{(x+3)(x+4)(x-3)(x-4)}$$

39. Simplify.

$$\left(1+\frac{2}{x}\right)\left(3-\frac{1}{x}\right)=\left(\frac{x}{x}+\frac{2}{x}\right)\left(\frac{3x}{x}-\frac{1}{x}\right)$$
$$=\left(\frac{x+2}{x}\right)\left(\frac{3x-1}{x}\right)$$
$$=\frac{(x+2)(3x-1)}{x^2}$$

41. Simplify the complex fraction.

$$\frac{4+\frac{1}{x}}{1-\frac{1}{x}}=\frac{\left(4+\frac{1}{x}\right)x}{\left(1-\frac{1}{x}\right)x}=\frac{4x+1}{x-1}$$

43. Simplify the complex fraction.

$$\frac{\frac{x}{y}-2}{y-x}=\frac{\left(\frac{x}{y}-2\right)y}{(y-x)y}=\frac{x-2y}{y(y-x)}$$

45. Simplify the complex fraction.

$$\frac{5-\frac{1}{x+2}}{1+\frac{3}{1+\frac{3}{x}}}=\frac{5-\frac{1}{x+2}}{1+\frac{3}{\frac{1(x)}{x}+\frac{3}{x}}}=\frac{5-\frac{1}{x+2}}{1+\frac{3}{\frac{x+3}{x}}}$$
$$=\frac{5-\frac{1}{x+2}}{1+3\div\frac{x+3}{x}}=\frac{5-\frac{1}{x+2}}{1+3\cdot\frac{x}{x+3}}$$
$$=\frac{5-\frac{1}{x+2}}{1+\frac{3x}{x+3}}=\frac{\frac{5(x+2)}{x+2}-\frac{1}{x+2}}{\frac{1(x+3)}{x+3}+\frac{3x}{x+3}}$$
$$=\frac{\frac{5(x+2)-1}{x+2}}{\frac{1(x+3)+3x}{x+3}}=\frac{\frac{5x+10-1}{x+2}}{\frac{x+3+3x}{x+3}}=\frac{\frac{5x+9}{x+2}}{\frac{4x+3}{x+3}}$$
$$=\frac{5x+9}{x+2}\div\frac{4x+3}{x+3}=\frac{5x+9}{x+2}\cdot\frac{x+3}{4x+3}$$
$$=\frac{(5x+9)(x+3)}{(x+2)(4x+3)}$$

47. Simplify the complex fraction.

$$\frac{1+\frac{1}{b-2}}{1-\frac{1}{b+3}}=\frac{\left(1+\frac{1}{b-2}\right)}{\left(1-\frac{1}{b+3}\right)}\cdot\frac{(b-2)(b+3)}{(b-2)(b+3)}$$
$$=\frac{1(b-2)(b+3)+1(b+3)}{1(b-2)(b+3)-1(b-2)}$$
$$=\frac{b^2+b-6+b+3}{b^2+b-6-b+2}=\frac{b^2+2b-3}{b^2-4}$$
$$=\frac{(b+3)(b-1)}{(b-2)(b+2)}$$

49. Simplify the complex fraction.

$$\frac{1-\frac{1}{x^2}}{1+\frac{1}{x}}=\frac{\left(1-\frac{1}{x^2}\right)}{\left(1+\frac{1}{x}\right)}\cdot\frac{x^2}{x^2}=\frac{x^2-1}{x^2+x}=\frac{(x-1)(x+1)}{x(x+1)}=\frac{x-1}{x}$$

51. Simplify the complex fraction.

$$2-\frac{m}{1-\frac{1-m}{-m}}=2-\frac{m}{1+\frac{1-m}{m}}$$
$$=2-\frac{m}{\frac{1(m)}{m}+\frac{1-m}{m}}=2-\frac{m}{\frac{m+1-m}{m}}$$
$$=2-\frac{m}{\frac{1}{m}}=2-\frac{(m)}{\left(\frac{1}{m}\right)}\cdot\frac{m}{m}$$
$$=2-m^2$$

53. Simplify the complex fraction.

$$\frac{\left(\frac{1}{x}-\frac{x-4}{x+1}\right)}{\frac{x}{x+1}}\cdot\frac{x(x+1)}{x(x+1)}=\frac{x+1-x(x-4)}{x(x)}$$
$$=\frac{x+1-x^2+4x}{x^2}$$
$$=\frac{-x^2+5x+1}{x^2}$$

55. Simplify the complex fraction.

$$\frac{\left(\frac{1}{x+3}-\frac{2}{x-1}\right)}{\left(\frac{x}{x-1}+\frac{3}{x+3}\right)}\cdot\frac{(x+3)(x-1)}{(x+3)(x-1)}=\frac{1(x-1)-2(x+3)}{x(x+3)+3(x-1)}$$
$$=\frac{x-1-2x-6}{x^2+3x+3x-3}$$
$$=\frac{-x-7}{x^2+6x-3}$$

57. Simplify the complex fraction.

$$\frac{\frac{x^2+3x-10}{x^2+x-6}}{\frac{x^2-x-30}{2x^2-15x+18}}=\frac{\frac{(x-2)(x+5)}{(x-2)(x+3)}}{\frac{(x+5)(x-6)}{(2x-3)(x-6)}}=\frac{\frac{x+5}{x+3}}{\frac{x+5}{2x-3}}$$
$$=\frac{x+5}{x+3}\div\frac{x+5}{2x-3}=\frac{x+5}{x+3}\cdot\frac{2x-3}{x+5}$$
$$=\frac{2x-3}{x+3}$$

59. Simplify and write answer with positive exponents.

$$\frac{a^{-1}+b^{-1}}{a-b}=\frac{\frac{1}{a}+\frac{1}{b}}{a-b}=\frac{\frac{1b}{ab}+\frac{1a}{ab}}{a-b}=\frac{\frac{b+a}{ab}}{a-b}$$
$$=\frac{b+a}{ab}\div(a-b)=\frac{b+a}{ab}\cdot\frac{1}{a-b}$$
$$=\frac{a+b}{ab(a-b)}$$

61. Simplify and write answer with positive exponents.

$$\frac{a^{-1}b-ab^{-1}}{a^2+b^2}=\frac{\frac{b}{a}-\frac{a}{b}}{a^2+b^2}=\frac{\frac{(b)b}{(b)a}-\frac{a(a)}{b(a)}}{a^2+b^2}$$
$$=\frac{\frac{b^2}{ab}-\frac{a^2}{ab}}{a^2+b^2}=\frac{\frac{b^2-a^2}{ab}}{a^2+b^2}$$
$$=\frac{b^2-a^2}{ab}\div a^2+b^2$$
$$=\frac{b^2-a^2}{ab}\cdot\frac{1}{a^2+b^2}$$
$$=\frac{b^2-a^2}{ab(a^2+b^2)}$$
$$=\frac{(b-a)(b+a)}{ab(a^2+b^2)}$$

63. a. $$\frac{2}{\frac{1}{180}+\frac{1}{110}}$$
$$=\frac{2}{\frac{110+180}{180(110)}}$$
$$=2\div\frac{290}{180(110)}=2\cdot\frac{(180)(110)}{290}$$

≈ 136.55 mph (to the nearest hundredth)

b. $$\frac{2}{\frac{1}{v_1}+\frac{1}{v_2}}=\frac{2}{\frac{v_2+v_1}{v_1v_2}}=\frac{2v_1v_2}{v_2+v_1}=\frac{2v_1v_2}{v_1+v_2}$$

65. $$\frac{1}{x}+\frac{1}{x+1}=\frac{x+1+x}{x(x+1)}=\frac{2x+1}{x(x+1)}$$

67. $$\frac{1}{x-2}+\frac{1}{x}+\frac{1}{x+2}$$
$$=\frac{x(x+2)+(x-2)(x+2)+x(x-2)}{x(x-2)(x+2)}$$
$$=\frac{x^2+2x+x^2-4+x^2-2x}{x(x-2)(x+2)}$$
$$=\frac{3x^2-4}{x(x-2)(x+2)}$$

Prepare for Section P.6

P1. Simplify.

$$(2-3x)(4-5x)=8-10x-12x+15x^2$$
$$=15x^2-22x+8$$

P3. Simplify.

$$\sqrt{96}=\sqrt{16\cdot 6}=4\sqrt{6}$$

P5. Simplify.

$$\frac{5+\sqrt{2}}{3-\sqrt{2}}=\frac{5+\sqrt{2}}{3-\sqrt{2}}\cdot\frac{3+\sqrt{2}}{3+\sqrt{2}}=\frac{15+8\sqrt{2}+2}{9-2}=\frac{17+8\sqrt{2}}{7}$$

Section P.6 Exercises

1. Write the complex number in standard form.

$\sqrt{-81}=i\sqrt{81}=9i$

3. Write the complex number in standard form.

$\sqrt{-98}=i\sqrt{98}=7i\sqrt{2}$

5. Write the complex number in standard form.

$\sqrt{16}+\sqrt{-81}=4+i\sqrt{81}=4+9i$

7. Write the complex number in standard form.

$5+\sqrt{-49}=5+i\sqrt{49}=5+7i$

9. Write the complex number in standard form.

$8-\sqrt{-18}=8-i\sqrt{18}=8-3i\sqrt{2}$

11. Simplify and write in standard form.

$$\begin{aligned}(5+2i)+(6-7i)&=5+2i+6-7i\\&=(5+6)+(2i-7i)\\&=11-5i\end{aligned}$$

13. Simplify and write in standard form.

$$\begin{aligned}(-2-4i)-(5-8i)&=-2-4i-5+8i\\&=(-2-5)+(-4i+8i)\\&=-7+4i\end{aligned}$$

15. Simplify and write in standard form.

$$\begin{aligned}(1-3i)+(7-2i)&=1-3i+7-2i\\&=(1+7)+(-3i-2i)\\&=8-5i\end{aligned}$$

17. Simplify and write in standard form.

$$\begin{aligned}(-3-5i)-(7-5i)&=-3-5i-7+5i\\&=(-3-7)+(-5i+5i)\\&=-10\end{aligned}$$

19. Simplify and write in standard form.

$$\begin{aligned}8i-(2-8i)&=8i-2+8i\\&=-2+(8i+8i)\\&=-2+16i\end{aligned}$$

21. Simplify and write in standard form.

$5i\cdot 8i=40i^2=40(-1)=-40$

23. Simplify and write in standard form.

$$\begin{aligned}\sqrt{-50}\cdot\sqrt{-2}&=i\sqrt{50}\cdot i\sqrt{2}=5i\sqrt{2}\cdot i\sqrt{2}\\&=5i^2(\sqrt{2})^2=5(-1)(2)\\&=-10\end{aligned}$$

25. Simplify and write in standard form.

$$\begin{aligned}3(2+5i)-2(3-2i)&=6+15i-6+4i\\&=(6-6)+(15i+4i)\\&=19i\end{aligned}$$

27. Simplify and write in standard form.

$$\begin{aligned}(4+2i)(3-4i)&=4(3-4i)+(2i)(3-4i)\\&=12-16i+6i-8i^2\\&=12-16i+6i-8(-1)\\&=12-16i+6i+8\\&=(12+8)+(-16i+6i)\\&=20-10i\end{aligned}$$

29. Simplify and write in standard form.

$$\begin{aligned}(-3-4i)(2+7i)&=-3(2+7i)-4i(2+7i)\\&=-6-21i-8i-28i^2\\&=-6-21i-8i-28(-1)\\&=-6-21i-8i+28\\&=(-6+28)+(-21i-8i)\\&=22-29i\end{aligned}$$

31. Simplify and write in standard form.

$$\begin{aligned}(4-5i)(4+5i)&=4(4+5i)-5i(4+5i)\\&=16+20i-20i-25i^2\\&=16+20i-20i-25(-1)\\&=16+20i-20i+25\\&=(16+25)+(20i-20i)\\&=41\end{aligned}$$

33. Simplify and write in standard form.

$$\begin{aligned}(3+\sqrt{-4})(2-\sqrt{-9})&=(3+i\sqrt{4})(2-i\sqrt{9})\\&=(3+2i)(2-3i)\\&=3(2-3i)+2i(2-3i)\\&=6-9i+4i-6i^2\\&=6-9i+4i-6(-1)\\&=6-9i+4i+6\\&=(6+6)+(-9i+4i)\\&=12-5i\end{aligned}$$

35. Simplify and write in standard form.

$$\begin{aligned}
&(3+2\sqrt{-18})(2+2\sqrt{-50})\\
&=(3+2i\sqrt{18})(2+2i\sqrt{50})\\
&=[3+2i(3\sqrt{2})][2+2i(5\sqrt{2})]\\
&=(3+6i\sqrt{2})(2+10i\sqrt{2})\\
&=3(2+10i\sqrt{2})+6i\sqrt{2}(2+10i\sqrt{2})\\
&=6+30i\sqrt{2}+12i\sqrt{2}+60i^2(\sqrt{2})^2\\
&=6+30i\sqrt{2}+12i\sqrt{2}+60(-1)(2)\\
&=6+30i\sqrt{2}+12i\sqrt{2}-120\\
&=(6-120)+(30i\sqrt{2}+12i\sqrt{2})\\
&=-114+42i\sqrt{2}
\end{aligned}$$

37. Write as a complex number in standard form.

$$\frac{6}{i}=\frac{6}{i}\cdot\frac{i}{i}=\frac{6i}{i^2}=\frac{6i}{-1}=-6i$$

39. Write as a complex number in standard form.

$$\begin{aligned}
\frac{6+3i}{i}&=\frac{6+3i}{i}\cdot\frac{i}{i}=\frac{6i+3i^2}{i^2}\\
&=\frac{6i+3(-1)}{-1}=\frac{6i-3}{-1}\\
&=3-6i
\end{aligned}$$

41. Write as a complex number in standard form.

$$\begin{aligned}
\frac{1}{7+2i}&=\frac{1}{7+2i}\cdot\frac{7-2i}{7-2i}=\frac{1(7-2i)}{(7+2i)(7-2i)}\\
&=\frac{7-2i}{49-4i^2}=\frac{7-2i}{49-4(-1)}=\frac{7-2i}{49+4}\\
&=\frac{7-2i}{53}=\frac{7}{53}-\frac{2}{53}i
\end{aligned}$$

43. Write as a complex number in standard form.

$$\begin{aligned}
\frac{2i}{1+i}&=\frac{2i}{1+i}\cdot\frac{1-i}{1-i}=\frac{2i(1-i)}{(1+i)(1-i)}\\
&=\frac{2i-2i^2}{1-i^2}=\frac{2i-2(-1)}{1-(-1)}\\
&=\frac{2i+2}{1+1}=\frac{2+2i}{2}=\frac{2}{2}+\frac{2}{2}i\\
&=1+i
\end{aligned}$$

45. Write as a complex number in standard form.

$$\begin{aligned}
\frac{5-i}{4+5i}&=\frac{5-i}{4+5i}\cdot\frac{4-5i}{4-5i}=\frac{(5-i)(4-5i)}{(4+5i)(4-5i)}\\
&=\frac{5(4-5i)-i(4-5i)}{4(4-5i)+5i(4-5i)}=\frac{20-25i-4i+5i^2}{16-20i+20i-25i^2}\\
&=\frac{20-25i-4i+5(-1)}{16-25(-1)}=\frac{20-25i-4i-5}{16+25}\\
&=\frac{(20-5)+(-25i-4i)}{16+25}\\
&=\frac{15-29i}{41}=\frac{15}{41}-\frac{29}{41}i
\end{aligned}$$

47. Write as a complex number in standard form.

$$\begin{aligned}
\frac{3+2i}{3-2i}&=\frac{3+2i}{3-2i}\cdot\frac{3+2i}{3+2i}=\frac{(3+2i)^2}{(3-2i)(3+2i)}\\
&=\frac{3^2+2(3)(2i)+(2i)^2}{3^2-(2i)^2}=\frac{9+12i+4i^2}{9-4i^2}\\
&=\frac{9+12i+4(-1)}{9-4(-1)}\\
&=\frac{9+12i-4}{9+4}=\frac{5+12i}{13}\\
&=\frac{5}{13}+\frac{12}{13}i
\end{aligned}$$

49. Write as a complex number in standard form.

$$\begin{aligned}
\frac{-7+26i}{4+3i}&=\frac{-7+26i}{4+3i}\cdot\frac{4-3i}{4-3i}=\frac{(-7+26i)(4-3i)}{(4+3i)(4-3i)}\\
&=\frac{-7(4-3i)+26i(4-3i)}{4^2-(3i)^2}\\
&=\frac{-28+21i+104i-78i^2}{16-9i^2}\\
&=\frac{-28+21i+104i-78(-1)}{16-9(-1)}\\
&=\frac{-28+21i+104i+78}{16+9}=\frac{50+125i}{25}\\
&=\frac{50}{25}+\frac{125}{25}i\\
&=2+5i
\end{aligned}$$

51. Write as a complex number in standard form.

$$\begin{aligned}
(3-5i)^2&=3^3+2(3)(-5i)+(-5i)^2\\
&=9-30i+25i^2\\
&=9-30i+25(-1)\\
&=9-30i-25\\
&=-16-30i
\end{aligned}$$

53. Write as a complex number in standard form.

$$\begin{aligned}(1+2i)^3 &= (1+2i)(1+2i)^2\\ &= (1+2i)[1^2+2(1)(2i)+(2i)^2]\\ &= (1+2i)[1+4i+4i^2]\\ &= (1+2i)[1+4i-4]\\ &= (1+2i)(-3+4i)\\ &= 1(-3+4i)+2i(-3+4i)\\ &= -3+4i-6i+8i^2\\ &= -3+4i-6i+8(-1)\\ &= -3+4i-6i-8\\ &= -11-2i\end{aligned}$$

55. Evaluate the power of i.

Use the Powers of i Theorem.

The remainder of $15 \div 4$ is 3.

$$i^{15} = i^3 = -i$$

57. Evaluate the power of i.

Use the Powers of i Theorem.

The remainder of $40 \div 4$ is 0.

$$-i^{40} = -(i^0) = -1$$

59. Evaluate the power of i.

Use the Powers of i Theorem.

The remainder of $25 \div 4$ is 1.

$$\frac{1}{i^{25}} = \frac{1}{i} = \frac{1}{i}\cdot\frac{i}{i} = \frac{i}{i^2} = \frac{i}{-1} = -i$$

61. Evaluate the power of i.

Use the Powers of i Theorem.

The remainder of $34 \div 4$ is 2.

$$i^{-34} = \frac{1}{i^{34}} = \frac{1}{i^2} = \frac{1}{-1} = -1$$

63. Evaluate.

Use $a = 3$, $b = -3$, $c = 3$.

$$\begin{aligned}\frac{-b+\sqrt{b^2-4ac}}{2a} &= \frac{-(-3)+\sqrt{(-3)^2-4(3)(3)}}{2(3)}\\ &= \frac{3+\sqrt{9-36}}{6} = \frac{3+\sqrt{-27}}{6}\\ &= \frac{3+i\sqrt{27}}{6} = \frac{3+3i\sqrt{3}}{6}\\ &= \frac{3}{6}+\frac{3\sqrt{3}}{6}i = \frac{1}{2}+\frac{\sqrt{3}}{2}i\end{aligned}$$

65. Evaluate.

Use $a = 2$ $b = 6$, $c = 6$.

$$\begin{aligned}\frac{-b+\sqrt{b^2-4ac}}{2a} &= \frac{-(6)+\sqrt{(6)^2-4(2)(6)}}{2(2)}\\ &= \frac{-6+\sqrt{36-48}}{4} = \frac{-6+\sqrt{-12}}{4}\\ &= \frac{-6+i\sqrt{12}}{4} = \frac{-6+2i\sqrt{3}}{4}\\ &= \frac{-6}{4}+\frac{2i\sqrt{3}}{4} = -\frac{3}{2}+\frac{\sqrt{3}}{2}i\end{aligned}$$

67. Evaluate.

Use $a = 4$, $b = -4$, $c = 2$.

$$\begin{aligned}\frac{-b+\sqrt{b^2-4ac}}{2a} &= \frac{-(-4)+\sqrt{(-4)^2-4(4)(2)}}{2(4)}\\ &= \frac{4+\sqrt{16-32}}{8} = \frac{4+\sqrt{-16}}{8}\\ &= \frac{4+i\sqrt{16}}{8} = \frac{4+4i}{8}\\ &= \frac{4}{8}+\frac{4i}{8} = \frac{1}{2}+\frac{1}{2}i\end{aligned}$$

Exploring Concepts with Technology

1. Use a calculator to find the first 20 iterations.

Below is a chart of a selection of those iterations.

Iteration	$p+3p(1-p)$	$4p-3p^2$
1	0.5	0.5
3	0.95703125	0.95703125
5	0.8198110957	0.8198110957
10	0.3846309658	0.3846309658
15	0.5610061236	0.5610061236
19	1.218765181	1.218765181
20	0.4188950251	0.4188950245

Chapter P Review Exercises

1. Classify the number. [P.1]

3: integer, rational number, real number, prime number

3. Classify the number. [P.1]

$-\frac{1}{2}$: rational number, real number

5. List the four smallest elements of the set. [P.1]

Let $x = 0, 1, 2, 3$. (We could have used $x = -3, -2, -1, 0$.) Then $\{y \mid y = x^2,\ x \in \text{integers}\} = \{0, 1, 4, 9\}$

7. Perform the operation. [P.1]

$A \cup B = \{1, 2, 3, 5, 7, 11\}$

9. Graph the interval, write in set-builder notation. [P.1]

$[-3, 2)$

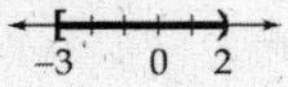

$-3 \quad 0 \quad 2$

$\{x | -3 \le x < 2\}$

11. Graph the set, write in interval notation. [P.1]

$\{x | -4 < x \le 2\}$

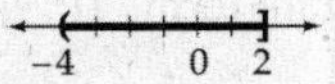

$-4 \quad 0 \quad 2$

$(-4, 2]$

13. Write without absolute value symbols. [P.1]

$|7| = 7$

15. Write without absolute value symbols. [P.1]

$|4 - \pi| = 4 - \pi$, because $4 > \pi$

17. Write without absolute value symbols. [P.1]

$|x-2| + |x+1|, \; -1 < x < 2$

$= 2 - x + x + 1$

$= 3$

19. Find the distance between –3 and 7. [P.1]

$|-3-7| = |-10| = 10$

21. Evaluate. [P.1]

$-4^4 = -4 \cdot 4 \cdot 4 \cdot 4 = -256$

23. Evaluate. [P.1]

$-5 \cdot 3^2 + 4\{5 - 2[-6 - (-4)]\}$

$= -5 \cdot 3^2 + 4\{5 - 2[-6 + 4]\}$

$= -5 \cdot 3^2 + 4\{5 - 2[-2]\}$

$= -5 \cdot 3^2 + 4\{5 + 4\}$

$= -5 \cdot 3 \cdot 3 + 4\{9\}$

$= -45 + 36$

$= -9$

25. Evaluate. [P.1]

$-3x^3 - 4xy - z^2, \quad x = -2, \; y = 3, \; z = -5$

$-3(-2)^3 - 4(-2)(3) - (-5)^2$

$= -3(-2)(-2)(-2) - 4(-2)(3) - (-5)(-5)$

$= 24 + 24 - 25$

$= 23$

27. Identify the property illustrated. [P.1]

Distributive property

29. Identify the property illustrated. [P.1]

Associative property of multiplication

31. Identify the property illustrated. [P.1]

Identity property of addition

33. Identify the property illustrated. [P.1]

Symmetric property of equality

35. Simplify the expression. [P.1]

$8 - 3(2x - 5) = 8 - 6x + 15 = -6x + 23$

37. Simplify the exponential expression. [P.2]

$-2^{-5} = -\dfrac{1}{2^5} = -\dfrac{1}{2 \cdot 2 \cdot 2 \cdot 2 \cdot 2} = -\dfrac{1}{32}$

39. Simplify the exponential expression. [P.2]

$\dfrac{2}{z^{-4}} = 2z^4$

41. Write the number in scientific notation. [P.2]

$620{,}000 = 6.2 \times 10^5$

43. Write the number in decimal form. [P.2]

$3.5 \times 10^4 = 35{,}000$

45. Evaluate the exponential expression. [P.2]

$25^{1/2} = \sqrt{25} = 5$

47. Evaluate the exponential expression. [P.2]

$36^{-1/2} = \dfrac{1}{\sqrt{36}} = \dfrac{1}{6}$

49. Simplify the expression. [P.2]

$(-4x^3y^2)(6x^4y^3) = -4(6)x^{3+4}y^{2+3} = -24x^7y^5$

51. Simplify the expression. [P.2]

$(-3x^{-2}y^3)^{-3} = (-3)^{-3}(x^{-2})^{-3}(y^3)^{-3}$

$= -\dfrac{1}{27}x^6y^{-9}$

$= -\dfrac{x^6}{27y^9}$

53. Simplify the expression. [P.2]

$$(-4x^{-3}y^2)^{-2}(8x^{-2}y^{-3})^2 = \left(\frac{1}{-4x^{-3}y^2}\right)^2(8x^{-2}y^{-3})^2$$
$$= \frac{1}{16x^{-6}y^4}\cdot 64x^{-4}y^{-6}$$
$$= 4x^{6-4}y^{-4-6} = 4x^2y^{-10}$$
$$= \frac{4x^2}{y^{10}}$$

55. Simplify the expression. [P.2]

$$(x^{-1/2})(x^{3/4}) = x^{-1/2+3/4} = x^{1/4}$$

57. Simplify the expression. [P.2]

$$\left(\frac{8x^{5/4}}{x^{1/2}}\right)^{2/3} = (8x^{5/4-1/2})^{2/3} = (8x^{5/4-2/4})^{2/3}$$
$$= (8x^{3/4})^{2/3} = 8^{2/3}x^{(3/4)(2/3)}$$
$$= (2^3)^{2/3}x^{(3/4)(2/3)} = 2^2x^{1/2}$$
$$= 4x^{1/2}$$

59. Simplify the radical expression. [P.2]

$$\sqrt{48a^2b^7} = \sqrt{16a^2b^6\cdot 3b} = 4ab^3\sqrt{3b}$$

61. Simplify the radical expression. [P.2]

$$\sqrt[3]{-135x^2y^7} = \sqrt[3]{-27y^6\cdot 5x^2y} = -3y^2\sqrt[3]{5x^2y}$$

63. Simplify the radical expression. [P.2]

$$b\sqrt{8a^4b^3} + 2a\sqrt{18a^2b^5}$$
$$= b\sqrt{4a^4b^2}\sqrt{2b} + 2a\sqrt{9a^2b^4}\sqrt{2b}$$
$$= 2a^2b^2\sqrt{2b} + 6a^2b^2\sqrt{2b}$$
$$= 8a^2b^2\sqrt{2b}$$

65. Simplify the radical expression. [P.2]

$$(3+2\sqrt{5})(7-3\sqrt{5}) = 21 - 9\sqrt{5} + 14\sqrt{5} - 30$$
$$= -9 + 5\sqrt{5}$$

67. Simplify the radical expression. [P.2]

$$(4-2\sqrt{7})^2 = 16 - 2(4\cdot 2\sqrt{7}) + 28 = 44 - 16\sqrt{7}$$

69. Simplify the radical expression. [P.2]

$$\frac{6}{\sqrt{8}} = \frac{6}{2\sqrt{2}} = \frac{3}{\sqrt{2}} = \frac{3}{\sqrt{2}}\cdot\frac{\sqrt{2}}{\sqrt{2}} = \frac{3\sqrt{2}}{2}$$

71. Simplify the radical expression. [P.2]

$$\frac{3+2\sqrt{7}}{9-3\sqrt{7}} = \frac{3+2\sqrt{7}}{9-3\sqrt{7}}\cdot\frac{9+3\sqrt{7}}{9+3\sqrt{7}}$$
$$= \frac{27+9\sqrt{7}+18\sqrt{7}+42}{81-63}$$
$$= \frac{69+27\sqrt{7}}{18} = \frac{3(23+9\sqrt{7})}{3\cdot 6}$$
$$= \frac{23+9\sqrt{7}}{6}$$

73. Write the polynomial in standard for, identify the degree, leading coefficient and constant term. [P.3]

$$-x^3 - 7x^2 + 4x + 5$$

degree: 3; leading coefficient: –1; constant: 5

75. Perform the operation. [P.3]

$$(2a^2+3a-7)+(-3a^2-5a+6)$$
$$= [2a^2+(-3a^2)]+[3a+(-5a)]+[-7+6]$$
$$= -a^2-2a-1$$

77. Perform the operation. [P.3]

$$(3x-2)(2x^2+4x-9)$$
$$= 3x(2x^2+4x-9)-2(2x^2+4x-9)$$
$$= 6x^3+12x^2-27x-4x^2-8x+18$$
$$= 6x^3+8x^2-35x+18$$

79. Perform the operation. [P.3]

$$(3x-4)(x+2) = 3x^2+6x-4x-8$$
$$= 3x^2+2x-8$$

81. Perform the operation. [P.3]

$$(2x+5)^2 = 4x^2+2(2x\cdot 5)+25$$
$$= 4x^2+20x+25$$

83. Factor out the GCF. [P.4]

$$12x^3y^4+10x^2y^3-34xy^2 = 2xy^2(6x^2y^2+5xy-17)$$

85. Factor out the GCF. [P.4]

$$(2x+7)(3x-y)-(3x+2)(3x-y)$$
$$= (3x-y)[2x+7-(3x+2)]$$
$$= (3x-7)(2x+7-3x-2)$$
$$= (3x-7)(-x+5)$$

87. Factor. [P.4]

$$x^2+7x-18=(x-2)(x+9)$$

89. Factor. [P.4]

$$2x^2+11x+12=(2x+3)(x+4)$$

91. Factor. [P.4]

$$\begin{aligned}6x^3y^2-12x^2y^2-144xy^2&=6xy^2(x^2-2x-24)\\&=6xy^2(x-6)(x+4)\end{aligned}$$

93. Factor. [P.4]

$$9x^2-100=(3x-10)(3x+10)$$

95. Factor. [P.4]

$$x^4-5x^2-6=(x^2)^2-5x^2-6=(x^2-6)(x^2+1)$$

97. Factor. [P.4]

$$x^3-27=(x-3)(x^2+3x+9)$$

99. Factor. [P.4]

$$\begin{aligned}&4x^4-x^2-4x^2y^2+y^2\\&=x^2(4x^2-1)-y^2(4x^2-1)\\&=(4x^2-1)(x^2-y^2)\\&=(2x+1)(2x-1)(x+y)(x-y)\end{aligned}$$

101. Factor. [P.4]

$$\begin{aligned}24a^2b^2-14ab^3-90b^4&=2b^2(12a^2-7ab-45b^2)\\&=2b^2(3a+5b)(4a-9b)\end{aligned}$$

103. Simplify the rational expression. [P.5]

$$\frac{6x^2-19x+10}{2x^2+3x-20}=\frac{(3x-2)(2x-5)}{(2x-5)(x+4)}=\frac{3x-2}{x+4}$$

105. Perform the operation and simplify. [P.5]

$$\begin{aligned}&\frac{10x^2+13x-3}{6x^2-13x-5}\cdot\frac{6x^2+5x+1}{10x^2+3x-1}\\&=\frac{(2x+3)(5x-1)}{(2x-5)(3x+1)}\cdot\frac{(2x+1)(3x+1)}{(2x+1)(5x-1)}\\&=\frac{2x+3}{2x-5}\end{aligned}$$

107. Perform the operation and simplify. [P.5]

$$\begin{aligned}\frac{x}{x^2-9}+\frac{2x}{x^2+x-12}&=\frac{x}{(x-3)(x+3)}+\frac{2x}{(x+4)(x-3)}\\&=\frac{x(x+4)+2x(x+3)}{(x-3)(x+3)(x+4)}\\&=\frac{x^2+4x+2x^2+6x}{(x-3)(x+3)(x+4)}\\&=\frac{3x^2+10x}{(x-3)(x+3)(x+4)}\\&=\frac{x(3x+10)}{(x-3)(x+3)(x+4)}\end{aligned}$$

109. Simplify the complex fraction. [P.5]

$$\begin{aligned}\frac{2+\frac{1}{x-5}}{3-\frac{2}{x-5}}&=\frac{\left(2+\frac{1}{x-5}\right)}{\left(3-\frac{2}{x-5}\right)}\cdot\frac{x-5}{x-5}\\&=\frac{2(x-5)+1}{3(x-5)-2}=\frac{2x-10+1}{3x-15-2}\\&=\frac{2x-9}{3x-17}\end{aligned}$$

111. Write the complex number in standard form. [P.6]

$$5+\sqrt{-64}=5+8i$$

113. Perform the operation and simplify. [P.6]

$$\begin{aligned}(2-3i)+(4+2i)&=2-3i+4+2i\\&=(2+4)+(-3i+2i)\\&=6-i\end{aligned}$$

115. Perform the operation and simplify. [P.6]

$$\begin{aligned}2i(3-4i)&=6i-8i^2\\&=6i-8(-1)\\&=6i+8\\&=8+6i\end{aligned}$$

117. Perform the operation and simplify. [P.6]

$$\begin{aligned}(3+i)^2&=3^2+2(3)(i)+i^2\\&=9+6i+(-1)\\&=8+6i\end{aligned}$$

119. Perform the operation and simplify. [P.6]

$$\begin{aligned}\frac{4-6i}{2i}&=\frac{2(2-3i)}{2i}=\frac{\cancel{2}(2-3i)}{\cancel{2}i}=\frac{2-3i}{i}\\&=\frac{2-3i}{i}\cdot\frac{i}{i}=\frac{2i-3i^2}{i^2}=\frac{2i-3(-1)}{-1}=\frac{2i+3}{-1}\\&=-3-2i\end{aligned}$$

Chapter P Test

1. Distributive property [P.1]

3. Simplify. [P.1]

$$|x+1|-|x-5|, \ -1<x<4$$
$$=x+1+x-5$$
$$=2x-4$$

5. Simplify. [P.2]

$$\frac{(2a^{-1}bc^{-2})^2}{(3^{-1}b)(2^{-1}ac^{-2})^3}=\frac{2^2a^{-2}b^2c^{-4}}{(3^{-1}b)(2^{-3}a^3c^{-6})}$$
$$=\frac{2^2\cdot 2^3\cdot 3^1\cdot b^2c^6}{ba^3a^2c^4}$$
$$=\frac{2^5\cdot 3\cdot bc^2}{a^5}$$
$$=\frac{96bc^2}{a^5}$$

7. Simplify. [P.2]

$$\frac{x^{1/3}y^{-3/4}}{x^{-1/2}y^{3/2}}=x^{1/3-(-1/2)}\,y^{-3/4-3/2}$$
$$=x^{1/3+1/2}\,y^{-3/4-3/2}=x^{2/6+3/6}\,y^{-3/4-6/4}$$
$$=x^{5/6}\,y^{-9/4}$$
$$=\frac{x^{5/6}}{y^{9/4}}$$

9. Simplify. [P.2]

$$(2\sqrt{3}-4)(5\sqrt{3}+2)=30+4\sqrt{3}-20\sqrt{3}-8$$
$$=22-16\sqrt{3}$$

11. Simplify. [P.2]

$$\frac{x}{\sqrt[4]{2x^3}}=\frac{x}{\sqrt[4]{2x^3}}\cdot\frac{\sqrt[4]{8x}}{\sqrt[4]{8x}}=\frac{x\sqrt[4]{8x}}{2x}=\frac{\sqrt[4]{8x}}{2}$$

13. Simplify. [P.2]

$$\frac{2+\sqrt{5}}{4-2\sqrt{5}}=\frac{2+\sqrt{5}}{4-2\sqrt{5}}\cdot\frac{4+2\sqrt{5}}{4+2\sqrt{5}}$$
$$=\frac{8+4\sqrt{5}+4\sqrt{5}+10}{16-20}$$
$$=\frac{18+8\sqrt{5}}{-4}=-\frac{2(9+4\sqrt{5})}{2\cdot 2}$$
$$=-\frac{9+4\sqrt{5}}{2}$$

15. Multiply. [P.3]

$$(3a+7b)(2a-9b)=6a^2-27ab+14ab-63b^2$$
$$=6a^2-13ab-63b^2$$

17. Factor. [P.4]

$$7x^2+34x-5=(7x-1)(x+5)$$

19. Factor. [P.4]

$$16x^4-2xy^3=2x(8x^3-y^3)$$
$$=2x(2x-y)(4x^2+2xy+y^2)$$

21. Simplify. [P.5]

$$\frac{x^2-2x-15}{25-x^2}=\frac{(x-5)(x+3)}{(5-x)(5+x)}=\left(\frac{x-5}{5-x}\right)\left(\frac{x+3}{x+5}\right)$$
$$=-1\cdot\left(\frac{x+3}{x+5}\right)=-\frac{x+3}{x+5}$$

23. Multiply. [P.5]

$$\frac{x^2-3x-4}{x^2+x-20}\cdot\frac{x^2+3x-10}{x^2+2x-8}$$
$$=\frac{(x+1)(x-4)}{(x+5)(x-4)}\cdot\frac{(x+5)(x-2)}{(x+4)(x-2)}$$
$$=\frac{x+1}{x+4}$$

25. Simplify. [P.5]

$$x-\frac{x}{x+\frac{1}{2}}=x-\frac{x}{\frac{2x}{2}+\frac{1}{2}}=x-\frac{x}{\frac{2x+1}{2}}$$
$$=x-x\div\frac{2x+1}{2}=x-x\cdot\frac{2}{2x+1}$$
$$=x-\frac{2x}{2x+1}=\frac{x(2x+1)}{2x+1}-\frac{2x}{2x+1}$$
$$=\frac{2x^2+x}{2x+1}-\frac{2x}{2x+1}=\frac{2x^2+x-2x}{2x+1}$$
$$=\frac{2x^2-x}{2x+1}=\frac{x(2x-1)}{2x+1}$$

27. Write in simplest form. [P.6]

$$(4-3i)-(2-5i)=4-3i-2+5i$$
$$=(4-2)+(-3i+5i)$$
$$=2+2i$$

29. Write in simplest form. [P.6]

$$\frac{3+4i}{5-i}=\frac{3+4i}{5-i}\cdot\frac{5+i}{5+i}=\frac{(3+4i)(5+i)}{(5-i)(5+i)}$$
$$=\frac{3(5+i)+4i(5+i)}{5^2-i^2}=\frac{15+3i+20i+4i^2}{25-i^2}$$
$$=\frac{15+3i+20i+4(-1)}{25-(-1)}$$
$$=\frac{15+3i+20i-4}{25+1}=\frac{(15-4)+(3i+20i)}{26}$$
$$=\frac{11+23i}{26}=\frac{11}{26}+\frac{23}{26}i$$

Chapter 1: Equations and Inequalities

Section 1.1 Exercises

1. Solve the equation.

$$\begin{aligned} 2x+10&=40\\ 2x&=40-10\\ 2x&=30\\ x&=15 \end{aligned}$$

3. Solve the equation.

$$\begin{aligned} 5x+2&=2x-10\\ 5x-2x&=-10-2\\ 3x&=-12\\ x&=-4 \end{aligned}$$

5. Solve the equation.

$$\begin{aligned} 2(x-3)-5&=4(x-5)\\ 2x-6-5&=4x-20\\ 2x-11&=4x-20\\ 2x-4x&=-20+11\\ -2x&=-9\\ x&=\frac{9}{2} \end{aligned}$$

7. Solve the equation.

$$\begin{aligned} 3x+5(1-2x)&=4-3(x+1)\\ 3x+5-10x&=4-3x-3\\ -7x+5&=1-3x\\ -4x&=-4\\ x&=1 \end{aligned}$$

9. Solve the equation.

$$\begin{aligned} 4(2r-17)+5(3r-8)&=0\\ 8r-68+15r-40&=0\\ 23r-108&=0\\ 23r&=108\\ r&=\frac{108}{23} \end{aligned}$$

11. Solve the equation.

$$\begin{aligned} \frac{3}{4}x+\frac{1}{2}&=\frac{2}{3}\\ 12\cdot\left(\frac{3}{4}x+\frac{1}{2}\right)&=12\cdot\left(\frac{2}{3}\right)\\ 9x+6&=8\\ 9x&=8-6\\ 9x&=2\\ x&=\frac{2}{9} \end{aligned}$$

13. Solve the equation.

$$\begin{aligned} \frac{2}{3}x-5&=\frac{1}{2}x-3\\ 6\cdot\left(\frac{2}{3}x-5\right)&=6\cdot\left(\frac{1}{2}x-3\right)\\ 4x-30&=3x-18\\ 4x-3x&=-18+30\\ x&=12 \end{aligned}$$

15. Solve the equation.

$$\begin{aligned} 0.2x+0.4&=3.6\\ 0.2x&=3.6-0.4\\ 0.2x&=3.2\\ x&=16 \end{aligned}$$

17. Solve the equation.

$$\begin{aligned} x+0.08(60)&=0.20(60+x)\\ x+4.8&=12+0.20x\\ x-0.20x&=12-4.8\\ 0.80x&=7.2\\ x&=9 \end{aligned}$$

19. Solve the equation.

$$\begin{aligned} 5[x-(4x-5)]&=3-2x\\ 5(x-4x+5)&=3-2x\\ 5(-3x+5)&=3-2x\\ -15x+25&=3-2x\\ -15x+2x&=3-25\\ -13x&=-22\\ x&=\frac{22}{13} \end{aligned}$$

21. Solve the equation.

$$\frac{40-3x}{5}=\frac{6x+7}{8}$$
$$40\cdot\left(\frac{40-3x}{5}\right)=40\cdot\left(\frac{6x+7}{8}\right)$$
$$8(40-3x)=5(6x+7)$$
$$320-24x=30x+35$$
$$-24x-30x=35-320$$
$$-54x=-285$$
$$x=\frac{95}{18}$$

23. Classify: contradiction, conditional equation, identity.

$$-3(x-5)=-3x+15$$
$$-3x+15=-3x+15$$

Identity

25. Classify: contradiction, conditional equation, identity.

$$2x+7=3(x-1)$$
$$2x+7=3x-3$$
$$2x-3x=-3-7$$
$$-x=-10$$
$$x=10$$

Conditional equation

27. Classify: contradiction, conditional equation, identity.

$$\frac{4x+8}{4}=x+8$$
$$4x+8=4(x+8)$$
$$4x+8=4x+32$$
$$8=32$$

Contradiction

29. Classify: contradiction, conditional equation, identity.

$$3[x-2(x-5)]-1=-3x+29$$
$$3[x-2x+10]-1=-3x+29$$
$$3[-x+10]-1=-3x+29$$
$$-3x+30-1=-3x+29$$
$$-3x+29=-3x+29$$

Identity

31. Classify: contradiction, conditional equation, identity.

$$2x-8=-x+9$$
$$3x=17$$
$$x=\frac{17}{3}$$

Conditional equation

33. Solve for x.

$$|x|=4$$
$$x=4 \quad \text{or} \quad x=-4$$

35. Solve for x.

$$|x-5|=2$$
$$x-5=2 \quad \text{or} \quad x-5=-2$$
$$x=7 \qquad x=3$$

37. Solve for x.

$$|2x-5|=11$$
$$2x-5=11 \quad \text{or} \quad 2x-5=-11$$
$$2x=16 \qquad 2x=-6$$
$$x=8 \qquad x=-3$$

39. Solve for x.

$$|2x+6|=10$$
$$2x+6=10 \quad \text{or} \quad 2x+6=-10$$
$$2x=4 \qquad 2x=-16$$
$$x=2 \qquad x=-8$$

41. Solve for x.

$$\left|\frac{x-4}{2}\right|=8$$
$$\frac{x-4}{2}=8 \quad \text{or} \quad \frac{x-4}{2}=-8$$
$$x-4=8(2) \qquad x-4=-8(2)$$
$$x-4=16 \qquad x-4=-16$$
$$x=20 \qquad x=-12$$

43. Solve for x.

$$|2x+5|=-8$$
$$|2x+5|\geq 0$$
$$-8\geq 0$$

Contradiction. There is no solution.

45. Solve for x.

$$2|x+3|+4=34$$
$$2|x+3|=30$$
$$|x+3|=15$$
$$x+3=15 \quad \text{or} \quad x+3=-15$$
$$x=12 \qquad x=-18$$

47. Solve for x.

$$|2x-a|=b, \quad b>0$$

$$2x-a=b \quad \text{or} \quad 2x-a=-b$$
$$2x=a+b \qquad 2x=a-b$$
$$x=\frac{a+b}{2} \qquad x=\frac{a-b}{2}$$

49. Estimate the volume of the pouch.

$$F=-5.5V+5400$$
$$550=-5.5V+5400$$
$$-4850=-5.5V$$
$$881.81\overline{81}=V$$
$$V\approx 882 \text{ cm}^3$$

51. Find the time.

$$d=|210-50t|$$

$$60=210-50t \quad \text{or} \quad -60=210-50t$$
$$-150=-50t \qquad -270=-50t$$
$$t=3 \qquad t=5.4$$

5.4 hours = 5 hours 24 minutes

Ruben will be exactly 60 miles from Barstow after 3 hours and after 5 hours and 24 minutes.

53. Find the number of square yards of carpet. Round to the nearest square yard.

$$45x+550=\text{Cost}$$
$$45x+550=3800$$
$$45x=3250$$
$$x\approx 72$$

Rounded to the nearest yard, 72 square yards can be carpeted for $3800.

55. Find the number of minutes. Round to the nearest tenth of a minute.

$$p=100-\frac{30}{N}t$$
$$25=100-\frac{30}{110}t$$
$$\frac{30}{110}t=75$$
$$t=275 \text{ seconds}$$
$$275 \text{ sec}\cdot\left(\frac{1 \text{ min}}{60 \text{ sec}}\right)\approx 4.6 \text{ min}$$

57. Find the maximum and minimum heart rates. Round to the nearest beat per minute.

$$\max=0.85(220-a) \qquad \min=0.65(220-a)$$
$$=0.85(220-25) \qquad =0.65(220-25)$$
$$=0.85(195) \qquad =0.65(195)$$
$$=165.75 \qquad =126.75$$

The maximum exercise heart rate for a person who is 25 years of age is about 166 beats per minute.

The minimum exercise heart rate for a person who is 25 years of age is about 127 beats per minute.

Prepare for Section 1.2

P1. Evaluate.

$$32-x$$
$$32-8\frac{1}{2}=23\frac{1}{2}$$

P3. Determine the property.

$$2l+2w=2(l+w)$$

Distributive property

P5. Add.

$$\frac{2}{5}x+\frac{1}{3}x=\frac{6}{15}x+\frac{5}{15}x=\frac{11}{15}x$$

Section 1.2 Exercises

1. Solve the formula.

$$V=\frac{1}{3}\pi r^2 h$$
$$3V=\pi r^2 h$$
$$\frac{3V}{\pi r^2}=h$$

3. Solve the formula.

$$I=Prt$$
$$\frac{I}{Pr}=t$$

5. Solve the formula.

$$F=\frac{Gm_1m_2}{d^2}$$
$$Fd^2=Gm_1m_2$$
$$\frac{Fd^2}{Gm_2}=m_1$$

7. Solve the formula.

$$a_n = a_1 + (n-1)d$$
$$a_n - a_1 = (n-1)d$$
$$\frac{a_n - a_1}{n-1} = d$$

9. Solve the formula.

$$S = \frac{a_1}{1-r}$$
$$S(1-r) = a_1$$
$$S - Sr = a_1$$
$$S - a_1 = Sr$$
$$\frac{S - a_1}{S} = r$$

11. Find Brees' quarterback rating.

$$\text{QB rating} = \frac{100}{6}[0.05(C-30) + 0.25(Y-3) + 0.2T + (2.375 - 0.25I)]$$
$$= \frac{100}{6}[0.05(65.0-30) + 0.25(7.98-3) + 0.2(5.4) + (2.375 - 0.25 \cdot 2.7)]$$
$$= 96.25$$

13. Estimate the reading level. Round to the nearest tenth.

$\text{SMOG} = \sqrt{42} + 3 = 9.5$

15. Estimate the reading level. Round to the nearest tenth.

$\text{GFI} = 0.4(14.8 + 15.1) \approx 12.0$

17. Find the width and length.

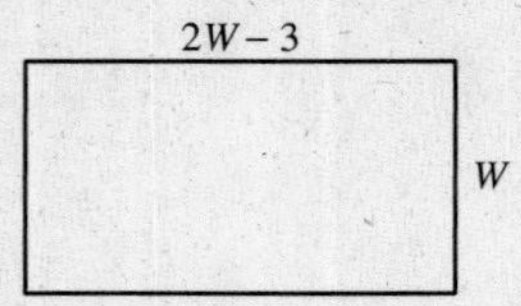

$$P = 2L + 2W$$
$$174 = 2(2W-3) + 2W$$
$$174 = 4W - 6 + 2W$$
$$180 = 6W$$
$$W = 30 \text{ ft}$$
$$L = 2W - 3 = 2(30) - 3 = 60 - 3 = 57 \text{ ft}$$

19. Find the length of each side.

3x 3x x

$$3x + 3x + x = 84$$
$$7x = 84$$
$$x = 12$$
$$3x = 3(12) = 36$$

The shortest side is 12 cm.

The longer sides are each 36 cm.

21. Use similar triangles to find the height of the tree.

Let h = the height of the tree.

$$\frac{h}{10} = \frac{6}{4}$$
$$h = 15 \text{ ft}$$

23. Find the distance from the building.

Let x = the distance from the building.

$$\frac{20}{50} = \frac{20-x}{6}$$
$$300\left(\frac{20}{50}\right) = 300\left(\frac{20-x}{6}\right)$$
$$120 = 1000 - 50x$$
$$50x = 880$$
$$x = 17.6 \text{ ft}$$

25. Find the number of pairs of sunglasses.

Let x = the number of sunglasses.

Profit = Revenue − Cost

$$17{,}884 = 29.99x - 8.95x$$
$$17{,}884 = 21.04x$$
$$x = 850$$

The manufacturer must sell 850 pairs of sunglasses to make a profit of $17,884.

27. Find the price of the book and the bookmark.

Let x = price of book

$10.10 - x$ = price of bookmark.

$$x = 10 + (10.10 - x)$$
$$2x = 20.10$$
$$x = 10.05$$
$$10.10 - 10.05 = 0.05$$

The price of the book is $10.05.

The price of the bookmark is $0.05.

29. Find the cost of the computer last year.

Let x = cost last year.

$$x - 0.20x = 750$$
$$0.80x = 750$$
$$x = 937.50$$

The cost of a computer last year was \$937.50.

31. Find how much was invested in each account.

Let x = amount invested at 8%.

$(14{,}000 - x) =$ amount invested at 6.5%.

$$0.08x + 0.065(14{,}000 - x) = 1024$$
$$0.08x + 910 - 0.065x = 1024$$
$$0.015x = 114$$
$$x = 7600$$
$$14{,}000 - x = 6400$$

\$7600 was invested at 8%.

\$6400 was invested at 6.5%.

33. Find the amount invested at 8%.

5.5%	2500
8%	x
7%	$2500 + x$

$$0.055(2500) + 0.08x = 0.07(2500 + x)$$
$$137.5 + 0.08x = 175 + 0.07x$$
$$0.01x = 37.5$$
$$x = 3750$$

\$3750 additional investment at 8%.

35. Find the length of the track.

$d = 6t$ →

← $d = 2(160 - t)$

Let t = the time to run to the end of the track.

Let $160 - t$ = the time in seconds to jog back.

$$6t = 2(160 - t)$$
$$6t = 320 - 2t$$
$$8t = 320$$
$$t = 40$$
$$d = 6(40) = 240 \text{ meters}$$

37. Find the time.

$d = 240(t + 3)$ →

← $d = 600t$

Let t = the time (in hours) of the second plane.

Let $t + 3$ = the time (in hours) of the first plane.

$$d = 240(t + 3)$$
$$d = 600t$$

$$240(t + 3) = 600t$$
$$240t + 720 = 600t$$
$$720 = 360t$$
$$2 = t$$
$$t = 2 \text{ hours}$$

39. Find the distance.

$d = 1865t$ →

← $d = 1100(2 - t)$

$$1865t = 1100(2 - t)$$
$$1865t = 2200 - 1100t$$
$$2965t = 2200$$
$$t \approx 0.741989$$
$$d = 1865t$$
$$d \approx 1383.81$$

The distance to the target is 1384 ft (to the nearest ft).

41. Find the speed of the second car.

30 seconds = 0.5 minutes = $\frac{0.5}{60}$ hour = $\frac{1}{120}$ hour.

500 meters = $\frac{1}{2}$ km.

	rate	time
Faster car	x	$\frac{1}{120}$
Slower car	80	$\frac{1}{120}$

$$x\left(\frac{1}{120}\right) - 80\left(\frac{1}{120}\right) = \frac{1}{2}$$
$$120\left[x\left(\frac{1}{120}\right) - 80\left(\frac{1}{120}\right)\right] = 120\left(\frac{1}{2}\right)$$
$$x - 80 = 60$$
$$x = 140 \text{ km/h}$$

43. Find the number of grams of pure silver.

1.00	x
0.45	$200 - x$
0.50	200

$$x + 0.45(200 - x) = 0.50(200)$$
$$x + 90 - 0.45x = 100$$
$$0.55x = 10$$
$$x = 18\frac{2}{11} \text{ g pure silver}$$

45. Find the number of liters of water.

0	x
0.12	160
0.20	$160-x$

$$0.12(160)-0=0.20(160-x)$$
$$19.2=32-0.20x$$
$$0.20x=12.8$$
$$x=64$$

64 liters of water

47. Find the number of grams of pure gold.

1.00	x
$\frac{14}{24}$ or $\frac{7}{12}$	15
$\frac{18}{24}$ or $\frac{3}{4}$	$15+x$

$$x+\frac{7}{12}(15)=\frac{3}{4}(15+x)$$
$$12\cdot\left[x+\frac{7}{12}(15)\right]=12\cdot\left[\frac{3}{4}(15+x)\right]$$
$$12x+7(15)=9(15+x)$$
$$12x+105=135+9x$$
$$3x=30$$
$$x=10$$

10 g of pure gold

49. Find the amount of each grade of tea.

\$6.50	x
\$4.25	$20-x$
\$5.60	20

$$6.50x+4.25(20-x)=5.60(20)$$
$$6.50x+85-4.25x=112$$
$$2.25x+85=112$$
$$2.25x=27$$
$$x=12 \text{ lbs of \$6.50 grade}$$
$$20-x=8 \text{ lbs of \$4.25 grade}$$

51. Find the number of pounds of each.

\$6	x
\$3	$25-x$
\$3.84	25

$$6x+3(25-x)=3.84(25)$$
$$6x+75-3x=96$$
$$3x=21$$
$$x=7 \text{ lbs of cranberries}$$
$$25-x=18 \text{ lb of granola}$$

53. Find the number of pounds of each kind of coffee.

\$8	x
\$4	$50-x$
\$5.50	50

$$8x+4(50-x)=5.50(50)$$
$$8x+200-4x=275$$
$$4x=75$$
$$x=18.75 \text{ lb of \$8 coffee}$$
$$50-x=31.25 \text{ lb of \$4 coffee}$$

55. Find the time working together.

Let t = the time it takes both electricians working together to wire the house.

The first electrician does $\frac{1}{14}$ of the job every hour.

The second electrician does $\frac{1}{18}$ of the job every hour.

$$\frac{1}{14}t+\frac{1}{18}t=1$$
$$126\left[\frac{1}{14}t+\frac{1}{18}t\right]=126\cdot 1$$
$$9t+7t=126$$
$$16t=126$$
$$t=7\frac{7}{8} \text{ hours}$$

57. Find the time working together.

Let t = the time it takes both painters working together to paint the kitchen.

The painter can paint $\frac{1}{10}$ of the kitchen every hour.

The apprentice can paint $\frac{1}{15}$ of the kitchen every hour.

$$\frac{1}{10}t+\frac{1}{15}t=1$$
$$30\left[\frac{1}{10}t+\frac{1}{15}t\right]=30\cdot 1$$
$$3t+2t=30$$
$$5t=30$$
$$t=6 \text{ hours}$$

59. Find the time for the older machine.

Let t = time it takes the older machine to finish the job.

The new machine does $\frac{1}{12}$ of the job every hour.

The old machine does $\frac{1}{16}$ of the job every hour.

The new machine works for 4 hours: $4\left(\frac{1}{12}\right)=\frac{1}{3}$.

The old machine completes the job.

$$\frac{1}{3}+\frac{1}{16}t=1$$
$$\frac{1}{16}t=\frac{2}{3}$$
$$t=10\frac{2}{3}\text{ hours}$$

Prepare for Section 1.3

P1. Factor.

$$x^2-x-42=(x+6)(x-7)$$

P3. Write in *a* + *bi* form.

$$3+\sqrt{-16}=3+4i$$

P5. Evaluate.

$$\frac{-(-3)+\sqrt{(-3)^2-4(2)(1)}}{2(2)}=\frac{3+\sqrt{1}}{4}=1$$

Section 1.3 Exercises

1. Solve: factor and apply the zero product principle.

$$x^2-2x-15=0$$
$$(x+3)(x-5)=0$$

$$x+3=0 \quad\text{or}\quad x-5=0$$
$$x=-3 \qquad\qquad x=5$$

3. Solve: factor and apply the zero product principle.

$$2x^2-x=1$$
$$2x^2-x-1=0$$
$$(2x+1)(x-1)=0$$

$$2x+1=0 \quad\text{or}\quad x-1=0$$
$$2x=-1 \qquad\qquad x=1$$
$$x=-\frac{1}{2}$$

5. Solve: factor and apply the zero product principle.

$$8x^2+189x-72=0$$
$$(8x-3)(x+24)=0$$

$$8x-3=0 \quad\text{or}\quad x+2=0$$
$$8x=3 \qquad\qquad x=-24$$
$$x=\frac{3}{8}$$

7. Solve: factor and apply the zero product principle.

$$3x^2-7x=0$$
$$x(3x-7)=0$$

$$x=0 \quad\text{or}\quad 3x-7=0$$
$$3x=7$$
$$x=\frac{7}{3}$$

9. Solve: factor and apply the zero product principle.

$$(x-5)^2-9=0$$
$$[(x-5)-3]\,[(x-5)+3]=0$$
$$(x-8)(x-2)=0$$

$$x-8=0 \quad\text{or}\quad x-2=0$$
$$x=8 \qquad\qquad x=2$$

11. Solve by the square root procedure.

$$x^2=81$$
$$x=\pm\sqrt{81}$$
$$x=\pm 9$$

13. Solve by the square root procedure.

$$y^2=24$$
$$y=\pm\sqrt{24}$$
$$y=\pm 2\sqrt{6}$$

15. Solve by the square root procedure.

$$z^2=-16$$
$$z=\pm\sqrt{-16}$$
$$z=\pm 4i$$

17. Solve by the square root procedure.

$$(x-5)^2=36$$
$$x-5=\pm\sqrt{36}$$
$$x-5=\pm 6$$
$$x=5\pm 6$$

$$x=5+6 \quad\text{or}\quad x=5-6$$
$$x=11 \qquad\qquad x=-1$$

19. Solve by the square root procedure.

$$(x+2)^2=27$$
$$x+2=\pm\sqrt{27}$$
$$x+2=\pm 3\sqrt{3}$$
$$x=-2\pm 3\sqrt{3}$$

$$x=-2+3\sqrt{3} \quad\text{or}\quad x=-2-3\sqrt{3}$$

21. Solve by the square root procedure.

$$(z-4)^2+25=0$$
$$(z-4)^2=-25$$
$$z-4=\pm\sqrt{-25}$$
$$z-4=\pm 5i$$
$$z=4\pm 5i$$
$$z=4+5i \quad \text{or} \quad z=4-5i$$

23. Solve by the square root procedure.

$$2(x+3)^2-18=0$$
$$2(x+3)^2=18$$
$$(x+3)^2=9$$
$$x+3=\pm\sqrt{9}$$
$$x+3=\pm 3$$
$$x=-3\pm 3$$
$$x=-3+3 \quad \text{or} \quad x=-3-3$$
$$x=0 \qquad\qquad x=-6$$

25. Solve by the square root procedure.

$$(y-6)^2-4=14$$
$$(y-6)^2=18$$
$$y-6=\pm\sqrt{18}$$
$$y-6=\pm 3\sqrt{2}$$
$$y=6\pm 3\sqrt{2}$$
$$y=6+3\sqrt{2} \quad \text{or} \quad y=6-3\sqrt{2}$$

27. Solve by the square root procedure.

$$5(x+6)^2+60=0$$
$$5(x+6)^2=-60$$
$$(x+6)^2=-12$$
$$x+6=\pm\sqrt{-12}$$
$$x+6=\pm 2i\sqrt{3}$$
$$x=-6\pm 2i\sqrt{3}$$
$$x=-6+2i\sqrt{3} \quad \text{or} \quad x=-6-2i\sqrt{3}$$

29. Solve by the square root procedure.

$$2(x+4)^2=9$$
$$(x+4)^2=\frac{9}{2}$$
$$x+4=\pm\sqrt{\frac{9}{2}}$$
$$x+4=\pm\frac{\sqrt{9}}{\sqrt{2}}\cdot\frac{\sqrt{2}}{\sqrt{2}}$$
$$x+4=\pm\frac{3\sqrt{2}}{2}$$
$$x=-4\pm\frac{3\sqrt{2}}{2}=\frac{-8\pm 3\sqrt{2}}{2}$$
$$x=\frac{-8+3\sqrt{2}}{2} \quad \text{or} \quad x=\frac{-8-3\sqrt{2}}{2}$$

31. Solve by the square root procedure.

$$4(x-2)^2+15=0$$
$$4(x-2)^2=-15$$
$$(x-2)^2=-\frac{15}{4}$$
$$x-2=\pm\sqrt{-\frac{15}{4}}$$
$$x-2=\pm\frac{i\sqrt{15}}{2}$$
$$x=2\pm\frac{i\sqrt{15}}{2}=\frac{4\pm i\sqrt{15}}{2}$$
$$x=\frac{4+i\sqrt{15}}{2} \quad \text{or} \quad x=\frac{4-i\sqrt{15}}{2}$$

33. Solve by completing the square.

$$x^2-2x-15=0$$
$$x^2-2x+1=15+1$$
$$(x-1)^2=16$$
$$x-1=\pm\sqrt{16}$$
$$x=1\pm 4$$
$$x=1+4 \quad \text{or} \quad x=1-4$$
$$x=5 \qquad\qquad x=-3$$

35. Solve by completing the square.

$$2x^2-5x-12=0$$
$$2x^2-5x=12$$
$$x^2-\frac{5}{2}x=6$$
$$x^2-\frac{5}{2}x+\frac{25}{16}=6+\frac{25}{16}$$
$$\left(x-\frac{5}{4}\right)^2=\frac{121}{16}$$
$$x-\frac{5}{4}=\pm\sqrt{\frac{121}{16}}$$
$$x=\frac{5}{4}\pm\frac{11}{4}$$
$$x=\frac{5}{4}+\frac{11}{4} \quad \text{or} \quad x=\frac{5}{4}-\frac{11}{4}$$
$$x=4 \qquad x=-\frac{3}{2}$$

37. Solve by completing the square.

$$x^2+6x+1=0$$
$$x^2+6x+9=-1+9$$
$$(x+3)^2=8$$
$$x+3=\pm\sqrt{8}$$
$$x=-3\pm2\sqrt{2}$$
$$x=-3+2\sqrt{2} \quad \text{or} \quad x=-3-2\sqrt{2}$$

39. Solve by completing the square.

$$x^2+3x-1=0$$
$$x^2+3x+\frac{9}{4}=1+\frac{9}{4}$$
$$\left(x+\frac{3}{2}\right)^2=\frac{13}{4}$$
$$x+\frac{3}{2}=\pm\sqrt{\frac{13}{4}}$$
$$x=-\frac{3}{2}\pm\frac{\sqrt{13}}{2}=\frac{-3\pm\sqrt{13}}{2}$$
$$x=\frac{-3+\sqrt{13}}{2} \quad \text{or} \quad x=\frac{-3-\sqrt{13}}{2}$$

41. Solve by completing the square.

$$2x^2+4x-1=0$$
$$2x^2+4x=1$$
$$x^2+2x=\frac{1}{2}$$
$$x^2+2x+1=\frac{1}{2}+1$$
$$(x+1)^2=\frac{3}{2}$$
$$x+1=\pm\sqrt{\frac{3}{2}}$$
$$x=-1\pm\sqrt{\frac{3}{2}}=-1\pm\sqrt{\frac{3}{2}\cdot\frac{2}{2}}$$
$$x=-1\pm\frac{\sqrt{6}}{\sqrt{4}}=-1\pm\frac{\sqrt{6}}{2}=\frac{-2\pm\sqrt{6}}{2}$$
$$x=\frac{-2+\sqrt{6}}{2} \quad \text{or} \quad x=\frac{-2-\sqrt{6}}{2}$$

43. Solve by completing the square.

$$3x^2-8x=-1$$
$$x^2-\frac{8}{3}x=-\frac{1}{3}$$
$$x^2-\frac{8}{3}x+\frac{16}{9}=-\frac{1}{3}+\frac{16}{9}$$
$$\left(x-\frac{4}{3}\right)^2=\frac{13}{9}$$
$$x-\frac{4}{3}=\pm\sqrt{\frac{13}{9}}$$
$$x=\frac{4}{3}\pm\frac{\sqrt{13}}{3}=\frac{4\pm\sqrt{13}}{3}$$
$$x=\frac{4+\sqrt{13}}{3} \quad \text{or} \quad x=\frac{4-\sqrt{13}}{3}$$

45. Solve by completing the square.

$$x^2+4x+5=0$$
$$x^2+4x+4=-5+4$$
$$(x+2)^2=-1$$
$$x+2=\pm\sqrt{-1}$$
$$x+2=\pm i$$
$$x=-2\pm i$$
$$x=-2-i \quad \text{or} \quad x=-2+i$$

47. Solve by completing the square.

$$4x^2+4x+2=0$$
$$4x^2+4x=-2$$
$$x^2+x=-\frac{1}{2}$$
$$x^2+x+\frac{1}{4}=-\frac{1}{2}+\frac{1}{4}$$
$$\left(x+\frac{1}{2}\right)^2=-\frac{1}{4}$$
$$x+\frac{1}{2}=\pm\sqrt{-\frac{1}{4}}$$
$$x=-\frac{1}{2}\pm\frac{1}{2}i \text{ or } \frac{-1\pm i}{2}$$

$$x=\frac{-1+i}{2} \text{ or } x=\frac{-1-i}{2}$$

49. Solve by completing the square.

$$3x^2+2x+1=0$$
$$3x^2+2x=-1$$
$$x^2+\frac{2}{3}x=-\frac{1}{3}$$
$$x^2+\frac{2}{3}x+\frac{1}{9}=-\frac{1}{3}+\frac{1}{9}$$
$$\left(x+\frac{1}{3}\right)^2=-\frac{2}{9}$$
$$x+\frac{1}{3}=\pm\sqrt{-\frac{2}{9}}$$
$$x=-\frac{1}{3}\pm\frac{\sqrt{2}}{3}i \text{ or } \frac{-1\pm i\sqrt{2}}{3}$$

$$x=\frac{-1+i\sqrt{2}}{3} \text{ or } x=\frac{-1-i\sqrt{2}}{3}$$

51. Solve by using the quadratic formula.

$$x^2-2x-15=0,\ a=1,\ b=-2,\ c=-15$$

$$x=\frac{-b\pm\sqrt{b^2-4ac}}{2a}$$

$$x=\frac{-(-2)\pm\sqrt{(-2)^2-4(1)(-15)}}{2(1)}$$

$$x=\frac{2\pm\sqrt{4+60}}{2}=\frac{2\pm\sqrt{64}}{2}$$

$$x=\frac{2\pm 8}{2}$$

$$x=\frac{2+8}{2}=\frac{10}{2}=5 \quad\text{or}\quad x=\frac{2-8}{2}=\frac{-6}{2}=-3$$

$$x=5 \text{ or } x=-3$$

53. Solve by using the quadratic formula.

$$12x^2-11x-15=0,\ a=12,\ b=-11,\ c=-15$$

$$x=\frac{-b\pm\sqrt{b^2-4ac}}{2a}$$

$$x=\frac{-(-11)\pm\sqrt{(-11)^2-4(12)(-15)}}{2(12)}$$

$$x=\frac{11\pm\sqrt{121+720}}{24}=\frac{11\pm\sqrt{841}}{24}$$

$$x=\frac{11\pm 29}{24}$$

$$x=\frac{11+29}{24}=\frac{40}{24}=\frac{5}{3} \quad\text{or}\quad x=\frac{11-29}{24}=\frac{-18}{24}=-\frac{3}{4}$$

$$x=\frac{5}{3} \text{ or } x=-\frac{3}{4}$$

55. Solve by using the quadratic formula.

$$x^2-2x=2$$
$$x^2-2x-2=0,\ a=1,\ b=-2,\ c=-2$$

$$x=\frac{-b\pm\sqrt{b^2-4ac}}{2a}$$

$$x=\frac{-(-2)\pm\sqrt{(-2)^2-4(1)(-2)}}{2(1)}$$

$$x=\frac{2\pm\sqrt{4+8}}{2}=\frac{2\pm\sqrt{12}}{2}$$

$$x=\frac{2\pm 2\sqrt{3}}{2}=1\pm\sqrt{3}$$

$$x=1+\sqrt{3} \quad\text{or}\quad x=1-\sqrt{3}$$

57. Solve by using the quadratic formula.

$$x^2=-x+1$$
$$x^2+x-1=0,\ a=1,\ b=1,\ c=-1$$

$$x=\frac{-1\pm\sqrt{1^2-4(1)(-1)}}{2(1)}$$

$$x=\frac{-1\pm\sqrt{1+4}}{2}=\frac{-1\pm\sqrt{5}}{2}$$

$$x=\frac{-1+\sqrt{5}}{2} \quad\text{or}\quad x=\frac{-1-\sqrt{5}}{2}$$

59. Solve by using the quadratic formula.

$$4x^2 = 41 - 8x$$

$4x^2 + 8x - 41 = 0,\ a = 4,\ b = 8,\ c = -41$

$$x = \frac{-8 \pm \sqrt{8^2 - 4(4)(-41)}}{2(4)}$$

$$x = \frac{-8 \pm \sqrt{64 + 656}}{8} = \frac{-8 \pm \sqrt{720}}{8}$$

$$x = \frac{-8 \pm 12\sqrt{5}}{8} = \frac{-2 \pm 3\sqrt{5}}{2}$$

$$x = \frac{-2 + 3\sqrt{5}}{2} \text{ or } x = \frac{-2 - 3\sqrt{5}}{2}$$

61. Solve by using the quadratic formula.

$$\frac{1}{2}x^2 + \frac{3}{4}x - 1 = 0$$

$$4\left(\frac{1}{2}x^2 + \frac{3}{4}x - 1\right) = 4(0)$$

$$2x^2 + 3x - 4 = 0$$

$$x = \frac{-3 \pm \sqrt{3^2 - 4(2)(-4)}}{2(2)}$$

$$x = \frac{-3 \pm \sqrt{9 + 32}}{4}$$

$$x = \frac{-3 \pm \sqrt{41}}{4}$$

$$x = \frac{-3 + \sqrt{41}}{4} \text{ or } x = \frac{-3 - \sqrt{41}}{4}$$

63. Solve by using the quadratic formula.

$x^2 + 6x + 13 = 0,\ a = 1,\ b = 6,\ c = 13$

$$x = \frac{-6 \pm \sqrt{6^2 - 4(1)(13)}}{2(1)}$$

$$x = \frac{-6 \pm \sqrt{36 - 52}}{2} = \frac{-6 \pm \sqrt{-16}}{2}$$

$$x = \frac{-6 \pm 4i}{2} = -3 \pm 2i$$

$$x = -3 - 2i \text{ or } x = -3 + 2i$$

65. Solve by using the quadratic formula.

$$2x^2 = 2x - 13$$

$2x^2 - 2x + 13 = 0,\ a = 1,\ b = -2,\ c = 13$

$$x = \frac{-(-2) \pm \sqrt{(-2)^2 - 4(2)(13)}}{2(2)}$$

$$x = \frac{2 \pm \sqrt{4 - 104}}{4} = \frac{2 \pm \sqrt{-100}}{4}$$

$$x = \frac{2 \pm 10i}{4} = \frac{1 \pm 5i}{2}$$

$$x = \frac{1 - 5i}{2} \text{ or } x = \frac{1 + 5i}{2}$$

67. Solve by using the quadratic formula.

$x^2 + 2x + 29 = 0,\ a = 1,\ b = 2,\ c = 29$

$$x = \frac{-2 \pm \sqrt{2^2 - 4(1)(29)}}{2(1)}$$

$$x = \frac{-2 \pm \sqrt{4 - 116}}{2} = \frac{-2 \pm \sqrt{-112}}{2}$$

$$x = \frac{-2 \pm 4i\sqrt{7}}{2} = -1 \pm 2i\sqrt{7}$$

$$x = -1 - 2i\sqrt{7} \text{ or } x = -1 + 2i\sqrt{7}$$

69. Solve by using the quadratic formula.

$4x^2 + 4x + 13 = 0,\ a = 4,\ b = 4,\ c = 13$

$$x = \frac{-4 \pm \sqrt{4^2 - 4(4)(13)}}{2(4)}$$

$$x = \frac{-4 \pm \sqrt{16 - 208}}{8} = \frac{-4 \pm \sqrt{-192}}{8}$$

$$x = \frac{-4 \pm 8i\sqrt{3}}{8} = \frac{-1 \pm 2i\sqrt{3}}{2}$$

$$x = \frac{-1 - 2i\sqrt{3}}{2} \text{ or } x = \frac{-1 + 2i\sqrt{3}}{2}$$

71. Determine the discriminant, state the number of real solutions. $2x^2 - 5x - 7 = 0$

$$b^2 - 4ac = (-5)^2 - 4(2)(-7)$$
$$= 25 + 56 = 81$$

Two real solutions

73. Determine the discriminant, state the number of real solutions. $3x^2 - 2x + 10 = 0$

$$b^2 - 4ac = (-2)^2 - 4(3)(10)$$
$$= 4 - 120 = -116$$

No real solutions

75. Determine the discriminant, state the number of real solutions. $x^2 - 20x + 100 = 0$

$$b^2 - 4ac = (-20)^2 - 4(1)(100)$$
$$= 400 - 400 = 0$$

One real solution

77. Determine the discriminant, state the number of real solutions.

$$24x^2 + 10x - 21 = 0$$

$$b^2 - 4ac = (10)^2 - 4(24)(-21)$$
$$= 100 + 2016 = 2116$$

Two real solutions

79. Determine the discriminant, state the number of real solutions. $12x^2 + 15x + 7 = 0$

$$b^2 - 4ac = (15)^2 - 4(12)(7)$$
$$= 225 - 336 = -111$$

No real solutions

81. Find the altitude of the triangle. Round to the nearest tenth of a centimeter.

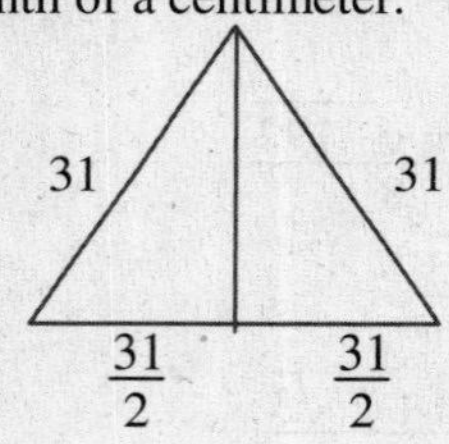

$$a^2 + \left(\frac{31}{2}\right)^2 = 31^2$$

$$a^2 = 31^2 - \left(\frac{31}{2}\right)^2$$

$$d = \sqrt{31^2 - \left(\frac{31}{2}\right)^2}$$

$$d \approx 26.8 \text{ cm}$$

83. Find the width and height.

$$\frac{a}{b} = \frac{4}{3}$$

$$a = \frac{4b}{3}$$

$$a^2 + b^2 = c^2$$

$$\left(\frac{4b}{3}\right)^2 + b^2 = 54^2$$

$$\frac{16b^2}{9} + b^2 = 2916$$

$$\frac{25b^2}{9} = 2916$$

$$b^2 = 1049.76$$

$$b = 32.4$$

$$a = \frac{4(32.4)}{3} = 43.2$$

The TV is 32.4 in. high and 43.2 in. wide.

85. Find the distance.

$$h = -0.0114x^2 + 1.732x$$
$$10 = -0.0114x^2 + 1.732x$$
$$0 = -0.0114x^2 + 1.732x - 10$$
$$a = -0.0114,\ b = 1.732,\ c = -10$$

$$x = \frac{-1.732 \pm \sqrt{1.732^2 - 4(-0.0114)(-10)}}{2(-0.0114)}$$

$$x = \frac{-1.732 \pm \sqrt{2.999824 - 0.456}}{-0.0228}$$

$$x = \frac{-1.732 \pm \sqrt{2.543824}}{-0.0228}$$

$$x = \frac{-1.732 + \sqrt{2.543824}}{-0.0228} \approx 6.0$$

$$x = \frac{-1.732 - \sqrt{2.543824}}{-0.0228} \approx 145.9$$

Since $x = 6.0$ ft is not realistic, it is not a solution.

Convert from feet to yards: $145.9 \text{ ft} \cdot \frac{1 \text{ yd}}{3 \text{ ft}} \approx 48.6 \text{ yd}$

The kicker can be up to 48.6 yd from the goalpost.

87. Find the number of racquets sold.

$$518{,}000 = -0.01x^2 + 168x - 120{,}000$$
$$0 = -0.01x^2 + 168x - 638{,}000$$
$$a = -0.01,\ b = 168,\ c = -638{,}000$$

$$x = \frac{-168 \pm \sqrt{168^2 - 4(-0.01)(-638{,}000)}}{2(-0.01)}$$

$$= \frac{-168 \pm \sqrt{2704}}{-0.02} = \frac{-168 \pm 52}{-0.02}$$

$$x=\frac{-168+52}{-0.02} \quad \text{or} \quad x=\frac{-168-52}{-0.02}$$
$$=\frac{-116}{-0.02}=5800 \qquad =\frac{-220}{-0.02}=11000$$

5800 or 11,000 racquets must be sold.

89. Find the dimensions.

Let w = width of region

Then $\frac{132-3w}{2}$ = length.

$$\text{Area} = \text{length(width)}$$
$$576=\frac{132-3w}{2}\cdot w$$
$$1152=132w-3w^2$$
$$3w^2-132w+1152=0$$
$$3(w^2-44w+384)=0$$
$$w^2-44w+384=0$$
$$(w-32)(w-12)=0$$
$$w-32=0 \quad \text{or} \quad w-12=0$$
$$w=32 \qquad w=12$$
$$\frac{132-3w}{2}=\frac{132-3(32)}{2}=18 \qquad \frac{132-3w}{2}=\frac{132-3(12)}{2}=48$$

The region is either 32 feet wide and 18 feet long, or 12 feet wide and 48 feet long.

91. Find the distances.

Solve $D=-45x^2+190x+200$ for x with $D = 250$.

$$250=-45x^2+190x+200$$
$$0=-45x^2+190x-50$$
$$a=-45,\ b=190,\ c=-50$$
$$x=\frac{-190\pm\sqrt{190^2-4(-45)(-50)}}{2(-45)}$$
$$=\frac{-190\pm\sqrt{27100}}{-90}\approx\frac{-190\pm164.2}{-90}$$
$$x\approx\frac{-190+164.2}{-90} \quad \text{or} \quad x\approx\frac{-190-164.2}{-90}$$
$$\approx 0.3 \text{ mile} \qquad \approx 3.9 \text{ miles}$$

93. Find the time in the air.

Solve $h=-16t^2+25.3t+20$ for t where $h = 17$.

$$17=-16t^2+25.3t+20$$
$$0=-16t^2+25.3t+3$$
$$t=\frac{-25.3\pm\sqrt{(25.3)^2-4(-16)(3)}}{2(-16)}=\frac{-25.3\pm\sqrt{832.09}}{-32}$$
$$t = 1.7 \quad \text{or} \quad t=-0.11$$

He was in the air for 1.7 s.

95. Find the times.

Solve $h=-16t^2+220t$ for t where $h = 350$.

$$350=-16t^2+220t$$
$$0=-16t^2+220t-350$$
$$a=-16,\ b=220,\ c=-350$$
$$t=\frac{-220\pm\sqrt{220^2-4(-16)(-350)}}{2(-16)}$$
$$=\frac{-220\pm\sqrt{26000}}{-32}\approx\frac{-220\pm161.245}{-32}$$
$$t\approx\frac{-220+161.245}{-32} \quad \text{or} \quad t\approx\frac{-220-161.245}{-32}$$
$$\approx 1.8 \text{ seconds} \qquad \approx 11.9 \text{ seconds}$$

97. Determine whether the baseball will clear the fence.

Solve $s=103.9t$ for t where $s = 360$ to find the time it takes the ball to reach the fence.

$$360=103.9t$$
$$t\approx 3.465 \text{ seconds}$$

Next, evaluate $h=-16t^2+50t+4.5$ where $t = 3.465$ to determine if the ball is at least 10 feet in the air when it reaches the fence.

$$h=-16(3.465)^2+50(3.465)+4.5$$
$$h\approx-14.3$$

No, the ball will not clear the fence.

99. Find the number of people.

Solve $h=\frac{1}{2}n(n-1)$ for n where $h = 36$.

$$36=\frac{1}{2}n(n-1)$$
$$72=n(n-1)=n^2-n$$
$$0=n^2-n-72=(n-9)(n+8)$$
$$n-9=0 \quad \text{or} \quad n+8=0$$
$$n=9 \text{ people} \qquad n=-8 \text{ (no)}$$

101. Find the year.

Solve $N = 0.3453x^2 - 9.417x + 164.1$ for x, where $N = 200$.

$$200 = 0.3453x^2 - 9.417x + 164.1$$
$$0 = 0.3453x^2 - 9.417x - 35.9$$
$$a = 0.3453,\ b = -9.417,\ c = -35.9$$
$$x = \frac{-(-9.417) \pm \sqrt{(-9.417)^2 - 4(0.3453)(-35.9)}}{2(0.3453)}$$
$$= \frac{9.417 \pm \sqrt{138.265}}{0.6906} \approx \frac{9.417 \pm 11.759}{0.6906}$$
$$x \approx \frac{9.417 + 11.759}{0.6906} \quad \text{or} \quad x \approx \frac{9.417 - 11.759}{0.6906}$$
$$\approx 30.663 \qquad\qquad \approx -3.39 \text{ (not in the future)}$$

There will be more than 200,000 centenarians living in the US in about 31 years from 200, or 2031.

103.a. If $t = 0$ represents the year 2000, then the year 2010 is represented by $t = 10$.

Evaluate $N = 0.004t^2 + 0.103t + 8.242$ for $t = 10$.

$$N = 0.004(10)^2 + 0.103(10) + 8.242$$
$$= 0.004(100)^2 + 0.103(10) + 8.242$$
$$= 0.4 + 1.03 + 8.242$$
$$= 9.672 \text{ thousand}$$

About 9700 objects

b. Solve $N = 0.004t^2 + 0.103t + 8.242$ for t where $N = 15$.

$$15 = 0.004t^2 + 0.103t + 8.242$$
$$0 = 0.004t^2 + 0.103t - 6.758$$
$$a = 0.004,\ b = 0.103,\ c = -6.758$$
$$t = \frac{-0.103 \pm \sqrt{0.103^2 - 4(0.004)(-6.758)}}{2(0.004)}$$
$$= \frac{-0.103 \pm \sqrt{0.010609 + 0.108128}}{0.008}$$
$$= \frac{-0.103 \pm \sqrt{0.118737}}{0.008} \approx \frac{-0.103 \pm 0.3446}{0.008}$$
$$t \approx \frac{-0.103 + 0.3446}{0.008} \quad \text{or} \quad t \approx \frac{-0.103 - 0.3446}{0.008}$$
$$\approx 30.2 \qquad\qquad \approx -56$$
$$(\text{not } 0 \le t \le 15)$$

30.2 years from 2000 will be in 2030.

Mid-Chapter 1 Quiz

1. Solve.

$$6 - 4(2x + 1) = 5(3 - 2x)$$
$$6 - 8x - 4 = 15 - 10x$$
$$2 - 8x = 15 - 10x$$
$$2x = 13$$
$$x = \frac{13}{2}$$

3. Solve by factoring.

$$x^2 - 5x = 6$$
$$x^2 - 5x - 6 = 0$$
$$(x + 1)(x - 6) = 0$$
$$x = -1 \text{ or } x = 6$$

5. Solve by quadratic formula.

$$x^2 - 6x + 12 = 0$$
$$a = 1,\ b = -6,\ c = 12$$
$$x = \frac{-(-6) \pm \sqrt{(-6)^2 - 4(1)(12)}}{2(1)}$$
$$x = \frac{6 \pm \sqrt{-12}}{2} = \frac{6 \pm 2i\sqrt{3}}{2} = 3 \pm i\sqrt{3}$$
$$x = 3 - i\sqrt{3} \text{ or } x = 3 + i\sqrt{3}$$

7. Find the amount of each solution.

9%	x
4%	$500 - x$
6%	500

$$0.09x + 0.04(500 - x) = 0.06(500)$$
$$0.09x + 20 - 0.04x = 30$$
$$0.05x = 10$$
$$x = 200$$
$$500 - x = 300$$

200 ml of 9% solution and 300 ml of 4% solution.

Prepare for Section 1.4

P1. Factor.

$$x^3 - 16x = x(x^2 - 16)$$
$$= x(x + 4)(x - 4)$$

P3. Evaluate.

$$8^{2/3} = \left(\sqrt[3]{8}\right)^2 = 2^2 = 4$$

P5. Multiply.

$$\begin{aligned}(1+\sqrt{x-5})^2 &= 1^2+2\sqrt{x-5}+(\sqrt{x-5})^2\\ &= 1+2\sqrt{x-5}+x-5\\ &= x+2\sqrt{x-5}-4\end{aligned}$$

Section 1.4 Exercises

1. Solve by factoring and using principle of zero products.

$$\begin{aligned}x^3-25x&=0\\ x(x^2-25)&=0\\ x(x-5)(x+5)&=0\end{aligned}$$

$x=0,\ x=5,$ or $x=-5$

3. Solve by factoring and using principle of zero products.

$$\begin{aligned}x^3-2x^2-x+2&=0\\ x^2(x-2)-(x-2)&=0\\ (x-2)(x^2-1)&=0\\ (x-2)(x-1)(x+1)&=0\end{aligned}$$

$x=2,\ x=1,$ or $x=-1$

5. Solve by factoring and using principle of zero products.

$$\begin{aligned}x^3-3x^2-5x+15&=0\\ x^2(x-3)-5(x-3)&=0\\ (x-3)(x^2-5)&=0\end{aligned}$$

$x-3=0$ or $x^2-5=0$

$x=3 \qquad x^2=5$

$x=\pm\sqrt{5}$

$x=3,\ x=-\sqrt{5},\ x=\sqrt{5}$

7. Solve by factoring and using principle of zero products.

$$\begin{aligned}3x^3+2x^2-27x-18&=0\\ x^2(3x+2)-9(3x+2)&=0\\ (3x+2)(x^2-9)&=0\\ (3x+2)(x+3)(x-3)&=0\end{aligned}$$

$x=-\frac{2}{3},\ x=-3,\ x=3$

9. Solve by factoring and using principle of zero products.

$$\begin{aligned}x^3-8&=0\\ (x-2)(x^2+2x+4)&=0\end{aligned}$$

$x=2,$ or $x^2+2x+4=0$

$x^2+2x=-4$ Use completing the square.

$$\begin{aligned}x^2+2x+1&=-4+1\\ (x+1)^2&=-3\\ x+1&=\pm\sqrt{-3}\\ x&=-1\pm i\sqrt{3}\end{aligned}$$

Thus the solutions are $2,\ -1+i\sqrt{3},\ -1-i\sqrt{3}$.

11. Solve by factoring and using principle of zero products.

$$\begin{aligned}x^4-2x^3+27x-54&=0\\ x^3(x-2)+27(x-2)&=0\\ (x-2)(x^3+27)&=0\\ (x-2)(x+3)(x^2-3x+9)&=0\end{aligned}$$

$$x=\frac{-(-3)\pm\sqrt{(-3)^2-4(1)(9)}}{2(1)}$$

$$x=\frac{3\pm\sqrt{-27}}{2}=\frac{3\pm 3i\sqrt{3}}{2}$$

The solutions are 2, –3, $\frac{3-3i\sqrt{3}}{2},\ \frac{3+3i\sqrt{3}}{2}$.

13. Solve the rational equation.

$$\begin{aligned}\frac{5}{x+4}-2&=\frac{7x+18}{x+4}\\ (x+4)\left(\frac{5}{x+4}-2\right)&=(x+4)\left(\frac{7x+18}{x+4}\right)\\ 5-2(x+4)&=7x+18\\ 5-2x-8&=7x+18\\ -2x-3&=7x+18\\ -9x&=21\\ x&=-\frac{7}{3}\end{aligned}$$

15. Solve the rational equation.

$$\begin{aligned}2+\frac{9}{r-3}&=\frac{3r}{r-3}\\ (r-3)\left(2+\frac{9}{r-3}\right)&=(r-3)\left(\frac{3r}{r-3}\right)\\ 2(r-3)+9&=3r\\ 2r-6+9&=3r\\ 2r+3&=3r\\ 2r-3r&=-3\\ -r&=-3\\ r&=3\end{aligned}$$

No solution because each side is undefined when $r=3$.

17. Solve the rational equation.

$$\frac{3}{x+2}=\frac{5}{2x-7}$$
$$3(2x-7)=5(x+2)$$
$$6x-21=5x+10$$
$$6x-5x=10+21$$
$$x=31$$

19. Solve the rational equation.

$$x-\frac{2x+3}{x+3}=\frac{2x+9}{x+3}$$
$$(x+3)\left(x-\frac{2x+3}{x+3}\right)=(x+3)\left(\frac{2x+9}{x+3}\right)$$
$$x(x+3)-(2x+3)=2x+9$$
$$x^2+3x-2x-3=2x+9$$
$$x^2-x-12=0$$
$$(x+3)(x-4)=0$$
$$x=-3 \text{ or } x=4$$

4 checks as a solution. –3 is not in the domain and does not check as a solution.

21. Solve the rational equation.

$$\frac{5}{x-3}-\frac{3}{x-2}=\frac{4}{x-3}$$
$$(x-3)(x-2)\left(\frac{5}{x-3}-\frac{3}{x-2}\right)=(x-3)(x-2)\left(\frac{4}{x-3}\right)$$
$$5(x-2)-3(x-3)=4(x-2)$$
$$5x-10-3x+9=4x-8$$
$$2x-1=4x-8$$
$$2x-4x=-8+1$$
$$-2x=-7$$
$$x=\frac{7}{2}$$

23. Solve the rational equation.

$$\frac{x}{x+1}-\frac{x+2}{x-1}=\frac{x-12}{x+1}$$
$$(x+1)(x-1)\left(\frac{x}{x+1}-\frac{x+2}{x-1}\right)=(x+1)(x-1)\left(\frac{x-12}{x+1}\right)$$
$$x(x-1)-(x+1)(x+2)=(x-1)(x-12)$$
$$x^2-x-x^2-3x-2=x^2-13x+12$$
$$-4x-2=x^2-13x+12$$
$$0=x^2-9x+14$$
$$0=(x-2)(x-7)$$
$$x=2 \text{ or } x=7$$

25. Solve the rational equation.

$$\frac{4-3x}{2x+1}+\frac{3x+2}{x+2}=\frac{4x-5}{2x+1}$$
$$(2x+1)(x+2)\left(\frac{4-3x}{2x+1}+\frac{3x+2}{x+2}\right)=(2x+1)(x+2)\left(\frac{4x-5}{2x+1}\right)$$
$$(x+2)(4-3x)+(2x+1)(3x+2)=(x+2)(4x-5)$$
$$-3x^2-2x+8+6x^2+7x+2=4x^2+3x-10$$
$$3x^2+5x+10=4x^2+3x-10$$
$$0=x^2-2x-20$$
$$20+1=x^2-2x+1$$
$$21=(x-1)^2$$
$$\pm\sqrt{21}=x-1$$
$$x=1\pm\sqrt{21}$$
$$x=1-\sqrt{21} \text{ or } x=1+\sqrt{21}$$

27. Solve the radical equation.

$$\sqrt{x-4}-6=0$$
$$\sqrt{x-4}=6$$
$$x-4=36$$
$$x=40$$

Check $\sqrt{40-4}-6=0$
$$\sqrt{36}-6=0$$
$$6-6=0$$
$$0=0$$

The solution is 40.

29. Solve the radical equation.

$$\sqrt{3x-5}-\sqrt{x+2}=1$$
$$(\sqrt{3x-5})^2=(1+\sqrt{x+2})^2$$
$$3x-5=1+2\sqrt{x+2}+x+2$$
$$2x-8=2\sqrt{x+2}$$
$$(x-4)^2=(\sqrt{x+2})^2$$
$$x^2-8x+16=x+2$$
$$x^2-9x+14=0$$
$$(x-7)(x-2)=0$$
$$x=7 \text{ or } x=2$$

Check $\sqrt{3(7)-5}-\sqrt{7+2}=1$
$$\sqrt{16}-\sqrt{9}=1$$
$$4-3=1$$
$$1=1$$

$$\sqrt{3(2)-5}-\sqrt{2+2}=1$$
$$\sqrt{1}-\sqrt{4}=1$$
$$1-2=1$$
$$-1=1 \text{ (No)}$$

The solution is 7.

31. Solve the radical equation.

$$\sqrt{2x+11}-\sqrt{2x-5}=2$$
$$(\sqrt{2x+11})^2=(2+\sqrt{2x-5})^2$$
$$2x+11=4+4\sqrt{2x-5}+2x-5$$
$$12=4\sqrt{2x-5}$$
$$(3)^2=(\sqrt{2x-5})^2$$
$$9=2x-5$$
$$14=2x$$
$$7=x$$

Check $\sqrt{2(7)+11}-\sqrt{2(7)-5}=2$
$$\sqrt{25}-\sqrt{9}=2$$
$$5-3=2$$
$$2=2$$

7 checks as the solution.

33. Solve the radical equation.

$$\sqrt{x-4}+\sqrt{x+1}=1$$
$$\sqrt{x-4}=1-\sqrt{x+1}$$
$$\left(\sqrt{x-4}\right)^2=\left(1-\sqrt{x+1}\right)^2$$
$$x-4=1-2\sqrt{x+1}+x+1$$
$$2\sqrt{x+1}=6$$
$$\left(2\sqrt{x+1}\right)^2=6^2$$
$$4(x+1)=36$$
$$4x+4=36$$
$$4x=32$$
$$x=8$$

Check $\sqrt{8-4}+\sqrt{8+1}=1$
$$\sqrt{4}+\sqrt{9}=1$$
$$2+3=1$$
$$5=1 \text{ (No)}$$

There is no solution.

35. Solve the radical equation.

$$\sqrt{9x-20}=x$$
$$\left(\sqrt{9x-20}\right)^2=x^2$$
$$9x-20=x^2$$
$$0=x^2-9x+20$$
$$0=(x-4)(x-5)$$
$$x=4 \text{ or } x=5$$

Check $\sqrt{9(4)-20}=4$ $\quad$ $\sqrt{9(5)-20}=5$
$$\sqrt{16}=4 \qquad \sqrt{25}=5$$
$$4=4 \qquad 5=5$$

4 and 5 check as solutions.

37. Solve the radical equation.

$$\sqrt{2x-1}-\sqrt{x-1}=1$$
$$\sqrt{2x-1}=1+\sqrt{x-1}$$
$$\left(\sqrt{2x-1}\right)^2=\left(1+\sqrt{x-1}\right)^2$$
$$2x-1=1+2\sqrt{x-1}+x-1$$
$$x-1=2\sqrt{x-1}$$
$$(x-1)^2=\left(2\sqrt{x-1}\right)^2$$
$$x^2-2x+1=4(x-1)$$
$$x^2-2x+1=4x-4$$
$$x^2-6x+5=0$$
$$(x-1)(x-5)=0$$
$$x=1 \text{ or } x=5$$

Check $\sqrt{2(1)-1}-\sqrt{(1)-1}=1$ $\quad$ $\sqrt{2(5)-1}-\sqrt{5-1}=1$
$$\sqrt{1}-\sqrt{0}=1 \qquad \sqrt{9}-\sqrt{4}=1$$
$$1=1 \qquad 3-2=1$$
$$1=1$$

1 and 5 check as solutions.

39. Solve the radical equation.

$$\sqrt{-7x+2}+x=2$$
$$\sqrt{-7x+2}=2-x$$
$$\left(\sqrt{-7x+2}\right)^2=(2-x)^2$$
$$-7x+2=4-4x+x^2$$
$$0=x^2+3x+2$$
$$0=(x+2)(x+1)$$
$$x=-2 \text{ or } x=-1$$

Check

$$\sqrt{-7(-2)+2}+(-2)=2 \qquad \sqrt{-7(-1)+2}+(-1)=2$$
$$\sqrt{16}-2=2 \qquad \sqrt{9}-1=2$$
$$4-2=2 \qquad 3-1=2$$
$$2=2 \qquad 2=2$$

–2 and –1 check as solutions.

41. Solve the radical equation.

$$\sqrt[3]{x^3-2x-13}=x-1$$
$$\left(\sqrt[3]{x^3-2x-13}\right)^3=(x-1)^3$$
$$x^3-2x-13=x^3-3x^2+3x-1$$
$$3x^2-5x-12=0$$
$$(3x+4)(x-3)=0$$
$$3x+4=0 \quad \text{or} \quad x-3=0$$
$$x=-\frac{4}{3} \qquad x=3$$

Check $\sqrt[3]{\left(-\frac{4}{3}\right)^3-2\left(-\frac{4}{3}\right)-13}=-\frac{4}{3}-1$

$$\sqrt[3]{-\frac{343}{27}}=-\frac{7}{3}$$
$$-\frac{7}{3}=-\frac{7}{3}$$
$$\sqrt[3]{(3)^3-2(3)-13}=3-1$$
$$\sqrt[3]{8}=2$$
$$2=2$$

Both $-\frac{4}{3}$ and 3 check as solutions.

43. Solve.

$$x^{1/3}=2$$
$$\left(x^{1/3}\right)^3=(2)^3$$
$$x=8$$

45. Solve.

$$x^{2/5}=9$$
$$\left(x^{2/5}\right)^{5/2}=(9)^{5/2}$$
$$|x|=243$$
$$x=-243,\ 243$$

47. Solve.

$$x^{3/2}=27$$
$$\left(x^{3/2}\right)^{2/3}=(27)^{2/3}$$
$$x=9$$

49. Solve.

$$3x^{2/3}-16=59$$
$$3x^{2/3}=75$$
$$x^{2/3}=25$$
$$\left(x^{2/3}\right)^{3/2}=(25)^{3/2}$$
$$|x|=125$$
$$x=-125,\ 125$$

51. Solve.

$$4x^{3/4}-31=77$$
$$4x^{3/4}=108$$
$$x^{3/4}=27$$
$$\left(x^{3/4}\right)^{4/3}=(27)^{4/3}$$
$$x=81$$

53. Find all real solutions.

$$x^4-9x^2+14=0$$

Let $u=x^2$.

$$u^2-9u+14=0$$
$$(u-7)(u-2)=0$$
$$u=7 \quad \text{or} \quad u=2$$
$$x^2=7 \qquad x^2=2$$
$$x=\pm\sqrt{7} \qquad x=\pm\sqrt{2}$$

The solutions are $\sqrt{7},\ -\sqrt{7},\ \sqrt{2},\ -\sqrt{2}$.

55. Find all real solutions.

$$2x^4-11x^2+12=0$$

Let $u=x^2$.

$$2u^2-11u+12=0$$
$$(2u-3)(u-4)=0$$
$$u=\frac{3}{2} \quad \text{or} \quad u=4$$
$$x^2=\frac{3}{2} \qquad x^2=4$$
$$x=\pm\sqrt{\frac{3}{2}}=\pm\frac{\sqrt{6}}{2} \qquad x=\pm 2$$

The solutions are $\frac{\sqrt{6}}{2},\ -\frac{\sqrt{6}}{2},\ 2,\ -2$.

57. Find all real solutions.

$x^6 + x^3 - 6 = 0$

Let $u = x^3$.

$$u^2 + u - 6 = 0$$
$$(u-2)(u+3) = 0$$

$u = 2$ or $u = -3$

$x^3 = 2$ $\quad x^3 = -3$

$x = \sqrt[3]{2}$ $\quad x = \sqrt[3]{-3} = -\sqrt[3]{3}$

The solutions are $\sqrt[3]{2}$ and $-\sqrt[3]{3}$.

59. Find all real solutions.

$x^{1/2} - 3x^{1/4} + 2 = 0$

Let $u = x^{1/4}$.

$$u^2 - 3u + 2 = 0$$
$$(u-1)(u-2) = 0$$

$u = 1$ or $u = 2$

$x^{1/4} = 1$ $\quad x^{1/4} = 2$

$x = 1$ $\quad x = 16$

The solutions are 1 and 16.

61. Find all real solutions.

$3x^{2/3} - 11x^{1/3} - 4 = 0$

Let $u = x^{1/3}$.

$$3u^2 - 11u - 4 = 0$$
$$(3u+1)(u-4) = 0$$

$u = -\frac{1}{3}$ or $u = 4$

$x^{1/3} = -\frac{1}{3}$ $\quad x^{1/3} = 4$

$x = -\frac{1}{27}$ $\quad x = 64$

The solutions are $-\frac{1}{27}$ and 64.

63. Find all real solutions.

$x^4 + 8x^2 - 9 = 0$

Let $u = x^2$.

$$u^2 + 8u - 9 = 0$$
$$(u+9)(u-1) = 0$$

$u = -9$ or $u = 1$

$x^2 = -9$ $\quad x^2 = 1$

$x = \pm\sqrt{-9}$ $\quad x = \pm 1$

$x = \pm 3i$

The solutions are –1, 1, $-3i$, $3i$.

65. Find all real solutions.

$x^{2/5} - x^{1/5} - 2 = 0$

Let $u = x^{1/5}$.

$$u^2 - u - 2 = 0$$
$$(u+1)(u-2) = 0$$

$u = -1$ or $u = 2$

$x^{1/5} = -1$ $\quad x^{1/5} = 2$

$x = -1$ $\quad x = 32$

The solutions are –1 and 32.

67. Find all real solutions.

$9x - 52\sqrt{x} + 64 = 0$

Let $\sqrt{x} = u$.

$$9u^2 - 52u + 64 = 0$$
$$(9u-16)(u-4) = 0$$

$u = \frac{16}{9}$ or $u = 4$

$\sqrt{x} = \frac{16}{9}$ $\quad \sqrt{x} = 4$

$x = \frac{256}{81}$ $\quad x = 16$

The solutions are $\frac{256}{81}$ and 16.

69. Find the speed of the boat.

Using the formula $t = \frac{d}{r}$, we get the equation.

$$\frac{24}{v+3} + \frac{24}{v-3} = 6$$
$$(v+3)(v-3)\left(\frac{24}{v+3} + \frac{24}{v-3}\right) = (v+3)(v-3)6$$
$$24(v-3) + 24(v+3) = (v^2 - 9)(6)$$
$$24v - 72 + 24v + 72 = 6v^2 - 54$$
$$48v = 6v^2 - 54$$
$$0 = 6v^2 - 48v - 54$$
$$0 = 6(v^2 - 8v - 9)$$
$$0 = 6(v+1)(v-9)$$

$v = -1$ (No) or $v = 9$

The speed is 9 mph.

71. Find the time it takes the assistant, working alone.

Let x = the number of hours the assistant would take to build the fence working alone.

The worker does $\frac{1}{8}$ of the job per hour; the assistant does $\frac{1}{x}$ of the job per hour.

worker	$\frac{1}{8}$	5
assistant	$\frac{1}{x}$	5

$$\left(\frac{1}{8}\right)(5)+\left(\frac{1}{x}\right)(5)=1$$
$$\frac{5}{8}+\frac{5}{x}=1$$
$$8x\left(\frac{5}{8}+\frac{5}{x}\right)=1(8x)$$
$$5x+40=8x$$
$$40=3x$$
$$\frac{40}{3}=x$$
$$x=13\frac{1}{3}\text{ hours}$$

73. Find the time for the experienced painter.

Let t = the time for the experienced painter.

$2t-5$ is the time for the apprentice painter.

$\frac{1}{t}$ is the rate for the experienced painter.

$\frac{1}{2t-5}$ is the rate for the apprentice painter.

$$\frac{1}{t}\cdot 6+\frac{1}{2t-5}\cdot 6=1$$
$$t(2t-5)\left(\frac{6}{t}+\frac{6}{2t-5}\right)=t(2t-5)\cdot 1$$
$$6(2t-5)+6t=2t^2-5t$$
$$12t-30+6t=2t^2-5t$$
$$0=2t^2-23t+30$$
$$0=(2t-3)(t-10)$$
$$t=\frac{3}{2}\text{ (No) or } t=10$$

It takes the experience painter 10 hours.

75. Find the time.

$$T=\frac{\sqrt{s}}{4}+\frac{s}{1100}$$
$$T=\frac{\sqrt{7100}}{4}+\frac{7100}{1100}$$
$$T\approx 27.5\text{ seconds}$$

77. Find the radius.

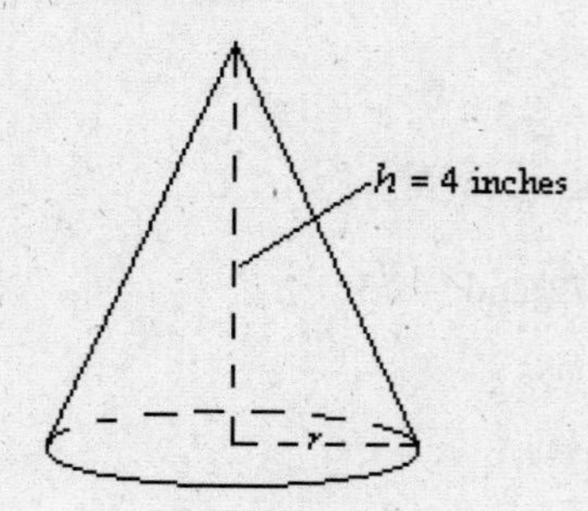

$$L=\pi r\sqrt{r^2+h^2}$$
$$15\pi=\pi r\sqrt{r^2+4^2}$$
$$15=r\sqrt{r^2+16}$$
$$225=r^2(r^2+16)$$
$$0=r^4+16r^2-225$$

Let $u=r^2$.

$$u^2+16u-225=0$$
$$(u+25)(u-9)=0$$
$$u=9 \quad\text{or}\quad u=-25\text{ (No)}$$
$$r^2=9$$
$$r=3$$

The radius is 3 in.

79. Find the length of the edge.

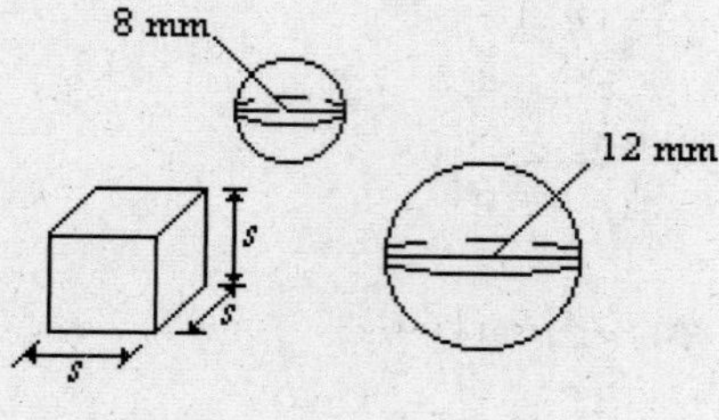

$d_1=8$ mm, $d_2=12$ mm

$V_c=s^3 \qquad V_s=\frac{4}{3}\pi r^3$

$$s^3=\frac{4}{3}\pi(4)^3+\frac{4}{3}\pi(6)^3=\frac{4}{3}\pi(64+216)\approx 1172.86$$
$$s\approx 10.5\text{ mm}$$

The side is approximately 10.5 mm.

81. Find the height.

$$d=\sqrt{1.5h}$$
$$14=\sqrt{1.5h}$$
$$196=1.5h$$
$$131\approx h$$

The height is approximately 131 ft.

83. Show that $\dfrac{AD}{AB}=\phi$.

Label the point G that is between B and C.

Note that $EG = ED$, by construction.

Let $x = FD$.

Then $EG = 1 + x$.

$$EG=\sqrt{EF^2+FG^2}=\sqrt{1^2+2^2}=\sqrt{5}$$
$$\sqrt{5}=1+x$$
$$x=\sqrt{5}-1$$
$$\frac{AD}{AB}=\frac{2+x}{2}=\frac{2+\sqrt{5}-1}{2}=\frac{1+\sqrt{5}}{2}=\phi$$

85. Find the distance.

$$\frac{16-x}{22}+\frac{\sqrt{16+x^2}}{7}=2$$
$$154\left(\frac{16-x}{22}+\frac{\sqrt{16+x^2}}{7}\right)=154(2)$$
$$7(16-x)+22\sqrt{16+x^2}=308$$
$$112-7x+22\sqrt{16+x^2}=308$$
$$22\sqrt{16+x^2}=7x+196$$
$$\left(22\sqrt{16+x^2}\right)^2=(7x+196)^2$$
$$484(16+x^2)=49x^2+2744x+38{,}416$$
$$7744+484x^2=49x^2+2744x+38{,}416$$
$$435x^2-2744x-30{,}672=0$$

Use quadratic formula to solve for x.

$$x=\frac{-(-2744)\pm\sqrt{(-2744)^2-4(435)(-30{,}672)}}{2(435)}$$
$$x=\frac{2744\pm\sqrt{60{,}898{,}816}}{870}$$
$$x=\frac{2744-\sqrt{60{,}898{,}816}}{870}\approx-5.8\text{ km (No)}$$
$$x=\frac{2744+\sqrt{60{,}898{,}816}}{870}\approx 12.1\text{ km}$$

The distance is about 12.1 km.

Prepare for Section 1.5

P1. Find $\{x|x>2\}\cap\{x|x>5\}$.

$\{x\,|\,x>5\}$

P3. Evaluate.

$$\frac{7+3}{7-2}=2$$

P5. Find the value of x where the expression is undefined.

$$\frac{x-3}{2x-7},\ 2x-7\neq 0$$

It is undefined for $x=\dfrac{7}{2}$.

Section 1.5 Exercises

1. Find the solution set in set builder notation and graph.

$$2x+3<11$$
$$2x<11-3$$
$$2x<8$$
$$x<4$$

$\{x|x<4\}$

-5 -4 -3 -2 -1 0 1 2 3 4 5

3. Find the solution set in set builder notation and graph.

$$x+4>3x+16$$
$$-2x>12$$
$$x<-6$$

$\{x|x<-6\}$

-7 -6 -5 -4 -3 -2 -1 0 1 2 3

5. Find the solution set in set builder notation and graph.

$$-3(x+2)\le 5x+7$$
$$-3x-6\le 5x+7$$
$$-8x\le 13$$
$$x\ge-\frac{13}{8}$$

$\left\{x\middle|x\ge-\dfrac{13}{8}\right\}$

-5 -4 -3 -2 -1 0 1 2 3 4 5

7. Find the solution set in set builder notation and graph.

$$-4(3x-5)>2(x-4)$$
$$-12x+20>2x-8$$
$$-14x>-28$$
$$x<2$$

$\{x|x<2\}$

-5 -4 -3 -2 -1 0 1 2 3 4 5

9. Find the solution set in set builder notation and graph.

$4x+1>-2$ and $4x+1\le 17$

$4x>-3$ and $4x\le 16$

$x>-\frac{3}{4}$ and $x\le 4$

$\left\{x\middle|x>-\frac{3}{4}\right\}\cap\{x|x\le 4\}=\left\{x\middle|-\frac{3}{4}<x\le 4\right\}$

-5 -4 -3 -2 -1 0 1 2 3 4 5

11. Find the solution set in set builder notation and graph.

$10\ge 3x-1\ge 0$

$11\ge 3x\ge 1$

$\frac{11}{3}\ge x\ge\frac{1}{3}$

$\left\{x\middle|\frac{1}{3}\le x\le\frac{11}{3}\right\}$

-5 -4 -3 -2 -1 0 1 2 3 4 5

13. Find the solution set in set builder notation and graph.

$x+2<-1$ or $x+3\ge 2$

$x<-3$ or $x\ge -1$

$\{x|x<-3\}\cup\{x|x\ge -1\}=\{x|x<-3 \text{ or } x\ge -1\}$

-5 -4 -3 -2 -1 0 1 2 3 4 5

15. Find the solution set in set builder notation and graph.

$-4x+5>9$ or $4x+1<5$

$-4x>4$ or $4x<4$

$x<-1$ or $x<1$

$\{x|x<-1\}\cup\{x|x<1\}=\{x|x<1\}$

-5 -4 -3 -2 -1 0 1 2 3 4 5

17. Find the solution set in interval notation

$|2x-1|>4$

$2x-1<-4$ or $2x-1>4$

$2x<-3$ or $2x>5$

$x<-\frac{3}{2}$ or $x>\frac{5}{2}$

$\left(-\infty, -\frac{3}{2}\right)\cup\left(\frac{5}{2}, \infty\right)$

19. Find the solution set in interval notation

$|x+3|\ge 5$

$x+3\le -5$ or $x+3\ge 5$

$x\le -8$ or $x\ge 2$

$(-\infty, -8]\cup[2, \infty)$

21. Find the solution set in interval notation

$|3x-10|\le 14$

$-14\le 3x-10\le 14$

$-4\le 3x\le 24$

$-\frac{4}{3}\le x\le 8$

$\left[-\frac{4}{3}, 8\right]$

23. Find the solution set in interval notation

$|4-5x|\ge 24$

$4-5x\le -24$ or $4-5x\ge 24$

$-5x\le -28$ or $-5x\ge 20$

$x\ge\frac{28}{5}$ or $x\le -4$

$(-\infty, -4]\cup\left[\frac{28}{5}, \infty\right)$

25. Find the solution set in interval notation

$|x-5|\ge 0$ (Note: The absolute value of *any* real number is greater than or equal to 0.)

$(-\infty, \infty)$

27. Find the solution set in interval notation

$|x-4|\le 0$

(Note: No absolute value is less than 0.)

$x-4=0$

$x=4$

$\{4\}$

29. Use the critical value method to solve. Write solution in interval notation.

$x^2+7x>0$

$x(x+7)>0$

The product $x(x+7)$ is positive.

$x = 0$ is a critical value.

$x + 7 = 0 \Rightarrow x = -7$ is a critical value.

$x(x+7)$ ++++|- - - - - - - -|++++
-7 0

$(-\infty, -7) \cup (0, \infty)$

31. Use the critical value method to solve. Write solution in interval notation.

$$x^2 - 16 \le 0$$
$$(x-4)(x+4) \le 0$$

The product $(x-4)(x+4)$ is negative or zero.

$x - 4 = 0 \Rightarrow x = 4$ is a critical value.

$x + 4 = 0 \Rightarrow x = -4$ is a critical value.

$(x-4)(x+4)$ ++++|- - - - - - - -|+++
-4 0 4

$[-4, 4]$

33. Use the critical value method to solve. Write solution in interval notation.

$$x^2 + 7x + 10 < 0$$
$$(x+5)(x+2) < 0$$

The product $(x+5)(x+2)$ is negative.

$x + 5 = 0 \Rightarrow x = -5$ is a critical value.

$x + 2 = 0 \Rightarrow x = -2$ is a critical value.

$(x+5)(x+2)$ ++++|- - -|++++++
-5 -2 0

$(-5, -2)$

35. Use the critical value method to solve. Write solution in interval notation.

$$x^2 - 3x \ge 28$$
$$x^2 - 3x - 28 \ge 0$$
$$(x-7)(x+4) \ge 0$$

The product $(x-7)(x+4)$ is positive or zero.

$x - 7 = 0 \Rightarrow x = 7$ is a critical value.

$x + 4 = 0 \Rightarrow x = -4$ is a critical value.

$(x-7)(x+4)$ ++|- - - - - - - - - - - -|++
-4 0 7

$(-\infty, -4] \cup [7, \infty)$

37. Use the critical value method to solve. Write solution in interval notation.

$$x^3 - x^2 - 16x + 16 < 0$$
$$x^2(x-1) - 16(x-1) < 0$$
$$(x-1)(x^2 - 16) < 0$$
$$(x-1)(x+4)(x-4) < 0$$

The product $(x-1)(x+4)(x-4)$ is negative.

$x - 1 = 0 \Rightarrow x = 1$ is a critical value.

$x - 4 = 0 \Rightarrow x = 4$ is a critical value.

$x + 4 = 0 \Rightarrow x = -4$ is a critical value.

The critical values are –4, 1, and 4.

$(x-1)(x+4)(x-4)$ - -|++++|- - - -|++
-4 1 4

$(-\infty, -4) \cup (1, 4)$

39. Use the critical value method to solve. Write solution in interval notation.

$$x^4 - 20x^2 + 64 \ge 0$$
$$(x^2 - 16)(x^2 - 4) \ge 0$$
$$(x+4)(x-4)(x+2)(x-2) \ge 0$$

The product is positive or zero.

$x + 4 = 0 \Rightarrow x = -4$ is a critical value.

$x - 4 = 0 \Rightarrow x = 4$ is a critical value.

$x + 2 = 0 \Rightarrow x = -2$ is a critical value.

$x - 2 = 0 \Rightarrow x = 2$ is a critical value.

The critical values are –4, –2, 2, and 4.

$(x+4)(x-4)(x+2)(x-2)$ ++|- - - -|+++|- - - -|++
-4 -2 2 4

$(-\infty, -4] \cup [-2, 2] \cup [4, \infty)$

41. Use the critical value method to solve. Write solution in interval notation.

$$\frac{x+4}{x-1} < 0$$

The quotient $\frac{x+4}{x-1}$ is negative.

$x + 4 = 0 \Rightarrow x = -4$

$x - 1 = 0 \Rightarrow x = 1$

The critical values are –4 and 1.

$\dfrac{x+4}{x-1}$ 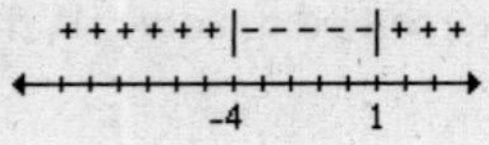

(–4, 1)

43. Use the critical value method to solve. Write solution in interval notation.

$$\frac{x-5}{x+8} \geq 3$$
$$\frac{x-5}{x+8} - 3 \geq 0$$
$$\frac{x-5-3(x+8)}{x+8} \geq 0$$
$$\frac{x-5-3x-24}{x+8} \geq 0$$
$$\frac{-2x-29}{x+8} \geq 0$$

The quotient $\dfrac{-2x-29}{x+8}$ is positive or zero.

$-2x-29=0 \Rightarrow x=-\dfrac{29}{2}$

$x+8=0 \Rightarrow \ x=-8$

The critical values are $-\dfrac{29}{2}$ and -8.

$\dfrac{-2x-29}{x+8}$ 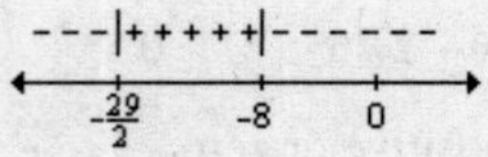

The denominator cannot equal zero $\Rightarrow x \neq -8$.

$\left[-\dfrac{29}{2}, \ -8\right)$

45. Use the critical value method to solve. Write solution in interval notation.

$$\frac{x}{2x+7} \geq 4$$
$$\frac{x}{2x+7} - 4 \geq 0$$
$$\frac{x-4(2x+7)}{2x+7} \geq 0$$
$$\frac{x-8x-28}{2x+7} \geq 0$$
$$\frac{-7x-28}{2x+7} \geq 0$$

The quotient $\dfrac{-7x-28}{2x+7}$ is positive or zero.

$-7x-28=0 \Rightarrow x=-4$

$2x+7=0 \Rightarrow \ x=-\dfrac{7}{2}$

The critical values are -4 and $-\dfrac{7}{2}$.

$\dfrac{-7x-28}{2x+7}$ 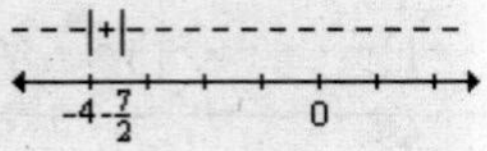

The denominator cannot equal zero $\Rightarrow x \neq -\dfrac{7}{2}$

$\left[-4, \ -\dfrac{7}{2}\right)$

47. Use the critical value method to solve. Write solution in interval notation.

$$\frac{(x+1)(x-4)}{x-2} < 0$$

The quotient $\dfrac{(x+1)(x-4)}{x-2}$ is negative.

$x+1=0 \Rightarrow x=-1$
$x-4=0 \Rightarrow x=4$
$x-2=0 \Rightarrow x=2$

The critical values are –1, 4, and 2.

$\dfrac{(x+1)(x-4)}{x-2}$ 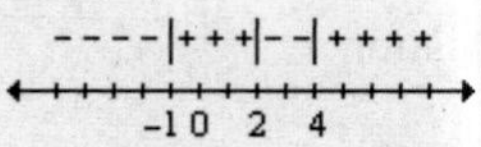

$(-\infty, \ -1) \cup (2, \ 4)$

49. Use the critical value method to solve. Write solution in interval notation.

$$\frac{x+2}{x-5} \leq 2$$
$$\frac{x+2}{x-5} - 2 \leq 0$$
$$\frac{x+2-2(x-5)}{x-5} \leq 0$$
$$\frac{x+2-2x+10}{x-5} \leq 0$$
$$\frac{-x+12}{x-5} \leq 0$$

The quotient $\dfrac{-x+12}{x-5}$ is negative or zero.

$-x+12=0 \Rightarrow x=12$
$x-5=0 \Rightarrow x=5$

The critical values are 12 and 5.

$\dfrac{-x+12}{x-5}$ 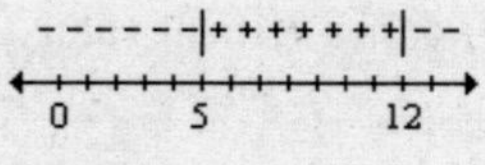

The denominator cannot equal zero $\Rightarrow x \neq 5$.

$(-\infty, \ 5) \cup [12, \ \infty)$

51. Use the critical value method to solve. Write solution in interval notation.

$$\frac{6x^2-11x-10}{x}>0$$
$$\frac{(3x+2)(2x-5)}{x}>0$$

The quotient $\frac{(3x+2)(2x-5)}{x}$ is positive.

$3x+2=0 \Rightarrow x=-\frac{2}{3}$

$2x-5=0 \Rightarrow x=\frac{5}{2}$

$x=0$

The critical values are $-\frac{2}{3}$, $\frac{5}{2}$, and 0.

$\frac{(3x+2)(2x-5)}{x}$ 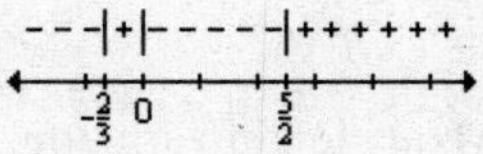

$\left(-\frac{2}{3},\ 0\right)\cup\left(\frac{5}{2},\ \infty\right)$

53. Use the critical value method to solve. Write solution in interval notation.

$$\frac{x^2-6x+9}{x-5}\le 0$$
$$\frac{(x-3)(x-3)}{x-5}\le 0$$

The quotient $\frac{(x-3)(x-3)}{x-5}$ is negative or zero.

$x-3=0 \Rightarrow x=3$

$x-5=0 \Rightarrow x=5$

The critical values are 3 and 5.

$\frac{(x-3)(x-3)}{x-5}$ 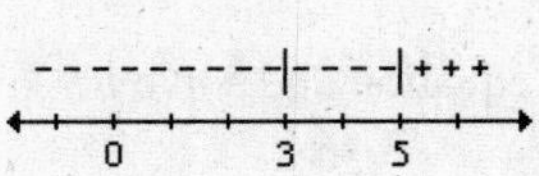

The denominator cannot equal zero $\Rightarrow x\ne 5$.

$(-\infty,\ 5)$

55. Find the conditions to use the LowCharge plan.

LowCharge: $5 + 0.01x$

FeeSaver: $1 + 0.08x$

$$5+0.01x<1+0.08x$$
$$4<0.07x$$
$$57.1<x$$

LowCharge is less expensive if you use more than 57 checks.

57. Find the range of heights, h.

Let h = the height of the package.

$$\text{length} + \text{girth} \le 130$$
$$\text{length} + 2(\text{width}) + 2(\text{height}) \le 130$$
$$34+2(22)+2h\le 130$$
$$34+44+2h\le 130$$
$$78+2h\le 130$$
$$2h\le 52$$
$$h\le 26$$

The height must be more than 0 but less than or equal to 26 inches.

59. Find the mileage if the value is \$25,000.

$$25{,}000=-181.14m+34{,}814$$
$$-9814=-181.14m$$
$$m\approx 54$$

The mileage is about 54,000 miles.

61. Find the corresponding temperature range.

$$68\le \quad F \quad \le 104$$
$$68\le \frac{9}{5}C+32\le 104$$
$$36\le \quad \frac{9}{5}C \quad \le 72$$
$$\frac{5}{9}(36)\le \frac{5}{9}\left(\frac{9}{5}C\right)\le \frac{5}{9}(72)$$
$$20^\circ\le \quad C \quad \le 40^\circ$$

63. Find the range of mean weights of men.

$$-2.575<\frac{190-\mu}{2.45}<2.575$$
$$-6.30875<190-\mu<6.30875$$
$$-196.30875< \ -\mu \ <-183.69125$$
$$196.30875> \ \mu \ >183.69125$$
$$183.7\text{ lb}< \ \mu \ <196.3\text{ lb}$$

65. Estimate the range of heights.

Solve $|h-(2.47f+54.10)|\le 3.72$ for h where $f = 32.24$.

$$|h-(2.47f+54.10)|\le 3.72$$
$$|h-[2.47(32.24)+54.10]|\le 3.72$$
$$|h-(79.6328+54.10)|\le 3.72$$
$$|h-133.7328|\le 3.72$$

$h-133.7328\le 3.72$ or $h-133.7328\ge -3.72$

$h\le 137.4528$ $\qquad$ $h\ge 130.0128$

The height, to the nearest 0.1 cm, is from 130.0 cm to 137.5 cm.

67. Find the interval where monthly revenue is greater than zero.

$$R = 420x - 2x^2$$
$$420x - 2x^2 > 0$$
$$2x(210 - x) > 0$$

The product is positive.

$$2x = 0 \Rightarrow x = 0$$
$$210 - x = 0 \Rightarrow x = 210$$

Critical values are 0 and 210.

$2x(210 - x)$ - - - -|+ + + + + + + + +|- - -

0 210

($0, $210)

69. Find the number of books.

$$\frac{14.25x + 350{,}000}{x} < 50$$
$$14.25x + 350{,}000 < 50x$$
$$-35.75x < -350{,}000$$
$$x > 9790.2$$

At least 9791 books must be published.

71. Find the time interval.

$$s = -16t^2 + v_0 t + s_0, \quad s > 48,\ v_0 = 64,\ s_0 = 0$$

$$-16t^2 + 64t > 48$$
$$-16t^2 + 64t - 48 > 0$$
$$-16(t^2 - 4t + 3) > 0$$
$$-16(t - 1)(t - 3) > 0$$

The product is positive.

The critical values are 1 and 3.

$(t-1)(t-3)$ - - - - - -|+ + + + + + +|- - -

1 3

1 second $< t <$ 3 seconds

The ball is higher than 48 ft between 1 and 3 seconds.

Prepare for Section 1.6

P1. Solve.

$$1820 = k(28)$$
$$65 = k$$

P3. Evaluate.

$$k \cdot \frac{3}{5^2}$$
$$225 \cdot \frac{3}{5^2} = 27$$

P5. The area becomes 4 times as large.

Section 1.6 Exercises

1. Write the equation of variation.

$$d = kt$$

3. Write the equation of variation.

$$y = \frac{k}{x}$$

5. Write the equation of variation.

$$m = knp$$

7. Write the equation of variation.

$$V = klwh$$

9. Write the equation of variation.

$$A = ks^2$$

11. Write the equation of variation.

$$F = \frac{km_1 m_2}{d^2}$$

13. Write the equation and solve for k.

$$y = kx$$
$$64 = k \cdot 48$$
$$\frac{64}{48} = k$$
$$\frac{4}{3} = k$$

15. Write the equation and solve for k.

$$r = kt^2$$
$$144 = k \cdot 108^2$$
$$\frac{144}{108^2} = k$$
$$\frac{2^4 \cdot 3^2}{2^4 \cdot 3^6} = k$$
$$\frac{1}{81} = k$$

17. Write the equation and solve for k.

$$T = krs^2$$
$$210 = k \cdot 30 \cdot 5^2$$
$$\frac{210}{30 \cdot 5^2} = k$$
$$\frac{7}{25} = k$$
$$0.28 = k$$

19. Write the equation and solve for k.

$$V = klwh$$
$$240 = k \cdot 8 \cdot 6 \cdot 5$$
$$\frac{240}{8 \cdot 6 \cdot 5} = k$$
$$1 = k$$

21. Find the volume of the balloon.

$$V = kT$$
$$0.85 = k \cdot 270$$
$$\frac{0.85}{270} = k$$
$$\frac{0.17}{54} = k$$

Thus $V = \frac{0.17}{54}T = \frac{0.17}{54} \cdot 324 = (0.17)6 = 1.02$ liters

23. Find the number of semester hours.

$$s = k \cdot q$$
$$34 = k \cdot 51$$
$$\frac{2}{3} = k$$
$$p = \frac{2}{3} \cdot 93$$
$$p = 62 \text{ semester hours}$$

25. Find the amount of juice. Round to the nearest tenth of a fluid ounce.

$$j = k \cdot d^3$$
$$6 = k \cdot (4)^3$$
$$\frac{3}{32} = k$$
$$p = \frac{3}{32} \cdot (5)^3$$
$$p \approx 11.7 \text{ fl oz}$$

27. $T = k\sqrt{l}$

$$1.8 = k\sqrt{3}$$
$$k = \frac{1.8}{\sqrt{3}} \approx 1.03923$$

a. $T = \frac{1.8}{\sqrt{3}}\sqrt{10} = \frac{1.8\sqrt{30}}{3} = 0.6\sqrt{30} \approx 3.3$ seconds

b.

$$T = k\sqrt{l}$$
$$\frac{T}{k} = \sqrt{l}$$
$$\frac{2}{1.03923} = \sqrt{l}$$
$$l = \frac{4}{1.03923^2} \approx 3.7 \text{ ft}$$

29. Find the speed of the gear with 48 teeth.

$$r = \frac{k}{t}$$
$$30 = \frac{k}{64}$$
$$1920 = k$$
$$r = \frac{1920}{48}$$
$$r = 40 \text{ revolutions per minute}$$

31. Find the sound intensity.

$$I = \frac{k}{d^2}$$
$$0.5 = \frac{k}{7^2}$$
$$24.5 = k$$
$$I = \frac{24.5}{d^2}$$
$$I = \frac{24.5}{10^2}$$
$$I = 0.245 \text{ W/m}^2$$

33. a. $V = kr^2h$

$$V_1 = k(3r)^2h$$
$$= 9(kr^2h)$$
$$= 9V$$

Thus the new volume is 9 times the original volume.

b. $V_2 = kr^2(3h)$

$$= 3(kr^2h)$$
$$= 3V$$

Thus the new volume is 3 times the original volume.

c. $V_3 = k(3r)^2(3h)$

$$= k9r^2 \cdot 3 \cdot h$$
$$= 27(kr^2h)$$
$$= 27V$$

Thus the new volume is 27 times the original volume.

35. Find what happens to the volume.

$$V = \frac{knT}{P}$$

$$V_1 = \frac{k(3n)T}{\left(\frac{1}{2}p\right)}$$

$$= 6\left(\frac{knT}{p}\right)$$

$$= 6V$$

Thus the new volume is 6 times larger than the original volume.

37. Find Glavine's ERA.

For Randy Johnson,

$$\text{ERA} = \frac{kr}{i}$$

$$2.32 = \frac{k(67)}{(260)}$$

$$9.00 = k$$

For Tom Glavine,

$$\text{ERA} = \frac{9(74)}{(224.2)} = 2.97$$

39. Find the force. Round to the nearest 10 lb.

$$F = \frac{kws^2}{r}$$

$$2800 = \frac{k \cdot 1800 \cdot 45^2}{425}$$

$$\frac{2800 \cdot 425}{1800 \cdot 45^2} = k$$

$$\frac{14 \cdot 425}{9 \cdot 45^2} = k$$

$$0.3264746 \approx k$$

$$\text{Thus } F = \frac{(0.3264746) \cdot 1800 \cdot 55^2}{450}$$

$$\approx 3950 \text{ pounds}$$

41. Find the distance. Round to the nearest million miles.

$$T = kd^{3/2}$$

$$365 = k \cdot 93^{3/2}$$

$$\frac{365}{93^{3/2}} = k$$

$$\text{Thus} \quad 686 = \frac{365}{93^{3/2}} \cdot d^{3/2}$$

$$\frac{686 \cdot 93^{3/2}}{365} = d^{3/2}$$

$$\left(\frac{686 \cdot 93^{3/2}}{365}\right)^{2/3} = d$$

$$93\left(\frac{686}{365}\right)^{2/3} = d$$

$$142 \text{ million miles} \approx d$$

Chapter 1 Review Exercises

1. Solve. [1.1]

$$4 - 5x = 3x + 14$$

$$-8x = 10$$

$$x = -\frac{5}{4}$$

3. Solve. [1.1]

$$\frac{4x}{3} - \frac{4x-1}{6} = \frac{1}{2}$$

$$6\left(\frac{4x}{3} - \frac{4x-1}{6}\right) = 6\left(\frac{1}{2}\right)$$

$$2(4x) - (4x - 1) = 3$$

$$8x - 4x + 1 = 3$$

$$4x + 1 = 3$$

$$4x = 2$$

$$x = \frac{1}{2}$$

5. Solve. [1.1]

$$|x - 3| = 2$$

$$x - 3 = 2 \quad \text{or} \quad x - 3 = -2$$

$$x = 5 \qquad\qquad x = 1$$

7. Solve. [1.1]

$$|2x + 1| = 5$$

$$2x + 1 = 5 \quad \text{or} \quad 2x + 1 = -5$$

$$2x = 4 \qquad\qquad 2x = -6$$

$$x = 2 \qquad\qquad x = -3$$

9. Solve. [1.2]

$$V = \pi r^2 h$$

$$\frac{V}{\pi r^2} = h$$

11. Solve. [1.2]

$$A=\frac{h}{2}(b_1+b_2)$$
$$2A=h(b_1+b_2)$$
$$2A=hb_1+hb_2$$
$$2A-hb_2=hb_1$$
$$\frac{2A-hb_2}{h}=b_1$$

13. Solve. [1.3]

$$x^2-5x+6=0$$
$$(x-2)(x-3)=0$$
$$x-2=0 \quad \text{or} \quad x-3=0$$
$$x=2 \qquad x=3$$

15. Solve. [1.3]

$$(x-2)^2=50$$
$$x-2=\pm\sqrt{50}$$
$$x=2\pm5\sqrt{2}$$
$$x=2+5\sqrt{2} \text{ or } x=2-5\sqrt{2}$$

17. Solve. [1.3]

$$x^2-6x-1=0$$
$$x^2-6x=1$$
$$x^2-6x+9=1+9$$
$$(x-3)^2=10$$
$$x-3=\pm\sqrt{10}$$
$$x=3\pm\sqrt{10}$$
$$x=3+\sqrt{10} \text{ or } x=3-\sqrt{10}$$

19. Solve. [1.3]

$3x^2-x-1=0$ Use the quadratic formula.

$$x=\frac{-(-1)\pm\sqrt{(-1)^2-4(3)(-1)}}{2(3)}$$
$$x=\frac{1\pm\sqrt{13}}{6}$$
$$x=\frac{1+\sqrt{13}}{6} \quad \text{or} \quad x=\frac{1-\sqrt{13}}{6}$$

21. Determine whether the equations has real or complex solutions. [1.3]

$$2x^2+4x=5$$
$$2x^2+4x-5=0$$
$$b^2-4ac=(4)^2-4(2)(-5)$$
$$=16+40=56>0$$

There are two real number solutions.

23. Solve the equation. [1.4]

$$3x^3-5x^2=0$$
$$x^2(3x-5)=0$$
$$x^2=0\Rightarrow x=0$$
$$3x-5=0\Rightarrow x=\frac{5}{3}$$
$$x=0 \quad \text{or} \quad x=\frac{5}{3}$$

25. Solve the equation. [1.4]

$$2x^3+3x^2-8x-12=0$$
$$x^2(2x+3)-4(2x+3)=0$$
$$(2x+3)(x^2-4)=0$$
$$(2x+3)(x+2)(x-2)=0$$
$$x=-\frac{3}{2},\ x=-2, \text{ or } x=2$$

27. Solve the equation. [1.4]

$$\frac{x}{x+2}+\frac{1}{4}=5$$
$$4(x+2)\left(\frac{x}{x+2}+\frac{1}{4}\right)=5(4)(x+2)$$
$$4x+x+2=20(x+2)$$
$$5x+2=20x+40$$
$$-15x=38$$
$$x=-\frac{38}{15}$$

29. Solve the equation. [1.4]

$$3x+\frac{2}{x-2}=\frac{4x-1}{x-2}$$
$$(x-2)\left(3x+\frac{2}{x-2}\right)=(x-2)\left(\frac{4x-1}{x-2}\right)$$
$$3x(x-2)+2=4x-1$$
$$3x^2-6x+2=4x-1$$
$$3x^2-10x+3=0$$
$$(3x-1)(x-3)=0$$

$3x-1=0$ or $x-3=0$

$x=\frac{1}{3}$ $\quad x=3$

31. Solve the equation. [1.4]

$$\sqrt{2x+6}-1=3$$
$$\sqrt{2x+6}=4$$
$$\left(\sqrt{2x+6}\right)^2=4^2$$
$$2x+6=16$$
$$2x=10$$
$$x=5$$

Check $\sqrt{2(5)+6}-1=3$
$$\sqrt{16}-1=3$$
$$4-1=3$$
$$3=3$$

5 checks as a solution.

33. Solve the equation. [1.4]

$$\sqrt{-2x-7}+2x=-7$$
$$\sqrt{-2x-7}=-2x-7$$
$$\left(\sqrt{-2x-7}\right)^2=(-2x-7)^2$$
$$-2x-7=4x^2+28x+49$$
$$0=4x^2+30x+56$$
$$0=2\left(2x^2+15x+28\right)$$
$$0=2(2x+7)(x+4)$$

$x=-\frac{7}{2}$ or $x=-4$

Check $\sqrt{-2\left(-\frac{7}{2}\right)-7}+2\left(-\frac{7}{2}\right)=-7$
$$\sqrt{7-7}-7=-7$$
$$-7=-7$$
$$\sqrt{-2(-4)-7}+2(-4)=-7$$
$$\sqrt{8-7}-8=-7$$
$$1-8=-7$$
$$-7=-7$$

$-\frac{7}{2}$ and -4 check as solutions.

35. Solve the equation. [1.4]

$$\sqrt{3x+4}+\sqrt{x-3}=5$$
$$\sqrt{3x+4}=5-\sqrt{x-3}$$
$$\left[\sqrt{3x+4}\right]^2=\left[5-\sqrt{x-3}\right]^2$$
$$3x+4=25-10\sqrt{x-3}+x-3$$
$$2x-18=-10\sqrt{x-3}$$
$$x-9=-5\sqrt{x-3}$$
$$(x-9)^2=\left[-5\sqrt{x-3}\right]^2$$
$$x^2-18x+81=25(x-3)$$
$$x^2-18x+81=25x-75$$
$$x^2-43x+156=0$$
$$(x-4)(x-39)=0$$

$x=4$ or $x=39$

Check $\sqrt{3(4)+4}+\sqrt{4-3}=5$
$$\sqrt{16}+\sqrt{1}=5$$
$$4+1=5$$
$$5=5$$
$$\sqrt{3(39)+4}+\sqrt{39-3}=5$$
$$\sqrt{121}+\sqrt{36}=5$$
$$11+6=5$$
$$17=5 \text{ (No)}$$

The solution is 4.

37. Solve the equation. [1.4]

$$x^{5/4}-32=0$$
$$x^{5/4}=32$$
$$\left(x^{5/4}\right)^{4/5}=(32)^{4/5}$$
$$x=16$$

39. Solve the equation. [1.4]

$$6x^4-23x^2+20=0$$

Let $u=x^2$.

$$6u^2-23u+20=0$$
$$(3u-4)(2u-5)=0$$

$a=\frac{4}{3}$ or $u=\frac{5}{2}$

$x^2=\frac{4}{3}$ $\qquad x^2=\frac{5}{2}$

$x=\pm\sqrt{\frac{4}{3}}$ $\qquad x=\pm\sqrt{\frac{5}{2}}$

$x=\pm\frac{2}{\sqrt{3}}\left(\frac{\sqrt{3}}{\sqrt{3}}\right)$ $\qquad x=\pm\frac{\sqrt{5}}{\sqrt{2}}\left(\frac{\sqrt{2}}{\sqrt{2}}\right)$

$x=\pm\frac{2\sqrt{3}}{3}$ $\qquad x=\pm\frac{\sqrt{10}}{2}$

41. Solve and write the solution in interval notation. [1.5]

$-3x+4\ge -2$

$-3x\ge -2-4$

$-3x\ge -6$

$x\le 2$

$(-\infty, 2]$

43. Solve and write the solution in interval notation. [1.5]

$3x+1>7$ or $3x+2<-7$

$3x>6$ $\qquad 3x<-9$

$x>2$ $\qquad x<-3$

$(-\infty,-3)\cup(2,\infty)$

45. Solve and write the solution in interval notation. [1.5]

$61\le \frac{9}{5}C+32\le 95$

$29\le \frac{9}{5}C \le 63$

$\frac{145}{9}\le C \le 35$

$\left[\frac{145}{9}, 35\right]$

47. Solve and write the solution in interval notation. [1.5]

$|3x-4|<2$

$-2<3x-4<2$

$2< 3x <6$

$\frac{2}{3}< x <2$

$\left(\frac{2}{3}, 2\right)$

49. Solve and write the solution in interval notation. [1.5]

$0<|x-2|<1$

If $x-2\ge 0$, then $2< x <3$.

If $x-2<0$, then $0<x-2<-1$

$2> x >1.$

$(1, 2)\cup(2, 3)$

51. Solve and write the solution in interval notation. [1.5]

$x^2+x-6\ge 0$

$(x+3)(x-2)\ge 0$

The product is positive or zero.

$x+3=0\Rightarrow x=-3$

$x-2=0\Rightarrow x=2$

Critical values are –3 and 2.

$(x+3)(x-2)$ 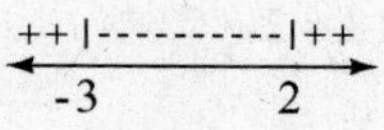

$(-\infty,-3]\cup[2,\infty)$

53. Solve and write the solution in interval notation. [1.5]

$\frac{x+3}{x-4}>0$

The quotient is positive.

$x+3=0\Rightarrow x=-3$

$x-4=0\Rightarrow x=4$

The critical values are –3 and 4.

$\frac{x+3}{x-4}$ 

$(-\infty, -3)\cup(4,\infty)$

55. Solve and write the solution in interval notation. [1.5]

$\frac{2x}{3-x}\le 10$

$\frac{2x}{3-x}-10\le 0$

$\frac{2x-10(3-x)}{3-x}\le 0$

$\frac{2x-30+10x}{3-x}\le 0$

$\frac{12x-30}{3-x}\le 0$

The quotient is negative or zero.

$12x-30=0\Rightarrow x=\frac{5}{2}$

$3-x=0\Rightarrow x=3$

The critical values are $\frac{5}{2}$ and 3.

$\dfrac{12x-30}{3-x}$ - - - - - - - |+| - - - - - -

Denominator $\neq 0 \Rightarrow x \neq 3$.

$\left(-\infty,\ \dfrac{5}{2}\right] \cup (3,\ \infty)$

57. Find the width and length. [1.2]

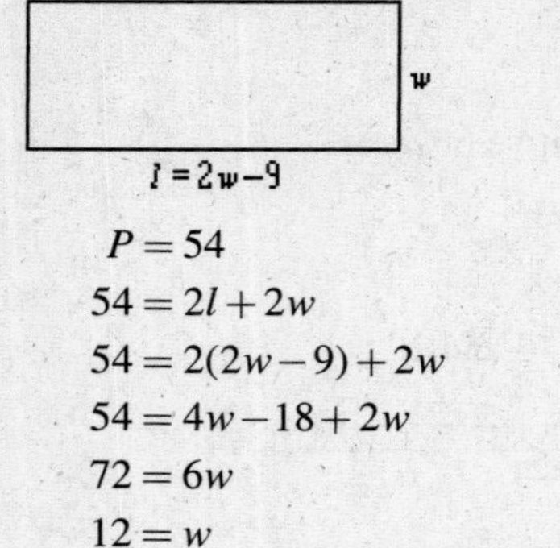

$$P = 54$$
$$54 = 2l + 2w$$
$$54 = 2(2w-9) + 2w$$
$$54 = 4w - 18 + 2w$$
$$72 = 6w$$
$$12 = w$$
$$2w - 9 = 2(12) - 9 = 24 - 9 = 15$$

width = 12 ft, length = 15 ft

59. Find the height of the tree. [1.2]

$$\frac{h}{15} = \frac{9}{6}$$
$$h = 22.5 \text{ ft}$$

61. Find the diameter. Round to the nearest foot. [1.2]

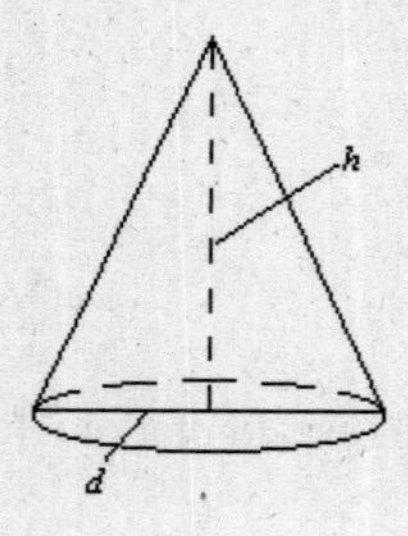

$$h = \frac{1}{4}d$$
$$V = 144$$
$$d = 2r$$
$$r = \frac{d}{2}$$
$$V = \frac{1}{3}\pi r^2 h$$
$$144 = \frac{1}{3}\pi\left(\frac{d}{2}\right)^2\left(\frac{1}{4}d\right)$$
$$144 = \frac{\pi d^3}{48}$$
$$\frac{144(48)}{\pi} = d^3$$
$$d \approx 13 \text{ ft}$$

63. Find the monthly maintenance cost. [1.2]

Let x = monthly maintenance cost per owner

$$18x = 24(x-12)$$
$$18x = 24x - 288$$
$$-6x = -288$$
$$x = 48$$
$$18x = 864$$

The total monthly maintenance cost is $864.

65. Find the distance. [1.2]

8t →

← 6(7 − t)

$$d = rt$$
$$d = 8t \qquad d = 6(7-t)$$
$$6(7-t) = 8t$$
$$42 - 6t = 8t$$
$$42 = 14t$$
$$3 = t$$

$d = 8(3) = 24$ nautical miles

67. Find the amount of each solution. [1.2]

5%	x
11%	$600-x$
7%	600

$$0.05x + 0.11(600-x) = 0.07(600)$$
$$0.05x + 66 - 0.11x = 42$$
$$-0.06x = -24$$
$$x = 400 \text{ ml of } 5\%$$
$$600 - x = 200 \text{ ml of } 11\%$$

69. Find the amount of gold alloy. [1.2]

$460	x
$220	25
$310	$x+25$

$$460x + 220(25) = 310(x+25)$$
$$460x + 5500 = 310x + 7750$$
$$150x = 2250$$
$$x = 15 \text{ oz}$$

71. Find the time it would take the apprentice, working alone. [1.4]

Let x = time for apprentice

$x - 9$ = time for mason

$\frac{1}{x}$ = rate for apprentice

$\frac{1}{x-9}$ = rate for mason

$$6\left(\frac{1}{x}+\frac{1}{x-9}\right)=1$$

$$6x(x-9)\left(\frac{1}{x}+\frac{1}{x-9}\right)=1x(x-9)$$

$$6(x-9)+6x=x^2-9x$$

$$6x-54+6x=x^2-9x$$

$$0=x^2-21x+54$$

$$0=(x-18)(x-3)$$

$$x=18 \quad \text{or} \quad x=3$$

Note: $x=3 \Rightarrow$ mason's time $=-6$ hours. Thus $x \neq 3$.

Apprentice takes 18 hours to build the wall.

73. Find the distance of AX. [1.3]

Let x = the distance of AX.

$$\sqrt{60^2+x^2}=\sqrt{40^2+(100-x)^2}$$

$$\left(\sqrt{60^2+x^2}\right)^2=\left(\sqrt{40^2+(100-x)^2}\right)^2$$

$$3600+x^2=1600+10{,}000-200x+x^2$$

$$200x=8000$$

$$x=40$$

The distance AX is 40 yards.

75. Find the time. [1.3]

$$5=-4.9t^2+7.5t+10$$

$$0=-4.9t^2+7.5t+5$$

$$t=\frac{-7.5\pm\sqrt{(7.5)^2-4(-4.9)(5)}}{2(-4.9)}$$

$$t=\frac{-7.5\pm\sqrt{154.25}}{-9.8}$$

$$t\approx 2.0 \quad \text{or} \quad t\approx -0.5 \text{ (No)}$$

In about 2.0 seconds.

77. Find the range of mean heights. [1.5]

$$-1.645<\frac{63.8-\mu}{0.45}<1.645$$

$$-0.74025<63.8-\mu<0.74025$$

$$-64.54025< -\mu <-63.05975$$

$$64.5> \mu >63.1$$

$$63.1< \mu <64.5 \text{ lb}$$

79. Find the range of diameters. [1.5]

Let C = the circumference, r = the radius, and d = the diameter.

$$C=2\pi r=\pi d$$

$$29.5\le C\le 30.0$$

$$29.5\le \pi d\le 30.0$$

$$\frac{29.5}{\pi}\le d\le \frac{30.0}{\pi}$$

$$9.39\le d\le 9.55$$

The diameter of the basketball is from 9.39 to 9.55 inches.

81. Find the acceleration. [1.6]

$$F=ka$$

$$10=k(2)$$

$$5=k$$

$$F=5a$$

$$15=5a$$

$$3=a$$

$$a=3 \text{ ft/s}^2$$

83. Find the number of players. [1.6]

$$N=\frac{k}{p}$$

$$5000=\frac{k}{150}$$

$$750{,}000=k$$

$$N=\frac{750{,}000}{p}$$

$$N=\frac{750{,}000}{125}$$

$$N=6000 \text{ players}$$

85. Find the acceleration. Round to the nearest hundredth of a meter per second squared. [1.6]

$$a=\frac{km}{r^2}$$

$$9.8=\frac{k(5.98\times 10^{26})}{(6{,}370{,}000)^2}$$

$$9.8(6{,}370{,}000)^2=k(5.98\times 10^{26})$$

$$\frac{9.8(6{,}370{,}000)^2}{5.98\times 10^{26}}=k$$

$$k\approx 6.6497\times 10^{-13}$$

$$a = \frac{6.6497\times10^{-13}m}{r^2}$$

$$a = \frac{(6.6497\times10^{-13})(7.46\times10^{24})}{(1,740,000)^2}$$

$$a \approx 1.64 \text{ meters/sec}^2$$

Chapter 1 Test

1. Solve. [1.1]

$$\frac{2x}{3}+\frac{1}{2}=\frac{x}{2}-\frac{3}{4}$$
$$12\left(\frac{2x}{3}+\frac{1}{2}\right)=12\left(\frac{x}{2}-\frac{3}{4}\right)$$
$$8x+6=6x-9$$
$$2x=-15$$
$$x=-\frac{15}{2}$$

3. Solve for x. [1.2]

$$ax-c=c(x-d)$$
$$ax-c=cx-cd$$
$$ax-cx=c-cd$$
$$x(a-c)=c-cd$$
$$x=\frac{c-cd}{a-c},\ a\neq c$$

5. Solve by completing the square. [1.3]

$$2x^2-8x+1=0$$
$$x^2-4x=-\frac{1}{2}$$
$$x^2-4x+4=-\frac{1}{2}+4$$
$$(x-2)^2=\frac{7}{2}$$
$$x-2=\pm\sqrt{\frac{7}{2}}=\pm\frac{\sqrt{14}}{2}$$
$$x=2\pm\frac{\sqrt{14}}{2}=\frac{4\pm\sqrt{14}}{2}$$

The solutions are $\frac{4-\sqrt{14}}{2}$ and $\frac{4+\sqrt{14}}{2}$.

7. Find the discriminant and number of solutions. [1.3]

$$2x^2+3x+1=0$$
$$a=2,\ b=3,\ c=1$$
$$b^2-4ac=(3)^2-4(2)(1)=9-8=1$$

The discriminant, 1, is a positive number. Therefore, there are two real solutions

9. Solve. [1.4]

$$\sqrt{3x+1}-\sqrt{x-1}=2$$
$$\sqrt{3x+1}=\sqrt{x-1}+2$$
$$\left(\sqrt{3x+1}\right)^2=\left(\sqrt{x-1}+2\right)^2$$
$$3x+1=x-1+4\sqrt{x-1}+4$$
$$2x-2=4\sqrt{x-1}$$
$$x-1=2\sqrt{x-1} \quad \text{Remove factor of 2.}$$
$$(x-1)^2=\left(2\sqrt{x-1}\right)^2$$
$$x^2-2x+1=4(x-1)$$
$$x^2-2x+1=4x-4$$
$$x^2-6x+5=0$$
$$(x-1)(x-5)=0$$
$$x=1 \text{ or } x=5$$

Check $\sqrt{3(1)+1}-\sqrt{1-1}=2$
$$\sqrt{4}-\sqrt{0}=2$$
$$2=2$$
$$\sqrt{3(5)+1}-\sqrt{5-1}=2$$
$$\sqrt{16}-\sqrt{4}=2$$
$$4-2=2$$
$$2=2$$

1 and 5 check as solutions.

11. Solve. [1.4]

$$\frac{3}{x+2}-\frac{3}{4}=\frac{5}{x+2}$$
$$4(x+2)\left(\frac{3}{x+2}-\frac{3}{4}\right)=4(x+2)\left(\frac{5}{x+2}\right)$$
$$4(3)-3(x+2)=4(5)$$
$$12-3x-6=20$$
$$-3x=14$$
$$x=-\frac{14}{3}$$

13. Solve. [1.4]

$$x^3-64=0$$
$$(x-4)(x^2+4x+16)=0$$
$$x-4=0 \text{ or } x^2+4x+16=0$$
$$x=4$$

$$x=\frac{-4\pm\sqrt{4^2-4(1)(16)}}{2(1)}$$

$$x=\frac{-4\pm\sqrt{-48}}{2}=\frac{-4\pm 4i\sqrt{3}}{2}=-2\pm 2i\sqrt{3}$$

The solutions are $4,\ -2-2i\sqrt{3},\ -2+2i\sqrt{3}$.

15. Solve and write in interval notation. [1.5]

$|3x-4|>5$

$3x-4>5$ or $3x-4<-5$

$3x>9$ $\quad$ $3x<-1$

$x>3$ $\quad$ $x<-\frac{1}{3}$

$\left(-\infty,-\frac{1}{3}\right)\cup(3,\infty)$

17. Solve and write in interval notation. [1.5]

$$\frac{x^2+x-12}{x+1}\geq 0$$

$$\frac{(x+4)(x-3)}{x+1}\geq 0$$

The quotient is positive or zero.

$x+4=0\Rightarrow x=-4$

$x-3=0\Rightarrow x=3$

$x+1=0\Rightarrow x=-1$

Critical values are –4, 3, and –1.

$\frac{(x+4)(x-3)}{x+1}$ - - - |+ + +|- - - -|+ + +

-4 -1 0 3

Denominator $\neq 0\Rightarrow x\neq -1$.

$[-4,\ -1)\cup[3,\ \infty)$

19. Find the time for the assistant working alone. [1.2]

Let x = number of hours the assistant needs to cover the parking lot.

$$6\left[\frac{1}{10}+\frac{1}{x}\right]=1$$

$$10x(6)\left[\frac{1}{10}+\frac{1}{x}\right]=10x(1)$$

$$6x+60=10x$$

$$-4x=-60$$

$$x=15$$

The assistant takes 15 hours to cover the parking lot.

21. Find the amount of each. [1.2]

\$3.45	x
\$2.70	$50-x$
\$3.15	50

$$3.45x+2.7(50-x)=3.15(50)$$

$$3.45x+135-2.70x=157.50$$

$$0.75x=22.5$$

$$x=30$$

$$50-x=20$$

30 lb of ground beef and 20 lb of sausage.

23. Find Zoey's speed. [1.4]

Let $x+4$ = Zoey's speed.

$$\frac{15}{x}-\frac{15}{x+4}=1$$

$$x(x+4)\left(\frac{15}{x}-\frac{15}{x+4}\right)=x(x+4)(1)$$

$$15(x+4)-15x=x(x+4)$$

$$15x+60-15x=x^2+4x$$

$$0=x^2+4x-60$$

$$0=(x+10)(x-6)$$

$$x=-10\text{ (No) or } x=6$$

$x+4=10$

Zoey's speed is 10 mph.

25. Find the velocity of the meteorite. [1.6]

$$v=\frac{k}{\sqrt{d}}$$

$$4=\frac{k}{\sqrt{3000}}$$

$$k=4\sqrt{3000}=40\sqrt{30}$$

$$v=\frac{40\sqrt{30}}{\sqrt{2500}}=\frac{40\sqrt{30}}{50}$$

$$v=\frac{4\sqrt{30}}{5}\approx 4.4\text{ miles/second}$$

Cumulative Review Exercises

1. Evaluate. [P.1]

$4+3(-5)=4-15=-11$

3. Perform the operations and simplify. [P.3]

$$(3x-5)^2-(x+4)(x-4)$$
$$=(9x^2-30x+25)-(x^2-16)$$
$$=9x^2-30x+25-x^2+16$$
$$=8x^2-30x+41$$

5. Simplify. [P.5]

$$\frac{7x-3}{x-4}-5=\frac{7x-3-5x+20}{x-4}=\frac{2x+17}{x-4}$$

7. Simplify. [P.6]

$$(2+5i)(2-5i)=4-25i^2=4+25=29$$

9. Solve using the quadratic formula. [1.3]

$$2x^2-4x=3$$
$$2x^2-4x-3=0$$
$$x=\frac{-(-4)\pm\sqrt{(-4)^2-4(2)(-3)}}{2(2)}=\frac{4\pm2\sqrt{10}}{4}$$
$$=\frac{2\pm\sqrt{10}}{2}$$

11. Solve. [1.4]

$$x=3+\sqrt{9-x}$$
$$x-3=\sqrt{9-x}$$
$$(x-3)^2=\left(\sqrt{9-x}\right)^2$$
$$x^2-6x+9=9-x$$
$$x^2-5x=0$$
$$x(x-5)=0$$
$$x=0 \text{ or } x=5$$

Check 0:	Check 5:
$0=3+\sqrt{9-0}$	$5=3+\sqrt{9-5}$
$0=3+\sqrt{9}$	$5=3+\sqrt{4}$
$0=3+3$	$5=3+2$
$0=6$	$5=5$
No	

The solution is 5.

13. Solve. [1.4]

$$2x^4-11x^2+15=0 \quad \text{Let } u=x^2.$$
$$2u^2-11u+15=0$$
$$(2u-5)(u-3)=0$$

$$2u-5=0 \quad \text{or} \quad u-3=0$$
$$u=x^2=\frac{5}{2} \qquad u=3$$
$$x^2=3$$
$$x=\pm\sqrt{\frac{5}{2}}=\pm\frac{\sqrt{10}}{2} \qquad x=\pm\sqrt{3}$$

The solutions are $-\frac{\sqrt{10}}{2}, \frac{\sqrt{10}}{2}, -\sqrt{3}, \sqrt{3}$.

15. Solve. Write the solution in interval notation. [1.5]

$$|x-6|\ge 2$$
$$x-6\ge 2 \quad \text{or} \quad x-6\le -2$$
$$x\ge 8 \qquad x\le 4$$

The solution is $(-\infty, 4]\cup[8, \infty)$.

17. Find the dimensions of the fence. [1.2]

w

$w+16$

$$\text{Perimeter} = 2(\text{Length})+2(\text{Width})$$
$$200=2(w+16)+2w$$
$$200=2w+32+2w$$
$$168=4w$$
$$42=w$$
$$w=42$$
$$w+16=58$$

The width is 42 feet; the length is 58 feet.

19. Find the range of scores. [1.5]

Let x = the score on the fourth test.

$$80\le\frac{86+72+94+x}{4}<90 \quad \text{and} \quad 0\le x\le 100$$
$$80\le\frac{252+x}{4}<90$$
$$320\le 252+x<360$$
$$68\le x<108$$

$$[68, 108)\cap[0, 100]=[68, 100]$$

The fourth test score must be from 68 to 100.

Chapter 2: Functions and Graphs

Section 2.1 Exercises

1. Plot the points:

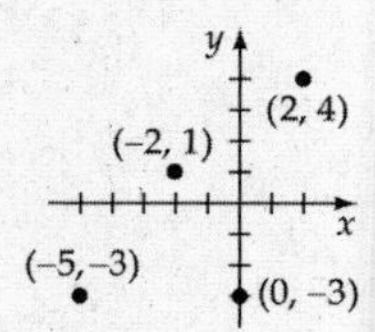

3. a. Find the decrease: The average debt decreased between 2001 and 2002.

b. Find the average debt in 2008:

Increase between 2006 to 2007: $22.7-22.3=0.4$

Then the increase from 2007 to 2008:

$22.7+0.4=23.1$, or \$23,100.

5. Find the distance: (6, 4), (–8, 11)

$$\begin{aligned} d &= \sqrt{(-8-6)^2+(11-4)^2} \\ &= \sqrt{(-14)^2+(7)^2} \\ &= \sqrt{196+49} \\ &= \sqrt{245} \\ &= 7\sqrt{5} \end{aligned}$$

7. Find the distance: (–4, –20), (–10, 15)

$$\begin{aligned} d &= \sqrt{(-10-(-4))^2+(15-(-20))^2} \\ &= \sqrt{(-6)^2+(35)^2} \\ &= \sqrt{36+1225} \\ &= \sqrt{1261} \end{aligned}$$

9. Find the distance: (5, –8), (0, 0)

$$\begin{aligned} d &= \sqrt{(0-5)^2+(0-(-8))^2} \\ &= \sqrt{(-5)^2+(8)^2} \\ &= \sqrt{25+64} \\ &= \sqrt{89} \end{aligned}$$

11. Find the distance: $(\sqrt{3}, \sqrt{8})$, $(\sqrt{12}, \sqrt{27})$

$$\begin{aligned} d &= \sqrt{(\sqrt{12}-\sqrt{3})^2+(\sqrt{27}-\sqrt{8})^2} \\ &= \sqrt{(2\sqrt{3}-\sqrt{3})^2+(3\sqrt{3}-2\sqrt{2})^2} \\ &= \sqrt{(\sqrt{3})^2+(3\sqrt{3}-2\sqrt{2})^2} \\ &= \sqrt{3+(27-12\sqrt{6}+8)} \\ &= \sqrt{3+27-12\sqrt{6}+8} \\ &= \sqrt{38-12\sqrt{6}} \end{aligned}$$

13. Find the distance: (a, b), $(-a, -b)$

$$\begin{aligned} d &= \sqrt{(-a-a)^2+(-b-b)^2} \\ &= \sqrt{(-2a)^2+(-2b)^2} \\ &= \sqrt{4a^2+4b^2} \\ &= \sqrt{4(a^2+b^2)} \\ &= 2\sqrt{a^2+b^2} \end{aligned}$$

15. Find the distance: $(x, 4x)$, $(-2x, 3x)$

$$\begin{aligned} d &= \sqrt{(-2x-x)^2+(3x-4x)^2} \text{ with } x<0 \\ &= \sqrt{(-3x)^2+(-x)^2} \\ &= \sqrt{9x^2+x^2} \\ &= \sqrt{10x^2} \\ &= -x\sqrt{10} \quad (\text{Note: since } x<0, \sqrt{x^2}=-x) \end{aligned}$$

17. Find the points:

$$\begin{aligned} \sqrt{(4-x)^2+(6-0)^2} &= 10 \\ \left(\sqrt{(4-x)^2+(6-0)^2}\right)^2 &= 10^2 \\ 16-8x+x^2+36 &= 100 \\ x^2-8x-48 &= 0 \\ (x-12)(x+4) &= 0 \end{aligned}$$

$$x=12 \quad \text{or} \quad x=-4$$

The points are (12, 0), (– 4, 0).

19. Find the midpoint: (1, –1), (5, 5)

$$M=\left(\frac{x_1+x_2}{2},\ \frac{y_1+y_2}{2}\right)$$
$$=\left(\frac{1+5}{2},\ \frac{-1+5}{2}\right)$$
$$=\left(\frac{6}{2},\ \frac{4}{2}\right)$$
$$=(3,\ 2)$$

21. Find the midpoint: (6, –3), (6, 11)

$$M=\left(\frac{6+6}{2},\ \frac{-3+11}{2}\right)$$
$$=\left(\frac{12}{2},\ \frac{8}{2}\right)$$
$$=(6,\ 4)$$

23. Find the midpoint: (1.75, 2.25), (–3.5, 5.57)

$$M=\left(\frac{1.75+(-3.5)}{2},\ \frac{2.25+5.57}{2}\right)$$
$$=\left(-\frac{1.75}{2},\ \frac{7.82}{2}\right)$$
$$=(-0.875,\ 3.91)$$

25. Graph the equation: $x-y=4$

x	y
0	−4
2	−2
4	0

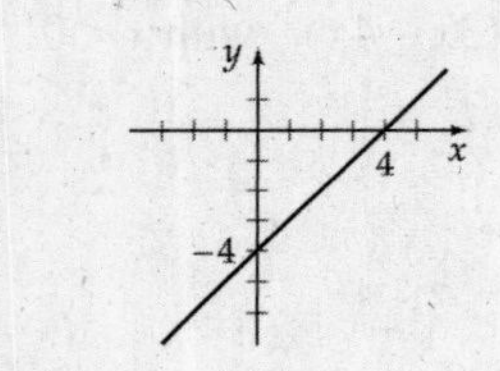

27. Graph the equation: $y=0.25x^2$

x	y
−4	4
−2	1
0	0
2	1
4	4

29. Graph the equation: $y=-2|x-3|$

x	y
0	−6
2	−2
3	0
4	−2
6	−6

31. Graph the equation: $y=x^2-3$

x	y
−2	1
−1	−2
0	−3
1	−2
2	1

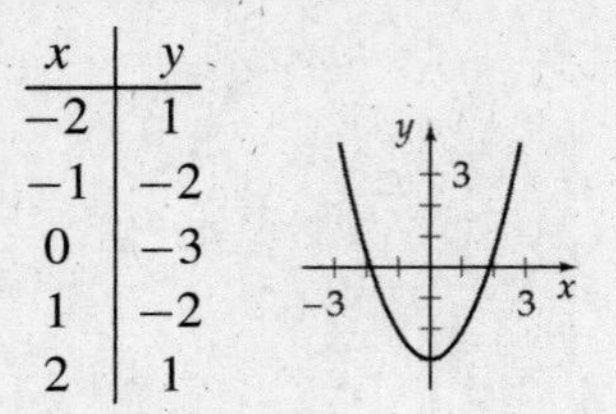

33. Graph the equation: $y=\frac{1}{2}(x-1)^2$

x	y
−1	2
0	0.5
1	0
2	0.5
3	2

35. Graph the equation: $y=x^2+2x-8$

x	y
−4	0
−2	−8
−1	−9
0	−8
2	0

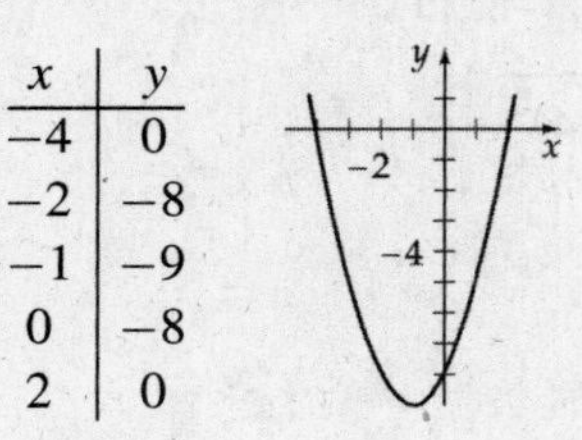

37. Graph the equation: $y=-x^2+2$

x	y
−2	−2
−1	1
0	2
1	1
2	−2

39. Find the x- and y-intercepts and graph: $2x+5y=12$

For the y-intercept, let $x = 0$ and solve for y.

$$2(0)+5y=12$$
$$y=\frac{12}{5},\quad y\text{-intercept: }\left(0,\ \frac{12}{5}\right)$$

For the x-intercept, let $y = 0$ and solve for x.

$$2x+5(0)=12$$
$$x=6,\quad x\text{-intercept: }(6,\ 0)$$

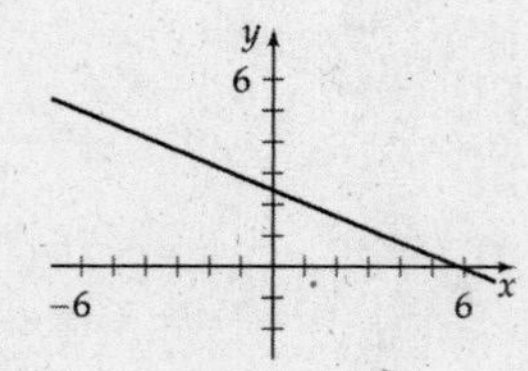

41. Find the x- and y-intercepts and graph: $x=-y^2+5$

For the y-intercept, let $x = 0$ and solve for y.

$$0=-y^2+5$$
$$y=\pm\sqrt{5},\ y\text{-intercepts: } (0,-\sqrt{5}),(0,\ \sqrt{5})$$

For the x-intercept, let $y = 0$ and solve for x.

$$x=-(0)^2+5$$
$$x=5,\ x\text{-intercept: } (5,\ 0)$$

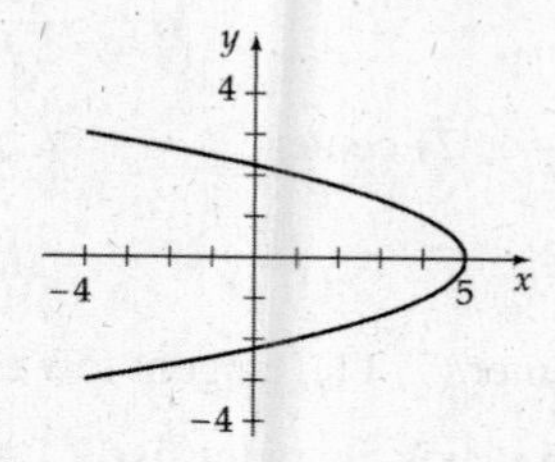

43. Find the x- and y-intercepts and graph: $x=|y|-4$

For the y-intercept, let $x = 0$ and solve for y.

$$0=|y|-4$$
$$y=\pm4,\ y\text{-intercepts: } (0,-4),(0,\ 4)$$

For the x-intercept, let $y = 0$ and solve for x.

$$x=|0|-4$$
$$x=-4,\ x\text{-intercept: } (-4,\ 0)$$

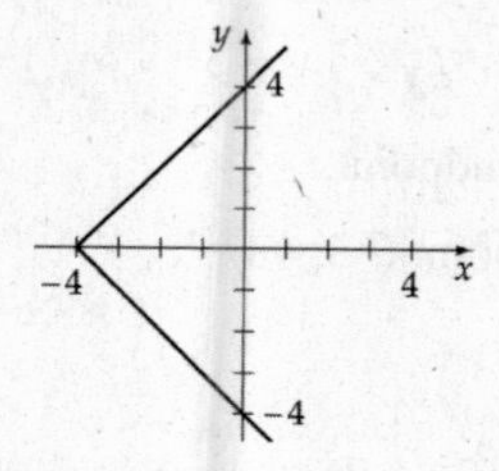

45. Find the x- and y-intercepts and graph: $x^2+y^2=4$

For the y-intercept, let $x = 0$ and solve for y.

$$(0)^2+y^2=4$$
$$y=\pm2,\ y\text{-intercepts: } (0,-2),(0,\ 2)$$

For the x-intercept, let $y = 0$ and solve for x.

$$x^2+(0)^2=4$$
$$x=\pm2,\ x\text{-intercepts: } (-2,\ 0),(2,\ 0)$$

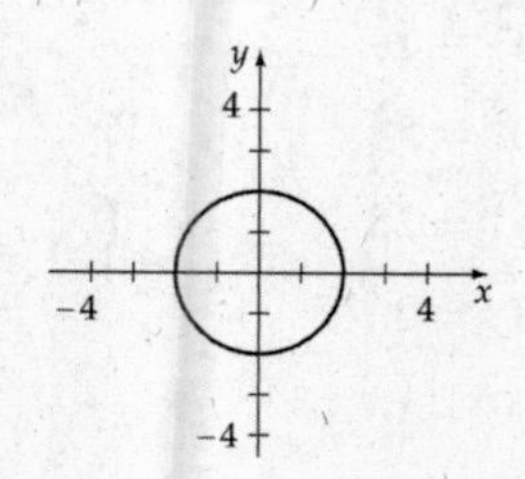

47. Find the x- and y-intercepts and graph: $|x|+|y|=4$

Intercepts: $(0,\ \pm4),\ (\pm4,\ 0)$

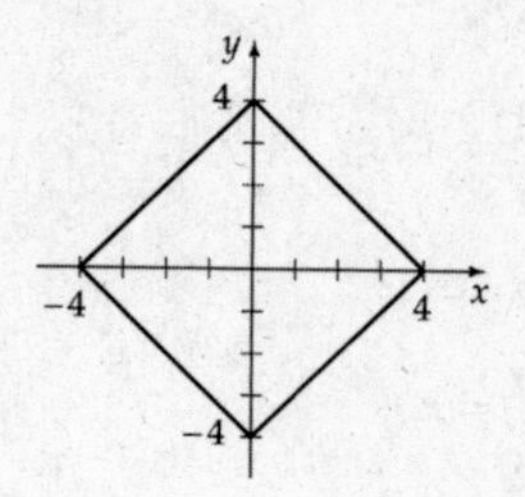

49. Find center and radius: $x^2+y^2=36$

center (0, 0), radius 6

51. Find center and radius: $(x-1)^2+(y-3)^2=49$

center (1, 3), radius 7

53. Find center and radius: $(x+2)^2+(y+5)^2=25$

center (−2, −5), radius 5

55. Find center and radius: $(x-8)^2+y^2=\frac{1}{4}$

center (8, 0), radius $\frac{1}{2}$

57. Find circle equation: center (4, 1), radius 2

$$(x-4)^2+(y-1)^2=2^2$$

59. Find circle equation: center $\left(\frac{1}{2},\ \frac{1}{4}\right)$, radius $\sqrt{5}$

$$\left(x-\frac{1}{2}\right)^2+\left(y-\frac{1}{4}\right)^2=\left(\sqrt{5}\right)^2$$

61. Find circle equation: center (0, 0), through (−3, 4)

$$(x-0)^2+(y-0)^2=r^2$$
$$(-3-0)^2+(4-0)^2=r^2$$
$$(-3)^2+4^2=r^2$$
$$9+16=r^2$$
$$25=5^2=r^2$$
$$(x-0)^2+(y-0)^2=5^2$$

63. Find circle equation: center (1, 3), through (4, –1)

$$(x+2)^2+(y-5)^2=r^2$$
$$(x-1)^2+(y-3)^2=r^2$$
$$(4-1)^2+(-1-3)^2=r^2$$
$$3^2+(-4)^2=r^2$$
$$9+16=r^2$$
$$25=5^2=r^2$$
$$(x-1)^2+(y-3)^2=5^2$$

65. Find center and radius: $x^2+y^2-6x+5=0$

$$x^2-6x \quad +y^2=-5$$
$$x^2-6x+9+y^2=-5+9$$
$$(x-3)^2+y^2=2^2$$

center (3, 0), radius 2

67. Find center and radius: $x^2+y^2-14x+8y+53=0$

$$x^2-14x \quad +y^2+8y \quad =-53$$
$$x^2-14x+49+y^2+8y+16=-53+49+16$$
$$(x-7)^2+(y+4)^2=12$$

center (7, –4), radius $\sqrt{12}=2\sqrt{3}$

69. Find center and radius: $x^2+y^2-x+3y-\frac{15}{4}=0$

$$x^2-x \quad +y^2+3y \quad =\frac{15}{4}$$
$$x^2-x+\frac{1}{4} \quad +y^2+3y+\frac{9}{4}=\frac{15}{4}+\frac{1}{4}+\frac{9}{4}$$
$$\left(x-\frac{1}{2}\right)^2+\left(y+\frac{3}{2}\right)^2=\left(\frac{5}{2}\right)^2$$

center $\left(\frac{1}{2},\ -\frac{3}{2}\right)$, radius $\frac{5}{2}$

71. Find center and radius: $x^2+y^2+3x-6y+2=0$

$$x^2+3x \quad +y^2-6y \quad =-2$$
$$x^2+3x+\frac{9}{4} \quad +y^2-6y+9=-2+\frac{9}{4}+9$$
$$\left(x+\frac{3}{2}\right)^2+(y-3)^2=\left(\frac{\sqrt{37}}{2}\right)^2$$

center $\left(-\frac{3}{2},\ 3\right)$, radius $\frac{\sqrt{37}}{2}$

73. Find circle equation: endpoints (2, 3) and (–4, 11)

$$d=\sqrt{(-4-2)^2+(11-3)^2}$$
$$=\sqrt{36+64}$$
$$=\sqrt{100}$$
$$=10$$

Since the diameter is 10, the radius is 5.

The center is the midpoint of the line segment from (2, 3) to (–4, 11).

$$\left(\frac{2+(-4)}{2},\ \frac{3+11}{2}\right)=(-1,\ 7) \text{ center}$$
$$(x+1)^2+(y-7)^2=5^2$$

75. Find circle equation: center (7, 11), tangent to x-axis

Since it is tangent to the x-axis, its radius is 11.

$$(x-7)^2+(y-11)^2=11^2$$

77. Find other endpoint: endpoint (5, 1), midpoint (9, 3)

$$\left(\frac{x+5}{2},\ \frac{y+1}{2}\right)=(9,\ 3)$$

therefore $\frac{x+5}{2}=9$ and $\frac{y+1}{2}=3$

$x+5=18$ $\quad$ $y+1=6$

$x=13$ $\quad$ $y=5$

Thus (13, 5) is the other endpoint.

79. Find other endpoint: endpoint (–3, –8), midpoint (2, –7)

$$\left(\frac{x+(-3)}{2},\ \frac{y+(-8)}{2}\right)=(2,\ -7)$$

therefore $\frac{x-3}{2}=2$ and $\frac{y-8}{2}=-7$

$x-3=4$ $\quad$ $y-8=-14$

$x=7$ $\quad$ $y=-6$

Thus (7, –6) is the other endpoint.

81. Find the formula:

$$\sqrt{(3-x)^2+(4-y)^2}=5$$
$$(3-x)^2+(4-y)^2=5^2$$
$$9-6x+x^2+16-8y+y^2=25$$
$$x^2-6x+y^2-8y=0$$

Prepare for Section 2.2

P1. x^2+3x-4

$(-3)^2+3(-3)-4=9-9-4=-4$

P3. $d=\sqrt{(3-(-4))^2+(-2-1)^2}=\sqrt{49+9}=\sqrt{58}$

P5.
$$x^2-x-6=0$$
$$(x+2)(x-3)=0$$
$$x+2=0 \quad x-3=0$$
$$x=-2 \quad x=3$$

$-2, 3$

Section 2.2 Exercises

1. Is y a function of x?

$$2x+3y=7$$
$$3y=-2x+7$$
$$y=-\frac{2}{3}x+\frac{7}{3},\ y \text{ is a function of } x.$$

3. Is y a function of x?

$$-x+y^2=2$$
$$y^2=x+2$$
$$y=\pm\sqrt{x+2},\ y \text{ is a not function of } x.$$

5. Is y a function of x?

$y=4\pm\sqrt{x}$, y is not a function of x since

for each $x>0$ there are two values of x.

7. Is y a function of x?

$y=\sqrt[3]{x}$, y is a function of x.

9. Is y a function of x?

$$y^2=x^2$$
$$y=\pm\sqrt{x^2},\ y \text{ is a not function of } x.$$

11. The set of ordered pairs defines y as a function of x since each x is paired with exactly one y.

13. The set of ordered pair does not define y as a function of x since 4 is paired with 4 and 5.

15. The set of ordered pairs defines y as a function of x since each x is paired with exactly one y.

17. Evaluate the function $f(x)=3x-1$,

a. $f(2)=3(2)-1$
$=6-1$
$=5$

b. $f(-1)=3(-1)-1$
$=-3-1$
$=-4$

c. $f(0)=3(0)-1$
$=0-1$
$=-1$

d. $f\left(\frac{2}{3}\right)=3\left(\frac{2}{3}\right)-1$
$=2-1$
$=1$

e. $f(k)=3(k)-1$
$=3k-1$

f. $f(k+2)=3(k+2)-1$
$=3k+6-1$
$=3k+5$

19. Evaluate the function $A(w)=\sqrt{w^2+5}$,

a. $A(0)=\sqrt{(0)^2+5}=\sqrt{5}$

b. $A(2)=\sqrt{(2)^2+5}=\sqrt{9}=3$

c. $A(-2)=\sqrt{(-2)^2+5}=\sqrt{9}=3$

d. $A(4)=\sqrt{4^2+5}=\sqrt{21}$

e. $A(r+1)=\sqrt{(r+1)^2+5}$
$=\sqrt{r^2+2r+1+5}$
$=\sqrt{r^2+2r+6}$

f. $A(-c)=\sqrt{(-c)^2+5}=\sqrt{c^2+5}$

21. Evaluate the function $f(x)=\frac{1}{|x|}$,

a. $f(2)=\frac{1}{|2|}=\frac{1}{2}$

b. $f(-2)=\frac{1}{|-2|}=\frac{1}{2}$

c. $f\left(-\frac{3}{5}\right)=\frac{1}{\left|-\frac{3}{5}\right|}$

$=\frac{1}{3/5}$

$=1\div\frac{3}{5}=1\cdot\frac{5}{3}$

$=\frac{5}{3}$

d. $f(2)+f(-2)=\frac{1}{2}+\frac{1}{2}=1$

e. $f(c^2+4)=\frac{1}{|c^2+4|}=\frac{1}{c^2+4}$

f. $f(2+h)=\frac{1}{|2+h|}$

23. Evaluate the function $s(x)=\frac{x}{|x|}$,

a. $s(4)=\frac{4}{|4|}=\frac{4}{4}=1$

b. $s(5)=\frac{5}{|5|}=\frac{5}{5}=1$

c. $s(-2)=\frac{-2}{|-2|}=\frac{-2}{2}=-1$

d. $s(-3)=\frac{-3}{|-3|}=\frac{-3}{3}=-1$

e. Since $t>0$, $|t|=t$.

$s(t)=\frac{t}{|t|}=\frac{t}{t}=1$

f. Since $t<0$, $|t|=-t$.

$s(t)=\frac{t}{|t|}=\frac{t}{-t}=-1$

25. a. Since $x=-4<2$, use $P(x)=3x+1$.

$P(-4)=3(-4)+1=-12+1=-11$

b. Since $x=\sqrt{5}\geq 2$, use $P(x)=-x^2+11$.

$P(\sqrt{5})=-(\sqrt{5})^2+11=-5+11=6$

c. Since $x=c<2$, use $P(x)=3x+1$.

$P(c)=3c+1$

d. Since $k\geq 1$, then $x=k+1\geq 2$,

so use $P(x)=-x^2+11$.

$P(k+1)=-(k+1)^2+11=-(k^2+2k+1)+11$

$=-k^2-2k-1+11$

$=-k^2-2k+10$

27. For $f(x)=3x-4$, the domain is the set of all real numbers.

29. For $f(x)=x^2+2$, the domain is the set of all real numbers.

31. For $f(x)=\frac{4}{x+2}$, the domain is $\{x|x\neq -2\}$.

33. For $f(x)=\sqrt{7+x}$, the domain is $\{x|x\geq -7\}$.

35. For $f(x)=\sqrt{4-x^2}$, the domain is $\{x|-2\leq x\leq 2\}$.

37. For $f(x)=\frac{1}{\sqrt{x+4}}$, the domain is $\{x|x>-4\}$.

39. To graph $f(x)=3x-4$, plot points and draw a smooth graph.

x	-1	0	1	2
$y=f(x)=3x-4$	-7	-4	-1	2

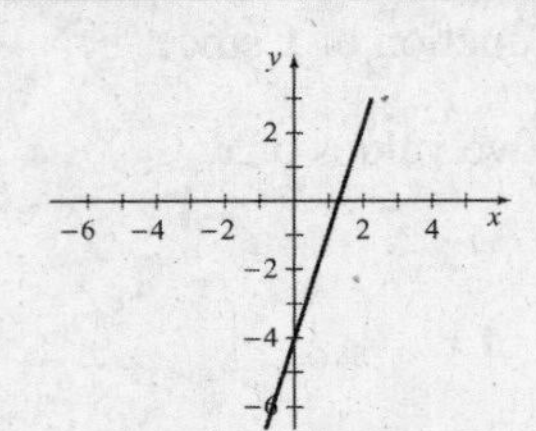

41. To graph $g(x)=x^2-1$, plot points and draw a smooth graph.

x	-2	-1	0	1	2
$y=g(x)=x^2-1$	3	0	-1	0	3

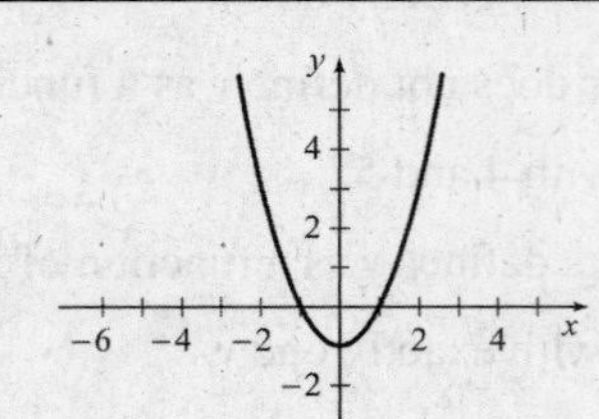

43. To graph $f(x)=\sqrt{x+4}$, plot points and draw a smooth graph.

x	-4	-2	0	2	5
$y=f(x)=\sqrt{x+4}$	0	$\sqrt{2}$	2	$\sqrt{6}$	3

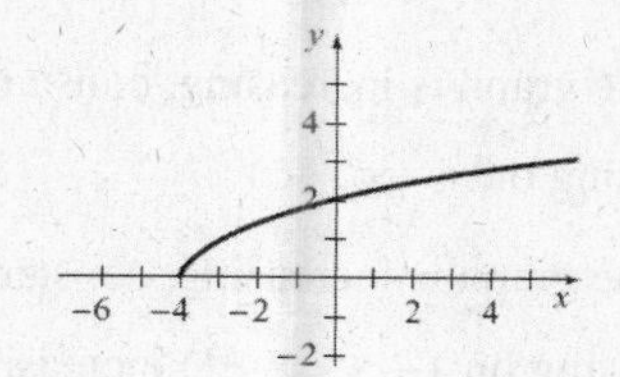

45. To graph $f(x)=\|x-2\|$, plot points and draw a smooth graph.

x	-3	0	2	4	6
$y=f(x)=\|x-2\|$	5	2	0	2	4

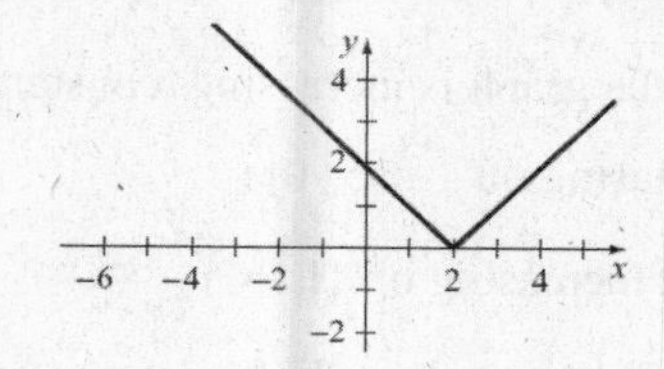

47. To graph $L(x)=\left[\!\left[\frac{1}{3}x\right]\!\right]$ for $-6\le x\le 6$, plot points and draw a smooth graph.

x	-6	-4	-3	-1	0	4	6
$y=L(x)=\left[\!\left[\frac{1}{3}x\right]\!\right]$	-2	-2	-1	-1	0	1	2

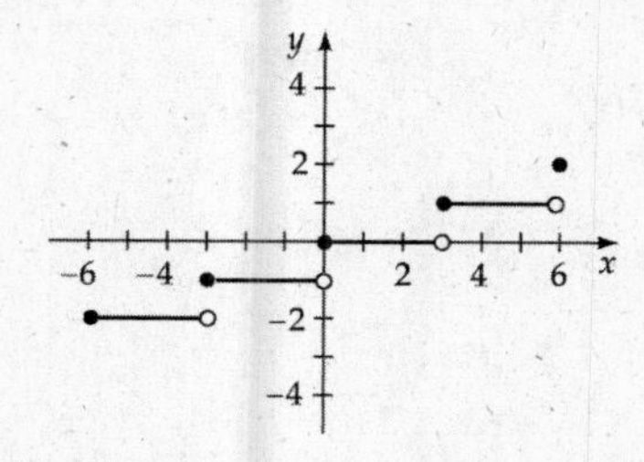

49. To graph $N(x)=\operatorname{int}(-x)$ for $-3\le x\le 3$, plot points and draw a smooth graph.

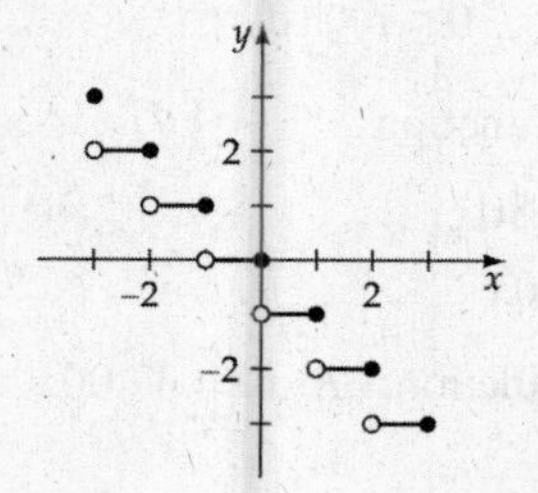

51. Find the value of a in the domain of $f(x)=3x-2$ for which $f(a)=10$.

$3a-2=10$ Replace $f(a)$ with $3a-2$

$3a=12$

$a=4$

53. Find the values of a in the domain of $f(x)=x^2+2x-2$ for which $f(a)=1$.

$a^2+2a-2=1$ Replace $f(a)$ with a^2+2a-2

$a^2+2a-3=0$

$(a+3)(a-1)=0$

$a+3=0 \quad a-1=0$

$a=-3 \quad a=1$

55. Find the values of a in the domain of $f(x)=|x|$ for which $f(a)=4$.

$|a|=4$ Replace $f(a)$ with $|a|$

$a=-4 \quad a=4$

57. Find the values of a in the domain of $f(x)=x^2+2$ for which $f(a)=1$.

$a^2+2=1$ Replace $f(a)$ with a^2+2

$a^2=-1$

There are no real values of a.

59. Find the zeros of f for $f(x)=3x-6$.

$f(x)=0$

$3x-6=0$

$3x=6$

$x=2$

61. Find the zeros of f for $f(x)=5x+2$.

$f(x)=0$

$5x+2=0$

$5x=-2$

$x=-\frac{2}{5}$

63. Find the zeros of *f* for $f(x)=x^2-4$.

$$f(x)=0$$
$$x^2-4=0$$
$$(x+2)(x-2)=0$$
$$x+2=0 \quad x-2=0$$
$$x=-2 \qquad x=2$$

65. Find the zeros of *f* for $f(x)=x^2-5x-24$.

$$f(x)=0$$
$$x^2-5x-24=0$$
$$(x+3)(x-8)=0$$
$$x+3=0 \quad x-8=0$$
$$x=-3 \qquad x=8$$

67. $\dfrac{\text{int}\left[10^2(2.3458)+0.5\right]}{10^2}=\dfrac{\text{int}[235.08]}{100}=\dfrac{235}{100}=2.35$

69. $\dfrac{\text{int}\left[10^3(34.05622)+0.5\right]}{10^3}=\dfrac{\text{int}[34,056.72]}{1000}=\dfrac{34,056}{1000}$
$=34.056$

71. $\dfrac{\text{int}\left[10^4(0.08951)+0.5\right]}{10^4}=\dfrac{\text{int}[895.6]}{10,000}=\dfrac{895}{10,000}$
$=0.0895$

73. a. $C(2.8)=0.44-0.17\text{int}(1-2.8)$
$=0.44-0.17\text{int}(-1.8)$
$=0.44-0.17(-2)$
$=0.44+0.34$
$=\$0.78$

b. $C(w)$

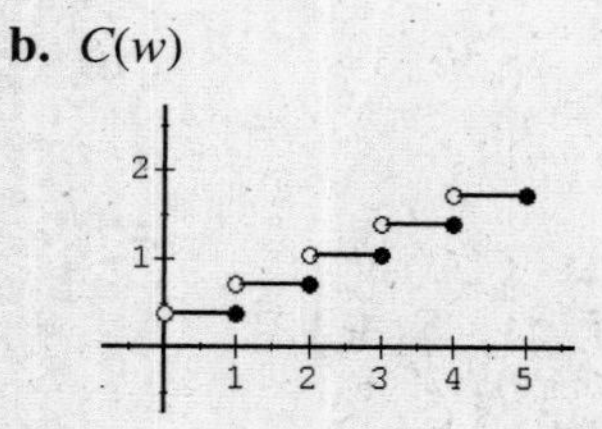

75. Determine which graphs are functions.

a. Yes; every vertical line intersects the graph in one point.

b. Yes; every vertical line intersects the graph in one point.

c. No; some vertical lines intersect the graph at more than one point.

d. Yes; every vertical line intersects the graph in one point.

77. Determine where the graph is increasing, constant, or decreasing. Decreasing on $(-\infty, 0]$; increasing on $[0, \infty)$

79. Determine where the graph is increasing, constant, or decreasing. Increasing on $(-\infty, \infty)$

81. Determine where the graph is increasing, constant, or decreasing. Decreasing on $(-\infty, -3]$; increasing on $[-3, 0]$; decreasing on $[0, 3]$; increasing on $[3, \infty)$

83. Determine where the graph is increasing, constant, or decreasing. Constant on $(-\infty, 0]$; increasing on $[0, \infty)$

85. Determine where the graph is increasing, constant, or decreasing. Decreasing on $(-\infty, 0]$; constant on [0, 1]; increasing on $[1, \infty)$

87. Determine which functions from 77-81 are one-to-one. *g* and *F* are one-to-one since every horizontal line intersects the graph at one point.
f, *V*, and *p* are not one-to-one since some horizontal lines intersect the graph at more than one point.

89. a. Write the width.

$$2l+2w=50$$
$$2w=50-2l$$
$$w=25-l$$

b. Write the area.

$$A=lw$$
$$A=l(25-l)$$
$$A=25l-l^2$$

91. Write the function.

$v(t)=80,000-6500t, \quad 0\le t\le 10$

93. a. Write the total cost function.

$$C(x)=5(400)+22.80x$$
$$=2000+22.80x$$

b. Write the revenue function. $R(x)=37.00x$

c. Write the profit function.

$$P(x) = 37.00x - C(x)$$
$$= 37.00 - [2000 + 22.80x]$$
$$= 37.00x - 2000 - 22.80x$$
$$= 14.20x - 2000$$

Note x is a natural number.

95. Write the function.

$$\frac{15}{3} = \frac{15-h}{r}$$
$$5 = \frac{15-h}{r}$$
$$5r = 15 - h$$
$$h = 15 - 5r$$
$$h(r) = 15 - 5r$$

97. Write the function.

$$d = \sqrt{(3t)^2 + (50)^2}$$
$$d = \sqrt{9t^2 + 2500} \text{ meters, } 0 \le t \le 60$$

99. Write the function.

$$d = \sqrt{(45 - 8t)^2 + (6t)^2} \text{ miles}$$

where t is the number of hours after 12:00 noon

101.a. Write the function.

Circle

$$C = 2\pi r$$
$$x = 2\pi r$$
$$r = \frac{x}{2\pi}$$
$$\text{Area} = \pi r^2 = \pi\left(\frac{x}{2\pi}\right)^2$$
$$= \frac{x^2}{4\pi}$$

Square

$$C = 4s$$
$$20 - x = 4s$$
$$s = 5 - \frac{x}{4}$$
$$\text{Area} = s^2 = \left(5 - \frac{x}{4}\right)^2$$
$$= 25 - \frac{5}{2}x + \frac{x^2}{16}$$

$$\text{Total Area} = \frac{x^2}{4\pi} + 25 - \frac{5}{2}x + \frac{x^2}{16}$$
$$= \left(\frac{1}{4\pi} + \frac{1}{16}\right)x^2 - \frac{5}{2}x + 25$$

b. Complete the table.

x	0	4	8	12	16	20
Total Area	25	17.27	14.09	15.46	21.37	31.83

c. Find the domain. Domain: [0, 20].

103.a. Write the function.

Left side triangle

$$c^2 = 20^2 + (40 - x)^2$$
$$c = \sqrt{400 + (40 - x)^2}$$

Right side triangle

$$c^2 = 30^2 + x^2$$
$$c = \sqrt{900 + x^2}$$

$$\text{Total length} = \sqrt{900 + x^2} + \sqrt{400 + (40 - x)^2}$$

b. Complete the table.

x	0	10	20	30	40
Total Length	74.72	67.68	64.34	64.79	70

c. Find the domain. Domain: [0, 40].

105. Complete the table.

x	5	10	12.5	15	20
$Y(x)$	275	375	385	390	394

Answers accurate to the nearest apple.

107. Find c.

$$f(c) = c^2 - c - 5 = 1$$
$$c^2 - c - 6 = 0$$
$$(c-3)(c+2) = 0$$
$$c - 3 = 0 \quad \text{or} \quad c + 2 = 0$$
$$c = 3 \qquad c = -2$$

109. Determine if 1 is in the range.

1 is not in the range of $f(x)$, since

$$1 = \frac{x-1}{x+1} \text{ only if } x + 1 = x - 1 \text{ or } 1 = -1.$$

111. Use a graphing utility to graph; set to "dot" mode.

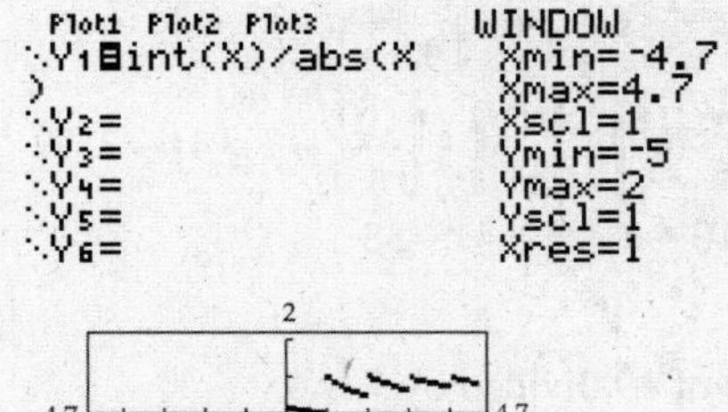

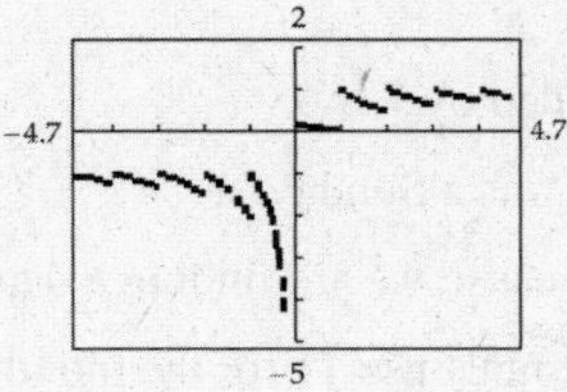

113. Use a graphing utility to graph; set to "dot" mode.

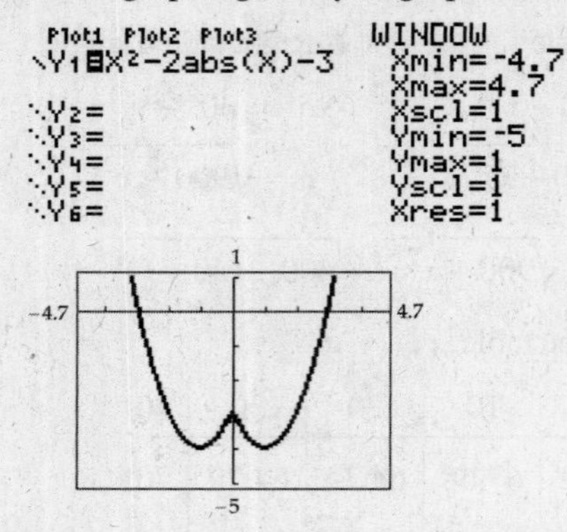

115. Use a graphing utility to graph; set to "dot" mode.

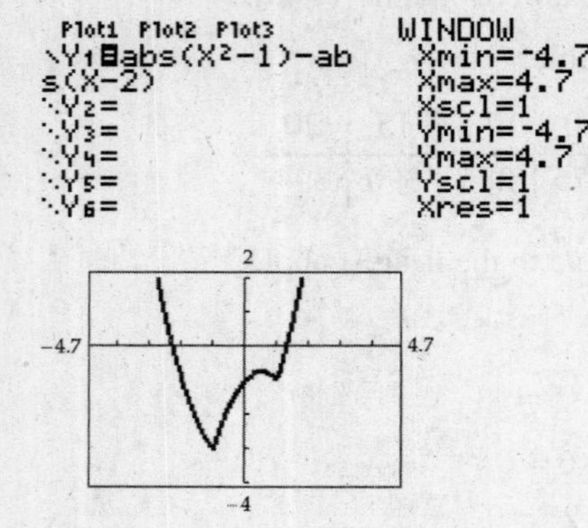

117. Find all fixed points.

$$a^2+3a-3=a$$
$$a^2+2a-3=0$$
$$(a-1)(a+3)=0$$
$$a=1 \quad \text{or} \quad a=-3$$

119.a. Let $m=10, d=7, c=19,$ and $y=41$. Then

$$\begin{aligned} z &= \left[\!\left[\frac{13m-1}{5}\right]\!\right]+\left[\!\left[\frac{y}{4}\right]\!\right]+\left[\!\left[\frac{c}{4}\right]\!\right]+d+y-2c \\ &= \left[\!\left[\frac{13\cdot 10-1}{5}\right]\!\right]+\left[\!\left[\frac{41}{4}\right]\!\right]+\left[\!\left[\frac{19}{4}\right]\!\right]+7+41-2\cdot 19 \\ &= 25+10+4+7+41-38 \\ &= 49 \end{aligned}$$

The remainder of 49 divided by 7 is 0.

Thus December 7, 1941, was a Sunday.

b. This one is tricky. Because we are finding a date in the month of January, we must use 11 for the month and we must use the previous year, which is 2019. Thus we let $m=11,\ d=1,\ c=20,$ and $y=19$.

Then

$$\begin{aligned} z &= \left[\!\left[\frac{13m-1}{5}\right]\!\right]+\left[\!\left[\frac{y}{4}\right]\!\right]+\left[\!\left[\frac{c}{4}\right]\!\right]+d+y-2c \\ &= \left[\!\left[\frac{13\cdot 11-1}{5}\right]\!\right]+\left[\!\left[\frac{19}{4}\right]\!\right]+\left[\!\left[\frac{20}{4}\right]\!\right]+1+19-2\cdot 20 \\ &= 28+4+5+1+19-40 \\ &= 17 \end{aligned}$$

The remainder of 17 divided by 7 is 3.

Thus January 1, 2020, will be a Wednesday.

c. Let $m=5,\ d=4,\ c=17,$ and $y=76$. Then

$$\begin{aligned} z &= \left[\!\left[\frac{13m-1}{5}\right]\!\right]+\left[\!\left[\frac{y}{4}\right]\!\right]+\left[\!\left[\frac{c}{4}\right]\!\right]+d+y-2c \\ &= \left[\!\left[\frac{13\cdot 5-1}{5}\right]\!\right]+\left[\!\left[\frac{76}{4}\right]\!\right]+\left[\!\left[\frac{17}{4}\right]\!\right]+4+76-2\cdot 17 \\ &= 12+19+4+4+76-34 \\ &= 81 \end{aligned}$$

The remainder of 81 divided by 7 is 4.

Thus July 4, 1776 was a Thursday.

d. Answers will vary.

Prepare for Section 2.3

P1. $d=5-(-2)=7$

P3. $\dfrac{-4-4}{2-(-3)}=\dfrac{-8}{5}$

P5. $3x-5y=15$

$$-5y=-3x+15$$
$$y=\frac{3}{5}x-3$$

Section 2.3 Exercises

1. Find the slope.

$$m=\frac{y_2-y_1}{x_2-x_1}=\frac{7-4}{1-3}=\frac{3}{-2}=-\frac{3}{2}$$

3. Find the slope.

$$m=\frac{2-0}{0-4}=-\frac{1}{2}$$

5. Find the slope.

$$m=\frac{-7-2}{3-3}=\frac{-9}{0} \quad \text{undefined}$$

7. Find the slope.

$$m=\frac{-2-4}{-4-(-3)}=\frac{-6}{-1}=6$$

9. Find the slope.

$$m = \frac{\frac{7}{2} - \frac{1}{2}}{\frac{7}{3} - (-4)} = \frac{\frac{6}{2}}{\frac{19}{3}} = 3 \cdot \frac{3}{19} = \frac{9}{19}$$

11. Find the slope.

$$m = \frac{f(3+h) - f(3)}{3+h-3} = \frac{f(3+h) - f(3)}{h}$$

13. Find the slope.

$$m = \frac{f(h) - f(0)}{h - 0} = \frac{f(h) - f(0)}{h}$$

15. Find the slope, y-intercept, and graph. $y = 2x - 4$

$m = 2$, y-intercept $(0, -4)$

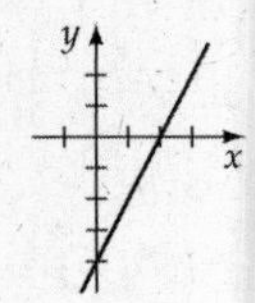

17. Find the slope, y-intercept, and graph. $y = \frac{3}{4}x + 1$

$m = \frac{3}{4}$, y-intercept $(0, 1)$

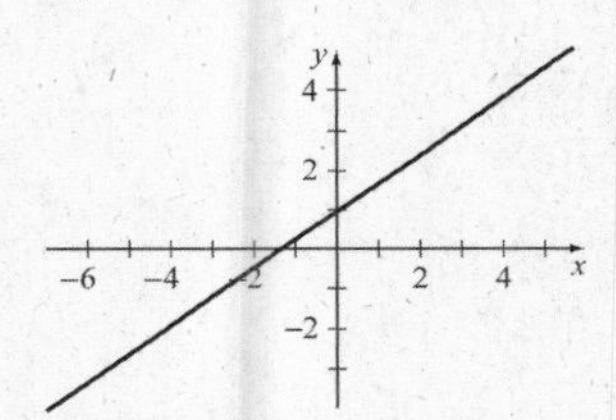

19. Find the slope, y-intercept, and graph. $y = -2x + 3$

$m = -2$, y-intercept $(0, 3)$

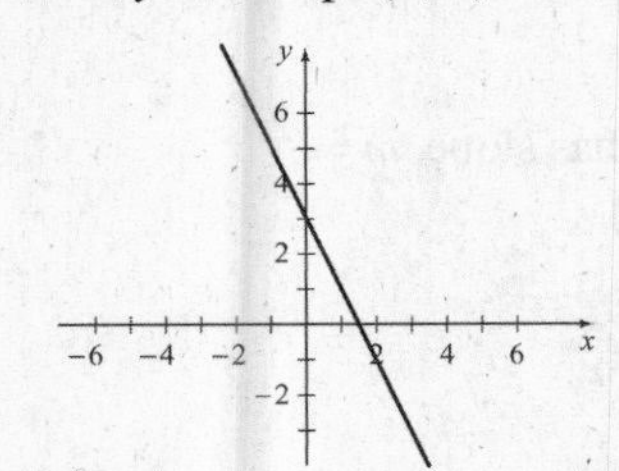

21. Find the slope, y-intercept, and graph. $y = 3$

$m = 0$, y-intercept $(0, 3)$

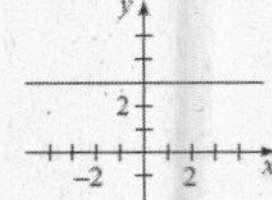

23. Find the slope, y-intercept, and graph. $y = 2x$

$m = 2$, y-intercept $(0, 0)$

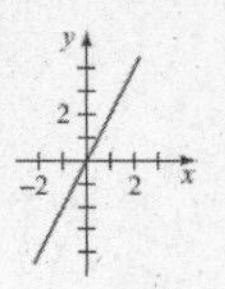

25. Find the slope, y-intercept, and graph. $y = x$

$m = 1$, y-intercept $(0, 0)$

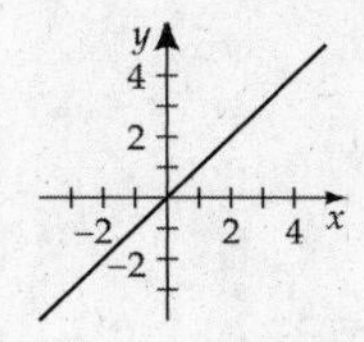

27. Write slope-intercept form, find intercepts, and graph.

$$2x + y = 5$$
$$y = -2x + 5$$

x-intercept $\left(\frac{5}{2}, 0\right)$, y-intercept $(0, 5)$

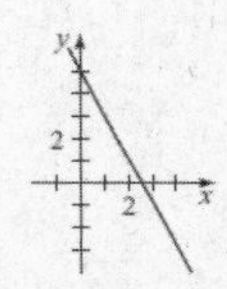

29. Write slope-intercept form, find intercepts, and graph.

$$4x + 3y - 12 = 0$$
$$3y = -4x + 12$$
$$y = -\frac{4}{3}x + 4$$

x-intercept $(3, 0)$, y-intercept $(0, 4)$

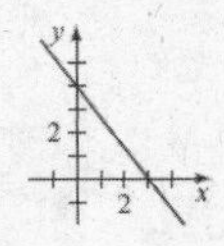

31. Write slope-intercept form, find intercepts, and graph.

$$2x - 5y = -15$$
$$-5y = -2x - 15$$
$$y = \frac{2}{5}x + 3$$

x-intercept $\left(-\frac{15}{2}, 0\right)$, y-intercept $(0, 3)$

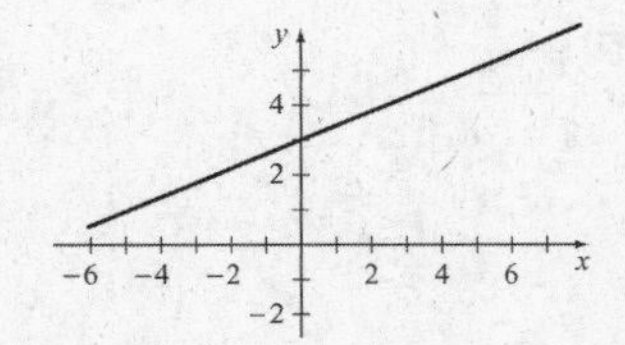

33. Write slope-intercept form, find intercepts, and graph.

$$x + 2y = 6$$
$$y = -\frac{1}{2}x + 3$$

x-intercept (6, 0), y-intercept (0, 3)

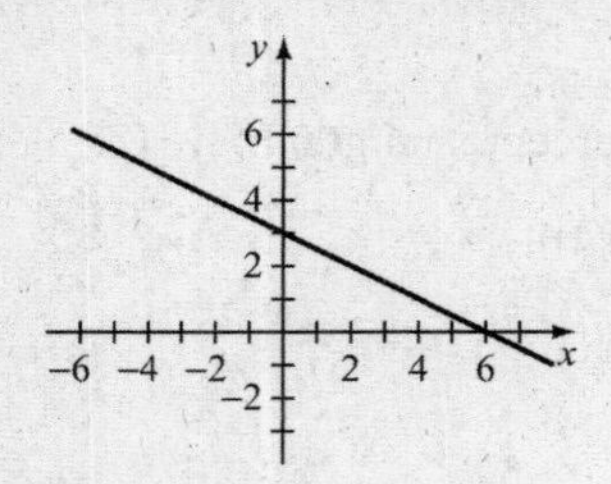

35. Find the equation.

Use $y = mx + b$ with $m = 1,\ b = 3$.

$$y = x + 3$$

37. Find the equation.

Use $y = mx + b$ with $m = \frac{3}{4},\ b = \frac{1}{2}$.

$$y = \frac{3}{4}x + \frac{1}{2}$$

39. Find the equation.

Use $y = mx + b$ with $m = 0,\ b = 4$.

$$y = 4$$

41. Find the equation.

$$y - 2 = -4(x - (-3))$$
$$y - 2 = -4x - 12$$
$$y = -4x - 10$$

43. Find the equation.

$$m = \frac{4-1}{-1-3} = \frac{3}{-4} = -\frac{3}{4}$$
$$y - 1 = -\frac{3}{4}(x - 3)$$
$$y = -\frac{3}{4}x + \frac{9}{4} + \frac{4}{4}$$
$$y = -\frac{3}{4}x + \frac{13}{4}$$

45. Find the equation.

$$m = \frac{-1-11}{2-7} = \frac{-12}{-5} = \frac{12}{5}$$
$$y - 11 = \frac{12}{5}(x - 7)$$
$$y - 11 = \frac{12}{5}x - \frac{84}{5}$$
$$y = \frac{12}{5}x - \frac{84}{5} + \frac{55}{5}$$
$$= \frac{12}{5}x - \frac{29}{5}$$

47. Find the equation.

$y = 2x + 3$ has slope $m = 2$.

$$y - y_1 = 2(x - x_1)$$
$$y + 4 = 2(x - 2)$$
$$y + 4 = 2x - 4$$
$$y = 2x - 8$$

49. Find the equation.

$y = -\frac{3}{4}x + 3$ has slope $m = -\frac{3}{4}$.

$$y - y_1 = -\frac{3}{4}(x - x_1)$$
$$y - 2 = -\frac{3}{4}(x + 4)$$
$$y - 2 = -\frac{3}{4}x - 3$$
$$y = -\frac{3}{4}x - 1$$

51. Find the equation.

$$2x - 5y = 2$$
$$-5y = -2x + 2$$

$y = \frac{2}{5}x - \frac{2}{5}$ has slope $m = \frac{2}{5}$.

$$y - y_1 = \frac{2}{5}(x - x_1)$$
$$y - 2 = \frac{2}{5}(x - 5)$$
$$y - 2 = \frac{2}{5}x - 2$$
$$y = \frac{2}{5}x$$

53. Find the equation.

$y = 2x - 5$ has perpendicular slope $m = -\frac{1}{2}$.

$$y - y_1 = -\frac{1}{2}(x - x_1)$$
$$y + 4 = -\frac{1}{2}(x - 3)$$
$$y + 4 = -\frac{1}{2}x + \frac{3}{2}$$
$$y = -\frac{1}{2}x - \frac{5}{2}$$

55. Find the equation.

$y = -\frac{3}{4}x + 1$ has perpendicular slope $m = \frac{4}{3}$.

$$y - y_1 = \frac{4}{3}(x - x_1)$$
$$y - 0 = \frac{4}{3}(x + 6)$$
$$y = \frac{4}{3}x + 8$$

57. Find the equation.

$$-x - 4y = 6$$
$$-4y = x + 6$$

$y = -\frac{1}{4}x - \frac{3}{2}$ has perpendicular slope $m = 4$.

$$y - y_1 = 4(x - x_1)$$
$$y - 2 = 4(x - 5)$$
$$y - 2 = 4x - 20$$
$$y = 4x - 18$$

59. Find *a*.

$$f(a) = 2a + 3 = -1$$
$$2a = -4$$
$$a = -2$$

61. Find *a*.

$$f(a) = 1 - 4a = 3$$
$$-4a = 2$$
$$a = -\frac{1}{2}$$

63. Find the zero of *f*.

$$f(x) = 3x - 12$$
$$3x - 12 = 0$$
$$3x = 12$$
$$x = 4$$

The x-intercept of the graph of $f(x)$ is (4, 0).

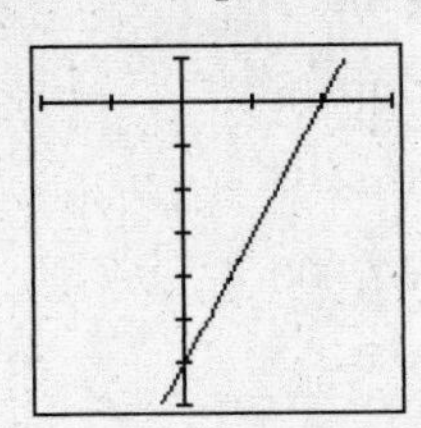

Xmin = – 4, Xmax = 6, Xscl=2,
Ymin = –12.2, Ymax = 2, Yscl = 2

65. Find the zero of *f*.

$$f(x) = \frac{1}{4}x + 5$$
$$\frac{1}{4}x + 5 = 0$$
$$\frac{1}{4}x = -5$$
$$x = -20$$

The x-intercept of the graph of $f(x)$ is $(-20, 0)$.

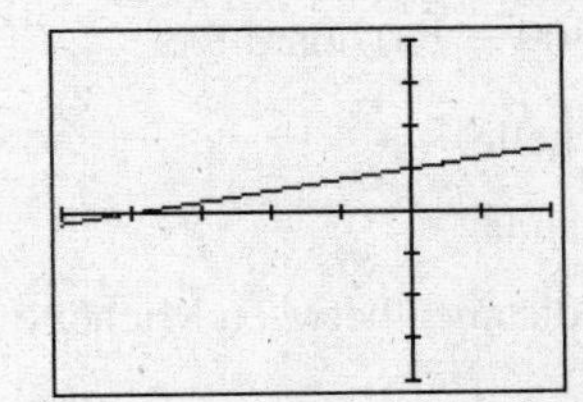

Xmin = –30, Xmax = 30, Xscl = 10,
Ymin = –10, Ymax = 10, Yscl = 1

67. Find the slope and explain the meaning.

$$m = \frac{1505 - 1482}{28 - 20} = 2.875$$

The value of the slope indicates that the speed of sound in water increases 2.875 m/s for a one-degree Celsius increase in temperature.

69. a. $m = \frac{30 - 20}{22 - 12} = 1$

$$H(c) - 30 = 1(c - 22)$$
$$H(c) = c + 8$$

b. $H(18) = (18) + 8 = 26$ mpg

71. a. $m=\dfrac{63,000-38,000}{2010-2000}=2500$

$N(t)-63,000=2500(t-2010)$

$N(t)=2500t-4,962,000$

b. $60,000=2500t-4,962,000$

$5,022,000=2500t$

$2008.8=t$

The number of jobs will exceed 60,000 in 2008.

73. a. $m=\dfrac{240-180}{18-16}=30$

$B(d)-180=30(d-16)$

$B(d)=30d-300$

b. The value of the slope means that a 1-inch increase in the diameter of a log 32 ft long results in an increase of 30 board-feet of lumber that can be obtained from the log.

c. $B(19)=30(19)-300=270$ board feet

75. Line A represents Michelle.

Line B represents Amanda.

Line C represents the distance between Michelle and Amanda.

77. a. $m=\dfrac{79.96-19.50}{0-65}\approx-0.93$

$y-79.96=-0.93(x-0)$

$y=-0.93x+79.96$

b. $y=-0.93(25)+79.96=56.71\approx 57$ years

79. Determine the profit function and break-even point.

$P(x)=92.50x-(52x+1782)$

$P(x)=92.50x-52x-1782$

$P(x)=40.50x-1782$

$40.50x-1782=0$

$40.50x=1782$

$x=\dfrac{1782}{40.50}$

$x=44$, the break-even point

81. Determine the profit function and break-even point.

$P(x)=259x-(180x+10,270)$

$P(x)=259x-180x-10,270$

$P(x)=79x-10,270$

$79x-10,270=0$

$79x=10,270$

$x=\dfrac{10,270}{79}$

$x=130$, the break-even point

83. a. $C(0)=8(0)+275=0+275=\$275$

b. $C(1)=8(1)+275=8+275=\$283$

c. $C(10)=8(10)+275=80+275=\$355$

d. The marginal cost is the slope of $C(x)=8x+275$, which is \$8 per unit.

85. a. $C(t)=19,500.00+6.75t$

b. $R(t)=55.00t$

c. $P(t)=R(t)-C(t)$

$P(t)=55.00t-(19,500.00+6.75t)$

$P(t)=55.00t-19,500.00-6.75t$

$P(t)=48.25t-19,500.00$

d. $48.25t=19,500.00$

$t=\dfrac{19,500.00}{48.25}$

$t=404.1451$ days $\approx$ 405 days

87. The equation of the line through (0,0) and $P(3,4)$ has slope $\frac{4}{3}$.

The path of the rock is on the line through $P(3,4)$ with slope $-\frac{3}{4}$, so $y-4=-\frac{3}{4}(x-3)$.

$y-4=-\frac{3}{4}x+\frac{9}{4}$

$y=-\frac{3}{4}x+\frac{9}{4}+4$

$y=-\frac{3}{4}x+\frac{25}{4}$

The point where the rock hits the wall at $y=10$ is the point of intersection of $y=-\frac{3}{4}x+\frac{25}{4}$ and $y=10$.

$$-\frac{3}{4}x+\frac{25}{4}=10$$
$$-3x+25=40$$
$$-3x=15$$
$$x=-5 \text{ feet}$$

Therefore the rock hits the wall at (–5, 10).

The x-coordinate is –5.

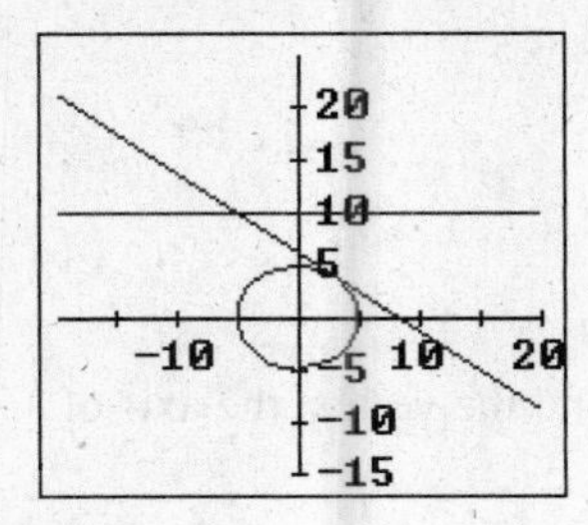

89. **a.** $h = 1$ so

$$Q(2+h,[2+h]^2+1)=Q(3,3^2+1)=Q(3,10)$$

$$m=\frac{10-5}{3-2}=\frac{5}{1}=5$$

b. $h = 0.1$ so

$$Q(2+h,[2+h]^2+1)=Q(2.1,2.1^2+1)=Q(2.1,5.41)$$

$$m=\frac{5.41-5}{2.1-2}=\frac{0.41}{0.1}=4.1$$

c. $h = 0.01$ so

$$Q(2+h,[2+h]^2+1)=Q(2.01,2.01^2+1)$$
$$=Q(2.01,5.0401)$$

$$m=\frac{5.0401-5}{2.01-2}=\frac{0.0401}{0.01}=4.01$$

d. As h approaches 0, the slope of PQ seems to be approaching 4.

e. $x_1=2,\ y_1=5, x_2=2+h,\ y_2=[2+h]^2+1$

$$m=\frac{y_2-y_1}{x_2-x_1}=\frac{[2+h]^2+1-5}{(2+h)-2}=\frac{(4+4h+h^2)+1-5}{h}$$
$$=\frac{4h+h^2}{h}=4+h$$

91. $$m=\frac{(x+h)^2-x^2}{x+h-x}=\frac{x^2+2xh+h^2-x^2}{h}$$
$$=\frac{2xh+h^2}{h}=\frac{h(2x+h)}{h}=2x+h$$

93. The slope of the line through (3, 9) and (x, y) is $\frac{15}{2}$, so $\frac{y-9}{x-3}=\frac{15}{2}$.

Therefore

$$2(y-9)=15(x-3)$$
$$2y-18=15x-45$$
$$2y-15x+27=0$$
$$2x^2-15x+27=0 \quad \text{Substituting } y=x^2$$
$$(2x-9)(x-3)=0$$
$$x=\frac{9}{2} \quad \text{or} \quad x=3$$

If $x=\frac{9}{2},\ y=x^2=\left(\frac{9}{2}\right)^2=\frac{81}{4}\Rightarrow\left(\frac{9}{2},\ \frac{81}{4}\right)$.

If $x=3,\ y=x^2=(3)^2=9\Rightarrow(3,\ 9)$, but this is the point itself. The point $\left(\frac{9}{2},\ \frac{81}{4}\right)$ is on the graph of $y=x^2$, and the slope of the line containing (3, 9) and $\left(\frac{9}{2},\ \frac{81}{4}\right)$ is $\frac{15}{2}$.

Mid-Chapter 2 Quiz

1. Find the midpoint and length.

$$M=\left(\frac{-3+1}{2},\ \frac{4-2}{2}\right)=\left(\frac{-2}{2},\ \frac{2}{2}\right)=(-1,1)$$

$$d=\sqrt{(1-(-3))^2+(-2-4)^2}=\sqrt{(4)^2+(-6)^2}$$
$$=\sqrt{16+36}=\sqrt{52}$$
$$=2\sqrt{13}$$

3. Evaluate.

$$f(x)=x^2-6x+1$$
$$f(-3)=(-3)^2-6(-3)+1=9+18+1=28$$

5. Find the zeros of f for $f(x)=x^2-x-12$.

$$f(x)=0$$
$$x^2-x-12=0$$
$$(x+3)(x-4)=0$$
$$x+3=0 \quad x-4=0$$
$$x=-3 \quad\quad x=4$$

7. Find the equation.

$$2x+3y=5$$
$$3y=-2x+5$$
$$y=-\frac{2}{3}x+\frac{5}{3} \text{ has slope } m=-\frac{2}{3}.$$

$$y-y_1=-\frac{2}{3}(x-x_1)$$
$$y-(-1)=-\frac{2}{3}(x-3)$$
$$y+1=-\frac{2}{3}x+2$$
$$y=-\frac{2}{3}x+1$$

Prepare for Section 2.4

P1. $3x^2+10x-8=(3x-2)(x+4)$

P3. $f(-3)=2(-3)^2-5(-3)-7$
$$=18+15-7$$
$$=26$$

P5. $x^2+3x-2=0$

$$x=\frac{-3\pm\sqrt{(3)^2-4(1)(-2)}}{2(1)}$$
$$=\frac{-3\pm\sqrt{17}}{2}$$

Section 2.4 Exercises

1. d

3. b

5. g

7. c

9. Write in standard form, find the vertex, the axis of symmetry and graph.

$$f(x)=(x^2+4x)+1$$
$$=(x^2+4x+4)+1-4$$
$$=(x+2)^2-3 \quad \text{standard form,}$$

vertex $(-2, -3)$, axis of symmetry $x = -2$

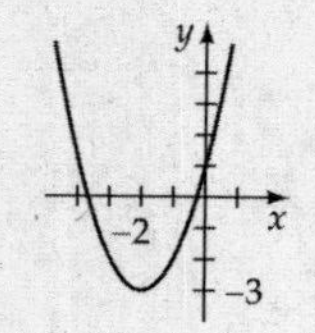

11. Write in standard form, find the vertex, the axis of symmetry and graph.

$$f(x)=(x^2-8x)+5$$
$$=(x^2-8x+16)+5-16$$
$$=(x-4)^2-11 \quad \text{standard form,}$$

vertex $(4, -11)$, axis of symmetry $x = 4$

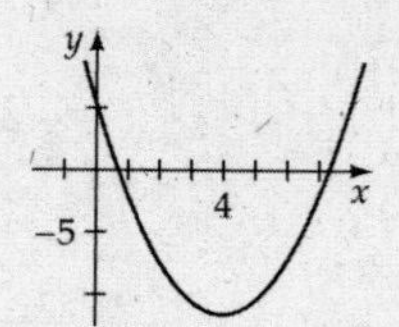

13. Write in standard form, find the vertex, the axis of symmetry and graph.

$$f(x)=(x^2+3x)+1$$
$$=\left(x^2+3x+\frac{9}{4}\right)+1-\frac{9}{4}$$
$$=\left(x+\frac{3}{2}\right)^2+\frac{4}{4}-\frac{9}{4}$$
$$=\left(x+\frac{3}{2}\right)^2-\frac{5}{4} \quad \text{standard form,}$$

vertex $\left(-\frac{3}{2}, -\frac{5}{4}\right)$, axis of symmetry $x = -\frac{3}{2}$

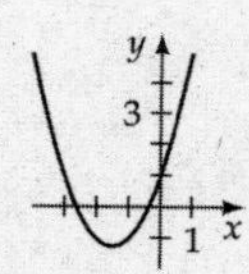

15. Write in standard form, find the vertex, the axis of symmetry and graph.

$$f(x)=-x^2+4x+2$$
$$=-(x^2-4x)+2$$
$$=-(x^2-4x+4)+2+4$$
$$=-(x-2)^2+6 \quad \text{standard form,}$$

vertex $(2, 6)$, axis of symmetry $x = 2$

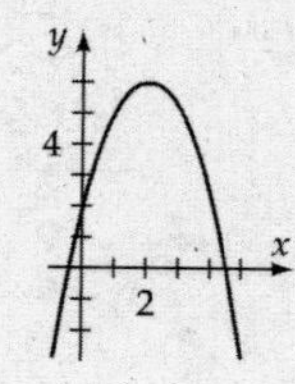

17. Write in standard form, find the vertex, the axis of symmetry and graph.

$$\begin{aligned} f(x) &= -3x^2 + 3x + 7 \\ &= -3(x^2 - 1x) + 7 \\ &= -3\left(x^2 - 1x + \frac{1}{4}\right) + 7 + \frac{3}{4} \\ &= -3\left(x - \frac{1}{2}\right)^2 + \frac{28}{4} + \frac{3}{4} \\ &= -3\left(x - \frac{1}{2}\right)^2 + \frac{31}{4} \quad \text{standard form,} \end{aligned}$$

vertex $\left(\frac{1}{2}, \frac{31}{4}\right)$, axis of symmetry $x = \frac{1}{2}$

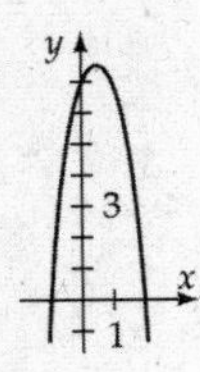

19. Find the vertex, write the function in standard form.

$$x = \frac{-b}{2a} = \frac{10}{2(1)} = 5$$

$$\begin{aligned} y = f(5) &= (5)^2 - 10(5) \\ &= 25 - 50 = -25 \end{aligned}$$

vertex $(5, -25)$

$f(x) = (x-5)^2 - 25$

21. Find the vertex, write the function in standard form.

$$x = \frac{-b}{2a} = \frac{0}{2(1)} = 0$$

$y = f(0) = (0)^2 - 10 = -10$

vertex $(0, -10)$

$f(x) = x^2 - 10$

23. Find the vertex, write the function in standard form.

$$x = \frac{-b}{2a} = \frac{-6}{2(-1)} = \frac{-6}{-2} = 3$$

$$\begin{aligned} y = f(3) &= -(3)^2 + 6(3) + 1 \\ &= -9 + 18 + 1 \\ &= 10 \end{aligned}$$

vertex $(3, 10)$

$f(x) = -(x-3)^2 + 10$

25. Find the vertex, write the function in standard form.

$$x = \frac{-b}{2a} = \frac{3}{2(2)} = \frac{3}{4}$$

$$\begin{aligned} y = f\left(\frac{3}{4}\right) &= 2\left(\frac{3}{4}\right)^2 - 3\left(\frac{3}{4}\right) + 7 \\ &= 2\left(\frac{9}{16}\right) - \frac{9}{4} + 7 \\ &= \frac{9}{8} - \frac{9}{4} + 7 = \frac{9}{8} - \frac{18}{8} + \frac{56}{8} \\ &= \frac{47}{8} \end{aligned}$$

vertex $\left(\frac{3}{4}, \frac{47}{8}\right)$

$f(x) = 2\left(x - \frac{3}{4}\right)^2 + \frac{47}{8}$

27. Find the vertex, write the function in standard form.

$$x = \frac{-b}{2a} = \frac{-1}{2(-4)} = \frac{1}{8}$$

$$\begin{aligned} y = f\left(\frac{1}{8}\right) &= -4\left(\frac{1}{8}\right)^2 + \left(\frac{1}{8}\right) + 1 \\ &= -4\left(\frac{1}{64}\right) + \frac{1}{8} + 1 \\ &= -\frac{1}{16} + \frac{1}{8} + 1 = -\frac{1}{16} + \frac{2}{16} + \frac{16}{16} \\ &= \frac{17}{16} \end{aligned}$$

vertex $\left(\frac{1}{8}, \frac{17}{16}\right)$

$f(x) = -4\left(x - \frac{1}{8}\right)^2 + \frac{17}{16}$

29. Find the range, find x.

$$\begin{aligned} f(x) &= x^2 - 2x - 1 \\ &= (x^2 - 2x) - 1 \\ &= (x^2 - 2x + 1) - 1 - 1 \\ &= (x-1)^2 - 2 \end{aligned}$$

vertex $(1, -2)$

The y-value of the vertex is -2.

The parabola opens up since $a = 1 > 0$.

Thus the range is $\{y \mid y \geq -2\}$.

$f(x) = 2 = x^2 - 2x - 1$
$0 = x^2 - 2x - 3$
$0 = (x+1)(x-3)$

$x+1=0$ or $x-3=0$
$x=-1$ $x=3$

31. Find the range, find x.

$$\begin{aligned} f(x) &= -2x^2 + 5x - 1 \\ &= -2\left(x^2 - \frac{5}{2}x\right) - 1 \\ &= -2\left(x^2 - \frac{5}{2}x + \frac{25}{16}\right) - 1 + 2\left(\frac{25}{16}\right) \\ &= -2\left(x - \frac{5}{4}\right)^2 - \frac{8}{8} + \frac{25}{8} \\ &= -2\left(x - \frac{5}{4}\right)^2 + \frac{17}{8} \end{aligned}$$

vertex $\left(\frac{5}{4}, \frac{17}{8}\right)$

The y-value of the vertex is $\frac{17}{8}$.

The parabola opens down since $a = -2 < 0$.

Thus the range is $\left\{y \middle| y \le \frac{17}{8}\right\}$.

$f(x) = 2 = -2x^2 + 5x - 1$
$2x^2 - 5x + 3 = 0$
$(x-1)(2x-3) = 0$

$x-1=0$ or $2x-3=0$
$x=1$ $x=\frac{3}{2}$

33. Find the real zeros and x-intercepts.

$$\begin{aligned} f(x) &= x^2 + 2x - 24 \\ &= (x+6)(x-4) \end{aligned}$$

$x+6=0$ $x-4=0$
$x=-6$ $x=4$
$(-6, 0)$ $(4, 0)$

35. Find the real zeros and x-intercepts.

$$\begin{aligned} f(x) &= 2x^2 + 11x + 12 \\ &= (x+4)(2x+3) \end{aligned}$$

$x+4=0$ $2x+3=0$
$x=-4$ $x=-\frac{3}{2}$
$(-4, 0)$ $\left(-\frac{3}{2}, 0\right)$

37. Find the minimum or maximum.

$$\begin{aligned} f(x) &= x^2 + 8x \\ &= (x^2 + 8x + 16) - 16 \\ &= (x+4)^2 - 16 \end{aligned}$$

minimum value of -16 when $x = -4$

39. Find the minimum or maximum.

$$\begin{aligned} f(x) &= -x^2 + 6x + 2 \\ &= -(x^2 - 6x) + 2 \\ &= -(x^2 - 6x + 9) + 2 + 9 \\ &= -(x-3)^2 + 11 \end{aligned}$$

maximum value of 11 when $x = 3$

41. Find the minimum or maximum.

$$\begin{aligned} f(x) &= 2x^2 + 3x + 1 \\ &= 2\left(x^2 + \frac{3}{2}x\right) + 1 \\ &= 2\left(x^2 + \frac{3}{2}x + \frac{9}{16}\right) + 1 - 2\left(\frac{9}{16}\right) \\ &= 2\left(x + \frac{3}{4}\right)^2 + \frac{8}{8} - \frac{9}{8} \\ &= 2\left(x + \frac{3}{4}\right)^2 - \frac{1}{8} \end{aligned}$$

minimum value of $-\frac{1}{8}$ when $x = -\frac{3}{4}$

43. Find the minimum or maximum.

$$\begin{aligned} f(x) &= 5x^2 - 11 \\ &= 5(x^2) - 11 \\ &= 5(x-0)^2 - 11 \end{aligned}$$

minimum value of -11 when $x = 0$

45. Find the minimum or maximum.

$$\begin{aligned} f(x) &= -\frac{1}{2}x^2 + 6x + 17 \\ &= -\frac{1}{2}(x^2 - 12x) + 17 \\ &= -\frac{1}{2}(x^2 - 12x + 36) + 17 + 18 \\ &= -\frac{1}{2}(x-6)^2 + 35 \end{aligned}$$

maximum value of 35 when $x = 6$

47. $A(t) = -4.9t^2 + 90t + 9000$

Microgravity begins and ends at a height of 9000 m.

$$9000 = -4.9t^2 + 90t + 9000$$
$$4.9t^2 - 90t = 0$$
$$t(4.9t - 90) = 0$$
$$t = 0 \quad 4.9t - 90 = 0$$
$$t = 0 \qquad t = \frac{90}{4.9} \approx 18.4$$

The time of microgravity is 18.4 seconds.

49. $h(x) = -\frac{3}{64}x^2 + 27 = -\frac{3}{64}(x - 0)^2 + 27$

a. The maximum height of the arch is 27 feet.

b.
$$h(10) = -\frac{3}{64}(10)^2 + 27$$
$$= -\frac{3}{64}(100) + 27$$
$$= -\frac{75}{16} + 27$$
$$= -\frac{75}{16} + \frac{432}{16}$$
$$= \frac{357}{16} = 22\frac{5}{16} \text{ feet}$$

c.
$$h(x) = 8 = -\frac{3}{64}x^2 + 27$$
$$8 - 27 = -\frac{3}{64}x^2$$
$$-19 = -\frac{3}{64}x^2$$
$$64(-19) = -3x^2$$
$$\frac{64(-19)}{-3} = x^2$$
$$\sqrt{\frac{64(-19)}{-3}} = x$$
$$8\sqrt{\frac{19}{3}} = x$$
$$\frac{8\sqrt{19}\sqrt{3}}{3} = x$$
$$\frac{8\sqrt{57}}{3} = x$$
$$20.1 \approx x$$

$h(x) = 8$ when $x \approx 20.1$ feet

51. a.
$$3w + 2l = 600$$
$$3w = 600 - 2l$$
$$w = \frac{600 - 2l}{3}$$

b.
$$A = w \times l$$
$$A = \left(\frac{600 - 2l}{3}\right) l$$
$$= 200l - \frac{2}{3}l^2$$

c.
$$A = -\frac{2}{3}(l^2 - 300l)$$
$$A = -\frac{2}{3}(l^2 - 300l + 150^2) + 15,000$$

In standard form,
$$A = -\frac{2}{3}(l - 150)^2 + 15,000$$

The maximum area of 15,000 ft^2 is produced when

$l = 150$ ft and the width $w = \frac{600 - 2(150)}{3} = 100$ ft .

53. a.
$$T(t) = -0.7t^2 + 9.4t + 59.3$$
$$= -0.7\left(t^2 - \frac{9.4}{0.7}\right) + 59.3$$
$$= -0.7\left(t^2 - \frac{94}{7}\right) + 59.3$$
$$= -0.7\left(t^2 - \frac{94}{7}t + \left(\frac{47}{7}\right)^2\right) + 59.3 + \left(0.7 \times \left(\frac{47}{7}\right)^2\right)$$
$$= -0.7\left(t - \frac{47}{7}\right)^2 + 90.857$$
$$= -0.7\left(t - 6\frac{5}{7}\right)^2 + 91$$

The temperature is a maximum when

$t = \frac{47}{7} = 6\frac{5}{7}$ hours after 6:00 A.M. Note $\frac{5}{7}$(60 min) $\approx$ 43 min.

Thus the temperature is a maximum at 12:43 P.M.

b. The maximum temperature is approximately 91°F.

55. $t = -\dfrac{b}{2a} = -\dfrac{82.86}{2(-279.67)} = 0.14814$

$E(0.14814) = -279.67(0.14814)^2 + 82.86(0.14814)$

≈ 6.1

The maximum energy is 6.1 joules.

57. $h(x) = -0.002x^2 - 0.03x + 8$

$h(39) = -0.002(39)^2 - 0.03(39) + 8 = 3.788 > 3$

Solve for x using quadratic formula.

$-0.002x^2 - 0.03x + 8 = 0$

$x^2 + 15x - 4000 = 0$

$x = \dfrac{-15 \pm \sqrt{(15)^2 - 4(1)(-4000)}}{2(1)}$

$= \dfrac{-15 \pm \sqrt{16,225}}{2}$, use positive value of x

$x \approx 56.2$

Yes, the conditions are satisfied.

59. a. $E(v) = -0.018v^2 + 1.476v + 3.4$

$= -0.018\left(v^2 - \dfrac{1.476}{0.018}v\right) + 3.4$

$= -0.018(v^2 - 82v) + 3.4$

$= -0.018\left(v^2 - 82v + 41^2\right) + 3.4 + 0.018(41)^2$

$= -0.018(v - 41)^2 + 33.658$

The maximum fuel efficiency is obtained at a speed of 41 mph.

b. The maximum fuel efficiency for this car, to the nearest mile per gallon, is 34 mpg.

61. $-\dfrac{b}{2a} = -\dfrac{296}{2(-0.2)} = 740$

$R(740) = 296(740) - 0.2(740)^2 = 109,520$

Thus, 740 units yield a maximum revenue of $109,520.

63. $-\dfrac{b}{2a} = -\dfrac{1.7}{2(-0.01)} = 85$

$P(85) = -0.01(85)^2 + 1.7(85) - 48 = 24.25$

Thus, 85 units yield a maximum profit of $24.25.

65. $P(x) = R(x) - C(x)$

$= x(102.50 - 0.1x) - (52.50x + 1840)$

$= -0.1x^2 + 50x - 1840$

The break-even points occur when $R(x) = C(x)$ or $P(x) = 0$.

Thus, $0 = -0.1x^2 + 50x - 1840$

$x = \dfrac{-50 \pm \sqrt{50^2 - 4(-0.1)(-1840)}}{2(-0.1)}$

$= \dfrac{-50 \pm \sqrt{1764}}{-0.2} = \dfrac{-50 \pm 42}{-0.2}$

$x = 40$ or $x = 460$

The break-even points occur when $x = 40$ or $x = 460$.

67. Let x = the number of people that take the tour.

a. $R(x) = x(15.00 + 0.25(60 - x))$

$= x(15.00 + 15 - 0.25x)$

$= -0.25x^2 + 30.00x$

b. $P(x) = R(x) - C(x)$

$= (-0.25x^2 + 30.00x) - (180 + 2.50x)$

$= -0.25x^2 + 27.50x - 180$

c. $-\dfrac{b}{2a} = -\dfrac{27.50}{2(-0.25)} = 55$

$P(55) = -0.25(55)^2 + 27.50(55) - 180 = \$576.25$

d. The maximum profit occurs when $x = 55$ tickets.

69. $h(t) = -16t^2 + 128t$

a. $-\dfrac{b}{2a} = -\dfrac{128}{2(-16)} = 4$ seconds

b. $h(4) = -16(4)^2 + 128(4) = 256$ feet

c. $0 = -16t^2 + 128t$

$0 = -16t(t - 8)$

$-16t = 0$ or $t - 8 = 0$

$t = 0 \qquad t = 8$

The projectile hits the ground at $t = 8$ seconds.

71. $y(x) = -0.014x^2 + 1.19x + 5$

$$-\frac{b}{2a} = -\frac{1.19}{2(-0.014)}$$
$$= 42.5$$

$$y(42.5) = -0.014(42.5)^2 + 1.19(42.5) + 5$$
$$= 30.2875$$
$$\approx 30 \text{ feet}$$

73. Find height and radius.

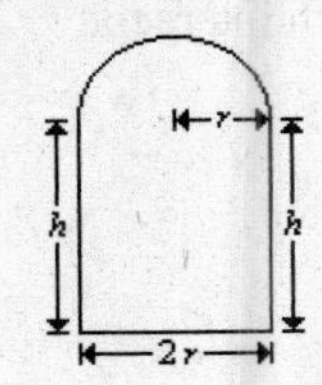

The perimeter is $48 = \pi r + h + 2r + h$.

Solve for h.

$$48 - \pi r - 2r = 2h$$
$$\frac{1}{2}(48 - \pi r - 2r) = h$$

Area = semicircle + rectangle

$$A = \frac{1}{2}\pi r^2 + 2rh$$
$$= \frac{1}{2}\pi r^2 + 2r\left(\frac{1}{2}\right)(48 - \pi r - 2r)$$
$$= \frac{1}{2}\pi r^2 + r(48 - \pi r - 2r)$$
$$= \frac{1}{2}\pi r^2 + 48r - \pi r^2 - 2r^2$$
$$= \left(\frac{1}{2}\pi - \pi - 2\right) r^2 + 48r$$
$$= \left(-\frac{1}{2}\pi - 2\right) r^2 + 48r$$

Graph the function A to find that its maximum occurs when $r \approx 6.72$ feet.

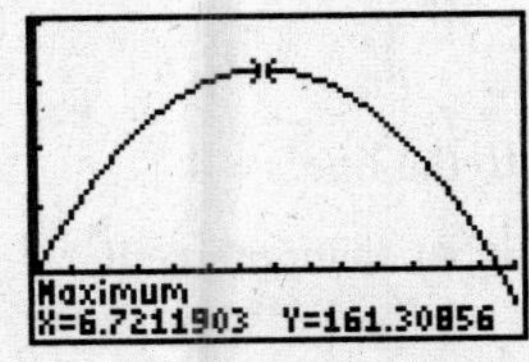

Xmin = 0, Xmax = 14, Xscl = 1

Ymin = −50, Ymax = 200, Yscl = 50

$$h = \frac{1}{2}(48 - \pi r - 2r)$$
$$\approx \frac{1}{2}(48 - \pi(6.72) - 2(6.72))$$
$$\approx 6.72 \text{ feet}$$

Hence the optimal window has its semicircular radius equal to its height.

Note: Using calculus it can be shown that the exact value of $r = h = \frac{48}{\pi + 4}$.

Prepare for Section 2.5

P1. $f(x) = x^2 + 4x - 6$

$$-\frac{b}{2a} = -\frac{4}{2(1)} = -2$$
$$x = -2$$

P3. $f(-2) = 2(-2)^3 - 5(-2) = -16 + 10 = -6$

$$-f(2) = -[2(2)^3 - 5(2)] = -[16 - 10] = -6$$
$$f(-2) = -f(2)$$

P5. $\frac{-a + a}{2} = 0, \quad \frac{b + b}{2} = b$

midpoint is $(0, b)$

Section 2.5 Exercises

1. Plot the points.

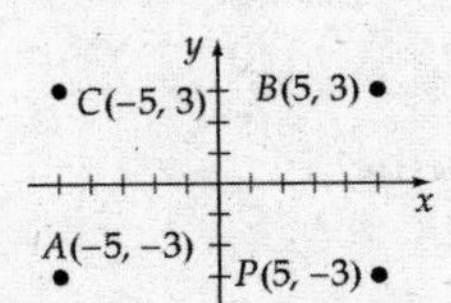

3. Plot the points.

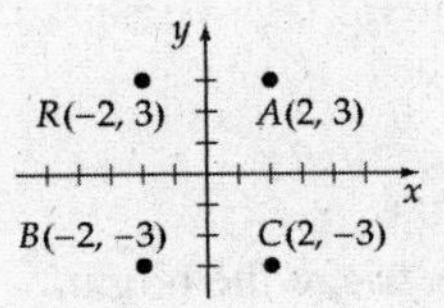

5. Plot the points.

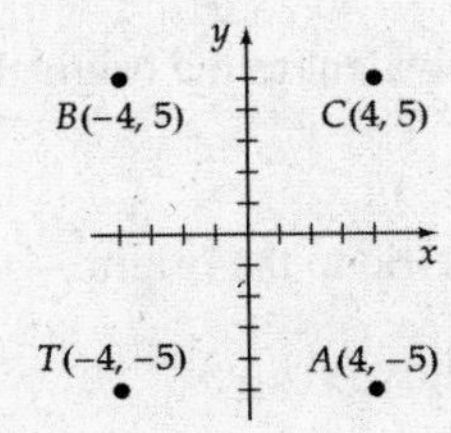

7. Sketch the graph symmetric to the x-axis.

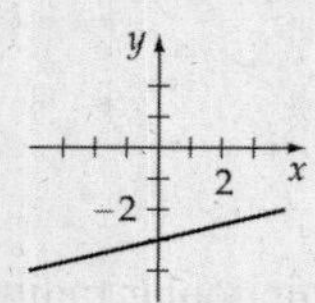

9. Sketch the graph symmetric to the y-axis.

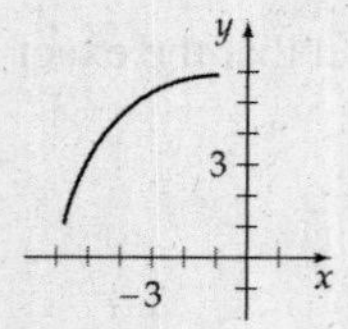

11. Sketch the graph symmetric to the origin.

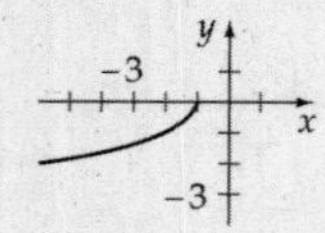

13. Determine if the graph is symmetric.

a. No

b. Yes

15. Determine if the graph is symmetric.

a. No

b. No

17. Determine if the graph is symmetric.

a. Yes

b. Yes

19. Determine if the graph is symmetric.

a. Yes

b. Yes

21. Determine if the graph is symmetric.

a. Yes

b. Yes

23. Determine if the graph is symmetric to the origin.

No, since $(-y) = 3(-x) - 2$ simplifies to

$(-y) = -3x - 2$, which is not equivalent to the original

equation $y = 3x - 2$.

25. Determine if the graph is symmetric to the origin.

Yes, since $(-y) = -(-x)^3$ implies

$-y = x^3$ or $y = -x^3$, which is the original equation.

27. Determine if the graph is symmetric to the origin.

Yes, since $(-x)^2 + (-y)^2 = 10$ simplifies to the

original equation.

29. Determine if the graph is symmetric to the origin.

Yes, since $-y = \frac{-x}{|-x|}$ simplifies to the original

equation.

31. Graph the equation and label each intercept.

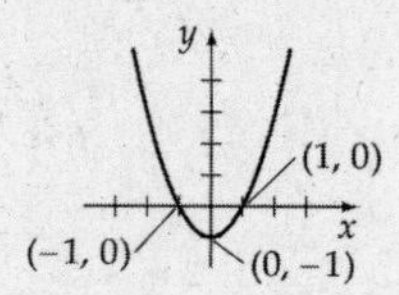

symmetric with respect to the y-axis

33. Graph the equation and label each intercept.

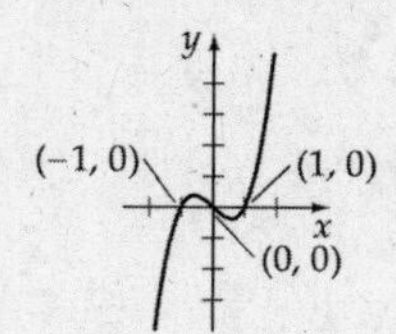

symmetric with respect to the origin

35. Graph the equation and label each intercept.

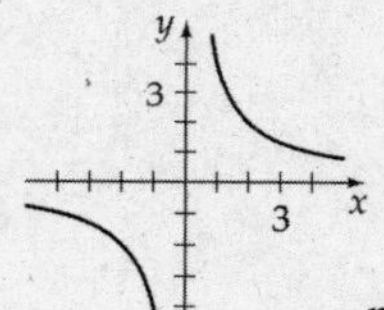

no intercepts

symmetric with respect to the origin

37. Graph the equation and label each intercept.

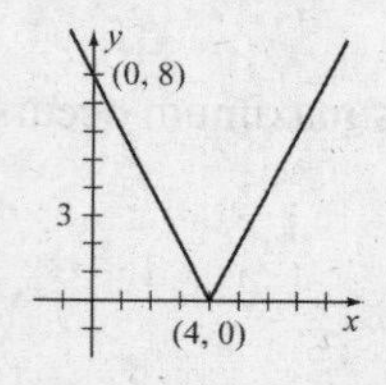

symmetric with respect to the line $x = 4$

39. Graph the equation and label each intercept.

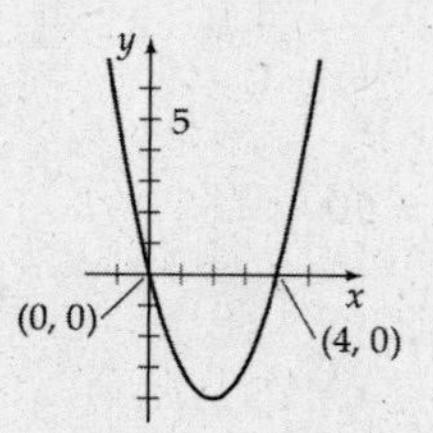

symmetric with respect to the line $x = 2$

41. Graph the equation and label each intercept.

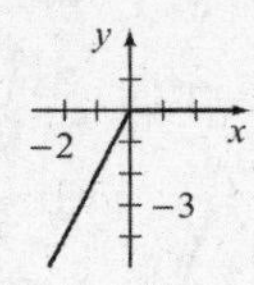

Intercept $(a, 0)$, $a \geq 0$

no symmetry

43. Determine if the function is odd, even or neither.

Even since $g(-x) = (-x)^2 - 7 = x^2 - 7 = g(x)$.

45. Determine if the function is odd, even or neither.

Odd, since $F(-x) = (-x)^5 + (-x)^3$
$= -x^5 - x^3$
$= -F(x)$.

47. Determine if the function is odd, even or neither.

Even

49. Determine if the function is odd, even or neither.

Even

51. Determine if the function is odd, even or neither.

Even

53. Determine if the function is odd, even or neither.

Even

55. Determine if the function is odd, even or neither.

Neither

57. Sketch the graphs.

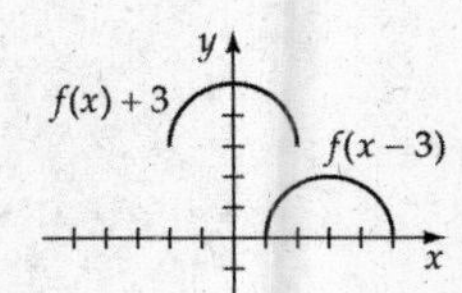

59. Sketch the graphs.

a. $f(x+2)$

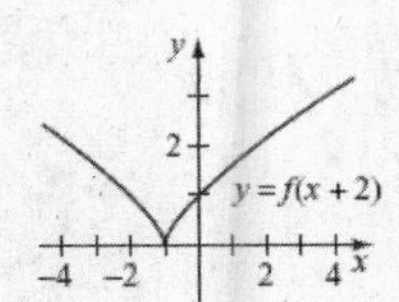

b. $f(x)+2$

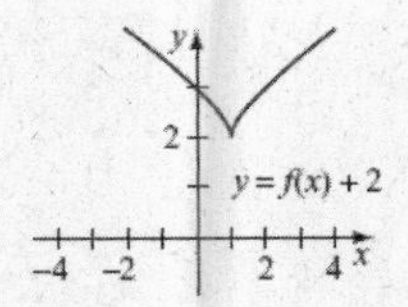

61. Sketch the graphs.

a. $y = f(x-2)+1$

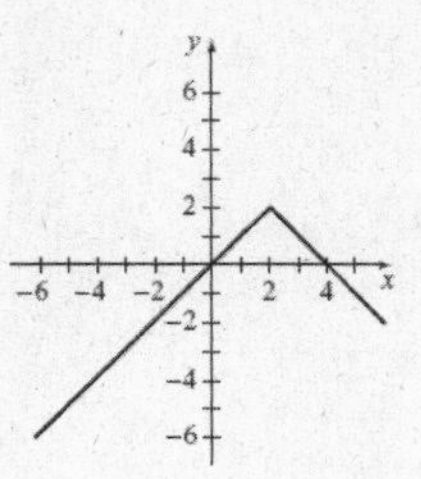

b. $y = f(x+3)-2$

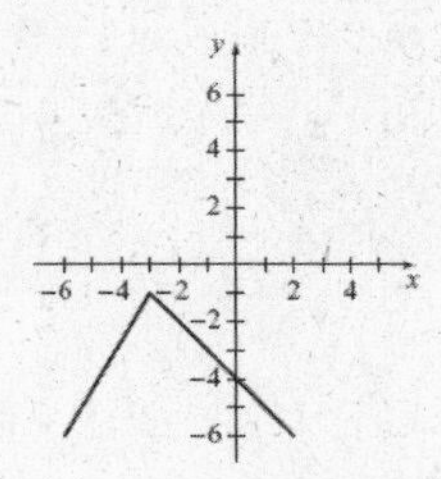

63. a. Give three points on the graph.

$f(x+3)$

$(-2-3, 5) = (-5, 5)$

$(0-3, -2) = (-3, -2)$

$(1-3, 0) = (-2, 0)$

b. Give three points on the graph.

$f(x)+1$

$(-2, 5+1) = (-2, 6)$

$(0, -2+1) = (0, -1)$

$(1, 0+1) = (1, 1)$

65. Sketch the graphs.

a. $f(-x)$

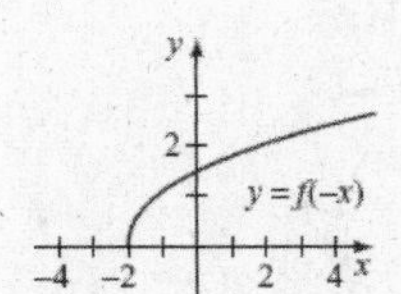

b. $-f(x)$

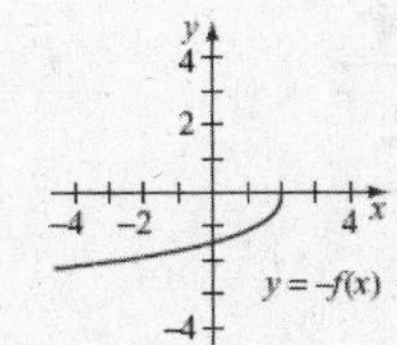

67. **a.** Give two points on the graph.

$f(-x)$

$(--1, 3) = (1, 3)$

$(-2, -4)$

b. Give two points on the graph.

$-f(x)$

$(-1, -3)$

$(2, --4) = (2, 4)$

69. Sketch the graphs.

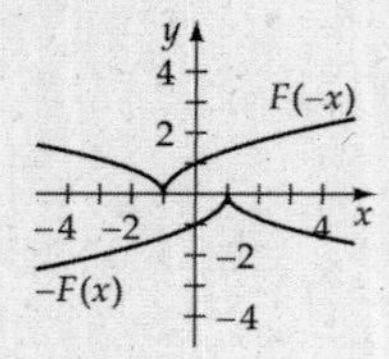

71. Sketch the graph.

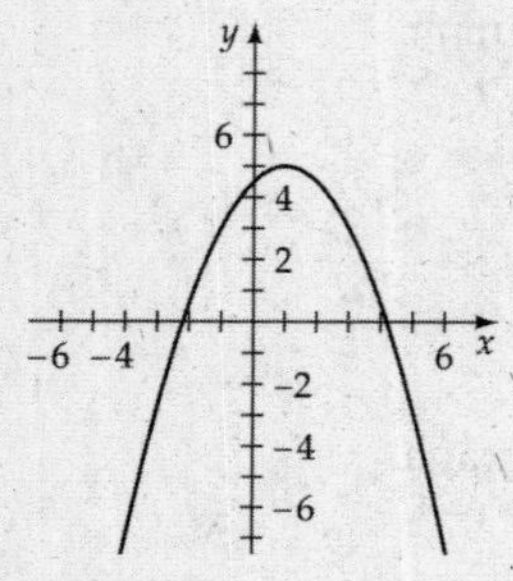

73. Sketch the graphs.

a.

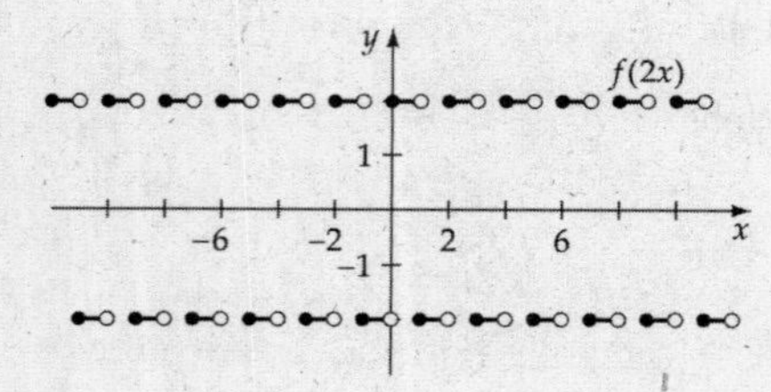

b.

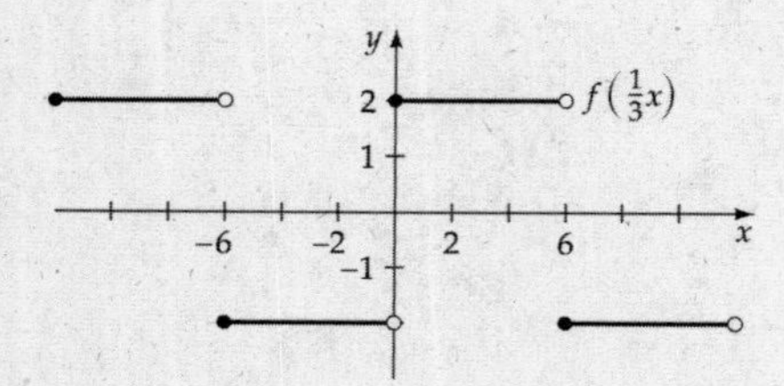

75. Sketch the graphs.

a.

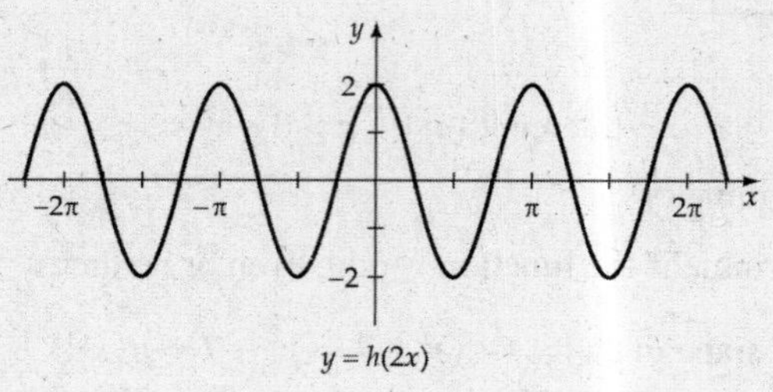

$y = h(2x)$

b.

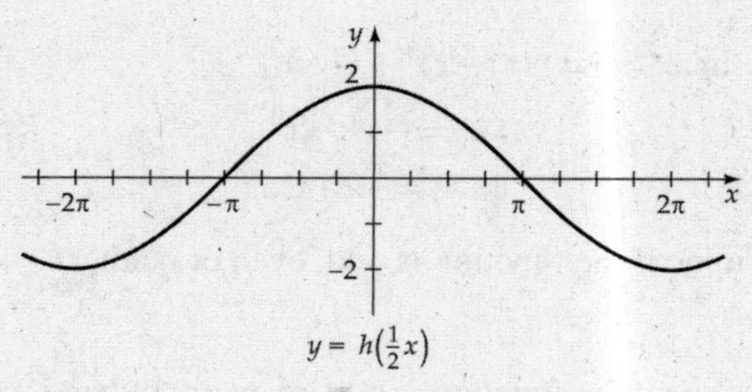

$y = h\left(\frac{1}{2}x\right)$

77. Graph using a graphing utility.

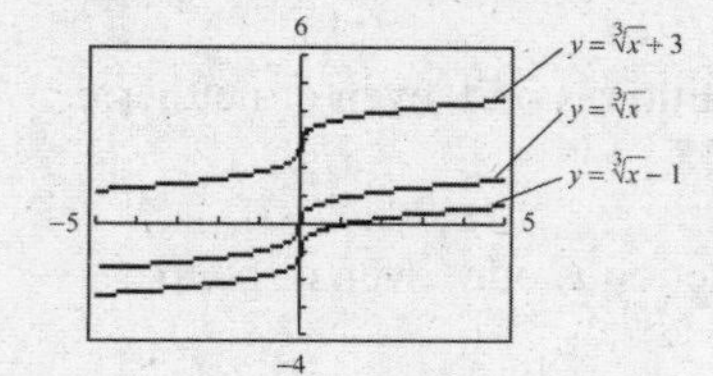

79. Graph using a graphing utility.

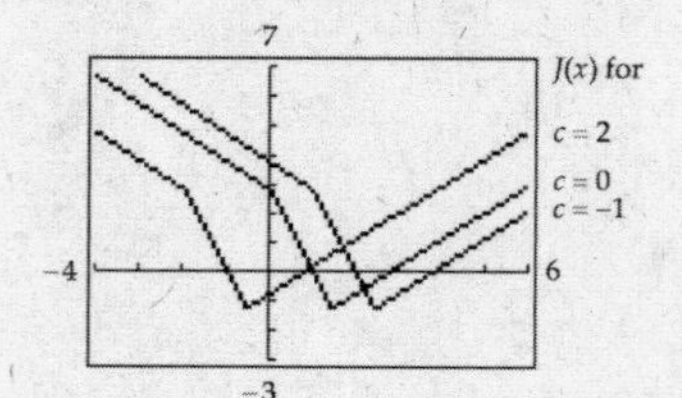

81. Graph using a graphing utility.

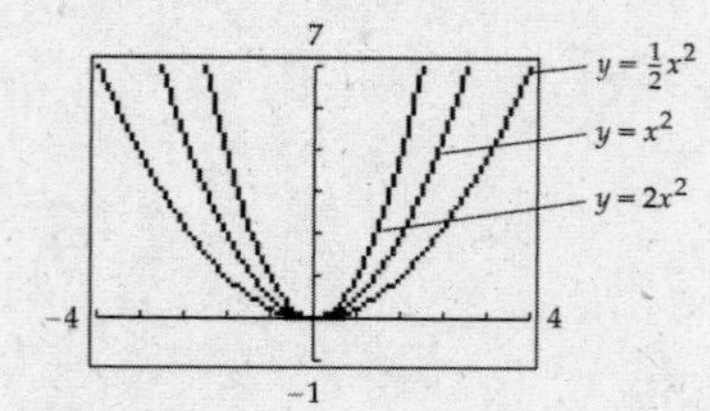

83. Graph using a graphing utility.

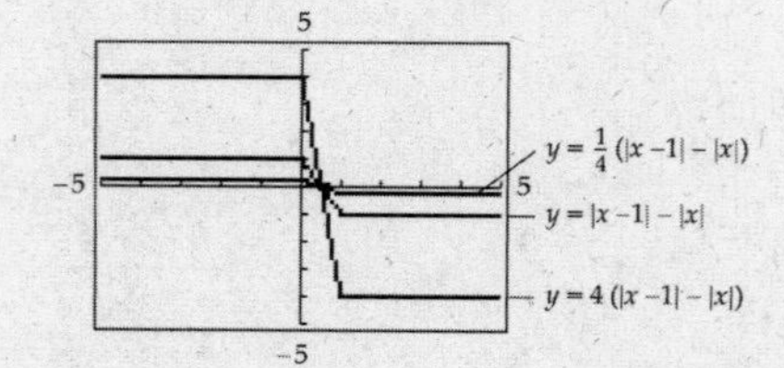

85. Graph for each value of c.

a. $c = 1$

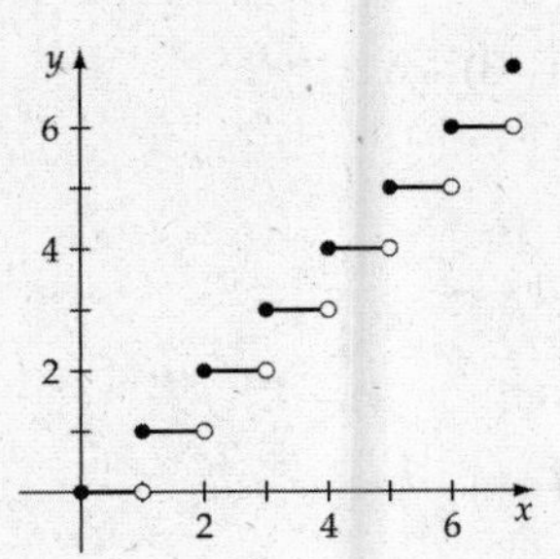

b. $c = \frac{1}{2}$

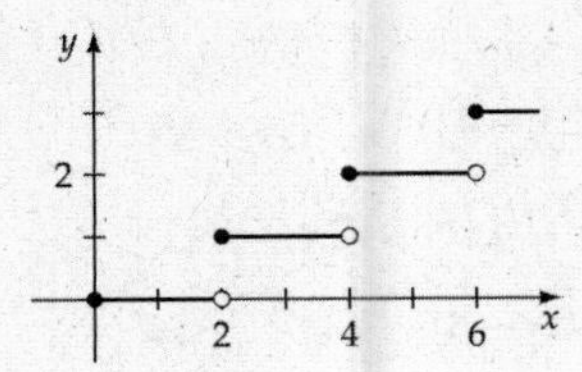

c. $c = 2$

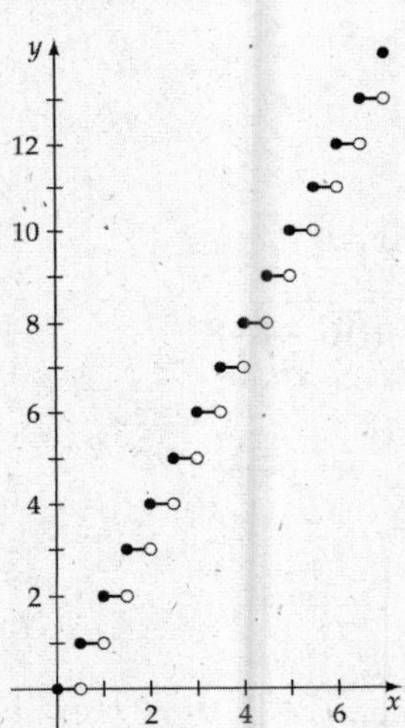

Prepare for Section 2.6

P1. $(2x^2+3x-4)-(x^2+3x-5)=x^2+1$

P3. $f(3a)=2(3a)^2-5(3a)+2$

$=18a^2-15a+2$

P5. Domain: all real numbers except $x = 1$

Section 2.6 Exercises

1. Perform the operations and find the domain.

$f(x)+g(x)=(x^2-2x-15)+(x+3)$

$=x^2-x-12$ Domain all real numbers

$f(x)-g(x)=(x^2-2x-15)-(x+3)$

$=x^2-3x-18$ Domain all real numbers

$f(x)g(x)=(x^2-2x-15)(x+3)$

$=x^3+x^2-21x-45$

Domain all real numbers

$f(x)/g(x)=(x^2-2x-15)/(x+3)$

$=x-5$ Domain $\{x \mid x \neq -3\}$

3. Perform the operations and find the domain.

$f(x)+g(x)=(2x+8)+(x+4)$

$=3x+12$ Domain all real numbers

$f(x)-g(x)=(2x+8)-(x+4)$

$=x+4$ Domain all real numbers

$f(x)g(x)=(2x+8)(x+4)$

$=2x^2+16x+32$ Domain all real numbers

$f(x)/g(x)=(2x+8)/(x+4)$

$=[2(x+4)]/(x+4)$

$=2$ Domain $\{x \mid x \neq -4\}$

5. Perform the operations and find the domain.

$f(x)+g(x)=(x^3-2x^2+7x)+x$

$=x^3-2x^2+8x$ Domain all real numbers

$f(x)-g(x)=(x^3-2x^2+7x)-x$

$=x^3-2x^2+6x$ Domain all real numbers

$f(x)g(x)=(x^3-2x^2+7x)x$

$=x^4-2x^3+7x^2$ Domain all real numbers

$f(x)/g(x)=(x^3-2x^2+7x)/x$

$=x^2-2x+7$ Domain $\{x \mid x \neq 0\}$

7. Perform the operations and find the domain.

$f(x)+g(x)=(4x-7)+(2x^2+3x-5)$

$=2x^2+7x-12$ Domain all real numbers

$f(x)-g(x)=(4x-7)-(2x^2+3x-5)$

$=-2x^2+x-2$ Domain all real numbers

$f(x)g(x)=(4x-7)(2x^2+3x-5)$

$=8x^3-14x^2+12x^2-20x-21x+35$

$=8x^3-2x^2-41x+35$

Domain all real numbers

$f(x)/g(x) = (4x-7)/(2x^2+3x-5)$

$$= \frac{4x-7}{2x^2+3x-5}$$

Domain $\left\{x \mid x \neq 1, x \neq -\frac{5}{2}\right\}$

9. Perform the operations and find the domain.

$f(x)+g(x) = \sqrt{x-3}+x$ Domain $\{x \mid x \geq 3\}$

$f(x)-g(x) = \sqrt{x-3}-x$ Domain $\{x \mid x \geq 3\}$

$f(x)g(x) = x\sqrt{x-3}$ Domain $\{x \mid x \geq 3\}$

$f(x)/g(x) = \dfrac{\sqrt{x-3}}{x}$ Domain $\{x \mid x \geq 3\}$

11. Perform the operations and find the domain.

$f(x)+g(x) = \sqrt{4-x^2}+2+x$

Domain $\{x \mid -2 \leq x \leq 2\}$

$f(x)-g(x) = \sqrt{4-x^2}-2-x$

Domain $\{x \mid -2 \leq x \leq 2\}$

$f(x)g(x) = \left(\sqrt{4-x^2}\right)(2+x)$

Domain $\{x \mid -2 \leq x \leq 2\}$

$f(x)/g(x) = \dfrac{\sqrt{4-x^2}}{2+x}$ Domain $\{x \mid -2 \leq x \leq 2\}$

13. Evaluate the function.

$$(f+g)(x) = x^2-x-2$$

$$\begin{aligned}(f+g)(5) &= (5)^2-(5)-2\\ &= 25-5-2\\ &= 18\end{aligned}$$

15. Evaluate the function.

$$(f+g)(x) = x^2-x-2$$

$$\begin{aligned}(f+g)\left(\frac{1}{2}\right) &= \left(\frac{1}{2}\right)^2-\left(\frac{1}{2}\right)-2\\ &= \frac{1}{4}-\frac{1}{2}-2\\ &= -\frac{9}{4}\end{aligned}$$

17. Evaluate the function.

$$(f-g)(x) = x^2-5x+6$$

$$\begin{aligned}(f-g)(-3) &= (-3)^2-5(-3)+6\\ &= 9+15+6\\ &= 30\end{aligned}$$

19. Evaluate the function.

$$(f-g)(x) = x^2-5x+6$$

$$\begin{aligned}(f-g)(-1) &= (-1)^2-5(-1)+6\\ &= 1+5+6\\ &= 12\end{aligned}$$

21. Evaluate the function.

$$\begin{aligned}(fg)(x) &= \left(x^2-3x+2\right)(2x-4)\\ &= 2x^3-6x^2+4x-4x^2+12x-8\\ &= 2x^3-10x^2+16x-8\end{aligned}$$

$$\begin{aligned}(fg)(7) &= 2(7)^3-10(7)^2+16(7)-8\\ &= 686-490+112-8\\ &= 300\end{aligned}$$

23. Evaluate the function.

$$(fg)(x) = 2x^3-10x^2+16x-8$$

$$\begin{aligned}(fg)\left(\frac{2}{5}\right) &= 2\left(\frac{2}{5}\right)^3-10\left(\frac{2}{5}\right)^2+16\left(\frac{2}{5}\right)-8\\ &= \frac{16}{125}-\frac{40}{25}+\frac{32}{5}-8\\ &= \frac{-384}{125} = -3.072\end{aligned}$$

25. Evaluate the function.

$$\left(\frac{f}{g}\right)(x) = \frac{x^2-3x+2}{2x-4}$$

$$\left(\frac{f}{g}\right)(x) = \frac{1}{2}x-\frac{1}{2}$$

$$\begin{aligned}\left(\frac{f}{g}\right)(-4) &= \frac{1}{2}(-4)-\frac{1}{2} = -2-\frac{1}{2}\\ &= -2\frac{1}{2} \text{ or } -\frac{5}{2}\end{aligned}$$

27. Evaluate the function.

$$\left(\frac{f}{g}\right)(x) = \frac{1}{2}x-\frac{1}{2}$$

$$\begin{aligned}\left(\frac{f}{g}\right)\left(\frac{1}{2}\right) &= \frac{1}{2}\left(\frac{1}{2}\right)-\frac{1}{2} = \frac{1}{4}-\frac{1}{2}\\ &= -\frac{1}{4}\end{aligned}$$

29. Find the difference quotient.

$$\frac{f(x+h)-f(x)}{h}=\frac{[2(x+h)+4]-(2x+4)}{h}$$
$$=\frac{2x+2(h)+4-2x-4}{h}$$
$$=\frac{2h}{h}$$
$$=2$$

31. Find the difference quotient.

$$\frac{f(x+h)-f(x)}{h}=\frac{[(x+h)^2-6]-(x^2-6)}{h}$$
$$=\frac{x^2+2x(h)+(h)^2-6-x^2+6}{h}$$
$$=\frac{2x(h)+h^2}{h}$$
$$=2x+h$$

33. Find the difference quotient.

$$\frac{f(x+h)-f(x)}{h}$$
$$=\frac{2(x+h)^2+4(x+h)-3-(2x^2+4x-3)}{h}$$
$$=\frac{2x^2+4xh+2h^2+4x+4h-3-2x^2-4x+3}{h}$$
$$=\frac{4xh+2h^2+4h}{h}$$
$$=4x+2h+4$$

35. Find the difference quotient.

$$\frac{f(x+h)-f(x)}{h}=\frac{-4(x+h)^2+6-(-4x^2+6)}{h}$$
$$=\frac{-4x^2-8xh-4h^2+6+4x^2-6}{h}$$
$$=\frac{-8xh-4h^2}{h}$$
$$=-8x-4h$$

37. Find the composite functions.

$$(g\circ f)(x)=g[f(x)]$$
$$=g[3x+5]$$
$$=2[3x+5]-7$$
$$=6x+10-7$$
$$=6x+3$$

$$(f\circ g)(x)=f[g(x)]$$
$$=f[2x-7]$$
$$=3[2x-7]+5$$
$$=6x-21+5$$
$$=6x-16$$

39. Find the composite functions.

$$(g\circ f)(x)=g[x^2+4x-1]$$
$$=[x^2+4x-1]+2$$
$$=x^2+4x+1$$

$$(f\circ g)(x)=f[x+2]$$
$$=[x+2]^2+4[x+2]-1$$
$$=x^2+4x+4+4x+8-1$$
$$=x^2+8x+11$$

41. Find the composite functions.

$$(g\circ f)(x)=g[f(x)]$$
$$=g[x^3+2x]$$
$$=-5[x^3+2x]$$
$$=-5x^3-10x$$

$$(f\circ g)(x)=f[g(x)]$$
$$=f[-5x]$$
$$=[-5x]^3+2[-5x]$$
$$=-125x^3-10x$$

43. Find the composite functions.

$$(g\circ f)(x)=g[f(x)]$$
$$=g\left[\frac{2}{x+1}\right]$$
$$=3\left[\frac{2}{x+1}\right]-5=\frac{6}{x+1}-\frac{5(x+1)}{x+1}$$
$$=\frac{6-5x-5}{x+1}$$
$$=\frac{1-5x}{x+1}$$

$$(f\circ g)(x)=f[g(x)]$$
$$=f[3x-5]$$
$$=\frac{2}{[3x-5]+1}$$
$$=\frac{2}{3x-4}$$

45. Find the composite functions.

$$(g\circ f)(x)=g[f(x)]=g\left[\frac{1}{x^2}\right]$$
$$=\sqrt{\left[\frac{1}{x^2}\right]-1}$$
$$=\sqrt{\frac{1-x^2}{x^2}}$$
$$=\frac{\sqrt{1-x^2}}{|x|}$$

$$(f\circ g)(x)=f[g(x)]$$
$$=f\left[\sqrt{x-1}\right]$$
$$=\frac{1}{\left[\sqrt{x-1}\right]^2}$$
$$=\frac{1}{x-1}$$

47. Find the composite functions.

$$(g\circ f)(x)=g\left[\frac{3}{|5-x|}\right]$$
$$=-\frac{2}{\left[\frac{3}{|5-x|}\right]}$$
$$=\frac{-2|5-x|}{3}$$

$$(f\circ g)(x)=f\left[-\frac{2}{x}\right]$$
$$=\frac{3}{\left|5-\left[-\frac{2}{x}\right]\right|}$$
$$=\frac{3}{\left|5+\frac{2}{x}\right|}$$
$$=\frac{3}{\frac{|5x+2|}{|x|}}$$
$$=\frac{3|x|}{|5x+2|}$$

49. Evaluate the composite function.

$$(g\circ f)(x)=4x^2+2x-6$$
$$(g\circ f)(4)=4(4)^2+2(4)-6$$
$$=64+8-6$$
$$=66$$

51. Evaluate the composite function.

$$(f\circ g)(x)=2x^2-10x+3$$
$$(f\circ g)(-3)=2(-3)^2-10(-3)+3$$
$$=18+30+3$$
$$=51$$

53. Evaluate the composite function.

$$(g\circ h)(x)=9x^4-9x^2-4$$
$$(g\circ h)(0)=9(0)^4-9(0)^2-4$$
$$=-4$$

55. Evaluate the composite function.

$$(f\circ f)(x)=4x+9$$
$$(f\circ f)(8)=4(8)+9$$
$$=41$$

57. Evaluate the composite function.

$$(h\circ g)(x)=-3x^4+30x^3-75x^2+4$$
$$(h\circ g)\left(\frac{2}{5}\right)=-3\left(\frac{2}{5}\right)^4+30\left(\frac{2}{5}\right)^3-75\left(\frac{2}{5}\right)^2+4$$
$$=-\frac{48}{625}+\frac{240}{125}-\frac{300}{25}+4$$
$$=\frac{-48+1200-7500+2500}{625}$$
$$=-\frac{3848}{625}$$

59. Evaluate the composite function.

$$(g\circ f)(x)=4x^2+2x-6$$
$$(g\circ f)(\sqrt{3})=4(\sqrt{3})^2+2(\sqrt{3})-6$$
$$=12+2\sqrt{3}-6$$
$$=6+2\sqrt{3}$$

61. Evaluate the composite function.

$$(g\circ f)(x)=4x^2+2x-6$$
$$(g\circ f)(2c)=4(2c)^2+2(2c)-6$$
$$=16c^2+4c-6$$

63. Evaluate the composite function.

$$(g\circ h)(x)=9x^4-9x^2-4$$
$$(g\circ h)(k+1)$$
$$=9(k+1)^4-9(k+1)^2-4$$
$$=9(k^4+4k^3+6k^2+4k+1)-9k^2-18k-9-4$$
$$=9k^4+36k^3+54k^2+36k+9-9k^2-18k-13$$
$$=9k^4+36k^3+45k^2+18k-4$$

65. Show $(g\circ f)(x)=(f\circ g)(x)$.

$$(g\circ f)(x)$$
$$=g[f(x)]$$
$$=g[2x+3]$$
$$=5(2x+3)+12$$
$$=10x+15+12$$
$$=10x+27$$

$$(f\circ g)(x)$$
$$=f[g(x)]$$
$$=f[5x+12]$$
$$=2(5x+12)+3$$
$$=10x+24+3$$
$$=10x+27$$

$$(g\circ f)(x)=(f\circ g)(x)$$

67. Show $(g\circ f)(x)=(f\circ g)(x)$.

$$(g\circ f)(x) = g[f(x)] = g\left[\frac{6x}{x-1}\right] = \frac{5\left(\frac{6x}{x-1}\right)}{\frac{6x}{x-1}-2} = \frac{\frac{30x}{x-1}}{\frac{6x-2x+2}{x-1}} = \frac{\frac{30x}{x-1}}{\frac{4x+2}{x-1}} = \frac{30x}{x-1}\cdot\frac{x-1}{2(2x+1)} = \frac{15x}{2x+1}$$

$$(f\circ g)(x) = f[g(x)] = f\left[\frac{5x}{x-2}\right] = \frac{6\left(\frac{5x}{x-2}\right)}{\frac{5x}{x-2}-1} = \frac{\frac{30x}{x-2}}{\frac{5x-x+2}{x-2}} = \frac{\frac{30x}{x-2}}{\frac{4x+2}{x-2}} = \frac{30x}{x-2}\cdot\frac{x-2}{2(2x+1)} = \frac{15x}{2x+1}$$

$(g\circ f)(x)=(f\circ g)(x)$

69. Show $(g\circ f)(x)=x$ and $(f\circ g)(x)=x$.

$$(g\circ f)(x)=g[f(x)] = g[2x+3] = \frac{[2x+3]-3}{2} = \frac{2x}{2} = x$$

$$(f\circ g)(x)=f[g(x)] = f\left[\frac{x-3}{2}\right] = 2\left[\frac{x-3}{2}\right]+3 = x-3+3 = x$$

71. Show $(g\circ f)(x)=x$ and $(f\circ g)(x)=x$.

$$(g\circ f)(x)=g[f(x)] = g\left[\frac{4}{x+1}\right] = \frac{4-\left[\frac{4}{x+1}\right]}{\left[\frac{4}{x+1}\right]} = \frac{\frac{4x+4-4}{x+1}}{\frac{4}{x+1}} = \frac{4x}{x+1}\cdot\frac{x+1}{4} = x$$

$$(f\circ g)(x)=f[g(x)] = f\left[\frac{4-x}{x}\right] = \frac{4}{\left[\frac{4-x}{x}\right]+1} = \frac{4}{\frac{4-x+x}{x}} = \frac{4}{\frac{4}{x}} = 4\cdot\frac{x}{4} = x$$

73. a. $r=1.5t$ and $A=\pi r^2$

$$\text{so } A(t)=\pi[r(t)]^2 = \pi(1.5t)^2$$

$$A(2)=2.25\pi(2)^2 = 9\pi \text{ square feet} \approx 28.27 \text{ square feet}$$

b. $r=1.5t$

$h=2r=2(1.5t)=3t$ and

$V=\frac{1}{3}\pi r^2h$ so

$$V(t)=\frac{1}{3}\pi(1.5t)^2[3t] = 2.25\pi t^3$$

Note: $V=\frac{1}{3}\pi r^2h=\frac{1}{3}(\pi r^2h)=\frac{1}{3}hA$

$$=\frac{1}{3}(3t)(2.25\pi t^2)=2.25\pi t^3$$

$$V(3)=2.25\pi(3)^3 = 60.75\pi \text{ cubic feet} \approx 190.85 \text{ cubic feet}$$

75. a. Since

$$d^2+4^2=s^2,$$
$$d^2=s^2-16$$
$$d=\sqrt{s^2-16}$$
$$d=\sqrt{(48-t)^2-16} \quad \text{Substitute } 48-t \text{ for } s$$
$$=\sqrt{2304-96t+t^2-16}$$
$$=\sqrt{t^2-96t+2288}$$

b. $s(35)=48-35=13$ ft

$$d(35)=\sqrt{35^2-96(35)+2288} = \sqrt{153}\approx 12.37 \text{ ft}$$

77. $(Y\circ F)(x)=Y(F(x))$ converts x inches to yards. F takes x inches to feet, and then Y takes feet to yards.

79. a. On $[0, 1]$, $a=0$

$\Delta t=1-0=1$

$C(a+\Delta t)=C(1)=99.8$ (mg/L)/h

$C(a)=C(0)=0$

$$\text{Average rate of change}=\frac{C(1)-C(0)}{1}=99.8-0=99.8$$

This is identical to the slope of the line through $(0, C(0))$ and $(1, C(1))$ since

$$m = \frac{C(1) - C(0)}{1 - 0} = C(1) - C(0)$$

b. On $[0, 0.5]$, $a = 0$, $\Delta t = 0.5$

Average rate of change

$$= \frac{C(0.5) - C(0)}{0.5} = \frac{78.1 - 0}{0.5} = 156.2 \text{ (mg/L)/h}$$

c. On $[1, 2]$, $a = 1$, $\Delta t = 2 - 1 = 1$

Average rate of change

$$= \frac{C(2) - C(1)}{1} = \frac{50.1 - 99.8}{1} = -49.7 \text{ (mg/L)/h}$$

d. On $[1, 1.5]$, $a = 1$, $\Delta t = 1.5 - 1 = 0.5$

Average rate of change

$$= \frac{C(1.5) - C(1)}{0.5} = \frac{84.4 - 99.8}{0.5} = \frac{-15.4}{0.5} = -30.8 \text{ (mg/L)/h}$$

e. On $[1, 1.25]$, $a = 1$, $\Delta t = 1.25 - 1 = 0.25$

Average rate of change

$$= \frac{C(1.25) - C(1)}{0.25} = \frac{95.7 - 99.8}{0.25} = \frac{-4.1}{0.25} = -16.4 \text{ (mg/L)/h}$$

f. On $[1, 1 + \Delta t]$,

$$\begin{aligned}
&Con(1 + \Delta t) \\
&= 25(1 + \Delta t)^3 - 150(1 + \Delta t)^2 + 225(1 + \Delta t) \\
&= 25(1 + 3\Delta t + 3(\Delta t)^2 + 1(\Delta t)^3) - 150(1 + 2(\Delta t) + (\Delta t)^2) \\
&\quad + 225(1 + \Delta t) \\
&= 25 + 75(\Delta t) + 75(\Delta t)^2 + 25(\Delta t)^3 \\
&\quad - 150 - 300(\Delta t) - 150(\Delta t)^2 + 225 + 225(\Delta t) \\
&= 100 - 75(\Delta t)^2 + 25(\Delta t)^3
\end{aligned}$$

$Con(1) = 100$

Average rate of change

$$\begin{aligned}
&= \frac{Con(1 + \Delta t) - Con(1)}{\Delta t} \\
&= \frac{100 - 75(\Delta t)^2 + 25(\Delta t)^3 - 100}{\Delta t} \\
&= \frac{-75(\Delta t)^2 + 25(\Delta t)^3}{\Delta t} \\
&= -75(\Delta t) + 25(\Delta t)^2
\end{aligned}$$

As Δt approaches 0, the average rate of change over $[1, 1 + \Delta t]$ seems to approach 0 (mg/L)/h.

Prepare for Section 2.7

P1. Slope: $-\frac{1}{3}$; y-intercept: $(0, 4)$

P3. $y = -0.45x + 2.3$

P5. $f(2) = 3(2)^2 + 4(2) - 1 = 12 + 8 - 1 = 19$

Section 2.7 Exercises

1. The scatter diagram suggests no relationship between x and y.

3. The scatter diagram suggests a linear relationship between x and y.

5. Figure A better approximates a graph that can be modeled by an equation than does Figure B. Thus Figure A has a coefficient of determination closer to 1.

7. Enter the data on your calculator. The technique for a TI-83 calculator is illustrated here. Press STAT.

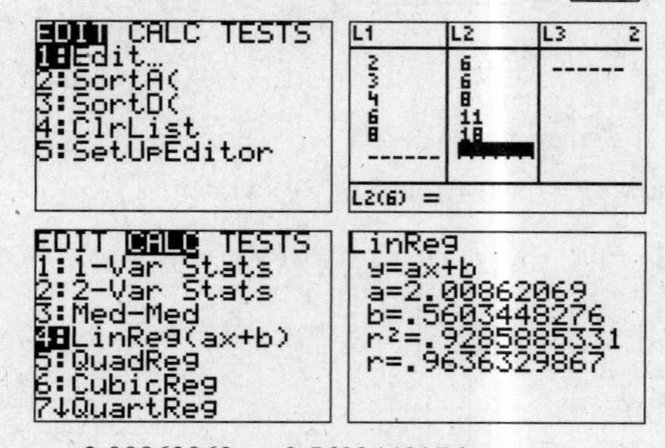

$y = 2.00862069x + 0.5603448276$

9. Enter the data on your calculator. The technique for a TI-83 calculator is illustrated here. Press STAT.

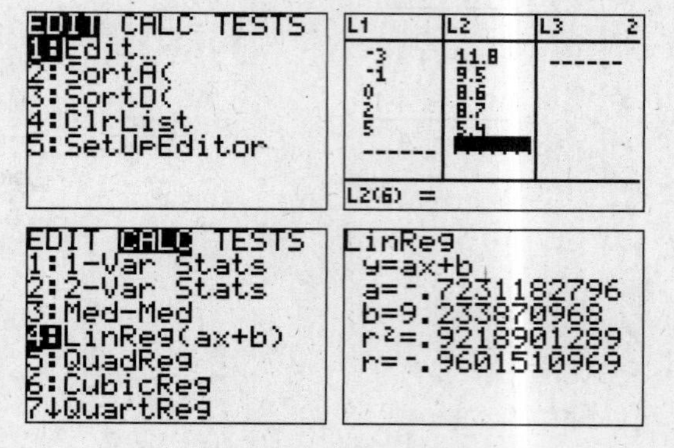

$y = -0.7231182796x + 9.233870968$

11. Enter the data on your calculator. The technique for a TI-83 calculator is illustrated here. Press STAT.

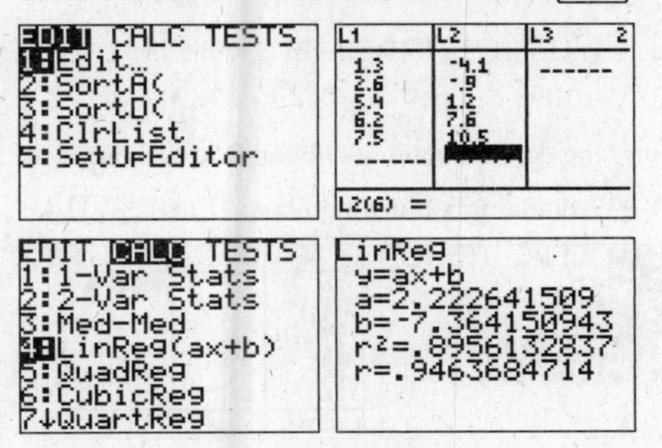

$y = 2.222641509x - 7.364150943$

13. Enter the data on your calculator. The technique for a TI-83 calculator is illustrated here. Press STAT.

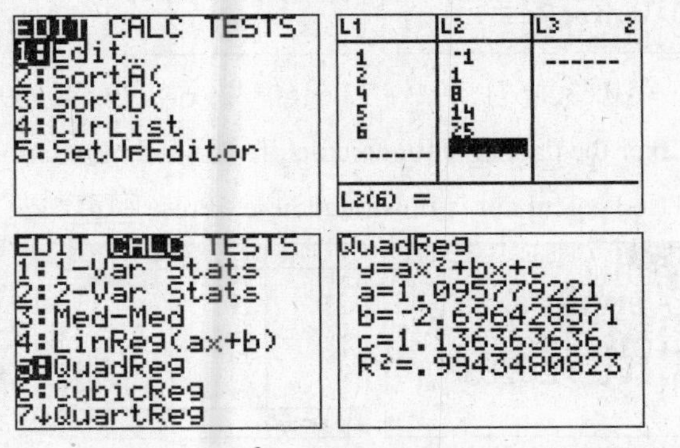

$y = 1.095779221x^2 - 2.696428571x + 1.136363636$

15. Enter the data on your calculator. The technique for a TI-83 calculator is illustrated here. Press STAT.

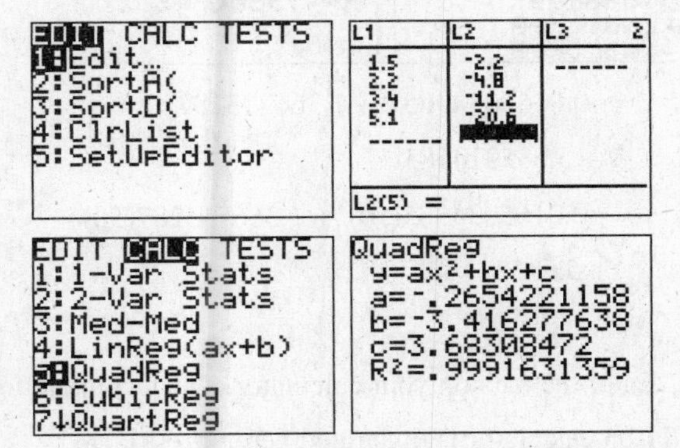

$y = -0.2987274717x^2 - 3.20998141x + 3.416463667$

17. Enter the data on your calculator. The technique for a TI-83 calculator is illustrated here. Press STAT.

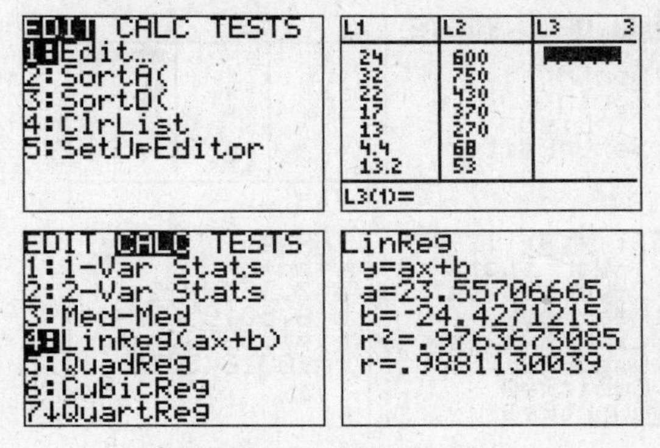

a. $y = 23.55706665x - 24.4271215$

b. $y = 23.55706665(54) - 24.4271215 \approx 1248$ cm

19. Enter the data on your calculator. The technique for a TI-83 calculator is illustrated here. Press STAT.

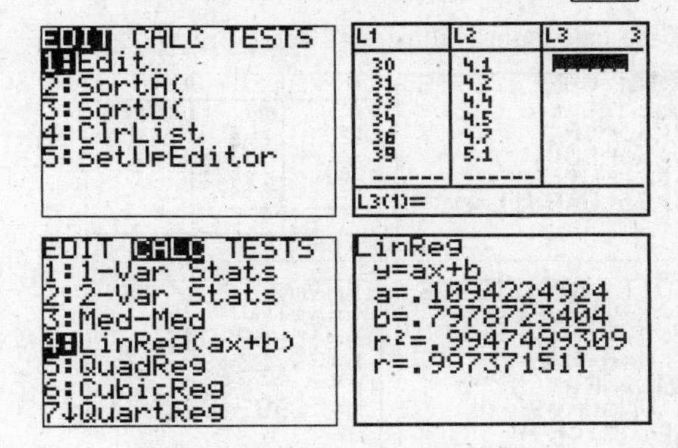

a. $y = 0.1094224924x + 0.7978723404$

b. $y = 0.1094224924(32) + 0.7978723404 \approx 4.3$ m/s

21. Enter the data on your calculator. The technique for a TI-83 calculator is illustrated here. Press STAT.

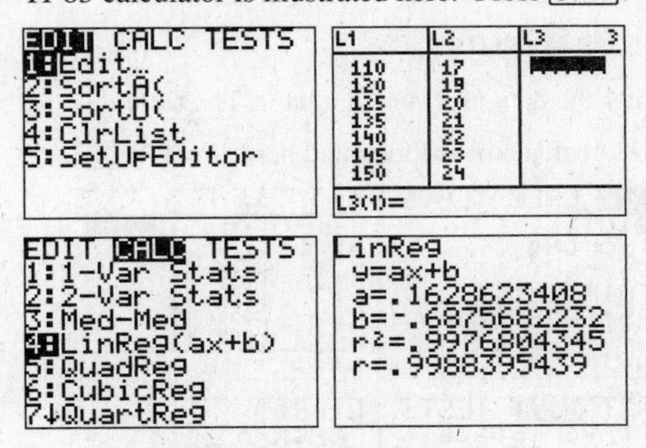

a. $y = 0.1628623408x - 0.6875682232$

b. $y = 0.1628623408(158) - 0.6875682232 \approx 25$

23. Enter the data on your calculator. The technique for a TI-83 calculator is illustrated here. Press STAT.

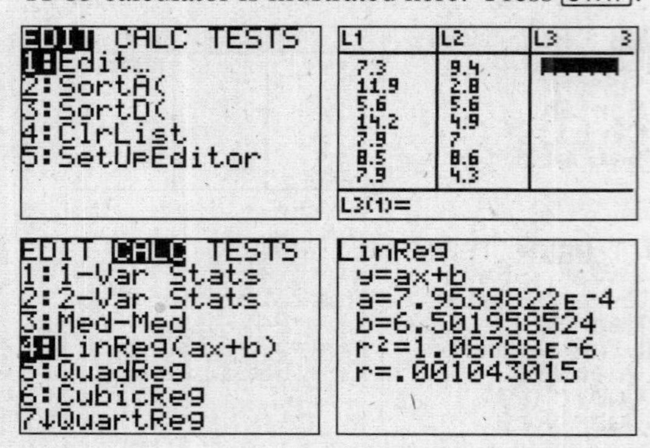

The value of r is close to 0. Therefore, no, there is not a strong linear relationship between the current and the torque.

25. Enter the data on your calculator. The technique for a TI-83 calculator is illustrated here. Press STAT.

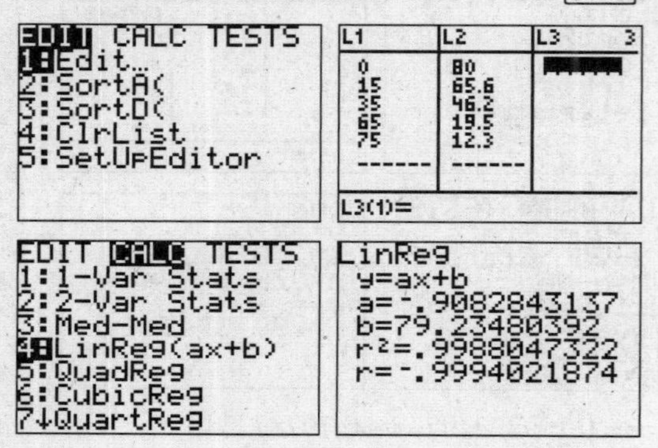

a. Yes, there is a strong linear correlation.

b. $y = -0.9082843137x + 79.23480392$

c. $y = -0.9082843137(25) + 79.23480392$

≈ 57 years

27. Enter the data on your calculator. The technique for a TI-83 calculator is illustrated here. Press STAT.

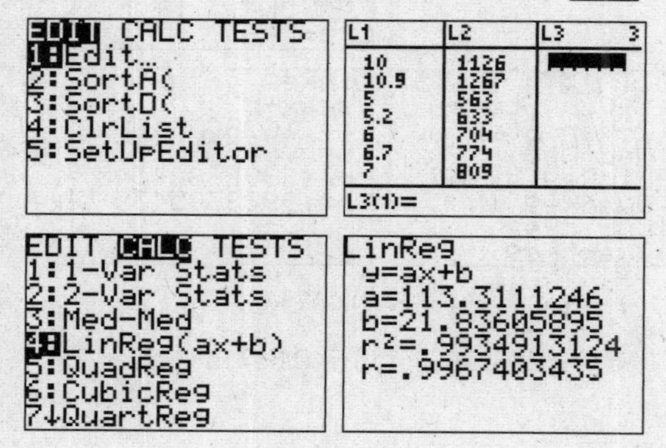

$y = 113.3111246x + 21.83605895$

a. Positively

b. $y = 113.3111246(9.5) + 21.83605895$

≈ 1098 calories

29. Enter the data on your calculator. The technique for a TI-83 calculator is illustrated here. Press STAT.

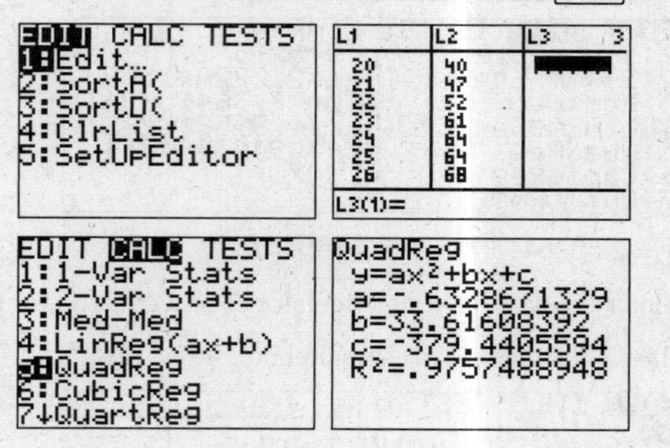

$y = -0.6328671329x^2 + 33.61608329x - 379.4405594$

31. Enter the data on your calculator. The technique for a TI-83 calculator is illustrated here. Press STAT.

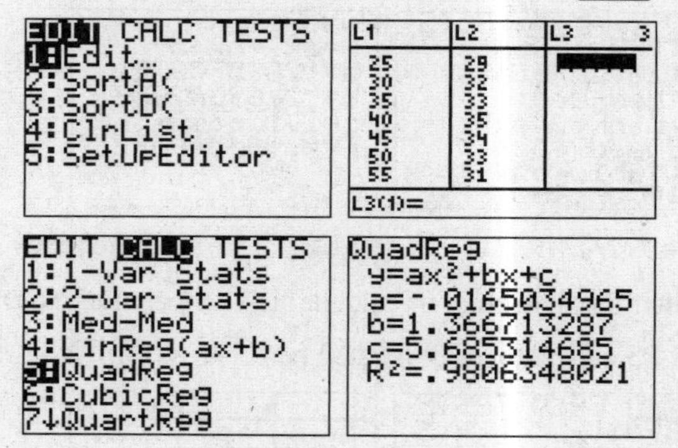

a. $y = -0.0165034965x^2 + 1.366713287x + 5.685314685$

b. $y = -0.0165034965(50)^2 + 1.366713287(50) + 5.685314685$

≈ 32.8 mpg

33. a. Enter the data on your calculator. The technique for a TI-83 calculator is illustrated here. Press STAT.

5-lb ball

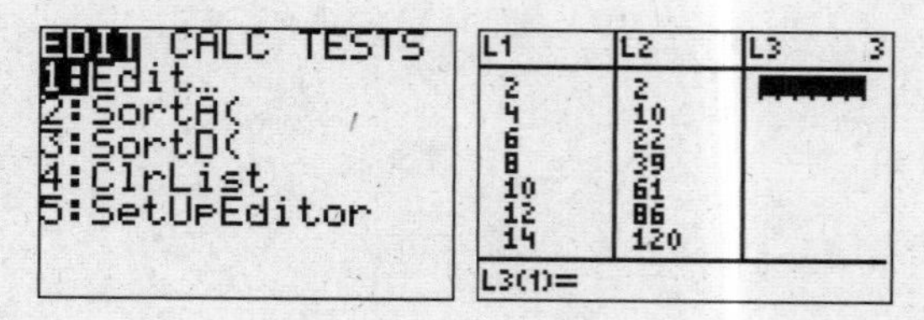

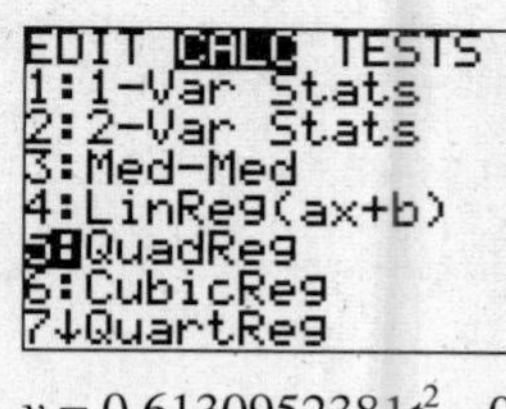

$y = 0.6130952381t^2 - 0.0714285714t + 0.1071428571$

10-lb ball

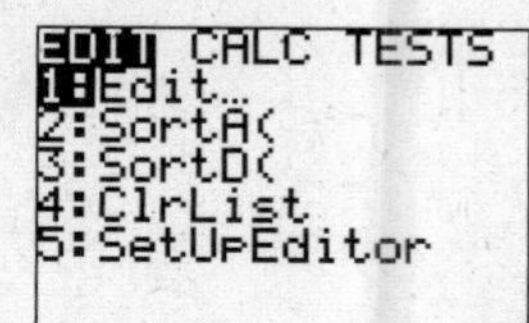

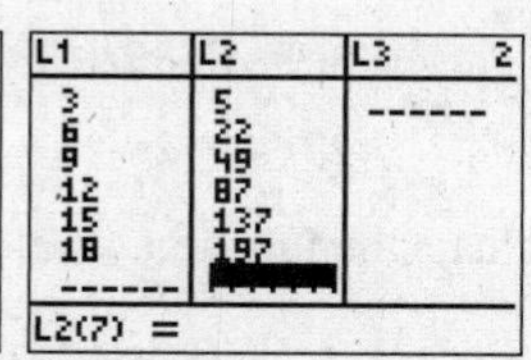

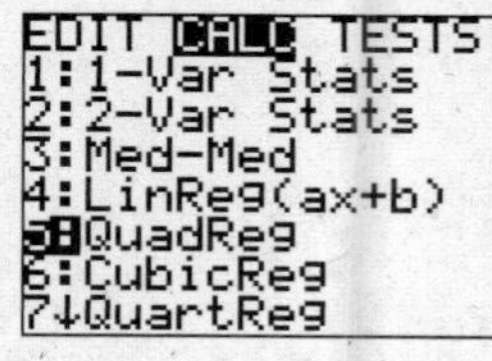

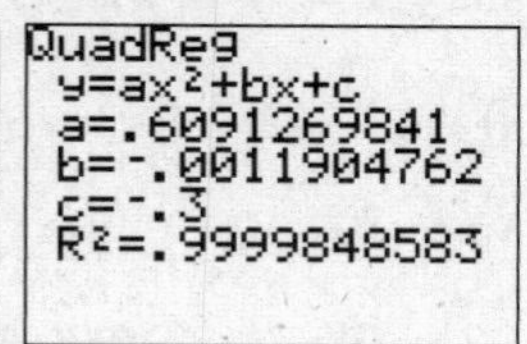

$y = 0.6091269841t^2 - 0.0011904762t - 0.3$

15-lb ball

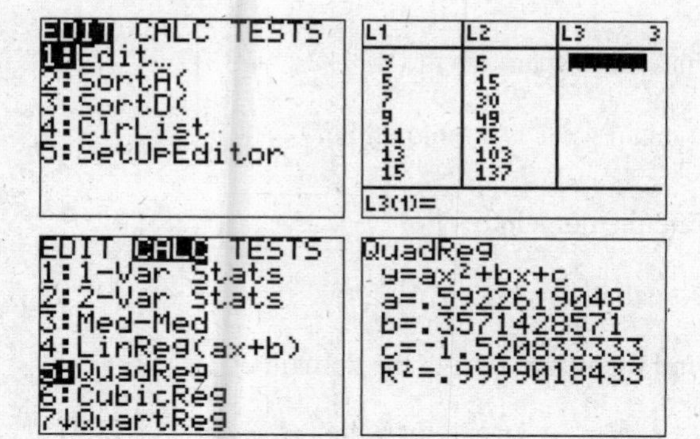

$y = 0.5922619048t^2 + 0.3571428571t - 1.520833333$

b. All the regression equations are approximately the same. Therefore, there is one equation of motion.

Exploring Concepts with Technology

1. Use Dot mode. Enter the function as

Y1=X²*(X<2)-X*(X≥2)

and graph this in the standard viewing window.

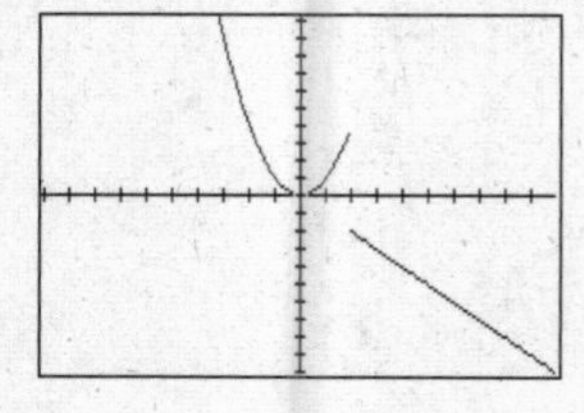

3. Use Dot mode. Enter the function as

Y1=(-X²+1)*(X<0)+(X²-1)*(X≥0)

and graph this in the standard viewing window.

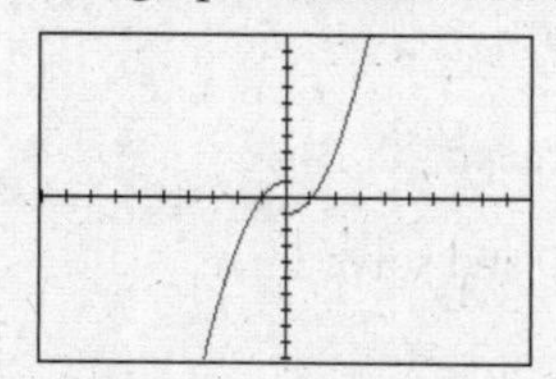

Chapter 2 Review Exercises

1. Finding the distance. [2.1]

$$d = \sqrt{(7-(-3))^2 + (11-2)^2}$$
$$= \sqrt{10^2 + 9^2} = \sqrt{100+81} = \sqrt{181}$$

3. Finding the midpoint: (2, 8), (–3, 12). [2.1]

$$M = \left(\frac{2+(-3)}{2}, \frac{8+12}{2}\right) = \left(-\frac{1}{2}, 10\right)$$

5. Graph the equation: $2x - y = -2$. [2.1]

x	y
–1	0
0	2
1	4

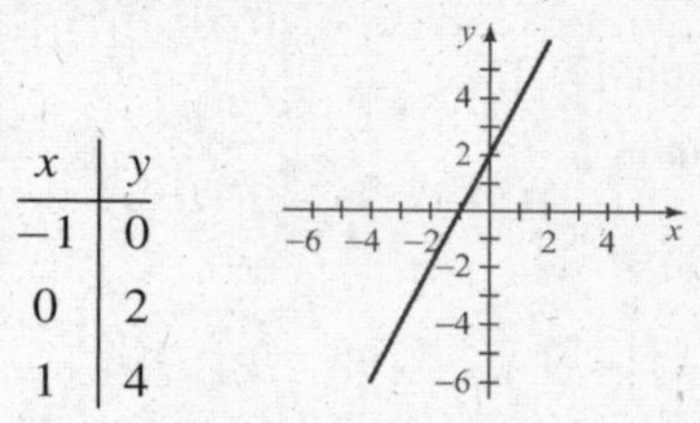

7. Graph the equation: $y = |x-2|+1$. [2.1]

x	y
–1	4
0	3
2	1
3	2
4	3

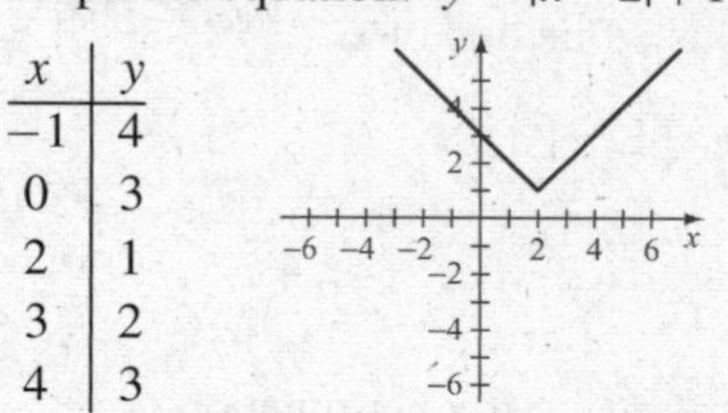

9. Finding x- and y-intercepts and graph: $x = y^2 - 1$ [2.1]

For the y-intercept, let $x = 0$ and solve for y.

$$0 = y^2 - 1$$
$$y = \pm 1, \text{ } y\text{-intercepts: } (0,-1),(0, 1)$$

For the x-intercept, let $y = 0$ and solve for x.

$$x = (0)^2 - 1$$
$$x = -1, \text{ } x\text{-intercept: } (-1, 0)$$

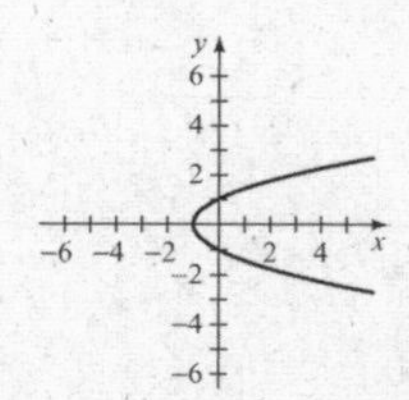

11. Finding x- and y-intercepts and graph: [2.1]

$3x+4y=12$

For the y-intercept, let $x = 0$ and solve for y.

$$3(0)+4y=12$$
$$y=3, \ y\text{-intercept: } (0, 3)$$

For the x-intercept, let $y = 0$ and solve for x.

$$3x+4(0)=12$$
$$x=4, \ x\text{-intercept: } (4, 0)$$

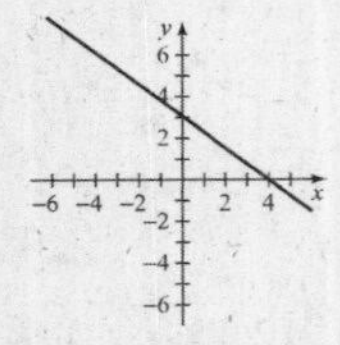

13. Finding the center and radius. [2.1]

$(x-3)^2+(y+4)^2=81$

center (3, –4), radius 9

15. Finding the equation. [2.1]

Center: (2, –3), radius 5

$(x-2)^2+(y+3)^2=5^2$

17. Is y a function of x? [2.2]

$$x-y=4$$
$$y=x-4, \ y \text{ is a function of } x.$$

19. Is y a function of x? [2.2]

$$|x|+|y|=4$$
$$|y|=-|x|+4$$
$$y=\pm(-|x|+4), \ y \text{ is a not function of } x.$$

21. Evaluate the function $f(x)=3x^2+4x-5$, [2.2]

a. $f(1)=3(1)^2+4(1)-5$
$$=3(1)+4-5=3+4-5$$
$$=2$$

b. $f(-3)=3(-3)^2+4(-3)-5$
$$=3(9)-12-5$$
$$=27-12-5$$
$$=10$$

c. $f(t)=3t^2+4t-5$

d. $f(x+h)=3(x+h)^2+4(x+h)-5$
$$=3(x^2+2xh+h^2)+4x+4h-5$$
$$=3x^2+6xh+3h^2+4x+4h-5$$

e. $3f(t)=3(3t^2+4t-5)$
$$=9t^2+12t-15$$

f. $f(3t)=3(3t)^2+4(3t)-5$
$$=3(9t^2)+12t-5$$
$$=27t^2+12t-5$$

23. Evaluate the function. [2.2]

a. Since $x=3\geq 0$, use $f(x)=x^2-3$.

$f(3)=(3)^2-3=9-3=6$

b. Since $x=-2<0$, use $f(x)=3x+2$.

$f(-2)=3(-2)+2=-6+2=-4$

c. Since $x=0\geq 0$, use $f(x)=x^2-3$.

$f(0)=(0)^2-3=0-3=-3$

25. Find the domain of $f(x)=-2x^2+3$. [2.2]

Domain $\{x|x \text{ is a real number}\}$

27. Find the domain of $f(x)=\sqrt{25-x^2}$. [2.2]

Domain $\{x|-5\leq x\leq 5\}$

29. Find the values of a in the domain of

$f(x)=x^2+2x-4$ for which $f(a)=-1$. [2.2]

$$a^2+2a-4=-1 \quad \text{Replace } f(a) \text{ with } a^2+2a-4$$
$$a^2+2a-3=0$$
$$(a+3)(a-1)=0$$
$$a+3=0 \quad a-1=0$$
$$a=-3 \quad a=1$$

31. Graph $f(x)=|x-1|-1$ [2.2]

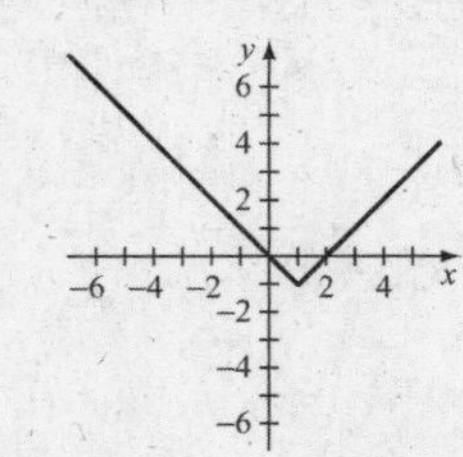

33. Find the zeros of f for $f(x)=2x+6$. [2.2]

$$\begin{aligned} f(x)&=0\\ 2x+6&=0\\ 2x&=-6\\ x&=-3 \end{aligned}$$

35. Evaluate the function $g(x)=[\![2x]\!]$. [2.2]

a. $g(\pi)=[\![2\pi]\!]\approx[\![6.283185307]\!]=6$

b. $g\left(-\frac{2}{3}\right)=\left[\!\!\left[2\left(-\frac{2}{3}\right)\right]\!\!\right]=\left[\!\!\left[-\frac{4}{3}\right]\!\!\right]\approx[\![-1.333333]\!]=-2$

c. $g(-2)=[\![2(-2)]\!]=-4$

37. Find the slope. [2.3]

$$m=\frac{-1-6}{4+3}=\frac{-7}{7}=-1$$

39. Find the slope. [2.3]

$$m=\frac{-2+2}{-3-4}=\frac{0}{-7}=0$$

41. Graph $f(x)=-\frac{3}{4}x+2$. [2.3]

$m=-\frac{3}{4}$, y-intercept (0, 2)

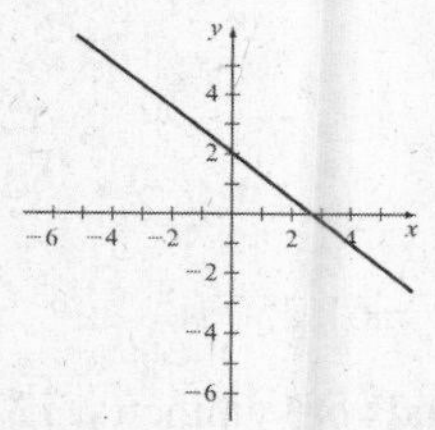

43. Graph $3x-4y=8$. [2.3]

$$\begin{aligned} 3x-4y&=8\\ -4y&=-3x+8\\ y&=\frac{3}{4}x-2 \end{aligned}$$

x-intercept $\left(\frac{8}{3},\ 0\right)$, y-intercept (0, –2)

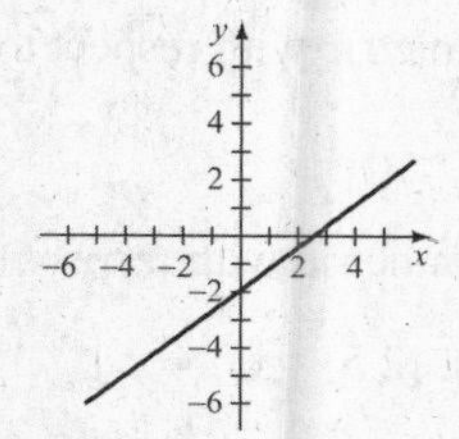

45. Find the equation. [2.3]

$$\begin{aligned} y-2&=-\frac{2}{3}(x+3)\\ y-2&=-\frac{2}{3}x-2\\ y&=-\frac{2}{3}x \end{aligned}$$

47. Find the equation. [2.3]

$$m=\frac{6-3}{1+2}=\frac{3}{3}=1$$

$$\begin{aligned} y-6&=1(x-1)\\ y-6&=x-1\\ y&=x+5 \end{aligned}$$

49. Find the equation. [2.3]

$y=\frac{2}{3}x-1$ has slope $m=\frac{2}{3}$.

$$\begin{aligned} y-y_1&=\frac{2}{3}(x-x_1)\\ y-(-5)&=\frac{2}{3}(x-3)\\ y+5&=\frac{2}{3}x-2\\ y&=\frac{2}{3}x-7 \end{aligned}$$

51. Find the equation. [2.3]

$y=-\frac{3}{2}x-2$ has perpendicular slope $m=\frac{2}{3}$.

$$\begin{aligned} y-y_1&=\frac{2}{3}(x-x_1)\\ y-(-1)&=\frac{2}{3}(x-3)\\ y+1&=\frac{2}{3}x-2\\ y&=\frac{2}{3}x-3 \end{aligned}$$

53. Find the function. [2.3]

$$m=\frac{175-155}{118-106}=\frac{20}{12}=\frac{5}{3}$$

$$\begin{aligned} f(x)-175&=\frac{5}{3}(x-118)\\ f(x)-175&=\frac{5}{3}x-\frac{590}{3}\\ f(x)&=\frac{5}{3}x-\frac{65}{3} \end{aligned}$$

55. Write the quadratic equation in standard form. [2.4]

$$f(x)=(x^2+6x)+10$$
$$f(x)=(x^2+6x+9)+10-9$$
$$f(x)=(x+3)^2+1$$

57. Write the quadratic equation in standard form. [2.4]

$$f(x)=-x^2-8x+3$$
$$f(x)=-(x^2+8x)+3$$
$$f(x)=-(x^2+8x+16)+3+16$$
$$f(x)=-(x+4)^2+19$$

59. Write the quadratic equation in standard form. [2.4]

$$f(x)=-3x^2+4x-5$$
$$f(x)=-3\left(x^2-\tfrac{4}{3}x\right)-5$$
$$f(x)=-3\left(x^2-\tfrac{4}{3}x+\tfrac{4}{9}\right)-5+\tfrac{4}{3}$$
$$f(x)=-3\left(x-\tfrac{2}{3}\right)^2-\tfrac{11}{3}$$

61. Find the vertex. [2.4]

$$\frac{-b}{2a}=\frac{-(-6)}{2(3)}=\frac{6}{6}=1$$

$$\begin{aligned} f(1)&=3(1)^2-6(1)+11\\ &=3(1)-6+11\\ &=3-6+11\\ &=8 \end{aligned}$$

Thus the vertex is (1, 8).

63. Find the vertex. [2.4]

$$\frac{-b}{2a}=\frac{-(60)}{2(-6)}=\frac{-60}{-12}=5$$

$$\begin{aligned} f(5)&=-6(5)^2+60(5)+11\\ &=-6(25)+300+11\\ &=-150+300+11\\ &=161 \end{aligned}$$

Thus the vertex is (5, 161).

65. Find the value. [2.4]

$$\begin{aligned} f(x)&=-x^2+6x-3\\ &=-(x^2-6x)-3\\ &=-(x^2-6x+9)-3+9\\ &=-(x-3)^2+6 \end{aligned}$$

maximum value of 6

67. Find the maximum height. [2.4]

$$h(t)=-16t^2+50t+4$$

$$-\frac{b}{2a}=-\frac{50}{2(-16)}=\frac{25}{16}$$

$$h\left(\frac{25}{16}\right)=-16\left(\frac{25}{16}\right)^2+50\left(\frac{25}{16}\right)+4=43.0625$$

The ball reaches a maximum height of 43.0625 ft.

69. Find the maximum area. [2.4]

Let x be the width. Using the formula for perimeter for three sides, $P=2w+l \Rightarrow 700=2x+l$

$$l=700-2x$$

Using the formula for area, $A=lw$. Then

$$A(x)=x(700-2x)$$
$$A(x)=-2x^2+700x$$
$$-\frac{b}{2a}=-\frac{700}{2(-2)}=175$$

$$A(175)=-2(175)^2+700(175)=61,250\text{ ft}^2$$

71. Sketch a graph with different kinds of symmetry. [2.5]

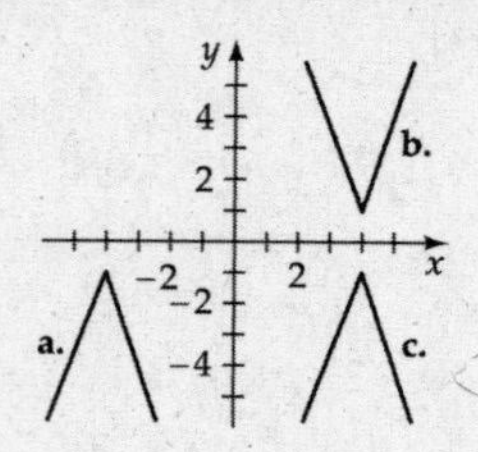

73. The graph of $x=y^2+3$ is symmetric with respect to the x-axis. [2.5]

75. The graph of $y^2=x^2+4$ is symmetric with respect to the x-axis, y-axis, and the origin. [2.5]

77. The graph of $xy=8$ is symmetric with respect to the origin. [2.5]

79. The graph of $|x+y|=4$ is symmetric with respect to the origin. [2.5]

81. Sketch the graph $g(x)=-2x-4$. [2.5]

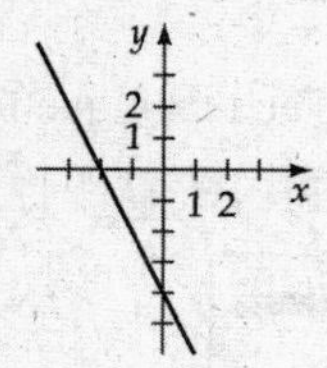

a. Domain all real numbers

Range all real numbers

b. g is neither even nor odd

83. Sketch the graph $g(x)=\sqrt{16-x^2}$. [2.5]

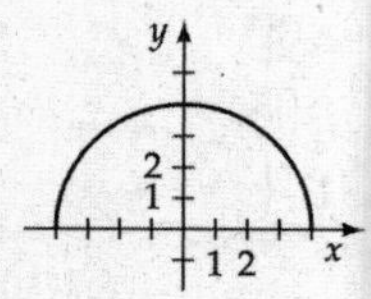

a. Domain $\{x|-4\le x\le 4\}$

Range $\{y|0\le y\le 4\}$

b. g is an even function

85. Sketch the graph $g(x)=2[\![x]\!]$. [2.5]

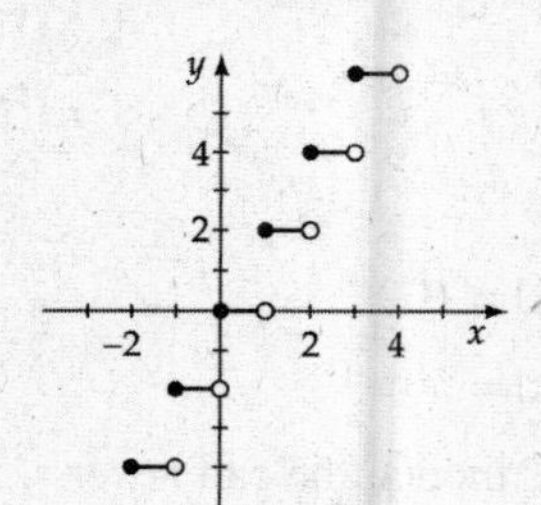

a. Domain all real numbers

Range $\{y|y \text{ is an even integer}\}$

b. g is neither even nor odd

87. $g(x)=f(x+3)$ [2.5]

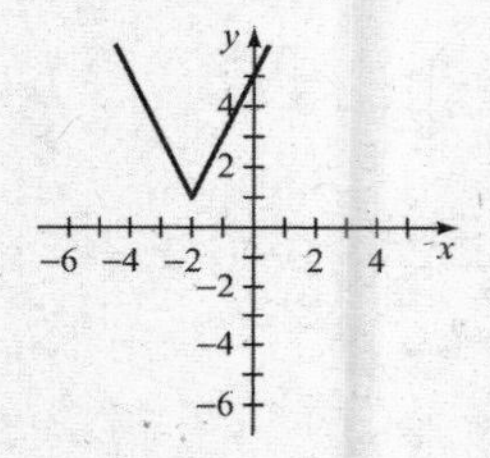

89. $g(x)=f(x+2)-1$ [2.5]

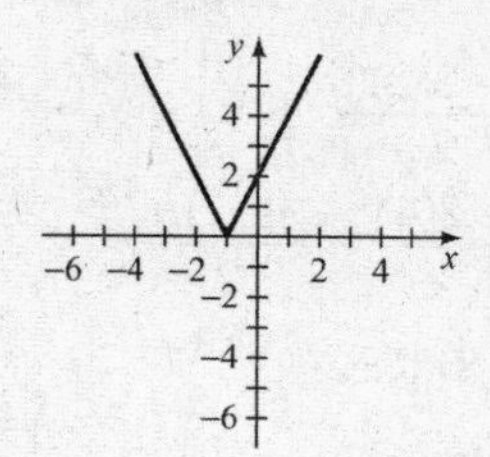

91. $g(x)=-f(x)$ [2.5]

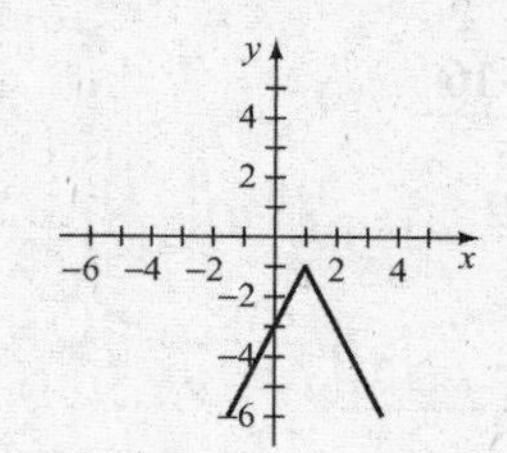

93. $g(x)=\frac{1}{2}f(x)$ [2.5]

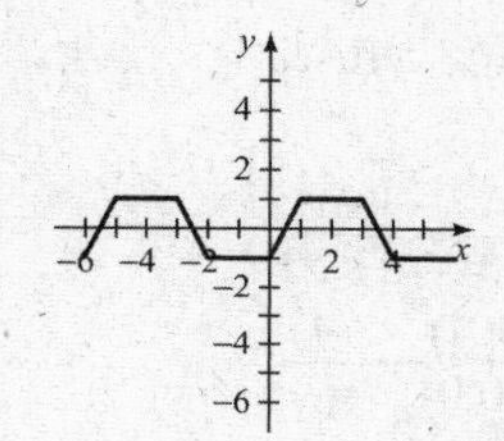

95. $g(x)=f\left(\frac{1}{2}x\right)$ [2.5]

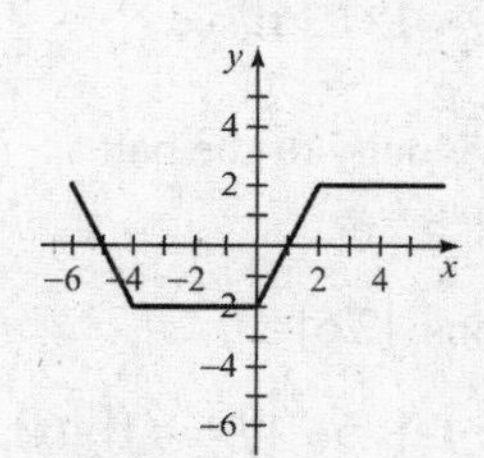

97. Find the difference quotient. [2.6]

$$\frac{f(x+h)-f(x)}{h}$$
$$=\frac{4(x+h)^2-3(x+h)-1-(4x^2-3x-1)}{h}$$
$$=\frac{4x^2+8xh+4h^2-3x-3h-1-4x^2+3x+1}{h}$$
$$=\frac{8xh+4h^2-3h}{h}$$
$$=8x+4h-3$$

99. $s(t) = 3t^2$ [2.4]

a. $\text{Average velocity} = \dfrac{3(4)^2 - 3(2)^2}{4-2}$

$= \dfrac{3(16) - 3(4)}{2} = \dfrac{48-12}{2}$

$= \dfrac{36}{2} = 18 \text{ ft/sec}$

b. $\text{Average velocity} = \dfrac{3(3)^2 - 3(2)^2}{3-2}$

$= \dfrac{3(9) - 3(4)}{1}$

$= \dfrac{27-12}{1} = 15 \text{ ft/sec}$

c. $\text{Average velocity} = \dfrac{3(2.5)^2 - 3(2)^2}{2.5-2}$

$= \dfrac{3(6.25) - 3(4)}{0.5} = \dfrac{18.75-12}{0.5}$

$= \dfrac{6.75}{0.5} = 13.5 \text{ ft/sec}$

d. $\text{Average velocity} = \dfrac{3(2.01)^2 - 3(2)^2}{2.01-2}$

$= \dfrac{3(4.0401) - 3(4)}{0.01}$

$= \dfrac{12.1203-12}{0.01}$

$= \dfrac{0.1203}{0.01} = 12.03 \text{ ft/sec}$

e. It appears that the average velocity of the ball approaches 12 ft/sec.

101. Evaluate the composite functions. [2.6]

a. $(f \circ g)(-5) = f(g(-5)) = f(|-5-1|) = f(|-6|)$

$= f(6) = 2(6)^2 + 7$

$= 72 + 7 = 79$

b. $(g \circ f)(-5) = g(f(-5)) = g(2(-5)^2 + 7)$

$= g(57) = |57 - 1|$

$= 56$

c. $(f \circ g)(x) = f(g(x))$

$= 2|x-1|^2 + 7$

$= 2x^2 - 4x + 2 + 7$

$= 2x^2 - 4x + 9$

d. $(g \circ f)(x) = g(f(x))$

$= |2x^2 + 7 - 1|$

$= |2x^2 + 6|$

$= 2x^2 + 6$

103.a. Enter the data on your calculator. The technique for a TI-83 calculator is illustrated here. Press STAT. [2.7]

```
EDIT CALC TESTS
1:Edit
2:SortA(
3:SortD(
4:ClrList
5:SetUpEditor
```

```
L1    L2    L3   3
0     180
10    163
20    147
30    133
40    118
50    105
L3(1)=
```

```
EDIT CALC TEST
1:1-Var Stats
2:2-Var Stats
3:Med-Med
4:LinReg (ax+b)
5:QuadReg
6:CubicReg
7↓QuartReg
```

```
QuadReg
 y=ax²+bx+c
 a=.0047952048
 b=-1.756843157
 c=180.4065934
 R²=.999144351
```

$h = 0.0047952048t^2 - 1.756843157t + 180.4065934$

b. Empty $\Rightarrow y = 0 \Rightarrow$ the graph intersects the x-axis. Graph the equation, and notice that it never intersects the x-axis.

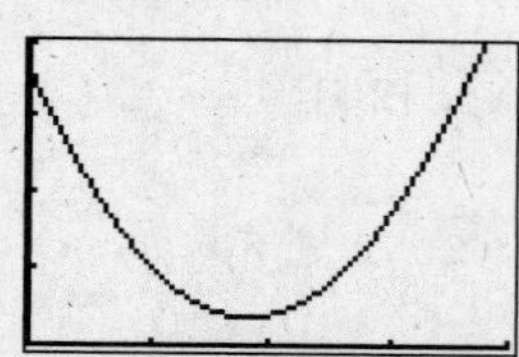

Xmin = 0, Xmax = 400, Xscl = 100

Ymin = 0, Ymax = 200, Xscl = 50

Thus, no, on the basis of this model, the can never empties.

c. The regression line is a model of the data and is not based on physical principles.

Chapter 2 Test

1. Finding the midpoint and length. [2.1]

$\text{midpoint} = \left(\dfrac{x_1 + x_2}{2}, \dfrac{y_1 + y_2}{2}\right)$

$= \left(\dfrac{-2+4}{2}, \dfrac{3+(-1)}{2}\right)$

$= \left(\dfrac{2}{2}, \dfrac{2}{2}\right) = (1, 1)$

$$\text{length} = d = \sqrt{(x_1 - x_2)^2 + (y_1 - y_2)^2}$$
$$= \sqrt{(-2-4)^2 + (3-(-1))^2}$$
$$= \sqrt{(-6)^2 + 4^2} = \sqrt{36+16} = \sqrt{52}$$
$$= 2\sqrt{13}$$

3. Graphing $y = |x+2|+1$. [2.1]

x	y
−4	3
−3	2
−2	1
−1	2
0	3

5. Determining the domain of the function. [2.2]

$$x^2 - 16 \geq 0$$
$$(x-4)(x+4) \geq 0$$

The product is positive or zero.

The critical values are 4 and −4.

+++ |−−−−−−−−|+++

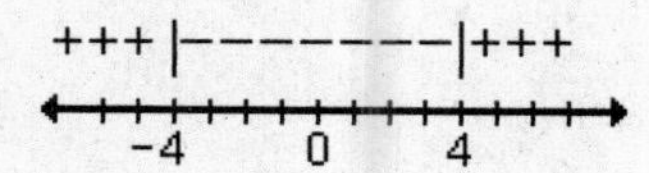

The domain is $\{x | x \geq 4 \text{ or } x \leq -4\}$.

7. Find the slope. [2.3]

$$m = \frac{3-(-2)}{-1-5} = \frac{5}{-6} = -\frac{5}{6}$$

9. Finding the equation in slope-intercept form. [2.3]

$$3x - 2y = 4$$
$$-2y = -3x + 4$$
$$y = \frac{3}{2}x - 2$$

Slope of perpendicular line is $-\frac{2}{3}$.

$$y - y_1 = m(x - x_1)$$
$$y + 2 = -\frac{2}{3}(x-4)$$
$$y + 2 = -\frac{2}{3}x + \frac{8}{3}$$
$$y = -\frac{2}{3}x + \frac{8}{3} - \frac{6}{3}$$
$$y = -\frac{2}{3}x + \frac{2}{3}$$

11. Finding the maximum or minimum value. [2.4]

$$-\frac{b}{2a} = -\frac{-4}{2(1)} = 2$$
$$f(2) = 2^2 - 4(2) - 8 = 4 - 8 - 8$$
$$= -12$$

The minimum value of the function is −12.

13. Identify the type of symmetry. [2.5]

a. $(-y)^2 = x+1$

$y^2 = x+1$ symmetric with respect to x-axis

b. $-y = 2(-x)^3 + 3(-x)$

$y = 2x^3 + 3x$ symmetric with respect to origin

c. $y = 3(-x)^2 - 2$

$y = 3x^2 - 2$ symmetric with respect to y-axis

15. $g(x) = f\left(\frac{1}{2}x\right)$ [2.5]

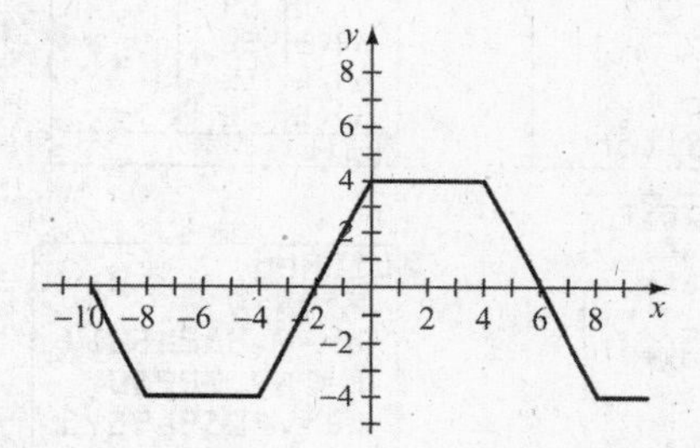

17. $g(x) = f(x-1) + 3$ [2.5]

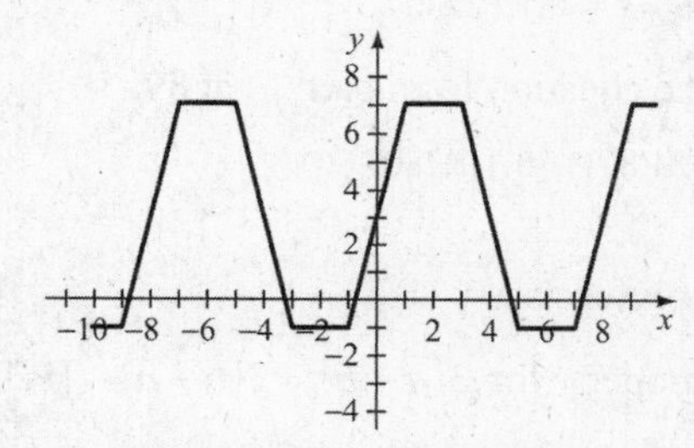

19. Perform the operations. [2.6]

a. $(f-g)(x) = (x^2 - x + 2) - (2x - 1) = x^2 - 3x + 3$

b. $(f \cdot g)(-2) = ((-2)^2 - (-2) + 2)(2(-2) - 1)$

$= (8)(-5) = -40$

c. $(f \circ g)(3) = f(g(3)) = f(2(3) - 1)$

$= f(5) = 5^2 - 5 + 2$

$= 22$

d. $(g \circ f)(x) = g(f(x)) = 2(x^2 - x + 2) - 1$

$= 2x^2 - 2x + 3$

21. Find the maximum area. [2.4]

Using the formula for perimeter for three sides,

$$P = 2w + l \Rightarrow 80 = 2x + y$$
$$y = 80 - 2x$$

Using the formula for area, $A = xy$. Then

$$A(x) = x(80 - 2x)$$
$$A(x) = -2x^2 + 80x$$
$$-\frac{b}{2a} = -\frac{80}{2(-2)} = 20$$
$$y = 80 - 2(20) = 40$$

$x = 20$ ft and $y = 40$ ft

23. a. Enter the data on your calculator. The technique for a TI-83 calculator is illustrated here. Press STAT. [2.7]

```
EDIT CALC TESTS
1:Edit
2:SortA(
3:SortD(
4:ClrList
5:SetUpEditor
```

```
L1    |L2  |L3   3
 93.2 | 28 |
 92.3 | 26 |
 91.9 | 39 |
 89.5 | 56 |
 89.6 | 56 |
 90.5 | 36 |
L3(1)=
```

```
EDIT CALC TEST
1:1-Var Stats
2:2-Var Stats
3:Med-Med
4:LinReg (ax+b)
5:QuadReg
6:CubicReg
7↓QuartReg
```

```
LinReg
 y=ax+b
 a=-7.98245614
 b=767.122807
 r2=.805969575
 r=-.8977580826
```

$y = -7.98245614x + 767.122807$

b. Evaluating the equation from part (a) at 89.

$$y = -7.98245614(89) + 767.122807$$
$$\approx 57 \text{ calories}$$

Cumulative Review Exercises

1. Determine the property for $3(a+b) = 3(b+a)$. [P.1]

Commutative Property of Addition

3. Simplify. [P.1]

$$3 + 4(2x - 9) = 3 + 8x - 36$$
$$= 8x - 33$$

5. Simplify. [P.2]

$$\frac{24a^4b^3}{18a^4b^5} = \frac{4a^{4-4}b^{3-5}}{3} = \frac{4b^{-2}}{3} = \frac{4}{3b^2}$$

7. Simplifying. [P.5]

$$\frac{x^2 + 6x - 27}{x^2 - 9} = \frac{(x+9)(x-3)}{(x+3)(x-3)} = \frac{x+9}{x+3}$$

9. Solving for x. [1.1]

$$6 - 2(2x - 4) = 14$$
$$6 - 4x + 8 = 14$$
$$-4x = 0$$
$$x = 0$$

11. Solving for x. [1.3]

$$(2x - 1)(x + 3) = 4$$
$$2x^2 + 5x - 3 = 4$$
$$2x^2 + 5x - 7 = 0$$
$$(2x + 7)(x - 1) = 0$$
$$x = -\frac{7}{2} \text{ or } x = 1$$

13. Solving for x. [1.4]

$$x^4 - x^2 - 2 = 0$$

Let $u = x^2$.

$$u^2 - u - 2 = 0$$
$$(u - 2)(u + 1) = 0$$
$$u - 2 = 0 \quad \text{or} \quad u + 1 = 0$$
$$u = 2 \qquad u = -1$$
$$x^2 = 2 \qquad x^2 = -1$$
$$x = \pm\sqrt{2} \qquad x = \pm i$$

15. Finding the distance. [2.1]

$$\text{distance} = \sqrt{[-2-2]^2 + [-4-(-3)]^2}$$
$$= \sqrt{(-4)^2 + (-1)^2} = \sqrt{16+1}$$
$$= \sqrt{17}$$

17. Finding the equation of the line. [2.3]

The slope is $m = \frac{-1-(-3)}{-2-2} = \frac{-1+3}{-2-2} = \frac{2}{-4} = -\frac{1}{2}$

The equation is $y - (-3) = -\frac{1}{2}(x - 2)$

$$y = -\frac{1}{2}x - 2$$

19. Evaluating a quadratic function. [2.4]

$$h(x) = -0.002x^2 - 0.03x + 8$$
$$h(39) = -0.002(39)^2 - 0.03(39) + 8 = 3.788 \text{ ft}$$

Yes.

Chapter 3: Polynomial and Rational Functions

Section 3.1 Exercises

1. Use long division to divide.

$$\begin{array}{r} 5x^2 - 9x + 10 \\ x+3 \overline{)\,5x^3 + 6x^2 - 17x + 20} \\ \underline{5x^3 + 15x^2} \\ -9x^2 - 17x \\ \underline{-9x^2 - 27x} \\ 10x + 20 \\ \underline{10x + 30} \\ -10 \end{array}$$

Answer: $5x^2 - 9x + 10 - \dfrac{10}{x+3}$

3. Use long division to divide.

$$\begin{array}{r} x^3 + 2x^2 - x + 1 \\ x-2 \overline{)\,x^4 + 0x^3 - 5x^2 + 3x - 1} \\ \underline{x^4 - 2x^3} \\ 2x^3 - 5x^2 \\ \underline{2x^3 - 4x^2} \\ -x^2 + 3x \\ \underline{-x^2 + 2x} \\ x - 1 \\ \underline{x - 2} \\ 1 \end{array}$$

Answer: $x^3 + 2x^2 - x + 1 + \dfrac{1}{x-2}$

5. Use long division to divide.

$$\begin{array}{r} x^2 + 4x + 10 \\ x-3 \overline{)\,x^3 + x^2 - 2x - 5} \\ \underline{x^3 - 3x^2} \\ 4x^2 - 2x \\ \underline{4x^2 - 12x} \\ 10x - 5 \\ \underline{10x - 30} \\ 25 \end{array}$$

Answer: $x^2 + 4x + 10 + \dfrac{25}{x-3}$

7. Use long division to divide.

$$\begin{array}{r} x^2 + 3x - 2 \\ 2x^2 - x + 1 \overline{)\,2x^4 + 5x^3 - 6x^2 + 4x + 3} \\ \underline{2x^4 - x^3 + x^2} \\ 6x^3 - 7x^2 + 4x \\ \underline{6x^3 - 3x^2 + 3x} \\ -4x^2 + x + 3 \\ \underline{-4x^2 + 2x - 2} \\ -x + 5 \end{array}$$

Answer: $x^2 + 3x - 2 + \dfrac{-x+5}{2x^2 - x + 1}$

9. Use long division to divide.

$$\begin{array}{r} x^3 - x^2 + 2x - 1 \\ 2x^2 + 2x - 3 \overline{)\,2x^5 - x^3 + 5x^2 - 9x + 6} \\ \underline{2x^5 + 2x^4 - 3x^3} \\ -2x^4 + 2x^3 + 5x^2 \\ \underline{-2x^4 - 2x^3 + 3x^2} \\ 4x^3 + 2x^2 - 9x \\ \underline{4x^3 + 4x^2 - 6x} \\ -2x^2 - 3x + 6 \\ \underline{-2x^2 - 2x + 3} \\ -x + 3 \end{array}$$

Answer: $x^3 - x^2 + 2x - 1 + \dfrac{-x+3}{2x^2 + x - 3}$

11. Use synthetic division to divide.

$$\begin{array}{r|rrrr} 2 & 4 & -5 & 6 & -7 \\ & & 8 & 6 & 24 \\ \hline & 4 & 3 & 12 & 17 \end{array}$$

Answer: $4x^2 + 3x + 12 + \dfrac{17}{x-2}$

13. Use synthetic division to divide.

$$\begin{array}{r|rrrr} -1 & 4 & 0 & -2 & 3 \\ & & -4 & 4 & -2 \\ \hline & 4 & -4 & 2 & 1 \end{array}$$

Answer: $4x^2 - 4x + 2 + \dfrac{1}{x+1}$

15. Use synthetic division to divide.

$$\begin{array}{r|rrrrrr} 4 & 1 & 0 & -10 & 0 & 5 & -1 \\ & & 4 & 16 & 24 & 96 & 404 \\ \hline & 1 & 4 & 6 & 24 & 101 & 403 \end{array}$$

Answer: $x^4+4x^3+6x^2+24x+101+\dfrac{403}{x-4}$

17. Use synthetic division to divide.

$$\begin{array}{r|rrrrrr} 1 & 1 & 0 & 0 & 0 & 0 & -1 \\ & & 1 & 1 & 1 & 1 & 1 \\ \hline & 1 & 1 & 1 & 1 & 1 & 0 \end{array}$$

Answer: $x^4+x^3+x^2+x+1$

19. Use synthetic division to divide.

$$\begin{array}{r|rrrr} \frac{1}{2} & 8 & -4 & 6 & -3 \\ & & 4 & 0 & 3 \\ \hline & 8 & 0 & 6 & 0 \end{array}$$

Answer: $8x^2+6$

21. Use synthetic division to divide.

$$\begin{array}{r|rrrrrrrrr} 2 & 1 & 0 & 1 & 0 & 1 & 0 & 1 & 0 & 4 \\ & & 2 & 4 & 10 & 20 & 42 & 84 & 170 & 340 \\ \hline & 1 & 2 & 5 & 10 & 21 & 42 & 85 & 170 & 344 \end{array}$$

Answer: $x^7+2x^6+5x^5+10x^4+21x^3+42x^2$

$+85x+170+\dfrac{344}{x-2}$

23. Use synthetic division to divide.

$$\begin{array}{r|rrrrrrr} -3 & 1 & 0 & 0 & 0 & 0 & 1 & -10 \\ & & -3 & 9 & -27 & 81 & -243 & 726 \\ \hline & 1 & -3 & 9 & -27 & 81 & -242 & 716 \end{array}$$

Answer: $x^5-3x^4+9x^3-27x^2+81x-242+\dfrac{716}{x+3}$

25. Find $P(c)$ using synthetic division and Remainder Theorem.

$$\begin{array}{r|rrrr} 2 & 3 & 1 & 1 & -5 \\ & & 6 & 14 & 30 \\ \hline & 3 & 7 & 15 & 25 \end{array}$$

$P(c)=P(2)=25$

27. Find $P(c)$ using synthetic division and Remainder Theorem.

$$\begin{array}{r|rrrrr} -2 & 4 & 0 & -6 & 0 & 5 \\ & & -8 & 16 & -20 & 40 \\ \hline & 4 & -8 & 10 & -20 & 45 \end{array}$$

$P(c)=P(-2)=45$

29. Find $P(c)$ using synthetic division and Remainder Theorem.

$$\begin{array}{r|rrrr} 10 & -2 & -2 & -1 & -20 \\ & & -20 & -220 & -2210 \\ \hline & -2 & -22 & -221 & -2230 \end{array}$$

$P(c)=P(10)=-2230$

31. Find $P(c)$ using synthetic division and Remainder Theorem.

$$\begin{array}{r|rrrrr} 3 & -1 & 0 & 0 & 0 & 1 \\ & & -3 & -9 & -27 & -81 \\ \hline & -1 & -3 & -9 & -27 & -80 \end{array}$$

$P(c)=P(3)=-80$

33. Find $P(c)$ using synthetic division and Remainder Theorem.

$$\begin{array}{r|rrrrr} 3 & 1 & -10 & 0 & 0 & 2 \\ & & 3 & -21 & -63 & -189 \\ \hline & 1 & -7 & -21 & -63 & -187 \end{array}$$

$P(c)=P(3)=-187$

35. Determine whether the binomial is a factor using synthetic division and the Factor Theorem.

$$\begin{array}{r|rrrr} 2 & 1 & 2 & -5 & -6 \\ & & 2 & 8 & 6 \\ \hline & 1 & 4 & 3 & 0 \end{array}$$

A remainder of 0 implies that $x-2$ is a factor of $P(x)$.

37. Determine whether the binomial is a factor using synthetic division and the Factor Theorem.

$$\begin{array}{r|rrrr} -1 & 2 & 1 & -3 & -1 \\ & & -2 & 1 & 2 \\ \hline & 2 & -1 & -2 & 1 \end{array}$$

A remainder of 1 implies that $x+1$ is not a factor of $P(x)$.

39. Determine whether the binomial is a factor using synthetic division and the Factor Theorem.

$$\begin{array}{r|rrrrr} -3 & 1 & 0 & -25 & 0 & 144 \\ & & -3 & 9 & 48 & -144 \\ \hline & 1 & -3 & -16 & 48 & 0 \end{array}$$

A remainder of 0 implies that $x+3$ is a factor of $P(x)$.

41. Determine whether the binomial is a factor using synthetic division and the Factor Theorem.

$$\begin{array}{r|rrrrrr} 5 & 1 & 2 & -22 & -50 & -75 & 0 \\ & & 5 & 35 & 65 & 75 & 0 \\ \hline & 1 & 7 & 13 & 15 & 0 & 0 \end{array}$$

A remainder of 0 implies that $x - 5$ is a factor of $P(x)$.

43. Determine whether the binomial is a factor using synthetic division and the Factor Theorem.

$$\begin{array}{r|rrrrr} \frac{1}{4} & 16 & -8 & 9 & 14 & 4 \\ & & 4 & -1 & 2 & 4 \\ \hline & 16 & -4 & 8 & 16 & 8 \end{array}$$

A remainder of 8 implies that $x - 1/4$ is not a factor of $P(x)$.

45. Use synthetic division to show c is a zero.

$$\begin{array}{r|rrrr} 2 & 3 & -8 & -10 & 28 \\ & & 6 & -4 & -28 \\ \hline & 3 & -2 & -14 & 0 \end{array}$$

47. Use synthetic division to show c is a zero.

$$\begin{array}{r|rrrrrr} 1 & 1 & 0 & 0 & 0 & 0 & -1 \\ & & 1 & 1 & 1 & 1 & 1 \\ \hline & 1 & 1 & 1 & 1 & 1 & 0 \end{array}$$

49. Use synthetic division to show c is a zero.

$$\begin{array}{r|rrrrr} -2 & 3 & 8 & 10 & 2 & -20 \\ & & -6 & -4 & -12 & 20 \\ \hline & 3 & 2 & 6 & -10 & 0 \end{array}$$

51. Use synthetic division to show c is a zero.

$$\begin{array}{r|rrrr} 11 & 2 & -18 & -50 & 66 \\ & & 22 & 44 & -66 \\ \hline & 2 & 4 & -6 & 0 \end{array}$$

53. Verify the given binomial is a factor.

$$\begin{array}{r|rrrr} 2 & 1 & 1 & 1 & -14 \\ & & 2 & 6 & 14 \\ \hline & 1 & 3 & 7 & 0 \end{array}$$

A remainder of 0 implies that $x - 2$ is a factor of $P(x)$.

$$P(x) = (x-2)(x^2 + 3x + 7)$$

55. Verify the given binomial is a factor.

$$\begin{array}{r|rrrrr} 4 & 1 & -1 & -9 & -11 & -4 \\ & & 4 & 12 & 12 & 4 \\ \hline & 1 & 3 & 3 & 1 & 0 \end{array}$$

A remainder of 0 implies that $x - 4$ is a factor of $P(x)$.

$$P(x) = (x-4)(x^3 + 3x^2 + 3x + 1)$$

57. a. Find how many different ways to select 3 cards.

$$\begin{array}{r|rrrr} 8 & 1 & -3 & 2 & 0 \\ & & 8 & 40 & 336 \\ \hline & 1 & 5 & 42 & 336 \end{array}$$

336 ways

b. Evaluate for $n = 8$ and compare to part a.

$$\begin{aligned} P(8) &= 8^3 - 3(8)^2 + 2(8) \\ &= 512 - 3(64) + 2(8) \\ &= 512 - 192 + 16 \\ &= 336 \text{ ways} \end{aligned}$$

They are the same.

59. a. Find the number of cards needed for 8 rows.

$$\begin{array}{r|rrr} 8 & 1.5 & 0.5 & 0 \\ & & 12 & 100 \\ \hline & 1.5 & 12.5 & 100 \end{array}$$

100 cards

b. Find the number of cards needed for 20 rows.

$$\begin{array}{r|rrr} 20 & 1.5 & 0.5 & 0 \\ & & 30 & 610 \\ \hline & 1.5 & 30.5 & 610 \end{array}$$

610 cards

61. a. Find the number of ways officers can be selected from 12 students.

$$\begin{array}{r|rrrrr} 12 & 1 & -6 & 11 & -6 & 0 \\ & & 12 & 72 & 996 & 11880 \\ \hline & 1 & 6 & 83 & 990 & 11880 \end{array}$$

11,880 ways

b. Find the number of ways officers can be selected from 24 students.

$$\begin{array}{r|rrrrr} 24 & 1 & -6 & 11 & -6 & 0 \\ & & 24 & 432 & 10632 & 255024 \\ \hline & 1 & 18 & 443 & 10626 & 225024 \end{array}$$

255,024 ways

63. a. Find the volume for $x = 6$ in.

$$\begin{array}{r|rrrr} 6 & 1 & 1 & 10 & -8 \\ & & 6 & 42 & 312 \\ \hline & 1 & 7 & 52 & 304 \end{array}$$

304 cubic inches

b. Find the volume for $x = 9$ in.

$$\begin{array}{r|rrrr} 9 & 1 & 1 & 10 & -8 \\ & & 9 & 90 & 900 \\ \hline & 1 & 10 & 100 & 892 \end{array}$$

892 cubic inches

65. Divide each by $x - 1$ and write as the quotient.

$$\begin{array}{r|rrrr} 1 & 1 & 0 & 0 & -1 \\ & & 1 & 1 & 1 \\ \hline & 1 & 1 & 1 & 0 \end{array}$$

So, $(x^3-1)\div(x-1)=x^2+x+1$

$$\begin{array}{r|rrrrrr} 1 & 1 & 0 & 0 & 0 & 0 & -1 \\ & & 1 & 1 & 1 & 1 & 1 \\ \hline & 1 & 1 & 1 & 1 & 1 & 0 \end{array}$$

So, $(x^5-1)\div(x-1)=x^4+x^3+x^2+x+1$

$$\begin{array}{r|rrrrrrrr} 1 & 1 & 0 & 0 & 0 & 0 & 0 & 0 & -1 \\ & & 1 & 1 & 1 & 1 & 1 & 1 & 1 \\ \hline & 1 & 1 & 1 & 1 & 1 & 1 & 1 & 0 \end{array}$$

So, $(x^7-1)\div(x-1)=x^6+x^5+x^4+x^3+x^2+x+1$

$(x^9-1)\div(x-1)$
$=x^8+x^7+x^6+x^5+x^4+x^3+x^2+x+1$

67. Find k.

$$\begin{array}{r|rrrr} 3 & 2 & 1 & -25 & k \\ & & 6 & 21 & -12 \\ \hline & 2 & 7 & -4 & 0 \end{array}$$

$k = 12$

69. Find k.

$$\begin{array}{r|rrrrr} -4 & 1 & 3 & -8 & k & 16 \\ & & -4 & 4 & 16 & -4k-64 \\ \hline & 1 & -1 & -4 & k+16 & 0 \end{array}$$

$$\begin{aligned} 16-4k-64&=0 \\ -4k&=48 \\ k&=-12 \end{aligned}$$

71. Find the remainder by using the Remainder Theorem.

$5(1)^{48}+6(1)^{10}-5(1)+7=5+6-5+7=13$

73. Determine whether i is a zero of the polynomial.

$$\begin{array}{r|rrrr} i & 1 & -3 & 1 & -3 \\ & & i & -1-3i & 3 \\ \hline & 1 & -3+i & -3i & 0 \end{array}$$

A remainder of 0 implies that $x - i$ is a factor of

x^3-3x^2+x-3.

Prepare for Section 3.2

P1. Find the minimum value.

$P(x)=x^2-4x+6$

$-\dfrac{b}{2a}=-\dfrac{-4}{2(1)}=-\dfrac{-4}{2}=2$

$P(2)=(2)^2-4(2)+6=4-8+6=2$

The minimum value is 2.

P3. Find the interval where the function is increasing.

$P(x)=x^2+2x+7$ is a parabola that opens up.

The x-value of the vertex is

$-\dfrac{b}{2a}=-\dfrac{2}{2(1)}=-\dfrac{2}{2}=-1$

The graph decreases from the left until it reaches the vertex, and then it increases.

$P(x)$ increases on the interval $[-1, \infty)$.

P5. Factor.

x^4-5x^2+4

$(x^2-4)(x^2-1)$

$(x+2)(x-2)(x+1)(x-1)$

Section 3.2 Exercises

1. Determine the far-left and far-right behavior.

Since $a_n = 3$ is positive and $n = 4$ is even, the graph of P goes up to its far left and up to its far right.

3. Determine the far-left and far-right behavior.

Since $a_n = 5$ is positive and $n = 5$ is odd, the graph of P goes down to its far left and up to its far right.

5. Determine the far-left and far-right behavior.

$P(x)=-4x^2-3x+2$

Since $a_n = -4$ is negative and $n = 2$ is even, the graph of P goes down to its far left and down to its far right.

7. Determine the far-left and far-right behavior.

$P(x)=\dfrac{1}{2}x^3+\dfrac{5}{2}x^2-1$

Since $a_n=\dfrac{1}{2}$ is positive and $n = 3$ is odd, the graph of

P goes down to its far left and up to its far right.

9. Determine the leading coefficient from the far-left and far-right behavior of the graph.

Up to the far left and down to the far right. $a < 0$.

11. Use a graphing utility to graph, estimate any relative maxima or minima.

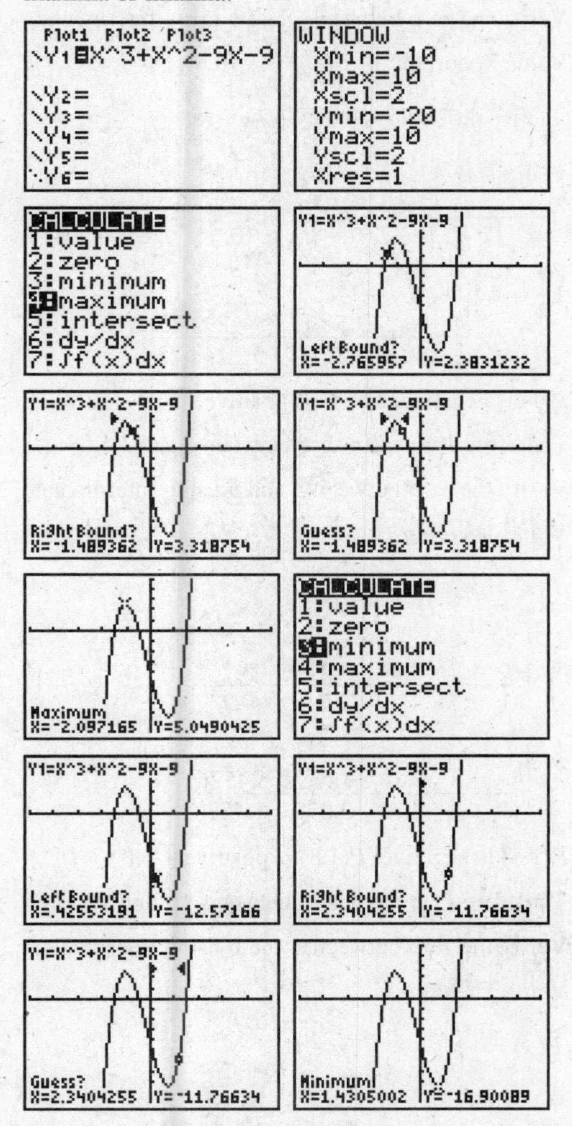

On a TI-83 calculator, the CALC feature is located above the TRACE key.

Relative maximum of $y \approx 5.0$ at $x \approx -2.1$.

Relative minimum of $y \approx -16.9$ at $x \approx 1.4$.

13. Use a graphing utility to graph, estimate any relative maxima or minima.

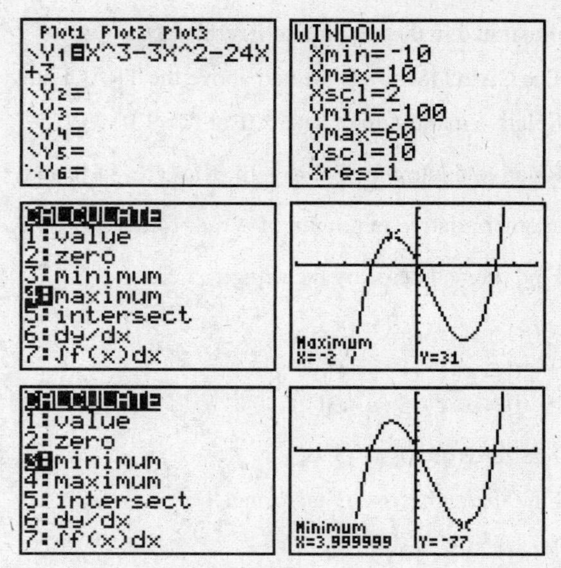

The step-by-step technique for a TI-83 calculator is illustrated in the solution to Exercise **11.**

The CALC feature is located above the TRACE key.

Relative maximum of $y \approx 31.0$ at $x \approx -2.0$.

Relative minimum of $y \approx -77.0$ at $x \approx 4$.

15. Use a graphing utility to graph, estimate any relative maxima or minima.

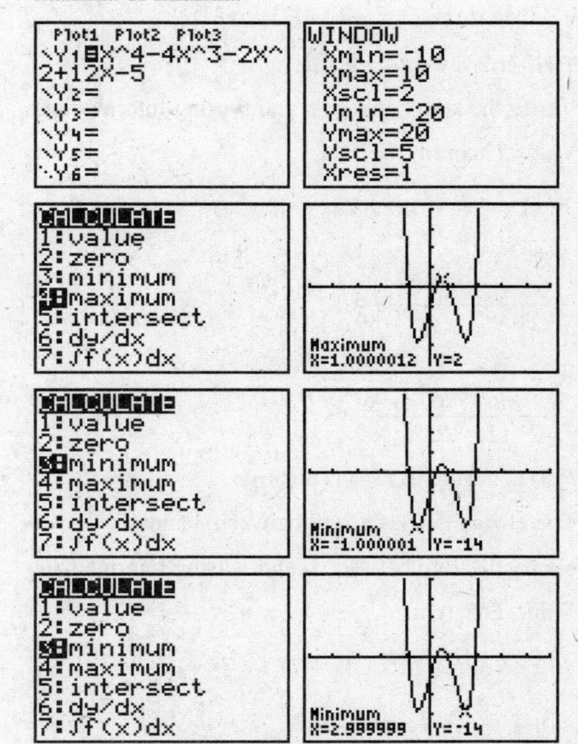

The step-by-step technique for a TI-83 calculator is

illustrated in the solution to Exercise **11.**

The CALC feature is located above the TRACE key.

Relative maximum of $y \approx 2.0$ at $x \approx 1.0$.

Relative minimum of $y \approx -14.0$ at $x \approx -1.0$, and

another relative minimum of $y \approx -14.0$ at $x \approx 3.0$.

17. Find the real zeros by factoring.

$$\begin{aligned} P(x) &= x^3 - 2x^2 - 15x \\ 0 &= x(x^2 - 2x - 15) \\ 0 &= x(x-5)(x+3) \end{aligned}$$

The zeros are 0, 5, –3.

19. Find the real zeros by factoring.

$$\begin{aligned} P(x) &= x^4 - 13x^2 + 36 \\ 0 &= (x^2 - 9)(x^2 - 4) \\ 0 &= (x+3)(x-3)(x+2)(x-2) \end{aligned}$$

The zeros are –3, 3, –2, 2.

21. Find the real zeros by factoring.

$$\begin{aligned} P(x) &= x^5 - 5x^3 + 4x \\ 0 &= x(x^4 - 5x^2 + 4) \\ 0 &= x(x^2 - 4)(x^2 - 1) \\ 0 &= x(x+2)(x-2)(x+1)(x-1) \end{aligned}$$

The zeros are 0, –2, 2; –1, 1.

23. Verify the zero between a and b using Intermediate Value Theorem.

$$P(x) = 2x^3 + 3x^2 - 23x - 42$$

$$\begin{array}{r|rrrr} 3 & 2 & 3 & -23 & -42 \\ & & 6 & 27 & 12 \\ \hline & 2 & 9 & 4 & -30 \end{array}$$

$$\begin{array}{r|rrrr} 4 & 2 & 3 & -23 & -42 \\ & & 8 & 44 & 84 \\ \hline & 2 & 11 & 21 & 42 \end{array}$$

$P(3)$ is negative; $P(4)$ is positive.

Therefore $P(x)$ has a zero between 3 and 4.

25. Verify the zero between a and b using Intermediate Value Theorem.

$$P(x) = 3x^3 + 7x^2 + 3x + 7$$

$$\begin{array}{r|rrrr} -3 & 3 & 7 & 3 & 7 \\ & & -9 & 6 & -27 \\ \hline & 3 & -2 & 9 & -20 \end{array}$$

$$\begin{array}{r|rrrr} -2 & 3 & 7 & 3 & 7 \\ & & -6 & -2 & -2 \\ \hline & 3 & 1 & 1 & 5 \end{array}$$

$P(-3)$ is negative; $P(-2)$ is positive.

Therefore $P(x)$ has a zero between –3 and –2.

27. Verify the zero between a and b using Intermediate Value Theorem.

$$P(x) = 4x^4 + 7x^3 - 11x^2 + 7x - 15$$

$$\begin{array}{r|rrrrr} 1 & 4 & 7 & -11 & 7 & -15 \\ & & 4 & 11 & 0 & 7 \\ \hline & 4 & 11 & 0 & 7 & -8 \end{array}$$

$$\begin{array}{r|rrrrr} 1\frac{1}{2} = 1.5 & 4 & 7 & -11 & 7 & -15 \\ & & 6 & 19.5 & 12.75 & 29.625 \\ \hline & 4 & 13 & 8.5 & 19.75 & 14.625 \end{array}$$

$P(1)$ is negative; $P(1.5)$ is positive.

Therefore $P(x)$ has a zero between 1 and 1.5.

29. Verify the zero between a and b using Intermediate Value Theorem.

$$P(x) = x^4 - x^2 - x - 4$$

$$\begin{array}{r|rrrrr} 1.7 & 1 & 0 & -1 & -1 & -4 \\ & & 1.7 & 2.89 & 3.213 & 3.7621 \\ \hline & 1 & 1.7 & 1.89 & 2.213 & -0.2379 \end{array}$$

$$\begin{array}{r|rrrrr} 1.8 & 1 & 0 & -1 & -1 & -4 \\ & & 1.8 & 3.24 & 4.032 & 5.4576 \\ \hline & 1 & 1.8 & 2.24 & 3.032 & 1.4526 \end{array}$$

$P(1.7)$ is negative; $P(1.8)$ is positive.

Therefore $P(x)$ has a zero between 1.7 and 1.8.

31. Verify the zero between a and b using Intermediate Value Theorem.

$$P(x) = -x^4 + x^3 + 5x - 1$$

$$\begin{array}{r|rrrrr} 0.1 & -1 & 1 & 0 & 5 & -1 \\ & & -0.1 & 0.09 & 0.009 & 0.5009 \\ \hline & -1 & 0.9 & 0.09 & 5.009 & -0.4991 \end{array}$$

$$\begin{array}{r|rrrrr} 0.2 & -1 & 1 & 0 & 5 & -1 \\ & & -0.2 & 0.16 & 0.032 & 1.0064 \\ \hline & -1 & 0.8 & 0.16 & 5.032 & 0.0064 \end{array}$$

$P(0.1)$ is negative; $P(0.2)$ is positive.

Therefore $P(x)$ has a zero between 0.1 and 0.2.

33. Determine x-intercepts. State which cross x-axis.

$$P(x) = (x-1)(x+1)(x-3)$$
$$0 = (x-1)(x+1)(x-3)$$

$x-1=0$ or $x+1=0$ or $x-3=0$

$x=1$ $\quad x=-1$ $\quad x=3$

The exponents on $(x + 1)$, $(x - 1)$, and $(x - 3)$ are odd integers. Therefore the graph of $P(x)$ will cross the x-axis at $(-1, 0)$, $(1, 0)$, and $(3, 0)$.

35. Determine x-intercepts. State which cross x-axis.

$$P(x) = -(x-3)^2(x-7)^5$$
$$0 = -(x-3)^2(x-7)^5$$

$x-3=0$ or $x-7=0$

$x=3$ $\quad x=7$

The exponent on $(x - 7)$ is an odd integer. Therefore the graph of $P(x)$ will cross the x-axis at $(7, 0)$.
The exponent on $(x - 3)$ is an even integer. Therefore the graph of $P(x)$ will intersect but not cross the x-axis at $(3, 0)$.

37. Determine x-intercepts. State which cross x-axis.

$$P(x) = (2x-3)^4(x-1)^{15}$$
$$0 = (2x-3)^4(x-1)^{15}$$

$2x-3=0$ or $x-1=0$

$x=\frac{3}{2}$ $\quad x=1$

The exponent on $(x-1)$ is an odd integer. Therefore the graph of $P(x)$ will cross the x-axis at $(1, 0)$.
The exponent on $(2x - 3)$ is an even integer. Therefore the graph of $P(x)$ will intersect but not cross the x-axis at $\left(\frac{3}{2}, 0\right)$.

39. Determine x-intercepts. State which cross x-axis.

$$P(x) = x^3 - 6x^2 + 9x$$
$$0 = x(x^2 - 6x + 9)$$
$$0 = x(x-3)^2$$

$x=0$ or $x-3=0$

$x=3$

The exponent on x is an odd integer. Therefore the graph of $P(x)$ will cross the x-axis at $(0, 0)$.
The exponent on $(x - 3)$ is an even integer. Therefore the graph of $P(x)$ will intersect but not cross the x-axis at $(3, 0)$.

41. Sketch the graph of the function.
Let $P(x) = 0$.

$$x^3 - x^2 - 2x = 0$$
$$x(x^2 - x - 2) = 0$$
$$x(x-2)(x+1) = 0$$

$x = 0$, $x = 2$, $x = -1$

The graph crosses the x-axis at $(0, 0)$, $(2, 0)$, and $(-1, 0)$.

Let $x = 0$. $P(0) = 0^3 - 0^2 - 2(0) = 0$

The y-intercept is $(0, 0)$.

x^3 has a positive coefficient and an odd exponent. Therefore, the graph goes down to the far left and up to the far right.

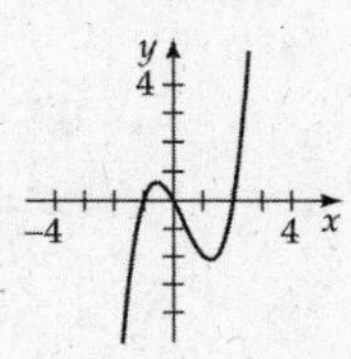

43. Sketch the graph of the function.
Let $P(x) = 0$.

$(x-2)(x+3)(x+1) = 0$

$x = 2$, $x = -3$, $x = -1$

The graph crosses the x-axis at $(2, 0)$, $(-3, 0)$, and $(-1, 0)$.

Let $x = 0$. $P(0) = -(0)^3 - 2(0)^2 + 5(0) + 6 = 6$

The y-intercept is $(0, 6)$.

$-x^3$ has a negative coefficient and an odd exponent. Therefore, the graph goes up to the far left and down to the far right.

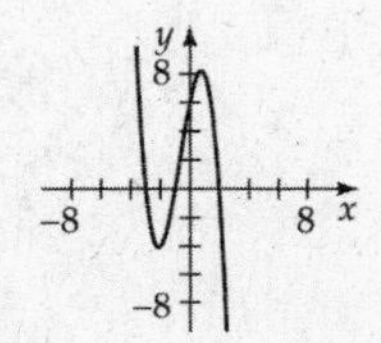

45. Sketch the graph of the function.

Let $P(x)=0$.

$(x-1)^2(x-3)(x+1)=0$

$x=1, x=3, x=-1$

The graph intersects the x-axis but does not cross it at (1, 0).

The graph crosses the x-axis at (3, 0) and (–1, 0).

Let $x=0$.

$P(0)=(0)^4-4(0)^3+2(0)^2+4(0)-3=-3$

The y-intercept is (0, –3).

x^4 has a positive coefficient and an even exponent.

Therefore, the graph goes up to the far left and up to the far right.

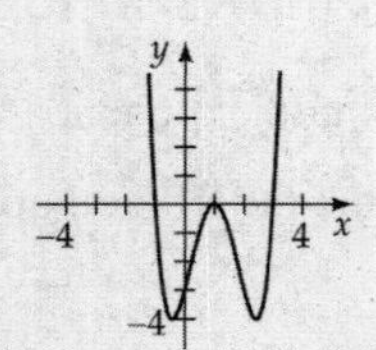

47. Sketch the graph of the function.

Let $P(x)=0$.

$x^3+6x^2+5x-12=0$

$(x-1)(x+3)(x+4)=0$

$x=1, x=-3, x=-4$

The graph crosses the x-axis at (1, 0), (–3, 0), and (–4, 0).

Let $x=0$. $P(0)=(0)^3+6(0)^2+5(0)-12=-12$

The y-intercept is (0, –12).

x^3 has a positive coefficient and an odd exponent.

Therefore, the graph goes down to the far left and up to the far right.

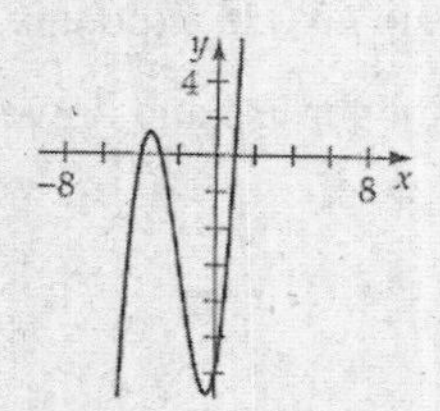

49. Sketch the graph of the function.

Let $P(x)=0$.

$$\begin{array}{r|rrrr} 1 & -1 & 0 & 7 & -6 \\ & & -1 & -1 & 6 \\ \hline & -1 & -1 & 6 & 0 \end{array}$$

$-x^3+7x-6=0$

$(x-1)(-x^2-x+6)=0$

$(x-1)(-x-3)(x-2)=0$

$x=1, x=-3, x=2$

The graph crosses the x-axis at (1, 0), (–3, 0), and (2, 0).

Let $x=0$. $P(0)=-(0)^3+7(0)-6=-6$

The y-intercept is (0, –6).

$-x^3$ has a negative coefficient and an odd exponent.

Therefore, the graph goes up to the far left and down to the far right.

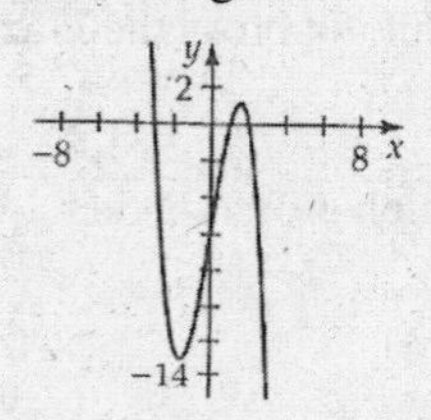

51. Sketch the graph of the function.

Let $P(x)=0$.

$-x^3+4x^2-4x=0$

$-x(x-2)^2=0$

$x=0, x=2$

The graph crosses the x-axis at (0, 0), and (2, 0).

Let $x=0$. $P(0)=-(0)^3+4(0)^2-4(0)=0$

The y-intercept is (0, 0).

$-x^3$ has a negative coefficient and an odd exponent.

Therefore, the graph goes up to the far left and down to the far right.

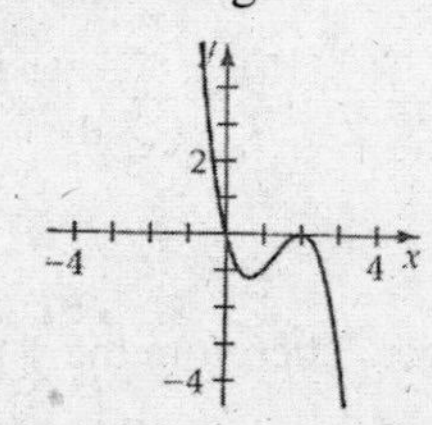

53. Sketch the graph of the function.

Let $P(x) = 0$.

$$-x^4 + 3x^3 + x^2 - 3x = 0$$
$$-x\left[x^3 - 3x^2 - x + 3\right] = 0$$
$$-x\left[x^2(x-3) - 1(x-3)\right] = 0$$
$$-x\left[(x-3)(x^2-1)\right] = 0$$
$$-x(x-3)(x+1)(x-1) = 0$$

$x = 0, x = 3, x = -1, x = 1$

The graph crosses the x-axis at (0, 0), (3, 0), (–1, 0) and (1, 0).

Let $x = 0$.

$P(0) = -(0)^4 + 3(0)^3 + 0^2 - 3(0) = 0$

The y-intercept is (0, 0).

$-x^4$ has a negative coefficient and an even exponent. Therefore, the graph goes down to the far left and down to the far right.

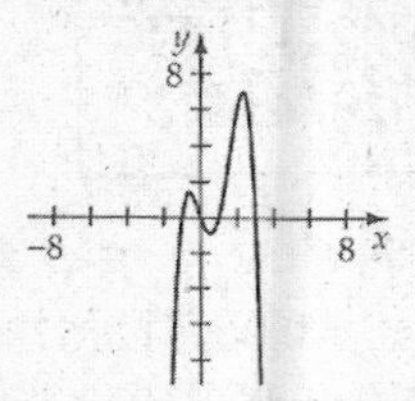

55. Sketch the graph of the function.

Let $P(x) = 0$.

$$x^5 - x^4 - 5x^3 + x^2 + 8x + 4 = 0$$
$$(x+1)^3(x-2)^2 = 0$$

$x = -1, x = 2$

The graph intersects the x-axis but does not cross it at (2, 0).

The graph crosses the x-axis at (–1, 0).

Let $x = 0$.

$P(0) = 0^5 - 0^4 - 5(0)^3 + 0^2 + 8(0) + 4 = 4$

The y-intercept is (0, 4).

x^5 has a positive coefficient and an odd exponent. Therefore, the graph goes down to the far left and up to the far right.

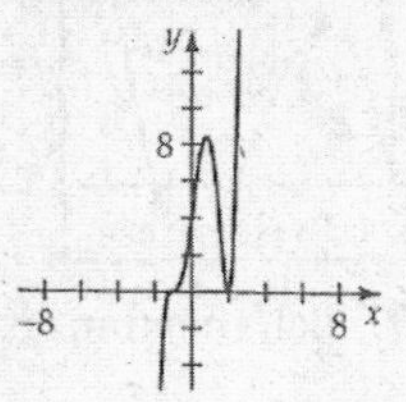

57. Explain how to produce graph P from graph Q.

Shift the graph of P vertically upward 2 units.

59. Explain how to produce graph P from graph Q.

Shift the graph of P horizontally 1 unit to the right.

61. Explain how to produce graph P from graph Q.

Shift the graph of P horizontally 2 units to the right and reflect this graph about the x-axis. Then shift the resulting graph vertically upward 3 units.

63. Use a graphing utility to find the maximum.

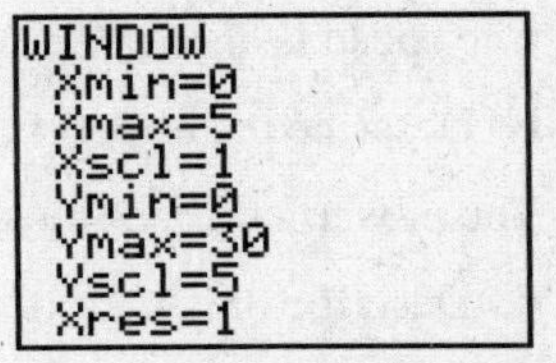

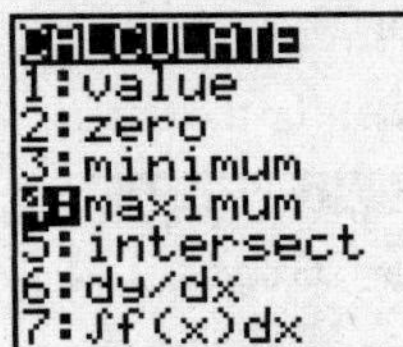

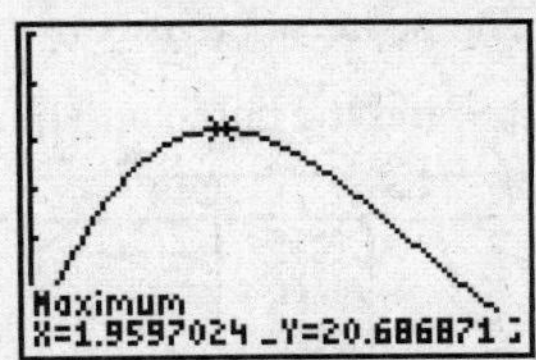

a. 20.69 milligrams

b. 1.968 hours $\times 60$ minutes per hour

≈ 118 minutes after taking the medication

65. a. Express the volume V as a function of x.

Volume = length × width × height

$$V(x) = (15-2x)(10-2x)x$$
$$= [15(10-2x) - 2x(10-2x)]x$$
$$= [150 - 30x - 20x + 4x^2]x$$
$$= [4x^2 - 50x + 150]x$$
$$= 4x^3 - 50x^2 + 150x$$

b. Find the value of x that maximizes the volume.

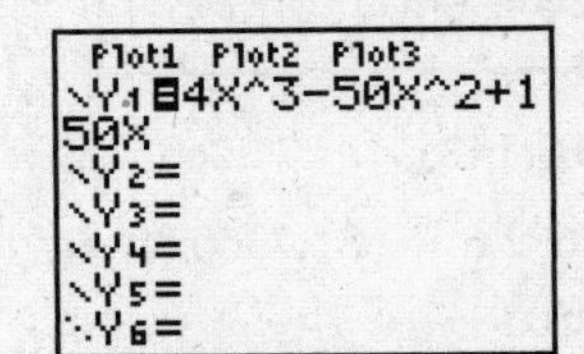

WINDOW
Xmin=-5
Xmax=10
Xscl=1
Ymin=-50
Ymax=200
Yscl=10
Xres=1

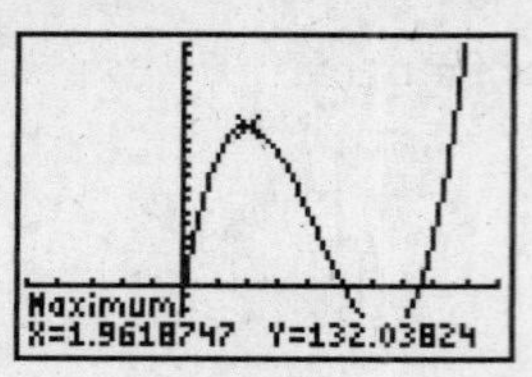

$x = 1.96$ inches (to the nearest 0.01 inch) maximizes the volume of the box.

67. a. Find the power.

$P(v) = 4.95v^3$

$P(8) = 4.95(8)^3 = 2534.4 \approx 2530$

The power is about 2530 W.

b. Find the wind speed.

$$10{,}000 = 4.95v^3$$
$$2020.20202 = v^3$$
$$12.6 \approx v$$

The speed is about 12.6 m/s.

c. Describe the effect if the wind speed is doubled.

The power increases by a factor of 8.

d. Describe the effect if the wind speed is tripled.

The power increases by a factor of 27.

69. a. Find the cubic and quartic model for the data.

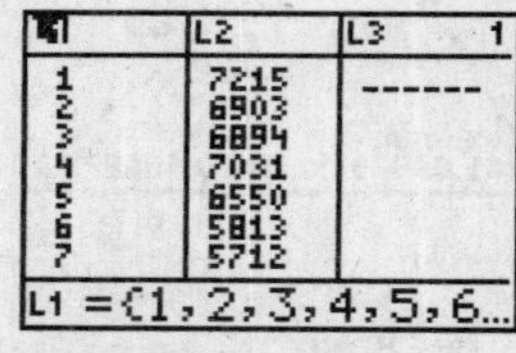

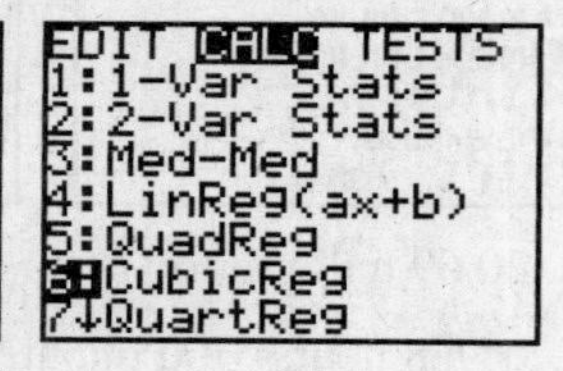

Cubic: $f(x) = 3.325304325x^3 - 51.92834943x^2$
$+13.21021571x + 7228.323232$

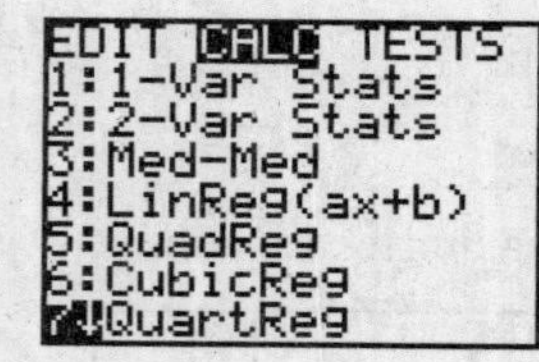

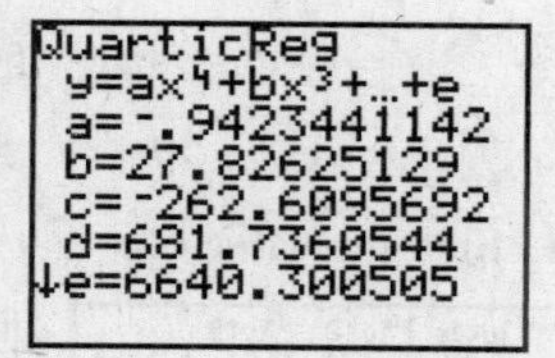

Quartic: $f(x) = -0.9423441142x^4 + 27.82625129x^3$
$-262.6095692x^2 + 681.7360544x$
$+6640.300505$

b. Answers will vary.

Cubic: $f(16) \approx 7766$, Quartic: $f(16) \approx 2539$

Neither model seems to accurately predict the number of sites in 2011 since both numbers are extreme.

71. a. Find the cubic and quartic model for the data.

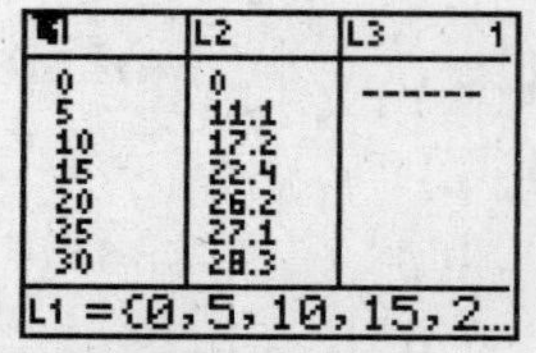

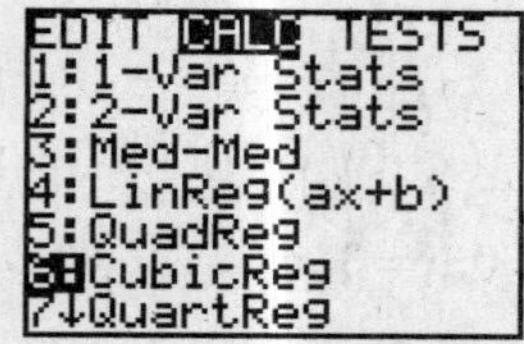

Cubic: $f(x) = 0.00015385409x^3 - 0.0297742717x^2$
$+1.674968246x + 2.136506708$

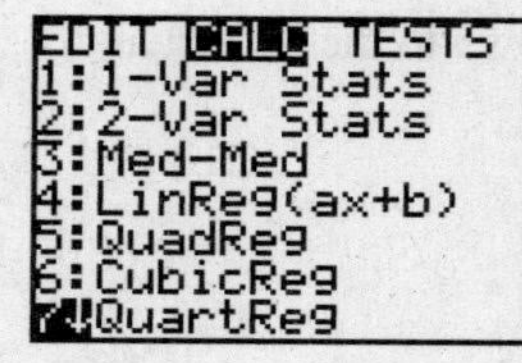

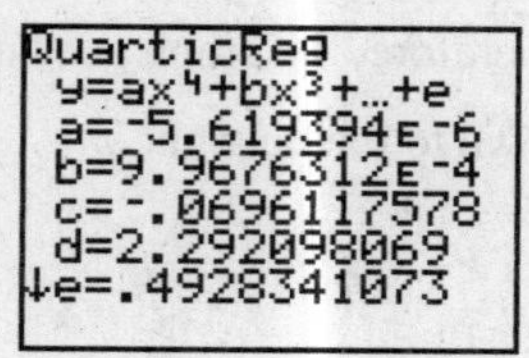

Quartic: $f(x) = -0.000005619394x^4$
$+0.00099676312x^3 - 0.0696117578x^2$
$+2.292098069x + 0.4928341073$

b. Use the cubic and quartic models to predict the fuel efficiency for a car traveling at 80 mph.

Cubic:

$$f(80) = 0.00015385409(80)^3 - 0.0297742717(80)^2$$
$$+1.674968246(80) + 2.136506708$$
$$\approx 24.4 \text{ mpg}$$

Quartic:

$$f(80) = -0.000005619394(80)^4$$
$$+0.00099676312(80)^3 - 0.0696117578(80)^2$$
$$+2.292098069(80) + 0.4928341073$$
$$\approx 18.5 \text{ mpg}$$

c. Determine which value from part b is more realistic.

Answers will vary; however, the downward trend for speeds greater than 50 mph suggests that 18.5 mpg is the more realistic value.

73. $P(x-3)$ shifts the graph horizontally three points to the right.

$(2+3, 0) = (5, 0)$

75. Shift the graph of $y = x^3$ horizontally two units to the right and vertically upward 1 unit.

Prepare for Section 3.3

P1. Find the zeros.

$$P(x) = 6x^2 - 25x + 14$$
$$0 = 6x^2 - 25x + 14$$
$$0 = (3x-2)(2x-7)$$

$$3x - 2 = 0 \quad \text{or} \quad 2x - 7 = 0$$
$$x = \frac{2}{3} \qquad\qquad x = \frac{7}{2}$$

P3. Use synthetic division to divide.

$$\begin{array}{r|rrrrr} 3 & 3 & 0 & -21 & -3 & -5 \\ & & 9 & 27 & 18 & 45 \\ \hline & 3 & 9 & 6 & 15 & 40 \end{array}$$

$$3x^3 + 9x^2 + 6x + 15 + \frac{40}{x-3}$$

P5. List all integer factors of 27.

$\pm1, \pm3, \pm9, \pm27$

Section 3.3 Exercises

1. Find the zeros and state the multiplicity.

$$P(x) = (x-3)^2(x+5)$$

The zeros are:

-5 (multiplicity 1), 3 (multiplicity 2).

3. Find the zeros and state the multiplicity.

$$P(x) = x^2(3x+5)^2$$

The zeros are:

$-\frac{5}{3}$ (multiplicity 2), 0 (multiplicity 2).

5. Find the zeros and state the multiplicity.

$$P(x) = (x^2-4)(x+3)^2$$
$$= (x+2)(x-2)(x+3)^2$$

The zeros are:

-3 (multiplicity 2), -2 (multiplicity 1), 2 (multiplicity 1).

7. Find the zeros and state the multiplicity.

$$P(x) = (3x-5)^2(2x-7)$$

The zeros are:

$\frac{5}{3}$ (multiplicity 2), $\frac{7}{2}$ (multiplicity 1).

9. List possible rational zeros using Rational Zero Theorem.

$$P(x) = x^3 + 3x^2 - 6x - 8$$
$$p = \pm \text{ factors of } 8 = \pm1, \pm2, \pm4, \pm8$$
$$q = \pm \text{ factors of } 1 = \pm1$$
$$\frac{p}{q} = \text{possible rational zeros} = \pm1, \pm2, \pm4, \pm8$$

11. List possible rational zeros using Rational Zero Theorem.

$$P(x) = 2x^3 + x^2 - 25x + 12$$
$$p = \pm \text{ factors of } 12 = \pm1, \pm2, \pm3, \pm4, \pm6, \pm12$$
$$q = \pm \text{ factors of } 2 = \pm1, \pm2$$
$$\frac{p}{q} = \text{possible rational zeros} = \pm1, \pm2, \pm3, \pm4, \pm6, \pm12, \pm\frac{1}{2}, \pm\frac{3}{2}$$

13. List possible rational zeros using Rational Zero Theorem.

$$P(x) = 6x^4 + 23x^3 + 19x^2 - 8x - 4$$
$$p = \pm \text{ factors of } 4 = \pm1, \pm2, \pm4$$
$$q = \pm \text{ factors of } 6 = \pm1, \pm2, \pm3, \pm6$$
$$\frac{p}{q} = \text{possible rational zeros} = \pm1, \pm2, \pm4, \pm\frac{1}{2}, \pm\frac{1}{3}, \pm\frac{1}{6}, \pm\frac{2}{3}, \pm\frac{4}{3}$$

15. List possible rational zeros using Rational Zero Theorem.

$$P(x) = 4x^4 - 12x^3 - 3x^2 + 12x - 7$$
$$p = \pm \text{ factors of } 7 = \pm1, \pm7$$
$$q = \pm \text{ factors of } 4 = \pm1, \pm2, \pm4$$
$$\frac{p}{q} = \text{possible rational zeros} = \pm1, \pm7, \pm\frac{1}{2}, \pm\frac{7}{2}, \pm\frac{1}{4}, \pm\frac{7}{4}$$

17. List possible rational zeros using Rational Zero Theorem.

$P(x) = x^5 - 32$

$p = \pm$ factors of $32 = \pm 1,\ \pm 2,\ \pm 4,\ \pm 8,\ \pm 16,\ \pm 32$

$q = \pm$ factors of $1 = \pm 1$

$\frac{p}{q} =$ possible rational zeros $= \pm 1,\ \pm 2,\ \pm 4,\ \pm 8,\ \pm 16,\ \pm 32$

19. Find the upper and lower bounds.

$$\begin{array}{r|rrrr} 1 & 1 & 3 & -6 & -6 \\ & & 1 & 4 & \\ \hline & 1 & 4 & -2 & \end{array}$$

Don't finish dividing. 1 is not an upper bound.

$$\begin{array}{r|rrrr} 2 & 1 & 3 & -6 & -6 \\ & & 2 & 10 & 8 \\ \hline & 1 & 5 & 4 & 2 \end{array}$$

The smallest integer that is an upper bound is 2.

$$\begin{array}{r|rrrr} -1 & 1 & 3 & -6 & -6 \\ & & -1 & & \\ \hline & 1 & 2 & & \end{array}$$

Don't finish dividing. –1 is not a lower bound.

$$\begin{array}{r|rrrr} -2 & 1 & 3 & -6 & -6 \\ & & -2 & & \\ \hline & 1 & 1 & & \end{array}$$

Don't finish dividing. –2 is not a lower bound.

$$\begin{array}{r|rrrr} -3 & 1 & 3 & -6 & -6 \\ & & -3 & 0 & \\ \hline & 1 & 0 & -6 & \end{array}$$

Don't finish dividing. –3 is not a lower bound.

$$\begin{array}{r|rrrr} -4 & 1 & 3 & -6 & -6 \\ & & -4 & 4 & \\ \hline & 1 & -1 & -2 & \end{array}$$

Don't finish dividing. –4 is not a lower bound.

$$\begin{array}{r|rrrr} -5 & 1 & 3 & -6 & -6 \\ & & -5 & 10 & -20 \\ \hline & 1 & -2 & 4 & -26 \end{array}$$

The largest integer that is a lower bound is –5.

21. Find the upper and lower bounds.

$$\begin{array}{r|rrrr} 3 & 2 & 1 & -25 & 10 \\ & & 6 & 21 & \\ \hline & 2 & 7 & -4 & \end{array}$$

Don't finish dividing. 3 is not an upper bound.

$$\begin{array}{r|rrrr} 4 & 2 & 1 & -25 & 10 \\ & & 8 & 36 & 44 \\ \hline & 2 & 9 & 11 & 54 \end{array}$$

The smallest integer that is an upper bound is 4.

$$\begin{array}{r|rrrr} -3 & 2 & 1 & -25 & 10 \\ & & -6 & 15 & \\ \hline & 2 & -5 & -10 & \end{array}$$

Don't finish dividing. –3 is not a lower bound.

$$\begin{array}{r|rrrr} -4 & 2 & 1 & -25 & 10 \\ & & -8 & 28 & -12 \\ \hline & 2 & -7 & 3 & -2 \end{array}$$

The largest integer that is a lower bound is –4.

23. Find the upper and lower bounds.

$$\begin{array}{r|rrrrr} 1 & 6 & 23 & 19 & -8 & -4 \\ & & 6 & 29 & 48 & 40 \\ \hline & 6 & 29 & 48 & 40 & 36 \end{array}$$

The smallest integer that is an upper bound is 1.

$$\begin{array}{r|rrrrr} -3 & 6 & 23 & 19 & -8 & -4 \\ & & -18 & & & \\ \hline & 6 & 5 & & & \end{array}$$

Don't finish dividing. –3 is not a lower bound.

$$\begin{array}{r|rrrrr} -4 & 6 & 23 & 19 & -8 & -4 \\ & & -24 & 4 & -92 & 400 \\ \hline & 6 & -1 & 23 & -100 & 396 \end{array}$$

The largest integer that is a lower bound is –4.

25. Find the upper and lower bounds.

$$\begin{array}{r|rrrrr} 3 & -4 & 12 & 3 & -12 & 7 \\ & & -12 & 0 & & \\ \hline & -4 & 0 & 3 & & \end{array}$$

Don't finish dividing. 3 is not an upper bound.

$$\begin{array}{r|rrrrr} 4 & -4 & 12 & 3 & -12 & 7 \\ & & -16 & -16 & -52 & -256 \\ \hline & -4 & -4 & -13 & -64 & -249 \end{array}$$

The smallest integer that is an upper bound is 4.

$$\begin{array}{r|rrrrr} -1 & -4 & 12 & 3 & -12 & 7 \\ & & 4 & -16 & 13 & -1 \\ \hline & -4 & 16 & -13 & 1 & 6 \end{array}$$

–1 is not a lower bound.

$$\begin{array}{r|rrrrr} -2 & -4 & 12 & 3 & -12 & 7 \\ & & 8 & -40 & 74 & -124 \\ \hline & -4 & 20 & -37 & 62 & -117 \end{array}$$

The largest integer that is a lower bound is –2.

27. Find the upper and lower bounds.

$$\begin{array}{r|rrrrrr} 1 & 1 & 0 & 0 & 0 & 0 & -32 \\ & & 1 & 1 & 1 & 1 & 1 \\ \hline & 1 & 1 & 1 & 1 & 1 & -31 \end{array}$$

1 is not an upper bound.

$$\begin{array}{r|rrrrrr} 2 & 1 & 0 & 0 & 0 & 0 & -32 \\ & & 2 & 4 & 8 & 16 & 32 \\ \hline & 1 & 2 & 4 & 8 & 16 & 0 \end{array}$$

The smallest integer that is an upper bound is 2.

$$\begin{array}{r|rrrrrr} -1 & 1 & 0 & 0 & 0 & 0 & -32 \\ & & -1 & 1 & -1 & 1 & -1 \\ \hline & 1 & -1 & 1 & -1 & 1 & -33 \end{array}$$

The largest integer that is a lower bound is –1.

29. State the number of positive and negative real zeros using Descartes' Rule of Signs.

$P(x) = x^3 + 3x^2 - 6x - 8$ has 1 change in sign.

Therefore there is one positive zero.

$P(-x) = -x^3 + 3x^2 + 6x - 8$ has 2 changes in sign.

Therefore there are two or no negative zeros.

31. State the number of positive and negative real zeros using Descartes' Rule of Signs.

$P(x) = 2x^3 + x^2 - 25x + 12$ has 2 changes in sign.

Therefore there are two or no positive zeros.

$P(-x) = -2x^3 + x^2 + 25x + 12$ has 1 change in sign.

Therefore there is one negative zero.

33. State the number of positive and negative real zeros using Descartes' Rule of Signs.

$P(x) = 6x^4 + 23x^3 + 19x^2 - 8x - 4$ has 1 change in sign. Therefore there is one positive zero.

$P(-x) = 6x^4 - 23x^3 + 19x^2 + 8x - 4$ has 3 changes in sign. Therefore there are three or one negative zero.

35. State the number of positive and negative real zeros using Descartes' Rule of Signs.

$P(x) = 4x^4 - 12x^3 - 3x^2 + 12x - 7$ has 3 changes in sign. Therefore there are three or one positive zeros.

$P(-x) = 4x^4 + 12x^3 - 3x^2 - 12x - 7$ has 1 change in sign. Therefore there is one negative zero.

37. State the number of positive and negative real zeros using Descartes' Rule of Signs.

$P(x) = x^5 - 32$ has 1 change in sign.

Therefore there is one positive zero.

$P(-x) = -x^5 - 32$ has no changes in sign.

Therefore there are no negative zeros.

39. State the number of positive and negative real zeros using Descartes' Rule of Signs.

$P(x) = 10x^6 - 9x^5 - 14x^4 - 8x^3 - 18x^2 + x + 6$ has 2 changes in sign. Therefore there are two or no positive zeros.

$P(-x) = 10x^6 + 9x^5 - 14x^4 + 8x^3 - 18x^2 - x + 6$ has 4 change in sign. Therefore there are four, two or no negative zeros.

41. State the number of positive and negative real zeros using Descartes' Rule of Signs.

$$P(x) = 12x^7 - 112x^6 + 421x^5 - 840x^4 + 1038x^3 - 938x^2 + 629x - 210$$

has 7 changes in sign. Therefore there are seven, five, three or one positive zeros.

$$P(-x) = -12x^7 - 112x^6 - 421x^5 - 840x^4 - 1038x^3 - 938x^2 - 629x - 210$$

has no changes in sign. Therefore there are no negative zeros.

43. Find the zeros and state multiplicity if more than one.

$P(x) = x^3 + 3x^2 - 6x - 8$

one positive and two or no negative real zeros

$\frac{p}{q} = \pm 1, \pm 2, \pm 4, \pm 8$

$$\begin{array}{r|rrrr} 2 & 1 & 3 & -6 & -8 \\ & & 2 & 10 & 8 \\ \hline & 1 & 5 & 4 & 0 \end{array}$$

$x^2 + 5x + 4 = 0$
$(x+4)(x+1) = 0$
$x = -4, -1$

The zeros of $P(x)$ are 2, –4, and –1.

45. Find the zeros and state multiplicity if more than one.

$P(x) = 2x^3 + x^2 - 25x + 12$ has two or no positive and one negative real zero.

$\frac{p}{q} = \pm 1, \pm 2, \pm 3, \pm 4, \pm 6, \pm 12, \pm \frac{1}{2}, \pm \frac{3}{2}$

$$\begin{array}{r|rrrr} 3 & 2 & 1 & -25 & 12 \\ & & 6 & 21 & -12 \\ \hline & 2 & 7 & -4 & 0 \end{array}$$

$2x^2 + 7x - 4 = 0$
$(2x-1)(x+4) = 0$
$x = \frac{1}{2}, -4$

The zeros of $P(x)$ are 3, $\frac{1}{2}$, –4.

47. Find the zeros and state multiplicity if more than one.

$P(x) = 6x^4 + 23x^3 + 19x^2 - 8x - 4$

one positive and three or one negative real zero

$$\begin{array}{r|rrrrr} -2 & 6 & 23 & 19 & -8 & -4 \\ & & -12 & -22 & 6 & 4 \\ \hline & 6 & 11 & -3 & -2 & 0 \end{array}$$

$$\begin{array}{r|rrrr} -2 & 6 & 11 & -3 & -2 \\ & & -12 & 2 & 2 \\ \hline & 6 & -1 & -1 & 0 \end{array}$$

$6x^2 - x - 1 = 0$
$(3x+1)(2x-1) = 0$
$x = -\frac{1}{3}, \frac{1}{2}$

The zeros of $P(x)$ are –2 (multiplicity 2), $-\frac{1}{3}$, $\frac{1}{2}$.

49. Find the zeros and state multiplicity if more than one.

$P(x) = 2x^4 - 9x^3 - 2x^2 + 27x - 12$

three or one positive and one negative real zero

$\frac{p}{q} = \pm 1, \pm 2, \pm 3, \pm 4, \pm 6, \pm 12, \pm \frac{1}{2}, \pm \frac{3}{2}$

$$\begin{array}{r|rrrrr} 4 & 2 & -9 & -2 & 27 & -12 \\ & & 8 & -4 & -24 & 12 \\ \hline & 2 & -1 & -6 & 3 & 0 \end{array}$$

$$\begin{array}{r|rrrr} \frac{1}{2} & 2 & -1 & -6 & 3 \\ & & 1 & 0 & -3 \\ \hline & 2 & 0 & -6 & 0 \end{array}$$

$2x^2 - 6 = 0$
$2(x^2 - 3) = 0$
$x = \pm\sqrt{3}$

The zeros of $P(x)$ are 4, $\frac{1}{2}$, $\sqrt{3}$, $-\sqrt{3}$.

51. Find the zeros and state multiplicity if more than one.

$P(x) = x^3 - 8x^2 + 8x + 24$

two or no positive and one negative real zero

$\frac{p}{q} = \pm 1, \pm 2, \pm 3, \pm 4, \pm 6, \pm 8, \pm 12, \pm 24$

$$\begin{array}{r|rrrr} 6 & 1 & -8 & 8 & 24 \\ & & 6 & -12 & -24 \\ \hline & 1 & -2 & -4 & 0 \end{array}$$

$x^2 - 2x - 4 = 0$

$$x = \frac{-(-2) \pm \sqrt{(-2)^2 - 4(1)(-4)}}{2(1)}$$
$$= \frac{2 \pm \sqrt{20}}{2} = \frac{2 \pm 2\sqrt{5}}{2} = 1 \pm \sqrt{5}$$

The zeros of $P(x)$ are 6, $1+\sqrt{5}$, $1-\sqrt{5}$.

53. Find the zeros and state multiplicity if more than one.

$P(x) = 2x^4 - 19x^3 + 51x^2 - 31x + 5$

four, two or no positive and no negative real zeros

$\frac{p}{q} = \pm 1, \pm 5, \pm \frac{1}{2}, \pm \frac{5}{2}$

$$\begin{array}{r|rrrrr} 5 & 2 & -19 & 51 & -31 & 5 \\ & & 10 & -45 & 30 & -5 \\ \hline & 2 & -9 & 6 & -1 & 0 \end{array}$$

$$\begin{array}{r|rrrr} \frac{1}{2} & 2 & -9 & 6 & -1 \\ & & 1 & -4 & 1 \\ \hline & 2 & -8 & 2 & 0 \end{array}$$

$2x^2 - 8x + 2 = 2(x^2 - 4x + 1) = 0$

$$x = \frac{-(-4) \pm \sqrt{(-4)^2 - 4(1)(1)}}{2(1)} = \frac{4 \pm \sqrt{12}}{2} = \frac{4 \pm 2\sqrt{3}}{2}$$

$= 2 \pm \sqrt{3}$

The zeros of $P(x)$ are 5, $\frac{1}{2}$, $2+\sqrt{3}$, $2-\sqrt{3}$.

55. Find the zeros and state multiplicity if more than one.

$P(x) = 3x^6 - 10x^5 - 29x^4 + 34x^3 + 50x^2 - 24x - 24$

three or one positive and three or one negative real zeros

$\frac{p}{q} = \pm 1, \pm 2, \pm 3, \pm 4, \pm 6, \pm 8, \pm 12, \pm 24,$
$\pm \frac{1}{3}, \pm \frac{2}{3}, \pm \frac{4}{3}, \pm \frac{3}{2}, \pm \frac{8}{3}$

$$\begin{array}{r|rrrrrrr} 1 & 3 & -10 & -29 & 34 & 50 & -24 & -24 \\ & & 3 & -7 & -36 & -2 & 48 & 24 \\ \hline & 3 & -7 & -36 & -2 & 48 & 24 & 0 \end{array}$$

$$\begin{array}{r|rrrrrr} -1 & 3 & -7 & -36 & -2 & 48 & 24 \\ & & -3 & 10 & 26 & -24 & -24 \\ \hline & 3 & -10 & -26 & 24 & 24 & 0 \end{array}$$

$$\begin{array}{r|rrrrr} -2 & 3 & -10 & -26 & 24 & 24 \\ & & -6 & 32 & -12 & -24 \\ \hline & 3 & -16 & 6 & 12 & 0 \end{array}$$

$$\begin{array}{r|rrrr} -\frac{2}{3} & 3 & -16 & 6 & 12 \\ & & -2 & 12 & -12 \\ \hline & 3 & -18 & 18 & 0 \end{array}$$

$3x^2 - 18x + 18 = 0$

$3(x^2 - 6x + 6) = 0$

$x^2 - 6x + 6 = 0$

$$x = \frac{-(-6) \pm \sqrt{(-6)^2 - 4(1)(6)}}{2(1)} = \frac{6 \pm \sqrt{12}}{2} = \frac{6 \pm 2\sqrt{3}}{2}$$

$= 3 \pm \sqrt{3}$

The zeros of $P(x)$ are 1, –1, –2, $-\frac{2}{3}$, $3+\sqrt{3}$, $3-\sqrt{3}$.

57. Find the zeros and state multiplicity if more than one.

$P(x) = x^3 - 3x - 2$

one positive and two or no negative real zeros

$\frac{p}{q} = \pm 1, \pm 2$

$$\begin{array}{r|rrrr} 2 & 1 & 0 & -3 & -2 \\ & & 2 & 4 & 2 \\ \hline & 1 & 2 & 1 & 0 \end{array}$$

$x^2 + 2x + 1 = 0$

$(x+1)^2 = 0$

$x = -1$

The zeros of $P(x)$ are 2, –1 (multiplicity 2).

59. Find the zeros and state multiplicity if more than one.

$P(x) = x^4 - 5x^2 - 2x = x(x^3 - 5x - 2)$

one positive and two or no negative real zeros

$\frac{p}{q} = \pm 1, \pm 2$

$$\begin{array}{r|rrrr} -2 & 1 & 0 & -5 & -2 \\ & & -2 & 4 & 2 \\ \hline & 1 & -2 & -1 & 0 \end{array}$$

$x^2 - 2x - 1 = 0$

$$x = \frac{-(-2) \pm \sqrt{(-2)^2 - 4(1)(-1)}}{2(1)}$$

$$= \frac{2 \pm \sqrt{8}}{2} = \frac{2 \pm 2\sqrt{2}}{2} = 1 \pm \sqrt{2}$$

The zeros of $P(x)$ are 0, – 2, $1+\sqrt{2}$, $1-\sqrt{2}$.

61. Find the zeros and state multiplicity if more than one.

$P(x) = x^4 + x^3 - 3x^2 - 5x - 2$

one positive and three or one negative real zeros

$\frac{p}{q} = \pm 1, \pm 2$

$$\begin{array}{r|rrrrr} -1 & 1 & 1 & -3 & -5 & -2 \\ & & -1 & 0 & 3 & 2 \\ \hline & 1 & 0 & -3 & -2 & 0 \end{array}$$

$$\begin{array}{r|rrrr} -1 & 1 & 0 & -3 & -2 \\ & & -1 & 1 & 2 \\ \hline & 1 & -1 & -2 & 0 \end{array}$$

$x^2 - x - 2 = 0$

$(x-2)(x+1) = 0$

$x = 2, -1$

The zeros of $P(x)$ are 2, –1 (multiplicity 3).

63. Find the zeros and state multiplicity if more than one.

$P(x) = 2x^4 - 17x^3 + 4x^2 + 35x - 24$

three or one positive and one negative real zeros

$\frac{p}{q} = \pm 1,\ \pm 2,\ \pm 3,\ \pm 4,\ \pm 6,\ \pm 8,\ \pm 12,\ \pm 24,$

$\pm\frac{1}{2},\ \pm\frac{3}{2}$

$$\begin{array}{r|rrrrr} 1 & 2 & -17 & 4 & 35 & -24 \\ & & 2 & -15 & -11 & 24 \\ \hline & 2 & -15 & -11 & 24 & 0 \end{array}$$

$$\begin{array}{r|rrrr} 1 & 2 & -15 & -11 & 24 \\ & & 2 & -13 & -24 \\ \hline & 2 & -13 & -24 & 0 \end{array}$$

$2x^2 - 13x - 24 = 0$

$(2x+3)(x-8) = 0$

$x = -\frac{3}{2},\ 8$

The zeros of $P(x)$ are 1(multiplicity 2), $-\frac{3}{2}$, 8.

65. Find the zeros and state multiplicity if more than one.

$P(x) = x^3 - 16x = x(x^2 - 16)$

one positive and one negative real zeros

$\frac{p}{q} = \pm 1,\ \pm 2,\ \pm 4,\ \pm 8,\ \pm 16$

$x(x^2 - 16) = 0$

$x(x+4)(x-4) = 0$

$x = -4,\ 0,\ 4$

The zeros of $P(x)$ are –4, 0, and 4.

67. Find n.

The original cube's dimensions are $n \times n \times n$.

The resulting solid measures $n \cdot n \cdot (n-2)$.

$n \cdot n \cdot (n-2) = 567$

$n^2(n-2) = 567$

$n^3 - 2n^2 - 567 = 0$

$$\begin{array}{r|rrrr} 9 & 1 & -2 & 0 & -567 \\ & & 9 & 63 & 567 \\ \hline & 1 & 7 & 63 & 0 \end{array}$$

$n = 9$ inches.

69. Find the value of x.

$[(x)(x+1)(x+2)] - [(2)(1)(x)] = 112$

$x(x^2 + 3x + 2) - 2x = 112$

$x^3 + 3x^2 + 2x - 2x = 112$

$x^3 + 3x^2 - 112 = 0$

$$\begin{array}{r|rrrr} 4 & 1 & 3 & 0 & -112 \\ & & 4 & 28 & 112 \\ \hline & 1 & 7 & 28 & 0 \end{array}$$

$x = 4$ inches

71. a. Find the maximum number of pieces.

$P(5) = \frac{5^3 + 5(5) + 6}{6} = \frac{125 + 25 + 6}{6}$

$= \frac{156}{6} = 26$ pieces

b. Find the fewest number of straight cuts.

$\frac{n^3 + 5n + 6}{6} = 64$

$n^3 + 5n + 6 = 384$

$n^3 + 5n - 378 = 0$

$$\begin{array}{r|rrrr} 7 & 1 & 0 & 5 & -378 \\ & & 7 & 49 & 378 \\ \hline & 1 & 7 & 54 & 0 \end{array}$$

At least 7 cuts are needed to produce 64 pieces.

73. Find the number of rows of the pyramid.

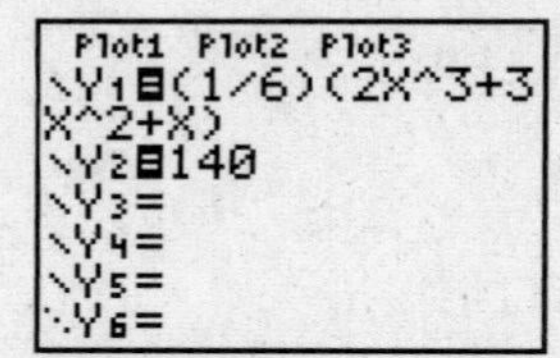

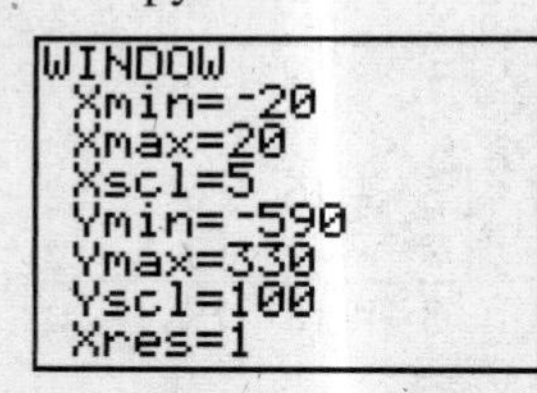

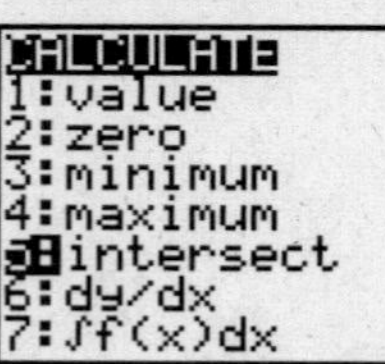

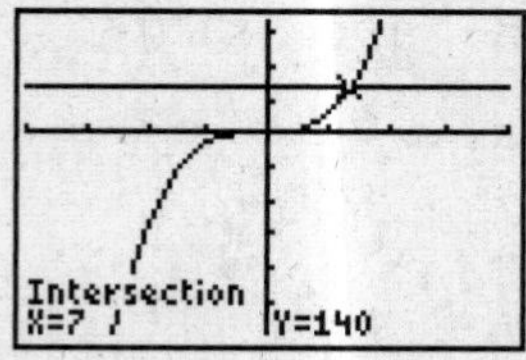

If 140 cannonballs are used, there are 7 rows in the pyramid.

75. Find x.

Plot1 Plot2 Plot3
\Y1=(5-X)(1-X)(.75-X)
\Y2=(5)(.75)(1)-.75
\Y3=
\Y4=
\Y5=

WINDOW
Xmin=-.5
Xmax=.75
Xscl=.25
Ymin=-1
Ymax=5
Yscl=1
Xres=1

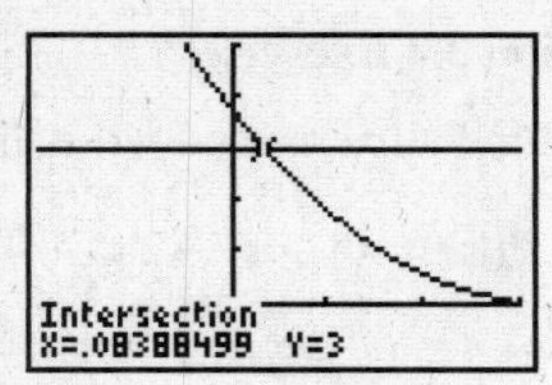

The company should decrease each dimension by 0.084 inch.

77. Find the possible lengths of l.

$$4w+l=81$$
$$l=81-4w$$

Volume = $w^2l=4900$

$$w^2(81-4w)=4900$$

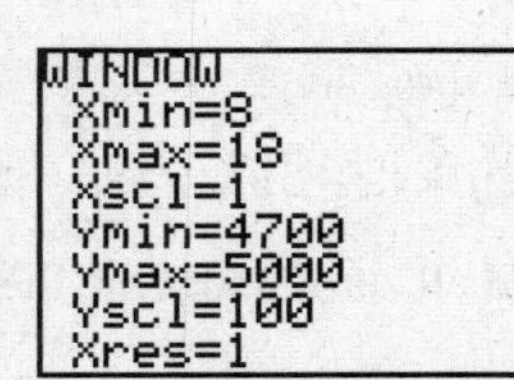

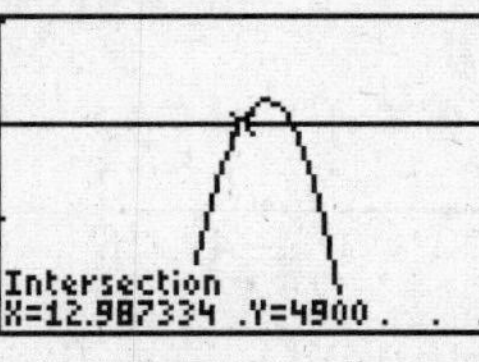

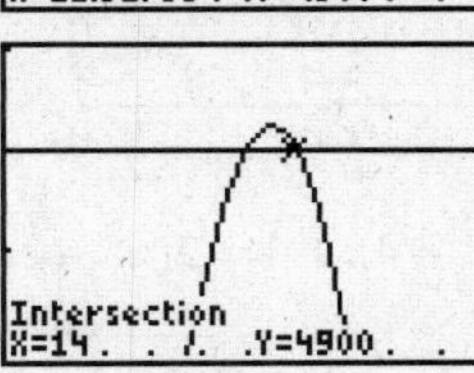

When w = 12.9875, then

$l=81-4w=81-4(12.9875)=29.05$ in.

When w = 14, then $l=81-4w=81-4(14)=25$ in.

Thus, the lengths can be 25 in. or 29.05 in.

79. Find the giraffe's height, rounding to the nearest tenth of a foot.

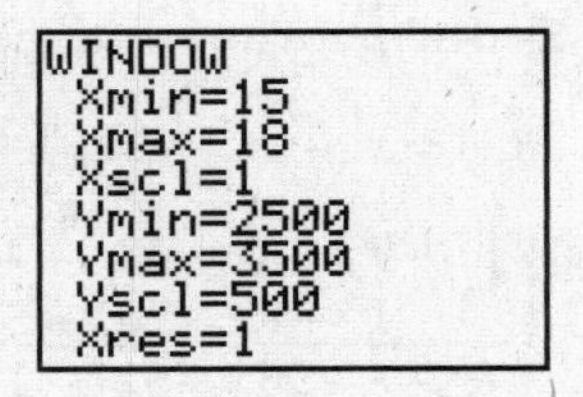

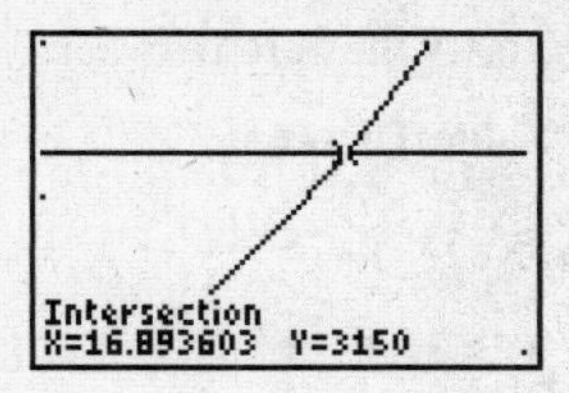

The giraffe is 16.9 feet tall.

81. Find B and verify the absolute value is less than B.

$$B=\left(\frac{\text{max of }(|-5|,|-28|,|15|)}{|2|}+1\right)$$

$$B=\left(\frac{28}{2}+1\right)$$

$$B=15$$

$$|-3|=3<15$$

$$\left|\frac{1}{2}\right|=\frac{1}{2}<15$$

$$|5|=5<15$$

The absolute value of each zero is less than B.

83. Find B and verify the absolute value is less than B.

$$B=\left(\frac{\text{max of }(|-2|,|9|,|2|,|-10|)}{|1|}+1\right)$$

$$B=\left(\frac{10}{1}+1\right)$$

$$B=11$$

$$|1+3i|=\sqrt{10}<11$$

$$|1-3i|=\sqrt{10}<11$$

$$|1|=1<11$$

$$|-1|=1<11$$

The absolute value of each zero is less than B.

Mid-Chapter 3 Quiz

1. Use the Remainder Theorem to find $P(5)$.

$$\begin{array}{r|rrrr} 5 & 3 & 7 & -2 & -5 \\ & & 15 & 110 & 540 \\ \hline & 3 & 22 & 108 & 535 \end{array}$$

$$P(5)=535$$

3. Determine the far-right and far-left behavior.

Since $a_n=4$ is positive and n = 3 is odd, the graph of P goes down to the far left and up to the far right.

5. Verify the zero between 3 and 4 using Intermediate Value Theorem.

$P(x) = 2x^4 - 5x^3 + x^2 - 20x - 28$

$$\begin{array}{r|rrrrr} 3 & 2 & -5 & 1 & -20 & -28 \\ & & 6 & 3 & 12 & -24 \\ \hline & 2 & 1 & 4 & -8 & -52 \end{array}$$

$$\begin{array}{r|rrrrr} 4 & 2 & -5 & 1 & -20 & -28 \\ & & 8 & 12 & 52 & 128 \\ \hline & 2 & 3 & 13 & 32 & 100 \end{array}$$

$P(3)$ is negative; $P(4)$ is positive.

Therefore $P(x)$ has a zero between 3 and 4.

7. Find the relative maximum. [3.2]

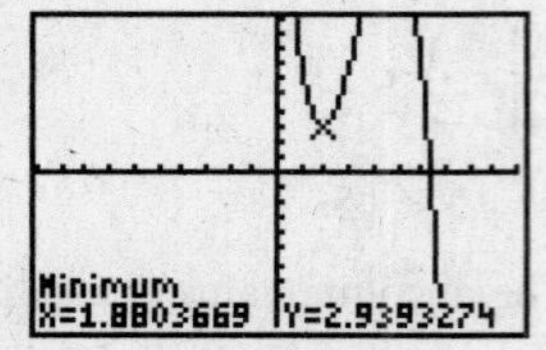

The relative minimum is $y \approx 2.94$ at $x \approx 1.88$.

Prepare for Section 3.4

P1. Find the conjugate of $3 - 2i$.

$3 + 2i$

P3. Multiply.

$$\begin{aligned} &(x-1)(x-3)(x-4) \\ &= (x-1)(x^2 - 7x + 12) \\ &= x(x^2 - 7x + 12) - 1(x^2 - 7x + 12) \\ &= x^3 - 7x^2 + 12x - x^2 + 7x - 12 \\ &= x^3 - 8x^2 + 19x - 12 \end{aligned}$$

P5. Solve.

$$\begin{aligned} x^2 + 9 &= 0 \\ x^2 &= -9 \\ x &= \pm\sqrt{-9} \\ x &= \pm 3i \end{aligned}$$

The solutions are $3i$ and $-3i$.

Section 3.4 Exercises

1. Find all zeros and write the polynomial as a product.

$P(x) = x^4 + x^3 - 2x^2 + 4x - 24$

$$\begin{array}{r|rrrrr} 2 & 1 & 1 & -2 & 4 & -24 \\ & & 2 & 6 & 8 & 24 \\ \hline & 1 & 3 & 4 & 12 & 0 \end{array}$$

$$\begin{array}{r|rrrr} -3 & 1 & 3 & 4 & 12 \\ & & -3 & 0 & -12 \\ \hline & 1 & 0 & 4 & 0 \end{array}$$

$$\begin{aligned} x^2 + 4 &= 0 \\ x^2 &= -4 \\ x &= \pm\sqrt{-4} = \pm 2i \end{aligned}$$

The zeros are $2,\ -3,\ 2i,\ -2i$.

$P(x) = (x-2)(x+3)(x-2i)(x+2i)$

3. Find all zeros and write the polynomial as a product.

$P(x) = 2x^4 - x^3 - 4x^2 + 10x - 4$

$$\begin{array}{r|rrrrr} \frac{1}{2} & 2 & -1 & -4 & 10 & -4 \\ & & 1 & 0 & -2 & 4 \\ \hline & 2 & 0 & -4 & 8 & 0 \end{array}$$

$$\begin{array}{r|rrrr} -2 & 2 & 0 & -4 & 8 \\ & & -4 & 8 & -8 \\ \hline & 2 & -4 & 4 & 0 \end{array}$$

$2x^2 - 4x + 4 = 2(x^2 - 2x + 2) = 0$

$$x = \frac{-(-2) \pm \sqrt{(-2)^2 - 4(1)(2)}}{2(1)}$$

$$= \frac{2 \pm \sqrt{-4}}{2} = \frac{2 \pm 2i}{2} = 1 \pm i$$

The zeros are $\frac{1}{2},\ -2,\ 1+i,\ 1-i$.

$P(x) = 2\left(x - \frac{1}{2}\right)(x+2)(x-1-i)(x-1+i)$

5. Find all zeros and write the polynomial as a product.

$P(x) = x^5 - 9x^4 + 34x^3 - 58x^2 + 45x - 13$

$$\begin{array}{r|rrrrrr} 1 & 1 & -9 & 34 & -58 & 45 & -13 \\ & & 1 & -8 & 26 & -32 & 13 \\ \hline & 1 & -8 & 26 & -32 & 13 & 0 \end{array}$$

$$\begin{array}{r|rrrrr} 1 & 1 & -8 & 26 & -32 & 13 \\ & & 1 & -7 & 19 & -13 \\ \hline & 1 & -7 & 19 & -13 & 0 \end{array}$$

$$\begin{array}{r|rrrr} 1 & 1 & -7 & 19 & -13 \\ & & 1 & -6 & 13 \\ \hline & 1 & -6 & 13 & 0 \end{array}$$

$x^2-6x+13=0$

$$x=\frac{-(-6)\pm\sqrt{(-6)^2-4(1)(13)}}{2(1)}=\frac{6\pm\sqrt{36-52}}{2}$$
$$=\frac{6\pm\sqrt{-16}}{2}=\frac{6\pm 4i}{2}=3\pm 2i$$

The zeros are 1 (multiplicity 3), $3+2i$, $3-2i$.

$P(x)=(x-1)^3(x-3-2i)(x-3+2i)$

7. Find all zeros and write the polynomial as a product.

$P(x)=2x^4-x^3-15x^2+23x+15$

$$\begin{array}{r|rrrrr} -3 & 2 & -1 & -15 & 23 & 15 \\ & & -6 & 21 & -18 & -15 \\ \hline & 2 & -7 & 6 & 5 & 0 \end{array}$$

$$\begin{array}{r|rrrr} -\frac{1}{2} & 2 & -7 & 6 & 5 \\ & & -1 & 4 & -5 \\ \hline & 2 & -8 & 10 & 0 \end{array}$$

$2x^2-8x+10=2(x^2-4x+5)=0$

$$x=\frac{-(-4)\pm\sqrt{(-4)^2-4(1)(5)}}{2(1)}$$
$$=\frac{4\pm\sqrt{-4}}{2}=\frac{4\pm 2i}{2}=2\pm i$$

The zeros are -3, $-\frac{1}{2}$, $2+i$, $2-i$.

$P(x)=2(x+3)\left(x+\frac{1}{2}\right)(x-2-i)(x-2+i)$

9. Find all zeros and write the polynomial as a product.

$P(x)=2x^4-14x^3+33x^2-46x+40$

$$\begin{array}{r|rrrrr} 4 & 2 & -14 & 33 & -46 & 40 \\ & & 8 & -24 & 36 & -40 \\ \hline & 2 & -6 & 9 & -10 & 0 \end{array}$$

$$\begin{array}{r|rrrr} 2 & 2 & -6 & 9 & -10 \\ & & 4 & -4 & 10 \\ \hline & 2 & -2 & 5 & 0 \end{array}$$

$2x^2-2x+5=0$

$$x=\frac{-(-2)\pm\sqrt{(-2)^2-4(2)(5)}}{2(2)}$$
$$=\frac{2\pm\sqrt{-36}}{4}=\frac{2\pm 6i}{4}=\frac{1}{2}\pm\frac{3}{2}i$$

The zeros are 4, 2, $\frac{1}{2}+\frac{3}{2}i$, $\frac{1}{2}-\frac{3}{2}i$.

$P(x)=(x-4)(x-2)\left(x-\frac{1}{2}-\frac{3}{2}i\right)\left(x-\frac{1}{2}+\frac{3}{2}i\right)$

11. Find all zeros and write the polynomial as a product.

$P(x)=2x^3-9x^2+18x-20$

$$\begin{array}{r|rrrr} \frac{5}{2} & 2 & -9 & 18 & -20 \\ & & 5 & -10 & 20 \\ \hline & 2 & -4 & 8 & 0 \end{array}$$

$2x^2-4x+8=2(x^2-2x+4)=0$

$$x=\frac{-(-2)\pm\sqrt{(-2)^2-4(1)(4)}}{2(1)}$$
$$=\frac{2\pm\sqrt{-12}}{2}=\frac{2\pm 2i\sqrt{3}}{2}=1\pm i\sqrt{3}$$

The zeros are $\frac{5}{2}$, $1+i\sqrt{3}$, $1-i\sqrt{3}$.

$P(x)=2\left(x-\frac{5}{2}\right)(x-1-i\sqrt{3})(x-1+i\sqrt{3})$

13. Find all zeros and write the polynomial as a product.

$P(x)=2x^4-x^3-2x^2+13x-6$

$$\begin{array}{r|rrrrr} -2 & 2 & -1 & -2 & 13 & -6 \\ & & -4 & 10 & -16 & 6 \\ \hline & 2 & -5 & 8 & -3 & 0 \end{array}$$

$$\begin{array}{r|rrrr} \frac{1}{2} & 2 & -5 & 8 & -3 \\ & & 1 & -2 & 3 \\ \hline & 2 & -4 & 6 & 0 \end{array}$$

$2x^2-4x+6=2(x^2-2x+3)=0$

$$x=\frac{-(-2)\pm\sqrt{(-2)^2-4(1)(3)}}{2(1)}$$
$$=\frac{2\pm\sqrt{-8}}{2}=\frac{2\pm 2i\sqrt{2}}{2}=1\pm i\sqrt{2}$$

The zeros are -2, $\frac{1}{2}$, $1+i\sqrt{2}$, $1-i\sqrt{2}$.

$P(x)=2(x+2)\left(x-\frac{1}{2}\right)(x-1-i\sqrt{2})(x-1+i\sqrt{2})$

15. Find all zeros and write the polynomial as a product.

$P(x) = 3x^5 + 2x^4 + 10x^3 + 6x^2 - 25x - 20$

$$\begin{array}{r|rrrrrr} -1 & 3 & 2 & 10 & 6 & -25 & -20 \\ & & -3 & 1 & -11 & 5 & 20 \\ \hline & 3 & -1 & 11 & -5 & -20 & 0 \end{array}$$

$$\begin{array}{r|rrrrr} -1 & 3 & -1 & 11 & -5 & -20 \\ & & -3 & 4 & -15 & 20 \\ \hline & 3 & -4 & 15 & -20 & 0 \end{array}$$

$$\begin{array}{r|rrrr} \frac{4}{3} & 3 & -4 & 15 & -20 \\ & & 4 & 0 & 20 \\ \hline & 3 & 0 & 15 & 0 \end{array}$$

$3x^2 + 15 = 0$

$3(x^2 + 5) = 0$

$x^2 + 5 = 0$

$x^2 = -5$

$x = \pm\sqrt{-5} = \pm\, i\sqrt{5}$

The zeros are -1, (multiplicity 2), $\frac{4}{3}$, $i\sqrt{5}$, $-i\sqrt{5}$.

$P(x) = 3(x+1)^2\left(x - \frac{4}{3}\right)(x - i\sqrt{5})(x + i\sqrt{5})$

17. Find the remaining zeros.

$$\begin{array}{r|rrrr} 1+i & 2 & -5 & 6 & -2 \\ & & 2+2i & -5-i & 2 \\ \hline & 2 & -3+2i & 1-i & 0 \end{array}$$

$$\begin{array}{r|rrr} 1-i & 2 & -3+2i & 1-i \\ & & 2-2i & -1+i \\ \hline & 2 & -1 & 0 \end{array}$$

$2x - 1 = 0$

$x = \frac{1}{2}$

The remaining zeros are $1 - i$, $\frac{1}{2}$.

19. Find the remaining zeros.

$$\begin{array}{r|rrrr} -i & 1 & 3 & 1 & 3 \\ & & -i & -1-3i & -3 \\ \hline & 1 & 3-i & -3i & 0 \end{array}$$

$$\begin{array}{r|rrr} i & 1 & 3-i & -3i \\ & & i & 3i \\ \hline & 1 & 3 & 0 \end{array}$$

$x + 3 = 0$

$x = -3$

The remaining zeros are i, -3.

21. Find the remaining zeros.

$$\begin{array}{r|rrrrr} 2-3i & 1 & -4 & 14 & -4 & 13 \\ & & 2-3i & -13 & 2-3i & -13 \\ \hline & 1 & -2-3i & 1 & -2-3i & 0 \end{array}$$

$$\begin{array}{r|rrrr} 2+3i & 1 & -2-3i & 1 & -2-3i \\ & & 2+3i & 0 & 2+3i \\ \hline & 1 & -2-3i & 1 & 0 \end{array}$$

$x^2 + 1 = 0$

$x^2 = -1$

$x = \pm i$

The remaining zeros are $2 + 3i$, i, $-i$.

23. Find the remaining zeros.

$$\begin{array}{r|rrrrr} 1+3i & 1 & -4 & 19 & -30 & 50 \\ & & 1+3i & -12-6i & 25+15i & -50 \\ \hline & 1 & -3+3i & 7-6i & -5+15i & 0 \end{array}$$

$$\begin{array}{r|rrrr} 1-3i & 1 & -3+3i & 7-6i & -5+15i \\ & & 1-3i & -2+6i & 5-15i \\ \hline & 1 & -2 & 5 & 0 \end{array}$$

$x^2 - 2x + 5 = 0$

$$x = \frac{-(-2) \pm \sqrt{(-2)^2 - 4(1)(5)}}{2(1)} = \frac{2 \pm \sqrt{-16}}{2} = \frac{2 \pm 4i}{2}$$

$= 1 \pm 2i$

The remaining zeros are $1 - 3i$, $1 + 2i$, $1 - 2i$.

25. Find the remaining zeros.

$$\begin{array}{r|rrrrrr} -2i & 1 & -3 & 7 & -13 & 12 & -4 \\ & & -2i & -4+6i & 12-6i & -12+2i & 4 \\ \hline & 1 & -3-2i & 3+6i & -1-6i & 2i & 0 \end{array}$$

$$\begin{array}{r|rrrrr} 2i & 1 & -3-2i & 3+6i & -1-6i & 2i \\ & & 2i & -6i & 6i & -2i \\ \hline & 1 & -3 & 3 & -1 & 0 \end{array}$$

$$\begin{array}{r|rrrr} 1 & 1 & -3 & 3 & -1 \\ & & 1 & -2 & 1 \\ \hline & 1 & -2 & 1 & 0 \end{array}$$

$x^2 - 2x + 1 = 0$

$(x-1)^2 = 0$

$x = 1$

The remaining zeros are $2i$, 1 (multiplicity 3).

27. Find the remaining zeros.

$$\begin{array}{r|rrrrr} 5+2i & 1 & -17 & 112 & -333 & 377 \\ & & 5+2i & -64-14i & 268+26i & -377 \\ \hline & 1 & -12+2i & 48-14i & -65+26i & 0 \end{array}$$

$$\begin{array}{r|rrrr} 5-2i & 1 & -12+2i & 48-14i & -65+26i \\ & & 5-2i & -35+14i & 65-26i \\ \hline & 1 & -7 & 13 & 0 \end{array}$$

$x^2-7x+13=0$

$$x=\frac{-(-7)\pm\sqrt{(-7)^2-4(1)(13)}}{2(1)}=\frac{7\pm\sqrt{-3}}{2}=\frac{7\pm i\sqrt{3}}{2}$$

$$=\frac{7}{2}\pm\frac{\sqrt{3}}{2}i$$

The remaining zeros are $5-2i,\ \frac{7}{2}+\frac{\sqrt{3}}{2}i,\ \frac{7}{2}-\frac{\sqrt{3}}{2}i.$

29. Determine the exact values of the solutions.

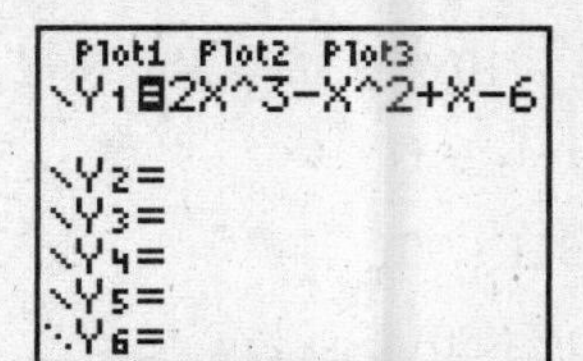

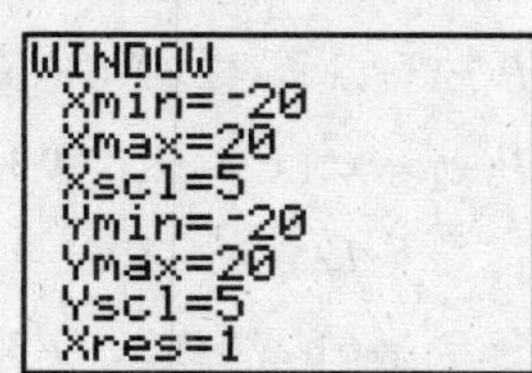

CALCULATE
1:value
2:zero
3:minimum
4:maximum
5:intersect
6:dy/dx
7:∫f(x)dx

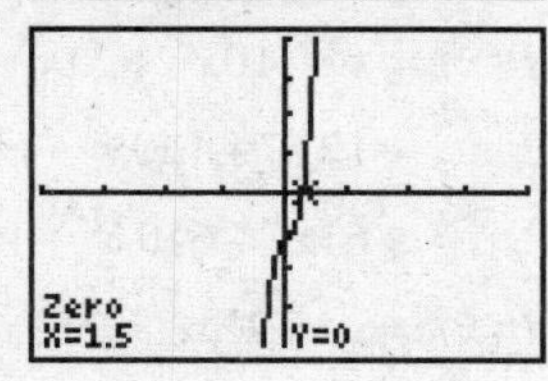

$$\begin{array}{r|rrrr} 1.5 & 2 & -1 & 1 & -6 \\ & & 3 & 3 & 6 \\ \hline & 2 & 2 & 4 & 0 \end{array}$$

$2x^2+2x+4=0$

$2(x^2+x+2)=0$

$$x=\frac{-1\pm\sqrt{(1)^2-4(1)(2)}}{2(1)}=\frac{-1\pm\sqrt{-7}}{2}=\frac{-1\pm i\sqrt{7}}{2}$$

$$=-\frac{1}{2}\pm\frac{\sqrt{7}}{2}i$$

The solutions are $1.5,\ -\frac{1}{2}+\frac{\sqrt{7}}{2}i,\ -\frac{1}{2}-\frac{\sqrt{7}}{2}i.$

31. Determine the exact values of the solutions.

Plot1 Plot2 Plot3
\Y1=24X^3-62X^2-7X+30
\Y2=
\Y3=
\Y4=
\Y5=
\Y6=

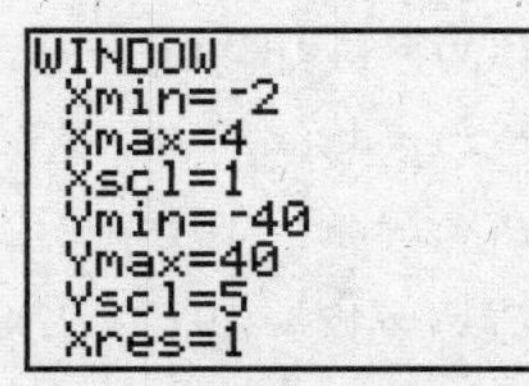

CALCULATE
1:value
2:zero
3:minimum
4:maximum
5:intersect
6:dy/dx
7:∫f(x)dx

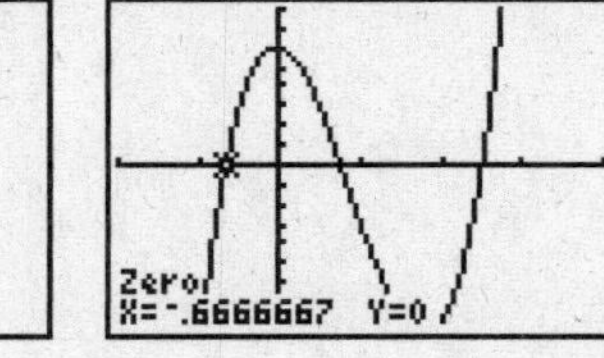

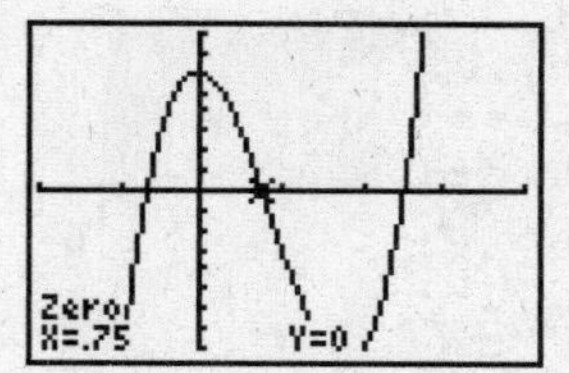

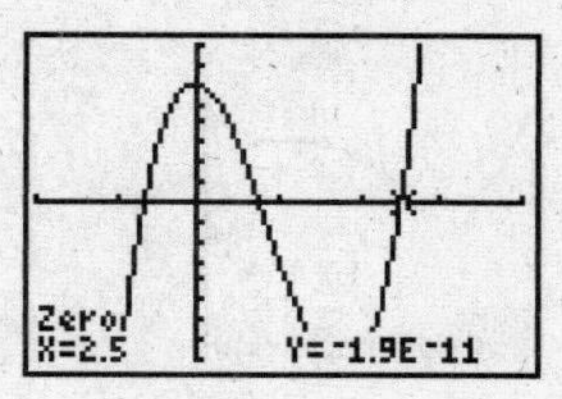

The solutions are $-0.\overline{6},\ 0.75,\ 2.5$ or $-\frac{2}{3},\ \frac{3}{4},\ \frac{5}{2}.$

33. Determine the exact values of the solutions.

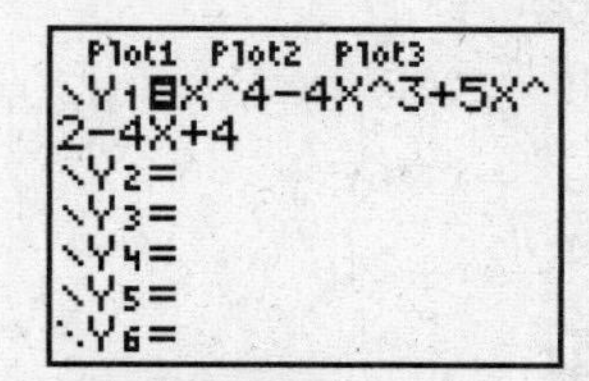

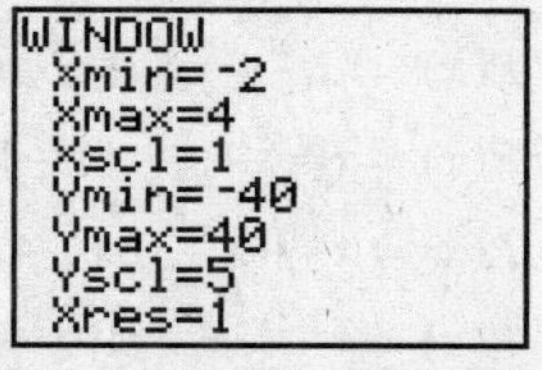

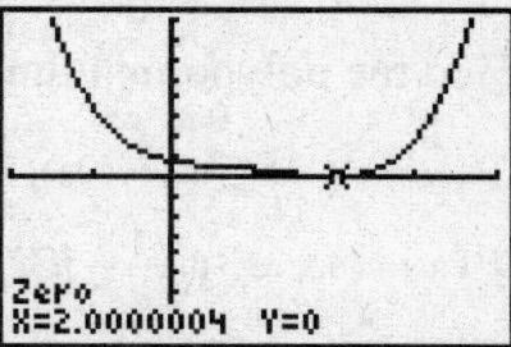

$$\begin{array}{r|rrrrr} 2 & 1 & -4 & 5 & -4 & 4 \\ & & 2 & -4 & 2 & -4 \\ \hline & 1 & -2 & 1 & -2 & 0 \end{array}$$

$$\begin{array}{r|rrrr} 2 & 1 & -2 & 1 & -2 \\ & & 2 & 0 & 2 \\ \hline & 1 & 0 & 1 & 0 \end{array}$$

$x^2+1=0$

$x^2=-1$

$x=\pm\sqrt{-1}=\pm i$

The solutions are 2 (multiplicity 2), i, $-i$.

35. Determine the exact values of the solutions.

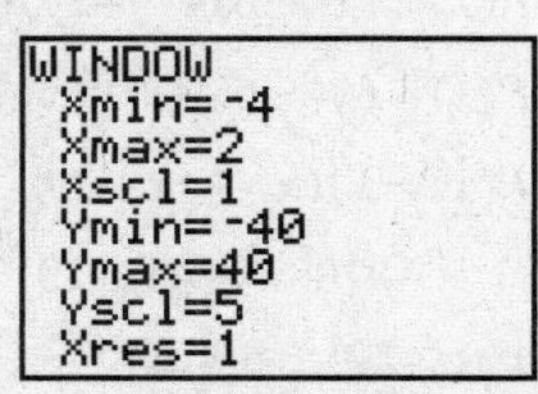

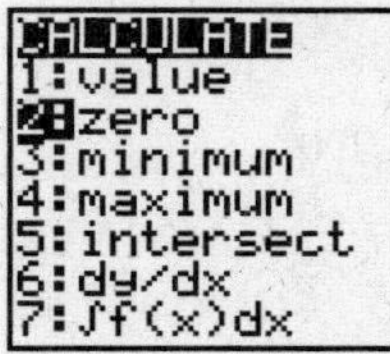

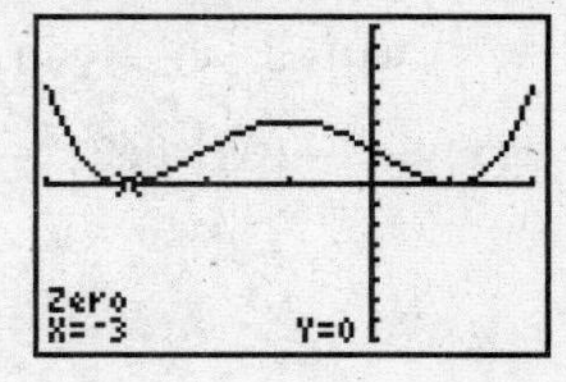

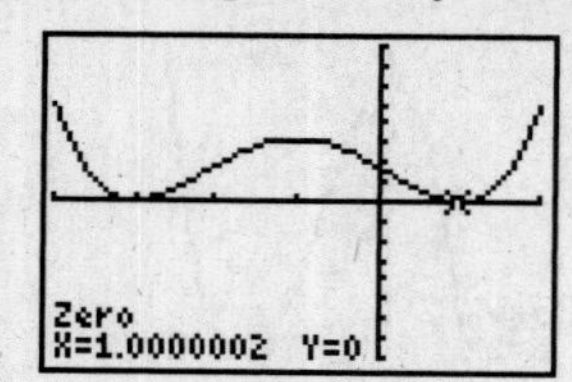

The solutions are –3 (multiplicity 2), 1 (multiplicity 2).

37. Find the polynomial function from the given zeros.

$$P(x)=(x-4)(x+3)(x-2)$$
$$P(x)=(x-4)(x^2+x-6)$$
$$P(x)=x(x^2+x-6)-4(x^2+x-6)$$
$$P(x)=x^3+x^2-6x-4x^2-4x+24$$
$$P(x)=x^3-3x^2-10x+24$$

39. Find the polynomial function from the given zeros.

$$P(x)=(x-3)(x-2i)(x+2i)$$
$$P(x)=(x-3)(x^2-[2i]^2)$$
$$P(x)=(x-3)(x^2-4i^2)$$
$$P(x)=(x-3)(x^2-4[-1])$$
$$P(x)=(x-3)(x^2+4)$$
$$P(x)=x^3-3x^2+4x-12$$

41. Find the polynomial function from the given zeros.

$$P(x)=[x-(3+i)][x-(3-i)][x-(2+5i)][x-(2-5i)]$$
$$P(x)=(x-3-i)(x-3+i)(x-2-5i)(x-2+5i)$$
$$P(x)=[(x-3)^2-i^2][(x-2)^2-25i^2]$$
$$P(x)=[(x^2-6x+9)-(-1)][(x^2-4x+4)-(25[-1])]$$
$$P(x)=[(x^2-6x+9)+1][(x^2-4x+4)+25]$$
$$P(x)=(x^2-6x+9+1)(x^2-4x+4+25)$$
$$P(x)=(x^2-6x+10)(x^2-4x+29)$$
$$P(x)=x^2(x^2-4x+29)-6x(x^2-4x+29)$$
$$+10(x^2-4x+29)$$
$$P(x)=x^4-4x^3+29x^2-6x^3+24x^2-174x$$
$$+10x^2-40x+290$$
$$P(x)=x^4-10x^3+63x^2-214x+290$$

43. Find the polynomial function from the given zeros.

$$P(x)=[x-(6+5i)][x-(6-5i)](x-2)(x-3)(x-5)$$
$$P(x)=[x-6-5i][x-6+5i](x-2)(x^2-8x+15)$$
$$P(x)=[(x-6)^2-(5i)^2]$$
$$\times[x(x^2-8x+15)-2(x^2-8x+15)]$$
$$P(x)=[(x^2-12x+36)-(25i^2)]$$
$$\times(x^3-8x^2+15x-2x^2+16x-30)$$
$$P(x)=[(x^2-12x+36)-(25[-1])]$$
$$\times(x^3-10x^2+31x-30)$$
$$P(x)=(x^2-12x+36+25)(x^3-10x^2+31x-30)$$
$$P(x)=(x^2-12x+61)(x^3-10x^2+31x-30)$$
$$P(x)=x^2(x^3-10x^2+31x-30)$$
$$-12x(x^3-10x^2+31x-30)$$
$$+61(x^3-10x^2+31x-30)$$
$$P(x)=x^5-10x^4+31x^3-30x^2$$
$$-12x^4+120x^3-372x^2+360x$$
$$+61x^3-610x^2+1891x-1830$$
$$P(x)=x^5-22x^4+212x^3-1012x^2+2251x-1830$$

45. Find the polynomial function from the given zeros.

Note: $4x-3=0$ if and only if $x=\frac{3}{4}$. Therefore, if $x=\frac{3}{4}$ (that is, $\frac{3}{4}$ is a zero), then $4x-3=0$.

$$P(x)=\left(x-\frac{3}{4}\right)[x-(2+7i)][x-(2-7i)]$$
$$P(x)=(4x-3)[x-2-7i][x-2+7i]$$
$$P(x)=(4x-3)[(x-2)^2-(7i)^2]$$
$$P(x)=(4x-3)[(x^2-4x+4)-49i^2]$$
$$P(x)=(4x-3)[(x^2-4x+4)-49(-1)]$$
$$P(x)=(4x-3)(x^2-4x+4+49)$$
$$P(x)=(4x-3)(x^2-4x+53)$$
$$P(x)=4x(x^2-4x+53)-3(x^2-4x+53)$$
$$P(x)=4x^3-16x^2+212x-3x^2+12x-159$$
$$P(x)=4x^3-19x^2+224x-159$$

47. Find the polynomial function.

If 2 – 5i is a zero, then 2 + 5i is also a zero.

$P(x)=[x-(2-5i)][x-(2+5i)](x+4)$
$P(x)=[x-2+5i][x-2-5i](x+4)$
$P(x)=[(x-2)^2-(5i)^2](x+4)$
$P(x)=[x^2-4x+4-25i^2](x+4)$
$P(x)=[x^2-4x+4-25(-1)](x+4)$
$P(x)=[x^2-4x+4+25](x+4)$
$P(x)=(x^2-4x+29)(x+4)$
$P(x)=x^2(x+4)-4x(x+4)+29(x+4)$
$P(x)=x^3+4x^2-4x^2-16x+29x+116$
$P(x)=x^3+13x+116$

49. Find the polynomial function.

If 4 + 3$i$and 5 - i are zeros, then 4 – 3i and 5 + i are also zeros.

$P(x)=[x-(4+3i)][x-(4-3i)]$
$\quad\times[x-(5-i)][x-(5+i)]$
$P(x)=(x^2-8x+25)(x^2-10x+26)$
$P(x)=x^4-10x^3+26x^2-8x^3+80x^2-208x$
$\quad+25x^2-250x+650$
$P(x)=x^4-18x^3+131x^2-458x+650$

51. Find the polynomial function.

$P(x)=(x+2)(x-1)(x-3)[x-(1+4i)][x-(1-4i)]$
$P(x)=(x^2+x-2)(x-3)[x-1-4i][x-1+4i]$
$P(x)=[x^3-2x^2-5x+6][(x-1)^2-(4i)^2]$
$P(x)=(x^3-2x^2-5x+6)[x^2-2x+1-16i^2]$
$P(x)=(x^3-2x^2-5x+6)[x^2-2x+1+16]$
$P(x)=(x^3-2x^2-5x+6)[x^2-2x+17]$
$P(x)=x^5-4x^4+16x^3-18x^2-97x+102$

53. Find the polynomial function.

$P(x)=a(x+1)(x-2)(x-3)$
$P(1)=a(1+1)(1-2)(1-3)$
$12=a(2)(-1)(-2)$
$12=4a$
$3=a$
$P(x)=3(x+1)(x-2)(x-3)$
$P(x)=(3x+3)(x^2-5x+6)$
$P(x)=3x(x^2-5x+6)+3(x^2-5x+6)$
$P(x)=3x^3-15x^2+18x+3x^2-15x+18$
$P(x)=3x^3-12x^2+3x+18$

55. Find the polynomial function.

$P(x)=a(x-3)(x+5)[x-(2+i)][x-(2-i)]$
$P(x)=a(x^2+2x-15)[x-2-i][x-2+i]$
$P(x)=a(x^2+2x-15)[(x-2)^2-i^2]$
$P(x)=a(x^2+2x-15)[x^2-4x+4-(-1)]$
$P(x)=a(x^2+2x-15)[x^2-4x+4+1]$
$P(x)=a(x^2+2x-15)(x^2-4x+5)$
$P(1)=a(1^2+2[1]-15)(1^2-4[1]+5)$
$48=a(1+2-15)(1-4+5)$
$48=a(-12)(2)$
$48=-24a$
$-2=a$
$P(x)=-2(x^2+2x-15)(x^2-4x+5)$
$P(x)=-2[x^2(x^2-4x+5)+2x(x^2-4x+5)$
$\quad-15(x^2-4x+5)]$
$P(x)=-2[x^4-4x^3+5x^2+2x^3-8x^2+10x$
$\quad-15x^2+60x-75)]$
$P(x)=-2(x^4-2x^3-18x^2+70x-75)$
$P(x)=-2x^4+4x^3+36x^2-140x+150$

57. Verify and explain.

$P(x)=x^3-x^2-ix^2-9x+9+9i$
$\quad=x^3+(-1-i)x^2-9x+(9+9i)$

$$\begin{array}{r|rrrr} 1+i & 1 & -1-i & -9 & 9+9i \\ & & 1+i & 0 & -9-9i \\ \hline & 1 & 0 & -9 & 0 \end{array}$$

Zero remainder implies 1 + i is a zero.

$$\begin{array}{r|rrr} 1-i & 1 & 0 & -9 \\ & & 1-i & 0 \\ \hline & 1 & 1-i & -9 \end{array}$$

Non-zero remainder implies $1-i$ is not a zero.

The Conjugate Pair Theorem does not apply because some of the coefficients of the polynomial function are not real numbers.

Prepare for Section 3.5

P1. Simplify.

$$\frac{x^2-9}{x^2-2x-15}=\frac{(x+3)(x-3)}{(x+3)(x-5)}=\frac{x-3}{x-5},\ x\neq-3$$

P3. Evaluate.

$$\frac{2(-3)^2+4(-3)-5}{-3+6}=\frac{2(9)-12-5}{3}=\frac{18-12-5}{3}=\frac{1}{3}$$

P5. Find the degree of the numerator and denominator.

The degree of the numerator, x^3+3x^2-5, is 3;

the degree of the denominator, x^2-4, is 2.

Section 3.5 Exercises

1. Find the domain.

$F(x)=\frac{1}{x}$ Then $x\neq 0$.

The domain is all real numbers except 0.

3. Find the domain.

$F(x)=\frac{x^2-3}{x^2+1}$ Then no restrictions on denominator.

The domain is all real numbers.

5. Find the domain.

$$F(x)=\frac{2x-1}{2x^2-15x+18}=\frac{2x-1}{(2x-3)(x-6)}$$

Then $x\neq\frac{3}{2},\ 6$.

The domain is all real numbers except $\frac{3}{2}$ and 6.

7. Find the domain.

$$F(x)=\frac{2x^2}{x^3-4x^2-12x}=\frac{2x^2}{x(x-6)(x+2)}$$

Then $x\neq 0,\ 6,-2$.

The domain is all real numbers except 0, 6 and –2.

9. Find all vertical asymptotes.

$$x^2+3x=0$$
$$x(x+3)=0$$
$$x=0 \quad\text{or}\quad x+3=0$$
$$x=-3$$

Vertical asymptotes: $x=0,\ x=-3$

11. Find all vertical asymptotes.

$$6x^2-5x-4=0$$
$$(3x-4)(2x+1)=0$$
$$3x-4=0 \quad\text{or}\quad 2x+1=0$$
$$x=\frac{4}{3} \qquad x=-\frac{1}{2}$$

Vertical asymptotes: $x=-\frac{1}{2},\ x=\frac{4}{3}$

13. Find all vertical asymptotes.

$$4x^3-25x^2+6x=0$$
$$x(4x^2-25x+6)=0$$
$$x(4x-1)(x-6)=0$$
$$x=0 \quad\text{or}\quad 4x-1=0 \quad\text{or}\quad x-6=0$$
$$x=\frac{1}{4} \qquad x=6$$

Vertical asymptotes: $x=0,\ x=\frac{1}{4}, x=6$

15. Find the horizontal asymptote.

Horizontal asymptote: $y=\frac{4}{1}=4$

17. Find the horizontal asymptote.

Horizontal asymptote: $y=\frac{15{,}000}{500}=30$

19. Find the horizontal asymptote.

Horizontal asymptote: $y=\frac{4}{\frac{1}{3}}=12$

21. Find all vertical and horizontal asymptotes. Sketch the graph and label intercepts and asymptotes.

$$F(x)=\frac{1}{x+4}$$

Vertical asymptote: $x=-4$

Horizontal asymptote: $y=0$

x-intercepts: none

y-intercept: $\left(0, \frac{1}{4}\right)$

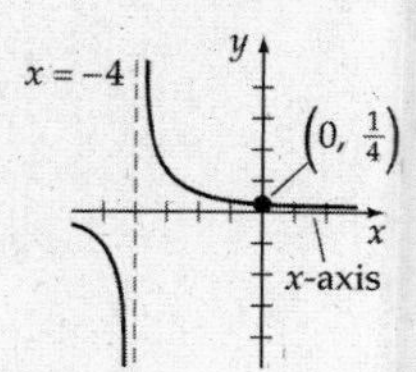

23. Find all vertical and horizontal asymptotes. Sketch the graph and label intercepts and asymptotes.

$F(x) = \frac{-4}{x-3}$

Vertical asymptote: $x = 3$

Horizontal asymptote: $y = 0$

x-intercepts: none

y-intercept: $\left(0, \frac{4}{3}\right)$

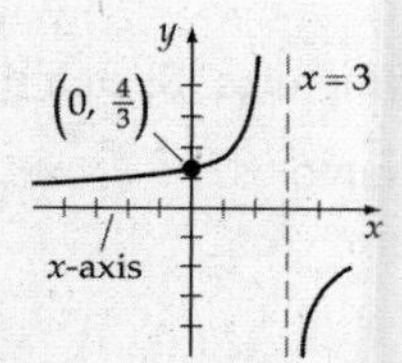

25. Find all vertical and horizontal asymptotes. Sketch the graph and label intercepts and asymptotes.

$F(x) = \frac{4}{x}$

Vertical asymptote: $x = 0$

Horizontal asymptote: $y = 0$

x-intercepts: none

y-intercept: none

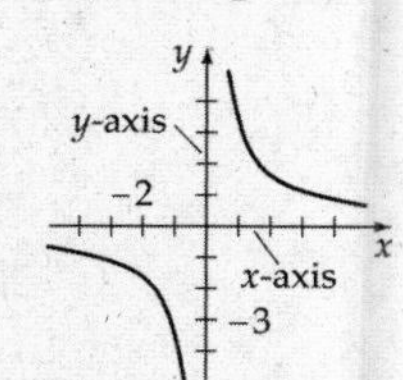

27. Find all vertical and horizontal asymptotes. Sketch the graph and label intercepts and asymptotes.

$F(x) = \frac{x}{x+4}$

Vertical asymptote: $x = -4$

Horizontal asymptote: $y = 1$

x-intercept: (0, 0)

y-intercept: (0, 0)

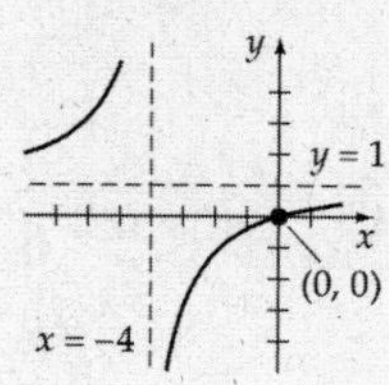

29. Find all vertical and horizontal asymptotes. Sketch the graph and label intercepts and asymptotes.

$F(x) = \frac{x+4}{2-x}$

Vertical asymptote: $x = 2$

Horizontal asymptote: $y = -1$

x-intercept: (–4, 0)

y-intercept: (0, 2)

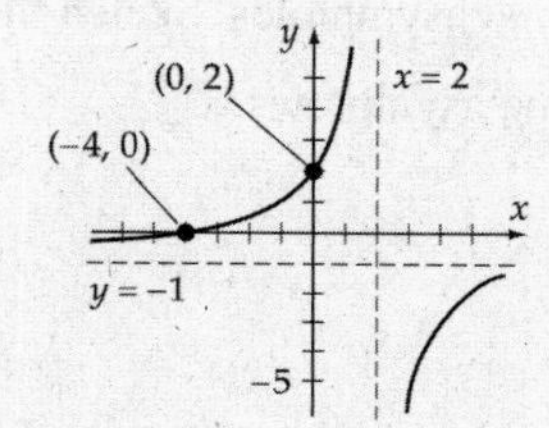

31. Find all vertical and horizontal asymptotes. Sketch the graph and label intercepts and asymptotes.

$F(x) = \frac{1}{x^2-9}$

Vertical asymptotes: $x = 3, x = -3$

Horizontal asymptote: $y = 0$

x-intercept: none

y-intercept: $\left(0, -\frac{1}{9}\right)$

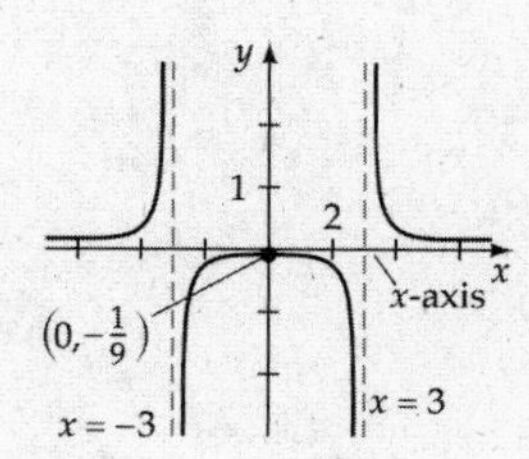

33. Find all vertical and horizontal asymptotes. Sketch the graph and label intercepts and asymptotes.

$$F(x)=\frac{1}{x^2+2x-3}$$

Vertical asymptotes: $x=-3, x=1$

Horizontal asymptote: $y=0$

x-intercept: none

y-intercept: $\left(0,-\frac{1}{3}\right)$

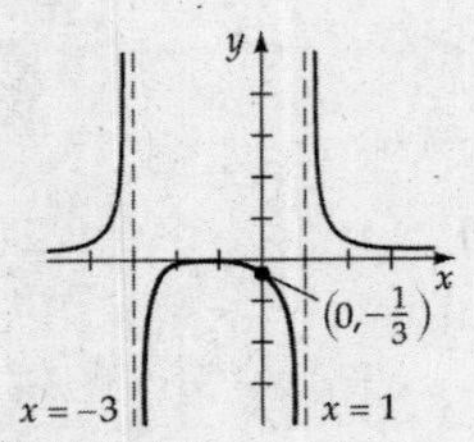

35. Find all vertical and horizontal asymptotes. Sketch the graph and label intercepts and asymptotes.

$$F(x)=\frac{x^2}{x^2+4x+4}$$

Vertical asymptote: $x=-2$

Horizontal asymptote: $y=1$

x-intercept: (0, 0)

y-intercept: (0, 0)

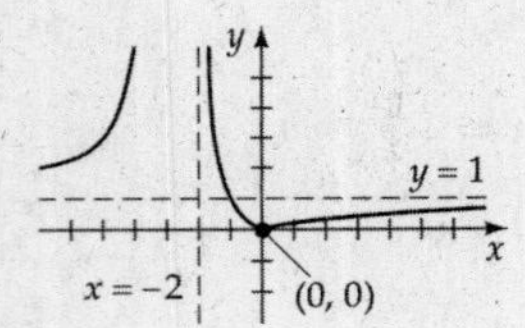

37. Find all vertical and horizontal asymptotes. Sketch the graph and label intercepts and asymptotes.

$$F(x)=\frac{10}{x^2+2}$$

Vertical asymptote: none

Horizontal asymptote: $y=0$

x-intercept: none

y-intercept: none

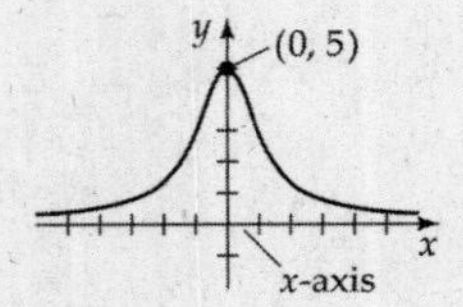

39. Find all vertical and horizontal asymptotes. Sketch the graph and label intercepts and asymptotes.

$$F(x)=\frac{2x^2-2}{x^2-9}$$

Vertical asymptotes: $x=3, x=-3$

Horizontal asymptote: $y=2$

x-intercepts: (–1, 0) and (1, 0)

y-intercept: $\left(0,\ \frac{2}{9}\right)$

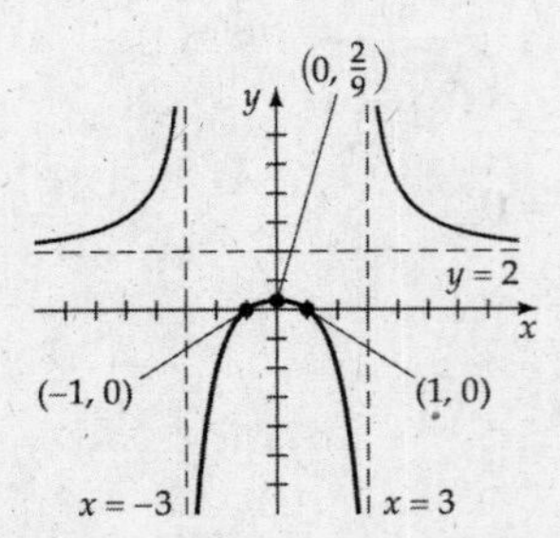

41. Find all vertical and horizontal asymptotes. Sketch the graph and label intercepts and asymptotes.

$$F(x)=\frac{x^2+x+4}{x^2+2x-1}$$

Vertical asymptotes: $x=-1+\sqrt{2}, x=-1-\sqrt{2}$

Horizontal asymptote: $y=1$

x-intercept: none

y-intercept: (0, –4)

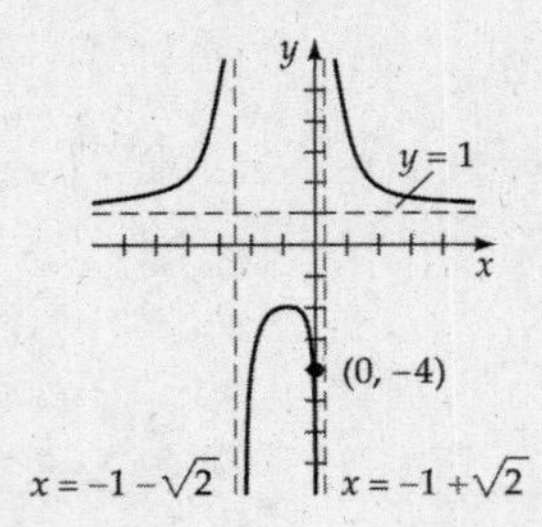

43. Find the slant asymptote.

$$F(x)=\frac{3x^2+5x-1}{x+4}$$

$$\begin{array}{r|rrr} -4 & 3 & 5 & -1 \\ & & -12 & 28 \\ \hline & 3 & -7 & 27 \end{array}$$

$$F(x)=3x-7+\frac{27}{x+4}$$

Slant asymptote: $y=3x-7$

45. Find the slant asymptote.

$$F(x)=\frac{x^3-1}{x^2}=\frac{x^3}{x^2}-\frac{1}{x^2}=x-\frac{1}{x^2}$$

Slant asymptote: $y=x$

47. Find the slant asymptote.

$$\begin{array}{r|rrr} 5 & -4 & 15 & 8 \\ & & -20 & -25 \\ \hline & -4 & -5 & -17 \end{array}$$

$$F(x)=-4x-5+\frac{-17}{x-5}$$

Slant asymptote: $y=-4x-5$

49. Find the vertical and slant asymptotes and graph.

$$F(x)=\frac{x^2-4}{x}=x-\frac{4}{x}$$

Slant asymptote: $y=x$

Vertical asymptote: $x=0$

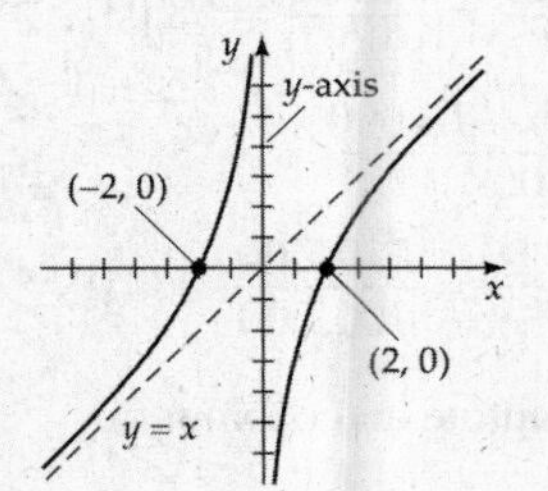

51. Find the vertical and slant asymptotes and graph.

$$\begin{array}{r|rrr} -3 & 1 & -3 & -4 \\ & & -3 & 18 \\ \hline & 1 & -6 & 14 \end{array}$$

$$F(x)=\frac{x^2-3x-4}{x+3}=x-6+\frac{14}{x+3}$$

Slant asymptote: $y=x-6$

Vertical asymptote: $x=-3$

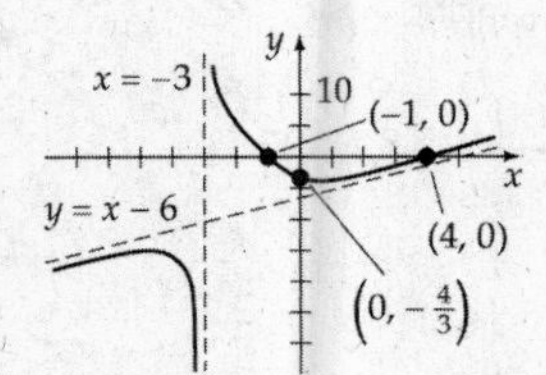

53. Find the vertical and slant asymptotes and graph.

$$\begin{array}{r|rrr} 4 & 2 & 5 & 3 \\ & & 8 & 52 \\ \hline & 2 & 13 & 55 \end{array}$$

$$F(x)=\frac{2x^2+5x+3}{x-4}=2x+13+\frac{55}{x-4}$$

Slant asymptote: $y=2x+13$

Vertical asymptote: $x=4$

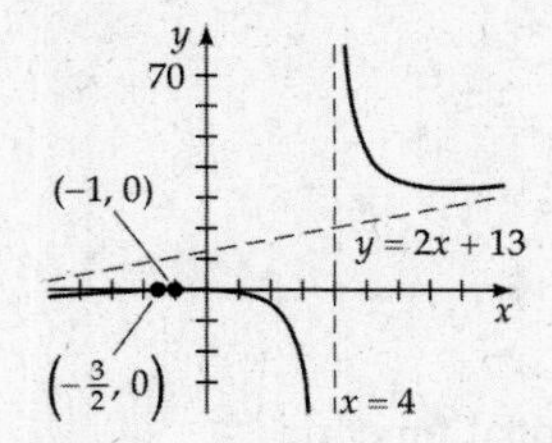

55. Find the vertical and slant asymptotes and graph.

$$\begin{array}{r|rrr} -2 & 1 & -1 & 0 \\ & & -2 & 6 \\ \hline & 1 & -3 & 6 \end{array}$$

$$F(x)=\frac{x^2-x}{x+2}=x-3+\frac{6}{x+2}$$

Slant asymptote: $y=x-3$

Vertical asymptote: $x=-2$

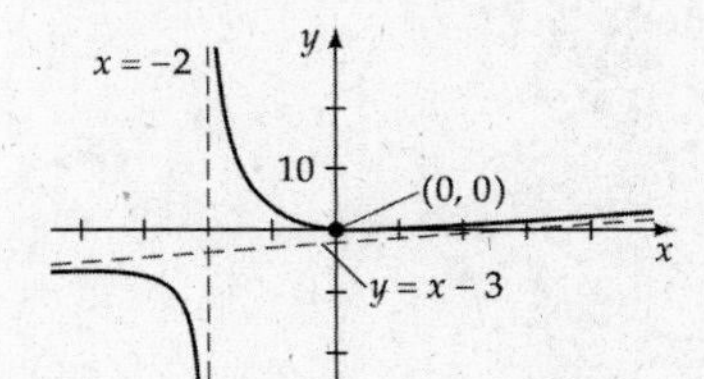

57. Find the vertical and slant asymptotes and graph.

$$\begin{array}{r} x \\ x^2-4\overline{)x^3 \qquad +1} \\ \underline{x^3-4x} \\ 4x+1 \end{array}$$

$$F(x)=\frac{x^3+1}{x^2-4}=x+\frac{4x+1}{x^2-4}$$

Slant asymptote: $y=x$

Vertical asymptotes: $x=2, x=-2$

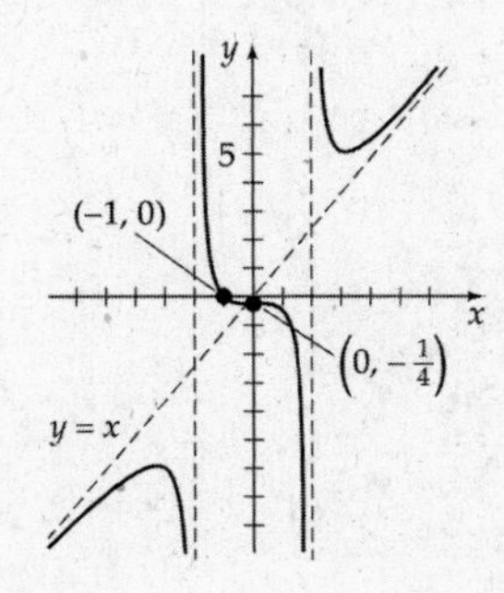

59. Sketch the graph of the rational function.

$$F(x)=\frac{x^2+x}{x+1}$$

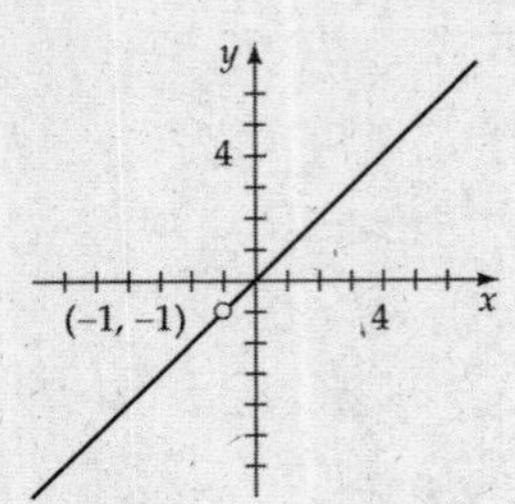

61. Sketch the graph of the rational function.

$$F(x)=\frac{2x^3+4x^2}{2x+4}$$

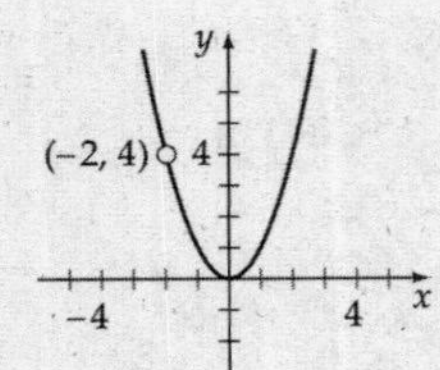

63. Sketch the graph of the rational function.

$$F(x)=\frac{-2x^3+6x}{2x^2-6x}$$

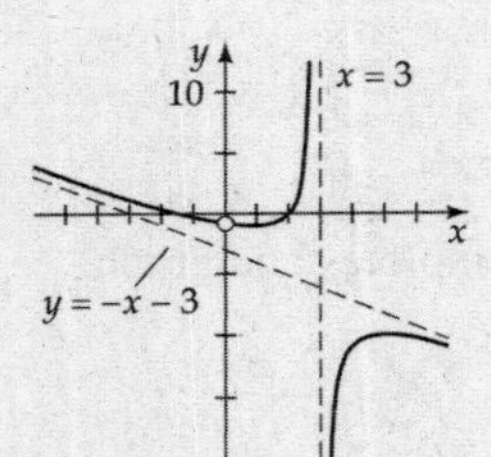

65. Sketch the graph of the rational function.

$$F(x)=\frac{x^2-3x-10}{x^2+4x+4}$$

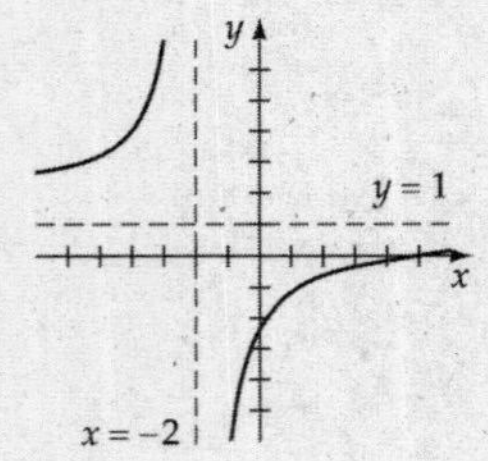

67. a. Find the current.

$$I(3)=\frac{9}{3+4.5}=1.2 \text{ amps}$$

b. Find the resistance.

$$0.24=\frac{9}{x+4.5}$$

$$0.24x+1.08=9$$

$$0.24x=7.92$$

$$x=33 \text{ ohms}$$

c. Find the horizontal asymptote and explain the meaning.

Horizontal asymptote: $y = 0$.

The current approaches 0 amps as the resistance of the variable resistor increases without bound.

69. a. Find the average costs per golf ball.

$$\overline{C}(1000)=\frac{0.43(1000)+76,000}{1000}=\frac{430+76,000}{1000}$$

$$=\frac{76,430}{1000}=\$76.43$$

$$\overline{C}(10,000)=\frac{0.43(10,000)+76,000}{10,000}$$

$$=\frac{4,300+76,000}{10,000}=\frac{80,300}{10,000}=\$8.03$$

$$\overline{C}(100,00)=\frac{0.43(100,000)+76,000}{100,000}$$

$$=\frac{43,000+76,000}{100,000}=\frac{119,000}{100,000}=\$1.19$$

b. Find the horizontal asymptote and explain the meaning.

Horizontal asymptote: $y = 0.43$.

As the number of golf balls that are produced increases, the average cost per golf ball approaches \$0.43.

71. a. Find the cost of removing 40% salt.

$$C(40)=\frac{2000(40)}{100-40}=\frac{80,000}{60}=\$1,333.33$$

b. Find the cost of removing 80% salt.

$$C(80)=\frac{2000(80)}{100-80}=\frac{160,000}{20}=\$8,000$$

c. Sketch the graph.

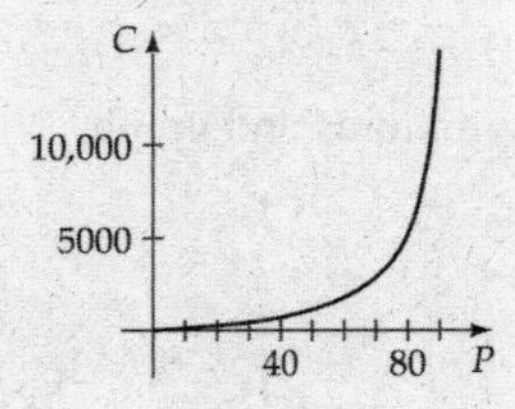

73. **a.** Find the monthly sales, round to nearest 100.

$$S(2) = \frac{150(2)}{1.5(2)^2 + 80} = \frac{300}{86} \approx 3.488 \text{ thousand} \approx 3500$$

$$S(4) = \frac{150(4)}{1.5(4)^2 + 80} = \frac{600}{104}$$

$$\approx 5.769 \text{ thousand} \approx 5800$$

$$S(10) = \frac{150(10)}{1.5(10)^2 + 80} = \frac{1500}{230}$$

$$\approx 6.522 \text{ thousand} \approx 6500$$

b. Find the month where there is a maximum sales.

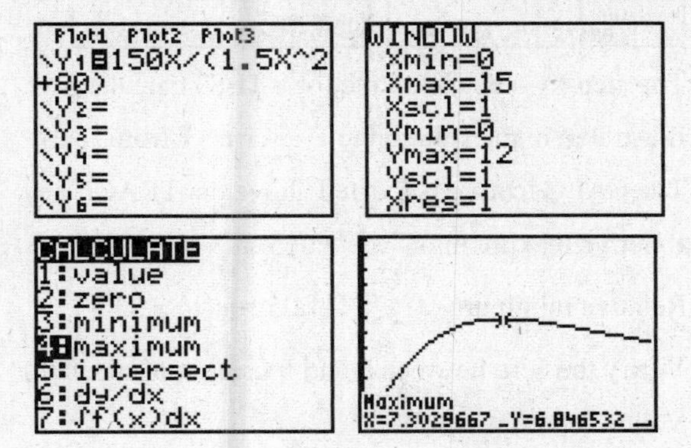

In the seventh month is the maximum.

c. Find the far-right behavior.

$(n < m)$ The sales will approach zero.

75. **a.** Graph, estimate r that makes a minimum.

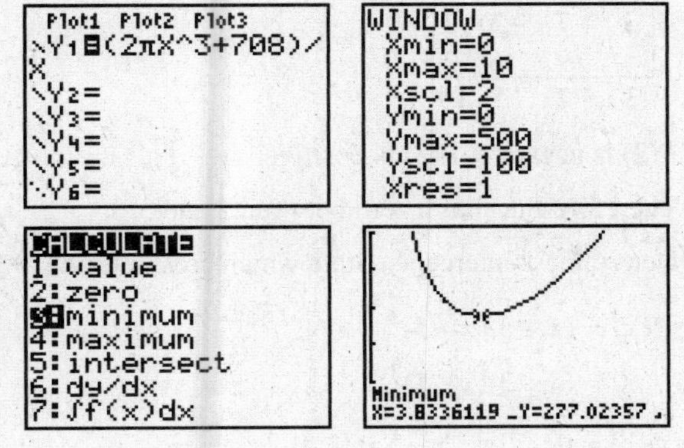

$r \approx 3.8$ centimeters

b. Determine whether there is a slant asymptote.

No. The degree of the numerator is not exactly one more than the degree of the denominator.

c. Explain the statement.

As the radius r increases without bound, the surface area approaches twice the area of a circle with radius r.

Since $V = \pi r^2 h$, then $h = \frac{V}{\pi r^2}$. $h \to 0$ as $r \to \infty$ so as the radius increases without bound, the surface area of the can approaches the area of the top and bottom of the can, two circles with radius r.

77. Determine where the graph intersects the horizontal asymptote.

Horizontal asymptote: $y = 2$

$$2 = \frac{2x^2 + 3x + 4}{x^2 + 4x + 7}$$

$$2x^2 + 8x + 14 = 2x^2 + 3x + 4$$

$$5x + 10 = 0$$

$$x = -2$$

The graph of F intersects its horizontal asymptote at $(-2, 2)$.

79. Create a rational function with given characteristics.

One possible function: $f(x) = \frac{2x^2}{x^2 - 9}$

Exploring Concepts with Technology

1.
```
In(2):=
NSolve[x^4-3x^3+x-5==0]
Out[2]=
{{x -> -1.14039},
{x -> 0.536692 - 1.06842 I},
{x -> 0.536692 + 1.06842 I},
{x -> 3.067}}
```

3.
```
In(4):=
NSolve[4x^5-3x^3+2x^2-x+2==0]
Out[4]=
{{x -> -1.25095},,
(x -> - 0.156173 - 0.685216 I},
{x -> -0.156173 + 0.685216 I},
{x -> 0.781647 - 0.445283 I},
{x -> 0.781647 + 0.445283 I}}
```

Chapter 3 Review Exercises

1. Use synthetic division to divide. [3.1]

$$\begin{array}{r|rrrr} 3 & 4 & -11 & 5 & -2 \\ & & 12 & 3 & 24 \\ \hline & 4 & 1 & 8 & 22 \end{array}$$

Answer: $4x^2+x+8+\dfrac{22}{x-3}$

3. Find $P(c)$ using the Remainder Theorem. [3.1]

$$\begin{array}{r|rrrr} 4 & 1 & 2 & -5 & 1 \\ & & 4 & 24 & 76 \\ \hline & 1 & 6 & 19 & 77 \end{array}$$

$P(4)=77$

5. Find $P(c)$ using the Remainder Theorem. [3.1]

$$\begin{array}{r|rrrrr} -2 & 6 & 0 & -12 & 8 & 1 \\ & & -12 & 24 & -24 & 32 \\ \hline & 6 & -12 & 12 & -16 & 33 \end{array}$$

$P(-2)=33$

7. Use synthetic division to show c is a zero. [3.1]

$$\begin{array}{r|rrrr} 3 & 1 & 2 & -26 & 33 \\ & & 3 & 15 & -33 \\ \hline & 1 & 5 & -11 & 0 \end{array}$$

9. Use synthetic division to show c is a zero. [3.1]

$$\begin{array}{r|rrrrrr} 1 & 1 & -1 & 0 & -2 & 1 & 1 \\ & & 1 & 0 & 0 & -2 & -1 \\ \hline & 1 & 0 & 0 & -2 & -1 & 0 \end{array}$$

11. Determine whether the binomial is a factor using the Factor Theorem. [3.1]

$$\begin{array}{r|rrrr} 5 & 1 & -11 & 39 & -45 \\ & & 5 & -30 & 45 \\ \hline & 1 & -6 & 9 & 0 \end{array}$$

A remainder of 0 implies that $x-5$ is a factor of $P(x)$.

13. Determine the far-left and far-right behavior. [3.2]

$P(x)=-2x^3-5x^2+6x-3$

Since $a_n=-2$ is negative and $n=3$ is odd, the graph of P goes up to its far left and down to its far right.

15. Use a graphing utility to graph, estimate any relative maxima or minima. [3.2]

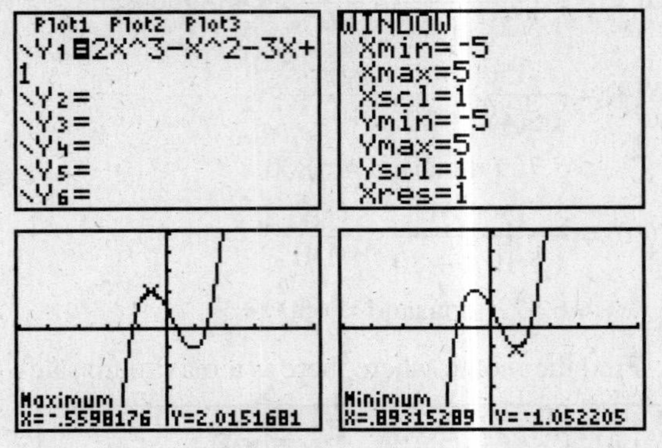

The step-by-step technique for a TI-83 calculator is illustrated in the solution to Exercise **11** from [3.2]. The CALC feature is located above the TRACE key.

Relative maximum of $y\approx 2.015$ at $x\approx -0.560$.

Relative minimum of $y\approx -1.052$ at $x\approx 0.893$.

17. Verify the zero between a and b using Intermediate Value Theorem. [3.2]

$P(x)=3x^3-7x^2-3x+7$

$$\begin{array}{r|rrrr} 2 & 3 & -7 & -3 & 7 \\ & & 6 & -2 & -10 \\ \hline & 3 & -1 & -5 & -3 \end{array}$$

$$\begin{array}{r|rrrr} 3 & 3 & -7 & -3 & 7 \\ & & 9 & 6 & 9 \\ \hline & 3 & 2 & 3 & 16 \end{array}$$

$P(2)$ is negative; $P(3)$ is positive.

Therefore $P(x)$ has a zero between 2 and 3.

19. Determine x-intercepts. State which cross x-axis. [3.2]

$P(x)=(x+3)(x-5)^2$

$0=(x+3)(x-5)^2$

$x+3=0$ or $x-5=0$

$x=-3$ $\quad$ $x=5$

The exponent on $(x+3)$ is an odd integer. Therefore the graph of $P(x)$ will cross the x-axis at $(-3, 0)$.

The exponent on $(x-5)$ is an even integer. Therefore the graph of $P(x)$ will intersect but not cross the x-axis at $(5, 0)$.

21. Sketch the graph of the function. [3.2]

Let $P(x)=0$.

$$x^3-x=0$$
$$x(x^2-1)=0$$
$$x(x+1)(x-1)=0$$

$x=0, x=-1, x=1$

The graph crosses the x-axis at (0, 0), (–1, 0), and (1, 0).

Let $x=0$. $P(0)=0^3-(0)=0$

The y-intercept is (0, 0).

x^3 has a positive coefficient and an odd exponent. Therefore, the graph goes down to the far left and up to the far right.

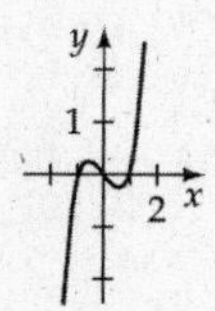

23. Sketch the graph of the function. [3.2]

Let $P(x)=0$.

$$x^4-6=0$$
$$\left(x^2+\sqrt{6}\right)\left(x^2-\sqrt{6}\right)=0$$
$$\left(x^2+\sqrt{6}\right)\left(x+\sqrt{\sqrt{6}}\right)\left(x-\sqrt{\sqrt{6}}\right)=0$$

$\sqrt{\sqrt{6}}\approx 1.565$

$x=-1.565, x=1.565$

The graph crosses the x-axis at (–1.565, 0), (1.565, 0).

Let $x=0$.

$P(0)=(0)^4+6=6$

The y-intercept is (0, 6).

x^4 has a positive coefficient and an even exponent. Therefore, the graph goes up to the far left and up to the far right.

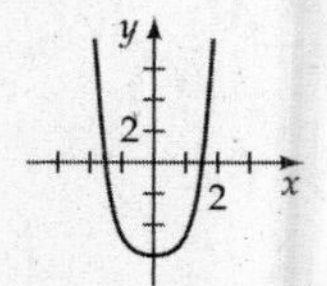

25. Sketch the graph of the function. [3.2]

Let $P(x)=0$.

$$x^4-10x^2+9=0$$
$$\left(x^2-1\right)\left(x^2-9\right)=0$$
$$(x+1)(x-1)(x+3)(x-3)=0$$

$x=-1, x=1, x=-3, x=3$

The graph crosses the x-axis at (–3, 0), (–1, 0), (1, 0) and (3, 0).

Let $x=0$.

$P(0)=(0)^4-10(0)^2+9=9$

The y-intercept is (0, 9).

x^4 has a positive coefficient and an even exponent. Therefore, the graph goes up to the far left and up to the far right.

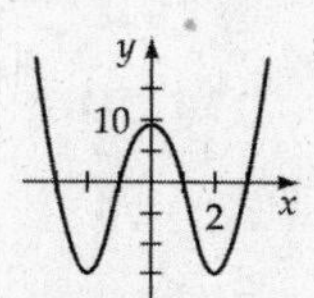

27. List the possible rational zeros. [3.3]

$P(x)=x^3-7x-6$

$p=\pm 1, \pm 2, \pm 3, \pm 6$

$q=\pm 1$

$\frac{p}{q}=\pm 1, \pm 2, \pm 3, \pm 6$

29. List the possible rational zeros. [3.3]

$P(x)=15x^3-91x^2+4x+12$

$p=\pm 1, \pm 2, \pm 3, \pm 4, \pm 6, \pm 12$

$q=\pm 1, \pm 3, \pm 5, \pm 15$

$\frac{p}{q}=\pm 1, \pm 2, \pm 3, \pm 4, \pm 6, \pm 12,$

$\pm\frac{1}{3}, \pm\frac{2}{3}, \pm\frac{4}{3}, \pm\frac{1}{5}, \pm\frac{2}{5}, \pm\frac{3}{5}, \pm\frac{4}{5}, \pm\frac{6}{5},$

$\pm\frac{12}{5}, \pm\frac{1}{15}, \pm\frac{2}{15}, \pm\frac{4}{15}$

31. List the possible rational zeros. [3.3]

$P(x)=x^3+x^2-x-1$

$p=\pm 1$

$q=\pm 1$

$\frac{p}{q}=\pm 1$

33. State the number of positive and negative real zeros using Descartes' Rule of Signs. [3.3]

$P(x)=x^3+3x^2+x+3$ has no change in sign.

Therefore there are no positive zeros.

$P(-x)=-x^3+3x^2-x+3$ has 3 changes in sign.

Therefore there are three or one negative zeros.

35. State the number of positive and negative real zeros using Descartes' Rule of Signs. [3.3]

$P(x)=x^4-x-1$ has 1 change in sign. Therefore there is one positive zero.

$P(-x)=x^4+x-1$ has 1 change in sign. Therefore there is one negative zero.

37. Find the zeros. [3.3]

$$\begin{array}{r|rrrr} 1 & 1 & 6 & 3 & -10 \\ & & 1 & 7 & 10 \\ \hline & 1 & 7 & 10 & 0 \end{array}$$

$$\begin{aligned} x^2+7x+10&=0 \\ (x+5)(x+2)&=0 \\ x=-5 \quad &\text{or} \quad x=-2 \end{aligned}$$

The zeros of $x^3+6x^2+3x-10$ are 1, –2, and –5.

39. Find the zeros. [3.3]

$$\begin{array}{r|rrrrr} -2 & 6 & 35 & 72 & 60 & 16 \\ & & -12 & -46 & -52 & -16 \\ \hline & 6 & 23 & 26 & 8 & 0 \end{array}$$

$$\begin{array}{r|rrrr} -2 & 6 & 23 & 26 & 8 \\ & & -12 & -22 & -8 \\ \hline & 6 & 11 & 4 & 0 \end{array}$$

$$\begin{aligned} 6x^2+11x+4&=0 \\ (3x+4)(2x+1)&=0 \\ x=-\frac{4}{3} \quad &\text{or} \quad x=-\frac{1}{2} \end{aligned}$$

The zeros of $6x^4+35x^3+72x^2+60x+16$ are –2 (multiplicity 2), $-\frac{4}{3}$, and $-\frac{1}{2}$.

41. Find the zeros. [3.3]

$$\begin{array}{r|rrrrr} 1 & 1 & -4 & 6 & -4 & 1 \\ & & 1 & -3 & 3 & -1 \\ \hline & 1 & -3 & 3 & -1 & 0 \end{array}$$

$$\begin{array}{r|rrrr} 1 & 1 & -3 & 3 & -1 \\ & & 1 & -2 & 1 \\ \hline & 1 & -2 & 1 & 0 \end{array}$$

$$\begin{aligned} x^2-2x+1&=0 \\ (x-1)(x-1)&=0 \\ x=1 \quad &\text{or} \quad x=1 \end{aligned}$$

The zero of $x^4-4x^3+6x^2-4x+1$ is 1 (multiplicity 4).

43. Find all zeros and write as a product. [3.4]

$P(x)=2x^4-9x^3+22x^2-29x+10$

$$\begin{array}{r|rrrrr} \frac{1}{2} & 2 & -9 & 22 & -29 & 10 \\ & & 1 & -4 & 9 & -10 \\ \hline & 2 & -8 & 18 & -20 & 0 \end{array}$$

$$\begin{array}{r|rrrr} 2 & 2 & -8 & 18 & -20 \\ & & 4 & -8 & 20 \\ \hline & 2 & -4 & 10 & 0 \end{array}$$

$$\begin{aligned} 2x^2-4x+10&=0 \\ 2(x^2-2x+5)&=0 \end{aligned}$$

$$x=\frac{-(-2)\pm\sqrt{(-2)^2-4(1)(5)}}{2(1)}=\frac{2\pm\sqrt{4-20}}{2}$$

$$=\frac{2\pm\sqrt{-16}}{2}=\frac{2\pm 4i}{2}=1\pm 2i$$

The zeros are $\frac{1}{2}$, 2, 1 + 2i, 1 – 2i.

$$P(x)=2\left(x-\frac{1}{2}\right)(x-2)(x-1-2i)(x-1+2i)$$

45. Find the remaining zeros. [3.4]

$$\begin{array}{r|rrrrr} 1-2i & 1 & -4 & 6 & -4 & -15 \\ & & 1-2i & -7+4i & 7+6i & 15 \\ \hline & 1 & -3-2i & -1+4i & 3+6i & 0 \end{array}$$

$$\begin{array}{r|rrrr} 1+2i & 1 & -3-2i & -1+4i & 3+6i \\ & & 1+2i & -2-4i & -3-6i \\ \hline & 1 & -2 & -3 & 0 \end{array}$$

$$\begin{aligned} x^2-2x-3&=0 \\ (x-3)(x+1)&=0 \\ x=3,\ x&=-1 \end{aligned}$$

The remaining zeros are 1 + 2i, 3, and –1.

47. Find the polynomial function. [3.4]

$$\begin{aligned} P(x) &= (x-4)(x+3)(2x-1) \\ &= (x^2-x-12)(2x-1) \\ &= 2x^3-x^2-2x^2+x-24x+12 \\ &= 2x^3-3x^2-23x+12 \end{aligned}$$

49. Find the polynomial function. [3.4]

$$\begin{aligned} P(x) &= (x-1)(x-2)(x-5i)(x+5i) \\ &= (x^2-3x+2)(x^2+25) \\ &= x^4+25x^2-3x^3-75x+2x^2+50 \\ &= x^4-3x^3+27x^2-75x+50 \end{aligned}$$

51. Find the domain. [3.5]

$F(x)=\dfrac{x^2}{x^2+7}$ There are no restrictions on the denominator.

The domain is all real numbers.

53. Find all vertical asymptotes. [3.5]

$$\begin{aligned} x^3-x^2-12x&=0 \\ x(x^2-x-12)&=0 \\ x(x+3)(x-4)&=0 \end{aligned}$$

The vertical asymptotes are $x = 0$, $x = -3$, $x = 4$.

55. Find the horizontal asymptote. [3.5]

$$y=\frac{3}{\frac{1}{2}}=6$$

57. Find the slant asymptote. [3.5]

$$f(x)=\frac{2x^3-3x^2-x+5}{x^2-x+1}$$

$$\begin{array}{r} 2x-1 \\ x^2-x+1\overline{\big)\,2x^3-3x^2-x+5} \\ \underline{2x^3-2x^2+2x} \\ -x^2-3x+5 \\ \underline{-x^2+x-1} \\ -4x+6 \end{array}$$

$$f(x)=2x-1+\frac{-4x+6}{x^2-x+1}$$

Slant asymptote: $y=2x-1$

59. Graph the rational function. [3.5]

$$f(x)=\frac{3x-2}{x}$$

Vertical asymptote: $x = 0$

Horizontal asymptote: $y = 3$

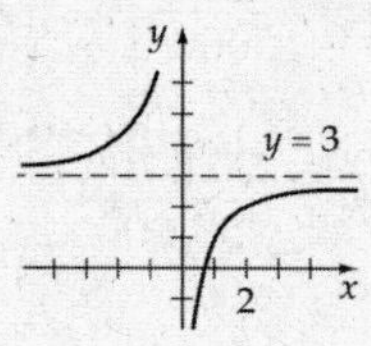

61. Graph the rational function. [3.5]

$$f(x)=\frac{12x-24}{x^2-4}$$

Vertical asymptotes: $x = -2$, $x = 2$

Horizontal asymptote: $y = 0$

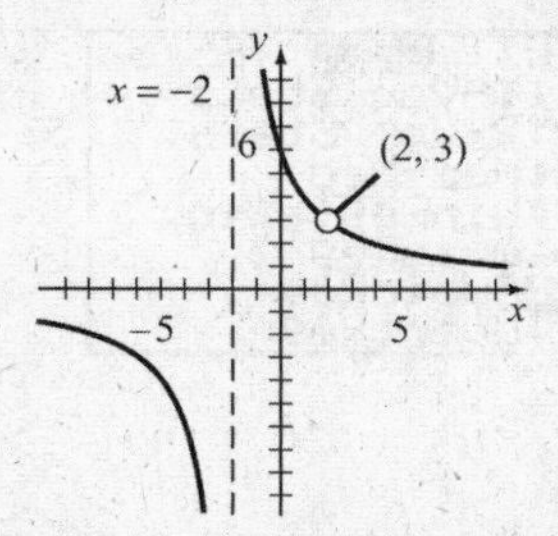

63. Graph the rational function. [3.5]

$$f(x)=\frac{2x^3-4x+6}{x^2-4}$$

Vertical asymptote: $x = -2$, $x = 2$

Horizontal asymptote: none

Slant asymptote: $y = 2x$

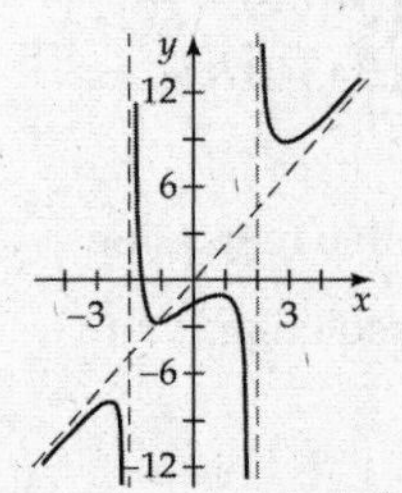

65. Graph the rational function. [3.5]

$$f(x)=\frac{3x^2-6}{x^2-9}$$

Vertical asymptotes: $x = -3$, $x = 3$

Horizontal asymptote: $y = 3$

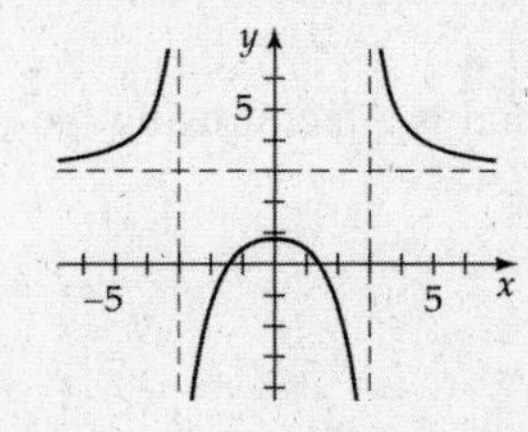

67. a. Find the average cost per skateboard. [3.5]

$$C(5000)=\frac{5.75(5000)+34,200}{5000}=\frac{62,950}{5000}$$
$$=\$12.59$$
$$C(50,000)=\frac{5.75(50,000)+34,200}{50,000}=\frac{321,700}{50,000}$$
$$\approx\$6.43$$

b. Find the horizontal asymptote and explain.

$y = 5.75$. As the number of skateboards produced increases, the average cost per skateboard approaches \$5.75.

69. a. Find the cubic regression function. [3.2]

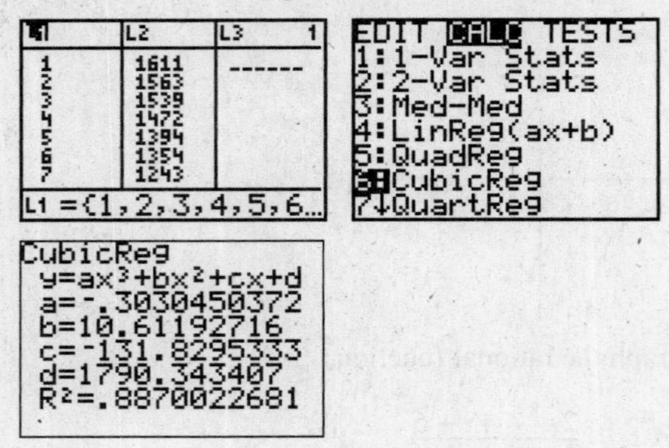

$$f(x)=-0.3030450372x^3+10.61192716x^2$$
$$-131.8295333x+1790.343407$$

b. Use the cubic regression function.

$$f(21)=-0.3030450372(21)^3+10.61192716(21)^2$$
$$-131.8295333(21)+1790.343407$$
$$\approx 895,000 \text{ thefts}$$

c. Answers will vary; however, cubic regression functions are often unreliable at predicting future results.

Chapter 3 Test

1. Use synthetic division to divide. [3.1]

$$\begin{array}{r|rrrr} -2 & 3 & 5 & 4 & -1 \\ & & -6 & 2 & -12 \\ \hline & 3 & -1 & 6 & -13 \end{array}$$

Answer: $3x^2-x+6+\frac{-13}{x+2}$

3. Determine whether the binomial is a factor using the Factor Theorem. [3.1]

$$\begin{array}{r|rrrrr} 1 & 1 & -4 & 7 & -6 & 2 \\ & & 1 & -3 & 4 & -2 \\ \hline & 1 & -3 & 4 & -2 & 0 \end{array}$$

A remainder of 0 implies that $x-1$ is a factor of

$x^4-4x^3+7x^2-6x+2$.

5. Find the real zeros. [3.2]

$$3x^3+7x^2-6x=0$$
$$x(3x^2+7x-6)=0$$
$$x(3x-2)(x+3)=0$$
$$x=0 \quad 3x-2=0 \quad \text{or} \quad x+3=0$$
$$x=\frac{2}{3} \qquad\qquad x=-3$$

The zeros of $3x^3+7x^2-6x=0$ are 0, $\frac{2}{3}$, and -3.

7. Find the zeros and state multiplicity. [3.3]

$$P(x)=(x^2-4)^2(2x-3)(x+1)^3$$
$$P(x)=(x-2)^2(x+2)^2(2x-3)(x+1)^3$$

The zeros of P are 2 (multiplicity 2),

–2 (multiplicity 2), $\frac{3}{2}$ (multiplicity 1), and

–1 (multiplicity 3).

9. State the number of positive and negative real zeros using Descartes' Rule of Signs. [3.3]

$P(x)=x^4-3x^3+2x^2-5x+1$ has 4 changes in sign.

Therefore there is four, two or zero positive zeros.

$P(-x)=x^4+3x^3+2x^2+5x+1$ has no change in sign. Therefore there are no negative zeros.

11. Find the remaining zeros. [3.4]

$$\begin{array}{r|rrrrr} 2+3i & 6 & -5 & 12 & 207 & 130 \\ & & 12+18i & -40+57i & -227+30i & -130 \\ \hline & 6 & 7+18i & -28+57i & -20+30i & 0 \end{array}$$

$$\begin{array}{r|rrrr} 2-3i & 6 & 7+18i & -28+57i & -20+30i \\ & & 12-18i & 38-57i & 20-30i \\ \hline & 6 & 19 & 10 & 0 \end{array}$$

$$6x^2+19x+10=0$$
$$(3x+2)(2x+5)=0$$
$$3x+2=0 \qquad 2x+5=0$$
$$x=-\frac{2}{3} \qquad x=-\frac{5}{2}$$

The remaining zeros are $2-3i$, $-\frac{2}{3}$ and $-\frac{5}{2}$.

13. Find the polynomial function. [3.4]

$$\begin{aligned} P(x) &= [x-(1+i)][x-(1-i)](x-3)(x) \\ &= (x^2-2x+2)(x-3)(x) \\ &= (x^3-5x^2+8x-6)(x) \\ &= x^4-5x^3+8x^2-6x \end{aligned}$$

15. Graph. [3.2]

$$P(x) = x^3-6x^2+9x+1=0$$

Let $x = 0$. $P(0) = 0^3-6(0)^2+9(0)+1=1$

The y-intercept is (0, 1).

x^3 has a positive coefficient and an odd exponent.

Therefore, the graph goes down to the far left and up to the far right.

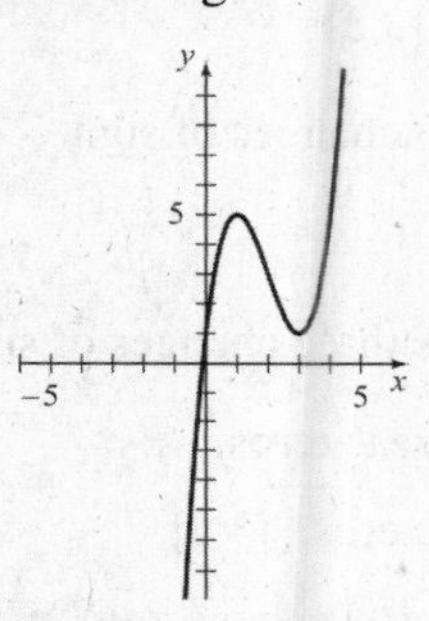

17. Graph. [3.5]

$$f(x) = \frac{2x^2+2x+1}{x+1}$$

Vertical asymptote: $x = -1$

Horizontal asymptote: none

Slant asymptote: $y = 2x$

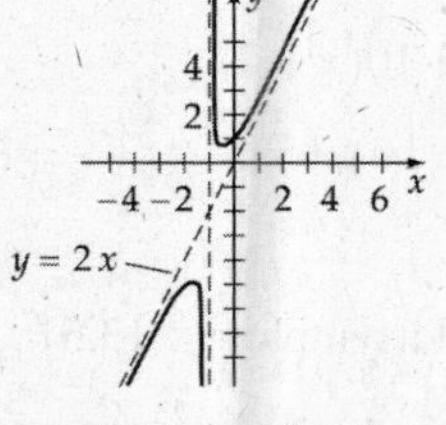

19. a. Evaluate the function at the values. [3.5]

$$w(t) = \frac{70t+120}{t+40}$$

$$w(1) = \frac{70(1)+120}{1+40} = \frac{70+120}{41} = \frac{190}{41}$$

≈ 5 words per minute

$$w(10) = \frac{70(10)+120}{10+40} = \frac{700+120}{50} = \frac{820}{50}$$

≈ 16 words per minute

$$w(20) = \frac{70(20)+120}{20+40} = \frac{1400+120}{60} = \frac{1520}{60}$$

≈ 25 words per minute

b. Find the number of hours.

$$\begin{aligned} 60 &= \frac{70t+120}{t+40} \\ 60t+2400 &= 70t+120 \\ 2280 &= 10t \\ t &= 228 \text{ hours} \end{aligned}$$

c. Explain the far-right behavior.

As $t \to \infty$, $w(t) \to \frac{70}{1} = 70$ words per minute

Cumulative Review Exercises

1. Write in $a + bi$ form. [P.6]

$$\begin{aligned} \frac{3+4i}{1-2i} &= \frac{3+4i}{1-2i} \cdot \frac{1+2i}{1+2i} = \frac{3+10i+8i^2}{1^2-4i^2} \\ &= \frac{3+10i+8(-1)}{1-4(-1)} = \frac{3+10i-8}{1+4} \\ &= \frac{-5+10i}{5} = -1+2i \end{aligned}$$

3. Solve. [1.4]

$$\begin{aligned} \sqrt{2x+5}-\sqrt{x-1} &= 2 \\ \sqrt{2x+5} &= 2+\sqrt{x-1} \\ \left(\sqrt{2x+5}\right)^2 &= \left(2+\sqrt{x-1}\right)^2 \\ 2x+5 &= 4+4\sqrt{x-1}+x-1 \\ 2x+5 &= 3+4\sqrt{x-1}+x \\ x+2 &= 4\sqrt{x-1} \\ (x+2)^2 &= \left(4\sqrt{x-1}\right)^2 \\ x^2+4x+4 &= 16(x-1) \\ x^2+4x+4 &= 16x-16 \\ x^2-12x+20 &= 0 \\ (x-2)(x-10) &= 0 \end{aligned}$$

$x = 2$, $x = 10$

Check 2:

$$\begin{aligned} \sqrt{2(2)+5}-\sqrt{(2)-1} &= 2 \\ \sqrt{9}-\sqrt{1} &= 2 \\ 3-1 &= 2 \\ 2 &= 2 \end{aligned}$$

Yes

Check 10:

$$\sqrt{2(10)+5}-\sqrt{(10)-1}=2$$
$$\sqrt{20+5}-\sqrt{10-1}=2$$
$$\sqrt{25}-\sqrt{9}=2$$
$$5-3=2$$
$$2=2$$

Yes

The solutions are $x = 2$, $x = 10$.

5. Find the distance between the points. [2.1]

$$d=\sqrt{(2-7)^2+[5-(-11)]^2}$$
$$=\sqrt{(2-7)^2+(5+11)^2}$$
$$=\sqrt{(-5)^2+(16)^2}$$
$$=\sqrt{25+256}$$
$$=\sqrt{281}$$

7. Find the difference quotient. [2.6]

$$P(x)=x^2-2x-3$$

$$\frac{P(x+h)-P(x)}{h}$$
$$=\frac{[(x+h)^2-2(x+h)-3]-(x^2-2x-3)}{h}$$
$$=\frac{x^2+2xh+h^2-2x-2h-3-x^2+2x+3}{h}$$
$$=\frac{2xh+h^2-2h}{h}$$
$$=2x+h-2$$

9. Find $(f-g)(x)$. [2.6]

$$(f-g)(x)=f(x)-g(x)$$
$$=x^3-2x+7-(x^2-3x-4)$$
$$=x^3-2x+7-x^2+3x+4$$
$$=x^2-x^2+x+11$$

11. Find $P(3)$ using Remainder Theorem. [3.1]

$$P(x)=2x^4-3x^2+4x-6$$

3	2	0	−3	4	−6
		6	18	45	147
	2	6	15	49	141

$P(3) = 141$

13. Find the relative maximum. [3.2]

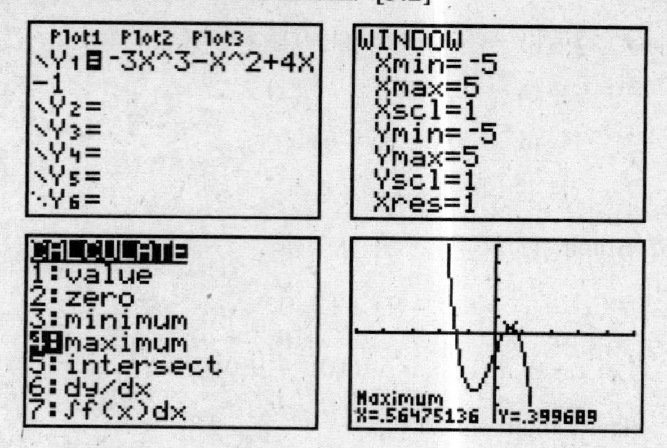

The relative maximum (to the nearest 0.0001) is 0.3997.

15. State the number of positive and negative real zeros using Descartes' Rule of Signs. [3.3]

$P(x)=x^3+x^2+2x+4$ has no changes of sign.

There are no positive real zeros.

$P(-x)=-x^3+x^2-2x+4$ has three changes of sign.

There are three or one negative real zeros.

17. Find the polynomial with given zeros. [3.4]

If $3 + i$ is a zero of $P(x)$, then $3 - i$ is also a zero.

$$P(x)=[x-(3+i)][x-(3-i)](x+2)$$
$$=[x-3-i][x-3+i](x+2)$$
$$=[(x-3)^2-i^2](x+2)$$
$$=[x^2-6x+9-(-1)](x+2)$$
$$=[x^2-6x+9+1](x+2)$$
$$=(x^2-6x+10)(x+2)$$
$$=x^2(x+2)-6x(x+2)+10(x+2)$$
$$=x^3+2x^2-6x^2-12x+10x+20$$
$$=x^3-4x^2-2x+20$$

19. Find the vertical and horizontal asymptotes. [3.5]

$$F(x)=\frac{4x^2}{x^2+x-6}$$

Vertical asymptotes:

$$x^2+x-6=0$$
$$(x+3)(x-2)=0$$
$$x=-3,\ x=2$$

Horizontal asymptote: $y=\frac{4}{1}$ Then $y=4$.

Chapter 4: Exponential and Logarithmic Functions

Section 4.1 Exercises

1. If $f(3)=7$, then $f^{-1}(7)=3$.

3. If $h^{-1}(-3)=-4$, then $h(-4)=-3$.

5. If 3 is in the domain of f^{-1}, then $f[f^{-1}(3)]=3$.

7. The domain of the inverse function f^{-1} is the range of *f*.

9. Draw the graph of the inverse relation, determine if it is a function.

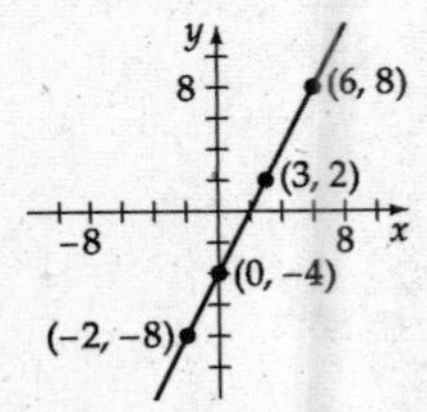

Yes, the inverse is a function.

11. Draw the graph of the inverse relation, determine if it is a function.

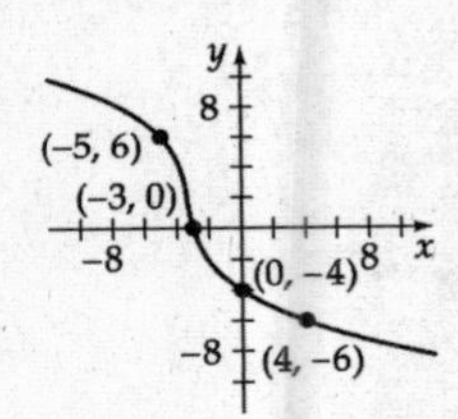

Yes, the inverse is a function.

13. Draw the graph of the inverse relation, determine if it is a function.

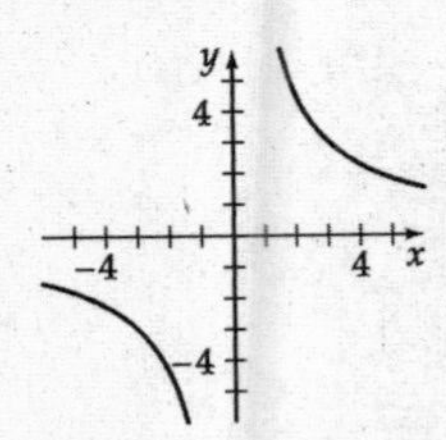

Yes, the inverse is a function.

15. Draw the graph of the inverse relation, determine if it is a function.

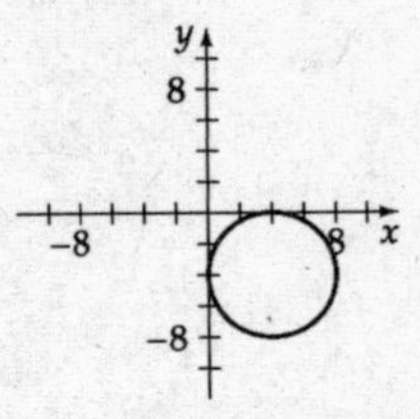

No, the inverse relation is not a function.

17. Use composition to determine whether *f* and *g* are inverses.

$$f(x)=4x;\ g(x)=\frac{x}{4}$$

$$f[g(x)]=f\left(\frac{x}{4}\right)=4\left(\frac{x}{4}\right)=x$$

$$g[f(x)]=g(4x)=\frac{4x}{4}=x$$

Yes, *f* and *g* are inverses of each other.

19. Use composition to determine whether *f* and *g* are inverses.

$$f(x)=4x-1;\ g(x)=\frac{1}{4}x+\frac{1}{4}$$

$$\begin{aligned}f[g(x)]&=f\left(\frac{1}{4}x+\frac{1}{4}\right)\\&=4\left(\frac{1}{4}x+\frac{1}{4}\right)-1=x+1-1\\&=x\end{aligned}$$

$$\begin{aligned}g[f(x)]&=g(4x-1)\\&=\frac{1}{4}(4x-1)+\frac{1}{4}=x-\frac{1}{4}+\frac{1}{4}\\&=x\end{aligned}$$

Yes, *f* and *g* are inverses of each other.

21. Use composition to determine whether *f* and *g* are inverses.

$$f(x)=-\frac{1}{2}x-\frac{1}{2};\ g(x)=-2x+1$$

$$\begin{aligned}f[g(x)]&=f(-2x+1)\\&=-\frac{1}{2}(-2x+1)-\frac{1}{2}=x-\frac{1}{2}-\frac{1}{2}\\&=x-1\\&\neq x\end{aligned}$$

No, *f* and *g* are not inverses of each other.

23. Use composition to determine whether *f* and *g* are inverses.

$$f(x)=\frac{5}{x-3};\ g(x)=\frac{5}{x}+3$$

$$f[g(x)]=f\left(\frac{5}{x}+3\right)$$

$$=\frac{5}{\frac{5}{x}+3-3}=\frac{5}{\frac{5}{x}}=5\cdot\frac{x}{5}$$

$$=x$$

$$g[f(x)]=g\left(\frac{5}{x-3}\right)$$

$$=\frac{5}{\frac{5}{x-3}}+3=x-3+3$$

$$=x$$

Yes, *f* and *g* are inverses of each other.

25. Use composition to determine whether *f* and *g* are inverses.

$$f(x)=x^3+2;\ g(x)=\sqrt[3]{x-2}$$

$$f[g(x)]=f\left(\sqrt[3]{x-2}\right)$$

$$=\left(\sqrt[3]{x-2}\right)^3+2=x-2+2$$

$$=x$$

$$g[f(x)]=g\left(x^3+2\right)$$

$$=\sqrt[3]{x^3+2-2}=\sqrt[3]{x^3}$$

$$=x$$

Yes, *f* and *g* are inverses of each other.

27. Find the inverse.

The inverse of {(−3, 1), (−2, 2), (1, 5), (4, −7)} is {(1, −3), (2, −2), (5, 1), (−7, 4)}.

29. Find the inverse.

The inverse of {(0, 1), (1, 2), (2, 4), (3, 8), (4, 16)} is {(1, 0), (2, 1), (4, 2), (8, 3), (16, 4)}.

31. Find $f^{-1}(x)$.

$$f(x)=2x+4$$

$$x=2y+4$$

$$x-4=2y$$

$$\frac{1}{2}x-2=y$$

$$f^{-1}(x)=\frac{1}{2}x-2$$

33. Find $f^{-1}(x)$.

$$f(x)=3x-7$$

$$x=3y-7$$

$$x+7=3y$$

$$\frac{1}{3}x+\frac{7}{3}=y$$

$$f^{-1}(x)=\frac{1}{3}x+\frac{7}{3}$$

35. Find $f^{-1}(x)$.

$$f(x)=-2x+5$$

$$x=-2y+5$$

$$x-5=-2y$$

$$-\frac{1}{2}x+\frac{5}{2}=y$$

$$f^{-1}(x)=-\frac{1}{2}x+\frac{5}{2}$$

37. Find $f^{-1}(x)$.

$$f(x)=\frac{2x}{x-1},\ x\neq 1$$

$$x=\frac{2y}{y-1}$$

$$x(y-1)=xy-x=2y$$

$$xy-2y=y(x-2)=x$$

$$y=\frac{x}{x-2}$$

$$f^{-1}(x)=\frac{x}{x-2},\ x\neq 2$$

39. Find $f^{-1}(x)$.

$$f(x)=\frac{x-1}{x+1},\ x\neq -1$$

$$x=\frac{y-1}{y+1}$$

$$x(y+1)=xy+x=y-1$$

$$xy-y=-x-1$$

$$y-xy=y(1-x)=x+1$$

$$y=\frac{x+1}{1-x}$$

$$f^{-1}(x)=\frac{x+1}{1-x},\ x\neq 1$$

41. Find $f^{-1}(x)$.

$$\begin{aligned} f(x) &= x^2+1,\ x \ge 0 \\ x &= y^2+1 \\ x-1 &= y^2 \\ \sqrt{x-1} &= y \\ f^{-1}(x) &= \sqrt{x-1},\ x \ge 1 \end{aligned}$$

Note: Do not use $\pm$ with the radical because the domain of f, and thus the range of f^{-1}, is nonnegative.

43. Find $f^{-1}(x)$.

$$\begin{aligned} f(x) &= \sqrt{x-2},\ x \ge 2 \\ x &= \sqrt{y-2} \\ x^2 &= y-2 \\ x^2+2 &= y \\ f^{-1}(x) &= x^2+2,\ x \ge 0 \end{aligned}$$

Note: The range of f, is nonnegative, therefore the domain of f^{-1} is also nonnegative.

45. Find $f^{-1}(x)$.

$$\begin{aligned} f(x) &= x^2+4x,\ x \ge -2 \\ x &= y^2+4y \\ x+4 &= y^2+4y+4 \\ x+4 &= (y+2)^2 \\ \sqrt{x+4} &= y+2 \\ y &= \sqrt{x+4}-2 \\ f^{-1}(x) &= \sqrt{x+4}-2,\ x \ge -4 \end{aligned}$$

47. Find $f^{-1}(x)$.

$$\begin{aligned} f(x) &= x^2+4x-1,\ x \le -2 \\ x &= y^2+4y-1 \\ x+1 &= y^2+4y \\ x+1+4 &= y^2+4y+4 \\ x+5 &= (y+2)^2 \\ -\sqrt{x+5} &= y+2 \\ -\sqrt{x+5}-2 &= y \\ f^{-1}(x) &= -\sqrt{x+5}-2,\ x \ge -5 \end{aligned}$$

49. Find f^{-1} and explain how it is used.

$$\begin{aligned} f(x) &= \frac{5}{9}(x-32) \\ x &= \frac{5}{9}(y-32) \\ \frac{9}{5}x &= y-32 \\ \frac{9}{5}x+32 &= y \\ f^{-1}(x) &= \frac{9}{5}x+32 \end{aligned}$$

$f^{-1}(x)$ is used to convert x degrees Celsius to an equivalent Fahrenheit temperature.

51. Find $s^{-1}(x)$.

$$\begin{aligned} s(x) &= 2x+24 \\ x &= 2y+24 \\ x-24 &= 2y \\ \frac{1}{2}x-12 &= y \\ s^{-1}(x) &= \frac{1}{2}x-12 \end{aligned}$$

53. **a.** Find $c(30)$.

$$c(x) = \frac{300+12x}{x}$$

$$c(30) = \frac{300+12(30)}{30} = \$22$$

The company charges \$22 per person to cater a dinner for 30 people.

b. Find $c^{-1}(x)$.

$$\begin{aligned} c(x) &= \frac{300+12x}{x} \\ x &= \frac{300+12y}{y} \\ xy &= 300+12y \\ xy-12y &= 300 \\ y(x-12) &= 300 \\ y &= \frac{300}{x-12} \\ c^{-1}(x) &= \frac{300}{x-12} \end{aligned}$$

c. $c^{-1}(\$15.00) = \dfrac{300}{15.00-12} = 100$ people

55. Find and explain how it could be used.

$$E(s) = 0.05s + 2500$$
$$s = 0.05y + 2500$$
$$s - 2500 = 0.05y$$
$$\frac{1}{0.05}s - \frac{2500}{0.05} = y$$
$$20s - 50,000 = y$$
$$E^{-1}(s) = 20s - 50,000$$

The executive can use the inverse function to determine the value of the software that must be sold in order to achieve a given monthly income.

57. a. $p(10) \approx 0.12 = 12\%$; $p(30) \approx 0.71 = 71\%$

b. The graph of p, for $1 \le n \le 60$, is an increasing function. Thus p has an inverse that is a function.

c. $p^{-1}(0.223)$ represents the number of people required to be in the group for a 22.3% probability that at least two of the people will share a birthday.

59. a. Encode the message.

D $f(13) = 2(13) - 1 = 25$
O $f(24) = 2(24) - 1 = 47$
(space) $f(36) = 2(36) - 1 = 71$
Y $f(34) = 2(34) - 1 = 67$
O $f(24) = 47$
U $f(30) = 2(30) - 1 = 59$
R $f(27) = 2(27) - 1 = 53$
(space) $f(36) = 71$
H $f(17) = 2(17) - 1 = 33$
O $f(24) = 47$
M $f(22) = 2(22) - 1 = 43$
E $f(14) = 2(14) - 1 = 27$
W $f(32) = 2(32) - 1 = 63$
O $f(24) = 47$
R $f(27) = 53$
K $f(20) = 2(20) - 1 = 39$

The code is
25 47 71 67 47 59 53 71 33 47 43 27 63 47 53 39.

b. Decode the message.

$f^{-1}(49) = \frac{49+1}{2} = 25$ P
$f^{-1}(33) = \frac{33+1}{2} = 17$ H
$f^{-1}(47) = \frac{47+1}{2} = 24$ O
$f^{-1}(45) = \frac{45+1}{2} = 23$ N
$f^{-1}(27) = \frac{27+1}{2} = 14$ E
$f^{-1}(71) = \frac{71+1}{2} = 36$ (space)
$f^{-1}(33) = 17$ H
$f^{-1}(47) = 24$ O
$f^{-1}(43) = \frac{43+1}{2} = 22$ M
$f^{-1}(27) = 14$ E

The message is PHONE HOME.

c. Answers will vary.

61. $f(2) = 7$, $f(5) = 12$, and $f(4) = c$. Because f is an increasing linear function, and 4 is between 2 and 5, then $f(4)$ is between $f(2)$ and $f(5)$. Thus, c is between 7 and 12.

63. f is a linear function, therefore f^{-1} is a linear function.

$$f(2) = 3 \Rightarrow f^{-1}(3) = 2$$
$$f(5) = 9 \Rightarrow f^{-1}(9) = 5$$

Since 6 is between 3 and 9, $f^{-1}(6)$ is between 2 and 5.

65. A horizontal line intersects the graph of the function at more than one point. Thus, the function is not a one-to-one function.

67. A horizontal line intersects the graph of the function at more than one point. Thus, the function is not a one-to-one function.

69. Find the slope and y-intercept of the inverse.

$f(x) = mx + b, \quad m \neq 0$

$$y = mx + b$$
$$x = my + b$$
$$x - b = my$$
$$\frac{x-b}{m} = y$$
$$f^{-1}(x) = \frac{1}{m}x - \frac{b}{m}$$

The slope is $\frac{1}{m}$ and the y-intercept is $\left(0, -\frac{b}{m}\right)$.

71. The reflection of f across the line given by $y=x$ yields f. Thus f is its own inverse.

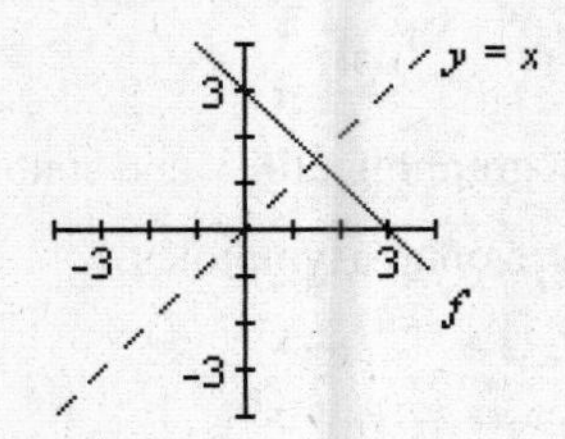

Prepare for Section 4.2

P1. Evaluate.

$2^3 = 2 \cdot 2 \cdot 2 = 8$

P3. Evaluate.

$$\frac{2^2 + 2^{-2}}{2} = \frac{4 + \frac{1}{4}}{2} = \frac{16+1}{8} = \frac{17}{8}$$

P5. Evaluate.

$$f(x) = 10^x$$
$$f(-1) = 10^{-1} = \frac{1}{10}$$
$$f(0) = 10^0 = 1$$
$$f(1) = 10^1 = 10$$
$$f(2) = 10^2 = 100$$

Section 4.2 Exercises

1. Evaluate.

$$f(0) = 3^0 = 1$$
$$f(4) = 3^4 = 81$$

3. Evaluate.

$$g(-2) = 10^{-2} = \frac{1}{100}$$
$$g(3) = 10^3 = 1000$$

5. Evaluate.

$$h(2) = \left(\frac{3}{2}\right)^2 = \frac{9}{4}$$
$$h(-3) = \left(\frac{3}{2}\right)^{-3} = \frac{8}{27}$$

7. Evaluate.

$$j(-2) = \left(\frac{1}{2}\right)^{-2} = 4$$
$$j(4) = \left(\frac{1}{2}\right)^4 = \frac{1}{16}$$

9. Evaluate using a calculator. Round to the hundredth.

$f(3.2) = 2^{3.2} \approx 9.19$

11. Evaluate using a calculator. Round to the hundredth.

$g(2.2) = e^{2.2} \approx 9.03$

13. Evaluate using a calculator. Round to the hundredth.

$h(\sqrt{2}) = 5^{\sqrt{2}} \approx 9.74$

15. $f(x) = 5^x$ is a basic exponential graph.

$g(x) = 1 + 5^{-x}$ is the graph of $f(x)$ reflected across the y-axis and moved up 1 unit.

$h(x) = 5^{x+3}$ is the graph of $f(x)$ moved to the left 3 units.

$k(x) = 5^x + 3$ is the graph of $f(x)$ moved up 3 units.

a. $k(x)$

b. $g(x)$

c. $h(x)$

d. $f(x)$

17. Sketch the graph.

$f(x) = 3^x$

19. Sketch the graph.

$f(x) = 10^x$

21. Sketch the graph.

$f(x) = \left(\frac{3}{2}\right)^x$

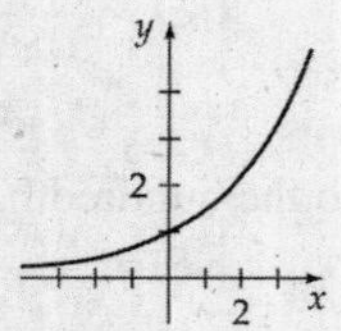

23. Sketch the graph.

$f(x) = \left(\frac{1}{3}\right)^x$

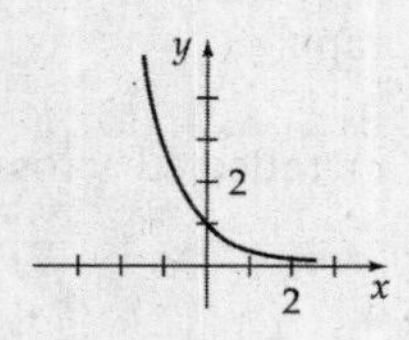

25. Explain how to get the second function from the first.
Shift the graph of f vertically upward 2 units.

27. Explain how to get the second function from the first.
Shift the graph of f horizontally to the right 2 units.

29. Explain how to get the second function from the first.
Reflect the graph of f across the y-axis.

31. Explain how to get the second function from the first.
Stretch the graph of f vertically away from the x-axis by a factor of 2.

33. Explain how to get the second function from the first.
Reflect the graph of f across the y-axis, and then shift this graph vertically upward 2 units.

35. Explain how to get the second function from the first.
Shift the graph of f horizontally 4 units to the right, and then reflect this graph across the x-axis.

37. Explain how to get the second function from the first.
Reflect the graph of f across the y-axis, and then shift this graph vertically upward 3 units.

39. Graph the function using a graphing utility and state the equation of existing horizontal asymptotes.

$f(x) = \dfrac{3^x + 3^{-x}}{2}$

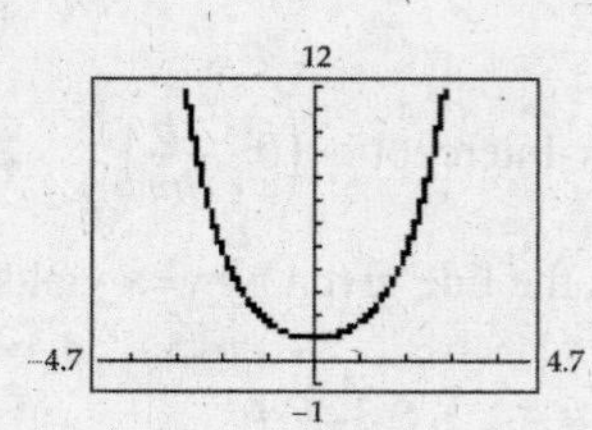

No horizontal asymptote

41. Graph the function using a graphing utility and state the equation of existing horizontal asymptotes.

$f(x) = \dfrac{e^x - e^{-x}}{2}$

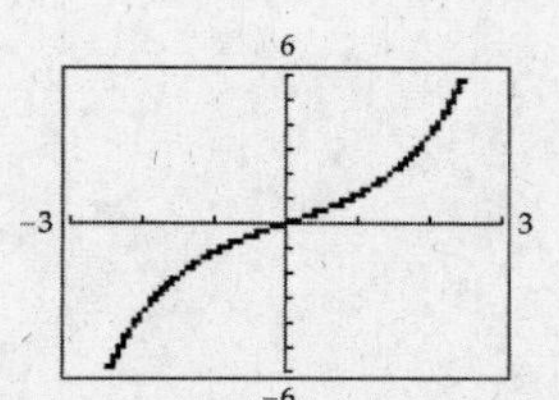

No horizontal asymptote

43. Graph the function using a graphing utility and state the equation of existing horizontal asymptotes.

$f(x) = -e^{(x-4)}$

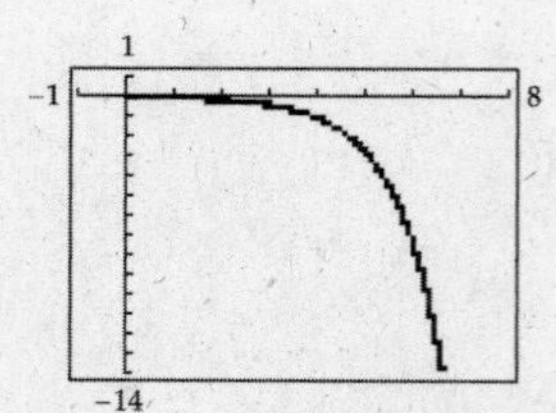

Horizontal asymptote: $y = 0$

45. Graph the function using a graphing utility and state the equation of existing horizontal asymptotes.

$f(x) = \dfrac{10}{1 + 0.4e^{-0.5x}}$, with $x \geq 0$

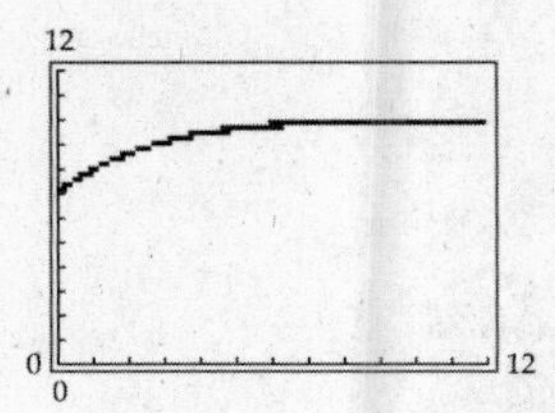

Horizontal asymptote: $y = 10$

47. a. $P(t) = 100 \cdot 2^{2t}$

$$P(3) = 100 \cdot 2^{2(3)} = 100 \cdot 2^{6} = 100 \cdot 64$$
$$= 6400 \text{ bacteria}$$

$$P(t) = 100 \cdot 2^{2t}$$

$$P(6) = 100 \cdot 2^{2(6)} = 100 \cdot 2^{12} = 100 \cdot 4096$$
$$= 409{,}600 \text{ bacteria}$$

b. Use a graphing utility to solve.

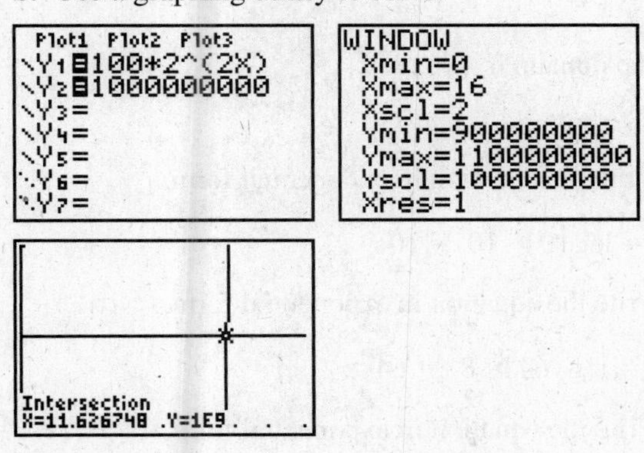

11.6 hours

49. a. $d(p) = 880e^{-0.18p}$

$$d(10) = 880e^{-0.18(10)} \approx 145 \text{ items per month}$$

$$d(p) = 880e^{-0.18p}$$

$$d(18) = 880e^{-0.18(18)} \approx 34 \text{ items per month}$$

b. As $p \to \infty$, $d(p) \to 0$. The demand will approach 0 items per month.

51. a. $P(x) = (0.9)^{x}$

$$P(3.5) = (0.9)^{3.5} \approx 69.2\%$$

b. Use a graphing utility to solve.

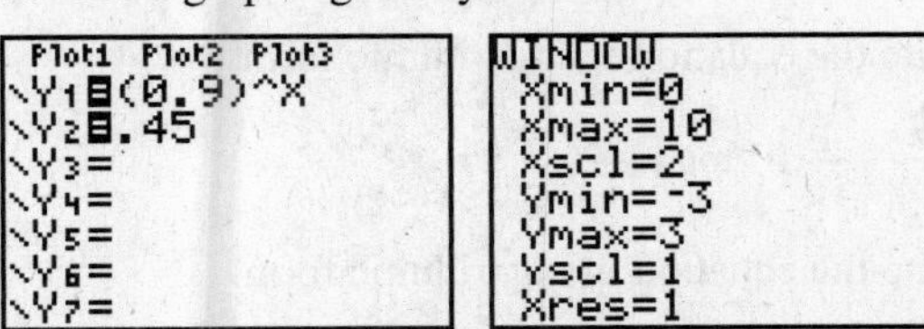

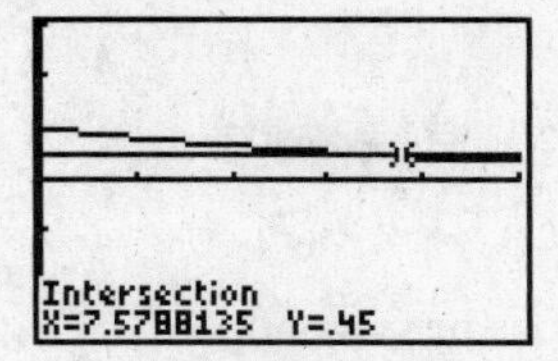

For a transparency of 45%, the UV index is 7.6.

53. a. $B(n) = \dfrac{3^{n+1} - 3}{2}$

$$B(5) = \frac{3^{5+1} - 3}{2} = \frac{3^{6} - 3}{2} = \frac{729 - 3}{2} = \frac{726}{2}$$
$$= 363 \text{ beneficiaries}$$

$$B(n) = \frac{3^{n+1} - 3}{2}$$

$$B(10) = \frac{3^{10+1} - 3}{2} = \frac{3^{11} - 3}{2} = \frac{177147 - 3}{2} = \frac{177144}{2}$$
$$= 88{,}572 \text{ beneficiaries}$$

b. Use a graphing utility to solve.

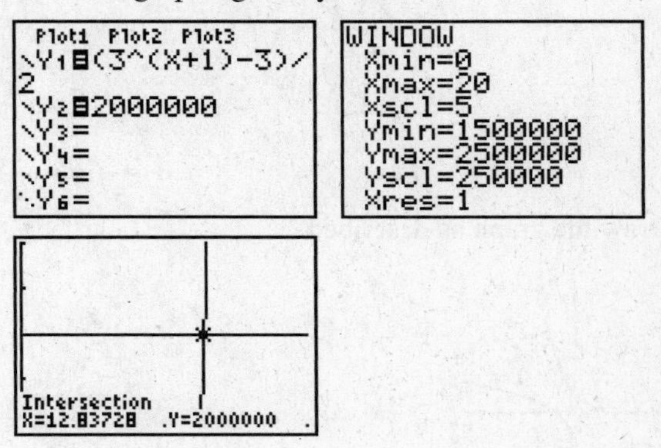

13 rounds

55. a. $T(t) = 65 + 115e^{-0.042t}$

$$T(10) = 65 + 115e^{-0.042(10)} = 65 + 115e^{-0.42}$$
$$\approx 141^{\circ} \text{ F}$$

b. Use a graphing utility to solve.

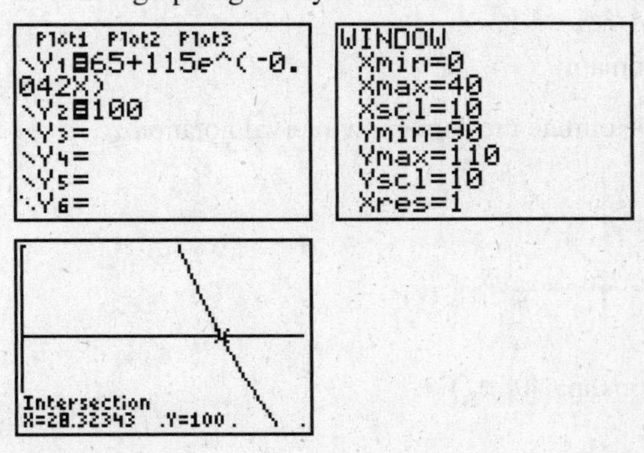

28.3 minutes

57. a. $f(n) = (27.5)2^{(n-1)/12}$

$f(40) = (27.5)2^{(40-1)/12} = (27.5)2^{39/12}$

$= (27.5)2^{3.25}$

≈ 261.63 vibrations per second

b. No. The function $f(n)$ is not a linear function. Therefore, the graph of $f(n)$ does not increase at a constant rate.

59. $f(x) = \dfrac{e^x - e^{-x}}{2}$ is an odd function. That is, prove $f(-x) = -f(x)$.

Proof: $f(x) = \dfrac{e^x - e^{-x}}{2}$

$f(-x) = \dfrac{e^{-x} - e^x}{2}$

$= \dfrac{-e^{-x} + e^x}{2}$

$= \dfrac{(e^x - e^{-x})}{2}$

$= -f(x)$

61. Draw the graph as described.

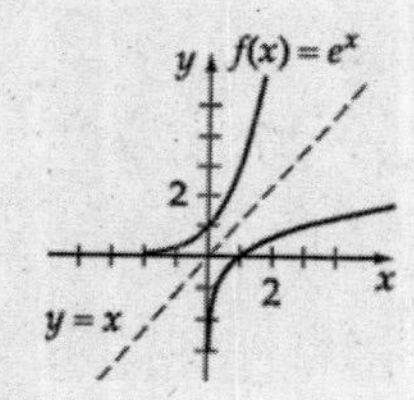

63. Determine the domain in interval notation.

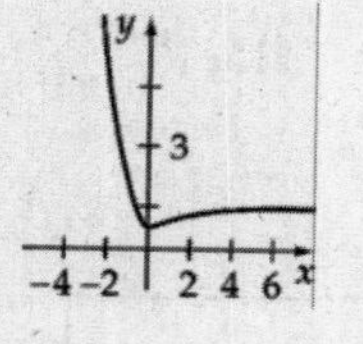

domain: $(-\infty, \infty)$

65. Determine the domain in interval notation.

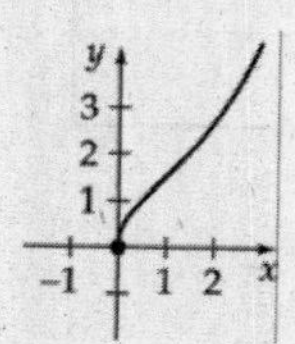

domain: $[0, \infty)$

Prepare for Section 4.3

P1. Determine the value of x.

$2^x = 16$

$2^x = 2^4$

$x = 4$

P3. Determine the value of x.

$x^4 = 625$

$x^4 = 5^4$

$x = 5$

P5. State the domain.

$g(x) = \sqrt{x - 2}$

$x - 2 \geq 0$

$x \geq 2$

The domain is $\{x \mid x \geq 2\}$.

Section 4.3 Exercises

1. Write the equation in exponential form.

$1 = \log 10 \Rightarrow 10^1 = 10$

3. Write the equation in exponential form.

$2 = \log_8 64 \Rightarrow 8^2 = 64$

5. Write the equation in exponential form.

$0 = \log_7 x \Rightarrow 7^0 = x$

7. Write the equation in exponential form.

$\ln x = 4 \Rightarrow e^4 = x$

9. Write the equation in exponential form.

$\ln 1 = 0 \Rightarrow e^0 = 1$

11. Write the equation in exponential form.

$2 = \log(3x + 1) \Rightarrow 10^2 = 3x + 1$

13. Write the equation in logarithmic form.

$3^2 = 9 \Rightarrow \log_3 9 = 2$

15. Write the equation in logarithmic form.

$4^{-2} = \dfrac{1}{16} \Rightarrow \log_4 \dfrac{1}{16} = -2$

17. Write the equation in logarithmic form.

$b^x = y \Rightarrow \log_b y = x$

19. Write the equation in logarithmic form.

$y = e^x \Rightarrow \ln y = x$

21. Write the equation in logarithmic form.

$100 = 10^2 \Rightarrow \log 100 = 2$

23. Write the equation in logarithmic form.

$e^2 = x + 5 \Rightarrow 2 = \ln(x+5)$

25. Evaluate without using a calculator.

$\log_4 16 = 2$ because $4^2 = 16$

27. Evaluate without using a calculator.

$\log_3 \frac{1}{243} = -5$ because $3^{-5} = \left(\frac{1}{3}\right)^5 = \frac{1}{243}$

29. Evaluate without using a calculator.

$\ln e^3 = 3$ because $e^3 = e^3$

31. Evaluate without using a calculator.

$\log \frac{1}{100} = -2$ because $10^{-2} = \frac{1}{10^2} = \frac{1}{100}$

33. Evaluate without using a calculator.

$\log_{0.5} 16 = \log_{1/2} 16 = -4$ because $\left(\frac{1}{2}\right)^{-4} = 2^4 = 16$

35. Evaluate without using a calculator.

$4 \log 1000 = 12 \Rightarrow \log 1000^4 = 12$

because $10^{12} = \left(10^3\right)^4 = (1000)^4$

37. Evaluate without using a calculator.

$2 \log_7 2401 = 8 \Rightarrow \log_7 2401^2 = 8$

because $7^8 = \left(7^4\right)^2 = (2401)^2$

39. Evaluate without using a calculator.

$\log_3 \sqrt[5]{9} = \frac{2}{5} \Rightarrow \log_3 9^{1/5} = \frac{2}{5}$

because $3^{2/5} = \left(3^2\right)^{1/5} = (9)^{1/5}$

41. Evaluate without using a calculator.

$5 \log_{13} \sqrt[3]{169} = \frac{10}{3} \Rightarrow \log_{13} 169^{5/3} = \frac{10}{3}$

because $13^{10/3} = \left(13^2\right)^{5/3} = (169)^{5/3}$

43. Graph the function using its exponential form.

$y = \log_4 x$

$x = 4^y$

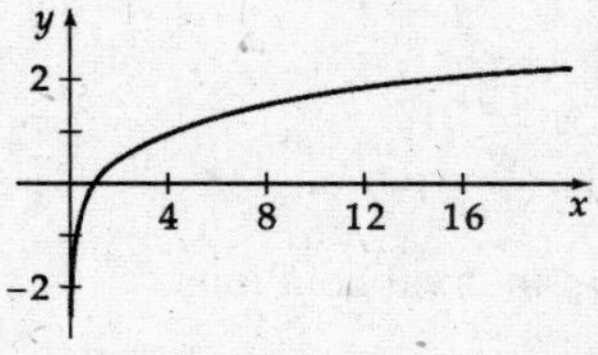

45. Graph the function using its exponential form.

$y = \log_{12} x$

$x = 12^y$

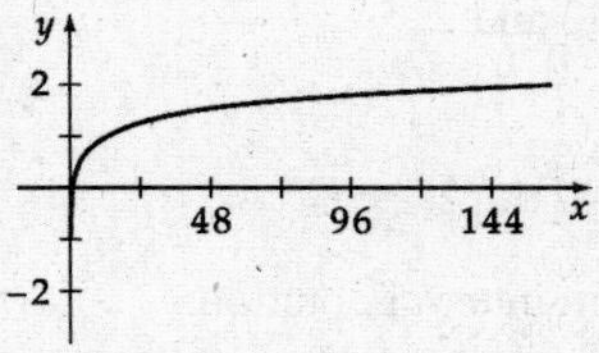

47. Graph the function using its exponential form.

$y = \log_{1/2} x$

$x = (1/2)^y$

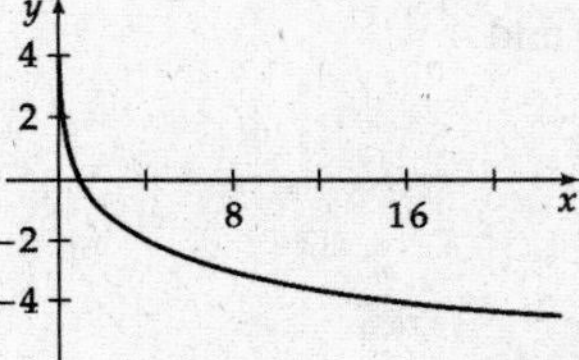

49. Graph the function using its exponential form.

$y = \log_{5/2} x$

$x = (5/2)^y$

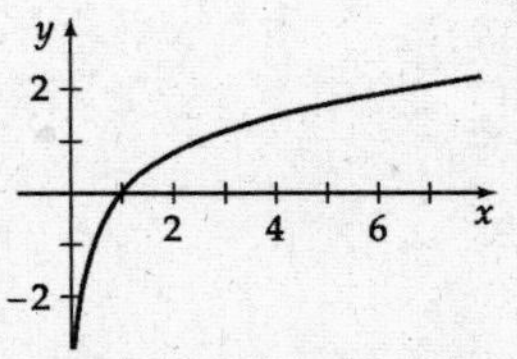

51. Find the domain, write in interval notation.

$f(x) = \log_5(x-3)$

$x - 3 > 0$

$x > 3$

The domain is $(3, \infty)$.

53. Find the domain, write in interval notation.

$k(x) = \log_{2/3}(11 - x)$

$11 - x > 0$

$-x > -11$

$x < 11$

The domain is $(-\infty, 11)$.

55. Find the domain, write in interval notation.

$P(x) = \ln(x^2 - 4)$

$x^2 - 4 > 0$

$(x+2)(x-2) > 0$

The critical values are –2 and 2.

The product is positive.

The domain is $(-\infty, -2) \cup (2, \infty)$.

57. Find the domain, write in interval notation.

$h(x) = \ln\left(\frac{x^2}{x-4}\right)$

$\frac{x^2}{x-4} > 0$

The critical values are 0 and 4.

The quotient is positive.

$x > 4$

The domain is $(4, \infty)$.

59. Find the domain, write in interval notation.

$x^3 - x > 0$

$x(x^2 - 1) > 0$

$x(x+1)(x-1) > 0$

Critical values are 0, ‾1 and 1.

Product is positive.

$-1 < x < 0$ or $x > 1$

$(-1, 0) \cup (1, \infty)$

61. Find the domain, write in interval notation.

$2x - 11 > 0$

$2x > 11$

$x > \frac{11}{2}$

The domain is $\left(\frac{11}{2}, \infty\right)$.

63. Find the domain, write in interval notation.

$3x - 7 > 0$

$3x > 7$

$x > \frac{7}{3}$

The domain is $\left(\frac{7}{3}, \infty\right)$.

65. Use translations to graph.

Shift 3 units to the right.

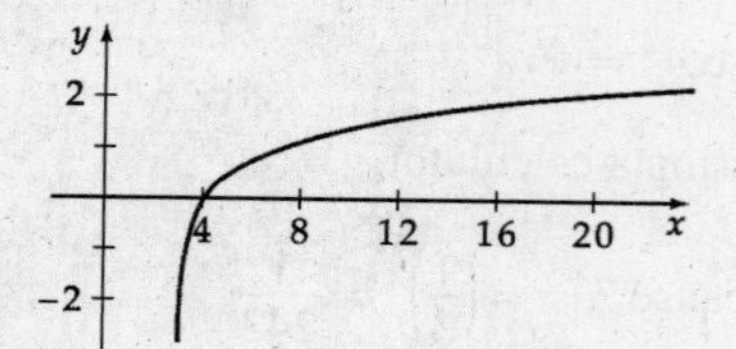

67. Use translations to graph.

Shift 2 units up.

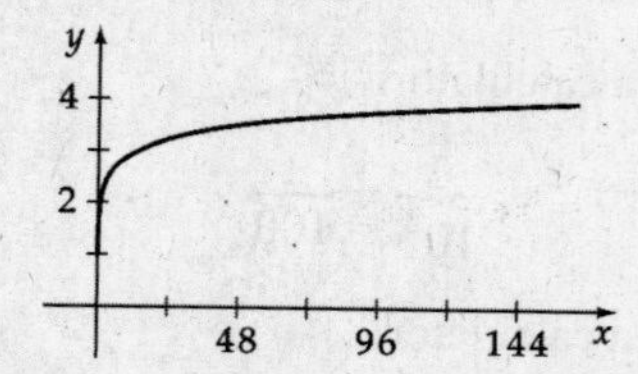

69. Use translations to graph.

Shift 3 units up.

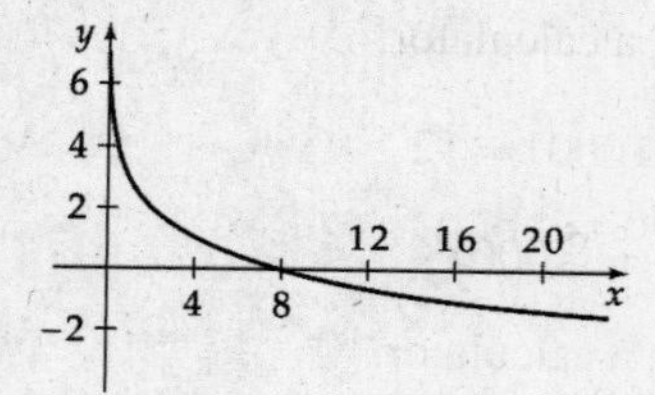

71. Use translations to graph.

Shift 4 units to the right and 1 unit up

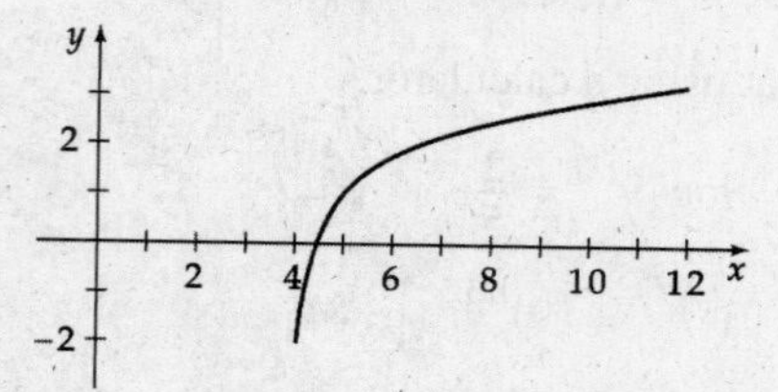

73. The graph of $f(x) = \log_5(x-2)$ is the graph of

$y = \log_5 x$ shifted 2 units to the right.

The graph of $g(x) = 2 + \log_5 x$ is the graph of

$y = \log_5 x$ shifted 2 units up.

The graph of $h(x) = \log_5(-x)$ is the graph of

$y = \log_5 x$ reflected across the y-axis.

The graph of $k(x) = -\log_5(x+3)$ is the graph of

$y = \log_5 x$ reflected across the x-axis and shifted

left 3 units.

a. $k(x)$

b. $f(x)$

c. $g(x)$

d. $h(x)$

75. Use a graphing utility to graph the function.

$f(x) = -2\ln x$

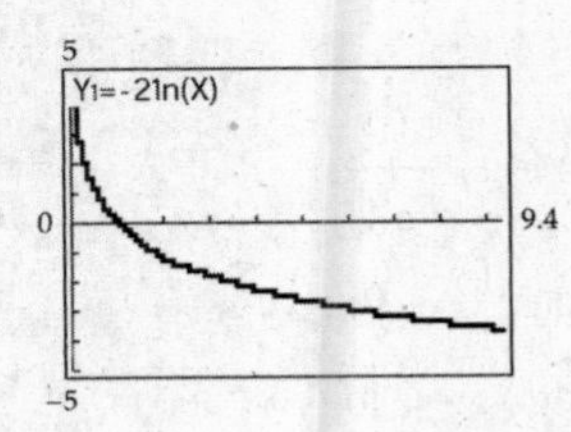

77. Use a graphing utility to graph the function.

$f(x) = |\ln x|$

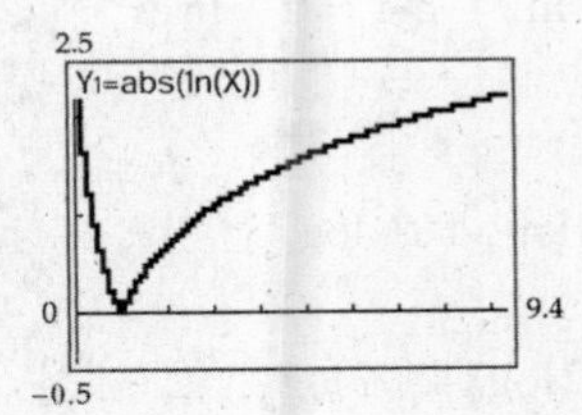

79. Use a graphing utility to graph the function.

$f(x) = \log \sqrt[3]{x}$

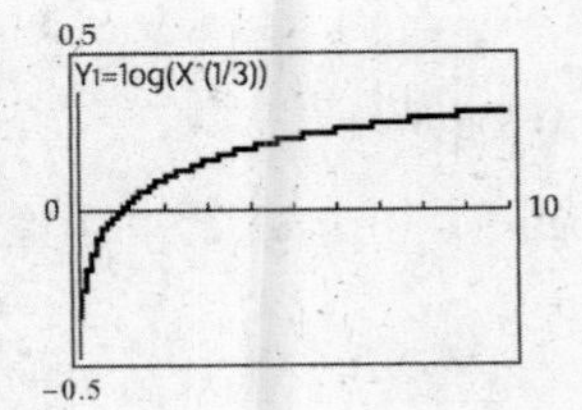

81. Use a graphing utility to graph the function.

$f(x) = \log(x+10)$

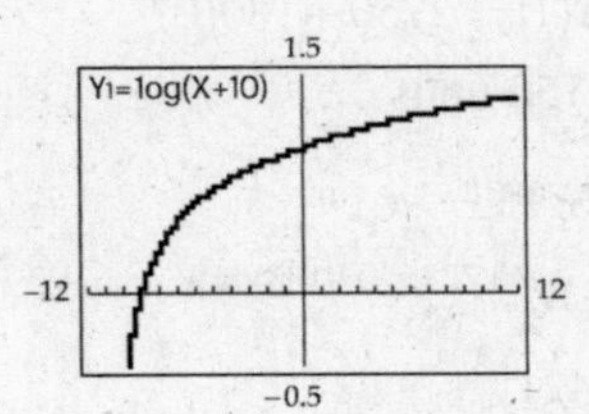

83. Use a graphing utility to graph the function.

$f(x) = 3\log|2x+10|$

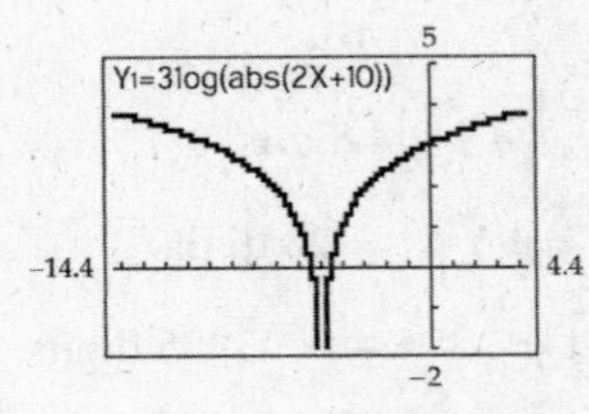

85. a. $r(t) = 0.69607 + 0.60781\ln t$

$r(9) = 0.69607 + 0.60781\ln 9 \approx 2.0\%$

b. Use a graphing calculator to solve.

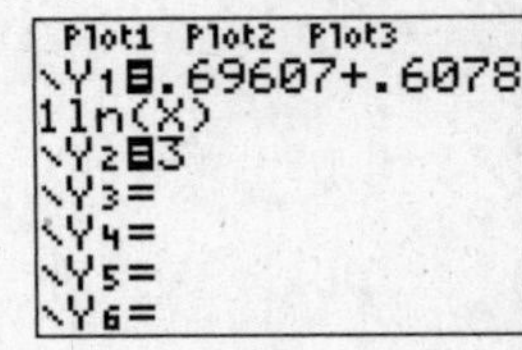

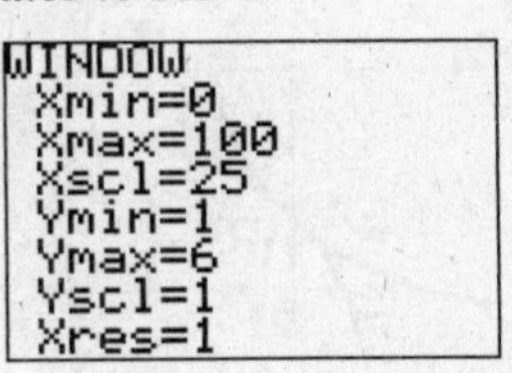

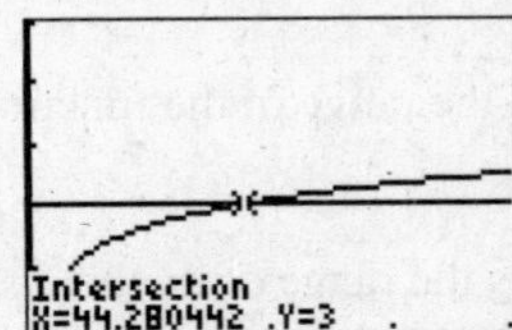

45 months

87. $N(x) = 2750 + 180\ln\left(\dfrac{x}{1000} + 1\right)$

a. $N(20{,}000) = 2750 + 180\ln\left(\dfrac{20{,}000}{1000} + 1\right)$

$= 2750 + 180\ln(21) \approx 3298$ units

$N(40{,}000) = 2750 + 180\ln\left(\dfrac{40{,}000}{1000} + 1\right)$

$= 2750 + 180\ln(41) \approx 3418$ units

$N(60{,}000) = 2750 + 180\ln\left(\dfrac{60{,}000}{1000} + 1\right)$

$= 2750 + 180\ln(61) \approx 3490$ units

b. $N(0) = 2750 + 180\ln\left(\frac{0}{1000}+1\right)$
$= 2750 + 180\ln(1) = 2750 + 180(0)$
$= 2750 + 0 = 2750$ units

89. Estimate the body surface area.

$BSA = 0.0003207 \cdot H^{0.3} \cdot W^{(0.7285-0.0188\log W)}$

$BSA = 0.0003207 \cdot (185.42)^{0.3}$
$\cdot (81,646.6)^{(0.7285-0.0188\log 81,646.6)}$

≈ 2.05 square meters

91. $N = \text{int}(x\log b) + 1$

a. $N = \text{int}(10\log 2) + 1 = 3 + 1 = 4$ digits

b. $N = \text{int}(200\log 3) + 1 = 95 + 1 = 96$ digits

c. $N = \text{int}(4005\log 7) + 1 = 3384 + 1 = 3385$ digits

d. $N = \text{int}(2,0996,001\log 2) + 1 = 6,320,429 + 1$
$= 6,320,430$ digits

93. $f(x)$ and $g(x)$ are inverse functions

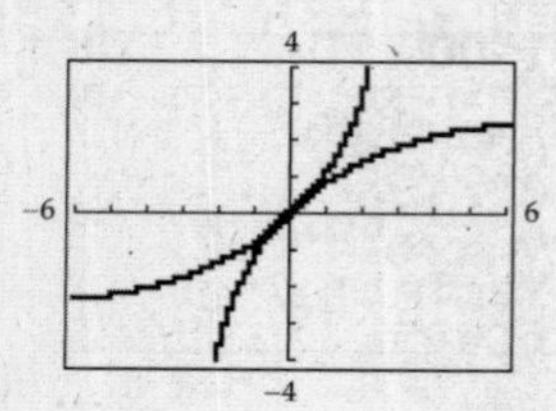

95. The domain of the inverse is the range of the function.

Range of f: $\{y \mid -1 < y \le 1\}$.

The domain of the function is the range of the inverse.

Range of g: all real numbers.

Prepare for Section 4.4

P1. Use a calculator to compare the values.

$\log 3 + \log 2 \approx 0.77815$
$\log 6 \approx 0.77815$

P3. Use a calculator to compare the values.

$3\log 4 \approx 1.80618$

$\log(4^3) \approx 1.80618$

P5. Use a calculator to compare the values.

$\ln 5 \approx 1.60944$

$\frac{\log 5}{\log e} \approx 1.60944$

Section 4.4 Exercises

1. Expand the expression.

$\log_b(xyz) = \log_b x + \log_b y + \log_b z$

3. Expand the expression.

$\ln\frac{x}{z^4} = \ln x - \ln z^4$
$= \ln x - 4\ln z$

5. Expand the expression.

$\log_2\frac{\sqrt{x}}{y^3} = \log_2\sqrt{x} - \log_2 y^3$
$= \log_2 x^{1/2} - \log_2 y^3$
$= \frac{1}{2}\log_2 x - 3\log_2 y$

7. Expand the expression.

$\log_7\frac{\sqrt{xz}}{y^2} = \log_7\frac{(xz)^{1/2}}{y^2} = \log_7\frac{x^{1/2}z^{1/2}}{y^2}$
$= \log_7 x^{1/2} + \log_7 z^{1/2} - \log_7 y^2$
$= \frac{1}{2}\log_7 x + \frac{1}{2}\log_7 z - 2\log_7 y$

9. Expand the expression.

$\ln(e^2 z) = \ln e^2 + \ln z = 2\ln e + \ln z = 2 + \ln z$

11. Expand the expression.

$\log_4\left(\frac{\sqrt[3]{z}}{16y^3}\right) = \log_4 z^{1/3} - \log_4 4^2 - \log_4 y^3$
$= \frac{1}{3}\log_4 z - 2\log_4 4 - 3\log_4 y$
$= \frac{1}{3}\log_4 z - 2 - 3\log_4 y$

13. Expand the expression.

$\log\sqrt{x\sqrt{z}} = \log\left(xz^{1/2}\right)^{1/2} = \log x^{1/2}z^{1/4}$
$= \frac{1}{2}\log x + \frac{1}{4}\log z$

15. Expand the expression.

$\ln\left(\sqrt[3]{z\sqrt{e}}\right) = \ln\left(ze^{1/2}\right)^{1/3} = \ln z^{1/3}e^{1/6}$
$= \ln z^{1/3} + \ln e^{1/6} = \frac{1}{3}\ln z + \frac{1}{6}\ln e$
$= \frac{1}{3}\ln z + \frac{1}{6}$

17. Write as a single expression.

$$\log(x+5)+2\log x=\log(x+5)+\log x^2 = \log[x^2(x+5)]$$

19. Write as a single expression.

$$\ln(x^2-y^2)-\ln(x-y)=\ln\frac{x^2-y^2}{x-y} = \ln\frac{(x+y)(x-y)}{x-y} = \ln(x+y)$$

21. Write as a single expression.

$$3\log x+\frac{1}{3}\log y+\log(x+1) = \log x^3+\log y^{1/3}+\log(x+1) = \log x^3+\log\sqrt[3]{y}+\log(x+1) = \log\left[x^3\cdot\sqrt[3]{y}(x+1)\right]$$

23. Write as a single expression.

$$\log\left(xy^2\right)-\log z=\log\left(\frac{xy^2}{z}\right)$$

25. Write as a single expression.

$$2\left(\log_6 x+\log_6 y^2\right)-\log_6(x+2) = \log_6 x^2+\log_6 y^4-\log_6(x+2) = \log_6\left(\frac{x^2y^4}{x+2}\right)$$

27. Write as a single expression.

$$2\ln(x+4)-\ln x-\ln\left(x^2-3\right) = \ln(x+4)^4-\ln x-\ln\left(x^2-3\right) = \ln\left[\frac{(x+4)^4}{x\left(x^2-3\right)}\right]$$

29. Write as a single expression.

$$\ln(2x+5)-\ln y-2\ln z+\frac{1}{2}\ln w = \ln(2x+5)-\ln y-\ln z^2+\ln w^{1/2} = \ln\left[\frac{(2x+5)\sqrt{w}}{yz^2}\right]$$

31. Write as a single expression.

$$\ln\left(x^2-9\right)-2\ln(x-3)+3\ln y = \ln(x+3)(x-3)-\ln(x-3)^2+\ln y^3 = \ln\left[\frac{(x+3)(x-3)y^3}{(x-3)^2}\right] = \ln\left[\frac{(x+3)y^3}{x-3}\right]$$

33. Approximate to the nearest ten-thousandth.

$$\log_7 20=\frac{\log 20}{\log 7}\approx 1.5395$$

35. Approximate to the nearest ten-thousandth.

$$\log_{11} 8=\frac{\log 8}{\log 11}\approx 0.8672$$

37. Approximate to the nearest ten-thousandth.

$$\log_6\frac{1}{3}=\frac{\log\frac{1}{3}}{\log 6}\approx -0.6131$$

39. Approximate to the nearest ten-thousandth.

$$\log_9\sqrt{17}=\frac{\log\sqrt{17}}{\log 9}\approx 0.6447$$

41. Approximate to the nearest ten-thousandth.

$$\log_{\sqrt{2}} 17=\frac{\log 17}{\log\sqrt{2}}\approx 8.1749$$

43. Approximate to the nearest ten-thousandth.

$$\log_\pi e=\frac{\log e}{\log\pi}\approx 0.8735$$

45. Graph the function using a graphing utility.

$$f(x)=\log_4 x=\frac{\log x}{\log 4}$$

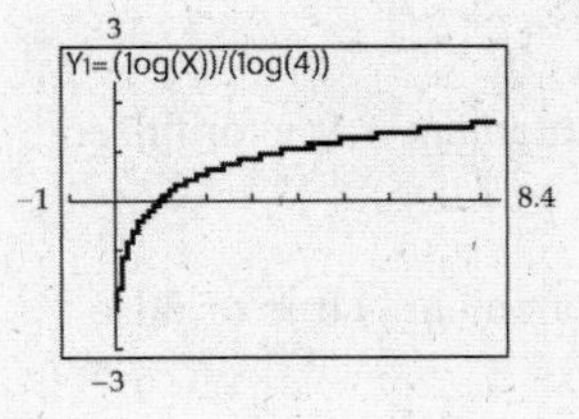

47. Graph the function using a graphing utility.

$$g(x) = \log_8(x-3) = \frac{\log(x-3)}{\log 8}$$

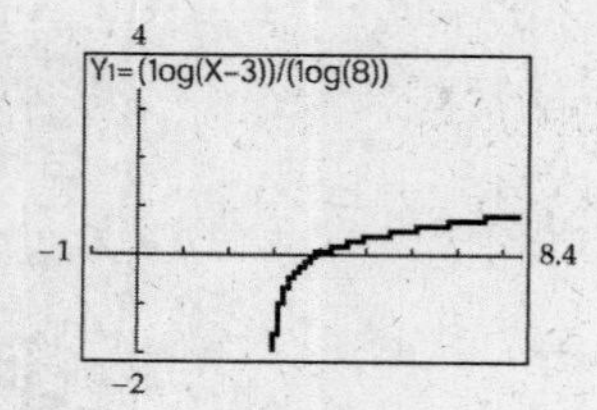

49. Graph the function using a graphing utility.

$$h(x) = \log_3(x-3)^2$$
$$= \frac{\log(x-3)^2}{\log 3}$$
$$= \frac{2\log(x-3)}{\log 3}$$

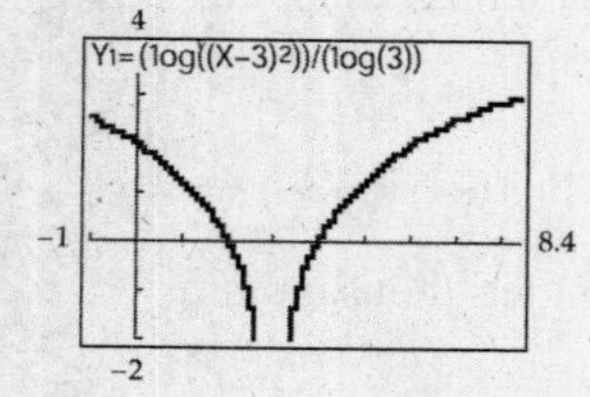

51. Graph the function using a graphing utility.

$$F(x) = -\log_5|x-2| = -\frac{\log|x-2|}{\log 5}$$

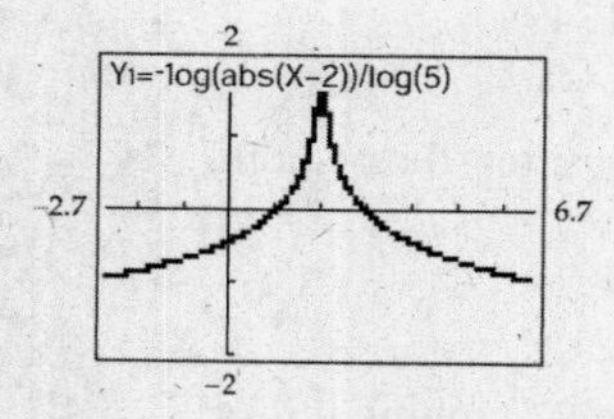

53. Determine whether the statement is true or false.

False. $\log 10 + \log 10 = 1 + 1 = 2$

but $\log(10+10) = \log 20 \neq 2$.

55. Determine whether the statement is true or false.

True.

57. Determine whether the statement is true or false.

False. $\log 100 - \log 10 = 2 - 1 = 1$

but $\log(100-10) = \log 90 \neq 1$

59. Determine whether the statement is true or false.

False. $\dfrac{\log 100}{\log 10} = \dfrac{2}{1} = 2$

but $\log 100 - \log 10 = 2 - 1$
$= 1$

61. Determine whether the statement is true or false.

False. $(\log 10)^2 = 1^2 = 1$ but $2\log 10 = 2(1) = 2$

63. Evaluate the expression without a calculator.

$$\log_3 5 \cdot \log_5 7 \cdot \log_7 9 = \frac{\log 5}{\log 3} \cdot \frac{\log 7}{\log 5} \cdot \frac{\log 9}{\log 7}$$
$$= \frac{\cancel{\log 5}}{\log 3} \cdot \frac{\cancel{\log 7}}{\cancel{\log 5}} \cdot \frac{\log 9}{\cancel{\log 7}}$$
$$= \frac{\log 9}{\log 3} = \frac{\log 3^2}{\log 3} = \frac{2\log 3}{\log 3}$$
$$= \frac{2\cancel{\log 3}}{\cancel{\log 3}} = 2$$

65. $\ln 500^{501} = 501 \ln 500 \approx 3113.52$

$\ln 506^{500} = 500 \ln 506 \approx 3113.27$

$\ln 500^{501}$ is larger.

67. $M = \log\left(\dfrac{I}{I_0}\right)$

$$M = \log\left(\frac{101,400 I_0}{I_0}\right)$$
$$= \log 101,400$$
$$\approx 5.0$$

69. $\log\left(\dfrac{I}{I_0}\right) = M$

$$\log\left(\frac{I}{I_0}\right) = 6.5$$
$$\frac{I}{I_0} = 10^{6.5}$$
$$I = 10^{6.5} I_0$$
$$I \approx 3,162,277.7 I_0$$

71. $M = \log\left(\frac{I}{I_0}\right)$

$M_5 = \log\left(\frac{I_5}{I_0}\right)$ $M_3 = \log\left(\frac{I_3}{I_0}\right)$

$5 = \log\left(\frac{I_5}{I_0}\right)$ $3 = \log\left(\frac{I_3}{I_0}\right)$

$10^5 = \frac{I_5}{I_0}$ $10^3 = \frac{I_3}{I_0}$

$10^5 I_0 = I_5$ $10^3 I_0 = I_3$

$$\frac{I_5}{I_3} = \frac{10^5 I_0}{10^3 I_0} = \frac{10^5}{10^3} = 10^{5-3} = 10^2 = 100 \text{ to } 1$$

73. $\frac{10^{8.9}}{10^{7.1}} = \frac{10^{8.9-7.1}}{1} = 10^{1.8} \text{ to } 1 \approx 63 \text{ to } 1$

75. $M = \log A + 3\log 8t - 2.92$
$= \log 18 + 3\log[8(31)] - 2.92 \approx 5.5$

77. $\text{pH} = -\log[H^+]$
$\text{pH} = -\log[3.97\times10^{-11}]$
$\text{pH} = 10.4$

10.4 > 7; milk of magnesia is a base

79. $\text{pH} = -\log[H^+]$
$9.5 = -\log[H^+]$
$-9.5 = \log[H^+]$
$10^{-9.5} = 10^{\log[H^+]}$
$[H^+] = 3.16\times10^{-10}$ mole per liter

81. $dB(I) = 10\log\left(\frac{I}{I_0}\right)$

a. $dB(1.58\times10^8 \cdot I_0) = 10\log\left(\frac{1.58\times10^8 \cdot I_0}{I_0}\right)$
$= 10\log(1.58\times10^8)$
≈ 82.0 decibels

b. $dB(10{,}800 \cdot I_0) = 10\log\left(\frac{10{,}800 \cdot I_0}{I_0}\right)$
$= 10\log(10{,}800)$
≈ 40.3 decibels

c. $dB(3.16\times10^{11} \cdot I_0) = 10\log\left(\frac{3.16\times10^{11} \cdot I_0}{I_0}\right)$
$= 10\log(3.16\times10^{11})$
≈ 115.0 decibels

d. $dB(1.58\times10^{15} \cdot I_0) = 10\log\left(\frac{1.58\times10^{15} \cdot I_0}{I_0}\right)$
$= 10\log(1.58\times10^{15})$
≈ 152.0 decibels

83. $dB(I) = 10\log\left(\frac{I}{I_0}\right)$

$120 = 10\log\left(\frac{I_{120}}{I_0}\right)$ $110 = 10\log\left(\frac{I_{110}}{I_0}\right)$

$12 = \log\left(\frac{I_{120}}{I_0}\right)$ $11 = \log\left(\frac{I_{110}}{I_0}\right)$

$10^{12} = \frac{I_{120}}{I_0}$ $10^{11} = \frac{I_{110}}{I_0}$

$10^{12} \cdot I_0 = I_{120}$ $10^{11} \cdot I_0 = I_{110}$

$$\frac{I_{120}}{I_{110}} = \frac{10^{12} \cdot I_0}{10^{11} \cdot I_0}$$
$$= \frac{10^{12}}{10^{11}} = 10^{12-11} = 10^1$$
= 10 times more intense

85. Determine the number of scales for each stage.

$S_n = S_0 \cdot 10^{\frac{n}{N}(\log S_f - \log S_0)}$

$S_1 = 1{,}000{,}000 \cdot 10^{\frac{1}{5}(\log 500{,}000 - \log 1{,}000{,}000)}$
$= 870{,}551$

$S_2 = 1{,}000{,}000 \cdot 10^{\frac{2}{5}(\log 500{,}000 - \log 1{,}000{,}000)}$
$= 757{,}858$

$S_3 = 1{,}000{,}000 \cdot 10^{\frac{3}{5}(\log 500{,}000 - \log 1{,}000{,}000)}$
$= 659{,}754$

$S_4 = 1{,}000{,}000 \cdot 10^{\frac{4}{5}(\log 500{,}000 - \log 1{,}000{,}000)}$
$= 574{,}349$

$S_5 = 1{,}000{,}000 \cdot 10^{\frac{5}{5}(\log 500{,}000 - \log 1{,}000{,}000)}$
$= 500{,}000$

The scales are 1:870,551; 1:757,858; 1:659,754; 1:574,349; 1:500,000.

87. Let $r = \log_b M$ and $s = \log_b N$.

Then $M = b^r$ and $N = b^s$.

Consider the quotient of M and N

$$\frac{M}{N} = \frac{b^r}{b^s}$$
$$\frac{M}{N} = b^{r-s}$$
$$\log_b \frac{M}{N} = r - s$$
$$\log_b \frac{M}{N} = \log_b M - \log_b N$$

Mid-Chapter 4 Quiz

1. Use composition to determine whether f and g are inverses.

$$f(x) = \frac{500+120x}{x};\ g(x) = \frac{500}{x-120}$$
$$f[g(x)] = f\left(\frac{500}{x-120}\right)$$
$$= \frac{500+120\left(\frac{500}{x-120}\right)}{\frac{500}{x-120}}$$
$$= \frac{500+120\left(\frac{500}{x-120}\right)}{\frac{500}{x-120}} \cdot \frac{x-120}{x-120}$$
$$= \frac{500(x-120)+120(500)}{500} = x - 120 + 120$$
$$= x$$
$$g[f(x)] = g\left(\frac{500+120x}{x}\right)$$
$$= \frac{500}{\frac{500+120x}{x} - 120} = \frac{500}{\frac{500+120x}{x} - \frac{120x}{x}}$$
$$= \frac{500}{\frac{500}{x}} = 500 \cdot \frac{x}{500}$$
$$= x$$

Yes, f and g are inverses of each other.

3. Evaluate.

$$f(x) = e^x$$
$$f(-2.4) = e^{-2.4} \approx 0.0907$$

5. Graph $f(x) = \log_3(x+3)$.

Shift 3 units to the left.

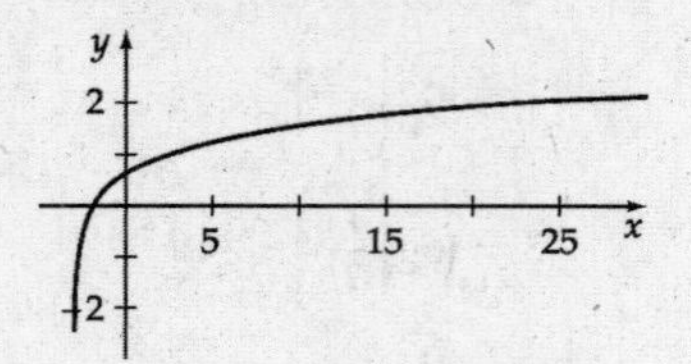

7. Write as a single logarithm.

$$\log_3 x^4 - 2\log_3 z + \log_3\left(xy^2\right)$$
$$= \log_3 x^4 - \log_3 z^2 + \log_3\left(xy^2\right)$$
$$= \log_3\left(\frac{x^5 y^2}{z^2}\right)$$

9. Find the magnitude.

$$M = \log\left(\frac{I}{I_0}\right)$$
$$M = \log\left(\frac{789,251 I_0}{I_0}\right)$$
$$= \log 789,251$$
$$\approx 5.9$$

Prepare for Section 4.5

P1. Write the expression in logarithmic form.

$$3^6 = 729 \Rightarrow \log_3 729 = 6$$

P3. Write the expression in logarithmic form.

$$a^{x+2} = b \Rightarrow \log_a b = x+2$$

P5. Solve for x.

$$165 = \frac{300}{1+12x}$$
$$165(1+12x) = 300$$
$$165 + 1980x = 300$$
$$1980x = 135$$
$$x = \frac{135}{1980} = \frac{3}{44}$$

Section 4.5 Exercises

1. Find the exact solution.

$$2^x = 64$$
$$2^x = 2^6$$
$$x = 6$$

3. Find the exact solution.

$$\begin{aligned} 49^x &= \frac{1}{343} \\ 7^{2x} &= 7^{-3} \\ 2x &= -3 \\ x &= -\frac{3}{2} \end{aligned}$$

5. Find the exact solution.

$$\begin{aligned} 2^{5x+3} &= \frac{1}{8} \\ 2^{5x+3} &= 2^{-3} \\ 5x+3 &= -3 \\ 5x &= -6 \\ x &= -\frac{6}{5} \end{aligned}$$

7. Find the exact solution.

$$\begin{aligned} \left(\frac{2}{5}\right)^x &= \frac{8}{125} \\ \left(\frac{2}{5}\right)^x &= \left(\frac{2}{5}\right)^3 \\ x &= 3 \end{aligned}$$

9. Find the exact solution.

$$\begin{aligned} 5^x &= 70 \\ \log(5^x) &= \log 70 \\ x\log 5 &= \log 70 \\ x &= \frac{\log 70}{\log 5} \end{aligned}$$

11. Find the exact solution.

$$\begin{aligned} 3^{-x} &= 120 \\ \log(3^{-x}) &= \log 120 \\ -x\log 3 &= \log 120 \\ -x &= \frac{\log 120}{\log 3} \\ x &= -\frac{\log 120}{\log 3} \end{aligned}$$

13. Find the exact solution.

$$\begin{aligned} 10^{2x+3} &= 315 \\ \log 10^{2x+3} &= \log 315 \\ (2x+3)\log 10 &= \log 315 \\ 2x+3 &= \log 315 \\ x &= \frac{\log 315 - 3}{2} \end{aligned}$$

15. Find the exact solution.

$$\begin{aligned} e^x &= 10 \\ \ln e^x &= \ln 10 \\ x &= \ln 10 \end{aligned}$$

17. Find the exact solution.

$$\begin{aligned} 2^{1-x} &= 3^{x+1} \\ \ln 2^{1-x} &= \ln 3^{x+1} \\ (1-x)\ln 2 &= (x+1)\ln 3 \\ \ln 2 - x\ln 2 &= x\ln 3 + \ln 3 \\ \ln 2 - x\ln 2 - x\ln 3 &= \ln 3 \\ -x\ln 2 - x\ln 3 &= \ln 3 - \ln 2 \\ -x(\ln 2 + \ln 3) &= \ln 3 - \ln 2 \\ x &= -\frac{(\ln 3 - \ln 2)}{(\ln 2 + \ln 3)} \\ x &= \frac{\ln 2 - \ln 3}{\ln 2 + \ln 3} \text{ or } \frac{\ln 2 - \ln 3}{\ln 6} \end{aligned}$$

19. Find the exact solution.

$$\begin{aligned} 2^{2x-3} &= 5^{-x-1} \\ \log 2^{2x-3} &= \log 5^{-x-1} \\ (2x-3)\log 2 &= (-x-1)\log 5 \\ 2x\log 2 - 3\log 2 &= -x\log 5 - \log 5 \\ 2x\log 2 + x\log 5 - 3\log 2 &= -\log 5 \\ 2x\log 2 + x\log 5 &= 3\log 2 - \log 5 \\ x(2\log 2 + \log 5) &= 3\log 2 - \log 5 \\ x &= \frac{3\log 2 - \log 5}{2\log 2 + \log 5} \end{aligned}$$

21. Find the exact solution.

$$\begin{aligned} \log(4x-18) &= 1 \\ 4x-18 &= 10^1 \\ 4x-18 &= 10 \\ 4x &= 28 \\ x &= 7 \end{aligned}$$

23. Find the exact solution.

$$\begin{aligned} \ln(x^2-12) &= \ln x \\ x^2 - 12 &= x \\ x^2 - x - 12 &= 0 \\ (x-4)(x+3) &= 0 \\ x = 4 \quad &\text{or} \quad x = -3 \text{ (No; not in domain.)} \\ x &= 4 \end{aligned}$$

25. Find the exact solution.

$$\begin{aligned}\log_2 x+\log_2(x-4)&=2\\ \log_2 x(x-4)&=2\\ \log_2(x^2-4x)&=2\\ 2^2&=x^2-4x\\ 0&=x^2-4x-4\\ x&=\frac{4\pm\sqrt{16-4(1)(-4)}}{2}\\ x&=\frac{4\pm4\sqrt{2}}{2}\\ x&=2\pm2\sqrt{2}\end{aligned}$$

$2-2\sqrt{2}$ is not a solution because the logarithm of a negative number is not defined. The solution is $x=2+2\sqrt{2}$.

27. Find the exact solution.

$$\begin{aligned}\log(5x-1)&=2+\log(x-2)\\ \log(5x-1)-\log(x-2)&=2\\ \log\frac{(5x-1)}{(x-2)}&=2\\ 10^2&=\frac{(5x-1)}{(x-2)}\\ 100(x-2)&=5x-1\\ 100x-200&=5x-1\\ 95x&=199\\ x&=\frac{199}{95}\end{aligned}$$

29. Find the exact solution.

$$\begin{aligned}\ln(1-x)+\ln(3-x)&=\ln 8\\ \ln[(1-x)(3-x)]&=\ln 8\\ (1-x)(3-x)&=8\\ 3-4x+x^2&=8\\ x^2-4x-5&=0\\ (x+1)(x-5)&=0\end{aligned}$$

$x=-1$ or $x=5$ (No; not in domain.)

The solution is $x=-1$.

31. Find the exact solution.

$$\begin{aligned}\log\sqrt{x^3-17}&=\frac{1}{2}\\ \frac{1}{2}\log\left(x^3-17\right)&=\frac{1}{2}\\ 10^1&=x^3-17\\ 27&=x^3\\ \sqrt[3]{27}&=\sqrt[3]{x^3}\\ 3&=x\end{aligned}$$

The solution is $x=3$.

33. Find the exact solution.

$$\begin{aligned}\log(\log x)&=1\\ 10^1&=\log x\\ 10^{10}&=x\end{aligned}$$

35. Find the exact solution.

$$\begin{aligned}\ln(e^{3x})&=6\\ 3x\ln e&=6\\ 3x(1)&=6\\ 3x&=6\\ x&=2\end{aligned}$$

37. Find the exact solution.

$$\begin{aligned}\log_7(5x)-\log_7 3&=\log_7(2x+1)\\ \log_7\left(\frac{5x}{3}\right)&=\log_7(2x+1)\\ \frac{5x}{3}&=2x+1\\ 5x&=6x+3\\ -3&=x\end{aligned}$$

$x=-3$ is not a solution because $\log_7(-15)$ is undefined.

No solution.

39. Find the exact solution.

$$\begin{aligned}e^{\ln(x-1)}&=4\\ \ln e^{\ln(x-1)}&=\ln 4\\ \ln(x-1)\ln e&=\ln 4\\ \ln(x-1)(1)&=\ln 4\\ (x-1)&=4\\ x&=5\end{aligned}$$

41. Find the exact solution.

$$\frac{10^x - 10^{-x}}{2} = 20$$
$$10^x(10^x - 10^{-x}) = 40(10^x)$$
$$10^{2x} - 1 = 40(10^x)$$
$$10^{2x} - 40(10)^x - 1 = 0$$

Let $u = 10^x$.

$$u^2 - 40u - 1 = 0$$
$$u = \frac{40 \pm \sqrt{40^2 - 4(1)(-1)}}{2}$$
$$= \frac{40 \pm \sqrt{1600 + 4}}{2}$$
$$= \frac{40 \pm \sqrt{1604}}{2}$$
$$= \frac{40 \pm 2\sqrt{401}}{2}$$
$$= 20 \pm \sqrt{401} \quad \text{Reject negative}$$
$$10^x = 20 + \sqrt{401}$$
$$\log 10^x = \log(20 + \sqrt{401})$$
$$x = \log(20 + \sqrt{401})$$

43. Find the exact solution.

$$\frac{10^x + 10^{-x}}{10^x - 10^{-x}} = 5$$
$$10^x + 10^{-x} = 5(10^x - 10^{-x})$$
$$10^x(10^x + 10^{-x}) = 5(10^x - 10^{-x})10^x$$
$$10^{2x} + 1 = 5(10^{2x} - 1)$$
$$4(10^{2x}) = 6$$
$$2(10^{2x}) = 3$$
$$(10^x)^2 = \frac{3}{2}$$
$$10^x = \sqrt{\frac{3}{2}}$$
$$x \log 10 = \log\sqrt{\frac{3}{2}}$$
$$x = \log\sqrt{\frac{3}{2}}$$
$$x = \frac{1}{2}\log\left(\frac{3}{2}\right)$$

45. Find the exact solution.

$$\frac{e^x + e^{-x}}{2} = 15$$
$$e^x(e^x + e^{-x}) = (30)e^x$$
$$e^{2x} + 1 = e^x(30)$$
$$e^{2x} - 30e^x + 1 = 0$$

Let $u = e^x$.

$$u^2 - 30u + 1 = 0$$
$$u = \frac{30 \pm \sqrt{900 - 4}}{2}$$
$$u = \frac{30 \pm \sqrt{896}}{2}$$
$$u = \frac{30 \pm 8\sqrt{14}}{2}$$
$$u = 15 \pm 4\sqrt{14}$$
$$e^x = 15 \pm 4\sqrt{14}$$
$$x \ln e = \ln(15 \pm 4\sqrt{14})$$
$$x = \ln(15 \pm 4\sqrt{14})$$

47. Find the exact solution.

$$\frac{1}{e^x - e^{-x}} = 4$$
$$1 = 4(e^x - e^{-x})$$
$$1(e^x) = 4(e^x)(e^x - e^{-x})$$
$$e^x = 4(e^{2x} - 1)$$
$$e^x = 4e^{2x} - 4$$
$$0 = 4e^{2x} - e^x - 4$$

Let $u = e^x$.

$$0 = 4u^2 - u - 4$$
$$u = \frac{1 \pm \sqrt{1 - 4(4)(-4)}}{8}$$
$$u = \frac{1 \pm \sqrt{65}}{8} \quad \text{Reject negative}$$
$$e^x = \frac{1 + \sqrt{65}}{8}$$
$$x \ln e = \ln\left(\frac{1 + \sqrt{65}}{8}\right)$$
$$x = \ln(1 + \sqrt{65}) - \ln 8$$

49. Use a graphing utility to approximate the solution.

$2^{-x+3} = x+1$

Graph $f = 2^{-x+3} - (x+1)$.

Its x-intercept is the solution.

$x \approx 1.61$

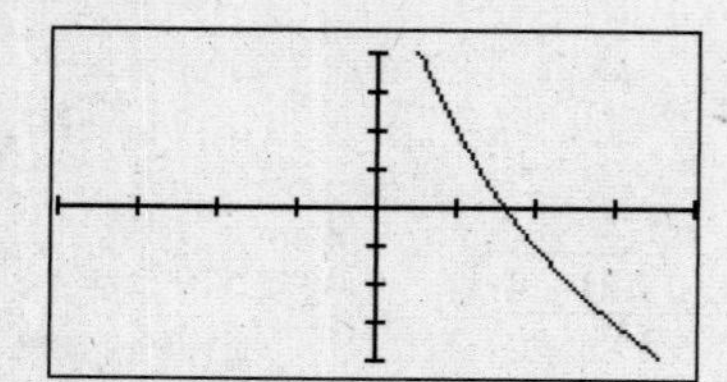

Xmin = −4, Xmax = 4, Xscl = 1,
Ymin = −4, Ymax = 4, Yscl = 1

51. Use a graphing utility to approximate the solution.

$e^{3-2x} - 2x = 1$

Graph $f = e^{3-2x} - 2x - 1$.

Its x-intercept is the solution.

$x \approx 0.96$

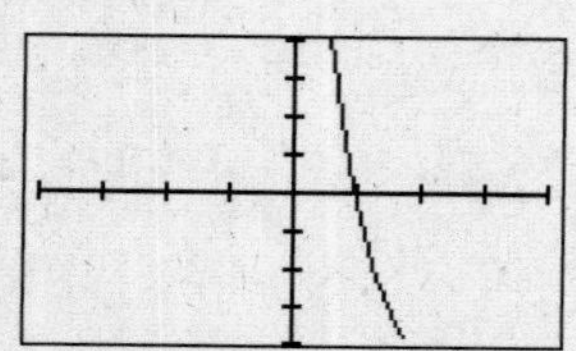

Xmin = −4, Xmax = 4, Xscl = 1,
Ymin = −4, Ymax = 4, Yscl = 1

53. Use a graphing utility to approximate the solution.

$3\log_2(x-1) = -x+3$

Graph $f = \dfrac{3\log(x-1)}{\log 2} + x - 3$.

Its x-intercept is the solution.

$x \approx 2.20$

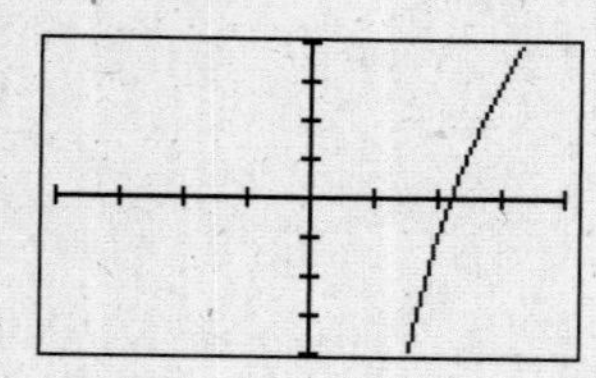

Xmin = −4, Xmax = 4, Xscl = 1,
Ymin = −4, Ymax = 4, Yscl = 1

55. Use a graphing utility to approximate the solution.

$\ln(2x+4) + \frac{1}{2}x = -3$

Graph $f = \ln(2x+4) + \frac{1}{2}x + 3$.

Its x-intercept is the solution.

$x \approx -1.93$

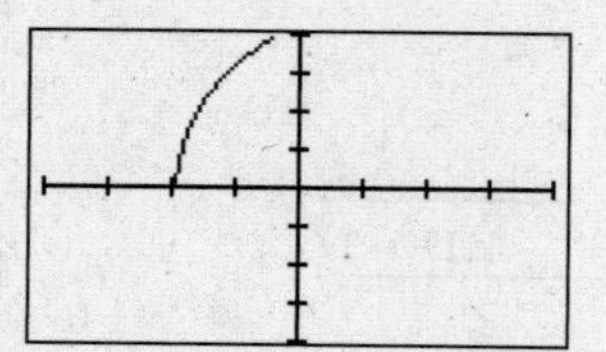

Xmin = −4, Xmax = 4, Xscl = 1,
Ymin = −4, Ymax = 4, Yscl = 1

57. Use a graphing utility to approximate the solution.

$2^{x+1} = x^2 - 1$

Graph $f = 2^{x+1} - x^2 + 1$.

Its x-intercept is the solution.

$x \approx -1.34$

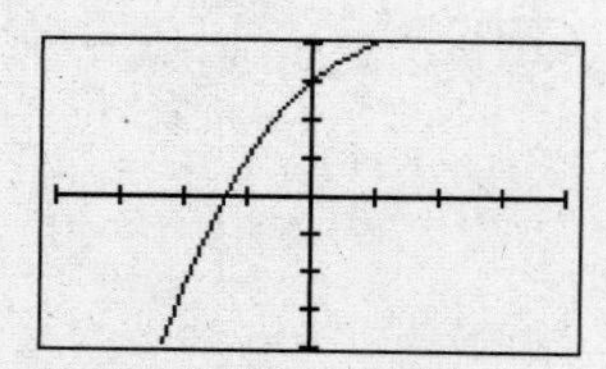

Xmin = −4, Xmax = 4, Xscl = 1,
Ymin = −4, Ymax = 4, Yscl = 1

59. a. $P(0) = 8500(1.1)^0 = 8500(1) = 8500$

$P(2) = 8500(1.1)^2 = 10,285$

b.

$$15,000 = 8500(1.1)^t$$
$$\ln 15,000 = 8500(1.1)^t$$
$$\ln 51,000 = \ln 8500 + t\ln(1.1)$$
$$\frac{\ln 15,000 - \ln 8500}{\ln(1.1)} = t$$
$$6 \approx t$$

The population will reach 15,000 in 6 years.

61. a. $T(10) = 36 + 43e^{-0.058(10)} = 36 + 43e^{-0.58}$

$T \approx 60° F$

b.
$$45 = 36 + 43e^{-0.058t}$$
$$\ln(45-36) = \ln 43 - 0.058t \ln e$$
$$\frac{\ln(45-36) - \ln 43}{-0.058} = t$$
$$t \approx 27 \text{ minutes}$$

63.
$$114 = 198 - (198 - 0.9)e^{-0.23x}$$
$$-84 = -197.1e^{-0.23x}$$
$$\frac{84}{197.1} = e^{-0.23x}$$
$$\ln\left(\frac{84}{197.1}\right) = -0.23x$$
$$x = \frac{\ln\left(\frac{84}{197.1}\right)}{-0.23}$$
$$x \approx 3.7 \text{ years}$$

65. Solve $5 + 29\ln(t+1) = 65$ for t.
$$5 + 29\ln(t+1) = 65$$
$$29\ln(t+1) = 60$$
$$\ln(t+1) = \frac{60}{29}$$
$$t + 1 = e^{60/29}$$
$$t = -1 + e^{60/29}$$
$$t \approx 6.9 \text{ months}$$

67. Consider the first function for time less than 10 seconds.
$$275 = -2.25x^2 + 56.26x - 0.28$$
$$0 = -2.25x^2 + 56.26x - 275.28$$
$$x = \frac{-56.26 \pm \sqrt{(56.26)^2 - 4(-2.25)(-275.28)}}{2(-2.25)}$$
$$x = 6.67 \text{ or } 18.33$$

18.33 s > 10 s, so it is not a solution. The solution is 6.67 s.

Consider the second function for time greater than 10 seconds.
$$275 = 8320(0.73)^x$$
$$\frac{275}{8320} = (0.73)^x$$
$$\ln\left(\frac{275}{8320}\right) = \ln 0.73^x$$
$$\ln\left(\frac{275}{8320}\right) = x\ln 0.73$$
$$x = \frac{\ln\left(\frac{275}{8320}\right)}{\ln 0.73} \approx 10.83$$

The solutions are 6.67 s and 10.83 s.

69. a. Use a graphing utility to graph.

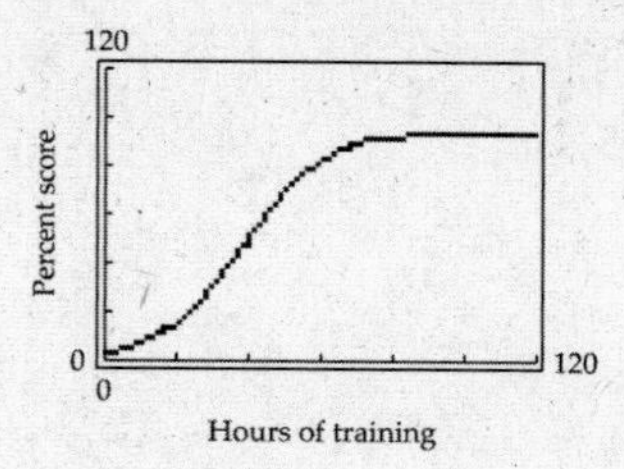

b. 48 hours

c. $P = 100$

d. As the number of hours of training increases, the test scores approach 100%.

71. a. Use a graphing utility to graph.

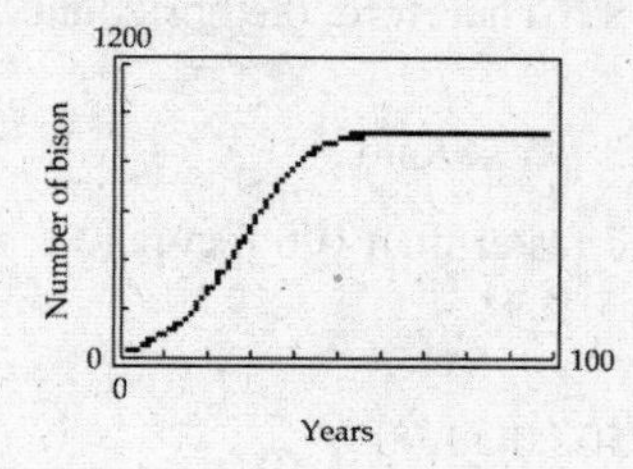

b. In 27 years or 2026

c. $B = 1000$

d. As the number of years increases, the bison population approaches but never exceeds 1000.

73. a. Use a graphing utility to graph.

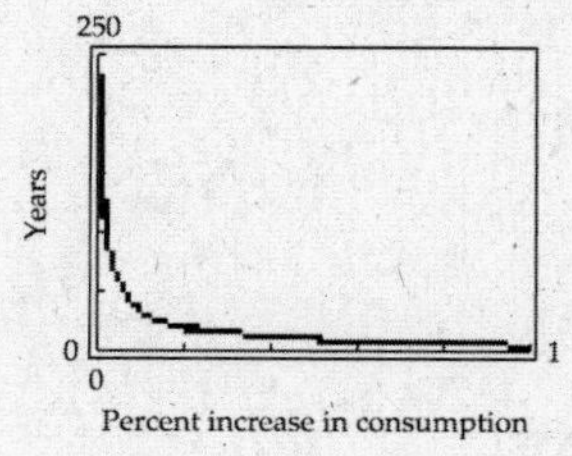

b. When $r = 3\%$, or 0.03, $T \approx 78$ years

c. When $T = 100$, $r \approx 0.019$, or 1.9%

75. a.

$$v = 100\left(\frac{e^{0.64t}-1}{e^{0.64t}+1}\right)$$

$$50 = 100\left(\frac{e^{0.64t}-1}{e^{0.64t}+1}\right)$$

$$\frac{50}{100} = \frac{e^{0.64t}-1}{e^{0.64t}+1}$$

$$0.5 = \frac{e^{0.64t}-1}{e^{0.64t}+1}$$

$$0.5(e^{0.64t}+1) = e^{0.64t}-1$$

$$0.5e0^{0.64t}+0.5 = e^{0.64t}-1$$

$$0.5e^{0.64t}-e^{0.64t} = -1.5$$

$$-0.5e^{0.64t} = -1.5$$

$$e^{0.64t} = 3$$

$$0.64t = \ln 3$$

$$t = \frac{\ln 3}{0.64}$$

$$t \approx 1.72$$

In approximately 1.72 seconds, the velocity will be 50 feet per second.

b. The horizontal asymptote is the value of $100\left[\frac{e^{0.64t}-1}{e^{0.64t}+1}\right]$ as $t \to \infty$. Therefore, the horizontal asymptote is v = 100 feet per second.

c. The object cannot fall faster than 100 feet per second.

77. Graph $V = 400{,}000 - 150{,}000(1.005)^x$ and $V = 100{,}000$.

They intersect when $x \approx 138.97$.

After 138 withdrawals, the account has $101,456.39.

After 139 withdrawals, the account has $99,963.67.

The designer can make at most 138 withdrawals and still have $100,000.

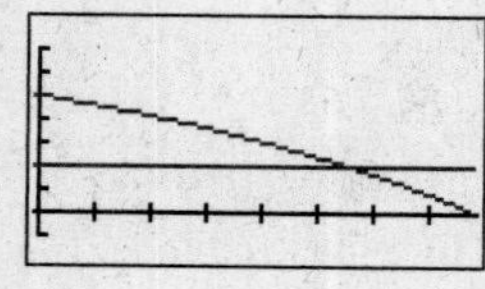

Xmin = 0, Xmax = 200, Xscl = 25
Ymin = −50000, Ymax = 350000, Yscl = 50000

79. The second step because $\log 0.5 < 0$. Thus the inequality sign must be reversed.

81. $\log(x+y) = \log x + \log y$

$\log(x+y) = \log xy$

Therefore $x + y = xy$

$$x - xy = -y$$

$$x(1-y) = -y$$

$$x = \frac{-y}{1-y}$$

$$x = \frac{y}{y-1}$$

83. Since $e^{0.336} \approx 1.4$,

$$F(x) = (1.4)^x \approx (e^{0.336})^x = e^{0.336x} = G(x)$$

Prepare for Section 4.6

P1. Evaluate. Round to the nearest hundredth.

$$A = 1000\left(1+\frac{0.1}{12}\right)^{12(2)} = 1220.39$$

P3. Solve. Round to the nearest ten-thousandth.

$$0.5 = e^{14k}$$

$$\ln 0.5 = \ln e^{14k}$$

$$\ln 0.5 = 14k$$

$$\frac{\ln 0.5}{14} = k$$

$$-0.0495 \approx k$$

P5. Solve. Round to the nearest thousandth.

$$6 = \frac{70}{5+9e^{-12k}}$$

$$6(5+9e^{-12k}) = 70$$

$$30+54e^{-12k} = 70$$

$$54e^{-12k} = 40$$

$$e^{-12k} = \frac{20}{27}$$

$$\ln e^{-12k} = \ln\frac{20}{27}$$

$$-12k = \ln\frac{20}{27}$$

$$k = -\frac{1}{12}\ln\frac{20}{27}$$

$$k \approx 0.025$$

Section 4.6 Exercises

1. a. $t=0$ hours, $N(0)=2200(2)^0=2200$ bacteria

b. $t=3$ hours, $N(3)=2200(2)^3=17{,}600$ bacteria

3. a. $N(t)=N_0e^{kt}$ where $N_0=24600$

$$N(5)=22{,}600e^{k(5)}$$
$$24{,}200=22{,}600e^{5k}$$
$$\frac{24{,}200}{22{,}600}=e^{5k}$$
$$\ln\left(\frac{24{,}200}{22{,}600}\right)=\ln\left(e^{5k}\right)$$
$$\ln\left(\frac{24{,}200}{22{,}600}\right)=5k$$
$$\frac{1}{5}\left[\ln\frac{24{,}200}{22{,}600}\right]=k$$
$$0.01368\approx k$$
$$N(t)=22{,}600e^{0.01368t}$$

b.
$$t=15$$
$$N(15)=22{,}600e^{0.01368(15)}$$
$$=22{,}600e^{0.2052}$$
$$\approx 27{,}700$$

5. $N(t)=N_0e^{kt}$ where $N_0=395{,}934$

$$N(4)=395{,}934e^{k(15)}$$
$$610{,}949=395{,}934e^{15k}$$
$$\frac{610{,}949}{395{,}934}=e^{15k}$$
$$\ln\left(\frac{610{,}949}{395{,}934}\right)=\ln\left(e^{15k}\right)$$
$$\ln\left(\frac{610{,}949}{395{,}934}\right)=15k$$
$$\frac{1}{15}\left[\ln\frac{610{,}949}{395{,}934}\right]=k$$
$$0.0289177304\approx k$$
$$N(t)=395{,}934e^{0.0289177304\,t}$$
$$t=22$$
$$N(22)=395{,}934e^{0.0289177304(22)}$$
$$=395{,}934e^{0.63619}$$
$$\approx 748{,}000$$

7. a. Graph the equation.

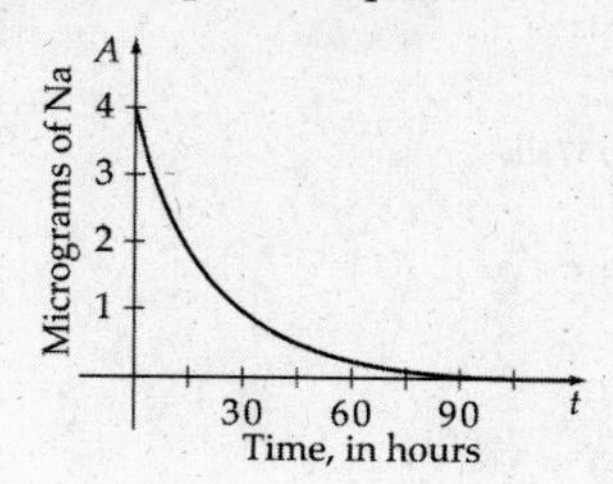

b. $A(5)=4e^{-0.23}\approx 3.18$ micrograms

c. Since $A = 4$ micrograms are present when $t = 0$, find the time t at which half remains—that is when $A = 2$.

$$2=4e^{-0.046t}$$
$$\frac{1}{2}=e^{-0.046t}$$
$$\ln\left(\frac{1}{2}\right)=-0.046t$$
$$\frac{\ln\left(\frac{1}{2}\right)}{-0.046}=t$$
$$15.07\approx t$$

The half-life of sodium-24 is about 15.07 hours.

d.
$$1=4e^{-0.046t}$$
$$\frac{1}{4}=e^{-0.046t}$$
$$\ln\left(\frac{1}{4}\right)=-0.046t$$
$$\frac{\ln\left(\frac{1}{4}\right)}{-0.046t}=t$$
$$30.14\approx t$$

The amount of sodium-24 will be 1 microgram after 30.14 hours.

9.
$$N(t)=N_0(0.5)^{t/5730}$$
$$N(t)=0.45N_0$$
$$0.45N_0=N_0(0.5)^{t/5730}$$
$$\ln(0.45)=\frac{t}{5730}\ln 0.5$$
$$5730\frac{\ln 0.45}{\ln 0.5}=t$$
$$6601\approx t$$

The bone is about 6601 years old.

11.
$$N(t) = N_0(0.5)^{t/5730}$$
$$N(t) = 0.75N_0$$
$$0.75N_0 = N_0(0.5)^{t/5730}$$
$$\ln 0.75 = \frac{t}{5730}\ln 0.5$$
$$5730\frac{\ln 0.75}{\ln 0.5} = t$$
$$2378 \approx t$$

The Rhind papyrus is about 2378 years old.

13. a. $P = 8000,\ r = 0.05,\ t = 4,\ n = 1$

$$B = 8000\left(1+\frac{0.05}{1}\right)^4 \approx \$9724.05$$

b. $t = 7,\ B = 8000\left(1+\frac{0.05}{1}\right)^7 \approx \$11,256.80$

15. a. $P = 38,000, r = 0.065, t = 4, n = 1$

$$B = 38,000\left(1+\frac{0.065}{1}\right)^4 \approx \$48,885.72$$

b. $n = 365$

$$B = 38,000\left(1+\frac{0.065}{365}\right)^{4(365)} \approx \$49,282.20$$

c. $n = 8760$

$$B = 38,000\left(1+\frac{0.065}{8760}\right)^{4(8760)} \approx \$49,283.30$$

17. $P = 15,000,\ r = 0.1,\ t = 5$

$$B = 15,000e^{5(0.1)} \approx \$24,730.82$$

19. $t = \frac{\ln 2}{r} \qquad r = 0.0784$

$$t = \frac{\ln 2}{0.0784}$$
$$t \approx 8.8 \text{ years}$$

21. $B = Pe^{rt} \qquad \text{Let } B = 3P$

$$3P = Pe^{rt}$$
$$3 = e^{rt}$$
$$\ln 3 = rt\ln e$$
$$t = \frac{\ln 3}{r}$$

23. $t = \frac{\ln 3}{r} \qquad r = 0.076$

$$t = \frac{\ln 3}{0.076}$$
$$t \approx 14 \text{ years}$$

25. a. Find the carrying capacity.

1900

b. Find the growth rate constant.

0.16

c. Find the initial population.

$$P(0) = \frac{1900}{1+8.5e^{-0.16(0)}} = 200$$

27. a. Find the carrying capacity.

157,500

b. Find the growth rate constant.

0.04

c. Find the initial population.

$$P(0) = \frac{157,500}{1+2.5e^{-0.04(0)}} = 45,000$$

29. a. Find the carrying capacity.

2400

b. Find the growth rate constant.

0.12

c. Find the initial population.

$$P(0) = \frac{2400}{1+7e^{-0.12(0)}} = 300$$

31. $a = \frac{c - P_0}{P_0} = \frac{5500-400}{400} = 12.75$

$$P(t) = \frac{c}{1+ae^{-bt}}$$
$$P(2) = \frac{5500}{1+12.75e^{-b(2)}}$$
$$780 = \frac{5500}{1+12.75e^{-2b}}$$
$$780\left(1+12.75e^{-2b}\right) = 5500$$
$$780 + 9945e^{-2b} = 5500$$
$$9945e^{-2b} = 4720$$
$$e^{-2b} = \frac{4720}{9945}$$
$$\ln e^{-2b} = \ln\frac{4720}{9945}$$
$$-2b = \ln\frac{4720}{9945}$$
$$b = -\frac{1}{2}\ln\frac{4720}{9945}$$
$$b \approx 0.37263$$

$$P(t) = \frac{5500}{1+12.75e^{-0.37263t}}$$

33. $a = \dfrac{c - P_0}{P_0} = \dfrac{100 - 18}{18} = 4.55556$

$$P(t) = \frac{c}{1 + ae^{-bt}}$$
$$P(3) = \frac{100}{1 + 4.55556e^{-b(3)}}$$
$$30 = \frac{100}{1 + 4.55556e^{-3b}}$$
$$30\left(1 + 4.55556e^{-3b}\right) = 100$$
$$30 + 136.67e^{-3b} = 100$$
$$136.67e^{-3b} = 70$$
$$e^{-3b} = \frac{70}{136.67}$$
$$\ln e^{-3b} = \ln\frac{70}{136.67}$$
$$-3b = \ln\frac{70}{136.67}$$
$$b = -\frac{1}{3}\ln\frac{70}{136.67}$$
$$b \approx 0.22302$$

$$P(t) = \frac{100}{1 + 4.55556e^{-0.22302t}}$$

35. a. $R(t) = \dfrac{625{,}000}{1 + 3.1e^{-0.045t}}$

$$R(1) = \frac{625{,}000}{1 + 3.1e^{-0.045(1)}} \approx \$158{,}000$$
$$R(2) = \frac{625{,}000}{1 + 3.1e^{-0.045(2)}} \approx \$163{,}000$$

b. $R(t) = \dfrac{625{,}000}{1 + 3.1e^{-0.045t}}$, as

$t \to \infty$, $R(t) \to \$625{,}000$

37. a. $a = \dfrac{c - P_0}{P_0} = \dfrac{1600 - 312}{312} = 4.12821$

$$P(t) = \frac{c}{1 + ae^{-bt}}$$
$$P(6) = \frac{1600}{1 + 4.12821e^{-b(6)}}$$
$$416 = \frac{1600}{1 + 4.12821e^{-6b}}$$
$$416\left(1 + 4.12821e^{-6b}\right) = 1600$$
$$416 + 1717.34e^{-6b} = 1600$$
$$1717.34e^{-6b} = 1184$$
$$e^{-6b} = \frac{1184}{1717.34}$$
$$\ln e^{-6b} = \ln\frac{1184}{1717.34}$$
$$-6b = \ln\frac{1184}{1717.34}$$
$$b = -\frac{1}{6}\ln\frac{1184}{1717.34}$$
$$b \approx 0.06198$$

$$P(t) = \frac{1600}{1 + 4.12821e^{-0.06198t}}$$

b. $P(10) = \dfrac{1600}{1 + 4.12821e^{-0.06198(10)}} \approx 497$ wolves

39. a. $a = \dfrac{c - P_0}{P_0} = \dfrac{8500 - 1500}{1500} = 4.66667$

$$P(t) = \frac{c}{1 + ae^{-bt}}$$
$$P(2) = \frac{8500}{1 + 4.66667e^{-b(2)}}$$
$$1900 = \frac{8500}{1 + 4.66667e^{-2b}}$$
$$1900\left(1 + 4.66667e^{-2b}\right) = 8500$$
$$1900 + 8866.673e^{-2b} = 8500$$
$$8866.673e^{-2b} = 6600$$
$$e^{-2b} = \frac{6600}{8866.673}$$
$$\ln e^{-2b} = \ln\frac{6600}{8866.673}$$
$$-2b = \ln\frac{6600}{8866.673}$$
$$b = -\frac{1}{2}\ln\frac{6600}{8866.673}$$
$$b \approx 0.14761$$

$$P(t) = \frac{8500}{1 + 4.66667e^{-0.14761t}}$$

b. $$4000 = \frac{8500}{1+4.66667e^{-0.14761t}}$$

$$4000\left(1+4.66667e^{-0.14761t}\right) = 8500$$
$$1+4.66667e^{-0.14761t} = 2.125$$
$$4.66667e^{-0.14761t} = 1.125$$
$$e^{-0.14761t} = \frac{1.125}{4.66667}$$
$$\ln e^{-0.14761t} = \ln\frac{1.125}{4.66667}$$
$$-0.14761t = \ln\frac{1.125}{4.66667}$$
$$t = -\frac{1}{0.14761}\ln\frac{1.125}{4.66667}$$
$$t \approx 9.6$$

The population will exceed 4000 in 2007 + 9 = 2016.

41. a. $A = 34^\circ\text{F},\ T_0 = 75^\circ\text{F},\ T_t = 65^\circ\text{F},\ t = 5$. Find k.

$$65 = 34 + (75-34)e^{-5k}$$
$$31 = 41e^{-5k}$$
$$\frac{31}{41} = e^{-5k}$$
$$\ln\left(\frac{31}{41}\right) = -5k$$
$$k = -\frac{1}{5}\ln\left(\frac{31}{41}\right)$$
$$k \approx 0.056$$

b. $A = 34^\circ\text{F},\ k = 0.056,\ T_0 = 75^\circ\text{F},\ t = 30$

$$T_t = 34 + (75-34)e^{-30(0.056)}$$
$$T_t = 34 + (41)e^{-1.68}$$
$$T_t \approx 42^\circ\text{F}$$

c. $T_t = 36^\circ\text{F},\ k = 0.056,\ T_t = 75^\circ\text{F},\ A = 34^\circ\text{F}$

$$36 = 34 + (75-34)e^{-0.056t}$$
$$2 = 41e^{-0.056t}$$
$$t \approx 54 \text{ minutes}$$

43. a. 10% of 80,000 is 8000.

$$8000 = 80{,}000\left(1-e^{-0.0005t}\right)$$
$$0.1 = 1-e^{-0.0005t}$$
$$-0.9 = -e^{-0.0005t}$$
$$0.9 = e^{-0.0005t}$$
$$\ln 0.9 = -0.0005t\ln e$$
$$\ln 0.9 = -0.0005t$$
$$\frac{\ln 0.9}{-0.0005} = t$$
$$211 \text{ h} \approx t$$

b. 50% of 80,000 is 40,000.

$$40{,}000 = 80{,}000\left(1-e^{-0.0005t}\right)$$
$$0.5 = 1-e^{-0.0005t}$$
$$-0.5 = -e^{-0.0005t}$$
$$0.5 = e^{-0.0005t}$$
$$\ln 0.5 = \ln\left(e^{-0.0005t}\right)$$
$$\ln 0.5 = -0.0005t$$
$$\frac{\ln 0.5}{-0.0005} = t$$
$$1386 \text{ h} \approx t$$

45. $V(t) = V_0(1-r)^t$

$$0.5V_0 = V_0(1-0.20)^t$$
$$0.5 = (1-0.20)^t$$
$$0.5 = 0.8^t$$
$$\ln 0.5 = \ln 0.8^t$$
$$\ln 0.5 = t\ln 0.8$$
$$\frac{\ln 0.5}{\ln 0.8} = t$$
$$3.1 \text{ years} \approx t$$

47. a. Graph the equation.

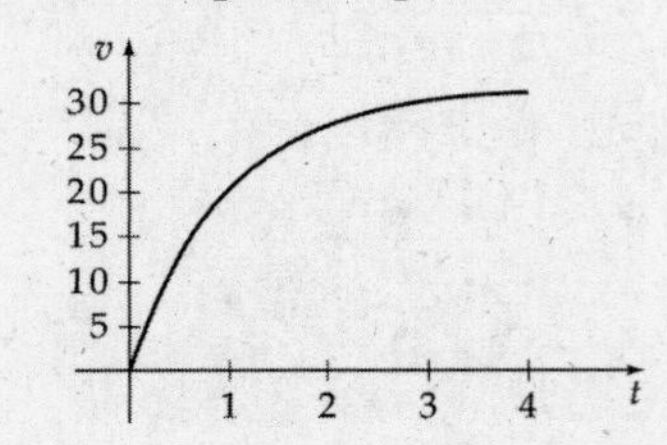

b. $20 = 32(1 - e^{-t})$

$0.625 = 1 - e^{-t}$

$e^{-t} = 0.375$

$-t = \ln 0.375$

$t \approx 0.98$ seconds

c. The horizontal asymptote is $v = 32$.

d. As time increases, the object's velocity approaches but never exceeds 32 ft/sec.

49. **a.** Graph the equation.

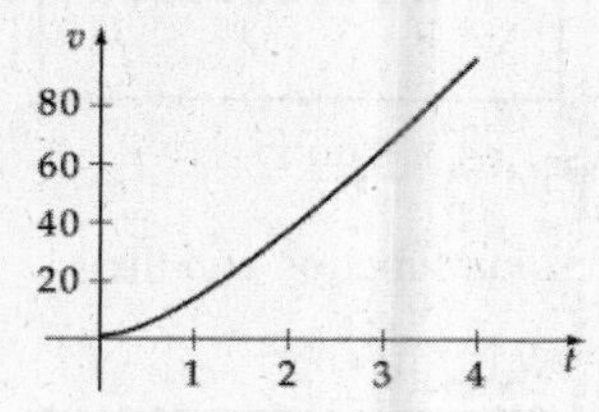

b. The graphs of $s = 32t + 32(e^{-t} - 1)$ and $s = 50$ intersect when $t \approx 2.5$ seconds.

c. The slope m of the secant line containing $(1, s(1))$ and $(2, s(2))$ is $m = \dfrac{s(2) - s(1)}{2 - 1} \approx 24.56$ ft/sec

d. The average speed of the object was 24.56 feet per second between $t = 1$ and $t = 2$.

51. Use the graph to solve.

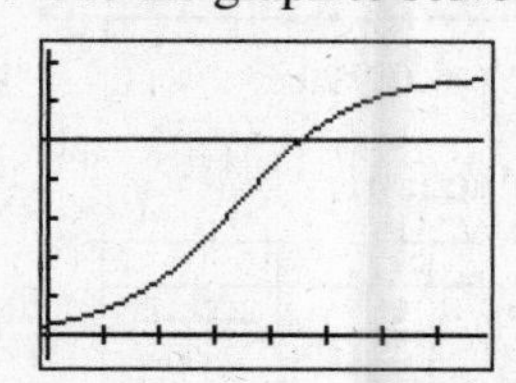

Xmin = 0, Xmax = 80, Xscl = 10,
Ymin = −10, Ymax = 110, Yscl = 15
When $P = 75\%$, $t \approx 45$ hours.

53. **a.** $A(1) = 0.5^{1/2}$

≈ 0.71 gram

b. $A(4) = 0.5^{4/2} + 0.5^{(4-3)/2}$

$= 0.5^2 + 0.5^{1/2}$

≈ 0.96 gram

c. $A(9) = 0.5^{9/2} + 0.5^{(9-3)/2} + 0.5^{(9-6)/2}$

$= 0.5^{4.5} + 0.5^3 + 0.5^{1.5}$

≈ 0.52 gram

55. $N(t) = 22{,}755e^{0.0287t}$

$= 22{,}755\left(e^{0.0287}\right)^t$

$\approx 22{,}755(1.0291)^t$

The annual growth rate is 2.91%.

57. **a.** $P_0 = \dfrac{1000}{1 + (-0.3333)} = \dfrac{1000}{0.6667} = 1500$ fish

b. As $t \to \infty$, $P(t) \to 1000$ fish.

59. **a.** $WR(107) = \dfrac{199.13}{1 + (-0.21726)(0.42536)}$

$= 219.41$ sec = 3 min, 39.41 sec

$WR(137) = \dfrac{199.13}{1 + (-0.21726)(0.33471)}$

$= 214.75$ sec = 3 min, 34.75 sec

b. As $t \to \infty$, $WR(t) \to 199.13$ sec = 3 min, 19.13 sec.

Prepare for Section 4.7

P1. This is a decreasing function.

P3. Evaluate.

$$P(0) = \frac{108}{1 + 2e^{-0.1(0)}} = \frac{108}{1 + 2} = 36$$

P5. Solve. Round to the nearest tenth.

$$10 = \frac{20}{1 + 2.2e^{-0.05t}}$$

$$10(1 + 2.2e^{-0.05t}) = 20$$

$$10 + 22e^{-0.05t} = 20$$

$$e^{-0.05t} = \frac{10}{22}$$

$$\ln e^{-0.05t} = \ln\frac{10}{22}$$

$$-0.05t = \ln\frac{10}{22}$$

$$t = -20\ln\frac{10}{22}$$

$$t \approx 15.8$$

Section 4.7 Exercises

1. Use a scatter plot to find the model function.

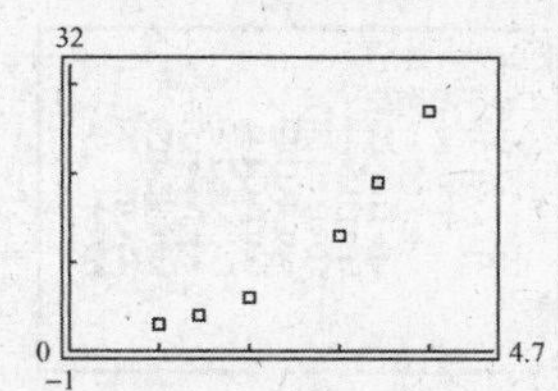

increasing exponential function

3. Use a scatter plot to find the model function.

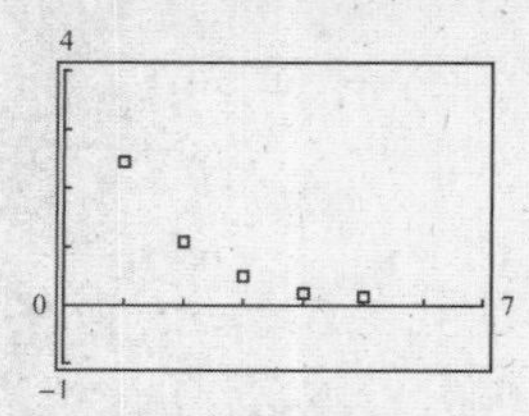

decreasing exponential function;
decreasing logarithmic function

5. Use a scatter plot to find the model function.

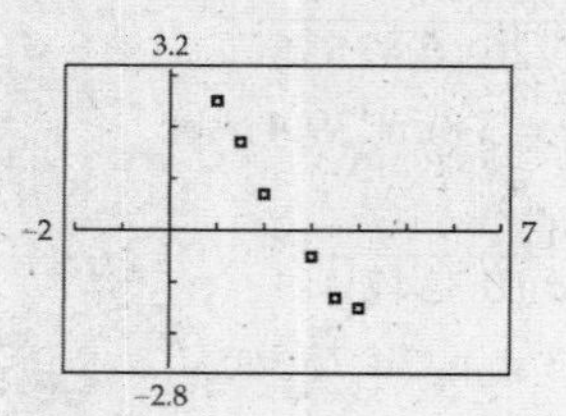

decreasing logarithmic function

7. Find the exponential regression function and the correlation coefficient.

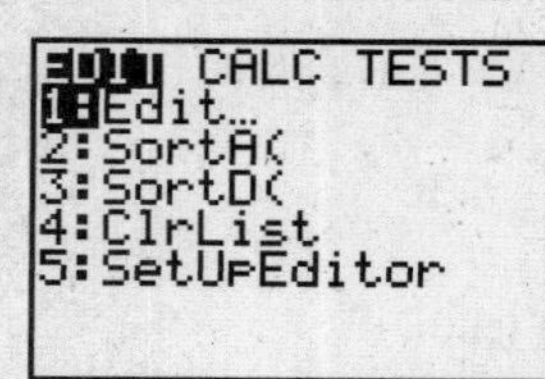

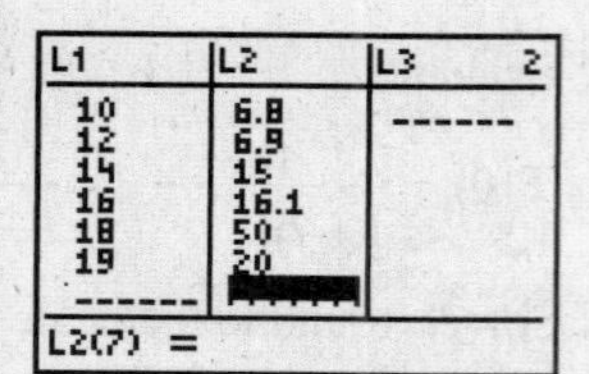

$y \approx 0.99628(1.20052)^x$; $r \approx 0.85705$

9. Find the exponential regression function and the correlation coefficient.

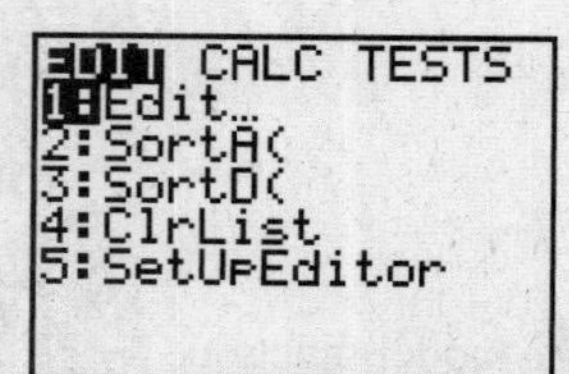

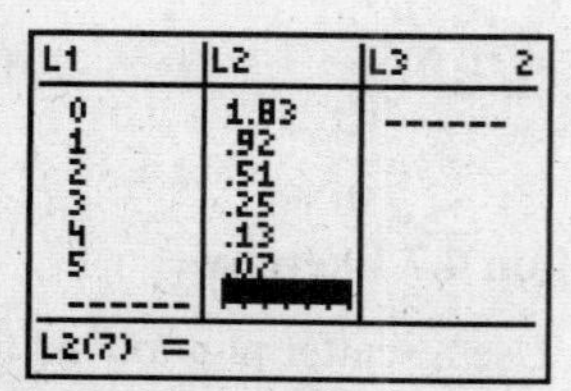

ExpReg
y=a*b^x
a=1.815049907
b=.519793213
r²=.9995676351
r=-.9997837942

$y \approx 1.81505(0.51979)^x$; $r \approx -0.99978$

11. Find the logarithmic regression function and the correlation coefficient.

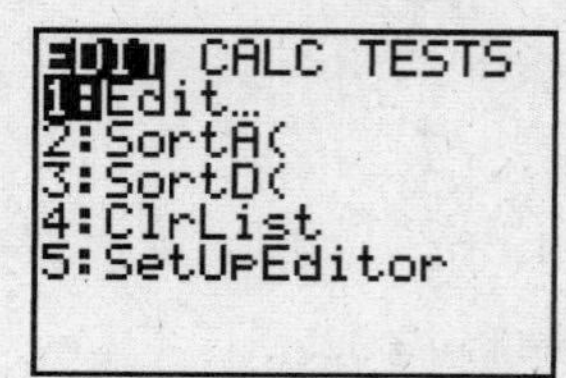

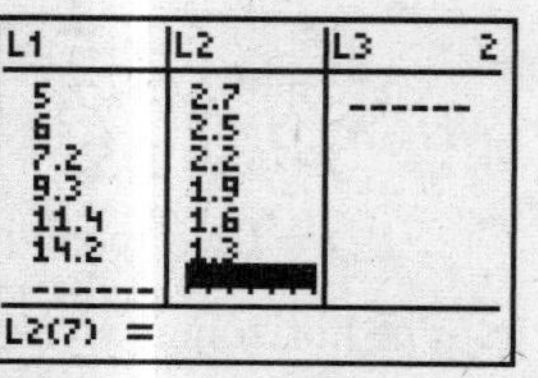

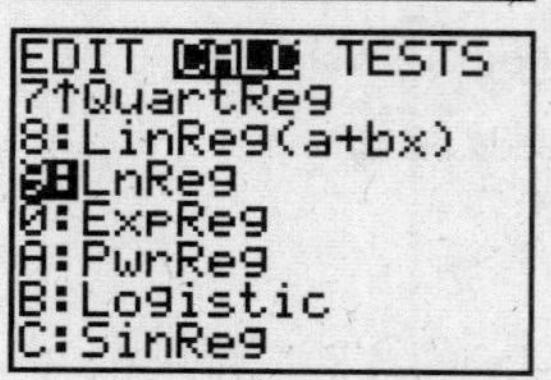

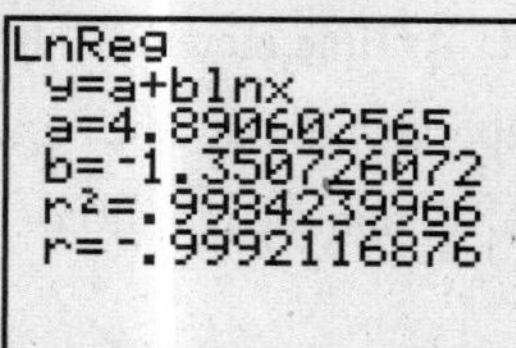

$y \approx 4.89060 - 1.35073 \ln x$; $r \approx -0.99921$

13. Find the logarithmic regression function and the correlation coefficient.

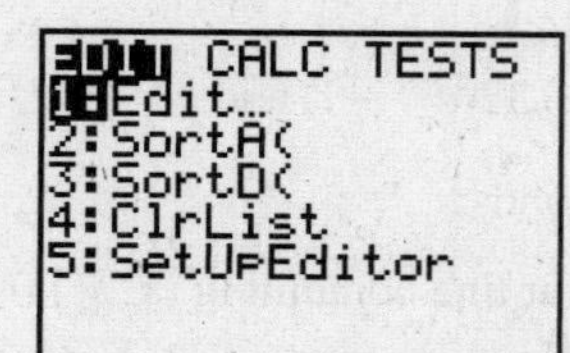

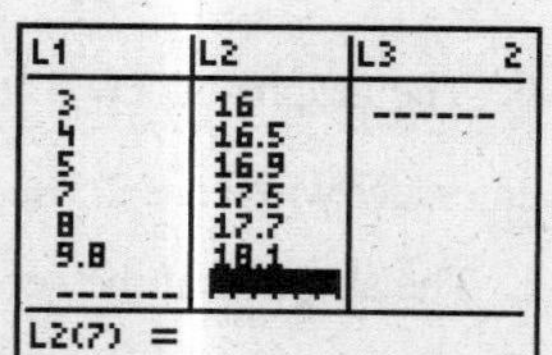

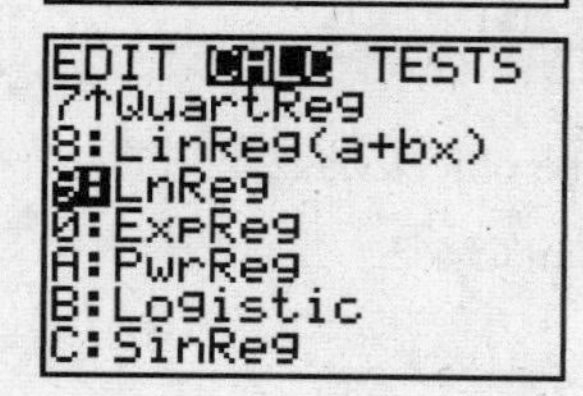

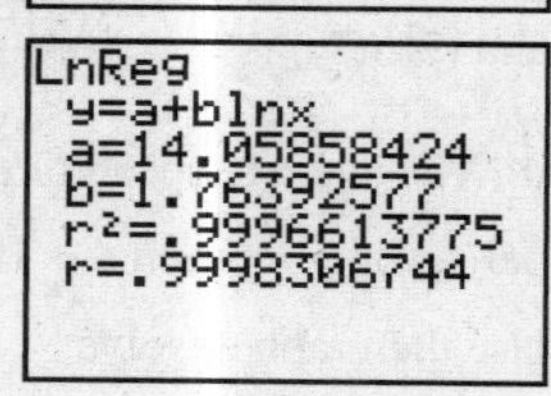

$y \approx 14.05858 + 1.76393 \ln x$; $r \approx 0.99983$

15. Find the logistic regression function.

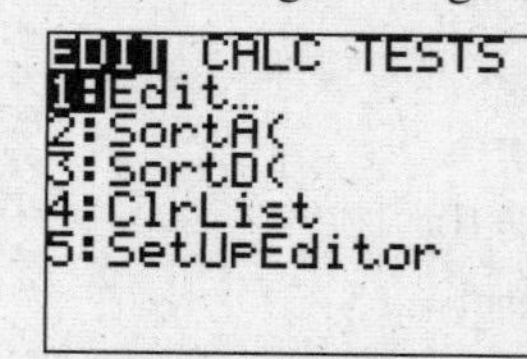

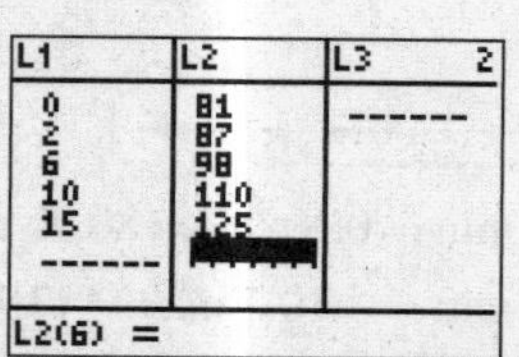

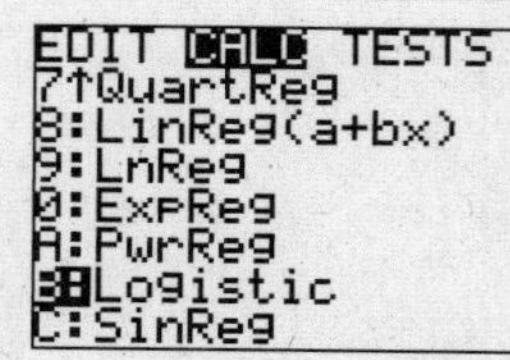

$$y \approx \frac{235.58598}{1 + 1.90188e^{-0.05101x}}$$

17. Find the logistic regression function.

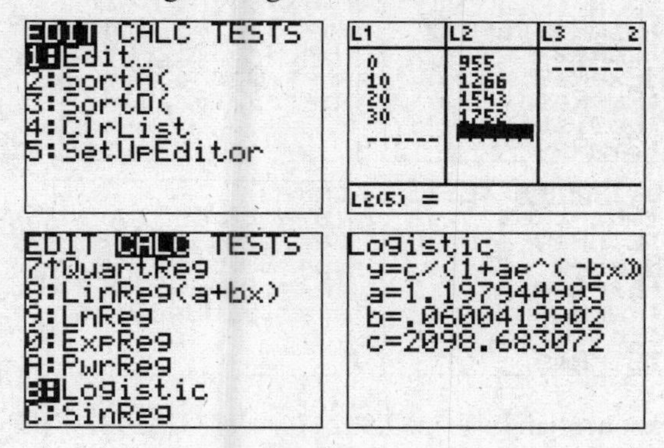

$$y \approx \frac{2098.68307}{1+1.19794e^{-0.06004x}}$$

19. a. Find the exponential regression model.

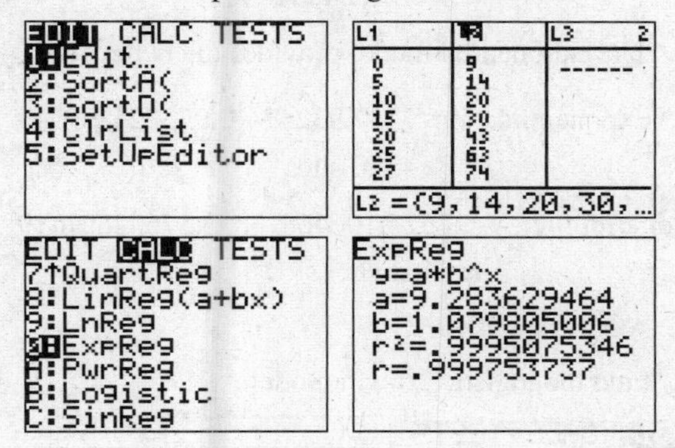

Exponential: $f(x) \approx 9.283629464(1.079805006)^x$

b. $f(34) \approx 9.283629464(1.079805006)^{34} \approx \$126$

21. a. Find the exponential regression model.

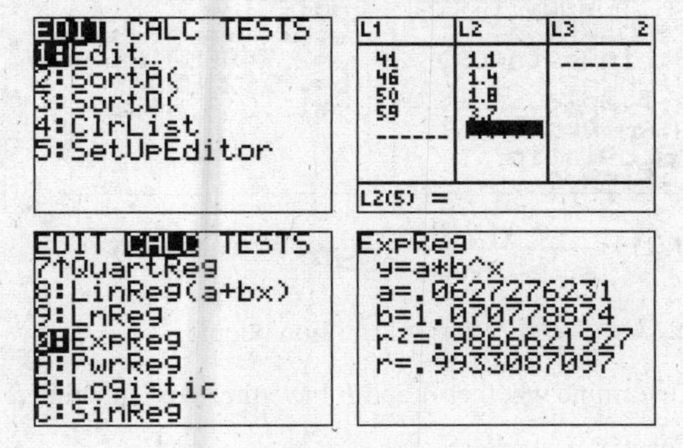

Exponential: $T \approx 0.06273(1.07078)^F$

b. $T \approx 0.06273(1.07078)^{65} \approx 5.3$ hours

23. a. Find the exponential regression model.

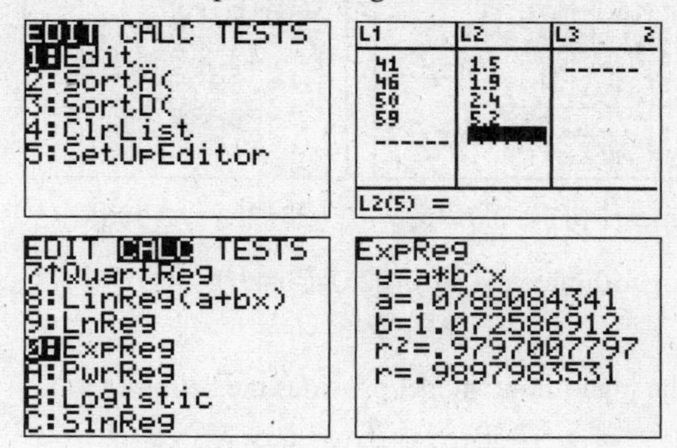

$T \approx 0.7881(1.07259)^F$

b. $T \approx 0.07881(1.07259)^{65} \approx 7.5$ hours

$7.5 - 5.3 = 2.2$ hours

25. An increasing logarithmic model provides a better fit because of the concave-downward nature of the graph.

27. a. Find the exponential regression model.

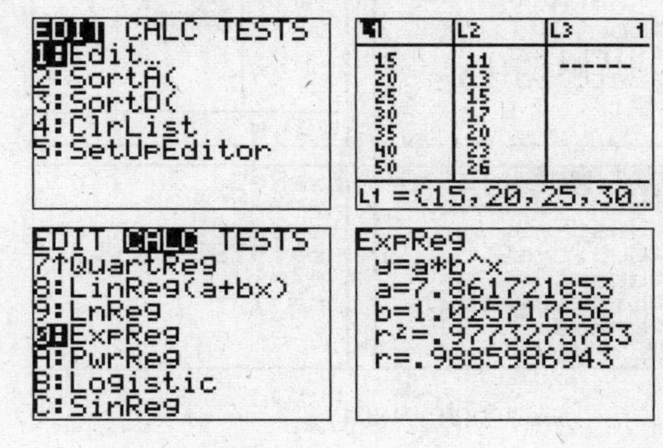

$p \approx 7.862(1.026)^y$

b. $p \approx 7.862(1.026)^{60} \approx 36$ cm

29. a. Find a linear and logarithmic model.

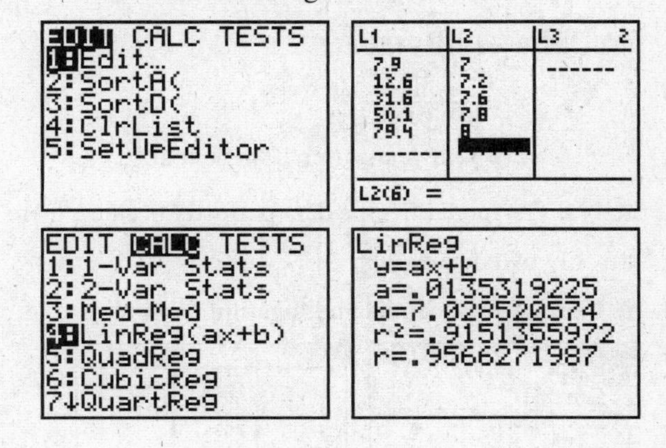

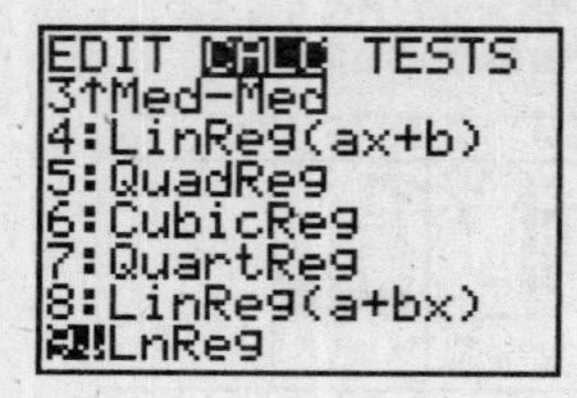

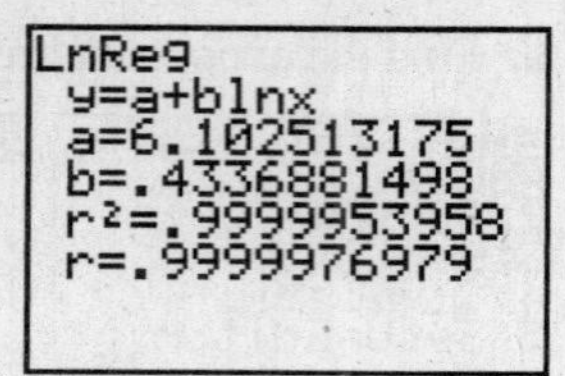

Linear: $\text{pH} \approx 0.01353q + 7.02852$; $r \approx 0.956627$.

Logarithmic: $\text{pH} \approx 6.10251 + 0.43369 \ln q$

$r \approx 0.999998$.

The logarithmic model provides the better fit.

b. $8.2 \approx 6.10251 + 0.43369 \ln q$

$2.09749 \approx 0.43369 \ln q$

$\dfrac{2.09749}{0.43369} \approx \ln q$

$q \approx e^{\frac{2.09749}{0.43369}} \approx 126.0$

31. a. Find the exponential model.

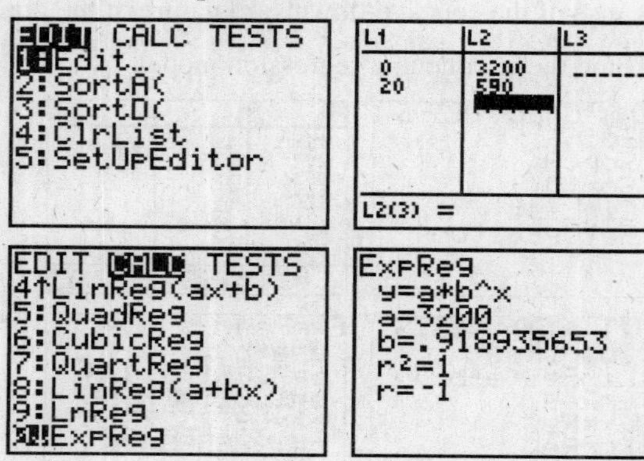

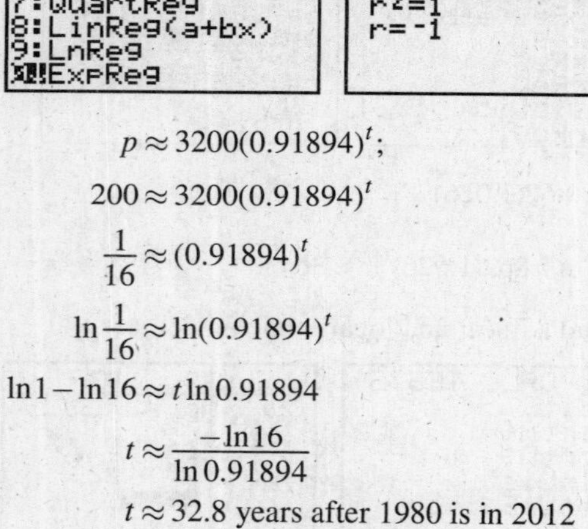

$p \approx 3200(0.91894)^t$;

$200 \approx 3200(0.91894)^t$

$\dfrac{1}{16} \approx (0.91894)^t$

$\ln \dfrac{1}{16} \approx \ln (0.91894)^t$

$\ln 1 - \ln 16 \approx t \ln 0.91894$

$t \approx \dfrac{-\ln 16}{\ln 0.91894}$

$t \approx 32.8$ years after 1980 is in 2012

b. No. The model fits the data perfectly because there are only two data points.

33. a. Find the exponential and logarithmic models.

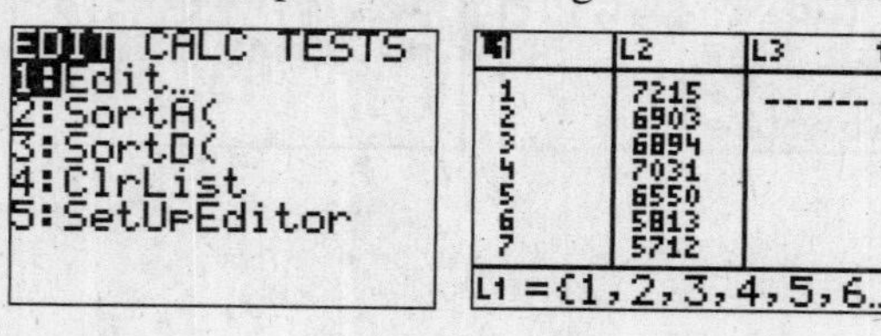

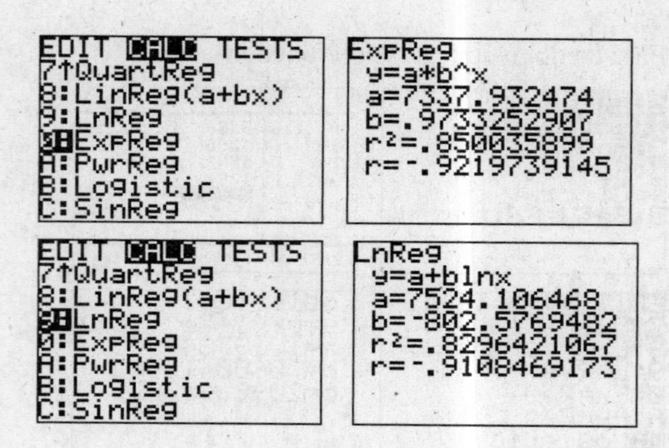

Exponential: $y \approx 7337.932474(0.9733252907)^x$;

$r \approx -0.9219739145$

Logarithmic: $y \approx 7524.106468 - 802.5769482 \ln x$

$r \approx -0.9108469173$

b. The exponential model provides the better fit.

c. Exponential: $y \approx 7337.932474(0.9733252907)^{19}$

≈ 4390 sites

Logarithmic: $y \approx 7524.106468 - 802.5769482 \ln 19$

≈ 5161 sites

d. Answers will vary.

35. a. Find the logistic growth model.

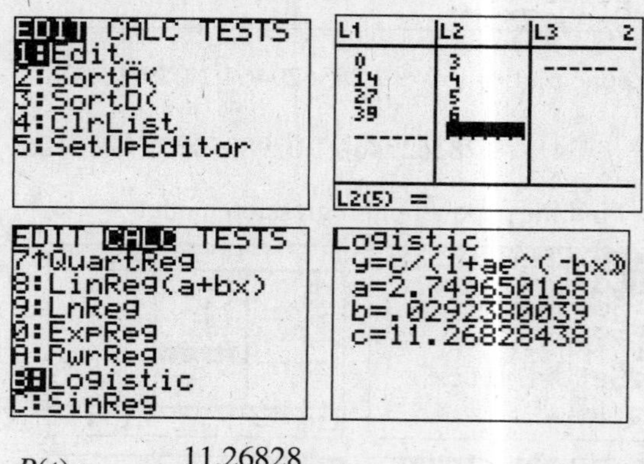

$P(t) \approx \dfrac{11.26828}{1 + 2.74965e^{-0.02924t}}$

b. As $t \to \infty$, $P(t) \to 11$ billion people

37. Determine whether A and B have the same function.

For A:

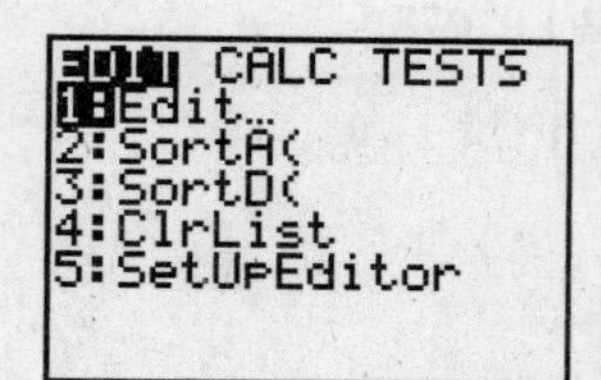

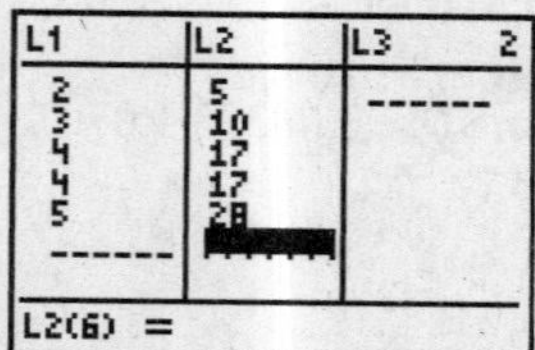

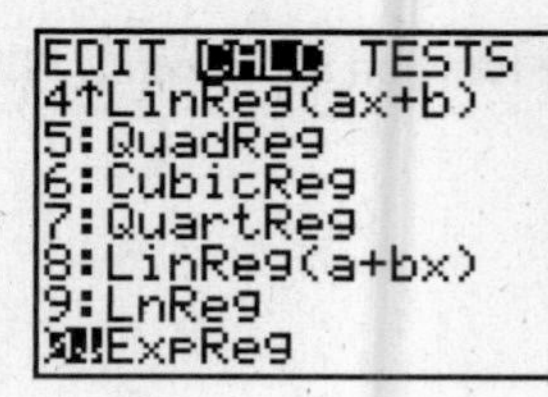

ExpReg
y=a*b^x
a=1.686257886
b=1.772013373
r²=.9936231263
r=.9968064638

For B:

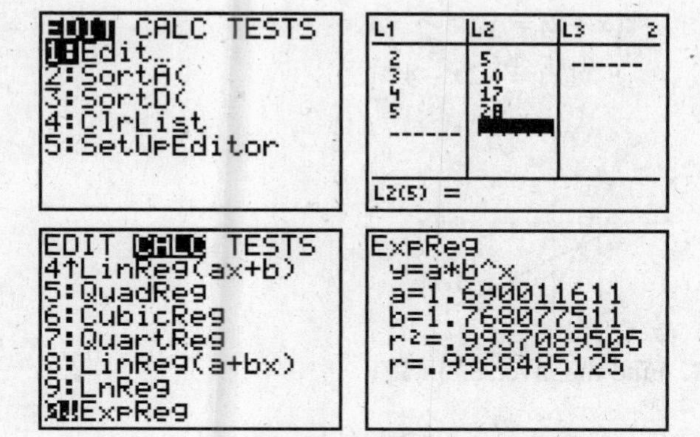

A and B have different exponential regression functions.

39. **a.** Find the exponential and power regression functions.

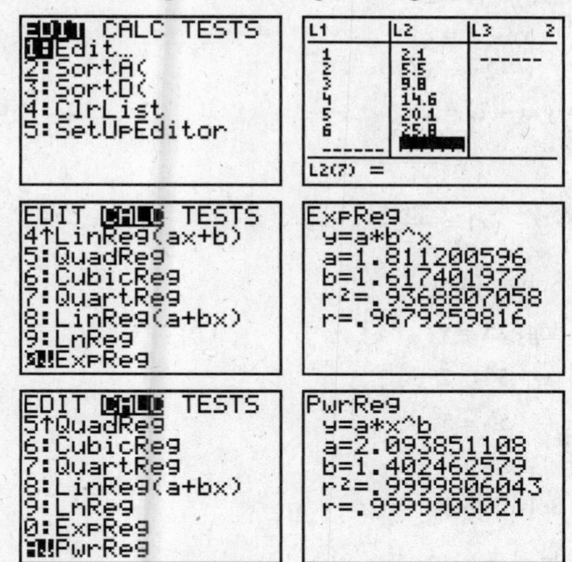

Exponential: $y \approx 1.81120(1.61740)^x$; $r \approx 0.96793$.

Power: $y \approx 2.09385(x^{1.40246})$; $r \approx 0.99999$.

b. The power regression provides the better fit.

Exploring Concepts with Technology

1. **a.** Graph the ordered pairs on a semilog scale.

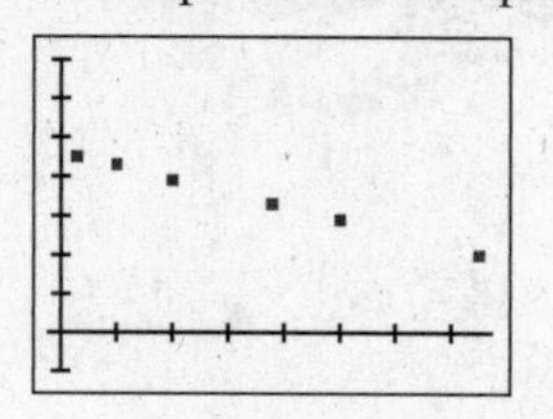

Xmin = −1, Xmax = 31, Xscl = 4,

Ymin = −1, Ymax = 7, Yscl = 1

b. $m = \dfrac{4.3-3.3}{4-15} \approx -0.0909$

c. $\ln A - 4.3 = -0.0909(t-4)$

$\ln A = -0.0909t + 4.664$

d. $e^{\ln A} = e^{-0.0909\,t+4.664}$

$A = e^{-0.0909\,t}e^{4.664}$

$A = e^{-0.0909\,t}e^{4.664}$

e. Graph equation in rectangular coordinate system.

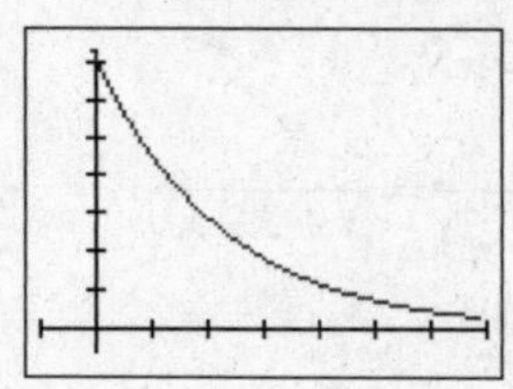

Xmin = −5, Xmax = 35, Xscl = 5,

Ymin = −10, Ymax = 110, Yscl = 15

f. $A(t) = 106e^{-0.0909t}$

At $t = 0$ there is $A = 106$ mg present

We must find t where $A = \frac{1}{2}(106) = 53$ mg.

$53 = 106e^{-0.0909\,t}$

$\frac{53}{106} = e^{-0.0909\,t}$

$\frac{1}{2} = e^{-0.0909\,t}$

$\ln 0.5 = -0.0909t$

$t = \dfrac{\ln 0.5}{-0.0909} \approx 7.6$ days

Chapter 4 Review Exercises

1. Draw the graph of the inverse. [4.1]

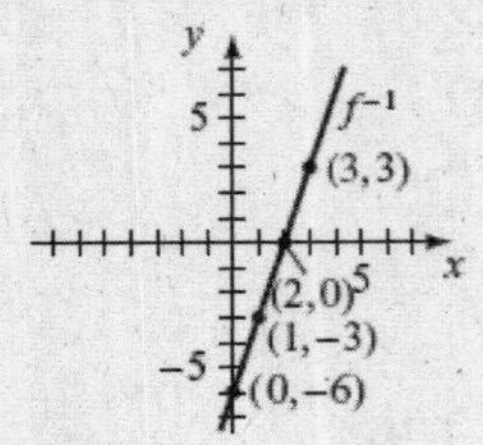

3. Determine whether the functions are inverses. [4.1]

$$F[G(x)] = F\left(\frac{x+5}{2}\right) = 2\left(\frac{x+5}{2}\right) - 5 = x + 5 - 5 = x$$

$$G[F(x)] = G(2x-5) = \frac{2x-5+5}{2} = \frac{2x}{2} = x$$

Yes, F and G are inverses.

5. Determine whether the functions are inverses. [4.1]

$$l[m(x)] = l\left(\frac{3}{x-1}\right) = \frac{\frac{3}{x-1}+3}{\frac{3}{x-1}} = \frac{3+3(x-1)}{3}$$

$$= \frac{3+3x-3}{3} = \frac{3x}{3} = x$$

$$m[l(x)] = m\left(\frac{x+3}{x}\right) = \frac{3}{\frac{x+3}{x}-1} = \frac{3x}{x+3-x}$$

$$= \frac{3x}{3} = x$$

Yes, l and m are inverses.

7. Find the inverse; sketch the function and inverse. [4.1]

$$y = 3x - 4$$
$$x = 3y - 4$$
$$x + 4 = 3y$$
$$\frac{x+4}{3} = y$$
$$f^{-1}(x) = \frac{1}{3}x + \frac{4}{3}$$

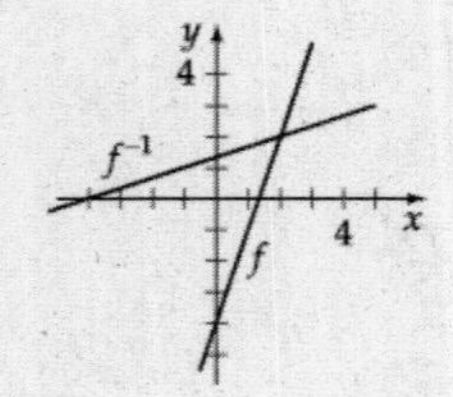

9. Find the inverse; sketch the function and inverse. [4.1]

$$y = -\frac{1}{2}x - 2$$
$$x = -\frac{1}{2}y - 2$$
$$x + 2 = -\frac{1}{2}y$$
$$-2(x+2) = y$$
$$h^{-1}(x) = -2x - 4$$

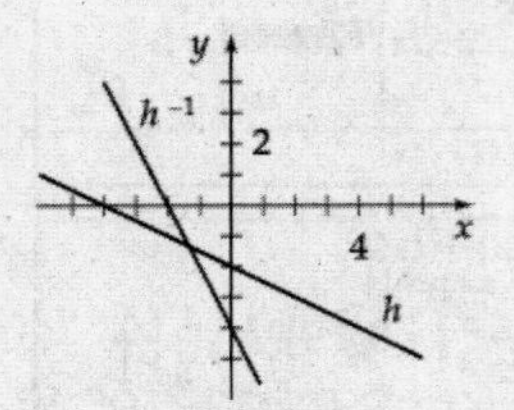

11. Find the inverse. [4.1]

$$f(x) = \frac{2x}{x-1},\ x > 1$$
$$x = \frac{2y}{y-1}$$
$$x(y-1) = 2y$$
$$xy - x = 2y$$
$$xy - 2y = x$$
$$y(x-2) = x$$
$$y = \frac{x}{x-2}$$
$$f^{-1}(x) = \frac{x}{x-2},\ x > 2$$

13. Solve. [4.3]

$$\log_5 25 = x$$
$$5^x = 25$$
$$5^x = 5^2$$
$$x = 2$$

15. Solve. [4.3]

$$\ln e^3 = x$$
$$e^x = e^3$$
$$x = 3$$

17. Solve. [4.5]

$$3^{2x+7} = 27$$
$$3^{2x+7} = 3^3$$
$$2x + 7 = 3$$
$$2x = -4$$
$$x = -2$$

19. Solve. [4.5]

$$2^x = \frac{1}{8}$$
$$2^x = 2^{-3}$$
$$x = -3$$

21. Solve. [4.5]

$$\log x^2 = 6$$
$$10^6 = x^2$$
$$1{,}000{,}000 = x^2$$
$$\pm\sqrt{1{,}000{,}000} = x$$
$$\pm 1000 = x$$

23. Solve. [4.5]

$$10^{\log 2x} = 14$$
$$2x = 14$$
$$x = 7$$

25. Sketch the graph of the function.

$f(x) = (2.5)^x$

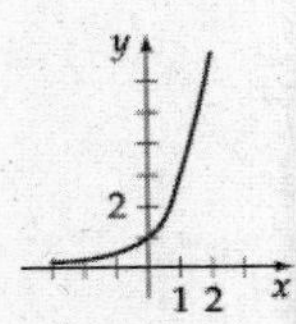

27. Sketch the graph of the function.

$f(x) = 3^{|x|}$

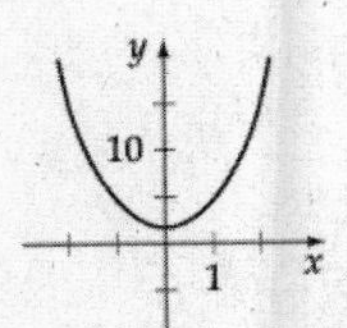

29. Sketch the graph of the function.

$f(x) = 2^x - 3$

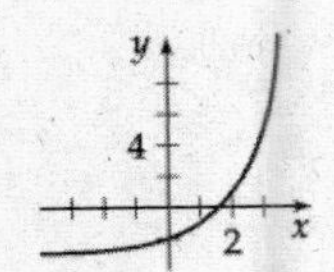

31. Sketch the graph of the function.

$f(x) = \log_5 x$

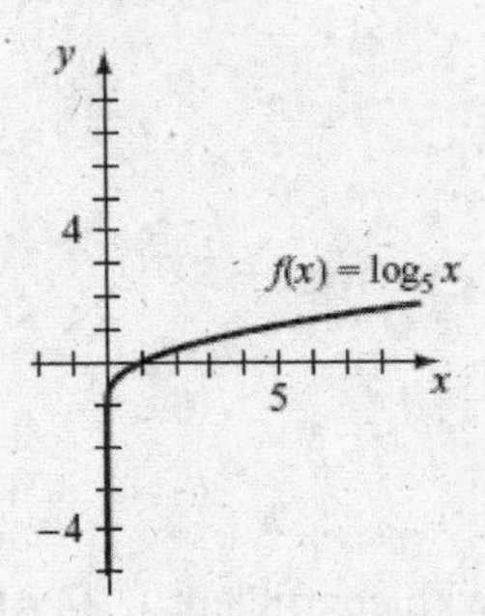

33. Sketch the graph of the function.

$f(x) = \frac{1}{3}\log x$

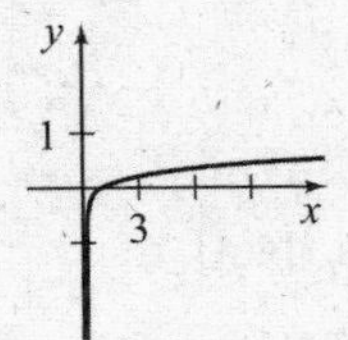

35. Sketch the graph of the function.

$f(x) = -\frac{1}{2}\ln x$

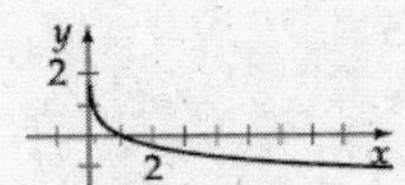

37. Use a graphing utility to graph the function.

$f(x) = \dfrac{4^x + 4^{-x}}{2}$

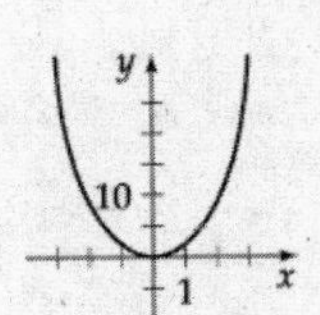

39. Change to exponential form. [4.3]

$$\log_4 64 = 3$$
$$4^3 = 64$$

41. Change to exponential form. [4.3]

$$\log_{\sqrt{2}} 4 = 4$$
$$(\sqrt{2})^4 = 4$$

43. Change to logarithmic form. [4.3]

$$5^3 = 125$$
$$\log_5 125 = 3$$

45. Change to logarithmic form. [4.3]

$$10^0 = 1$$
$$\log_{10} 1 = 0$$

47. Expand the expression. [4.4]

$$\log_b \frac{x^2 y^3}{z} = 2\log_b x + 3\log_b y - \log_b z$$

49. Expand the expression. [4.4]

$$\ln xy^3 = \ln x + 3\ln y$$

51. Write as a single logarithm with coefficient 1. [4.4]

$$2\log x + \frac{1}{3}\log(x+1) = \log\left(x^2\sqrt[3]{x+1}\right)$$

53. Write as a single logarithm with coefficient 1. [4.4]

$$\frac{1}{2}\ln 2xy - 3\ln z = \ln\frac{\sqrt{2xy}}{z^3}$$

55. Approximate to six significant digits. [4.4]

$$\log_5 101 = \frac{\log 101}{\log 5} \approx 2.86754$$

57. Approximate to six significant digits. [4.4]

$$\log_4 0.85 = \frac{\log 0.85}{\log 4} \approx -0.117233$$

59. Solve for x. [4.5]

$$4^x = 30$$
$$\log 4^x = \log 30$$
$$x\log 4 = \log 30$$
$$x = \frac{\log 30}{\log 4}$$

61. Solve for x. [4.5]

$$\ln(3x) - \ln(x-1) = \ln 4$$
$$\ln\frac{3x}{x-1} = \ln 4$$
$$\frac{3x}{x-1} = 4$$
$$3x = 4(x-1)$$
$$3x = 4x - 4$$
$$4 = x$$

63. Solve for x. [4.5]

$$e^{\ln(x+2)} = 6$$
$$(x+2) = 6$$
$$x + 2 = 6$$
$$x = 4$$

65. Solve for x. [4.5]

$$\frac{4^x + 4^{-x}}{4^x - 4^{-x}} = 2$$
$$4^x(4^x + 4^{-x}) = 2(4^x - 4^{-x})4^x$$
$$4^{2x} + 1 = 2(4^{2x} - 1)$$
$$4^{2x} + 1 = 2(4^{2x} - 1)$$
$$4^{2x} - 2\times 4^{2x} + 3 = 0$$
$$4^{2x} = 3$$
$$2^x \ln 4 = \ln 3$$
$$x = \frac{\ln 3}{2\ln 4}$$

67. Solve for x. [4.5]

$$\log(\log x) = 3$$
$$10^3 = \log x$$
$$10^{(10^3)} = x$$
$$10^{1000} = x$$

69. Solve for x. [4.5]

$$\log\sqrt{x-5} = 3$$
$$10^3 = \sqrt{x-5}$$
$$10^6 = x - 5$$
$$10^6 + 5 = x$$
$$x = 1{,}000{,}005$$

71. Solve for x. [4.5]

$$\log_4(\log_3 x) = 1$$
$$4 = \log_3 x$$
$$3^4 = x$$
$$81 = x$$

73. Solve for x. [4.5]

$$\log_5 x^3 = \log_5 16x$$
$$x^3 = 16x$$
$$x^2 = 16$$
$$x = 4 \quad \text{Reject negative}$$

75. Find the magnitude. [4.4]

$$m = \log\left(\frac{I}{I_0}\right) = \log\left(\frac{51,782,000 I_0}{I_0}\right) = \log 51,782,000 \approx 7.7$$

77. Find the ratio of larger to smaller intensities. [4.4]

$$\log\left(\frac{I_1}{I_0}\right) = 7.2 \qquad \text{and} \qquad \log\left(\frac{I_2}{I_0}\right) = 3.7$$
$$\frac{I_1}{I_0} = 10^{7.2} \qquad \frac{I_2}{I_0} = 10^{3.7}$$
$$I_1 = 10^{7.2} I_0 \qquad I_2 = 10^{3.7} I_0$$
$$\frac{I_1}{I_2} = \frac{10^{7.2} I_0}{10^{3.7} I_0} = \frac{10^{3.5}}{1} \approx \frac{3162}{1}$$

3162 to 1

79. Find the pH. [4.4]

$$\text{pH} = -\log\left[H_3O^+\right] = -\log\left[6.28 \times 10^{-5}\right] \approx 4.2$$

81. Find the balance. [4.6]

$P = 16,000,\ r = 0.08,\ t = 3$

a. $B = 16,000\left(1 + \frac{0.08}{12}\right)^{36} \approx \$20,323.79$

b. $B = 16,000e^{0.08(3)}$

$B = 16,000e^{0.24} \approx \$20,339.99$

83. Find the value. [4.6]

$S(n) = P(1-r)^n,\ P = 12,400,\ r = 0.29,\ t = 3$

$S(n) = 12,400(1-0.29)^3 \approx \$4438.10$

85. Find the exponential growth or decay function. [4.6]

$$N(0) = 1 \qquad N(2) = 5$$
$$1 = N_0 e^{k(0)} \qquad 5 = e^{2k}$$
$$1 = N_0 \qquad \ln 5 = 2k$$
$$k = \frac{\ln 5}{2} \approx 0.8047$$

Thus $N(t) = e^{0.8047t}$

87. Find the exponential growth or decay function. [4.6]

$4 = N(1) = N_0 e^k$ and thus $\frac{4}{N_0} = e^k$. Now, we also have $N(5) = 5 = N_0 e^{5k} = N_0\left(\frac{4}{N_0}\right)^5 = \frac{1024}{N_0^4}$.

$$N_0 = \sqrt[4]{\frac{1024}{5}} \approx 3.783$$

Thus $4 = 3.783e^k$.

$$\ln\left(\frac{4}{3.783}\right) = k$$
$$k \approx 0.0558$$

Thus $N_0 = 3.783e^{0.0558t}$.

89. a. Find the exponential growth function. [4.6]

$$N(1) = 25,200e^{k(1)} = 26,800$$
$$e^k = \frac{26,800}{25,200}$$
$$\ln e^k = \ln\left(\frac{26,800}{25,200}\right)$$
$$k \approx 0.061557893$$

$$N(t) = 25,200e^{0.061557893\,t}$$

b. $N(7) = 25,200e^{0.061557893(7)}$

$= 25,200e^{0.430905251}$

$\approx 38,800$

91. Answers will vary. [4.7]

93. a. Find the logistic model. [4.6]

$$P(t) = \frac{mP_0}{P_0 + (m - P_0)e^{-kt}}$$
$$P(3) = 360 = \frac{1400(210)}{210 + (1400 - 210)e^{-k(3)}}$$

$$360=\frac{294000}{210+1190e^{-3k}}$$
$$360\left(210+1190e^{-3k}\right)=294000$$
$$210+1190e^{-3k}=\frac{294000}{360}$$
$$1190e^{-3k}=\frac{29400}{36}-210$$
$$e^{-3k}=\frac{29400/36-210}{1190}$$
$$\ln e^{-3k}=\ln\left(\frac{29400/36-210}{1190}\right)$$
$$-3k=\ln\left(\frac{29400/36-210}{1190}\right)$$
$$k=-\frac{1}{3}\ln\left(\frac{29400/36-210}{1190}\right)$$
$$k\approx 0.2245763649$$

$$P(t)=\frac{294000}{210+1190e^{-0.22458t}}=\frac{1400}{1+\frac{17}{3}e^{-0.22458t}}$$

b. $$P(13)=\frac{294000}{210+1190e^{-0.22458(13)}}$$
$$=\frac{294000}{210+1190e^{-2.919492744}}$$
$$\approx 1070 \text{ coyotes}$$

Chapter 4 Test

1. Find the inverse; sketch the function and inverse. [4.1]

$$y=2x-3$$
$$x=2y-3$$
$$x+3=2y$$
$$\frac{1}{2}x+\frac{3}{2}=y$$
$$f^{-1}(x)=\frac{1}{2}x+\frac{3}{2}$$

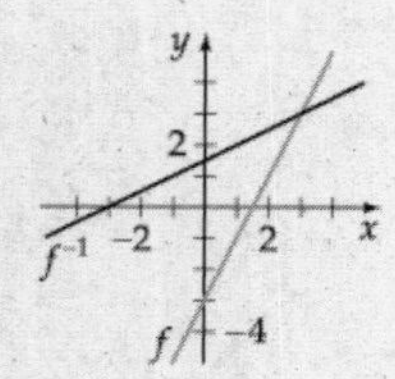

3. a. Write in exponential form. [4.3]

$$\log_b(5x-3)=c$$
$$b^c=5x-3$$

b. Write in logarithmic form.

$$3^{x/2}=y$$
$$\log_3 y=\frac{x}{2}$$

5. Write as a single logarithm with coefficient of 1. [4.4]

$$\log_{10}(2x+3)-3\log_{10}(x-2)$$
$$=\log_{10}(2x+3)-\log_{10}(x-2)^3$$
$$=\log_{10}\frac{2x+3}{(x-2)^3}$$

7. Sketch the graph of the function.

$$f(x)=3^{-x/2}$$

9. Solve. [4.5]

$$5^x=22$$
$$x\log 5=\log 22$$
$$x=\frac{\log 22}{\log 5}$$
$$x\approx 1.9206$$

11. Solve. [4.5]

$$\log(x+99)-\log(3x-2)=2$$
$$\log\frac{x+99}{3x-2}=2$$
$$\frac{x+99}{3x-2}=10^2$$
$$x+99=100(3x-2)$$
$$x+99=300x-200$$
$$-299x=-299$$
$$x=1$$

13. Find the balance. [4.6]

a. $$A=P\left(1+\frac{r}{n}\right)^{nt}$$
$$=20,000\left(1+\frac{0.078}{12}\right)^{12(5)}$$
$$=20,000(1.0065)^{60}$$
$$=\$29,502.36$$

b. $$A=Pe^{rt}$$
$$=20,000e^{0.078(5)}$$
$$=20,000e^{0.39}$$
$$=\$29,539.62$$

15. a. Find the magnitude. [4.4]

$$\begin{aligned} M &= \log\left(\frac{I}{I_0}\right) \\ &= \log\left(\frac{42,304,000 I_0}{I_0}\right) \\ &= \log 42,304,000 \\ &\approx 7.6 \end{aligned}$$

b. Compare the intensities.

$$\log\left(\frac{I_1}{I_0}\right) = 6.3 \qquad \text{and} \qquad \log\left(\frac{I_2}{I_0}\right) = 4.5$$

$$\frac{I_1}{I_0} = 10^{6.3} \qquad\qquad \frac{I_2}{I_0} = 10^{4.5}$$

$$I_1 = 10^{6.3} I_0 \qquad\qquad I_2 = 10^{4.5} I_0$$

$$\frac{I_1}{I_2} = \frac{10^{6.3} I_0}{10^{4.5} I_0} = \frac{10^{1.8}}{1} \approx \frac{63}{1}$$

Therefore the ratio is 63 to 1.

17. Find the age of the bone. [4.6]

$$\begin{aligned} P(t) = 0.5^{\,t/5730} &= 0.92 \\ \log 0.5^{\,t/5730} &= \log 0.92 \\ \frac{t}{5730}\log 0.5 &= \log 0.92 \\ \frac{t}{5730} &= \frac{\log 0.92}{\log 0.5} \\ t &= 5730\left(\frac{\log 0.92}{\log 0.5}\right) \\ t &\approx 690 \text{ years} \end{aligned}$$

19. a. Find the logarithmic and logistic models. [4.7]

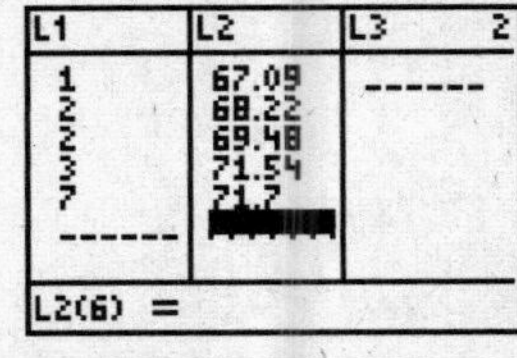

Logarithmic: $d \approx 67.35500994 + 2.540152486 \ln t$;

Logistic: $d \approx \dfrac{72.03782781}{1 + 0.1527878996 e^{-0.6775213733\,t}}$

b. Logarithmic: $d \approx 67.35500994 + 2.540152486 \ln(14)$
≈ 74.06 m

Logistic: $d \approx \dfrac{72.03782781}{1 + 0.1527878996 e^{-0.6775213733(14)}}$
≈ 72.04 m

Cumulative Review Exercises

1. Solve. [1.5]

$|x-4| \le 2 \Rightarrow -2 \le x-4 \le 2 \Rightarrow 2 \le x \le 6.$

The solution is [2, 6].

3. Find the distance. [2.1]

$$\begin{aligned} d &= \sqrt{(11-5)^2 + (7-2)^2} \\ &= \sqrt{6^2 + 5^2} = \sqrt{36+25} \\ &= \sqrt{61} \approx 7.8 \end{aligned}$$

5. Find $(g \circ f)(x)$ for $f(x) = 2x+1$, $g(x) = x^2 - 5$. [2.6]

$$\begin{aligned} (g \circ f)(x) &= g[f(x)] \\ &= g(2x+1) \\ &= (2x+1)^2 - 5 \\ &= 4x^2 + 4x + 1 - 5 \\ &= 4x^2 + 4x - 4 \end{aligned}$$

7. Find the weight. [1.6]

$$\begin{aligned} L &= kwd^2 \\ 1500 &= k(4)(8)^2 \\ 1500 &= 256k \\ \frac{1500}{256} &= \frac{375}{64} = k \\ L &= \frac{375}{64} wd^2 \\ L &= \frac{375}{64}(6)(10)^2 \\ L &\approx 3500 \text{ pounds} \end{aligned}$$

9. Find the zeros. [3.3]

$P(x) = x^4 - 5x^3 + x^2 + 15x - 12$ has three or one positive and one negative real zeros.

$\dfrac{p}{q} = \pm 1, \ \pm 2, \ \pm 3, \ \pm 4, \ \pm 6, \ \pm 12$ are the possible rational zeros.

$$\begin{array}{r|rrrrr} 1 & 1 & -5 & 1 & 15 & -12 \\ & & 1 & -4 & -3 & 12 \\ \hline & 1 & -4 & -3 & 12 & 0 \end{array}$$

$$\begin{array}{r|rrrr} 4 & 1 & -4 & -3 & 12 \\ & & 4 & 0 & -12 \\ \hline & 1 & 0 & -3 & 0 \end{array}$$

$$x^2 - 3 = 0$$
$$x^2 = 3$$
$$x = \pm\sqrt{3}$$

The zeros are 1, 4, $-\sqrt{3}$, $\sqrt{3}$.

11. Find the asymptotes. [3.5]

$$r(x) = \frac{3x-5}{x-4}$$

Vertical asymptote: $x - 4 = 0$

$$x = 4$$

Horizontal asymptote: $(n = m)$

$$y = \frac{3}{1}$$
$$y = 3$$

13. State if the function is increasing or decreasing. [4.2]

$f(x) = 0.4^x$ is a decreasing function since $0.4 < 1$.

15. Write in logarithmic form. [4.3]

$5^3 = 125 \Rightarrow \log_5 125 = 3$

17. Solve. Round to the nearest ten-thousandth. [4.5]

$$2e^x = 15$$
$$e^x = 7.5$$
$$\ln e^x = \ln 7.5$$
$$x = \ln 7.5$$
$$x \approx 2.0149$$

19. Solve. Round to the nearest ten-thousandth. [4.5]

$$\frac{e^x - e^{-x}}{2} = 12$$
$$e^x\left(e^x - e^{-x}\right) = (24)e^x$$
$$e^{2x} - 1 = e^x(24)$$
$$e^{2x} - 24e^x - 1 = 0$$

Let $u = e^x$.

$$u^2 - 24u - 1 = 0$$
$$u = \frac{24 \pm \sqrt{576 - 4(-1)}}{2}$$
$$u = \frac{24 \pm \sqrt{580}}{2}$$
$$u = \frac{24 \pm 2\sqrt{145}}{2}$$
$$u = 12 \pm \sqrt{145}$$
$$e^x = 12 \pm \sqrt{145}$$

$x \ln e = \ln(12 + \sqrt{145})$ cannot take ln of a negative

$$x = \ln(12 + \sqrt{145})$$
$$x \approx 3.1798$$

Chapter 5: Topics in Analytic Geometry

Section 5.1 Exercises

1. **a.** iii

b. i

c. iv

d. ii

3. Find the vertex, focus, and directrix; sketch the graph.

$x^2 = -4y$

$4p = -4$

$p = -1$

vertex $= (0,\ 0)$

focus $= (0,\ -1)$

directrix: $y = 1$

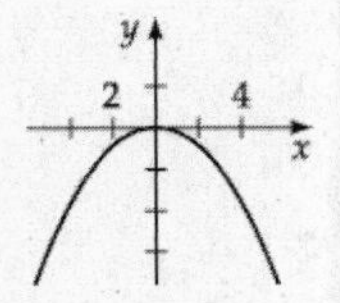

5. Find the vertex, focus, and directrix; sketch the graph.

$y^2 = \frac{1}{3}x$

$4p = \frac{1}{3}$

$p = \frac{1}{12}$

vertex $= (0,\ 0)$

focus $= \left(\frac{1}{12},\ 0\right)$

directrix: $x = -\frac{1}{12}$

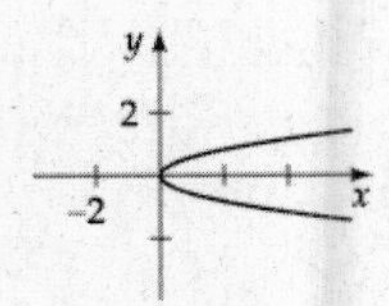

7. Find the vertex, focus, and directrix; sketch the graph.

$(x-2)^2 = 8(y+3)$

vertex $= (2,\ -3)$

$4p = 8, \quad p = 2$

$(h, k+p) = (2,\ -3+2) = (2,\ -1)$

focus $= (2,\ -1)$

$k - p = -3 - 2 = -5$

directrix: $y = -5$

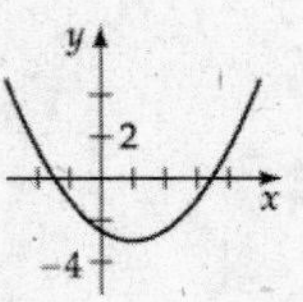

9. Find the vertex, focus, and directrix; sketch the graph.

$(y+4)^2 = -4(x-2)$

vertex $= (2,\ -4)$

$4p = -4 \quad p = -1$

$(h+p,\ k) = (2-1,\ -4) = (1,\ -4)$

focus $= (1,\ -4)$

$h - p = 2 + 1 = 3$

directrix: $x = 3$

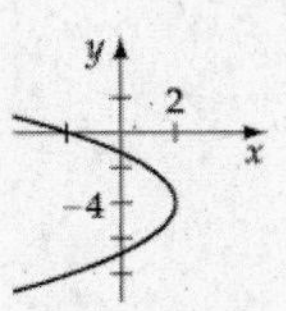

11. Find the vertex, focus, and directrix; sketch the graph.

$(y-1)^2 = 2x+8$

vertex $= (-4,\ 1)$

$4p = 2, \quad p = \frac{1}{2}$

$(h+p,\ k) = \left(-4+\frac{1}{2},\ 1\right) = \left(-\frac{7}{2},\ 1\right)$

focus $= \left(-\frac{7}{2},\ 1\right)$

$h - p = -4 - \frac{1}{2} = -\frac{9}{2}$

directrix: $x = -\frac{9}{2}$

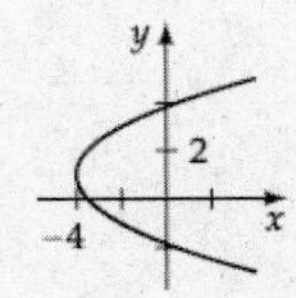

13. Find the vertex, focus, and directrix; sketch the graph.

$$(2x-4)^2 = 8y-16 \Rightarrow (x-2)^2 = 2(y-2)$$

vertex $= (2,\ 2)$

$4p = 2,\ p = \frac{1}{2}$

$(h,\ k+p) = \left(2,\ 2+\frac{1}{2}\right) = \left(2,\ \frac{5}{2}\right)$

focus $= \left(2,\ \frac{5}{2}\right)$

$k - p = 2 - \frac{1}{2} = \frac{3}{2}$

directrix: $y = \frac{3}{2}$

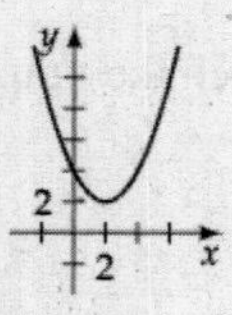

15. Find the vertex, focus, and directrix; sketch the graph.

$$x^2+8x-y+6=0$$
$$x^2+8x=y-6$$
$$x^2+8x+16=y-6+16$$
$$(x+4)^2=y+10$$

vertex $= (-4,\ -10)$

$4p = 1,\ p = \frac{1}{4}$

focus $= \left(-4,\ -\frac{39}{4}\right)$

directrix: $y = -\frac{41}{4}$

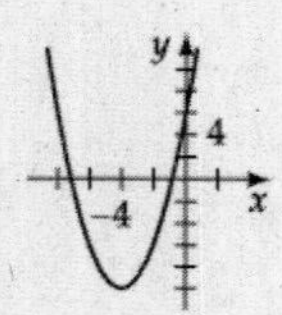

17. Find the vertex, focus, and directrix; sketch the graph.

$$x+y^2-3y+4=0$$
$$y^2-3y=-x-4$$
$$y^2-3y+9/4=-x-4+9/4$$
$$\left(y-\frac{3}{2}\right)^2=-\left(x+\frac{7}{4}\right)$$

vertex $= \left(-\frac{7}{4},\ \frac{3}{2}\right)$

$4p = -1,\ p = -\frac{1}{4}$

focus $= \left(-2,\ \frac{3}{2}\right)$

directrix: $x = -\frac{3}{2}$

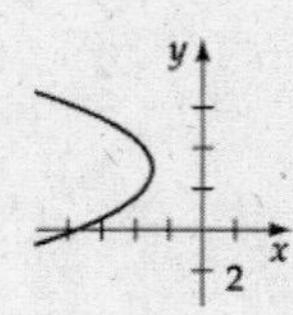

19. Find the vertex, focus, and directrix; sketch the graph.

$$2x-y^2-6y+1=0$$
$$-y^2-6y=-2x-1$$
$$y^2+6y=2x+1$$
$$y^2+6y+9=2x+1+9$$
$$(y+3)^2=2(x+5)$$

vertex $= (-5,\ -3)$

$4p = 2,\ p = \frac{1}{2}$

focus $= \left(-\frac{9}{2},\ -3\right)$

directrix: $x = -\frac{11}{2}$

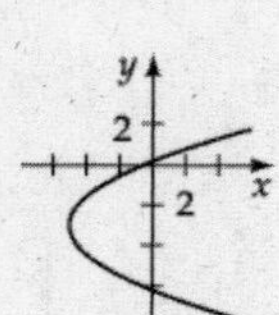

21. Find the vertex, focus, and directrix; sketch the graph.

$$x^2+3x+3y-1=0$$
$$x^2+3x=-3y+1$$
$$x^2+3x+\frac{9}{4}=-3y+1+\frac{9}{4}$$
$$\left(x+\frac{3}{2}\right)^2=-3\left(y-\frac{13}{12}\right)$$

vertex $\left(-\frac{3}{2},\ \frac{13}{12}\right)$

$4p = -3,\ p = -\frac{3}{4}$

focus $= \left(-\frac{3}{2},\ \frac{1}{3}\right)$

directrix: $y = \frac{11}{6}$

23. Find the vertex, focus, and directrix; sketch the graph.

$$2x^2-8x-4y+3=0$$
$$2(x^2-4x)=4y-3$$
$$2(x^2-4x+4)=4y-3+8$$
$$2(x-2)^2=4y+5$$
$$(x-2)^2=2y+\frac{5}{2}$$
$$(x-2)^2=2\left(y+\frac{5}{4}\right)$$

vertex $=\left(2,\ -\frac{5}{4}\right)$

$4p=2,\ p=\frac{1}{2}$

focus $=\left(2,\ -\frac{3}{4}\right)$

directrix $y=-\frac{7}{4}$

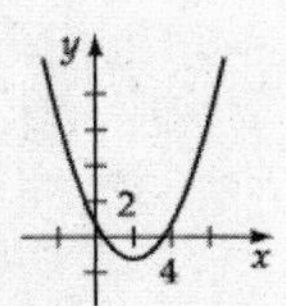

25. Find the vertex, focus, and directrix; sketch the graph.

$$2x+4y^2+8y-5=0$$
$$4y^2+8y=-2x+5$$
$$4(y^2+2y)=-2x+5$$
$$4(y^2+2y+1)=-2x+5+4$$
$$4(y+1)^2=-2x+9$$
$$(y+1)^2=-\frac{1}{2}x+\frac{9}{4}$$
$$(y+1)^2=-\frac{1}{2}\left(x-\frac{9}{2}\right)$$

vertex $=\left(\frac{9}{2},\ -1\right)$

$4p=-\frac{1}{2},\ p=-\frac{1}{8}$

focus $=\left(\frac{35}{8},\ -1\right)$

directrix $x=\frac{37}{8}$

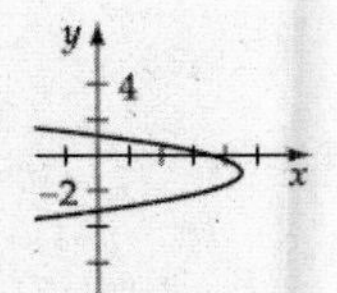

27. Find the vertex, focus, and directrix; sketch the graph.

$$3x^2-6x-9y+4=0$$
$$x^2-2x=3y-\frac{4}{3}$$
$$x^2-2x+1=3y-\frac{1}{3}$$
$$(x-1)^2=3\left(y-\frac{1}{9}\right)$$

vertex $=\left(1,\ \frac{1}{9}\right)$

$4p=3,\ p=\frac{3}{4}$

focus $=\left(1,\ \frac{31}{36}\right)$

directrix $y=-\frac{23}{36}$

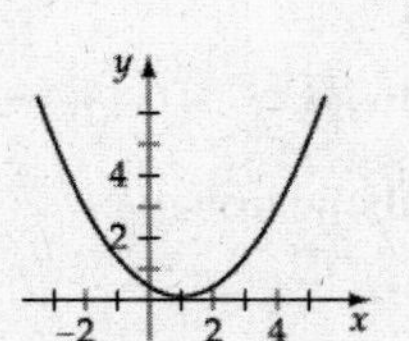

29. Find the equation of the parabola in standard form.

vertex $(0,\ 0)$, focus $(0,\ -4)$

$x^2=4py$

$p=-4$ since focus is $(0,p)$

$x^2=4(-4)y$

$x^2=-16y$

31. Find the equation of the parabola in standard form.

vertex $(-1,\ 2)$, focus $(-1,3)$

$(x-h)^2=4p(y-k)$

$h=-1,\ k=2.$

The distance p from the vertex to the focus is 1.

$(x+1)^2=4(1)(y-2)$

$(x+1)^2=4(y-2)$

33. Find the equation of the parabola in standard form.

focus $(3,\ -3)$, directrix $y=-5$

The vertex is the midpoint of the line segment joining $(3,-3)$ and the point $(3,-5)$ on the directrix.

$$(h,\ k)=\left(\frac{3+2}{2},\ \frac{-3+(-5)}{2}\right)=(3,\ -4)$$

The distance p from the vertex to the focus is 1.

$$4p = 4(1) = 4$$
$$(x-h)^2 = 4p(y-k)$$
$$(x-3)^2 = 4(y+4)$$

35. Find the equation of the parabola in standard form.

vertex $= (-4, 1)$, point: $(-2, 2)$ on the parabola.

Axis of symmetry $x = -4$.

If $P_1 = (-2, 2)$, $(x+4)^2 = 4p(y-1)$. Since $(-2, 2)$ is on the curve, we get

$$(-2+4)^2 = 4p(2-1)$$
$$4 = 4p \Rightarrow p = 1$$

Thus the equation in standard form is $(x+4)^2 = 4(y-1)$.

37. The equation of the mirrored parabolic trough is

$x^2 = 4py \quad -5 \le x \le 5$

Because (5, 1.5) is a point on the parabola, (5, 1.5) must be a solution of the equation. Thus

$$5^2 = 4p(1.5)$$
$$25 = 6p$$
$$4\frac{1}{6} = p$$

The focus is $4\frac{1}{6}$ ft, or 4 ft 2 inches above the vertex.

39. Place the satellite dish on an xy-coordinate system with its vertex at (0, −1) as shown.

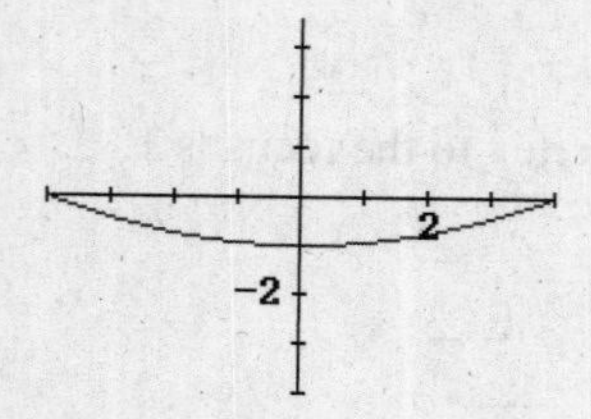

The equation of the parabola is

$x^2 = 4p(y+1) \quad -1 \le y \le 0$

Because (4, 0) is a point on this graph, (4, 0) must be a solution of the equation of the parabola. Thus,

$$16 = 4p(0+1)$$
$$16 = 4p$$
$$4 = p$$

Because p is the distance from the vertex to the focus, the focus is on the x-axis 4 feet above the vertex.

41. The focus of the parabola is $(p, 0)$ where $y^2 = 4px$.

Half of 18.75 inches is 9.375 inches.

Therefore, the point (3.66, 9.375) is on the parabola.

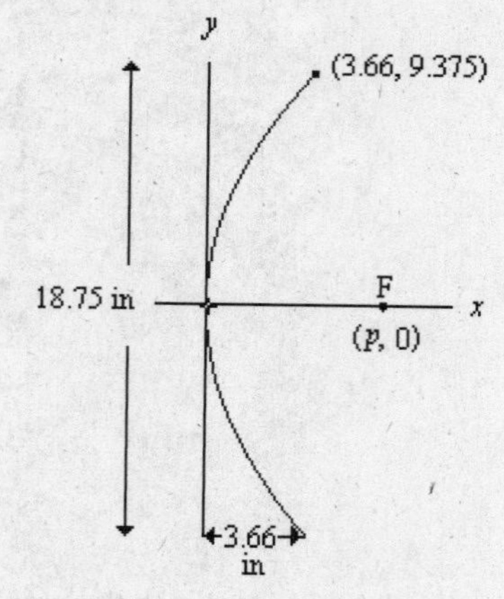

$$(9.375)^2 = 4p(3.66)$$
$$87.890625 = 14.64p$$
$$\frac{87.890625}{14.64} = p$$
$$p \approx 6.0 \text{ inches}$$

43. $S = \frac{\pi r}{6d^2}\left[\left(r^2 + 4d^2\right)^{3/2} - r^3\right]$

a. $r = 40.5$ feet

$d = 16$ feet

$$S = \frac{\pi(40.5)}{6(16)^2}\left[\left([40.5]^2 + 4[16]^2\right)^{3/2} - (40.5)^3\right]$$
$$= \frac{40.5\pi}{1536}\left[(2664.25)^{3/2} - 66430.125\right]$$
$$= \frac{40.5\pi}{1536}[137518.9228 - 66430.125]$$
$$= \frac{40.5\pi}{1536}[71088.79775]$$
$$\approx 5900 \text{ square feet}$$

b. $r = 125$ feet

$d = 52$ feet

$$S = \frac{\pi(125)}{6(52)^2}\left[\left([125]^2 + 4[52]^2\right)^{3/2} - (125)^3\right]$$
$$= \frac{125\pi}{16224}\left[(26441)^{3/2} - 1953125\right]$$
$$= \frac{125\pi}{16224}[4299488.724 - 1953125]$$
$$= \frac{125\pi}{16224}[2346363.724]$$
$$\approx 56,800 \text{ square feet}$$

45. The equation of the mirror is given by

$x^2 = 4py \quad -60 \le x \le 60$

Because p is the distance from the vertex to the focus and the coordinates of the focus are (0, 600), $p = 600$.

Therefore,

$$x^2 = 4(600)y$$
$$x^2 = 2400y$$

To determine a, substitute $(60, a)$ into the equation $x^2 = 2400y$ and solve for a.

$$x^2 = 2400y$$
$$60^2 = 2400a$$
$$3600 = 2400a$$
$$1.5 = a$$

The concave depth of the mirror is 1.5 inches.

47. Find the vertex.

$$x = -0.325y^2 + 13y + 120$$
$$x - 120 = -0.325y^2 + 13y$$
$$x - 120 = -0.325(y^2 - 40y)$$
$$x - 120 - 130 = -0.325(y^2 - 40y + 400)$$
$$x - 250 = -0.325(y - 20)^2$$
$$-\frac{40}{13}(x - 250) = (y - 20)^2$$

The vertex is (250, 20).

Find the focus.

$4p = -\frac{40}{13}, \quad p = -\frac{10}{13}$

The focus is $\left(\frac{3240}{13}, 20\right)$.

49. $x^2 = 4y$

$4p = 4, \quad p = 1$

focus = (0, 1)

Substituting the vertical coordinate of the focus for y to obtain x-coordinates of endpoints (x_1, y_1), (x_2, y_2), we have

$$x^2 = 4(1), \text{ or } x^2 = 4$$
$$x = \pm\sqrt{4}$$
$$x_1 = -2 \quad x_2 = 2$$

Length of latus rectum $= |x_2 - x_1| = 2 - (-2) = 4$.

51. $(x - h)^2 = 4p(y - k)$

focus $= (h, k + p)$

Substituting the vertical coordinate of the focus for y to obtain x-coordinates of endpoints (x_1, y_1), (x_2, y_2), we have

$$(x - h)^2 = 4p(k + p - k)$$
$$(x - h)^2 = 4p^2$$
$$x - h = \pm 2p$$
$$x_1 = h - 2p \quad x_2 = h + 2p$$

Solving for $|x_2 - x_1|$, we obtain

$\Delta x = |x_2 - x_1| = |h + 2p - h + 2p| = 4|p|$

or $(y - k)^2 = 4p(x - h)$

focus = $(h + p, k)$

Substituting the horizontal coordinate of the focus for x to obtain the y-coordinates of the endpoints (x_1, y_1), (x_2, y_2), we have

$$(y - k)^2 = 4p(h + p - h)$$
$$(y - k)^2 = 4p^2$$
$$y - k = \pm 2p$$

Solving for $|y_2 - y_1|$, we obtain

$\Delta y = |y_2 - y_1| = |k + 2p - k + 2p| = 4|p|$

Thus, the length of the latus rectum is $4|p|$.

53. Graph $(y+4)^2 = -(x-1)$.

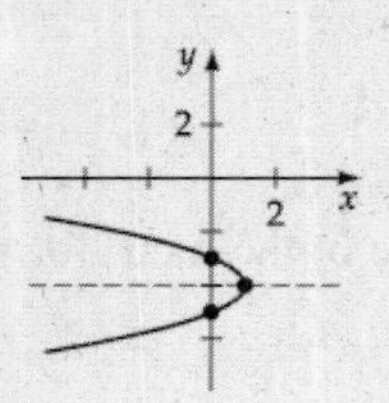

$4p = -1,\ p = -\frac{1}{4}$

focus $\left(\frac{3}{4},\ -4\right)$

one point: $\left(\frac{3}{4},\ k+2p\right) = \left(\frac{3}{4},\ -\frac{9}{2}\right)$

one point: $\left(\frac{3}{4},\ k-2p\right) = \left(\frac{3}{4},\ -\frac{7}{2}\right)$

55. Graph $y = \frac{7}{4} + \frac{1}{4}x|x|$.

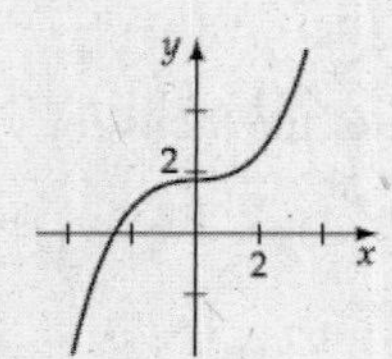

57. By definition, any point on the curve (x, y) will be equidistant from both the focus (1, 1) and the directrix, $(y_2 = -x_2 - 2)$.

If we let d_1 equal the distance from the focus to the point (x, y), we get $d_1 = \sqrt{(x-1)^2 + (y-1)^2}$.

To determine the distance d_2 from the point (x, y) to the line $y = -x - 2$, draw a line segment from (x, y) to the directrix so as to meet the directrix at a $90°$ angle. Now drop a line segment parallel to the y-axis from (x, y) to the directrix. This segment will meet the directrix at a 45° angle, thus forming a right isosceles triangle with the directrix and the line segment perpendicular to the directrix from (x, y). The length of this segment, which is the hypotenuse of the triangle, is the difference between y and the y-value of the directrix at x, or $-x - 2$. Thus, the hypotenuse has a length of $y + x + 2$, and since the right triangle is also isosceles, each leg has a length of $\frac{y+x+2}{\sqrt{2}}$.

But since d_2 is the length of the leg drawn from (x, y) to the directrix, $d_2 = \frac{y+x+2}{\sqrt{2}}$.

Thus, $d_1 = \sqrt{(x-1)^2 + (y-1)^2}$ and $d_2 = \frac{x+y+2}{\sqrt{2}}$.

By definition, $d_1 = d_2$. So by substitution,

$$\sqrt{(x-1)^2+(y-1)^2} = \frac{x+y+2}{\sqrt{2}}$$
$$\sqrt{2}\sqrt{(x-1)^2+(y-1)^2} = x+y+2$$
$$2\left[(x-1)^2+(y-1)^2\right] = x^2+y^2+4x+4y+2xy+4$$
$$2\left[x^2-2x+1+y^2-2y+1\right] = x^2+y^2+4x+4y+2xy+4$$
$$2x^2-4x+2y^2-4y+4 = x^2+y^2+4x+4y+2xy+4$$
$$x^2+y^2-8x-8y-2xy = 0$$

Prepare for Section 5.2

P1. Find the midpoint and length.

midpoint: $\left(\frac{x_1+x_2}{2},\ \frac{y_1+y_2}{2}\right) = \left(\frac{5+(-1)}{2},\ \frac{1+5}{2}\right)$

The midpoint is (2, 3).

length: $\sqrt{(x_2-x_1)^2+(y_2-y_1)^2}$

$\sqrt{(-1-5)^2+(5-1)^2} = \sqrt{36+16} = \sqrt{52}$

The length is $2\sqrt{13}$.

P3. Solve $x^2 - 2x = 2$.

$$x^2 - 2x = 2$$
$$x^2 - 2x + 1 = 2 + 1$$
$$(x-1)^2 = 3$$
$$x - 1 = \pm\sqrt{3}$$
$$x = 1 \pm \sqrt{3}$$

P5. Solve $(x-2)^2 + y^2 = 4$ for y.

$$(x-2)^2 + y^2 = 4$$
$$y^2 = 4 - (x-2)^2$$
$$y = \pm\sqrt{4-(x-2)^2}$$

Section 5.2 Exercises

1. **a.** iv

b. i

c. ii

d. iii

3. Find the center, vertices, and foci; sketch the graph.

$$\frac{x^2}{16}+\frac{y^2}{25}=1$$

$a^2=25\rightarrow a=5$

$b^2=16\rightarrow b=4$

$c=\sqrt{a^2-b^2}=\sqrt{25-16}=\sqrt{9}=3$

Center $(0, 0)$
Vertices $(0, \pm 5)$
Foci $(0, \pm 3)$

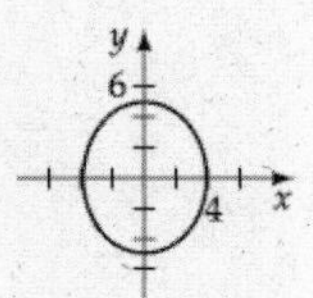

5. Find the center, vertices, and foci; sketch the graph.

$$\frac{x^2}{9}+\frac{y^2}{4}=1$$

$a^2=9\rightarrow a=3$

$b^2=4\rightarrow b=2$

$c=\sqrt{a^2-b^2}=\sqrt{9-4}=\sqrt{5}$

Center $(0, 0)$
Vertices $(\pm 3, 0)$
Foci $(\pm\sqrt{5}, 0)$

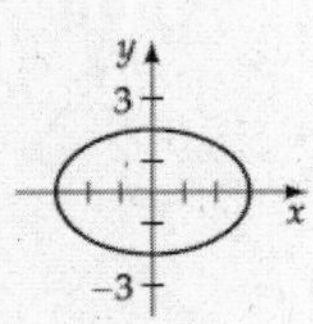

7. Find the center, vertices, and foci; sketch the graph.

$$\frac{x^2}{9}+\frac{y^2}{7}=1$$

$a^2=9\rightarrow a=3$

$b^2=7\rightarrow b=\sqrt{7}$

$c=\sqrt{a^2-b^2}=\sqrt{9-7}=\sqrt{2}$

Center $(0, 0)$
Vertices $(\pm 3, 0)$
Foci $(\pm\sqrt{2}, 0)$

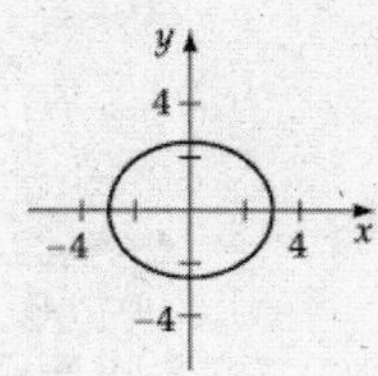

9. Find the center, vertices, and foci; sketch the graph.

$$\frac{4x^2}{9}+\frac{y^2}{16}=1$$

Rewrite as $\frac{x^2}{9/4}+\frac{y^2}{16}=1$

$a^2=16\rightarrow a=4$

$b^2=\frac{9}{4}\rightarrow b=\frac{3}{2}$

$c=\sqrt{a^2-b^2}=\sqrt{16-\frac{9}{4}}=\frac{\sqrt{55}}{2}$

Center $(0, 0)$
Vertices $(0, \pm 4)$
Foci $\left(0, \pm\frac{\sqrt{55}}{2}\right)$

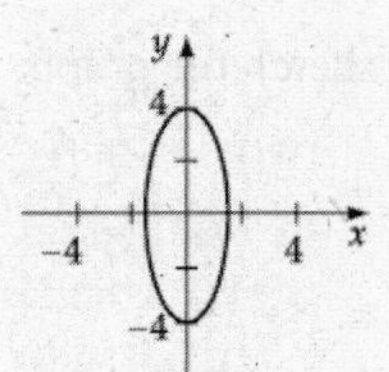

11. Find the center, vertices, and foci; sketch the graph.

$$\frac{(x-3)^2}{25}+\frac{(y+2)^2}{16}=1$$

Center $(3, -2)$
Vertices $(3\pm 5, -2)=(8, -2), (-2, -2)$
Foci $(3\pm 3, -2)=(6, -2), (0, -2)$

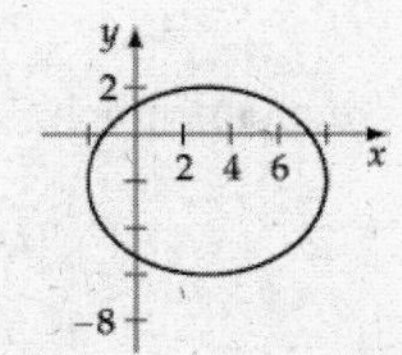

13. Find the center, vertices, and foci; sketch the graph.

$$\frac{(x+2)^2}{9}+\frac{y^2}{25}=1$$

Center $(-2,\ 0)$

Vertices $(-2,\ 5),\ (-2,\ -5)$

Foci $(-2,\ 4),\ (-2,\ -4)$

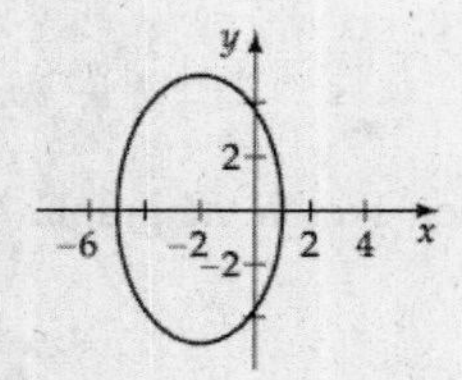

15. Find the center, vertices, and foci; sketch the graph.

$$\frac{(x-1)^2}{21}+\frac{(y-3)^2}{4}=1$$

Center $(1,\ 3)$

Vertices $\left(1\pm\sqrt{21},\ 3\right)$

Foci $\left(1\pm\sqrt{17},\ 3\right)$

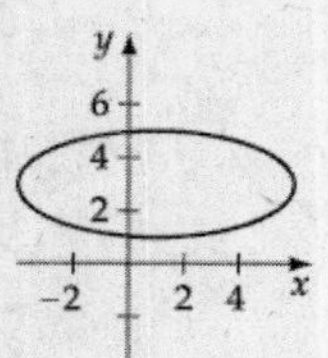

17. Find the center, vertices, and foci; sketch the graph.

$$\frac{9(x-1)^2}{16}+\frac{(y+1)^2}{9}=1$$

Center $(1,\ -1)$

Vertices $(1,\ -1\pm 3)=(1,\ 2),\ (1,\ -4)$

Foci $\left(1,\ -1\pm\frac{\sqrt{65}}{3}\right)$

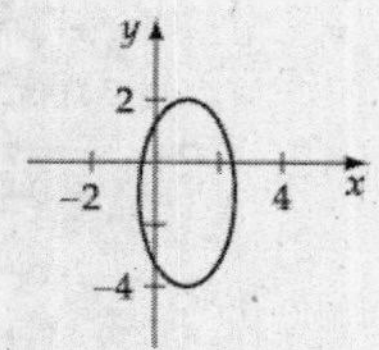

19. Find the center, vertices, and foci; sketch the graph.

$$3x^2+4y^2=12$$

$$\frac{x^2}{4}+\frac{y^2}{3}=1$$

Center $(0,\ 0)$

Vertices $(\pm 2,\ 0)$

Foci $(\pm 1,\ 0)$

21. Find the center, vertices, and foci; sketch the graph.

$$25x^2+16y^2=400$$

$$\frac{x^2}{16}+\frac{y^2}{25}=1$$

Center $(0,\ 0)$

Vertices $(0,\ \pm 5)$

Foci $(0,\ \pm 3)$

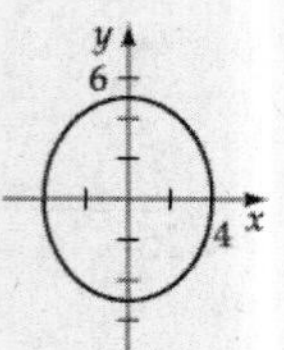

23. Find the center, vertices, and foci; sketch the graph.

$$64x^2+25y^2=400$$

$$\frac{x^2}{25/4}+\frac{y^2}{16}=1$$

Center $(0,\ 0)$

Vertices $(0,\ \pm 4)$

Foci $\left(0,\ \pm\frac{\sqrt{39}}{2}\right)$

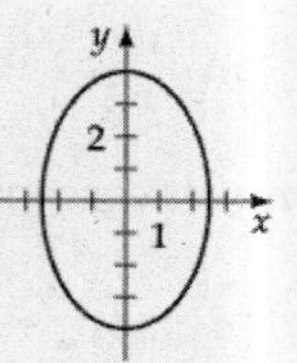

25. Find the center, vertices, and foci; sketch the graph.

$$4x^2+y^2-24x-8y+48=0$$

$$4\left(x^2-6x\right)+\left(y^2-8y\right)=-48$$

$$4\left(x^2-6x+9\right)+\left(y^2-8y+16\right)=-48+36+16$$

$$4(x-3)^3+(y-4)^2=4$$

$$\frac{(x-3)^2}{1}+\frac{(y-4)^2}{4}=1$$

Center $(3,\ 4)$

Vertices $(3,\ 4\pm 2)=(3,\ 6),\ (3,\ 2)$

Foci $\left(3,\ 4\pm\sqrt{3}\right)$

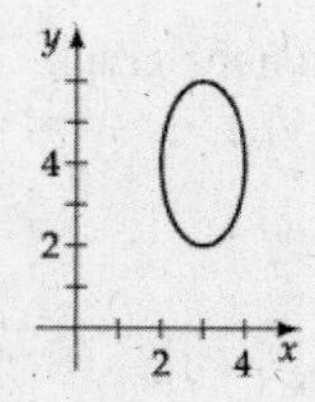

27. Find the center, vertices, and foci; sketch the graph.

$$5x^2+9y^2-20x+54y+56=0$$
$$5(x^2-4x)+9(y^2+6y)=-56$$
$$5(x^2-4x+4)+9(y^2+6y+9)=-56+20+81$$
$$5(x-2)^2+9(y+3)^2=45$$
$$\frac{(x-2)^2}{9}+\frac{(y+3)^2}{5}=1$$

Center $(2,\ -3)$

Vertices $(2\pm 3,\ -3)=(-1,\ -3),\ (5,\ -3)$

Foci $(2\pm 2,\ 3)=(0,\ -3),\ (4,\ -3)$

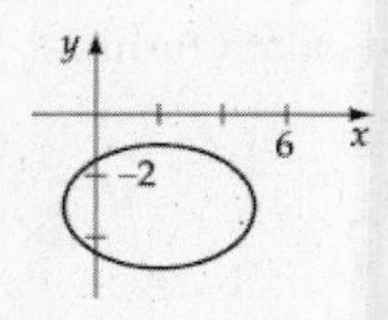

29. Find the center, vertices, and foci; sketch the graph.

$$16x^2+9y^2-64x-80=0$$
$$16(x^2-4x)+9y^2=80$$
$$16(x^2-4x+4)+9y^2=80+64$$
$$16(x-2)^2+9y^2=144$$
$$\frac{(x-2)^2}{9}+\frac{y^2}{16}=1$$

Center $(2, 0)$

Vertices $(2,\ \pm 4)=(2,\ 4),\ (2,\ -4)$

Foci $(2\pm\sqrt{7})$

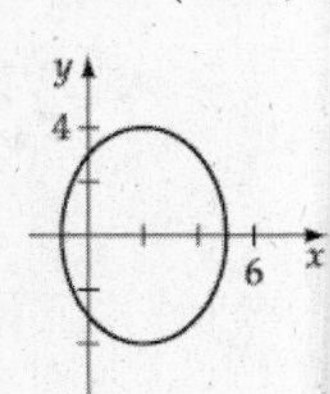

31. Find the center, vertices, and foci; sketch the graph.

$$25x^2+16y^2+50x-32y-359=0$$
$$25(x^2+2x)+16(y^2-2y)=359$$
$$25(x^2+2x+1)+16(y^2-2y+1)=359+25+16$$
$$25(x+1)^2+16(y-1)^2=400$$
$$\frac{(x+1)^2}{16}+\frac{(y-1)^2}{25}=1$$

Center $(-1,\ 1)$

Vertices $(-1,\ 1\pm 5)=(-1,\ 6),\ (-1,\ -4)$

Foci $(-1,\ 1\pm 3)=(-1,\ 4),\ (-1,\ -2)$

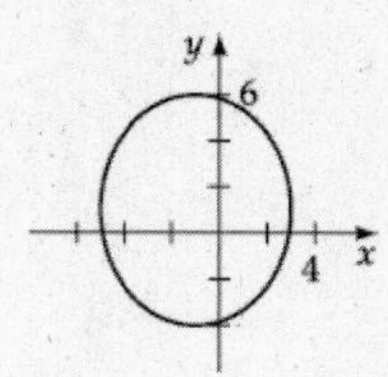

33. Find the center, vertices, and foci; sketch the graph.

$$8x^2+25y^2-48x+50y+47=0$$
$$8(x^2-6x)+25(y^2+2y)=-47$$
$$8(x^2-6x+9)+25(y^2+2y+1)=-47+72+25$$
$$8(x-3)^2+25(y+1)^2=50$$
$$\frac{(x-3)^2}{25/4}+\frac{(y+1)^2}{2}=1$$

Center: $(3,\ -1)$

Vertices: $\left(3\pm\frac{5}{2},\ -1\right)=\left(\frac{11}{2},\ -1\right),\ \left(\frac{1}{2},\ -1\right)$

Foci: $\left(3\pm\frac{\sqrt{17}}{2},\ -1\right)$

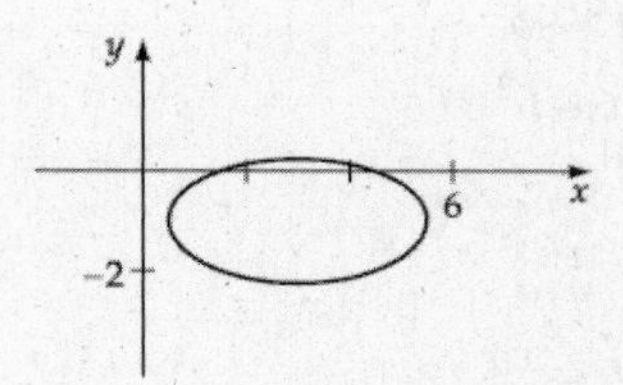

35. Find the equation of the ellipse in standard form.

$$2a=10,\ a=5,\ a^2=25$$
$$c=4$$
$$c^2=a^2-b^2$$
$$16=25-b^2$$
$$b^2=9$$
$$\frac{x^2}{25}+\frac{y^2}{9}=1$$

37. Find the equation of the ellipse in standard form.

$$a=6,\ a^2=36$$
$$b=4,\ b^2=16$$
$$\frac{x^2}{36}+\frac{y^2}{16}=1$$

39. Find the equation of the ellipse in standard form.

$$\begin{aligned} 2a &= 12 \\ a &= 6 \\ a^2 &= 36 \\ \frac{x^2}{36}+\frac{y^2}{b^2} &= 1 \\ \frac{(2)^2}{36}+\frac{(-3)^2}{b^2} &= 1 \\ \frac{4}{36}+\frac{9}{b^2} &= 1 \\ \frac{9}{b^2} &= \frac{8}{9} \\ 8b^2 &= 81 \\ b^2 &= \frac{81}{8} \\ \frac{x^2}{36}+\frac{y^2}{81/8} &= 1 \end{aligned}$$

41. Find the equation of the ellipse in standard form.

$$\begin{aligned} c &= 3 \\ 2a &= 8 \\ a &= 4 \\ a^2 &= 16 \\ c^2 &= a^2-b^2 \\ 9 &= 16-b^2 \\ b^2 &= 7 \\ \frac{(x+2)^2}{16}+\frac{(y-4)^2}{7} &= 1 \end{aligned}$$

43. Find the equation of the ellipse in standard form.

$$\begin{aligned} 2a &= 10 \\ a &= 5 \\ a^2 &= 25 \end{aligned}$$

Since the center of the ellipse is (2, 4) and the point (3, 3) is on the ellipse, we have

$$\begin{aligned} \frac{(x-2)^2}{b^2}+\frac{(y-4)^2}{a^2} &= 1 \\ \frac{(3-2)^2}{b^2}+\frac{(3-4)^2}{25} &= 1 \\ \frac{1}{b^2} &= 1-\frac{1}{25} \\ b^2 &= \frac{25}{24} \\ \frac{(x-2)^2}{25/24}+\frac{(y-4)^2}{25} &= 1 \end{aligned}$$

45. Find the equation of the ellipse in standard form.

center (5, 1)

$$\begin{aligned} c &= 3 \\ 2a &= 10 \\ a &= 5 \\ a^2 &= 25 \\ c^2 &= a^2-b^2 \\ 9 &= 25-b^2 \\ b^2 &= 16 \\ \frac{(x-5)^2}{16}+\frac{(y-1)^2}{25} &= 1 \end{aligned}$$

47. Find the equation of the ellipse in standard form.

$$\begin{aligned} 2a &= 10 \\ a &= 5 \\ a^2 &= 25 \\ \frac{c}{a} &= \frac{2}{5} \\ \frac{c}{5} &= \frac{2}{5} \\ c &= 2 \\ c^2 &= a^2-b^2 \\ 4 &= 25-b^2 \\ b^2 &= 21 \\ \frac{x^2}{25}+\frac{y^2}{21} &= 1 \end{aligned}$$

49. Find the equation of the ellipse in standard form.

center (0, 0)

$$\begin{aligned} c &= 4 \\ \frac{c}{a} &= \frac{2}{3} \\ \frac{4}{a} &= \frac{2}{3} \\ a &= 6 \\ c^2 &= a^2-b^2 \\ 16 &= 36-b^2 \\ b^2 &= 20 \\ \frac{x^2}{20}+\frac{y^2}{36} &= 1 \end{aligned}$$

51. Find the equation of the ellipse in standard form.

center (1, 3)

$$c = 2$$
$$\frac{c}{a} = \frac{2}{5}$$
$$\frac{2}{a} = \frac{2}{5}$$
$$a = 5$$
$$c^2 = a^2 - b^2$$
$$4 = 25 - b^2$$
$$b^2 = 21$$
$$\frac{(x-1)^2}{25} + \frac{(y-3)^2}{21} = 1$$

53. Find the equation of the ellipse in standard form.

$$2a = 24, \ a = 12$$
$$\frac{c}{a} = \frac{2}{3}$$
$$\frac{c}{12} = \frac{2}{3}$$
$$c = 8$$
$$c^2 = a^2 - b^2$$
$$64 = 144 - b^2$$
$$b^2 = 80$$
$$\frac{x^2}{80} + \frac{y^2}{144} = 1$$

55. $484 = 64 + c^2$

$$c^2 = 420$$
$$c = 20.494$$
$$2c = 40.9878 \approx 41$$

The emitter should be placed 41 cm away.

57. Aphelion $= 2a -$ perihelion

$$934.34 = 2a - 835.14$$
$$a = 884.74 \text{ million miles}$$

Aphelion $= a + c = 934.34$

$$884.74 + c = 934.34$$
$$c = 49.6 \text{ million miles}$$
$$b = \sqrt{a^2 - c^2} = \sqrt{884.74^2 - 49.6^2}$$
$$\approx 883.35 \text{ million miles}$$

An equation of the orbit of Saturn is

$$\frac{x^2}{884.74^2} + \frac{y^2}{883.35^2} = 1$$

59. $a =$ semimajor axis $= 50$ feet

$b =$ height $= 30$ feet

$$c^2 = a^2 - b^2$$
$$c^2 = 50^2 - 30^2$$
$$c = \sqrt{1600} = 40$$

The foci are located 40 feet to the right and to the left of center.

61. $2a = 36 \quad 2b = 9$

$$a = 18 \qquad b = \frac{9}{2}$$
$$c^2 = a^2 - b^2$$
$$c^2 = 18^2 - \left(\frac{9}{2}\right)^2$$
$$c^2 = 324 - \frac{81}{4}$$
$$c^2 = \frac{1215}{4}$$
$$c = \frac{9\sqrt{15}}{2}$$

Since one focus is at (0, 0), the center of the ellipse is at $\left(9\sqrt{15}/2, \ 0\right)$

(17.43, 0). The equation of the path of Halley's Comet in astronomical units is

$$\frac{\left(x - 9\sqrt{15}/2\right)^2}{324} + \frac{y^2}{81/4} = 1$$

63. $\frac{x^2}{75^2} + \frac{y^2}{34^2} = 1$

Solve for y, where $x = 55$.

$$\frac{55^2}{75^2} + \frac{y^2}{34^2} = 1$$
$$\frac{y^2}{34^2} = 1 - \frac{55^2}{75^2}$$
$$y^2 = 34^2\left(1 - \frac{55^2}{75^2}\right)$$
$$y = \sqrt{34^2\left(1 - \frac{55^2}{75^2}\right)}$$
$$y \approx 23 \text{ ft}$$
$$h = y + 1 = 23 + 1 = 24 \text{ ft}$$

65. a. $c^2 = a^2 - b^2$

$c^2 = 4^2 - 3^2$

$c^2 = 7$

$c = \sqrt{7}$

$\sqrt{7}$ ft to the right and left of O.

b. $2a = 2(4) = 8$ ft

67. $9y^2 + 36y + 16x^2 - 108 = 0$

$$y = \frac{-36 \pm \sqrt{36^2 - 4(9)(16x^2 - 108)}}{2(9)}$$
$$= \frac{-36 \pm \sqrt{1296 - 36(16x^2 - 108)}}{18}$$
$$= \frac{-36 \pm \sqrt{1296 - 576x^2 + 3888}}{18}$$
$$= \frac{-36 \pm \sqrt{-576x^2 + 5184}}{18}$$
$$= \frac{-36 \pm \sqrt{576(-x^2 + 9)}}{18}$$
$$= \frac{-36 \pm 24\sqrt{(-x^2 + 9)}}{18}$$
$$= \frac{-6 \pm 4\sqrt{(-x^2 + 9)}}{3}$$

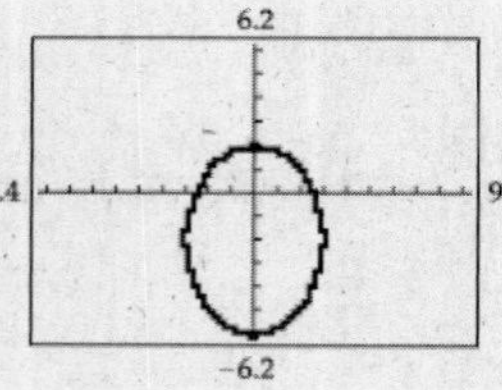

69. $9y^2 + 18y + 4x^2 + 24x + 44 = 0$

$$y = \frac{-18 \pm \sqrt{18^2 - 4(9)(4x^2 + 24x + 44)}}{2(9)}$$
$$= \frac{-18 \pm \sqrt{324 - 36(4x^2 + 24x + 44)}}{18}$$
$$= \frac{-18 \pm \sqrt{324 - 144x^2 - 864x - 1584}}{18}$$
$$= \frac{-18 \pm \sqrt{-144x^2 - 864x - 1260}}{18}$$
$$= \frac{-18 \pm \sqrt{36(-4x^2 - 24x - 35)}}{18}$$
$$= \frac{-18 \pm 6\sqrt{-4x^2 - 24x - 35}}{18}$$
$$= \frac{-3 \pm \sqrt{-4x^2 - 24x - 35}}{3}$$

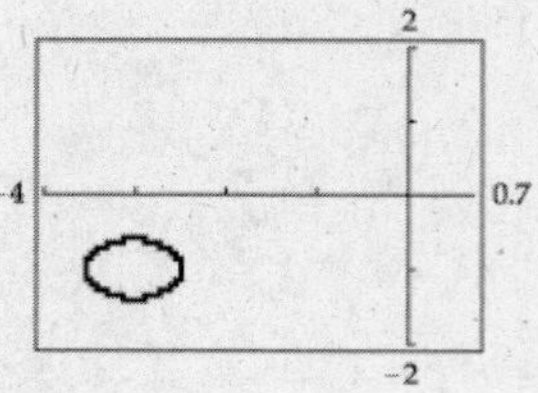

71. Find the area.

$$\frac{x^2}{5} + \frac{y^2}{3} = 1$$

$a^2 = 5,\ a = \sqrt{5}$

$b^2 = 3,\ b = \sqrt{3}$

$A = \pi ab = \pi(\sqrt{5})(\sqrt{3}) = \pi\sqrt{15}$ square units

73. The sum of the distances between the two foci and a point on the ellipse is $2a$.

$$2a = \sqrt{\left(\frac{9}{2} - 0\right)^2 + (3-3)^2} + \sqrt{\left(\frac{9}{2} - 0\right)^2 + (3+3)^2}$$
$$= \sqrt{\left(\frac{9}{2}\right)^2} + \sqrt{\frac{225}{4}}$$
$$= \frac{9}{2} + \frac{15}{2}$$
$$= 12$$

$a = 6$

$c = 3$

$c^2 = a^2 - b^2$

$9 = 36 - b^2$

$b^2 = 27$

$$\frac{x^2}{36} + \frac{y^2}{27} = 1$$

75. The sum of the distances between the two foci and a point on the ellipse is $2a$.

$$2a = \sqrt{(5-2)^2 + (3+1)^2} + \sqrt{(5-2)^2 + (3-3)^2}$$
$$= \sqrt{25} + \sqrt{3^2}$$
$$= 5 + 3$$
$$= 8$$

$$a = 4$$
$$c = 2$$
$$c^2 = a^2 - b^2$$
$$4 = 16 - b^2$$
$$b^2 = 12$$
$$\frac{(x-1)^2}{16} + \frac{(y-2)^2}{12} = 1$$

77. Center $(1, -1)$

$$c^2 = a^2 - b^2$$
$$c^2 = 16 - 9$$
$$c^2 = 7$$
$$c = \sqrt{7}$$

The latus rectum is on the graph of $y = -1 + \sqrt{7}$, or $y = -1 - \sqrt{7}$

$$\frac{(x-1)^2}{9} + \frac{(y+1)^2}{16} = 1$$
$$\frac{(x-1)^2}{9} + \frac{(-1+\sqrt{7}+1)^2}{16} = 1$$
or
$$\frac{(x-1)^2}{9} + \frac{(-1-\sqrt{7}+1)^2}{16} = 1$$
$$\frac{(x-1)^2}{9} + \frac{7}{16} = 1$$
$$\frac{(x-1)^2}{9} = \frac{9}{16}$$
$$16(x-1)^2 = 81$$
$$(x-1)^2 = \frac{81}{16}$$
$$x - 1 = \pm\sqrt{\frac{81}{16}}$$
$$x - 1 = \pm\frac{9}{4}$$
$$x = \frac{13}{4} \text{ and } -\frac{5}{4}$$

The x-coordinates of the endpoints of the latus rectum are $\frac{13}{4}$ and $-\frac{5}{4}$.

$$\left|\frac{13}{4} - \left(-\frac{5}{4}\right)\right| = \frac{9}{2}$$

The length of the latus rectum is $\frac{9}{2}$.

79. Let us transform the general equation of an ellipse into an $x'y'$ - coordinate system where the center is at the origin by replacing $(x-h)$ by x' and $(y-k)$ by y'.

We have $\frac{x'^2}{a^2} + \frac{y'^2}{b^2} = 1$.

Letting $x' = c$ and solving for y' yields

$$\frac{(c)^2}{a^2} + \frac{y'^2}{b^2} = 1$$
$$b^2c^2 + a^2y'^2 = a^2b^2$$
$$a^2y'^2 = a^2b^2 - b^2c^2$$
$$a^2y'^2 = b^2(a^2 - c^2)$$

But since $c^2 = a^2 - b^2$, $b^2 = a^2 - c^2$, we can substitute to obtain

$$a^2y'^2 = b^2(b^2)$$
$$y'^2 = \frac{b^4}{a^2}$$
$$y' = \pm\sqrt{\frac{b^4}{a^2}} = \pm\frac{b^2}{a}$$

The endpoints of the latus rectum, then, are

$$\left(c,\ \frac{b^2}{a}\right) \text{ and } \left(c,\ -\frac{b^2}{a}\right).$$

The distance between these points is $\frac{2b^2}{a}$.

Mid-Chapter 5 Quiz

1. Find the vertex, focus, and directrix; sketch the graph.

$$-2x + 3y^2 + 6y = 3$$
$$3y^2 + 6y = 2x + 3$$
$$y^2 + 2y = \frac{2}{3}x + 1$$
$$y^2 + 2y + 1 = \frac{2}{3}x + 1 + 1$$
$$(y+1)^2 = \frac{2}{3}(x+3)$$

vertex $= (-3, -1)$

$4p = \frac{2}{3}$, $p = \frac{1}{6}$

focus $= \left(-3 + \frac{1}{6}, -1\right) = \left(-\frac{17}{6}, -1\right)$

directrix: $y = -3 - \frac{1}{6} = -\frac{19}{6}$

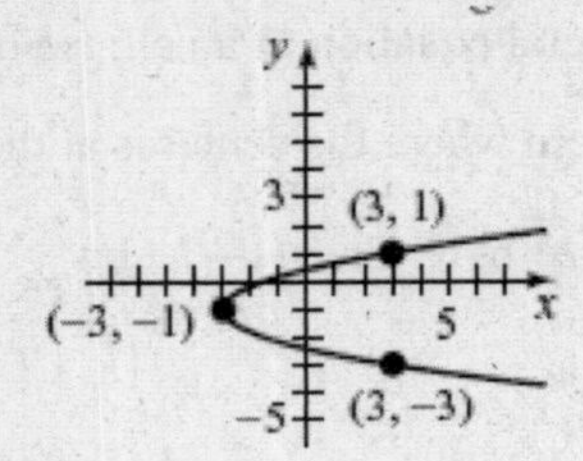

3. Find the equation of the ellipse in standard form.

$$2b = 2, \ b = 1$$
$$b^2 = 1$$
$$c = 2$$
$$c^2 = 4$$
$$c^2 = a^2 - b^2$$
$$4 = a^2 - 1$$
$$a^2 = 5$$
$$\frac{(x-5)^2}{1^2} + \frac{(y+3)^2}{(\sqrt{5})^2} = 1$$

5. Find the equation of the parabola in standard form.

vertex (6, 2), focus (4, 2)

$$(y-k)^2 = 4p(x-h)$$
$$h = 6, \ k = 2$$

Since the focus is $(h + p, k)$, $6 + p = 4$ and $p = -2$.

$$(y-2)^2 = 4(-2)(x-6)$$
$$(y-2)^2 = -8(x-6)$$

Prepare for Section 5.3

P1. Find the midpoint and length.

$$\frac{4+-2}{2} = 1$$
$$\frac{-3+1}{2} = -1$$

Midpoint: (1, –1)

$$\sqrt{(-2-4)^2 + (1--3)^2} = \sqrt{52} = 2\sqrt{13}$$

Length: $2\sqrt{13}$

P3. Simplify.

$$\frac{4}{\sqrt{8}} = \frac{4\sqrt{8}}{8} = \frac{8\sqrt{2}}{8} = \sqrt{2}$$

P5. Solve.

$$\frac{x^2}{4} - \frac{y^2}{9} = 1$$
$$-\frac{y^2}{9} = 1 - \frac{x^2}{4}$$
$$y^2 = \frac{9x^2}{4} - 9$$
$$y = \pm\sqrt{\frac{9x^2}{4} - 9}$$
$$y = \pm\frac{3}{2}\sqrt{x^2 - 4}$$

Section 5.3 Exercises

1. **a.** iii

b. ii

c. i

d. iv

3. Find the center, vertices, foci, and asymptotes; graph.

$$\frac{x^2}{16} - \frac{y^2}{25} = 1$$

Center $(0, 0)$

Vertices $(\pm 4, 0)$

Foci $(\pm\sqrt{41}\ 0)$

Asymptotes $y = \pm\frac{5}{4}x$

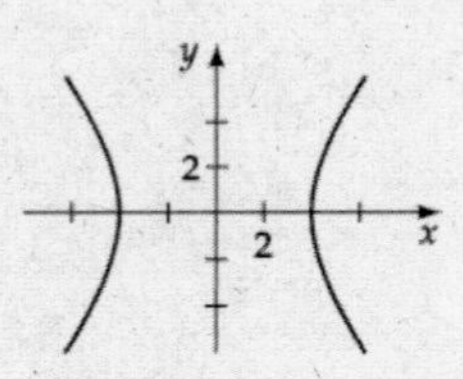

5. Find the center, vertices, foci, and asymptotes; graph.

$$\frac{y^2}{4} - \frac{x^2}{25} = 1$$

Center $(0, 0)$

Vertices $(0, \ \pm 2)$

Foci $(0, \ \pm\sqrt{29})$

Asymptotes $y = \pm\frac{2}{5}x$

7. Find the center, vertices, foci, and asymptotes; graph.

$$\frac{x^2}{7}-\frac{y^2}{9}=1$$

Center $(0, 0)$

Vertices $(\pm\sqrt{7}, 0)$

Foci $(\pm 4, 0)$

Asymptotes $y=\pm\frac{3\sqrt{7}}{7}x$

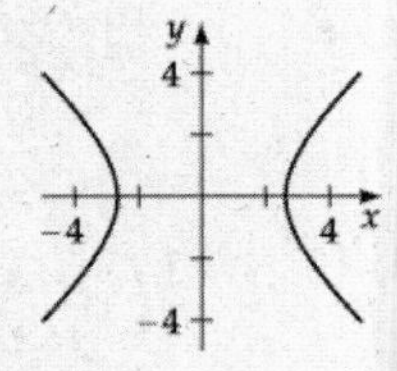

9. Find the center, vertices, foci, and asymptotes; graph.

$$\frac{4x^2}{9}-\frac{y^2}{16}=1$$

Center $(0, 0)$

Vertices $\left(\pm\frac{3}{2}, 0\right)$

Foci $\left(\pm\frac{\sqrt{73}}{2}, 0\right)$

Asymptotes $y=\pm\frac{8}{3}x$

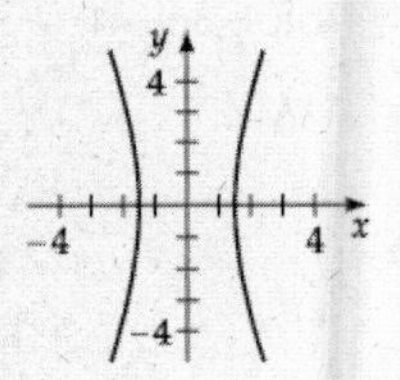

11. Find the center, vertices, foci, and asymptotes; graph.

$$\frac{(x-3)^2}{16}-\frac{(y+4)^2}{9}=1$$

Center $(3, -4)$

Vertices $(3\pm 4, -4)=(7, -4), (-1, -4)$

Foci $(3\pm 5, -4)=(8, -4), (-2, -4)$

Asymptotes $y+4=\pm\frac{3}{4}(x-3)$

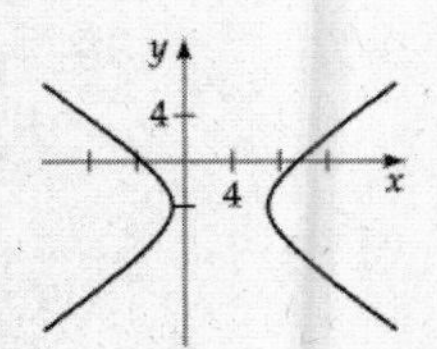

13. Find the center, vertices, foci, and asymptotes; graph.

$$\frac{(y+2)^2}{4}-\frac{(x-1)^2}{16}=1$$

Center $(1, -2)$

Vertices $(1, -2\pm 2)=(1, 0), (1, -4)$

Foci $(1,-2\pm 2\sqrt{5})=(1,-2+2\sqrt{5}), (1,-2-2\sqrt{5})$

Asymptotes $y+2=\pm\frac{1}{2}(x-1)$

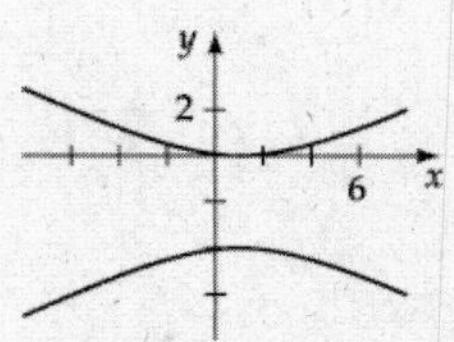

15. Find the center, vertices, foci, and asymptotes; graph.

$$\frac{(x+2)^2}{9}-\frac{y^2}{25}=1$$

Center $(-2, 0)$

Vertices $(-2\pm 3, 0)=(1, 0), (-5, 0)$

Foci $(-2\pm\sqrt{34}, 0)$

Asymptotes $y=\pm\frac{5}{3}(x+2)$

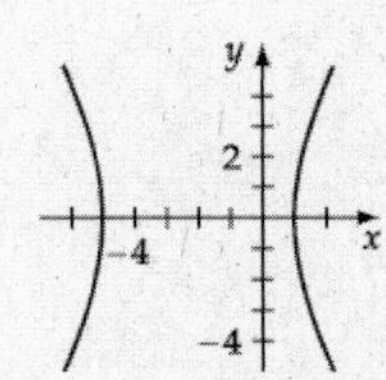

17. Find the center, vertices, foci, and asymptotes; graph.

$$\frac{9(x-1)^2}{16}-\frac{(y+1)^2}{9}=1$$

$$\frac{(x-1)^2}{16/9}-\frac{(y+1)^2}{9}=1$$

Center $(1, -1)$

Vertices $\left(1\pm\frac{4}{3}, -1\right)=\left(\frac{7}{3}, -1\right), \left(-\frac{1}{3}, -1\right)$

Foci $\left(1\pm\frac{\sqrt{97}}{3}, -1\right)$

Asymptotes $(y+1)=\pm\frac{9}{4}(x-1)$

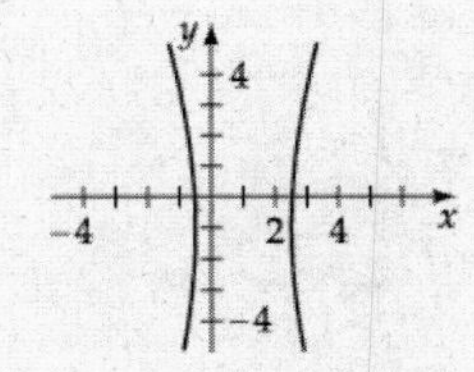

19. Find the center, vertices, foci, and asymptotes; graph.

$x^2 - y^2 = 9$

$\frac{x^2}{9} - \frac{y^2}{9} = 1$

Center $(0, 0)$

Vertices $(\pm 3, 0)$

Foci $(\pm 3\sqrt{2}, 0)$

Asymptotes $y = \pm x$

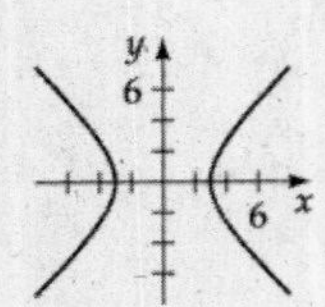

21. Find the center, vertices, foci, and asymptotes; graph.

$16y^2 - 9x^2 = 144$

$\frac{y^2}{9} - \frac{x^2}{16} = 1$

Center $(0, 0)$

Vertices $(0, \pm 3)$

Foci (0 ± 5)

Asymptotes $y = \pm \frac{3}{4}x$

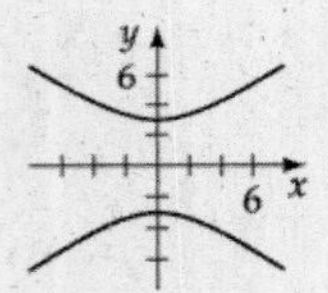

23. Find the center, vertices, foci, and asymptotes; graph.

$9y^2 - 36x^2 = 4$

$\frac{y^2}{4/9} - \frac{x^2}{1/9} = 1$

Center $(0, 0)$

Vertices $\left(0, \pm \frac{2}{3}\right)$

Foci $\left(0, \pm \frac{\sqrt{5}}{3}\right)$

Asymptotes $y = \pm 2x$

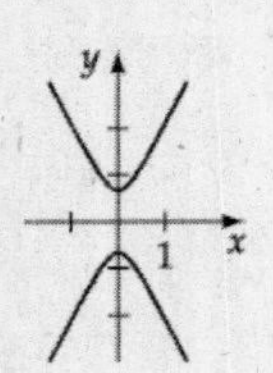

25. Find the center, vertices, foci, and asymptotes; graph.

$$x^2 - y^2 - 6x + 8y - 3 = 0$$
$$x^2 - y^2 - 6x + 8y = 3$$
$$(x^2 - 6x) - (y^2 - 8y) = 3$$
$$(x^2 - 6x + 9) - (y^2 - 8y + 16) = 3 + 9 - 16$$
$$(x-3)^2 - (y-4)^2 = -4$$
$$\frac{(y-4)^2}{4} - \frac{(x-3)^2}{4} = 1$$

Center $(3, 4)$

Vertices $(3, 4 \pm 2) = (3, 6), (3, 2)$

Foci $(3, 4 \pm 2\sqrt{2}) = (3, 4 + 2\sqrt{2}), (3, 4 - 2\sqrt{2})$

Asymptotes $y - 4 = \pm(x - 3)$

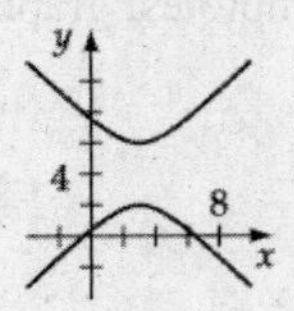

27. Find the center, vertices, foci, and asymptotes; graph.

$$9x^2 - 4y^2 + 36x - 8y + 68 = 0$$
$$9x^2 + 36x - 4y^2 - 8y = -68$$
$$9(x^2 + 4x) - 4(y^2 + 2y) = -68$$
$$9(x^2 + 4x + 4) - 4(y^2 + 2y + 1) = -68 + 36 - 4$$
$$9(x+2)^2 - 4(y+1)^2 = -36$$
$$\frac{(y+1)^2}{9} - \frac{(x+2)^2}{4} = 1$$

Center $(-2, -1)$

Vertices $(-2, -1 \pm 3) = (-2, 2), (-2, -4)$

Foci $(-2, -1 \pm \sqrt{13})$

$= (-2, -1 + \sqrt{13}), (-2, -1 - \sqrt{13})$

Asymptotes $y + 1 = \pm \frac{3}{2}(x + 2)$

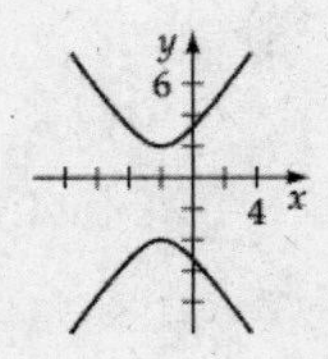

29. Solve for y and graph.

$$y=\frac{-6\pm\sqrt{6^2-4(-1)(4x^2+32x+39)}}{2(-1)}$$
$$=\frac{-6\pm\sqrt{36+4(4x^2+32x+39)}}{-2}$$
$$=\frac{-6\pm\sqrt{16x^2+128x+192}}{-2}$$
$$=\frac{-6\pm\sqrt{16(x^2+8x+12)}}{-2}$$
$$=\frac{-6\pm4\sqrt{x^2+8x+12}}{-2}$$
$$=3\pm2\sqrt{x^2+8x+12}$$

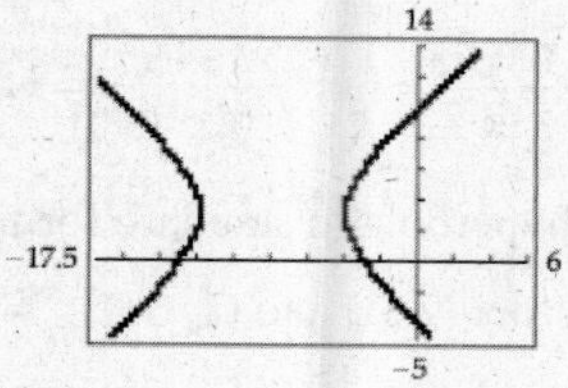

14

-17.5 6

-5

31. Solve for y and graph.

$$y=\frac{64\pm\sqrt{(-64)^2-4(-16)(9x^2-36x+116)}}{2(-16)}$$
$$=\frac{64\pm\sqrt{4096+64(9x^2-36x+116)}}{-32}$$
$$=\frac{64\pm\sqrt{64(9x^2-36x+116+64)}}{-32}$$
$$=\frac{64\pm8\sqrt{(9x^2-36x+180)}}{-32}$$
$$=\frac{64\pm8\sqrt{9(x^2-4x+20)}}{-32}$$
$$=\frac{64\pm24\sqrt{x^2-4x+20}}{-32}$$
$$=\frac{-8\pm3\sqrt{x^2-4x+20}}{4}$$

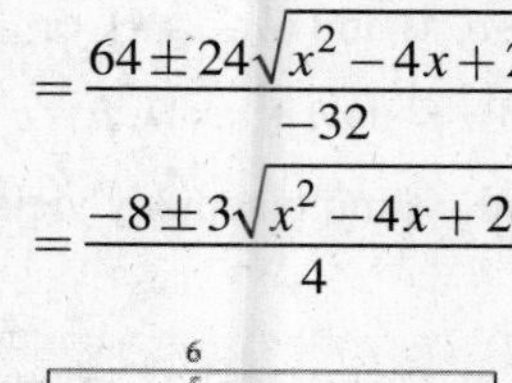

6

-7.5 16

-8

33. Solve for y and graph.

$$y=\frac{18\pm\sqrt{(-18)^2-4(-9)(4x^2+8x-6)}}{2(-9)}$$
$$=\frac{18\pm\sqrt{324+36(4x^2+8x-6)}}{-18}$$
$$=\frac{18\pm\sqrt{36(4x^2+8x-6+9)}}{-18}$$
$$=\frac{18\pm6\sqrt{4x^2+8x+3}}{-18}$$
$$=\frac{-3\pm\sqrt{4x^2+8x+3}}{3}$$

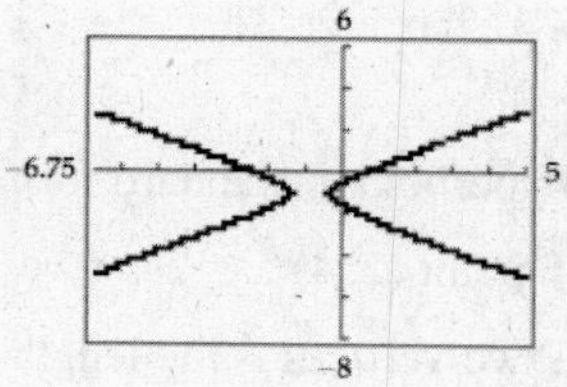

6

-6.75 5

-8

35. Find the equation of the hyperbola in standard form.

vertices $(3, 0)$ and $(-3, 0)$, foci $(4, 0)$ and $(-4, 0)$

Traverse axis is on x-axis. For a standard hyperbola, the vertices are at $(h + a, k)$ and $(h - a, k)$, $h + a = 3$, $h - a = -3$, and $k = 0$..

If $h + a = 3$ and $h - a = -3$, then $h = 0$ and $a = 3$.

The foci are located at $(4, 0)$ and $(-4, 0)$. Thus, $h = 0$ and $c = 4$.

Since $c^2=a^2+b^2$, $b^2=c^2-a^2$

$$b^2=(4)^2-(3)^2=16-9=7$$
$$\frac{(x-h)^2}{a^2}-\frac{(y-k)^2}{b^2}=1$$
$$\frac{(x-0)^2}{(3)^2}-\frac{(y-0)^2}{7}=1$$
$$\frac{x^2}{9}-\frac{y^2}{7}=1$$

37. Find the equation of the hyperbola in standard form.

foci $(0, 5)$ and $(0, -5)$, asymptotes $y = 2x$ and $y = -2x$

Transverse axis is on y-axis. Since foci are at $(h, k + c)$ and $(h, k-c)$, $k+c=5$, $k-c=-5$, and $h=0$.

Therefore, $k = 0$ and $c = 5$.

Since one of the asymptotes is $y=\frac{a}{b}x$, $\frac{a}{b}=2$ and

$a = 2b$.

$a^2 + b^2 = c^2$; then substituting $a = 2b$ and $c = 5$ yields

$(2b)^2 + b^2 = (5)^2$, or $5b^2 = 25$.

Therefore, $b^2 = 5$ and $b = \sqrt{5}$.

Since $a = 2b$, $a = 2(\sqrt{5}) = 2\sqrt{5}$.

$$\frac{(y-k)^2}{a^2} - \frac{(x-h)^2}{b^2} = 1$$

$$\frac{y^2}{(2\sqrt{5})^2} - \frac{x^2}{5} = 1$$

$$\frac{y^2}{20} - \frac{x^2}{5} = 1$$

39. Find the equation of the hyperbola in standard form.

vertices (0, 3) and (0, –3), point (2, 4)

The distance between the two vertices is the length of the transverse axis, which is $2a$.

$2a = |3-(-3)| = 6$ or $a = 3$.

Since the midpoint of the transverse axis is the center of the hyperbola, the center is given by

$\left(\frac{0+0}{2}, \frac{3+(-3)}{2}\right)$, or (0, 0)

Since both vertices lie on the y-axis, the transverse axis must be on the y-axis.

Taking the standard form of the hyperbola, we have

$$\frac{y^2}{a^2} - \frac{x^2}{b^2} = 1$$

Substituting the point (2, 4) for x and y, and 3 for a, we have $\frac{16}{9} - \frac{4}{b^2} = 1$

Solving for b^2 yields $b^2 = \frac{36}{7}$.

Therefore, the equation is $\frac{y^2}{9} - \frac{x^2}{36/7} = 1$.

41. Find the equation of the hyperbola in standard form.

vertices (0, 4) and (0, –4),

asymptotes $y = \frac{1}{2}x$ and $y = -\frac{1}{2}x$.

The length of the transverse axis, or the distance between the vertices, is equal to $2a$.

$2a = 4-(-4) = 8$, or $a = 4$

The center of the hyperbola, or the midpoint of the line segment joining the vertices, is

$\left(\frac{0+0}{2}, \frac{4+(-4)}{2}\right)$, or (0, 0).

Since both vertices lie on the y-axis, the transverse axis must lie on the y-axis. Therefore, the asymptotes are given by $y = \frac{a}{b}x$ and $y = -\frac{a}{b}x$. One asymptote is

$y = \frac{1}{2}x$. Thus $\frac{a}{b} = \frac{1}{2}$ or $b = 2a$.

Since $b = 2a$ and $a = 4$, $b = 2(4) = 8$.

Thus, the equation is $\frac{y^2}{4^2} - \frac{x^2}{8^2} = 1$ or $\frac{y^2}{16} - \frac{x^2}{64} = 1$.

43. Find the equation of the hyperbola in standard form.

vertices (6, 3) and (2, 3), foci (7, 3) and (1, 3)

Length of transverse axis = distance between vertices

$$2a = |6-2|$$
$$a = 2$$

The center of the hyperbola (h, k) is the midpoint of the line segment joining the vertices, or the point

$\left(\frac{6+2}{2}, \frac{3+3}{2}\right)$.

Thus, $h = \frac{6+2}{2}$, or 4, and $k = \frac{3+3}{3}$, or 3.

Since both vertices lie on the horizontal line $y = 3$, the transverse axis is parallel to the x-axis. The location of the foci is given by $(h + c, k)$ and $(h - c, k)$, or specifically (7, 3) and (1, 3). Thus $h + c = 7$, $h - c = 1$, and $k = 3$. Solving for h and c simultaneously yields $h = 4$ and $c = 3$.

Since $c^2 = a^2 + b^2$, $b^2 = c^2 - a^2$.

Substituting, we have $b^2 = 3^2 - 2^2 = 9 - 4 = 5$.

Substituting $a = 2$, $b^2 = 5$, $h = 4$, and $k = 3$ in the

standard equation $\frac{(x-h)^2}{a^2} - \frac{(y-k)^2}{b^2} = 1$

yields $\frac{(x-4)^2}{4} - \frac{(y-3)^2}{5} = 1$.

45. Find the equation of the hyperbola in standard form.

foci (1, –2) and (7, –2), slope of an asymptote $= \frac{5}{4}$

Both foci lie on the horizontal line $y = -2$; therefore, the transverse axis is parallel to the x-axis.

The foci are given by $(h + c, k)$ and $(h - c, k)$.

Thus, $h - c = 1$, $h + c = 7$, and $k = -2$. Solving simultaneously for h and c yields $h = 4$ and $c = 3$.

Since $y - k = \frac{b}{a}(x - h)$ is the equation for an asymptote, and the slope of an asymptote is given as

$\frac{5}{4}$, $\frac{b}{a} = \frac{5}{4}$, $b = \frac{5a}{4}$, and $b^2 = \frac{25a^2}{16}$.

Because $a^2 + b^2 = c^2$, substituting $c = 3$ and

$b^2 = \frac{25a^2}{16}$ yields $a^2 = \frac{144}{41}$.

Therefore, $b^2 = \frac{3600}{656} = \frac{225}{41}$.

Substituting in the standard equation for a hyperbola

yields $\frac{(x-4)^2}{144/41} - \frac{(y+2)^2}{225/41} = 1$

47. Find the equation of the hyperbola in standard form.

Because the transverse axis is parallel to the y-axis and the center is (7, 2), the equation of the hyperbola is

$$\frac{(y-2)^2}{a^2} - \frac{(x-7)^2}{b^2} = 1.$$

Because (9, 4) is a point on the hyperbola,

$$\frac{(4-2)^2}{a^2} - \frac{(9-7)^2}{b^2} = 1.$$

The slope of the asymptote is $\frac{1}{2}$. Therefore

$\frac{1}{2} = \frac{a}{b}$ or $b = 2a$.

Substituting, we have $\frac{4}{a^2} - \frac{4}{4a^2} = 1$

$\frac{4}{a^2} - \frac{1}{a^2} = 1$, or $a^2 = 3$.

Since $b = 2a$, $b^2 = 4a^2$, or $b^2 = 12$. The equation is

$$\frac{(y-2)^2}{3} - \frac{(x-7)^2}{12} = 1.$$

49. Find the equation of the hyperbola in standard form.

vertices (1, 6) and (1, 8), eccentricity = 2

Length of transverse axis = distance between vertices

$$2a = |6 - 8| = 2$$

$$a = 1 \text{ and } a^2 = 1$$

Center (midpoint of transverse axis) is $\left(\frac{1+1}{2}, \frac{6+8}{2}\right)$, or (1, 7).

Therefore, $h = 1$ and $k = 7$.

Since both vertices lie on the vertical line $x = 1$, the transverse axis is parallel to the y-axis.

Since $e = \frac{c}{a}$, $c = ae = (1)(2) = 2$.

Because $b^2 = c^2 - a^2$, $b^2 = (2)^2 - (1)^2 = 4 - 1 = 3$.

Substituting h, k, a^2, and b^2 into the standard equation

yields $\frac{(y-7)^2}{1} - \frac{(x-1)^2}{3} = 1$.

51. Find the equation of the hyperbola in standard form.

foci (4, 0) and (–4, 0), eccentricity = 2

Center (midpoint of line segment joining foci) is

$\left(\frac{4+(-4)}{2}, \frac{0+0}{2}\right)$, or (0, 0)

Thus, $h = 0$ and $k = 0$.

Since both foci lie on the horizontal line $y = 0$, the transverse axis is parallel to the x-axis. The locations of the foci are given by $(h + c, k)$ and $(h - c, k)$, or specifically (4, 0) and (–4, 0). Since $h = 0$, $c = 4$.

Because $e = \frac{c}{a}$, $a = \frac{c}{e} = \frac{4}{2} = 2$ and $a^2 = 4$.

Because $b^2 = c^2 - a^2$, $b^2 = 4^2 - 2^2 = 16 - 4 = 12$.

Substituting h, k, a^2 and b^2 into the standard formula

for a hyperbola yields $\frac{x^2}{4} - \frac{y^2}{12} = 1$.

53. Find the equation of the hyperbola in standard form. conjugate axis length = 4, center (4, 1), eccentricity = $\frac{4}{3}$

$2b$ = conjugate axis length = 4

$b = 2$ and $b^2 = 4$

Since $e = \frac{c}{a} = \frac{4}{3}$, $c = \frac{4a}{3}$ and $c^2 = \frac{16a^2}{9}$. Since

$a^2 + b^2 = c^2$, substituting $b^2 = 4$ and $c^2 = \frac{16a^2}{9}$ and

solving for a^2 yields $a^2 = \frac{36}{7}$.

Substituting into the two standard equations of a

hyperbola yields $\frac{(x-4)^2}{36/7} - \frac{(y-1)^2}{4} = 1$ and

$\frac{(y-1)^2}{36/7} - \frac{(x-4)^2}{4} = 1$.

55. a. Because the transmitters are 250 miles apart,

$2c = 250$ and $c = 125$.

$2a$ = rate × time

$2a = 0.186 \times 500 = 93$

Thus, $a = 46.5$ miles.

$b = \sqrt{c^2 - a^2} = \sqrt{125^2 - 46.5^2} = \sqrt{13{,}462.75}$ miles

The ship is located on the hyperbola given by

$$\frac{x^2}{2{,}162.25} - \frac{y^2}{13{,}462.75} = 1$$

b. $x = 100$

$$\frac{10{,}000}{2{,}162.25} - \frac{y^2}{13{,}462.75} = 1$$

$$\frac{-y^2}{13{,}462.75} \approx -3.6248121$$

$$y^2 \approx 48{,}799.939$$

$$y \approx 221$$

The ship is 221 miles from the coastline.

57. When the wave hits Earth, $z = 0$.

$$y^2 = x^2 + (0 - 10{,}000)^2$$

$$y^2 - x^2 = 10{,}000^2$$

It is a hyperbola.

59. a. Using the eccentricity, and $a = 2$,

$$\frac{c}{2} = \frac{\sqrt{17}}{4} \Rightarrow c = \frac{\sqrt{17}}{2}$$

Solve for b.

$$a^2 + b^2 = c^2$$

$$b^2 = c^2 - a^2$$

$$b^2 = \left(\frac{\sqrt{17}}{2}\right)^2 - 2^2 = \frac{17}{4} - \frac{16}{4}$$

$$b^2 = \frac{1}{4}$$

$$b = \frac{1}{2} = 0.5$$

$$\frac{x^2}{2^2} - \frac{y^2}{0.5^2} = 1$$

b. For FG, $y = 0.6$.

$$\frac{x^2}{2^2} - \frac{0.6^2}{0.5^2} = 1$$

$$\frac{x^2}{2^2} = 1 + \frac{0.6^2}{0.5^2}$$

$$x^2 = 2^2\left(1 + \frac{0.6^2}{0.5^2}\right)$$

$$x = \sqrt{2^2\left(1 + \frac{0.6^2}{0.5^2}\right)}$$

$$x \approx 3.1241$$

$$FG = 2x \approx 6.25 \text{ in.}$$

61. Identify the equation and graph.

$$4x^2 + 9y^2 - 16x - 36y + 16 = 0$$

$$4(x^2 - 4x) + 9(y^2 - 4y) = -16$$

$$4(x^2 - 4x + 4) + 9(y^2 - 4y + 4) = -16 + 16 + 36$$

$$4(x-2)^2 + 9(y-2)^2 = 36$$

$$\frac{(x-2)^2}{9} + \frac{(y-2)^2}{4} = 1$$

ellipse

center (2, 2)

vertices $(2 \pm 3, 2) = (5, 2), (-1, 2)$

foci $(2 \pm \sqrt{5}, 2) = (2 + \sqrt{5}, 2), (2 - \sqrt{5}, 2)$

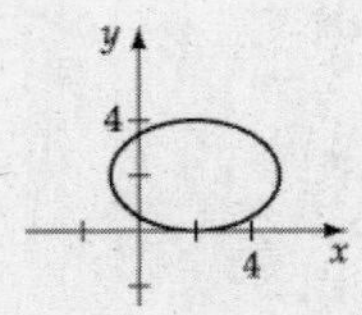

63. Identify the equation and graph.

$$5x-4y^2+24y-11=0$$
$$-4\left(y^2-6y\right)=-5x+11$$
$$-4\left(y^2-6y+9\right)=-5x+11-36$$
$$-4(y-3)^2=-5x-25$$
$$-4(y-3)^2=-5(x+5)$$
$$(y-3)^2=\frac{5}{4}(x+5)$$

parabola
vertex $(-5,\ 3)$
focus $\left(-5+\frac{5}{16},\ 3\right)=\left(-\frac{75}{16},\ 3\right)$
directrix $x=-5-\frac{5}{16}$, or $x=\frac{-85}{16}$

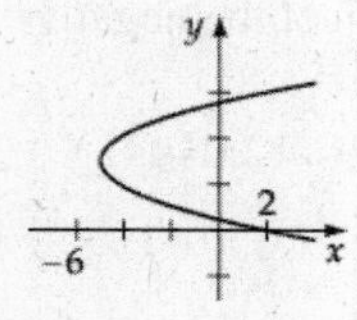

65. Identify the equation and graph.

$$x^2+2y-8x=0$$
$$x^2-8x=-2y$$
$$x^2-8x+16=-2y+16$$
$$(x-4)^2=-2(y-8)$$

parabola
vertex (4, 8)
foci $\left(4,\ 8-\frac{1}{2}\right)=\left(4,\ \frac{15}{2}\right)$
directrix $y=8+\frac{1}{2}$, or $y=\frac{17}{2}$

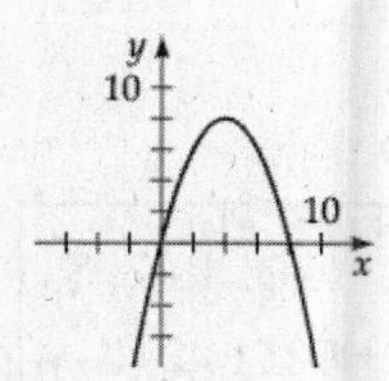

67. Identify the equation and graph.

$$25x^2+9y^2-50x-72y-56=0$$
$$25\left(x^2-2x\right)+9\left(y^2-8y\right)=56$$
$$25\left(x^2-2x+1\right)+9\left(y^2-8y+16\right)=56+25+144$$
$$25(x-1)^2+9(y-4)^2=225$$
$$\frac{(x-1)^2}{9}+\frac{(y-4)^2}{25}=1$$

ellipse
center (1, 4)
vertices $(1,\ 4\pm 5)=(1,\ 9),\ (1,\ -1)$
foci $(1,\ 4\pm 4)=(1,\ 8),\ (1,\ 0)$

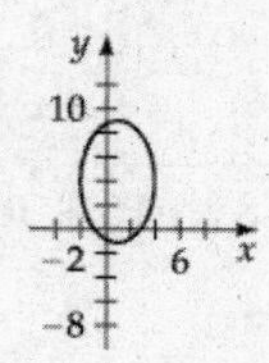

69. Find the equation of the hyperbola in standard form.

foci $F_1(2,\ 0),\ F_2(-2,\ 0)$ passing through $P_1(2,\ 3)$

$$d(P_1,\ F_2)-d(P_1,\ F_1)=\sqrt{(2+2)^2+3^2}-\sqrt{(2-2)^2+3^2}$$
$$=5-3=2$$

Let $P(x, y)$ be any point on the hyperbola. Since the difference between F_1P and F_2P is the same as the difference between F_1P_1 and F_2P_1, we have

$$\sqrt{(x-2)^2+y^2}-\sqrt{(x+2)^2+y^2}=2$$
$$\sqrt{(x-2)^2+y^2}=2+\sqrt{(x+2)^2+y^2}$$
$$x^2-4x+4+y^2=4+4\sqrt{(x+2)^2+y^2}+x^2+4x+4+y^2$$
$$-8x-4=4\sqrt{(x+2)^2+y^2}$$
$$-2x-1=\sqrt{(x+2)^2+y^2}$$
$$4x^2+4x+1=x^2+4x+4+y^2$$
$$3x^2-y^2=3$$
$$\frac{x^2}{1}-\frac{y^2}{3}=1$$

71. Find the equation of the hyperbola in standard form.

foci (0, 4) and (0, –4), point $\left(\frac{7}{3},\ 4\right)$

Difference in distances from (x, y) to foci = difference of distances from $\left(\frac{7}{3},\ 4\right)$ to foci

$$\sqrt{(x-0)^2+(y-4)^2}-\sqrt{(x-0)^2+(y+4)^2}$$
$$=\sqrt{\left(\frac{7}{3}-0\right)^2+(4-4)^2}-\sqrt{\left(\frac{7}{3}-0\right)^2+(4+4)^2}$$

$$\sqrt{x^2+y^2-8y+16}-\sqrt{x^2+y^2+8y+16}$$
$$=\frac{7}{3}-\frac{25}{3}=-6$$
$$\sqrt{x^2+y^2-8y+16}$$
$$=\sqrt{x^2+y^2+8y+16}-6$$
$$x^2+y^2-8y+16$$
$$=x^2+y^2+8y+16-12\sqrt{x^2+y^2+8y+16}+36$$
$$-16y-36$$
$$=-12\sqrt{x^2+y^2+8y+16}$$
$$4y+9$$
$$=3\sqrt{x^2+y^2+8y+16}$$
$$16y^2+72y+81$$
$$=9x^2+9y^2+72y+144$$
$$7y^2-9x^2=63$$

Thus, $\frac{y^2}{9}-\frac{x^2}{7}=1$

73.

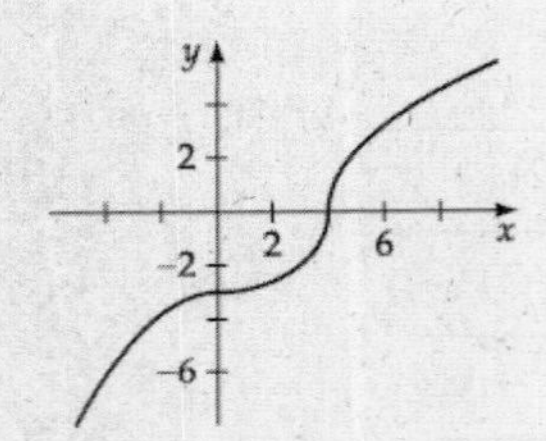

75. a. For the equation of the parabola:

$$x^2=4py, \quad p=16$$
$$x^2=64y$$

For the equation of the hyperbola: $V_1(0,-2), V_2(0,12)$

Halfway between the vertices is the point (0, 7).

$$\frac{(y-7)^2}{a^2}-\frac{x^2}{b^2}=1$$

a is the distance from the vertex to the center

$= 12 - 7 = 5.$

c is the midpoint between the foci = [16 – (–2)]/2 = 9.

$$b^2=c^2-a^2=9^2-5^2=81-25=56=\left(2\sqrt{14}\right)^2$$
$$\frac{(y-7)^2}{5^2}-\frac{x^2}{\left(2\sqrt{14}\right)^2}=1$$

b. The x-coordinate of D is 4, so set $x = 4$ and solve the parabolic equation for y.

$$(4)^2=64y$$
$$y=\frac{64}{16}=0.25$$
$$D=(4,\ 0.25)$$

The x-coordinate of P is 1, so set $x = 1$ and solve the hyperbolic equation for y.

$$\frac{(y-7)^2}{5^2}-\frac{1^2}{\left(2\sqrt{14}\right)^2}=1$$
$$\frac{(y-7)^2}{25}-\frac{1}{56}=1$$
$$\frac{(y-7)^2}{25}=\frac{57}{56}$$
$$(y-7)^2=25\left(\frac{57}{56}\right)$$
$$y-7=\sqrt{25\left(\frac{57}{56}\right)} \quad \text{Reject the negative}$$
$$y=7+\sqrt{25\left(\frac{57}{56}\right)}\approx 12.0444$$

$P = (1, 12.0444)$

c. The length BE is the same as the length of AR, where R is point where the y-axis and the line extended from EQ intersect.

The distance from A to the origin is 0.5 in.

From part b, we know the distance from the origin to the y-coordinate of P is 12.0444 in.

The distance between the y-coordinates of P and Q is 0.125 in.

Therefore, AR = 0.5 + 12.0444 + 0.125 = 12.6694 in.

Thus, BE = 12.6694 in.

Exploring Concepts with Technology

a. Using a graphing calculator,

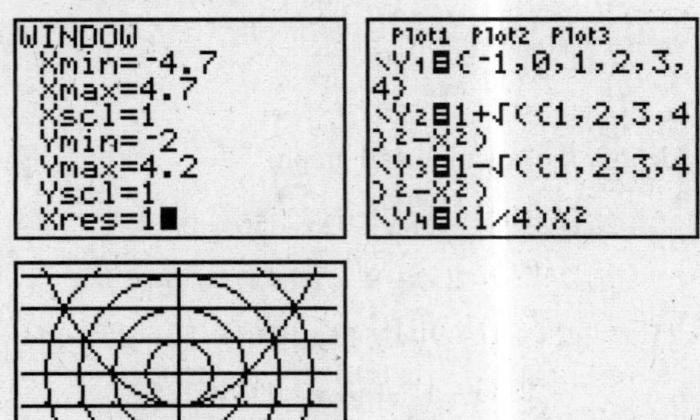

b. $y=3$ and $y=\frac{1}{4}x^2 \Rightarrow 3=\frac{1}{4}x^2 \Rightarrow 12=x^2 \Rightarrow x=2\sqrt{3}$.

Thus point A is located at $(2\sqrt{3},\ 3)$.

distance from A to (0, 1) = the radius of the circle = 4

distance from A to directrix ($y = -1$) = distance from $(2\sqrt{3},\ 3)$ to $(2\sqrt{3},\ -1)$ is

$$\sqrt{(2\sqrt{3}-2\sqrt{3})^2+(3-(-1))^2}=\sqrt{0+4^2}=4$$

c. $y=2$ and $y=\frac{1}{4}x^2 \Rightarrow 2=\frac{1}{4}x^2 \Rightarrow 8=x^2 \Rightarrow x=2\sqrt{2}$.

Thus, point B is located at $(2\sqrt{3},\ 3)$.

distance from B to (0, 1) = the radius of the circle = 3

distance from B to directrix ($y = -1$) = distance from $(2\sqrt{2},\ 2)$ to $(2\sqrt{2},\ -1)$ is

$$\sqrt{(2\sqrt{2}-2\sqrt{2})^2+(2-(-1))^2}=\sqrt{0+3^2}=3$$

d. For any point on a parabola, the distance from the point to the focus is the same as the distance from the point to the directrix.

Chapter 5 Review Exercises

1. Identify each conic section and parts, then graph. [5.3]

$$x^2-y^2=4 \Rightarrow \frac{x^2}{4}-\frac{y^2}{4}=1$$

hyperbola

center (0, 0); vertex $(\pm 2,0)$; foci $(\pm 2\sqrt{2},\ 0)$

asymptotes $y=\pm x$

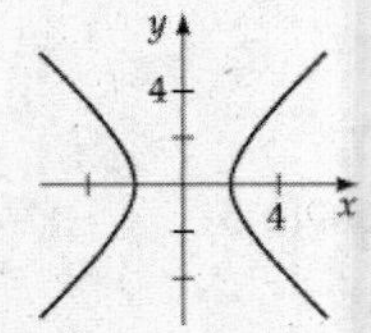

3. Identify each conic section and parts, then graph. [5.2]

$$x^2+4y^2-6x+8y-3=0$$
$$x^2-6x+4(y^2+2y)=3$$
$$(x^2-6x+9)+4(y^2+2y+1)=3+9+4$$
$$(x-3)^2+4(y+1)^2=16$$
$$\frac{(x-3)^2}{16}+\frac{(y+1)^2}{4}=1$$

ellipse

center (3, −1)

vertices $(3\pm 4,\ -1)=(7,\ -1),\ (-1,\ -1)$

foci $(3\pm 2\sqrt{3},\ -1)=(3+2\sqrt{3},\ -1),\ (3-2\sqrt{3},\ -1)$

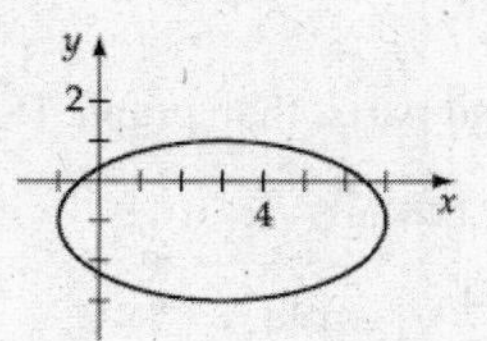

5. Identify each conic section and parts, then graph. [5.1]

$$3x-4y^2+8y+2=0$$
$$-4(y^2-2y)=-3x-2$$
$$-4(y^2-2y+1)=-3x-2-4$$
$$-4(y-1)^2=-3(x+2)$$
$$(y-1)^2=\frac{3}{4}(x+2)$$

parabola

vertex (−2, 1); focus $\left(-2+\frac{3}{16},\ 1\right)=\left(-\frac{29}{16},\ 1\right)$

directrix $x=-2-\frac{3}{16}$ or $x=-\frac{35}{16}$

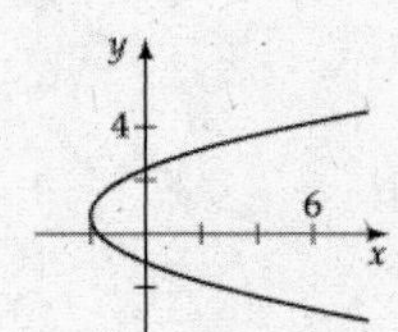

7. Identify each conic section and parts, then graph. [5.2]

$$9x^2+4y^2+36x-8y+4=0$$
$$9(x^2+4x)+4(y^2-2y)=-4$$
$$9(x^2+4x+4)+4(y^2-2x+1)=-4+36+4$$
$$9(x+2)^2+4(y-1)^2=36$$
$$\frac{(x+2)^2}{4}+\frac{(y-1)^2}{9}=1$$

ellipse

center (−2, 1)

vertices $(-2,\ 1\pm 3)=(-2,\ 4),\ (-2,\ -2)$

foci $\left(-2,\ 1\pm\sqrt{5}\right)=\left(-2,\ 1+\sqrt{5}\right),\ \left(-2,\ 1-\sqrt{5}\right)$

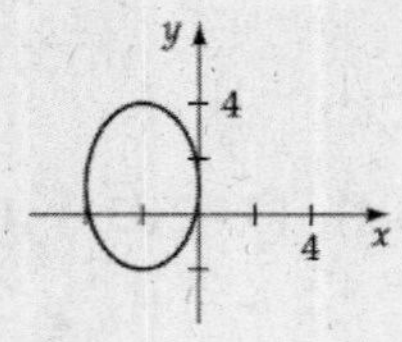

9. Identify each conic section and parts, then graph. [5.3]

$$4x^2-9y^2-8x+12y-144=0$$
$$4\left(x^2-2x\right)-9\left(y^2-\frac{4}{3}y\right)=144$$
$$4\left(x^2-2x+1\right)-9\left(y^2-\frac{4}{3}y+\frac{4}{9}\right)=144+4-4$$
$$4(x-1)^2-9\left(y-\frac{2}{3}\right)^2=144$$
$$\frac{(x-1)^2}{36}-\frac{(y-2/3)^2}{16}=1$$

hyperbola

center $\left(1,\ \frac{2}{3}\right)$

vertices $\left(1\pm 6,\ \frac{2}{3}\right)=\left(7,\ \frac{2}{3}\right),\ \left(-5,\ \frac{2}{3}\right)$

foci $\left(1\pm 2\sqrt{13},\ \frac{2}{3}\right)=\left(1+2\sqrt{13},\ \frac{2}{3}\right),\ \left(1-2\sqrt{13},\ \frac{2}{3}\right)$

asymptotes $y-\frac{2}{3}=\pm\frac{2}{3}(x-1)$

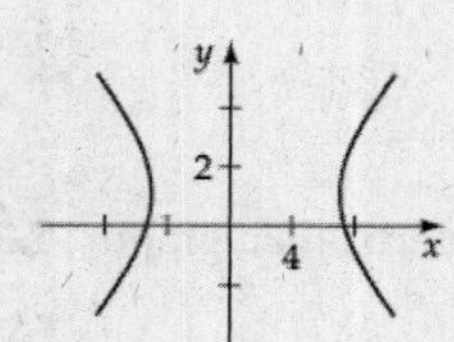

11. Identify each conic section and parts, then graph. [5.1]

$$4x^2+28x+32y+81=0$$
$$4\left(x^2+7x\right)=-32y-81$$
$$4\left(x^2+7x+\frac{49}{4}\right)=-32y-81+49$$
$$4\left(x+\frac{7}{2}\right)^2=-32(y+1)$$
$$\left(x+\frac{7}{2}\right)^2=-8(y+1)$$

parabola

vertex $\left(-\frac{7}{2},\ -1\right)$ $\quad 4p=-8,\ p=-2$

focus $\left(-\frac{7}{2},\ -3\right)$

directrix $y=1$

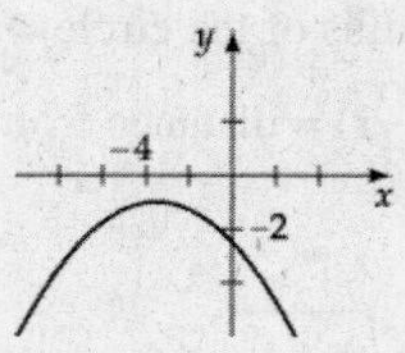

13. Find the eccentricity. [5.2]

$$4x^2+49y^2-48x-294y+389=0$$
$$4\left(x^2-12x\right)+49\left(y^2-6y\right)=-389$$
$$4\left(x^2-12x+36\right)+49\left(y^2-6y+9\right)=-389+144+441$$
$$4(x-6)^2+49(y-3)^2=196$$
$$\frac{(x-6)^2}{49}+\frac{(y-3)^2}{4}=1$$

$$c^2=a^2-b^2=49-4=45$$
$$c=\sqrt{45}=3\sqrt{5}$$
$$e=\frac{c}{a}=\frac{3\sqrt{5}}{7}$$

15. Find the equation of the conic section. [5.2]

$$2a=|7-(-3)|=10$$
$$a=5$$
$$a^2=25$$
$$2b=8$$
$$b=4$$
$$b^2=16$$

Center (2, 3)

$$\frac{(x-2)^2}{25}+\frac{(y-3)^2}{16}=1$$

17. Find the equation of the conic section. [5.3]

center (–2, 2), $c=3$

$$2a=4$$
$$a=2$$
$$a^2=4$$
$$c^2=a^2+b^2$$
$$9=4+b^2$$
$$b^2=5$$

$$\frac{(x+2)^2}{4}-\frac{(y-2)^2}{5}=1$$

19. Find the equation of the conic section. [5.1]

$$(x-h)^2 = 4p(y-k) \quad (y-k)^2 = 4p(x-h)$$
$$(3-0)^2 = 4p(4+2) \quad (4+2)^2 = 4p(3-0)$$
$$9 = 4p(6) \quad 36 = 4p(3)$$
$$p = \frac{3}{8} \quad p = 3$$

Thus, there are two parabolas that satisfy the given conditions:

$x^2 = \frac{3}{2}(y+2)$ or $(y+2)^2 = 12x$

21. Find the equation of the conic section. [5.2

$a = 6$ and the transverse axis is on the x-axis.

$$\pm\frac{b}{a} = \pm\frac{1}{9}$$
$$\frac{b}{6} = \frac{1}{9}$$
$$b = \frac{2}{3}$$
$$\frac{x^2}{36} - \frac{y^2}{4/9} = 1$$

23. Find the equation. [5.1]

focus (–2, 3), directrix $x = 2$

The vertex is the midpoint of the line segment joining (–2, 3), and (2, 3) on the directrix.

$$(h, k) = \left(\frac{-2+2}{2}, \frac{3+3}{2}\right) = (0, 3)$$

The directed distance p from the vertex to the focus is –2.

$$4p = 4(-2) = -8$$
$$(y-k)^2 = 4p(x-h)$$
$$(y-3)^2 = -8x$$

25. Find the equation. [5.2]

foci (–3, 1)and (5, 1), length of major axis 10

The center of the ellipse is the midpoint of the line segment joining the foci.

$$(h, k) = \left(\frac{-3+5}{2}, \frac{1+1}{2}\right) = (1, 1)$$

$$2a = 10$$
$$a = 5$$
$$a^2 = 25$$
$$c = 4$$
$$c^2 = a^2 - b^2$$
$$16 = 25 - b^2$$
$$b^2 = 9$$
$$\frac{(x-1)^2}{25} + \frac{(y-1)^2}{9} = 1$$

27. The equation of the mirror is [5.1]

$x^2 = 4py \quad -4 \le x \le 4$

Because (4, 0.1) is a point on the parabola, (4, 0.1) must be a solution of the equation. Thus

$$4^2 = 4p(0.1)$$
$$16 = 0.4p$$
$$40 = p$$

The focus is 40 inches above the vertex.

Chapter 5 Test

1. Find the vertex, focus, and directrix. [5.1]

$y = \frac{1}{8}x^2$ [5.1]

$$x^2 = 8y$$
$$4p = 8$$
$$p = 2$$

vertex: (0, 0)

focus: (0, 2)

directrix: $y = -2$

3. Write the equation in standard form. [5.1]

$$(h, k) = \left(\frac{-1+3}{2}, \frac{-2-2}{2}\right) = (1, -2)$$
$$p = -2$$
$$4p = -8$$
$$(y-k)^2 = 4p(x-h)$$
$$(y+2)^2 = -8(x-1)$$

5. Find the vertices and foci. [5.2]

$$a^2 = 64 \quad b^2 = 9 \quad c^2 = 55$$
$$a = 8 \quad b = 3 \quad c = \sqrt{55}$$

vertices: (0, 8), (0, −8)

foci: $\left(0, \sqrt{55}\right), \left(0, -\sqrt{55}\right)$

7. Find the vertices and foci. [5.2]

$$25x^2-150x+9y^2+18y+9=0$$
$$25\left(x^2-6x+9\right)+9\left(y^2+2y+1\right)=-9+225+9$$
$$25(x-3)^2+9(y-1)^2=225$$
$$\frac{(x-3)^2}{9}+\frac{(y+1)^2}{25}=1$$

$a=5 \quad b=3 \quad c=4$

vertices: (3, 4), (3, −6)

foci: (3, 3), (3, −5)

9. Find the eccentricity. [5.2]

$$9x^2+25y^2=81$$
$$\frac{9x^2}{81}+\frac{25y^2}{81}=1$$
$$\frac{x^2}{9}+\frac{y^2}{81/25}=1$$
$$c^2=a^2-b^2=9-\frac{81}{25}=\frac{225-81}{25}=\frac{144}{25}$$
$$c=\frac{12}{5}$$
$$e=\frac{c}{a}=\frac{12/5}{3}=\frac{4}{5}$$

11. Find the vertices, foci and asymptotes. [5.3]

$$\frac{x^2}{36}-\frac{y^2}{64}=1$$

vertices: (6, 0), (−6, 0), asymptotes: $y=\pm\frac{4}{3}x$

foci: (10, 0), (−10, 0)

13. Find the vertices and foci. [5.3]

$$\frac{(y-4)^2}{36}-\frac{(x+5)^2}{9}=1$$

vertices: $(-5,\ 4\pm 6)=(-5,\ 10),\ (-5,\ -2)$

$$c^2=36+9=45$$
$$c=3\sqrt{5}$$

foci: $\left(-5,\ 4\pm 3\sqrt{5}\right)$

15. Find the equation of the parabola. [5.1]

$$(y-4)^2=-16(x-2)$$

17. The equation of the mirror is [5.1]

$$y^2=4px \quad -4\le y\le 4$$

Because (4, 4) is a point on the parabola,

(4, 4) must be a solution of the equation. Thus

$$4^2=4p(4)$$
$$16=16p$$
$$1=p$$

The focus is 1 inch from the vertex.

Cumulative Review Exercises

1. Solve $x^4-2x^2-8=0$. [1.4]

Let $u=x^2$.

$$u^2-2u-8=0$$
$$(u-4)(u+2)=0$$
$$u=4 \quad or \quad u=-2$$
$$x^2=4 \qquad x^2=-2$$
$$x=\pm 2 \qquad x=\pm i\sqrt{2}$$

The solutions are $2,\ -2,\ i\sqrt{2},\ -i\sqrt{2}$.

3. Write the difference quotient. [2.6]

$$\frac{f(2+h)-f(2)}{h}=\frac{\left[1-(2+h)^2\right]-\left[1-(2)^2\right]}{h}$$
$$=\frac{1-4-4h-(h)^2-1+4}{h}$$
$$=\frac{-4h-h^2}{h}$$
$$=-4-h$$

5. Find the number of complex solutions.

By the Linear Factor Theorem, since the polynomial is of degree 6, there are 6 complex number solutions to $x^6+2x^4-3x^3-x^2+5x-7=0$.

7. Find the equations of the asymptotes. [2.3]

$x=-3,\ y=2$

9. Graph $f(x)=2^{-x+1}$. [4.2]

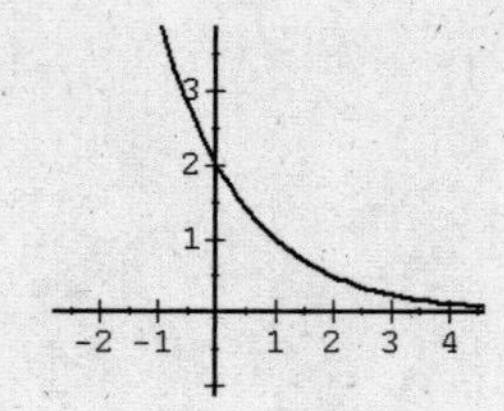

11. Sketch the graph of $y=-f(x)+2$. [2.5]

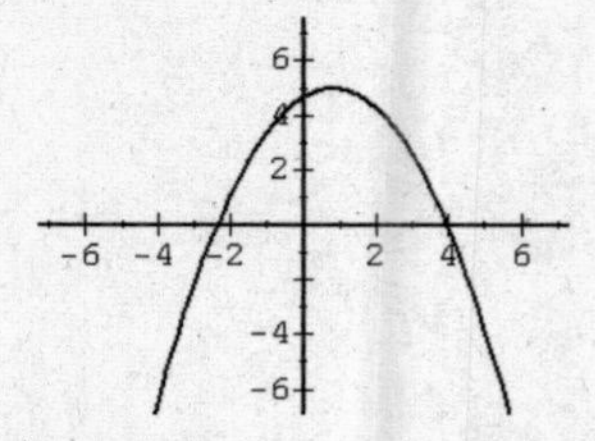

13. Find the remaining zeros. [3.4]

$$\begin{array}{r|rrrrr} 2i & 1 & 1 & -8 & 4 & -48 \\ & & 2i & -4+2i & -4-24i & 48 \\ \hline -2i & 1 & 1+2i & -12+2i & -24i & 0 \\ & & -2i & -2i & 24i & \\ \hline & 1 & 1 & -12 & 0 & \end{array}$$

$x^2+x-12=(x-3)(x+4)=0$

$x=3,\ x=-4$

The remaining zeros are $-2i$ and -4.

15. Is the graph symmetric with respect to x-axis, y-axis or origin? [2.5]

Not symmetric with respect to either axis.

Symmetric to the origin since $(-x)=(-y)^3-(-y)$ simplifies to $-x=-y^3+y$, which is equivalent to the original equation $x=y^3-y$.

17. Solve $x^2+3x-4<0$. [1.5]

$(x-1)(x+4)<0$

The product $(x-1)(x+4)$ is negative.

$x-1=0 \Rightarrow x=1$ is a critical value.

$x+4=0 \Rightarrow x=-4$ is a critical value.

$(x-1)(x+4)$ ++++++|-----|+++ (critical values -4, 1)

$(-4, 1)$

19. Express the area A in terms of d. [2.2]

Area of a square: $A=s^2$

Using the Pythagorean Theorem and solving for s^2:

$$s^2+s^2=d^2$$
$$s^2=\frac{d^2}{2}$$
$$A(d)=\frac{d^2}{2}$$

Chapter 6: Systems of Equations and Inequalities

Section 6.1 Exercises

1. Solve by substitution.

$$\begin{cases} 2x-3y=16 \\ x=2 \end{cases}$$

$$\begin{aligned} 2(2)-3y&=16 \\ -3y&=12 \\ y&=-4 \end{aligned}$$

The solution is (2, –4).

3. Solve by substitution.

$$\begin{cases} 3x+4y=18 \\ y=-2x+3 \end{cases}$$

$$\begin{aligned} 3x+4(-2x+3)&=18 \\ 3x-8x+12&=18 \\ -5x&=6 \\ x&=-\frac{6}{5} \end{aligned}$$

$$y=-2\left(-\frac{6}{5}\right)+3=\frac{27}{5}$$

The solution is $\left(-\frac{6}{5}, \frac{27}{5}\right)$.

5. Solve by substitution.

$$\begin{cases} -2x+3y=6 \\ x=2y-5 \end{cases}$$

$$\begin{aligned} -2(2y-5)+3y&=6 \\ -4y+10+3y&=6 \\ -y&=-4 \\ y&=4 \end{aligned}$$

$$x=2(4)-5=3$$

The solution is (3, 4).

7. Solve by substitution.

$$\begin{cases} 6x+5y=1 & (1) \\ x-3y=4 & (2) \end{cases}$$

Solve (2) for x: $x=3y+4$

$$\begin{aligned} 6(3y+4)+5y&=1 \\ 18y+24+5y&=1 \\ 23y&=-23 \\ y&=-1 \end{aligned}$$

$$x=3(-1)+4=1$$

The solution is (1, –1).

9. Solve by substitution.

$$\begin{cases} 7x+6y=-3 & (1) \\ y=\frac{2}{3}x-6 & (2) \end{cases}$$

$$\begin{aligned} 7x+6\left(\frac{2}{3}x-6\right)&=-3 \\ 7x+4x-36&=-3 \\ 11x&=33 \\ x&=3 \end{aligned}$$

$$y=\frac{2}{3}(3)-6=-4$$

The solution is (3, –4).

11. Solve by substitution.

$$\begin{cases} y=4x-3 \\ y=3x-1 \end{cases}$$

$$\begin{aligned} 4x-3&=3x-1 \\ x&=2 \end{aligned}$$

$$y=4(2)-3=5$$

The solution is (2, 5).

13. Solve by substitution.

$$\begin{cases} y=5x+4 \\ x=-3y-4 \end{cases}$$

$$\begin{aligned} y&=5(-3y-4)+4 \\ y&=-15y-20+4 \\ 16y&=-16 \\ y&=-1 \end{aligned}$$

$$x=-3(-1)-4=-1$$

The solution is (–1, –1).

15. Solve by substitution.

$$\begin{cases} 3x-4y=2 & (1) \\ 4x+3y=14 & (2) \end{cases}$$

Solve (1) for x and substitute into (2).

$$3x=4y+2 \rightarrow x=\frac{4y+2}{3}$$

$$4\left(\frac{4y+2}{3}\right)+3y=14$$
$$16y+8+9y=42$$
$$25y=34$$
$$y=\frac{34}{25}$$

$$x=\frac{4}{3}\left(\frac{34}{25}\right)+\frac{2}{3}=\frac{62}{25}$$

The solution is $\left(\frac{62}{25}, \frac{34}{25}\right)$.

17. Solve by substitution.

$$\begin{cases} 3x-3y=5 & (1) \\ 4x-4y=9 & (2) \end{cases}$$

Solve (1) for x and substitute into (2).

$$3x-3y=5 \rightarrow x=\frac{3y+5}{3}$$

$$4\left(\frac{3y+5}{3}\right)-4y=9$$
$$12y+20-12y=27$$
$$20=27$$

The system of equations is inconsistent

and has no solution.

19. Solve by substitution.

$$\begin{cases} 4x+3y=6 \\ y=-\frac{4}{3}x+2 \end{cases}$$

$$4x+3\left(-\frac{4}{3}x+2\right)=6$$
$$4x-4x+6=6$$
$$0=0$$

The system of equations is dependent.

Let $x=c$ and $y=-\frac{4}{3}c+2$.

The solutions are $\left(c, -\frac{4}{3}c+2\right)$.

21. Solve by elimination.

$$\begin{cases} 3x-y=10 & (1) \\ 4x+3y=-4 & (2) \end{cases}$$

$$\begin{aligned} 9x-3y&=30 \quad \text{3 times (1)} \\ 4x+3y&=-4 \quad \text{(2)} \\ \hline 13x \quad &=26 \\ x&=2 \end{aligned}$$

$$3(2)-y=10$$
$$6-y=10$$
$$y=-4$$

The solution is (2, –4).

23. Solve by elimination.

$$\begin{cases} 4x+7y=21 & (1) \\ 5x-4y=-12 & (2) \end{cases}$$

$$\begin{aligned} 20x+35y&=105 \quad \text{5 times (1)} \\ -20x+16y&=48 \quad -4 \text{ times (2)} \\ \hline 51y&=153 \\ y&=3 \end{aligned}$$

$$4x+7(3)=21$$
$$x=0$$

The solution is (0, 3).

25. Solve by elimination.

$$\begin{cases} 5x-3y=0 & (1) \\ 10x-6y=0 & (2) \end{cases}$$

$$\begin{aligned} -10x+6y&=0 \quad -2 \text{ times (1)} \\ 10x-6y&=0 \quad \text{(2)} \\ \hline 0&=0 \end{aligned}$$

$$5x-3c=0$$
$$x=\frac{3c}{5}$$

The solution is $\left(\frac{3c}{5}, c\right)$.

27. Solve by elimination.

$$\begin{cases} 6x+6y=1 & (1) \\ 4x+9y=4 & (2) \end{cases}$$

$$\begin{aligned} 12x+12y&=2 \quad \text{2 times (1)} \\ -12x-27y&=-12 \quad -3 \text{ times (2)} \\ \hline -15y&=-10 \\ y&=\frac{2}{3} \end{aligned}$$

$$6x+6\left(\frac{2}{3}\right)=1$$
$$6x=-3$$
$$x=-\frac{1}{2}$$

The solution is $\left(-\frac{1}{2}, \frac{2}{3}\right)$.

29. Solve by elimination.

$$\begin{cases} 3x+6y=11 & (1) \\ 2x+4y=9 & (2) \end{cases}$$

$$\begin{array}{rl} 6x+12y=22 & \text{2 times (1)} \\ -6x-12y=-27 & -3\text{ times (2)} \\ \hline 0=-5 & \end{array}$$

The system of equations is inconsistent and has no solution.

31. Solve by elimination.

$$\begin{cases} \frac{5}{6}x-\frac{1}{3}y=-6 & (1) \\ \frac{1}{6}x+\frac{2}{3}y=1 & (2) \end{cases}$$

$$\begin{array}{rl} \frac{5}{3}x-\frac{2}{3}y=-12 & \text{2 times (1)} \\ \frac{1}{6}x+\frac{2}{3}y=1 & \text{(2)} \\ \hline \frac{11}{6}x \quad =-11 & \end{array}$$

$$x=-6$$

$$\frac{1}{6}(-6)+\frac{2}{3}y=1$$

$$\frac{2}{3}y=2$$

$$y=3$$

The solution is (–6, 3).

33. Solve by elimination.

$$\begin{cases} \frac{3}{4}x+\frac{1}{3}y=1 & (1) \\ \frac{1}{2}x+\frac{2}{3}y=0 & (2) \end{cases}$$

$$\begin{array}{rl} 9x+4y=12 & \text{12 times (1)} \\ -3x-4y=0 & -6\text{ times (2)} \\ \hline 6x \quad =12 & \end{array}$$

$$x=2$$

$$3(2)+4y=0$$

$$4y=-6$$

$$y=-\frac{3}{2}$$

The solution is $\left(2,-\frac{3}{2}\right)$.

35. Solve by elimination.

$$\begin{cases} 2\sqrt{3}x-3y=3 & (1) \\ 3\sqrt{3}x+2y=24 & (2) \end{cases}$$

$$\begin{array}{rl} 6\sqrt{3}x-9y=9 & \text{3 times (1)} \\ -6\sqrt{3}x-4y=-48 & -2\text{ times (2)} \\ \hline -13y=-39 & \end{array}$$

$$y=3$$

$$2\sqrt{3}x-3(3)=3$$

$$2\sqrt{3}x=12$$

$$\sqrt{3}x=6$$

$$x=2\sqrt{3}$$

The solution is $(2\sqrt{3},\ 3)$.

37. Solve by elimination.

$$\begin{cases} 3\pi x-4y=6 & (1) \\ 2\pi x+3y=5 & (2) \end{cases}$$

$$\begin{array}{rl} 6\pi x-8y=12 & \text{2 times (1)} \\ -6\pi x-9y=-15 & -3\text{ times (2)} \\ \hline -17y=-3 & \end{array}$$

$$y=\frac{3}{17}$$

$$\begin{array}{rl} 9\pi x-12y=18 & \text{3 times (1)} \\ 8\pi x+12y=20 & \text{4 times (2)} \\ \hline 17\pi x \quad =38 & \end{array}$$

$$x=\frac{38}{17\pi}$$

The solution is $\left(\frac{38}{17\pi},\ \frac{3}{17}\right)$.

39. Solve by elimination.

$$\begin{cases} 3\sqrt{2}x-4\sqrt{3}y=-6 & (1) \\ 2\sqrt{2}x+3\sqrt{3}y=13 & (2) \end{cases}$$

$$\begin{array}{rl} 6\sqrt{2}x-8\sqrt{3}y=-12 & \text{2 times (1)} \\ -6\sqrt{2}x-9\sqrt{3}y=-39 & -3\text{ times (2)} \\ \hline -17\sqrt{3}y=-51 & \end{array}$$

$$y=\frac{3}{\sqrt{3}}$$

$$\begin{array}{rl} 9\sqrt{2}x-12\sqrt{3}y=-18 & \text{3 times (1)} \\ 8\sqrt{2}x+12\sqrt{3}y=52 & \text{4 times (2)} \\ \hline 17\sqrt{2}x=34 & \end{array}$$

$$x=\sqrt{2}$$

The solution is $(\sqrt{2},\ \sqrt{3})$.

41. Find the equilibrium price.

Solve the system by substitution.

$$20p - 2000 = -4p + 1000$$
$$24p = 3000$$
$$p = 125$$

The solution is \$125.

43. Find the rate of the plane in calm air and wind.

Rate of plane with the wind: $r + w$

Rate of plane against the wind: $r - w$

$$r \cdot t = d$$

$$\begin{cases} (r+w)\cdot 3 = 450 & (1) \\ (r-w)\cdot 5 = 450 & (2) \end{cases}$$

$$\begin{array}{r} r + w = 150 \\ r - w = 90 \\ \hline 2r \quad = 240 \\ r = 120 \end{array}$$

$$120 + w = 150$$
$$w = 30$$

Rate of plane = 120 mph.

Rate of wind = 30mph.

45. Find the rate of the boat in calm water and current.

Rate of boat with the current: $r + w$

Rate of boat against the wind: $r - w$

$$r \cdot t = d$$

$$\begin{cases} (r+w)\cdot 4 = 120 & (1) \\ (r-w)\cdot 6 = 120 & (2) \end{cases}$$

$$\begin{array}{r} r + w = 30 \\ r - w = 20 \\ \hline 2r \quad = 50 \\ r = 25 \end{array}$$

$$25 + w = 30$$
$$w = 5$$

Rate of boat = 25 mph.

Rate of current = 5 mph.

47. Find the cost for the iron and lead alloys.

$x =$ cost per kilogram of iron alloy

$y =$ cost per kilogram of lead alloy

$$\begin{cases} 30x + 45y = 1080 & (1) \\ 15x + 12y = 372 & (2) \end{cases}$$

$$\begin{array}{rl} 30x + 45y = 1080 & (1) \\ -30x - 24y = -744 & -2 \text{ times } (2) \\ \hline 21y = 336 & \\ y = 16 & \end{array}$$

$$15x + 12(16) = 372$$
$$15x = 180$$
$$x = 12$$

Cost of iron alloy: \$12 per kilogram

Cost of lead alloy: \$16 per kilogram

49. Find the amount of each alloy.

$x =$ amount of 40% gold

$y =$ amount of 60% gold

$$\begin{cases} x + y = 20 & (1) \\ 0.40x + 0.60y = (0.52)(20) & (2) \end{cases}$$

$$\begin{array}{rl} -0.40x - 0.40y = -8 & -0.40 \text{ times } (1) \\ 0.40x + 0.60y = 10.4 & (2) \\ \hline 0.20y = 2.4 & \\ y = 12 & \end{array}$$

$$x + 12 = 20$$
$$x = 8$$

Amount of 40% gold: 8 g

Amount of 60% gold: 12 g

51. Find the area.

Sketch a graph to visualize the right triangle.

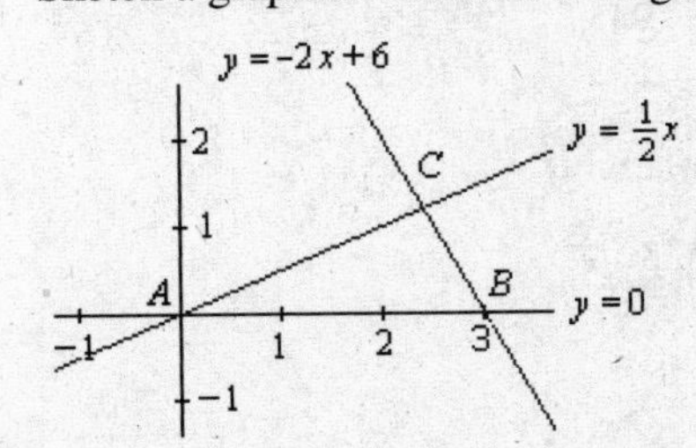

To find the coordinates of point A, solve the system

$$\begin{cases} y = 0 \\ y = \frac{1}{2}x \end{cases}$$

By substitution, $\frac{1}{2}x = 0$

$x = 0$ Thus A is (0, 0).

To find the coordinates of point B, solve the system

$$\begin{cases} y=0 \\ y=-2x+6 \end{cases}$$

By substitution, $-2x+6=0$

$$-2x=-6$$

$$x=3 \qquad \text{Thus } B \text{ is } (3, 0).$$

To find the coordinates of the point C, solve the system

$$\begin{cases} y=-2x+6 & (1) \\ y=\frac{1}{2}x & (2) \end{cases}$$

By substitution, $\frac{1}{2}x=-2x+6$

$$\frac{5}{2}x=6$$

$$x=\frac{12}{5}$$

Substituting $\frac{12}{5}$ for x in Equation (2), we have

$y=\frac{1}{2}\left(\frac{12}{5}\right)=\frac{6}{5}$. Thus C is $\left(\frac{12}{5}, \frac{6}{5}\right)$.

From the graph, $\angle C$ is the right angle.

Use the distance formula to find AC and BC.

$$AC=\sqrt{\left(\frac{12}{5}-0\right)^2+\left(\frac{6}{5}-0\right)^2}$$

$$=\sqrt{\frac{144}{25}+\frac{36}{25}}=\sqrt{\frac{180}{25}}=\frac{6}{5}\sqrt{5}$$

$$BC=\sqrt{\left(3-\frac{12}{5}\right)^2+\left(0-\frac{6}{5}\right)^2}$$

$$=\sqrt{\left(\frac{3}{5}\right)^2+\left(-\frac{6}{5}\right)^2}=\sqrt{\frac{45}{25}}=\frac{3}{5}\sqrt{5}$$

$$\text{Area}=\frac{1}{2}(\text{base})(\text{height})$$

$$=\frac{1}{2}\left(\frac{6}{5}\sqrt{5}\right)\left(\frac{3}{5}\sqrt{5}\right)$$

$$=\frac{9}{25}(5)$$

$$=\frac{9}{5} \text{ square units}$$

53. Find the largest possible value of Z.

$$\begin{array}{r} 5Z7 \\ +\ \underline{256} \\ XY3 \end{array}$$

Case 1: $Z+5+1\le 9$ Case 2: $Z+5+1>9$

$$\begin{cases} Z+5+1=Y \\ 5+2=X \end{cases} \qquad \begin{cases} Z+5+1=10+Y \\ 5+2+1=X \end{cases}$$

$$\begin{cases} Z+6=Y \\ 7=X \end{cases} \qquad \begin{cases} Z-4=Y \\ 8=X \end{cases}$$

$$X+Y=7+Z+6 \qquad X+Y=8+Z-4$$

$$X+Y=Z+13 \qquad X+Y=Z+4$$

$XY3$ is divisible by 3 $\Rightarrow X+Y$ is divisible by 3.

If $Z+13$ is divisible by 3, then $Z=2$, 5, or 8.

If $Z+4$ is divisible by 3, then $Z=2$, 5, or 8.

In both cases, the largest digit Z can be is 8.

55. Find the Pythagorean triples.

$$14=c-b \qquad 18=c-b$$

$$126=c+b \qquad 98=c+b$$

$$140=2c \qquad 116=2c$$

$$70=c, b=56 \qquad 58=c, b=40$$

$$294=c-b \qquad 2=c-b$$

$$6=c+b \qquad 882=c+b$$

$$300=2c \qquad 884=2c$$

$$150=c, b=144 \qquad 442=c, b=440$$

The Pythagorean triples are: 42, 56, 70; 42, 40, 58; 42, 144, 150; 42, 440, 442.

57. Find how many people like the skin cream.

x = people who like lip balm but do not like skin cream

y = people who like lip balm and skin cream

z = people who do not like lip balm but do like skin cream

w = people who do not like lip balm nor skin cream

$$\begin{cases} x+y+z+w=100 & (1) \\ 0.80(y+z)=y & (2) \\ 0.50(x+w)=w & (3) \\ x+y=77 & (4) \end{cases}$$

Rewrite the system by solving eq (2) for z, eq (3) for w, and eq (4) for x.

$$\begin{cases} x+y+z+w=100 & (1) \\ z=0.25y & (2) \\ w=x & (3) \\ x=-y+77 & (4) \end{cases}$$

Substitute the values from equations (2), (3), and (4) into equation (1) and solve for y.

$$\begin{aligned}(-y+77)+y+0.25y+(-y+77)&=100\\ -0.75y+154&=100\\ -0.75y&=-54\\ y&=72\end{aligned}$$

$z=0.25(72)=18$
$x=-72+77=5$
$w=5$

Find the number of people who like skin cream ($y+z$)

$y+z=72+18=90$ people like the skin cream

59. Find the pumping rate for each pump.

S = supply pump

A = outlet pump

$$\begin{cases}\frac{1}{2}S-\frac{1}{2}A=8750 & (1)\\ \frac{3}{4}S-2\left(\frac{3}{4}A\right)=11,250 & (2)\end{cases}$$

$$\begin{aligned}S-A&=17,500 \quad (1)\\ S-2A&=15,000 \quad \tfrac{4}{3}\text{ times (2)}\\ \hline A&=2500\\ S&=20,000\end{aligned}$$

The supply pump can pump 20,000 gal/h.

The outlet pump can pump 2500 gal/h.

Prepare for Section 6.2

P1. Solve.

$$\begin{aligned}2x-5y&=15\\ -5y&=-2x+15\\ y&=\frac{2}{5}x-3\end{aligned}$$

P3. Solve by substitution.

$$\begin{cases}5x-2y=10 & (1)\\ 2y=8 & (2)\end{cases}$$

$y=4$

$$\begin{aligned}5x-2(4)&=10\\ 5x&=18\\ x&=\frac{18}{5}\end{aligned}$$

The solution is $\left(\frac{18}{5},4\right)$.

P5. Solve by substitution.

$$\begin{cases}y=3x-4 & (1)\\ y=4x-2 & (2)\end{cases}$$

$$\begin{aligned}3x-4&=4x-2\\ x&=-2\end{aligned}$$

$y=3(-2)-4=-10$

The solution is $(-2,-10)$.

Section 6.2 Exercises

1. Solve the system of equations.

$$\begin{cases}2x-y+z=8 & (1)\\ 2y-3z=-11 & (2)\\ 3y+2z=3 & (3)\end{cases}$$

$$\begin{aligned}6y-9z&=-33 \quad \text{3 times (2)}\\ -6y-4z&=-6 \quad -2\text{ times (3)}\\ \hline -13z&=-39\\ z&=3 \quad (4)\end{aligned}$$

$$\begin{cases}2x-y+z=8\\ 2y-3z=-11\\ z=3 & (4)\end{cases}$$

$$\begin{aligned}2y-3(3)&=-11\\ y&=-1\end{aligned}$$

$$\begin{aligned}2x-(-1)+3&=8\\ x&=2\end{aligned}$$

The solution is $(2,-1,3)$.

3. Solve the system of equations.

$$\begin{cases}x+3y-2z=8 & (1)\\ 2x-y+z=1 & (2)\\ 3x+2y-3z=15 & (3)\end{cases}$$

$$\begin{aligned}-2x-6y+4z&=-16 \quad -2\text{ times (1)}\\ 2x-y+z&=1 \quad (2)\\ \hline -7y+5z&=-15 \quad (4)\end{aligned}$$

$$\begin{aligned}-3x-9y+6z&=-24 \quad -3\text{ times (1)}\\ 3x+2y-3z&=15 \quad (3)\\ \hline -7y+3z&=-9 \quad (5)\end{aligned}$$

$$\begin{cases}x+3y-2z=8 & (1)\\ -7y+5z=-15 & (4)\\ -7y+3z=-9 & (5)\end{cases}$$

$$\begin{array}{rl} -7y+5z=-15 & (4) \\ 7y-3z=9 & -1 \text{ times } (5) \\ \hline 2z=-6 & \\ z=-3 & (6) \end{array}$$

$$\begin{cases} x+3y-2z=8 & (1) \\ -7y+5z=-15 & (4) \\ z=-3 & (6) \end{cases}$$

$$-7y+5(-3)=-15 \qquad x+3(0)-2(-3)=8$$
$$y=0 \qquad x=2$$

The solution is (2, 0, –3).

5. Solve the system of equations.

$$\begin{cases} 3x+4y-z=-7 & (1) \\ x-5y+2z=19 & (2) \\ 5x+y-2z=5 & (3) \end{cases}$$

$$\begin{array}{rl} 3x+4y-z=-7 & (1) \\ -3x+15y-6z=-57 & -3 \text{ times } (2) \\ \hline 19y-7z=-64 & (4) \end{array}$$

$$\begin{array}{rl} -5x+25y-10z=-95 & -5 \text{ times } (2) \\ 5x+y-2z=5 & (3) \\ \hline 26y-12z=-90 & (5) \end{array}$$

$$\begin{cases} 3x+4y-z=-7 & (1) \\ 19y-7z=-64 & (4) \\ 26y-12z=-90 & (5) \end{cases}$$

$$\begin{array}{rl} 494y-182z=-1664 & 26 \text{ times } (4) \\ -494y+228z=1710 & -19 \text{ times } (3) \\ \hline 46z=46 & \\ z=1 & (6) \end{array}$$

$$\begin{cases} 3x+4y-z=-7 & (1) \\ 19y-7z=-64 & (4) \\ z=1 & (6) \end{cases}$$

$$19y-7(1)=-64$$
$$y=-3$$

$$3x+4(-3)-1=-7$$
$$x=2$$

The solution is (2, –3, 1).

7. Solve the system of equations.

$$\begin{cases} 2x-5y+3z=-18 & (1) \\ 3x+2y-z=-12 & (2) \\ x-3y-4z=-4 & (3) \end{cases}$$

$$\begin{array}{rl} 3x+2y-z=-12 & (2) \\ -3x+9y+12z=12 & -3 \text{ times } (3) \\ \hline 11y+11z=0 & \\ y+z=0 & (4) \end{array}$$

$$\begin{array}{rl} 2x-5y+3z=-18 & (1) \\ -2x+6y+8z=8 & -2 \text{ times } (3) \\ \hline y+11z=-10 & (5) \end{array}$$

$$\begin{cases} 2x-5y+3z=-18 & (1) \\ y+z=0 & (4) \\ y+11z=-10 & (5) \end{cases}$$

$$\begin{array}{rl} y+z=0 & (4) \\ -y-11z=10 & -1 \text{ times } (5) \\ \hline -10z=10 & \\ z=-1 & (6) \end{array}$$

$$\begin{cases} 2x-5y+3z=-18 & (1) \\ y+z=0 & (4) \\ z=-1 & (6) \end{cases}$$

$$y-1=0$$
$$y=1$$

$$2x-5(1)+3(-1)=-18$$
$$x=-5$$

The solution is (–5, 1, –1).

9. Solve the system of equations.

$$\begin{cases} x+2y-3z=-7 & (1) \\ 2x-y+4z=11 & (2) \\ 4x+3y-4z=-3 & (3) \end{cases}$$

$$\begin{array}{rl} -2x-4y+6z=14 & -2 \text{ times } (1) \\ 2x-y+4z=11 & (2) \\ \hline -5y+10z=25 & (4) \end{array}$$

$$\begin{array}{rl} -4x-8y+12z=28 & -4 \text{ times } (1) \\ 4x+3y-4z=-3 & (3) \\ \hline -5y+8z=25 & (5) \end{array}$$

$$\begin{cases} x+2y-3z=-7 & (1) \\ -y+2z=5 & (4) \\ -5y+8z=25 & (5) \end{cases}$$

$$\begin{array}{rl} 5y-10z=-25 & -1 \text{ times } (4) \\ \underline{-5y+8z=25} & (5) \\ -2z=0 & \\ z=0 & (6) \end{array}$$

$$\begin{cases} x+2y-3z=-7 & (1) \\ -y+2z=5 & (4) \\ z=0 & (6) \end{cases}$$

$$-y+2(0)=5$$
$$y=-5$$

$$x+2(-5)-3(0)=-7$$
$$x=3$$

The solution is (3, –5, 0).

11. Solve the system of equations.

$$\begin{cases} 2x-5y+2z=-4 & (1) \\ 3x+2y+3z=13 & (2) \\ 5x-3y-4z=-18 & (3) \end{cases}$$

$$\begin{array}{rl} 6x-15y+6z=-12 & 3 \text{ times } (1) \\ \underline{-6x-4y-6z=-26} & -2 \text{ times } (2) \\ -19y \quad =-38 & \\ y=2 & (4) \end{array}$$

$$\begin{array}{rl} 10x-25y+10z=-20 & 5 \text{ times } (1) \\ \underline{-10x+6y+8z=36} & -2 \text{ times } (3) \\ -19y+18z=16 & (5) \end{array}$$

$$\begin{cases} 2x-5y+2z=-4 & (1) \\ y=2 & (4) \\ -19y+18z=16 & (5) \end{cases}$$

$$\begin{array}{rl} 19y \quad =38 & 19 \text{ times } (4) \\ \underline{-19y+18z=16} & (5) \\ 18z=54 & \\ z=3 & (6) \end{array}$$

$$\begin{cases} 2x-5y+2z=-4 & (1) \\ y=2 & (4) \\ z=3 & (6) \end{cases}$$

$$2x-5(2)+2(3)=-4$$
$$x=0$$

The solution is (0, 2, 3).

13. Solve the system of equations.

$$\begin{cases} 2x+y-z=-2 & (1) \\ 3x+2y+3z=21 & (2) \\ 7x+4y+z=17 & (3) \end{cases}$$

$$\begin{array}{rl} 6x+3y-3z=-6 & 3 \text{ times } (1) \\ \underline{-6x-4y-6z=-42} & -2 \text{ times } (2) \\ -y-9z=-48 & (4) \end{array}$$

$$\begin{array}{rl} 14x+7y-7z=-14 & 7 \text{ times } (1) \\ \underline{-14x-8y-2z=-34} & -2 \text{ times } (3) \\ -y-9z=-48 & (5) \end{array}$$

$$\begin{cases} 2x+y-z=-2 & (1) \\ -y-9z=-48 & (4) \\ -y-9z=-48 & (5) \end{cases}$$

$$\begin{array}{rl} -y-9z=-48 & (4) \\ \underline{y+9z=48} & -1 \text{ times } (5) \\ 0=0 & (6) \end{array}$$

$$\begin{cases} 2x+y-z=-2 & (1) \\ -y-9z=-48 & (4) \\ 0=0 & (6) \end{cases}$$

The system of equations is dependent.

Let $z=c$. $\quad -y-9c=-48$
$$y=48-9c$$

$$2x+(48-9c)-c=-2$$
$$x=5c-25$$

The solution is $(5c-25,\ 48-9c,\ c)$.

15. Solve the system of equations.

$$\begin{cases} 3x-2y+3z=11 & (1) \\ 2x+3y+z=3 & (2) \\ 5x+14y-z=1 & (3) \end{cases}$$

$$\begin{array}{rl} 6x-4y+6z=22 & 2 \text{ times } (1) \\ \underline{-6x-9y-3z=-9} & -3 \text{ times } (2) \\ -13y+3z=13 & (4) \end{array}$$

$$\begin{array}{rl} 15x-10y+15z=55 & 5 \text{ times } (1) \\ \underline{-15x-42y+3z=-3} & -3 \text{ times } (3) \\ -52y+18z=52 & \\ -36y+9z=26 & (5) \end{array}$$

$$\begin{cases} 3x-2y+3z=11 & (1) \\ -13y+3z=13 & (4) \\ -26y+9z=26 & (5) \end{cases}$$

$$\begin{array}{rl} 26y-6z=-26 & -2 \text{ times } (4) \\ -26y+9z=26 & (5) \\ \hline z=0 & (6) \end{array}$$

$$\begin{cases} 3x-2y+3z=11 & (1) \\ -13y+3z=13 & (4) \\ z=0 & (6) \end{cases}$$

$$-13y+3(0)=13 \qquad 3x-2(-1)+3(0)=11$$
$$y=-1 \qquad x=3$$

The solution is $(3, -1, 0)$.

17. Solve the system of equations.

$$\begin{cases} 2x-3y+6z=3 & (1) \\ x+2y-4z=5 & (2) \\ 3x+4y-8z=7 & (3) \end{cases}$$

$$\begin{array}{rl} 2x-3y+6z=3 & (1) \\ -2x-4y+8z=-10 & -2 \text{ times } (2) \\ \hline -7y+14z=-7 & \\ -y+2z=-1 & (4) \end{array}$$

$$\begin{array}{rl} -3x-6y+12z=-15 & -3 \text{ times } (2) \\ 3x+4y-8z=7 & (3) \\ \hline -2y+4z=-8 & \\ -y+2z=-4 & (5) \end{array}$$

$$\begin{cases} 2x-3y+6z=3 & (1) \\ -y+2z=-1 & (4) \\ -y+2z=-4 & (5) \end{cases}$$

$$\begin{array}{rl} -y+2z=-1 & (4) \\ y-2z=4 & -1 \text{ times } (5) \\ \hline 0=3 & (6) \end{array}$$

$$\begin{cases} 2x-3y+6z=3 & (1) \\ -y+2z=-1 & (4) \\ 0=3 & (6) \end{cases}$$

The system of equations is inconsistent and has no solution.

19. Solve the system of equations.

$$\begin{cases} 2x-3y+5z=14 & (1) \\ x+4y-3z=-2 & (2) \end{cases}$$

$$\begin{array}{rl} 2x-3y+5z=14 & (1) \\ -2x-8y+6z=4 & -2 \text{ times } (2) \\ \hline -11y+11z=18 & (3) \end{array}$$

$$\begin{cases} 2x-3y+5z=14 & (1) \\ -11y+11z=18 & (3) \end{cases}$$

Let $z=c$. $\quad -11y+11c=18$

$$y=\frac{18-11c}{-11}$$
$$y=\frac{11c-18}{11}$$

$$2x-3\left(\frac{11c-18}{11}\right)+5c=14$$
$$2x=14-5c+\frac{33c-54}{11}$$
$$2x=\frac{154-55c+33c-54}{11}$$
$$x=\frac{50-11c}{11}$$

The solution is $\left(\frac{50-11c}{11}, \frac{11c-18}{11}, c\right)$.

21. Solve the system of equations.

$$\begin{cases} 6x-9y+6z=7 & (1) \\ 4x-6y+4z=9 & (2) \end{cases}$$

$$\begin{array}{rl} 24x-36y+24z=28 & 4 \text{ times } (1) \\ -24x+36y-24z=-54 & -6 \text{ times } (2) \\ \hline 0=-26 & (3) \end{array}$$

$$\begin{cases} 6x-9y+6z=7 & (1) \\ 0=-26 & (3) \end{cases}$$

The system of equations is inconsistent and has no solution.

23. Solve the system of equations.

$$\begin{cases} 5x+3y+2z=10 & (1) \\ 3x-4y-4z=-5 & (2) \end{cases}$$

$$\begin{array}{rl} 15x+9y+6z=30 & 3 \text{ times } (1) \\ -15x+20y+20z=25 & -5 \text{ times } (2) \\ \hline 29y+26z=55 & (3) \end{array}$$

$$\begin{cases} 5x+3y+2z=10 & (1) \\ 29y+26z=55 & (3) \end{cases}$$

Let $z=c$.

$$29y+26c=55$$
$$y=\frac{55-26c}{29}$$

$$5x+3\left(\frac{55-26c}{29}\right)+2c=10$$
$$5x=10-2c-\frac{165-78c}{29}$$
$$5x=\frac{290-58c-165+78c}{29}$$
$$x=\frac{25+4c}{29}$$

The solution is $\left(\frac{25+4c}{29}, \frac{55-26c}{29}, c\right)$.

25. Solve the homogeneous system of equations.

$$\begin{cases} x+3y-4z=0 & (1) \\ 2x+7y+z=0 & (2) \\ 3x-5y-2z=0 & (3) \end{cases}$$

$$\begin{array}{rl} -2x-6y+8z=0 & -2 \text{ times } (1) \\ 2x+7y+z=0 & (2) \\ \hline y+9z=0 & (4) \end{array}$$

$$\begin{array}{rl} -3x-9y+12z=0 & -3 \text{ times } (1) \\ 3x-5y-2z=0 & (3) \\ \hline -14y+10z=0 & \\ -7y+5z=0 & (5) \end{array}$$

$$\begin{cases} x+3y-4z=0 & (1) \\ y+9z=0 & (4) \\ -7y+5z=0 & (5) \end{cases}$$

$$\begin{array}{rl} 7y+63z=0 & 7 \text{ times } (4) \\ -7y+5z=0 & (5) \\ \hline 68z=0 & \\ z=0 & (6) \end{array}$$

$$\begin{cases} x+3y-4z=0 & (1) \\ y+9z=0 & (4) \\ z=0 & (6) \end{cases}$$

$$y+9(0)=0 \qquad x+3(0)-4(0)=0$$
$$y=0 \qquad x=0$$

The solution is (0, 0, 0).

27. Solve the homogeneous system of equations.

$$\begin{cases} 2x-3y+z=0 & (1) \\ 2x+4y-3z=0 & (2) \\ 6x-2y-z=0 & (3) \end{cases}$$

$$\begin{array}{rl} -2x+3y-z=0 & -1 \text{ times } (1) \\ 2x+4y-3z=0 & (2) \\ \hline 7y-4z=0 & (4) \end{array}$$

$$\begin{array}{rl} -6x+9y-3z=0 & -3 \text{ times } (1) \\ 6x-2y-z=0 & (3) \\ \hline 7y-4z=0 & (5) \end{array}$$

$$\begin{cases} 2x-3y+z=0 & (1) \\ 7y-4z=0 & (4) \\ 7y-4z=0 & (5) \end{cases}$$

$$\begin{array}{rl} -7y+4z=0 & -1 \text{ times } (4) \\ 7y-4z=0 & (5) \\ \hline 0=0 & (6) \end{array}$$

$$\begin{cases} 2x-3y+z=0 & (1) \\ 7y-4z=0 & (4) \\ 0=0 & (6) \end{cases}$$

Let $z=c$. Then $7y=4c$ or $y=\frac{4}{7}c$. Substitute for y and z in Eq. (1) and solve for x.

$$2x-3\left(\frac{4}{7}c\right)+c=0$$
$$2x=\frac{5}{7}c$$
$$x=\frac{5}{14}c$$

The solution is $\left(\frac{5}{14}c, \frac{4}{7}c, c\right)$.

29. Solve the homogeneous system of equations.

$$\begin{cases} 3x-5y+3z=0 & (1) \\ 2x-3y+4z=0 & (2) \\ 7x-11y+11z=0 & (3) \end{cases}$$

$$\begin{array}{rl} -6x+10y-6z=0 & -2 \text{ times } (1) \\ 6x-9y+12z=0 & 3 \text{ times } (2) \\ \hline y+6z=0 & (4) \end{array}$$

$$\begin{array}{rl} -21x+35y-21z=0 & -7 \text{ times } (1) \\ 21x-33y+33z=0 & 3 \text{ times } (3) \\ \hline 2y+12z=0 & (5) \end{array}$$

$$\begin{cases} 3x-5y+3z=0 & (1) \\ y+6z=0 & (4) \\ 2y+12z=0 & (5) \end{cases}$$

$$\begin{array}{rl} -2y-12z=0 & -2 \text{ times } (4) \\ 2y+12z=0 & (5) \\ \hline 0=0 & (6) \end{array}$$

$$\begin{cases} 3x-5y+3z=0 & (1) \\ y+6z=0 & (4) \\ 0=0 & (6) \end{cases}$$

From Eq. (4), $y=-6z$. Substitute into Eq. (1).

$$3x-5(-6z)+3z=0$$
$$3x=-33z$$
$$x=-11z$$

Let z be any real number c, then the solutions are $(-11c,\ -6c,\ c)$.

31. Solve the homogeneous system of equations.

$$\begin{cases} 4x-7y-2z=0 & (1) \\ 2x+4y+3z=0 & (2) \\ 3x-2y-5z=0 & (3) \end{cases}$$

$$\begin{array}{rl} 4x-7y-2z=0 & (1) \\ -4x-8y-6z=0 & -2 \text{ times } (2) \\ \hline -15y-8z=0 & (4) \end{array}$$

$$\begin{array}{rl} 6x+12y+9z=0 & 3 \text{ times } (2) \\ -6x+4y+10z=0 & -2 \text{ times } (3) \\ \hline 16y+19z=0 & (5) \end{array}$$

$$\begin{cases} 4x-7y-2z=0 & (1) \\ -15y-8z=0 & (4) \\ 16y+19z=0 & (5) \end{cases}$$

$$\begin{array}{rl} -240y-128z=0 & 16 \text{ times } (4) \\ 240y+285z=0 & 15 \text{ times } (5) \\ \hline 157z=0 & \\ z=0 & (6) \end{array}$$

$$\begin{cases} 4x-7y-2z=0 & (1) \\ -15y-8z=0 & (4) \\ z=0 & (6) \end{cases}$$

$z=0,\ y=0,\ x=0$. The solution is $(0, 0, 0)$.

33. Find the equation that passes through the points.

$$y=ax^2+bx+c$$
$$3=a(2)^2+b(2)+c$$
$$7=a(-2)^2+b(-2)+c$$
$$-2=a(1)^2+b(1)+c$$

$$\begin{cases} 4a+2b+c=3 & (1) \\ 4a-2b+c=7 & (2) \\ a+b+c=-2 & (3) \end{cases}$$

$$\begin{array}{rl} 4a+2b+c=3 & (1) \\ -4a+2b-c=-7 & -1 \text{ times } (2) \\ \hline 4b=-4 & (4) \end{array}$$

$$\begin{array}{rl} 4a+2b+c=3 & (1) \\ -4a-4b-4c=8 & -4 \text{ times } (3) \\ \hline -2b-3c=11 & (5) \end{array}$$

$$\begin{cases} 4a+2b+c=3 & (1) \\ 4b\quad=-4 & (4) \\ -2b-3c=11 & (5) \end{cases}$$

From (4): $4b=-4$
$$b=-1$$

From (5): $-2(-1)-3c=11$
$$c=-3$$

From (1): $4a+2(-1)-3=3$
$$a=2$$

The equation whose graph passes through the three points is $y=2x^2-x-3$.

35. Find the equation of the circle.

$$x^2+y^2+ax+by+c=0$$
$$5^2+3^2+a(5)+b(3)+c=0$$
$$(-1)^2+(-5)^2+a(-1)+b(-5)+c=0$$
$$(-2)^2+2^2+a(-2)+b(2)+c=0$$

$$\begin{cases} 5a+3b+c=-34 & (1) \\ -a-5b+c=-26 & (2) \\ -2a+2b+c=-8 & (3) \end{cases}$$

$$\begin{array}{rl} 5a+3b+c=-34 & (1) \\ a+5b-c=26 & -1 \text{ times } (2) \\ \hline 6a+8b=-8 & \\ 3a+4b=-4 & (4) \end{array}$$

$$\begin{array}{ll} 5a+3b+c=-34 & (1) \\ 2a-2b-c=8 & -1 \text{ times } (3) \\ \hline 7a+b=-26 & (5) \end{array}$$

$$\begin{cases} 5a+3b+c=-34 & (1) \\ 3a+4b=-4 & (4) \\ 7a+b=-26 & (5) \end{cases}$$

$$\begin{array}{ll} 3a+4b=-4 & (4) \\ -28a-4b=104 & -4 \text{ times } (5) \\ \hline -25a \quad =100 & \\ a=-4 & (6) \end{array}$$

$$\begin{cases} 5a+3b+c=-34 & (1) \\ 3a+4b=-4 & (4) \\ a=-4 & (6) \end{cases}$$

$$\begin{array}{ll} 3(-4)+4b=-4 & 5(-4)+3(2)+c=-34 \\ b=2 & c=-20 \end{array}$$

The equation whose graph passes through the three points is $x^2+y^2-4x+2y-20=0$.

37. Find the center and radius of the circle.

$$x^2+y^2+ax+by+c=0$$
$$(-2)^2+10^2+a(-2)+b(10)+c=0$$
$$(-12)^2+(-14)^2+a(-12)+b(-14)+c=0$$
$$5^2+3^2+a(5)+b(3)+c=0$$

$$\begin{cases} -2a+10b+c=-104 & (1) \\ -12a-14b+c=-340 & (2) \\ 5a+3b+c=-34 & (3) \end{cases}$$

$$\begin{array}{ll} -2a+10b+c=-104 & (1) \\ 12a+14b-c=340 & -1 \text{ times } (2) \\ \hline 10a+24b=236 & \\ 5a+12b=118 & (4) \end{array}$$

$$\begin{array}{ll} -2a+10b+c=-104 & (1) \\ -5a-3b-c=34 & -1 \text{ times } (3) \\ \hline -7a+7b \quad =-70 & \\ -a+b=-10 & (5) \end{array}$$

$$\begin{cases} -2a+10b+c=-104 & (1) \\ 5a+12b=118 & (4) \\ -a+b=-10 & (5) \end{cases}$$

$$\begin{array}{ll} 5a+12b=118 & (4) \\ -5a+5b=-50 & 5 \text{ times } (5) \\ \hline 17b=68 & \\ b=4 & (6) \end{array}$$

$$\begin{cases} -2a+10b+c=-104 & (1) \\ 5a+12b=118 & (4) \\ b=4 & (6) \end{cases}$$

$$\begin{array}{ll} 5a+12(4)=118 & -2(14)+10(4)+c=-104 \\ a=14 & c=-116 \end{array}$$

The equation whose graph passes through the three points is $x^2+y^2+14x+4y-116=0$.

$$(x^2+14x+49)+(y^2+4y+4)=116+49+4$$
$$(x+7)^2+(y+2)^2=169$$

The center is $(-7,\ -2)$ and radius is 13.

39. Find the traffic flow between B and C.

For intersection A, $275+225=x_1+x_2$
$$x_1+x_2=500$$

For intersection B, $x_2+90=x_3+150$
$$x_2-x_3=60$$

For intersection C, $x_1+x_3=240+200$
$$x_1+x_3=440$$

$$\begin{cases} x_1+x_2=500 & (1) \\ x_2-x_3=60 & (2) \\ x_1+x_3=440 & (3) \end{cases}$$

The equations are dependent.

Solve Eq. (2) for x_3 and substitute the inequality for x_2.

$x_3=x_2-60$

Because $150 \le x_2 \le 250$, then $90 \le x_3 \le 190$.

The flow between B and C is 90 to 190 cars per hour.

41. Find the estimated traffic flow between C and A, D and C, and B and D.

For intersection A, $256+x_4=389+x_1$
$$x_1-x_4=-133$$

For intersection B, $437+x_1=x_2+300$
$$x_1-x_2=-137$$

For intersection C, $298 + x_3 = 249 + x_4$

$$x_3 - x_4 = -49$$

For intersection D, $314 + x_2 = 367 + x_3$

$$x_2 - x_3 = 53$$

$$\begin{cases} x_1 - x_4 = -133 & (1) \\ x_1 - x_2 = -137 & (2) \\ x_3 - x_4 = -49 & (3) \\ x_2 - x_3 = 53 & (4) \end{cases}$$

The equations are dependent. Solving the system gives

$$x_1 = x_4 - 133$$
$$x_2 = x_4 + 4$$
$$x_3 = x_4 - 49$$

Because $125 \le x_1 \le 175$, then

$125 \le x_4 - 133 \le 175$ and $258 \le x_2 - 4 \le 308$

$258 \le x_4 \le 308$ $\quad 262 \le x_2 \le 312$

and $258 \le x_3 + 49 \le 308$

$209 \le x_3 \le 259$

The flow between C and A is 258 to 308 cars per hour

The flow between B and D is 262 to 312 cars per hour.

The flow between D and C is 209 to 259 cars per hour.

43. Find the position of the middle chime.

$w_1 d_1 + w_2 d_2 = w_3 d_3$ and $w_1 = 2$, $w_2 = 6$, $w_3 = 9$

$2d_1 + 6d_2 = 9d_3$ (1)

From the words in the exercise,

$d_1 + d_3 = 13$ (2)

$d_2 = \frac{1}{3} d_1$ (3)

$$\begin{array}{rll} 2d_1 + 6d_2 - 9d_3 = 0 & (1) \\ 2d_1 - 6d_2 \qquad = 0 & 6 \text{ times } (3) \\ \hline 4d_1 \qquad - 9d_3 = 0 & (4) \end{array}$$

$$\begin{array}{rll} 4d_1 - 9d_3 = 0 & (4) \\ 9d_1 + 9d_3 = 117 & 9 \text{ times } (2) \\ \hline 13d_1 \qquad = 117 & \\ d_1 = 9 & (5) \end{array}$$

Substitute into equation (3) and (2)

$d_2 = \frac{1}{3}(9) = 3$

$9 + d_3 = 13$

$d_3 = 4$

Therefore $d_1 = 9$ in., $d_2 = 3$ in., and $d_3 = 4$ in.

$d_2 + d_3 = 3 + 4 = 7$ in.

$d_1 - d_2 = 9 - 3 = 6$ in.

So the middle chime is 7 in. from the 9 ounce chime and 6 in. from the 2 ounce chime.

45. Find the equation of a plane.

$z = ax + by + c$

$$\begin{cases} a - b + c = 5 & (1) \\ 2a - 2b + c = 9 & (2) \\ -3a - b + c = -1 & (3) \end{cases}$$

$$\begin{array}{rll} 3a - 3b + 3c = 15 & 3 \text{ times } (1) \\ -3a - b + c = -1 & (3) \\ \hline -4b + 4c = 14 & (4) \end{array}$$

$$\begin{array}{rll} -2a + 2b - 2c = -10 & -2 \text{ times } (1) \\ 2a - 2b + c = 9 & (2) \\ \hline -c = -1 & \\ c = 1 & (5) \end{array}$$

Substitute 1 for c in Eq (4) and solve.

$-4b + 4(1) = 14$

$b = -\frac{5}{2}$

Substitute the value for b in Eq (1) and solve.

$a - \left(-\frac{5}{2}\right) + 1 = 5$

$a = \frac{3}{2}$

Thus, the equation of the plane is $z = \frac{3}{2}x - \frac{5}{2}y + 1$ or $3x - 5y - 2z = -2$.

47-48. These exercises follow the same steps until the end.

$$\begin{cases} x - 3y - 2z = A^2 & (1) \\ 2x - 5y + Az = 9 & (2) \\ 2x - 8y + z = 18 & (3) \end{cases}$$

$$\begin{array}{rll} -2x + 6y + 4z = -2A^2 & -2 \text{ times } (1) \\ 2x - 5y + Az = 9 & (2) \\ \hline y + (4 + A)z = -2A^2 + 9 & (4) \end{array}$$

$$\begin{array}{rl} -2x+6y+4z=-2A^2 & -2 \text{ times (1)} \\ 2x-8y+z=18 & (3) \\ \hline -2y+5z=-2A^2+18 & (5) \end{array}$$

$$\begin{cases} x-3y-2z=A^2 & (1) \\ y+(4+A)z=-2A^2+9 & (4) \\ -2y+5z=-2A^2+18 & (5) \end{cases}$$

$$\begin{array}{rl} 2y+(8+2A)z=-4A^2+18 & 2 \text{ times (5)} \\ -2y+5z=-2A^2+18 & (5) \\ \hline (13+2A)z=-6A^2+36 & (7) \end{array}$$

$$\begin{cases} x-3y-2z=A^2 & (1) \\ y+(4+A)z=-2A^2+9 & (4) \\ (13+2A)z=-6A^2+36 & (7) \end{cases}$$

For Exercise 47, the system of equations has no solution when $2A+13=0$ or $A=-\frac{13}{2}$.

49-50. These exercises follow the same steps until the end.

$$\begin{cases} x+2y+z=A^2 & (1) \\ -2x-3y+Az=1 & (2) \\ 7x+12y+A^2z=4A^2-3 & (3) \end{cases}$$

$$\begin{array}{rl} 2x+4y+2z=2A^2 & 2 \text{ times (1)} \\ -2x-3y+Az=1 & (2) \\ \hline y+(A+2)z=2A^2+1 & (4) \end{array}$$

$$\begin{array}{rl} -7x-14y-7z=-7A^2 & -7 \text{ times (1)} \\ 7x+12y+A^2z=4A^2-3 & (3) \\ \hline -2y+(A^2-7)z=-3A^2-3 & (5) \end{array}$$

$$\begin{cases} x+2y+z=A^2 & (1) \\ y+(A+2)z=2A^2+1 & (4) \\ -2y+(A^2-7)z=-3A^2-3 & (5) \end{cases}$$

$$\begin{array}{rl} 2y+(2A+4)z=4A^2+2 & 2 \text{ times (4)} \\ -2y+(A^2-7)z=-3A^2-3 & (5) \\ \hline (A^2+2A-3)z=A^2-1 & (6) \end{array}$$

$$\begin{cases} x+2y+z=A^2 & (1) \\ y+(A+2)z=2A^2+1 & (4) \\ (A^2+2A-3)z=A^2-1 & (6) \end{cases}$$

In Exercise 49, the system of equations will have a unique solution when $(A^2+2A-3)\neq 0$ in Eq. (6).

That is, $(A+3)(A-1)\neq 0$, or $A\neq -3, A\neq 1$

Prepare for Section 6.3

P1. Solve for x.

$$\begin{aligned} x^2+2x-2&=0 \\ x^2+2x+1&=2+1 \\ (x+1)^2&=3 \\ x+1&=\pm\sqrt{3} \\ x&=-1\pm\sqrt{3} \end{aligned}$$

P3. Name the graph of $(y+3)^2=8x$.

parabola

P5. Find the number of times the graphs intersect.

2

Section 6.3 Exercises

1. Solve the system of equations.

$$\begin{cases} y=x^2-x & (1) \\ y=2x-2 & (2) \end{cases}$$

Set the expressions for y equal to each other.

$$\begin{aligned} x^2-x&=2x-2 \\ x^2-3x+2&=0 \\ (x-2)(x-1)&=0 \end{aligned}$$

$$\begin{array}{ll} x-2=0 & x-1=0 \\ x=2 & x=1 \end{array}$$

When $x=2$, $y=2^2-2=2$ (From Eq. (1))

When $x=1$, $y=1^2-1=0$

The solutions are (1, 0) and (2, 2).

3. Solve the system of equations.

$$\begin{cases} y=2x^2-3x-3 & (1) \\ y=x-4 & (2) \end{cases}$$

Set the expressions for y equal to each other.

$$\begin{aligned} 2x^2-3x-3&=x-4 \\ 2x^2-4x+1&=0 \end{aligned}$$

$$x=\frac{4\pm\sqrt{16-4(2)(1)}}{2\cdot 2} \text{ (Quadratic Formula)}$$

$$=\frac{4\pm\sqrt{8}}{4}=\frac{4\pm 2\sqrt{2}}{4}=\frac{2\pm\sqrt{2}}{2}$$

Substitute for x in (1) and solve for y.

When $x=\frac{2+\sqrt{2}}{2}$, $y=\frac{2+\sqrt{2}}{2}-4=\frac{-6+\sqrt{2}}{2}$.

When $x=\frac{2-\sqrt{2}}{2}$, $y=\frac{2-\sqrt{2}}{2}-4=\frac{-6-\sqrt{2}}{2}$.

The solutions are

$$\left(\frac{2+\sqrt{2}}{2}, \frac{-6+\sqrt{2}}{2}\right) \text{ and } \left(\frac{2-\sqrt{2}}{2}, \frac{-6-\sqrt{2}}{2}\right).$$

5. Solve the system of equations.

$$\begin{cases} y=x^2-2x+3 & (1) \\ y=x^2-x-2 & (2) \end{cases}$$

Set the expressions for y equal to each other.

$$\begin{aligned} x^2-2x+3&=x^2-x-2 \\ -x&=-5 \\ x&=5 \end{aligned}$$

Substitute for x in Eq. (1).

$$y=5^2-2(5)+3=18$$

The solution is (5, 18).

7. Solve the system of equations.

$$\begin{cases} x+y=10 & (1) \\ xy=24 & (2) \end{cases}$$

Substitute y from Eq. (1) into Eq. (2).

$$\begin{aligned} x(10-x)&=24 \\ 10x-x^2&=24 \\ 0&=x^2-10x+24 \\ 0&=(x-4)(x-6) \end{aligned}$$

$x=4$ or $x=6$

Substitute for x in Eq. (1).

$$\begin{aligned} 4+y&=10 \qquad & 6+y&=10 \\ y&=6 & y&=4 \end{aligned}$$

The solutions are (4, 6) and (6, 4).

9. Solve the system of equations.

$$\begin{cases} 2x-y=1 & (1) \\ xy=6 & (2) \end{cases}$$

Solve Eq. (1) for y.

$$y=2x-1 \quad (3)$$

Substitute into Eq. (2).

$$\begin{aligned} x(2x-1)&=6 \\ 2x^2-x&=6 \\ 2x^2-x-6&=0 \\ (2x+3)(x-2)&=0 \end{aligned}$$

$$2x+3=0, \text{ or } x-2=0$$

$$x=-\frac{3}{2} \qquad x=2$$

Substitute for x in Eq. (3).

When $x=-\frac{3}{2}$, $y=2\left(-\frac{3}{2}\right)-1=-4$.

When $x=2$, $y=2(2)-1=3$.

The solutions are (–3/2, –4) and (2, 3).

11. Solve the system of equations.

$$\begin{cases} 3x^2-2y^2=1 & (1) \\ y=4x-3 & (2) \end{cases}$$

Substitute y from Eq. (2) into Eq. (1).

$$\begin{aligned} 3x^2-2(4x-3)^2&=1 \\ 3x^2-32x^2+48x-18&=1 \\ 29x^2-48x+19&=0 \\ (29x-19)(x-1)&=0 \end{aligned}$$

$$x=\frac{19}{29} \text{ or } x=1$$

Substitute for x in Eq. (2).

$$\begin{aligned} y&=4\left(\frac{19}{29}\right)-3 \qquad & y&=4(1)-3 \\ y&=\frac{76}{29}-\frac{87}{29} & y&=1 \\ y&=-\frac{11}{29} \end{aligned}$$

The solutions are (19/29, –11/29) and (1, 1).

13. Solve the system of equations.

$$\begin{cases} y=x^3+4x^2-3x-5 & (1) \\ y=2x^2-2x-3 & (2) \end{cases}$$

Set the expressions for y equal to each other.

$$\begin{aligned} x^3+4x^2-3x-5&=2x^2-2x-3 \\ x^3+2x^2-x-2&=0 \\ x^2(x+2)-(x+2)&=0 \\ (x+2)(x^2-1)&=0 \\ (x+2)(x-1)(x+1)&=0 \end{aligned}$$

$$x=-2, \ x=1, \text{ or } x=-1$$

Substitute for x in Eq. (2).

When $x=-2,\ y=2(-2)^2-2(-2)-3=9$

When $x=1,\ y=2(1)^2-2(1)-3=-3$

When $x=-1,\ y=2(-1)^2-2(-1)-3=1$

The solutions are (–2, 9), (1, –3) and (–1, 1).

15. Solve the system of equations.

$$\begin{cases} 2x^2+y^2=9 & (1) \\ x^2-y^2=3 & (2) \end{cases}$$

$$\begin{array}{rl} 2x^2+y^2=9 & (1) \\ x^2-y^2=3 & (2) \\ \hline 3x^2 \quad =12 & \\ x^2=4 & \\ x=\pm 2 & \end{array}$$

When $x=-2,\ (-2)^2-y^2=3$ From Eq. (2)

$$\begin{aligned} 4-y^2&=3 \\ -y^2&=-1 \\ y^2&=1 \\ y&=\pm 1 \end{aligned}$$

When $x=2,\ (2)^2-y^2=3$

$$\begin{aligned} 4-y^2&=3 \\ -y^2&=-1 \\ y^2&=1 \\ y&=\pm 1 \end{aligned}$$

The solutions are (–2, 1), (–2, –1), (2, 1), and (2, –1).

17. Solve the system of equations.

$$\begin{cases} x^2-2y^2=8 & (1) \\ x^2+3y^2=28 & (2) \end{cases}$$

Use the elimination method to eliminate x^2.

$$\begin{array}{rl} x^2-2y^2=8 & (1) \\ -x^2-3y^2=-28 & -1 \text{ times } (2) \\ \hline -5y^2=-20 & \\ y^2=4 & \\ y=\pm 2 & \end{array}$$

Substitute for y in Eq. (1).

$$\begin{array}{ll} x^2-2(2)^2=8 & x^2-2(-2)^2=8 \\ x^2=16 & x^2=16 \\ x=\pm 4 & x=\pm 4 \end{array}$$

The solutions are (4, 2), (–4, 2), (4, –2) and (–4, –2).

19. Solve the system of equations.

$$\begin{cases} 2x^2+4y^2=5 & (1) \\ 3x^2+8y^2=14 & (2) \end{cases}$$

Use the elimination method to eliminate y^2.

$$\begin{array}{rl} -4x^2-8y^2=-10 & -2 \text{ times } (1) \\ 3x^2+8y^2=14 & (2) \\ \hline -x^2=4 & \\ x^2=-4 & \end{array}$$

$x^2=-4$ has no real number solutions. The graphs of the equations do not intersect.

21. Solve the system of equations.

$$\begin{cases} x^2-2x+y^2=1 & (1) \\ 2x+y=5 & (2) \end{cases}$$

Substitute y from Eq. (2) into Eq. (1).

$$\begin{aligned} x^2-2x+(5-2x)^2&=1 \\ x^2-2x+25-20x+4x^2&=1 \\ 5x^2-22x+24&=0 \\ (5x-12)(x-2)&=0 \end{aligned}$$

$x=\frac{12}{5}$ or $x=2$

Substitute for x in Eq. (2).

$$\begin{array}{ll} 2\left(\frac{12}{5}\right)+y=5 & 2(2)+y=5 \\ y=\frac{1}{5} & y=1 \end{array}$$

The solutions are (12/5, 1/5) and (2, 1).

23. Solve the system of equations.

$$\begin{cases} (x-3)^2+(y+1)^2=5 & (1) \\ x-3y=7 & (2) \end{cases}$$

Substitute x from Eq. (2) into Eq. (1).

$$(3y+4)^2+(y+1)^2=5$$
$$9y^2+24y+16+y^2+2y+1=5$$
$$10y^2+26y+12=0$$
$$5y^2+13y+6=0$$
$$(5y+3)(y+2)=0$$
$$y=-\frac{3}{5} \text{ or } y=-2$$

Substitute for y in Eq. (2).

$$x=3\left(-\frac{3}{5}\right)+7 \qquad x=3(-2)+7$$
$$x=\frac{26}{5} \qquad x=1$$

The solutions are (26/5, –3/5) and (1, –2).

25. Solve the system of equations.

$$\begin{cases} x^2-3x+y^2=4 & (1) \\ 3x+y=11 & (2) \end{cases}$$

Substitute y from Eq. (2) into Eq. (1).

$$x^2-3x+(11-3x)^2=4$$
$$x^2-3x+121-66x+9x^2=4$$
$$10x^2-69x+117=0$$
$$(10x-39)(x-3)=0$$
$$x=\frac{39}{10} \text{ or } x=3$$

Substitute for x in Eq. (2).

$$3\left(\frac{39}{10}\right)+y=11 \qquad 3(3)+y=11$$
$$y=-\frac{7}{10} \qquad y=2$$

The solutions are (39/10, –7/10) and (3, 2).

27. Solve the system of equations.

$$\begin{cases} (x-1)^2+(y+2)^2=14 & (1) \\ (x+2)^2+(y-1)^2=2 & (2) \end{cases}$$

Expand the binomials and then subtract.

$$\begin{array}{rr} x^2-2x+1+y^2+4y+4=14 & (1) \\ x^2+4x+4+y^2-2y+1=2 & (2) \\ \hline -6x-3 \quad +6y+3=12 & \end{array}$$
$$-6x+6y=12$$
$$y=x+2$$

Substitute for y in Eq. (2).

$$(x+2)^2+(x+2-1)^2=2$$
$$x^2+4x+4+x^2+2x+1=2$$
$$2x^2+6x+3=0$$
$$x=\frac{-6\pm\sqrt{36-4\cdot2\cdot3}}{4}=\frac{-6\pm\sqrt{12}}{4}=\frac{-3\pm\sqrt{3}}{2}$$

Substitute for x in $y = x + 2$.

$$y=\frac{-3+\sqrt{3}}{2}+2 \qquad y=\frac{-3-\sqrt{3}}{2}+2$$
$$y=\frac{1+\sqrt{3}}{2} \qquad y=\frac{1-\sqrt{3}}{2}$$

The solutions are

$$\left(\frac{-3+\sqrt{3}}{2}, \frac{1+\sqrt{3}}{2}\right) \text{ and } \left(\frac{-3-\sqrt{3}}{2}, \frac{1-\sqrt{3}}{2}\right).$$

29. Solve the system of equations.

$$\begin{cases} (x+3)^2+(y-2)^2=20 & (1) \\ (x-2)^2+(y-3)^2=2 & (2) \end{cases}$$

Expand the binomials and then subtract.

$$\begin{array}{rr} x^2+6x+9+y^2-4y+4=20 & (1) \\ x^2-4x+4+y^2-6y+9=2 & (2) \\ \hline 10x+5 \quad +2y-5=18 & \end{array}$$
$$10x+2y=18$$
$$y=-5x+9$$

Find $y – 3$ and substitute in Eq. (2): $y-3=-5x+6$.

$$(x-2)^2+(-5x+6)^2=2$$
$$x^2-4x+4+25x^2-60x+36=2$$
$$26x^2-64x+38=0$$
$$13x^2-32x+19=0$$
$$(13x-19)(x-1)=0$$
$$x=\frac{19}{13} \text{ or } x=1$$

Substitute for x in $y = –5x + 9$.

$$y=-5\left(\frac{19}{13}\right)+9 \qquad y=-5(1)+9$$
$$y=\frac{22}{13} \qquad y=4$$

The solutions are (19/13, 22/13) and (1, 4).

31. Solve the system of equations.

$$\begin{cases}(x-1)^2+(y+1)^2=2 & (1)\\ (x+2)^2+(y-3)^2=3 & (2)\end{cases}$$

Expand the binomials and then subtract.

$$\begin{array}{rl} x^2-2x+1+y^2+2y+1=2 & (1)\\ x^2+4x+4+y^2-6y+9=3 & (2)\\ \hline -6x-3\quad +8y-8=-1 & \\ -6x+8y=10 & \\ y=\frac{3x+5}{4} & \end{array}$$

Find y + 1 and substitute in Eq. (1): $y+1=\frac{3x+9}{4}$.

$$(x-1)^2+\left(\frac{3x+9}{4}\right)^2=2$$

$$x^2-2x+1+\frac{9x^2+54x+81}{16}=2$$

$$16x^2-32x+16+9x^2+54x+81=32$$

$$25x^2+22x+65=0$$

$$x=\frac{-22\pm\sqrt{22^2-4(25)(65)}}{2(25)}=\frac{-22\pm\sqrt{-6016}}{50}$$

x is not a real number. There are no real solutions.

The curves do not intersect.

33. Find the width and height.

h = height

w = weight

$$2h+2w=25$$

$$wh=37.5$$

$$w=\frac{37.5}{h}$$

$$2h+2\left(\frac{37.5}{h}\right)=25$$

$$2h^2+75=25h$$

$$2h^2-25h+75=0$$

$$(2h-15)(h-5)=0$$

$$2h-15=0 \qquad h-5=0$$

$$h=7.5 \qquad h=5$$

Since the height is greater than the width, h = 7.5.

$$w=\frac{37.5}{7.5}=5$$

The width is 5 in. and the height is 7.5 in.

35. Find the dimensions of each carpet.

$$\begin{array}{r} x^2+y^2=865\\ x^2-y^2=703\\ \hline 2x^2=1568\\ x^2=784\\ x=28 \end{array}$$

$$28^2+y^2=865$$

$$y^2=81$$

$$y=9$$

The small carpet is 9 ft by 9 ft and the large carpet is 28 ft by 28 ft.

37. Find the radius of each globe.

r = radius of the small globe

R = radius of the large globe

$$V=\frac{4}{3}\pi r^3$$

$$\frac{4}{3}\pi R^3=8\left(\frac{4}{3}\pi r^3\right)$$

$$R^3=8r^3$$

$$\frac{4}{3}\pi R^3-\frac{4}{3}\pi r^3=15{,}012.62$$

$$\begin{array}{r} -\frac{4}{3}\pi R^3+\frac{32}{3}\pi r^3=0\\ \frac{4}{3}\pi R^3-\frac{4}{3}\pi r^3=15{,}012.62\\ \hline \frac{28}{3}\pi r^3=15{,}012.62 \end{array}$$

$$r^3=\frac{3(15{,}012.62)}{28\pi}$$

$$r=\sqrt[3]{\frac{3(15{,}012.62)}{28\pi}}$$

$$r\approx 8.0$$

$$R^3=8r^3$$

$$R^3=8(8.0)^3$$

$$R=16.0$$

The radius of the large globe is 16.0 in. and the radius of the small globe is 8.0 in.

39. Find the perimeter.

$$\begin{cases} x^2 = y & (1) \\ 18x - 22 = 3y + 5 & (2) \end{cases}$$

Substitute for y in Eq. (2).

$$18x - 22 = 3(x^2) + 5$$
$$0 = 3x^2 - 18x + 27$$
$$0 = 3(x^2 - 6x + 9)$$
$$0 = 3(x-3)^2$$

$x = 3$

$y = 3^2 = 9$

$$P = x^2 + 3y + 5 + y + 18x - 22$$
$$= 3^2 + 3(9) + 5 + 9 + 18(3) - 22$$
$$= 82 \text{ units}$$

41. Find values of r.

$$\begin{cases} x^2 + y^2 = r^2 & (1) \\ y = 2x + 1 & (2) \end{cases}$$

Substitute for y in Eq. (1)

$$x^2 + (2x+1)^2 = r^2$$
$$x^2 + 4x^2 + 4x + 1 = r^2$$
$$5x^2 + 4x + 1 = r^2$$

Minimize r^2 by completing the square.

$$r^2 = 5x^2 + 4x + 1 = 5\left(x^2 + \frac{4}{5}x + \frac{4}{25}\right) + 1 - \frac{4}{5}$$
$$= 5\left(x + \frac{2}{5}\right)^2 + \frac{1}{5}$$

Thus $\left(-\frac{2}{5}, \frac{1}{5}\right)$ is the point on both $x^2 + y^2 = r^2$ and $y = 2x + 1$ for which $x^2 + y^2 = r^2$ has the smallest radius.

Substitute for x in $r^2 = 5x^2 + 4x + 1$

$$r^2 = 5\left(-\frac{2}{5}\right)^2 + 4\left(-\frac{2}{5}\right) + 1 = \frac{1}{5}$$

$r = \sqrt{\frac{1}{5}}$ or $\frac{\sqrt{5}}{5}$ is the minimum radius.

Therefore $r \geq \frac{\sqrt{5}}{5}$.

43. Find the equilibrium price.

$$\begin{cases} x = \frac{p^2}{5} - 20 \\ x = \frac{17,710}{p+1} \end{cases}$$

Use a graphing calculator and INTERSECTION.

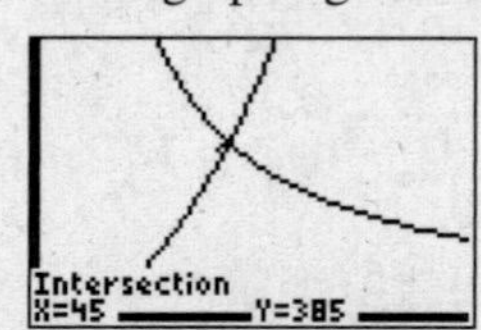

on [0, 100] by [0, 600]

The graphs intersect at (45, 385). The solution is $45.

45. Solve the system of equations.

$$\begin{cases} y = 2^x \\ y = x + 1 \end{cases}$$

Using a graphing calculator, graph the two equations on the same coordinate grid. Using the ZOOM feature, estimate the coordinates of the points where the graphs intersect. These coordinates are the solutions of the system of equations. For this system of equations, the solutions are (0, 1) and (1, 2).

47. Solve the system of equations.

$$\begin{cases} y = e^{-x} \\ y = x^2 \end{cases}$$

Using a graphing calculator, graph the two equations on the same coordinate grid. Using the ZOOM feature, estimate the coordinates of the points where the graphs intersect. These coordinates are the solutions of the system of equations. For this system of equations, the solution is approximately (0.7035, 0.4949).

49. Solve the system of equations.

$$\begin{cases} y = \sqrt{x} \\ y = \frac{1}{x-1} \end{cases}$$

Using a graphing calculator, graph the two equations on the same coordinate grid. Using the ZOOM feature, estimate the coordinates of the points where the graphs intersect. These coordinates are the solutions of the system of equations. For this system of equations, the solution is approximately (1.7549, 1.3247).

51. Solve the system of equations for rational-number ordered pairs.

$$\begin{cases} y = x^2 + 4 \\ x = y^2 - 24 \end{cases}$$

Solve by substitution.

$$x = (x^2+4)^2 - 24$$
$$x = x^4 + 8x^2 + 16 - 24$$
$$0 = x^4 + 8x^2 - x - 8$$
$$0 = (x-1)(x^3 + x^2 + 9x + 8)$$

$x^3 + x^2 + 9x + 8$ is not factorable over the rational numbers because the Rational Zero Theorem implies the only rational zeros are $\pm 1, \pm 2, \pm 4, \pm 8$. Thus, the only rational ordered-pair solution is (1, 5).

53. Solve the system of equations for rational-number ordered pairs.

$$x^2 - 3xy + y^2 = 5$$
$$x^2 - xy - 2y^2 = 0$$

Factor the second equation.

$(x-2y)(x+y) = 0$

Thus $x = 2y$ or $x = -y$. Substituting each expression into the first equation, we have

$$\begin{aligned} (2y)^2 - 3(2y)y + y^2 &= 5 \\ 4y^2 - 6y^2 + y^2 &= 5 \\ -y^2 &= 5 \\ y^2 &= -5 \end{aligned}$$

There are no rational solutions.

$$\begin{aligned} (-y)^2 - 3(-y)y + y^2 &= 5 \\ y^2 + 3y^2 + y^2 &= 5 \\ 5y^2 &= 5 \\ y^2 &= 1 \\ y &= \pm 1 \end{aligned}$$

Substituting into $x = -y$, we have $x = -1$ or $x = 1$. The rational ordered-pair solutions are (–1, 1) and (1, –1).

55. Solve the system of equations for rational-number ordered pairs.

$$\begin{cases} 2x^2 - 4xy - y^2 = 6 \\ 4x^2 - 3xy - y^2 = 6 \end{cases}$$

Subtract the two equations.

$$-2x^2 - xy = 0$$
$$-x(2x + y) = 0$$
$$x = 0 \text{ or } y = -2x$$

Substituting $x = 0$ into the first equation gives $-y^2 = 6$ or $y^2 = -6$. There are no rational solutions.

Substituting $y = -2x$ into the first equation gives

$$\begin{aligned} 2x^2 + 8x^2 - 4x^2 &= 6 \\ 6x^2 &= 6 \\ x^2 &= 1 \\ x &= \pm 1 \end{aligned}$$

The rational ordered-pair solutions are $(1, -2)$ and $(-1, 2)$.

Mid-Chapter 6 Quiz

1. Solve the system of equations.

$$\begin{cases} 2x - 3y = -15 & (1) \\ -3x + 4y = 19 & (2) \end{cases}$$

$$\begin{array}{rl} 6x - 9y = -45 & \text{3 times (1)} \\ -6x + 8y = 38 & \text{2 times (2)} \\ \hline -y = -7 & \\ y = 7 & \end{array}$$

$$2x - 3(7) = -15$$
$$x = 3$$

The solution is (3, 7).

3. Give an example of an inconsistent system of equation in two variables.

Answers will vary.

One example is $\begin{cases} x - 2y = 7 & (1) \\ -5x + 10y = 11 & (2) \end{cases}$

5. Solve the system of equations.

$$\begin{cases} 3x^2 + y^2 = 28 & (1) \\ x^2 - y^2 = 8 & (2) \end{cases}$$

$$\begin{array}{rl} 3x^2 + y^2 = 28 & (1) \\ x^2 - y^2 = 8 & (2) \\ \hline 4x^2 \quad = 36 & \end{array}$$

$x^2 = 9$

$x = \pm 3$

When $x = -3$, $(-3)^2 - y^2 = 8$

$-y^2 = -1$

$y^2 = 1$

$y = \pm 1$

When $x = 3$, $(3)^2 - y^2 = 8$

$-y^2 = -1$

$y^2 = 1$

$y = \pm 1$

The solutions are (–3, 1), (–3, –1), (3, 1), and (3, –1).

Prepare for Section 6.4

P1. Factor over the real numbers.

$x^4 + 14x^2 + 49 = (x^2 + 7)^2$

P3. Simplify.

$$\frac{7}{x} - \frac{6}{x-1} + \frac{10}{(x-1)^2}$$

$$= \frac{(x-1)^2}{(x-1)^2} \cdot \frac{7}{x} - \frac{x(x-1)}{x(x-1)} \cdot \frac{6}{x-1} + \frac{x}{x} \cdot \frac{10}{(x-1)^2}$$

$$= \frac{7x^2 - 14x + 7}{x(x-1)^2} - \frac{6x^2 - 6x}{x(x-1)^2} + \frac{10x}{x(x-1)^2}$$

$$= \frac{x^2 + 2x + 7}{x(x-1)^2}$$

P5. Solve.

$$\begin{cases} 0 = A + B & (1) \\ 3 = -2B + C & (2) \\ 16 = 7A - 2C & (3) \end{cases}$$

Solve Eq (1) for A and substitute into Eq (3).

$16 = -7B - 2C$ (4)

Multiply Eq (2) by 2 and add to Eq (4).

$$\begin{array}{l} 6 = -4B + 2C \\ 16 = -7B - 2C \\ \hline 22 = -11B \end{array}$$

$-2 = B$

$A = 2$

$C = 2B + 3$

$= 2(-2) + 3$

$= -1$

The solution is (2, –2, –1).

Section 6.4 Exercises

1. Determine the constants A and B.

$$\frac{x+15}{x(x-5)} = \frac{A}{x} + \frac{B}{x-5}$$

$x + 15 = A(x-5) + Bx$

$x + 15 = (A+B)x - 5A$

$$\begin{cases} 1 = A + B \\ 15 = -5A \end{cases} \quad A = -3 \quad \begin{array}{l} -3 + B = 1 \\ B = 4 \end{array}$$

3. Determine the constants A and B.

$$\frac{1}{(2x+3)(x-1)} = \frac{A}{2x+3} + \frac{B}{x-1}$$

$1 = A(x-1) + B(2x+3)$

$1 = Ax - A + 2Bx + 3B$

$1 = (A+2B)x + (-A+3B)$

$$\begin{array}{l} 0 = A + 2B \\ 1 = -A + 3B \\ \hline 1 = \quad 5B \end{array}$$

$B = \frac{1}{5} \qquad 0 = A + 2\left(\frac{1}{5}\right)$

$A = -\frac{2}{5}$

5. Determine the constants A, B and C.

$$\frac{x+9}{x(x-3)^2} = \frac{A}{x} + \frac{B}{x-3} + \frac{C}{(x-3)^2}$$

$x + 9 = A(x-3)^2 + Bx(x-3) + Cx$

$x + 9 = Ax^2 - 6Ax + 9A + Bx^2 - 3Bx + Cx$

$x + 9 = (A+B)x^2 + (-6A - 3B + C)x + 9A$

$$\begin{cases} 0 = A + B \\ 1 = -6A - 3B + C \\ 9 = 9A \end{cases}$$

$A = 1$ $\quad A + B = 0$ $\quad -6A - 3B + C = 1$

$1 + B = 0$ $\quad -6(1) - 3(-1) + C = 1$

$B = -1$ $\quad C = 4$

7. Determine the constants A, B and C.

$$\frac{4x^2+3}{(x-1)(x^2+x+5)} = \frac{A}{x-1} + \frac{Bx+C}{x^2+x+5}$$

$$4x^2 + 3 = A(x^2 + x + 5) + (Bx + C)(x - 1)$$
$$4x^2 + 3 = Ax^2 + Ax + 5A + Bx^2 - Bx + Cx - C$$
$$4x^2 + 3 = (A + B)x^2 + (A - B + C)x + (5A - C)$$

$$\begin{cases} 4 = A + B & (1) \\ 0 = A - B + C & (2) \\ 3 = 5A - C & (3) \end{cases}$$

From (3), $C = 5A - 3$. From (1), $B = 4 - A$.

Substitute C and B into Eq. (2).

$0 = A - (4 - A) + 5A - 3$ $\quad C = 5(1) - 3$ $\quad B = 4 - 1$

$0 = A - 4 + A + 5A - 3$ $\quad C = 2$ $\quad B = 3$

$7 = 7A$

$1 = A$

9. Determine the constants A, B, C, and D.

$$\frac{x^3+2x}{(x^2+1)^2} = \frac{Ax+B}{x^2+1} + \frac{Cx+D}{(x^2+1)^2}$$

$$x^3 + 2x = (Ax + B)(x^2 + 1) + (Cx + D)$$
$$x^3 + 2x = Ax^3 + Ax + Bx^2 + B + Cx + D$$
$$x^3 + 2x = Ax^3 + Bx^2 + (A + C)x + (B + D)$$

$$\begin{cases} 1 = A \\ 0 = B \\ 2 = A + C \\ 0 = B + D \end{cases}$$

$A = 1$ $\quad B = 0$ $\quad 1 + C = 2$ $\quad 0 + D = 0$

$C = 1$ $\quad D = 0$

11. Find the partial fraction decomposition.

$$\frac{8x+12}{x(x+4)} = \frac{A}{x} + \frac{B}{x+4}$$

$$8x + 12 = A(x + 4) + Bx$$
$$8x + 12 = Ax + 4A + Bx$$
$$8x + 12 = (A + B)x + 4A$$

$$\begin{cases} 8 = A + B \\ 12 = 4A \end{cases}$$

$A = 3$

$3 + B = 8$

$B = 5$

$$\frac{8x+12}{x(x+4)} = \frac{3}{x} + \frac{5}{x+4}$$

13. Find the partial fraction decomposition.

$$\frac{3x+50}{x^2-7x-18} = \frac{3x+50}{(x-9)(x+2)} = \frac{A}{x-9} + \frac{B}{x+2}$$

$$3x + 50 = A(x + 2) + B(x - 9)$$
$$3x + 50 = Ax + 2A + Bx - 9B$$
$$3x + 50 = (A + B)x + (2A - 9B)$$

$$\begin{cases} 3 = A + B \\ 50 = 2A - 9B \end{cases}$$

$-2A - 2B = -6$

$2A - 9B = 50$

$-11B = 44$

$B = -4$

$3 = A + (-4)$

$7 = A$

$$\frac{3x+50}{x^2-7x-18} = \frac{7}{x-9} + \frac{-4}{x+2}$$

15. Find the partial fraction decomposition.

$$\frac{16x+34}{4x^2+16x+15} = \frac{16x+34}{(2x+3)(2x+5)} = \frac{A}{2x+3} + \frac{B}{2x+5}$$

$$16x + 34 = A(2x + 5) + B(2x + 3)$$
$$16x + 34 = 2Ax + 5A + 2Bx + 3B$$
$$16x + 34 = (2A + 2B)x + (5A + 3B)$$

$$\begin{cases} 16 = 2A + 2B & (1) \\ 34 = 5A + 3B & (2) \end{cases}$$

$$\begin{array}{rl} 6A+6B=48 & 3\text{ times }(1) \\ -10A-6B=-68 & -2\text{ times }(2) \\ \hline -4A=-20 & \\ A=5 & \end{array}$$

$$2(5)+2B=16$$
$$B=3$$

$$\frac{16x+34}{4x^2+16x+15}=\frac{5}{2x+3}+\frac{3}{2x+5}$$

17. Find the partial fraction decomposition.

$$\frac{x-5}{(3x+5)(x-2)}=\frac{A}{3x+5}+\frac{B}{x-2}$$

$$x-5=A(x-2)+B(3x+5)$$
$$x-5=Ax-2A+3Bx+5B$$
$$x-5=(A+3B)x+(-2A+5B)$$

$$\begin{cases} 1=A+3B & (1) \\ -5=-2A+5B & (2) \end{cases}$$

$$\begin{array}{rl} 2A+6B=2 & 2\text{ times }(1) \\ -2A+5B=-5 & (2) \\ \hline 11B=-3 & \end{array}$$

$$B=-\frac{3}{11}$$

$$A+3\left(-\frac{3}{11}\right)=1$$
$$A=\frac{20}{11}$$

$$\frac{x-5}{(3x+5)(x-2)}=\frac{20}{11(3x+5)}+\frac{-3}{11(x-2)}$$

19. Find the partial fraction decomposition.

$$\begin{array}{r} x+3 \\ x^2-4\overline{)x^3+3x^2-4x-8} \\ \underline{x^3\qquad -4x} \\ 3x^2\qquad -8 \\ \underline{3x^2\qquad -12} \\ 4 \end{array}$$

$$\frac{x^3+3x^2-4x-8}{x^2-4}=x+3+\frac{4}{(x-2)(x+2)}$$

$$\frac{4}{(x-2)(x+2)}=\frac{A}{x-2}+\frac{B}{x+2}$$

$$4=A(x+2)+B(x-2)$$
$$4=Ax+2A+Bx-2B$$
$$4=(A+B)x+(2A-2B)$$

$$\begin{cases} 0=A+B & (1) \\ 4=2A-2B & (2) \end{cases}$$

$$\begin{array}{rl} 2A+2B=0 & 2\text{ times }(1) \\ 2A-2B=4 & (2) \\ \hline 4A\qquad =4 & \\ A=1 & \end{array}$$

$$1+B=0$$
$$B=-1$$

$$\frac{x^3+3x^2-4x-8}{x^2-4}=x+3+\frac{1}{x-2}+\frac{-1}{x+2}$$

21. Find the partial fraction decomposition.

$$\frac{3x^2+49}{x(x+7)^2}=\frac{A}{x}+\frac{B}{x+7}+\frac{C}{(x+7)^2}$$

$$3x^2+49=A(x+7)^2+Bx(x+7)+Cx$$
$$3x^2+49=Ax^2+14Ax+49A+Bx^2+7Bx+Cx$$
$$3x^2+49=(A+B)x^2+(14A+7B+C)x+49A$$

$$\begin{cases} 3=A+B \\ 0=14A+7B+C \\ 49=49A \end{cases}$$

$$A=1$$
$$1+B=3$$
$$B=2$$

$$14(1)+7(2)+C=0$$
$$C=-28$$

$$\frac{3x^2+49}{x(x+7)^2}=\frac{1}{x}+\frac{2}{x+7}+\frac{-28}{(x+7)^2}$$

23. Find the partial fraction decomposition.

$$\frac{5x^2-7x+2}{x^3-3x^2+x}=\frac{5x^2-7x+2}{x(x^2-3x+1)}=\frac{A}{x}+\frac{Bx+C}{x^2-3x+1}$$

$$5x^2-7x+2=A(x^2-3x+1)+(Bx+C)x$$
$$5x^2-7x+2=Ax^2-3Ax+A+Bx^2+Cx$$
$$5x^2-7x+2=(A+B)x^2+(-3A+C)x+A$$

$$\begin{cases} 5=A+B \\ -7=-3A+C \\ 2=A \end{cases}$$

$$A=2$$
$$2+B=5$$
$$B=3$$

$$-3(2)+C=-7$$
$$C=-1$$

$$\frac{5x^2-7x+2}{x^3-3x^2+x}=\frac{2}{x}+\frac{3x-1}{x^2-3x+1}$$

25. Find the partial fraction decomposition.

$$\frac{2x^3+9x^2+26x+41}{(x+3)^2(x^2+1)}=\frac{A}{x+3}+\frac{B}{(x+3)^2}+\frac{Cx+D}{x^2+1}$$

$$\begin{aligned}&2x^3+9x^2+26x+41\\&=A(x+3)(x^2+1)+B(x^2+1)+(Cx+D)(x+3)^2\\&=Ax^3+Ax+3Ax^2+3A+Bx^2+B\\&\quad+Cx^3+6Cx^2+9Cx+Dx^2+6Dx+9D\\&=(A+C)x^3+(3A+B+6C+D)x^2\\&\quad+(A+9C+6D)x+(3A+B+9D)\end{aligned}$$

$$\begin{cases}2=A+C & (1)\\9=3A+B+6C+D & (2)\\26=A+9C+6D & (3)\\41=3A+B+9D & (4)\end{cases}$$

$$\begin{array}{rl}3A+B+6C+D=9 & (2)\\-3A-B-9D=-41 & -1\text{ times }(4)\\\hline 6C-8D=-32 & (5)\end{array}$$

$$\begin{array}{rl}A+9C+6D=26 & (3)\\-A-C=-2 & -1\text{ times }(1)\\\hline 8C+6D=24 & (6)\end{array}$$

$$\begin{cases}2=A+C & (1)\\9=3A+B+6C+D & (2)\\-32=6C-8D & (5)\\24=8C+6D & (6)\end{cases}$$

$$\begin{array}{rl}36C-48D=-192 & 6\text{ times }(5)\\64C+48D=192 & 8\text{ times }(6)\\\hline 100C=0 & \end{array}$$

$$C=0$$
$$A+0=2$$
$$A=2$$
$$8(0)+6D=24$$
$$D=4$$
$$3(2)+B+9(4)=41$$
$$B=-1$$

$$\frac{2x^3+9x^2+26x+41}{(x+3)^2(x^2+1)}=\frac{2}{x+3}+\frac{-1}{(x+3)^2}+\frac{4}{x^2+1}$$

27. Find the partial fraction decomposition.

$$\frac{3x-7}{(x-4)^2}=\frac{A}{x-4}+\frac{B}{(x-4)^2}$$

$$3x-7=A(x-4)+B$$
$$3x-7=Ax-4A+B$$
$$3x-7=Ax+(-4A+B)$$

$$\begin{cases}3=A\\-7=-4A+B\end{cases}$$

$$B-4(3)=-7$$
$$B=5$$

$$\frac{3x-7}{(x-4)^2}=\frac{3}{x-4}+\frac{5}{(x-4)^2}$$

29. Find the partial fraction decomposition.

$$\frac{3x^3-x^2+34x-10}{(x^2+10)^2}=\frac{Ax+B}{x^2+10}+\frac{Cx+D}{(x^2+10)^2}$$

$$\begin{aligned}&3x^3-x^2+34x-10\\&=(Ax+B)(x^2+10)+Cx+D\\&=Ax^3+Bx^2+10Ax+10B+Cx+D\\&=Ax^3+Bx^2+(10A+C)x+(10B+D)\end{aligned}$$

$$\begin{cases}3=A\\-1=B\\34=10A+C\\-10=10B+D\end{cases}$$

$$10(3)+C=34 \qquad 10(-1)+D=-10$$
$$C=4 \qquad D=0$$

$$\frac{3x^3-x^2+34x-10}{(x^2+10)^2}=\frac{3x-1}{x^2+10}+\frac{4x}{(x^2+10)^2}$$

31. Find the partial fraction decomposition.

$$\frac{1}{k^2-x^2}=\frac{1}{(k-x)(k+x)}=\frac{A}{k-x}+\frac{B}{k+x}$$

$$1=A(k+x)+B(k-x)$$
$$1=Ak+Ax+Bk-Bx$$
$$1=(A-B)x+(Ak+Bk)$$

$$\begin{cases}0=A-B & (1)\\1=Ak+Bk & (2)\end{cases}$$

$$\begin{aligned} Ak - Bk &= 0 \quad k \text{ times (1)} \\ Ak + Bk &= 1 \quad (2) \\ \hline 2Ak &= 1 \\ A &= \frac{1}{2k} \end{aligned}$$

$$\frac{1}{2k} - B = 0$$

$$B = \frac{1}{2k}$$

$$\frac{1}{k^2 - x^2} = \frac{1}{2k(k-x)} + \frac{1}{2k(k+x)}$$

33. Find the partial fraction decomposition.

$$x^2 - x \overline{\big) x^3 - x^2 - x - 1} \quad \text{quotient } x$$

$$\frac{x^3 - x^2}{-x-1}$$

$$\frac{x^3 - x^2 - x - 1}{x^2 - x} = x + \frac{-x-1}{x^2 - x}$$

$$\frac{-x-1}{x(x-1)} = \frac{A}{x} + \frac{B}{x-1}$$

$$-x - 1 = Ax - A + Bx$$

$$-x - 1 = (A + B)x - A$$

$$\begin{cases} -1 = A + B \\ -1 = -A \end{cases}$$

$$A = 1$$

$$1 + B = -1$$

$$B = -2$$

$$\frac{x^3 - x^2 - x - 1}{x^2 - x} = x + \frac{1}{x} + \frac{-2}{x-1}$$

35. Find the partial fraction decomposition.

$$x^2 - x - 1 \overline{\big) 2x^3 - 4x^2 \quad + 5} \quad \text{quotient } 2x - 2$$

$$\frac{2x^3 - 2x^2 - 2x}{-2x^2 + 2x + 5}$$

$$\frac{-2x^2 + 2x + 2}{3}$$

$$\frac{2x^3 - 4x^2 + 5}{x^2 - x - 1} = 2x - 2 + \frac{3}{x^2 - x - 1}$$

37. Find the partial fraction decomposition.

$$\frac{x^2 - 1}{(x-1)(x+2)(x-3)} = \frac{(x-1)(x+1)}{(x-1)(x+2)(x-3)}$$

$$= \frac{x+1}{(x+2)(x-3)} = \frac{A}{x+2} + \frac{B}{x-3}$$

$$x + 1 = A(x-3) + B(x+2)$$

$$x + 1 = Ax - 3A + Bx + 2B$$

$$x + 1 = (A + B)x + (-3A + 2B)$$

$$\begin{cases} A + B = 1 & (1) \\ -3A + 2B = 1 & (2) \end{cases}$$

$$\begin{aligned} 3A + 3B &= 3 \quad 3 \text{ times (1)} \\ -3A + 2B &= 1 \quad (2) \\ \hline 5B &= 4 \\ B &= \frac{4}{5} \end{aligned}$$

$$A + \frac{4}{5} = 1$$

$$A = \frac{1}{5}$$

$$\frac{x^2 - 1}{(x-1)(x+2)(x-3)} = \frac{1}{5(x+2)} + \frac{4}{5(x-3)}$$

39. Find the partial fraction decomposition.

$$\frac{-x^4 - 4x^2 + 3x - 6}{x^4(x-2)} = \frac{A}{x} + \frac{B}{x^2} + \frac{C}{x^3} + \frac{D}{x^4} + \frac{E}{x-2}$$

$$\begin{aligned} &-x^4 - 4x^2 + 3x - 6 \\ &= Ax^3(x-2) + Bx^2(x-2) + Cx(x-2) \\ &\quad + D(x-2) + Ex^4 \\ &= Ax^4 - 2Ax^3 + Bx^3 - 2Bx^2 + Cx^2 - 2Cx \\ &\quad + Dx - 2D + Ex^4 \\ &= (A + E)x^4 + (-2A + B)x^3 + (-2B + C)x^2 \\ &\quad + (-2C + D)x + (-2D) \end{aligned}$$

$$\begin{cases} -1 = A + E & (1) \\ 0 = -2A + B & (2) \\ -4 = -2B + C & (3) \\ 3 = -2C + D & (4) \\ -6 = -2D & (5) \end{cases}$$

$$-2D = -6$$

$$D = 3$$

$$-2C + 3 = 3$$

$$C = 0$$

$$-2B+0=-4$$
$$B=2$$
$$-2A+2=0$$
$$A=1$$
$$1+E=-1$$
$$E=-2$$

$$\frac{-x^4-4x^2+3x-6}{x^4(x-2)}=\frac{1}{x}+\frac{2}{x^2}+\frac{3}{x^4}+\frac{-2}{x-2}$$

41. Find the partial fraction decomposition.

$$\frac{2x^2+3x-1}{(x^3-1)}=\frac{2x^2+3x-1}{(x-1)(x^2+x+1)}=\frac{A}{x-1}+\frac{Bx+C}{x^2+x+1}$$

$$2x^2+3x-1=A(x^2+x+1)+(Bx+C)(x-1)$$
$$2x^2+3x-1=Ax^2+Ax+A+Bx^2-Bx+Cx-C$$
$$2x^2+3x-1=(A+B)x^2+(A-B+C)x+(A-C)$$

$$\begin{cases} 2=A+B & (1) \\ 3=A-B+C & (2) \\ -1=A-C & (3) \end{cases}$$

Solve Eq. (1) for B and Eq. (3) for C and substitute into Eq. (2).

$$A+B=2 \qquad A-C=-1$$
$$B=2-A \qquad C=A+1$$

$$A-B+C=3$$
$$A-(2-A)+(A+1)=3$$
$$A-2+A+A+1=3$$
$$3A=4$$
$$A=\frac{4}{3}$$

$$B=2-A \qquad C=A+1$$
$$B=2-\frac{4}{3} \qquad C=\frac{4}{3}+1$$
$$B=\frac{2}{3} \qquad C=\frac{7}{3}$$

$$\frac{2x^2+3x-1}{x^3-1}=\frac{4}{3(x-1)}+\frac{2x+7}{3(x^2+x+1)}$$

43. Show the statement to be true.

$$\frac{1}{(b-a)(p(x)+a)}+\frac{1}{(a-b)(p(x)+b)}$$
$$=\frac{(a-b)(p(x)+b)+(b-a)(p(x)+a)}{(b-a)(a-b)(p(x)+a)(p(x)+b)}$$
$$=\frac{(a-b)p(x)+(a-b)b+(b-a)p(x)+(b-a)a}{(b-a)(a-b)(p(x)+a)(p(x)+b)}$$
$$=\frac{(a-b)p(x)+(a-b)b-(a-b)p(x)-(a-b)a}{(b-a)(a-b)(p(x)+a)(p(x)+b)}$$
$$=\frac{(a-b)b-(a-b)a}{(b-a)(a-b)(p(x)+a)(p(x)+b)}$$
$$=\frac{(a-b)(b-a)}{(b-a)(a-b)(p(x)+a)(p(x)+b)}$$
$$=\frac{1}{(p(x)+a)(p(x)+b)}$$

Prepare for Section 6.5

P1. Graph $y=-2x+3$.

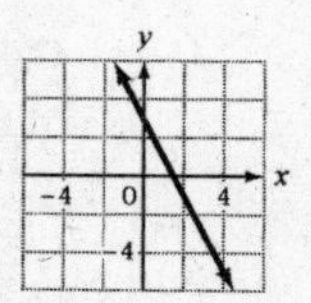

P3. Graph $y=|x|+1$.

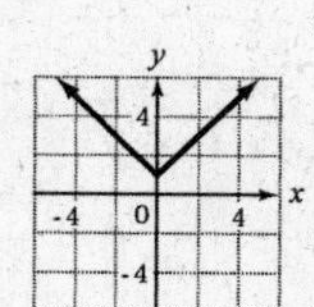

P5. Graph $\frac{x^2}{16}+\frac{y^2}{25}=1$.

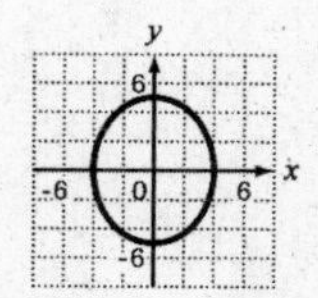

Section 6.5 Exercises

1. Sketch the graph of the inequality.

$y\le -2$ Use a solid line.

Test point (0, 0): not in region

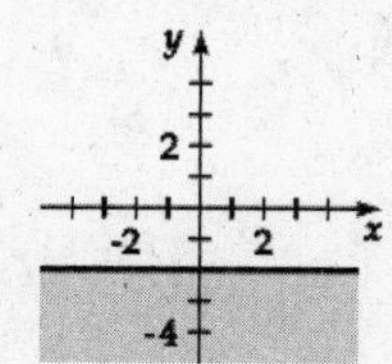

3. Sketch the graph of the inequality.

$y \geq 2x+3$ Use a solid line.

Test point (0, 0): not in region

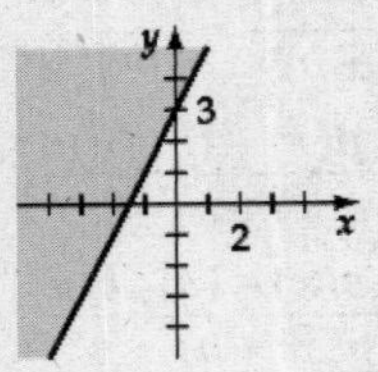

5. Sketch the graph of the inequality.

$2x-3y<6$

$y>\frac{2}{3}x-2$ Use a dashed line.

Test point (0, 0): is in region

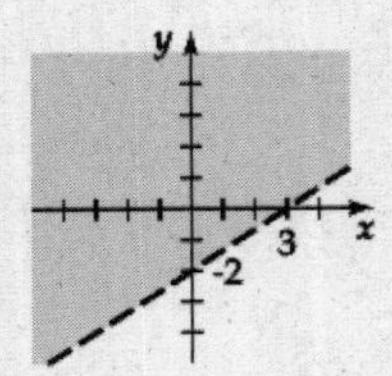

7. Sketch the graph of the inequality.

$4x+3y \leq 12$

$y \leq -\frac{4}{3}x+4$ Use a solid line.

Test point (0, 0): is in region

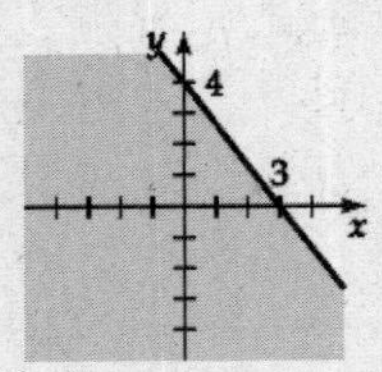

9. Sketch the graph of the inequality.

$y<x^2$ Use a dashed line.

Vertex: (0, 0)

Test point (1, 0): is in region

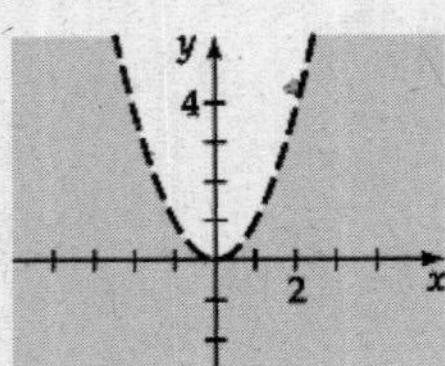

11. Sketch the graph of the inequality.

$y \geq x^2-2x-3$ Use a solid line.

Vertex: (1, –4)

Test point (0, 0): is in region

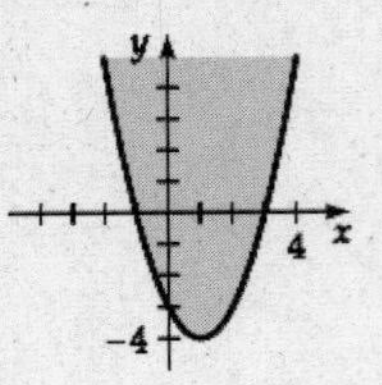

13. Sketch the graph of the inequality.

$(x-2)^2+(y-1)^2<16$ Use a dashed line.

Center: (2, 1), radius: 4

Test point (0, 0): is in region

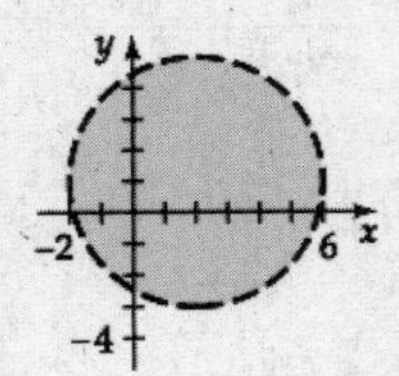

15. Sketch the graph of the inequality.

$\frac{(x-3)^2}{9}-\frac{(y+1)^2}{16}>1$ Use a dashed line.

Center: (3, –1), $a=3$, $b=4$

Test point (–1, 0): is in region

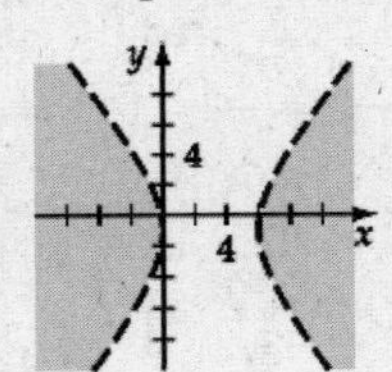

17. Sketch the graph of the inequality.

$$4x^2+9y^2-8x+18y \geq 23$$
$$4(x^2-2x+1)+9(y^2+2y+1) \geq 23+4+9$$
$$4(x-1)^2+9(y+1)^2 \geq 36$$
$$\frac{(x-1)^2}{9}+\frac{(y+1)^2}{4} \geq 1 \text{ Use a solid line.}$$

center (1, –1), $a=3$, $b=2$

Test point (0, 0): not in region

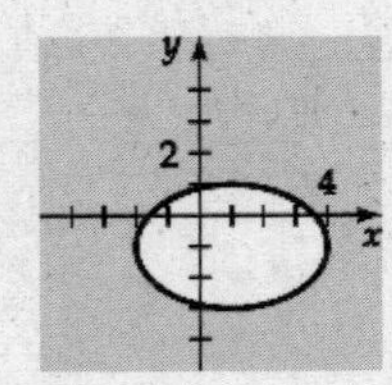

19. Sketch the graph of the inequality.

$y \geq |2x - 4|$ Use a solid line.

Test point (0, 0): not in region

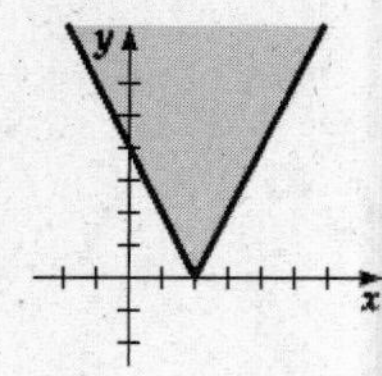

21. Sketch the graph of the inequality.

$y < 2^{x-1}$ Use a dashed line.

Test point (0, 0): is in region

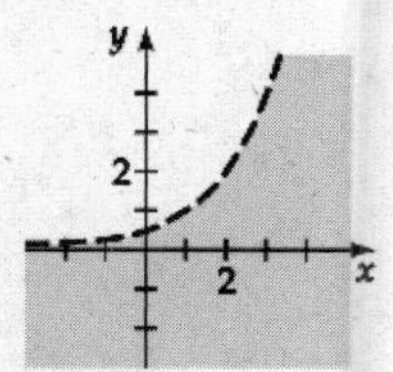

23. Sketch the graph of the solution set of the system of inequalities.

$$\begin{cases} 1 \leq x < 3 \\ -2 < y \leq 4 \end{cases}$$

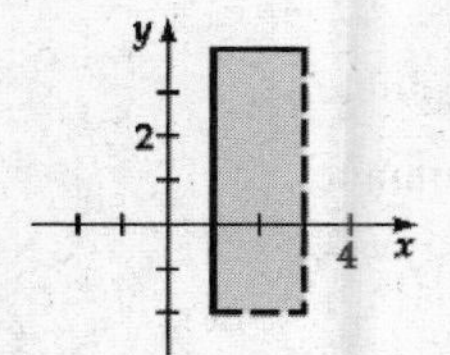

25. Sketch the graph of the solution set of the system of inequalities.

$$\begin{cases} 3x + 2y \geq 1 \\ x + 2y < -1 \end{cases}$$

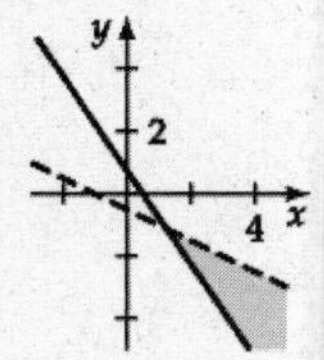

27. Sketch the graph of the solution set of the system of inequalities.

$$\begin{cases} 2x - y \geq -4 \\ 4x - 2y \leq -17 \end{cases}$$

The graphs of each of the two inequalities are shown.

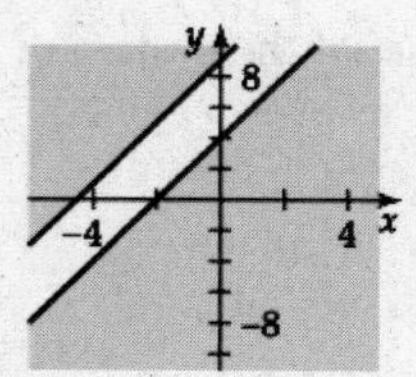

Because the solution sets of the inequalities do not intersect, the system has no solution and cannot be graphed.

29. Sketch the graph of the solution set of the system of inequalities.

$$\begin{cases} 4x - 3y < 14 \\ 2x + 5y \leq -6 \end{cases}$$

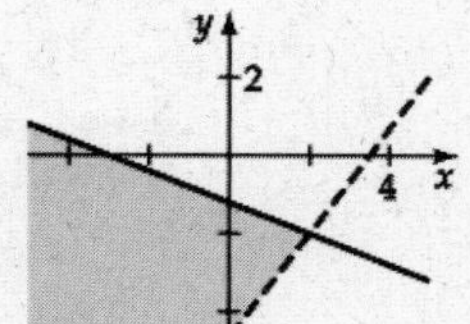

31. Sketch the graph of the solution set of the system of inequalities.

$$\begin{cases} y < 2x + 3 \\ y > 2x - 2 \end{cases}$$

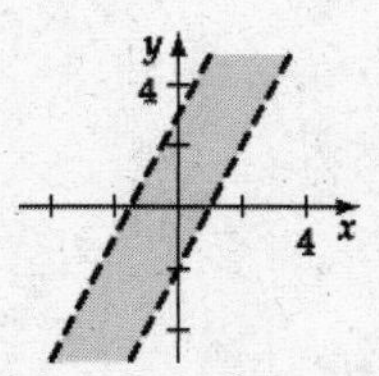

33. Sketch the graph of the solution set of the system of inequalities.

$$\begin{cases} y < 2x - 1 \\ y \geq x^2 + 3x - 7 \end{cases}$$

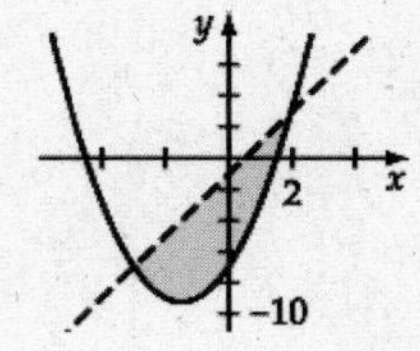

35. Sketch the graph of the solution set of the system of inequalities.

$$\begin{cases} x^2+y^2 \le 49 \\ 9x^2+4y^2 \ge 36 \end{cases}$$

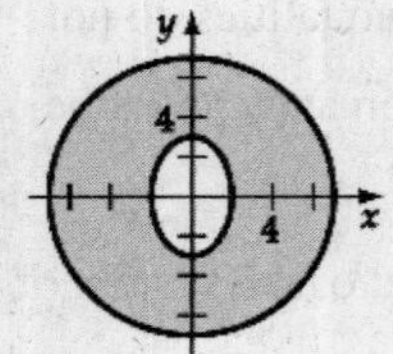

37. Sketch the graph of the solution set of the system of inequalities.

$$\begin{cases} (x-1)^2+(y+1)^2 \le 16 \\ (x-1)^2+(y+1)^2 \ge 4 \end{cases}$$

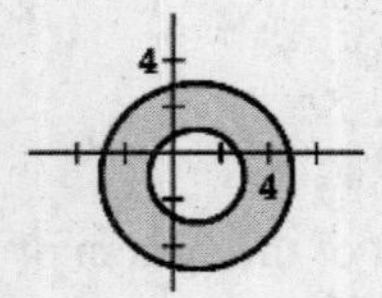

39. Sketch the graph of the solution set of the system of inequalities.

$$\begin{cases} \dfrac{(x-4)^2}{16}-\dfrac{(y+2)^2}{9}>1 \\ \dfrac{(x-4)^2}{25}+\dfrac{(y+2)^2}{9}<1 \end{cases}$$

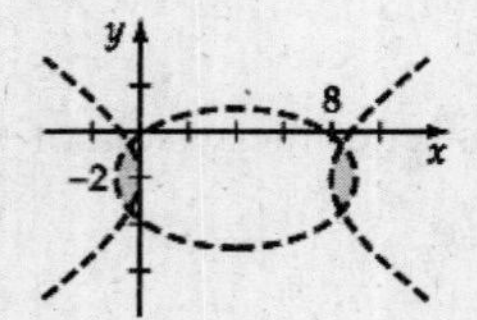

41. Sketch the graph of the solution set of the system of inequalities.

$$\begin{cases} 2x-3y \ge -5 \\ x+2y \le 7 \\ x \ge -1,\ y \ge 0 \end{cases}$$

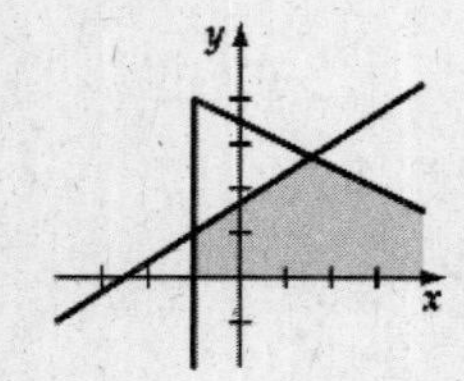

43. Sketch the graph of the solution set of the system of inequalities.

$$\begin{cases} 3x+2y \ge 14 \\ x+3y \ge 14 \\ x \le 10,\ y \le 8 \end{cases}$$

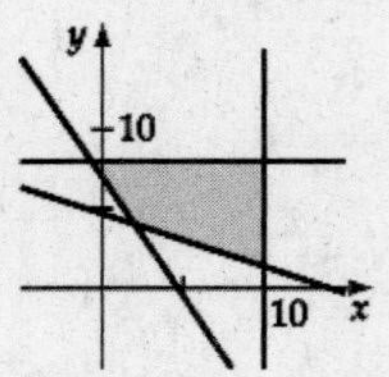

45. Find the heart rate range.

Substitute Ashley's age, 35, in the first inequality to find the minimum value.

$$y \ge 0.55(208-0.7x)$$
$$y \ge 0.55(208-0.7(35))$$
$$y \ge 100.925$$

Substitute Ashley's age in the second inequality to find the maximum value.

$$y \le 0.75(208-0.7x)$$
$$y \le 0.75(208-0.7(35))$$
$$y \le 137.625$$

The minimum is 101 beats per minute and maximum is 138 beats per minute.

47. Sketch the graph of the inequality.

$$|y| \ge |x|$$

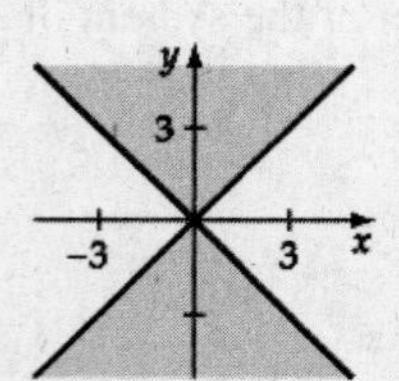

49. Sketch the graph of the inequality.

$$|x+y| \le 1$$

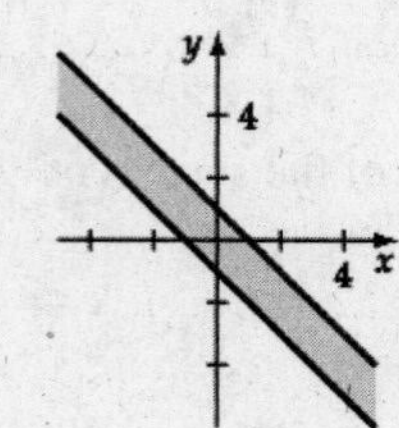

51. Sketch the graph of the inequality.

$|x|+|y|\le 1$

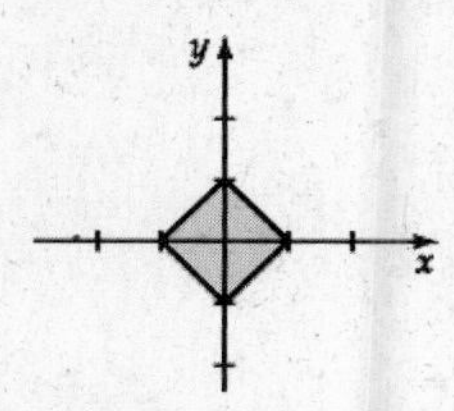

53. Sketch the graph of the inequalities. Explain.

If x is a negative number, then the inequality is reversed when multiplying both sides of the inequality by the negative number $\dfrac{1}{x}$.

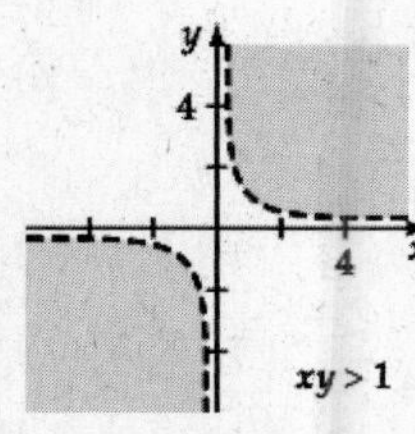

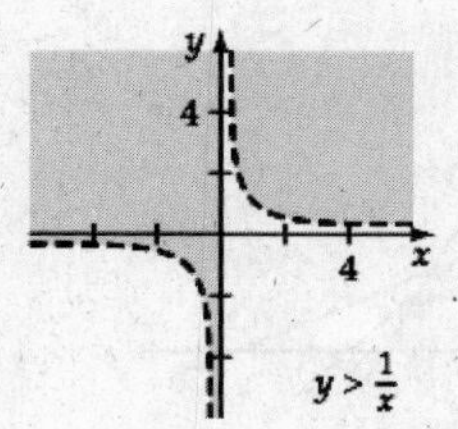

Prepare for Section 6.6

P1. Graph $2x+3y\le 12$.

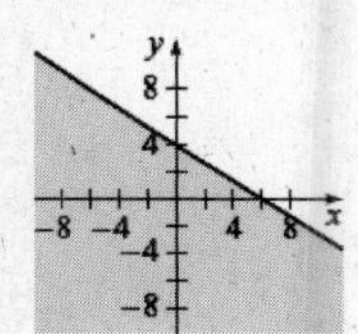

P3. Evaluate.

$$C=3x+4y$$
$$C(0,\ 5)=3(0)+4(5)=20$$
$$C(2,\ 3)=3(2)+4(3)=18$$
$$C(6,\ 1)=3(6)+4(1)=22$$
$$C(9,\ 0)=3(9)+4(0)=27$$

P5. Solve the system of equations.

$$\begin{cases}3x+y=6 & (1)\\ x+y=4 & (2)\end{cases}$$

Solve Eq (1) for y and substitute into Eq (2).

$$x+(-3x+6)=4$$
$$-2x+6=4$$
$$-2x=-2$$
$$x=1$$

$$y=-3(1)+6=3$$

The solution is (1, 3).

Section 6.6 Exercises

1. Find the minimum value and state where it is.

$$C(x,\ y)=3x+4y$$
$$C(0,\ 5)=3(0)+4(5)=20$$
$$C(2,\ 3)=3(2)+4(3)=18$$
$$C(6,\ 1)=3(6)+4(1)=22$$
$$C(9,\ 0)=3(9)+4(0)=27$$

The minimum of 18 occurs at (2, 3).

3. Find the maximum value and state where it is.

$$C(x,\ y)=2.5x+3y+5$$
$$C(0,\ 20)=2.5(0)+3(20)+5=65$$
$$C(5,\ 19)=2.5(5)+3(19)+5=74.5$$
$$C(20,\ 4)=2.5(20)+3(4)+5=67$$
$$C(22.5,\ 0)=2.5(22.5)+3(0)+5=61.25$$

The maximum of 74.5 occurs at (5, 19).

5. Solve the linear programming problem.

$$C=4x+2y$$

$$\begin{cases}x+y\ge 7\\ 4x+3y\ge 24\\ x\le 10,\ \ y\le 10\\ x\ge 0,\ \ y\ge 0\end{cases}$$

	$C=4x+2y$	
(0,10)	20	
(0,8)	16	minimum
(3,4)	20	
(7,0)	28	
(10,0)	40	
(10,10)	60	

(3, 4)

The minimum is 16 at (0, 8).

7. Solve the linear programming problem.

$$C=6x+7y$$

$$\begin{cases}x+2y\le 16\\ 5x+3y\le 45\\ x\ge 0,\ y\ge 0\end{cases}$$

$C=6x+7y$	
(0,8)	56
(0,0)	0
(9,0)	54
(6,5)	71 maximum

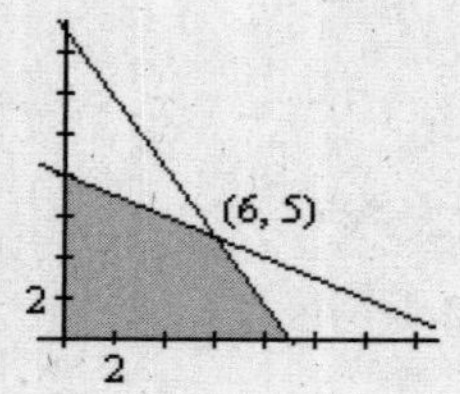

The maximum is 71 at (6, 5).

9. Solve the linear programming problem.

$C=x+6y$

$$\begin{cases} 5x+8y\le 120 \\ 7x+16y\le 192 \\ x\ge 0,\ y\ge 0 \end{cases}$$

$C=x+6y$	
(0,12)	72 maximum
(0,0)	0
(24,0)	24
(16,5)	46

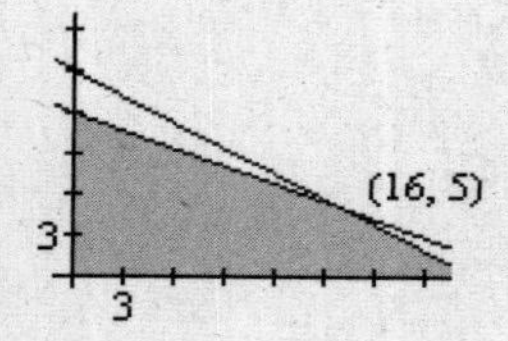

The maximum is 72 at (0, 12).

11. Solve the linear programming problem.

$C=4x+y$

$$\begin{cases} 3x+5y\ge 120 \\ x+y\ge 32 \\ x\ge 0,\ y\ge 0 \end{cases}$$

$C=4x+y$	
(40,0)	160
(0,32)	32 minimum
(20,12)	92

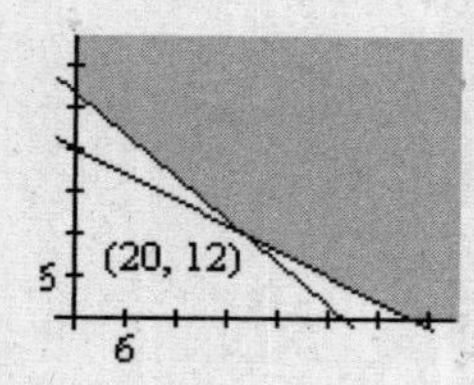

The minimum is 32 at (0, 32).

13. Solve the linear programming problem.

$C=2x+7y$

$$\begin{cases} x+y\le 10 \\ x+2y\le 16 \\ 2x+y\le 16 \\ x\ge 0,\ y\ge 0 \end{cases}$$

$C=2x+7y$	
(0,8)	56 maximum
(0,0)	0
(8,0)	16
(6,4)	40
(4,6)	50

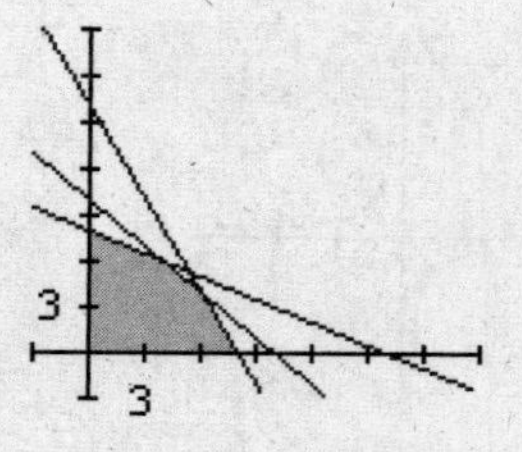

The maximum is 56 at (0, 8).

15. Solve the linear programming problem.

$C=3x+2y$

$$\begin{cases} 3x+y\ge 12 \\ 2x+7y\ge 21 \\ x+y\ge 8 \\ x\ge 0,\ y\ge 0 \end{cases}$$

$C=3x+2y$	
(0,12)	24
(2,6)	18 minimum
(7,1)	23
(10.5,0)	31.5

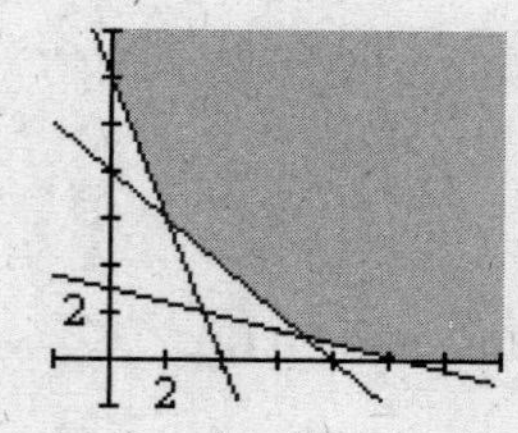

The minimum is 18 at (2, 6).

17. Solve the linear programming problem.

$C = 3x + 4y$

$$\begin{cases} 2x + y \le 10 \\ 2x + 3y \le 18 \\ x - y \le 2 \\ x \ge 0,\ y \ge 0 \end{cases}$$

$C = 3x + 4y$		
$(0,6)$	24	
$(3,4)$	25	maximum
$(4,2)$	20	
$(2,0)$	6	
$(0,0)$	0	

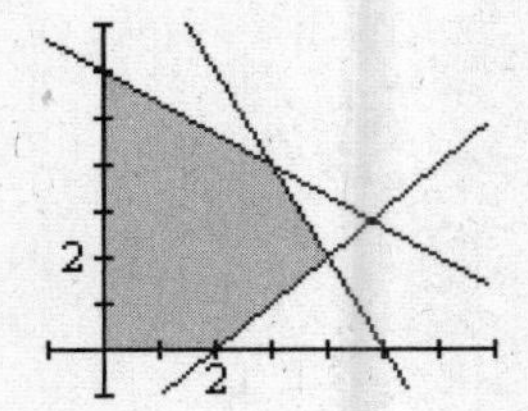

The maximum is 25 at (3, 4).

19. Solve the linear programming problem.

$C = 3x + 2y$

$$\begin{cases} x + 2y \ge 8 \\ 3x + y \ge 9 \\ x + 4y \ge 12 \\ x \ge 0,\ y \ge 0 \end{cases}$$

$C = 3x + 2y$		
$(0,9)$	18	
$(2,3)$	12	maximum
$(4,2)$	16	
$(12,0)$	36	

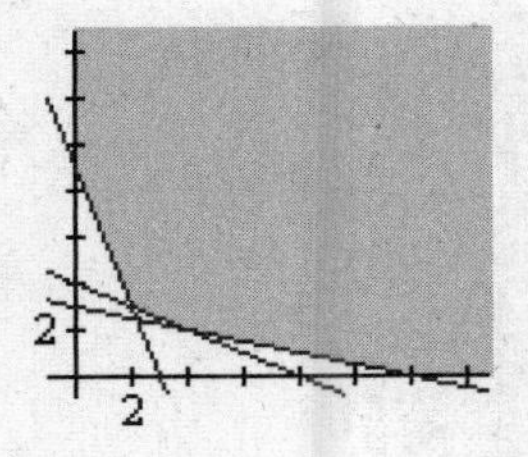

The minimum is 12 at (2, 3).

21. Solve the linear programming problem.

$C = 6x + 7y$

$$\begin{cases} x + 2y \le 900 \\ x + y \le 500 \\ 3x + 2y \le 1200 \\ x \ge 0,\ y \ge 0 \end{cases}$$

$C = 6x + 7y$		
$(0,450)$	3150	
$(100,400)$	3400	maximum
$(200,300)$	3300	
$(400,0)$	2400	
$(0,0)$	0	

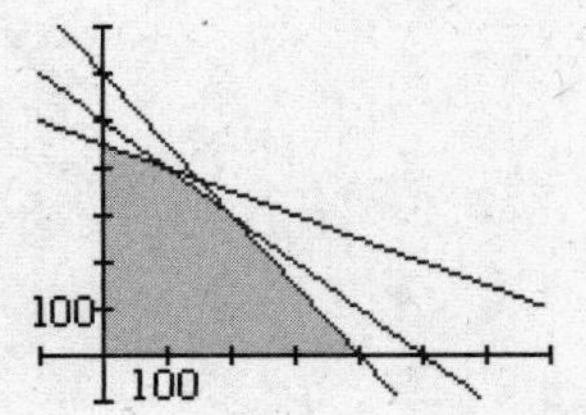

The maximum is 3400 at (100, 400).

23. Find the amount of each cereal to minimize the cost and find the minimum cost.

x = number of cups of Oat Flakes

y = number of cups of Crunchy O's

$C = 0.38x + 0.32y$

Constraints:

$$\begin{cases} 6x + 3y \ge 210 \\ 30x + 40y \ge 1200 \\ x \ge 0 \\ y \ge 0 \end{cases}$$

$C = 0.38x + 0.32y$		
$(0,70)$	\$22.40	
$(32,6)$	\$14.08	minimum
$(40,0)$	\$15.20	

The minimum cost of \$14.08 is achieved by mixing 32 cups of Oat Flakes and 6 cups of Crunchy O's.

25. Find the acres of each crop to maximize the profit.

W = acres of wheat to plant

B = acres of barley to plant

$P = 50W + 70B$

Constraints:

$$\begin{cases} 4W + 3B \le 200 \\ W + 2B \le 100 \\ W \ge 0,\ B \ge 0 \end{cases}$$

$P = 50W + 70B$		
(0,50)	3500	
(20,40)	3800	maximum
(50,0)	2500	
(0,0)	0	

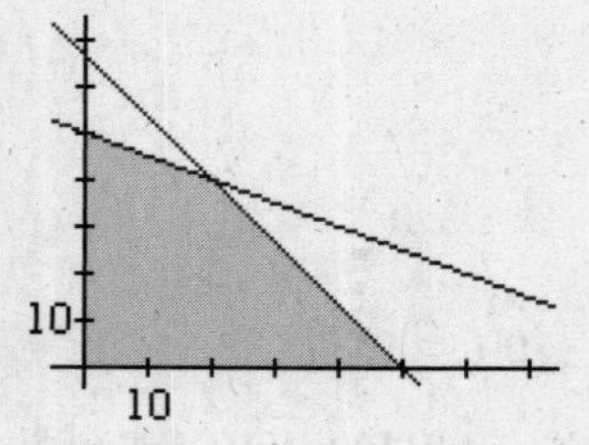

The maximum profit is achieved by planting 20 acres of wheat and 40 acres of barley.

27. Find the amount of each to maximize profit and find the maximum weekly profit.

x = number of economy boards

y = number of superior boards

Profit $= 26x + 42y$

Constraints:

$$\begin{cases} 2x + 2.5y \le 240 \\ 2x + 4y \le 312 \\ x \ge 0,\ y \ge 0 \end{cases}$$

Profit $= 26x + 42y$		
(0,78)	3276	
(120,0)	3120	
(60,48)	3576	maximum
(0,0)	0	

The maximized profit of $3576 is achieved by producing 60 economy boards and 48 superior boards.

29. Find the amount of each to minimize cost and find the minimum cost.

A = ounces of food group A

B = ounces of food group B

Cost $= 40A + 10B$

Constraints:

$$\begin{cases} 3A + B \ge 24 \\ A + B \ge 16 \\ A + 3B \ge 30 \\ A \ge 0,\ B \ge 0 \end{cases}$$

Cost $= 40A + 10B$		
(0,24)	240	minimum
(4,12)	280	
(9,7)	430	
(30,0)	1200	

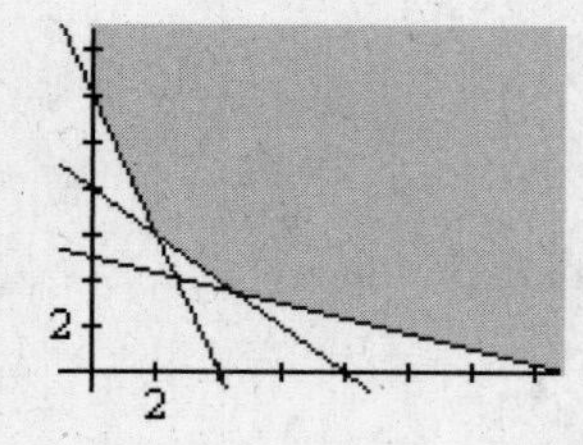

To minimize cost, use 24 ounces of food group B and zero ounces of food group A.

The minimum cost is $2.40.

31. Find the number of each to maximize profit and find the maximum profit.

x = number of 4-cylinder engines

y = number of 6-cylinder engines

Profit $= 150x + 250y$

Constraints:

$$\begin{cases} x + y \le 9 \\ 5x + 10y \le 80 \\ 3x + 2y \le 24 \\ x \ge 0,\ y \ge 0 \end{cases}$$

Profit $= 150x + 250y$		
(0,8)	2000	
(2,7)	2050	maximum
(6,3)	1650	
(8,0)	1200	
(0,0)	0	

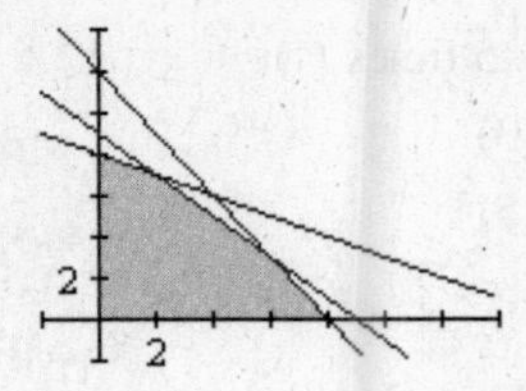

To achieve the maximum profit of \$2050, produce two 4-cylinder engines and seven 6-cylinder engines.

Exploring Concepts with Technology

System 1

When the coefficients are approximated to:

a. the nearest 0.01 the solution is (266.6666667, –1433.333333, 1366,666667).

b. the nearest 0.001 the solution is (30.50502049, –209.9900227, 226.6966693).

c. the nearest 0.0001 the solution is (27.32317303, –193.65719, 211.5374196).

d. the limit of most calculators (nearest 0.00000000001) the solution is (27.00000003, –192.0000002, 210.0000002).

System 1

When the coefficients are approximated to:

a. the nearest 0.01 the solution is (55.98194131, –295.2595937, 190.9706546, 118.510158).

b. the nearest 0.001 the solution is (81.29464132, –708.9158728, 1335.912871, –682.7827636).

c. the nearest 0.0001 the solution is (–90.6440169, 1188.443724, –3201.68217, 2258.171289).

d. the limit of most calculators (nearest 0.00000000001) the solution is (–64.00000245, 900.0000269, –2520.000064, 1820.000042)

Chapter 6 Review Exercises

1. Solve the system of equations. [6.1]

$$\begin{cases} 2x-4y=-3 & (1) \\ 3x+8y=-12 & (2) \end{cases}$$

$$\begin{array}{rl} 4x-8y=-6 & 2 \text{ times } (1) \\ \underline{3x+8y=-12} & (2) \\ 7x=-18 & \\ x=-\frac{18}{7} & \end{array}$$

$$2\left(-\frac{18}{7}\right)-4y=-3$$

$$y=-\frac{15}{28}$$

The solution is $\left(-\frac{18}{7},\ -\frac{15}{28}\right)$

3. Solve the system of equations. [6.1]

$$\begin{cases} 3x-4y=-5 & (1) \\ y=\frac{2}{3}x+1 & (2) \end{cases}$$

$$3x-4\left(\frac{2}{3}x+1\right)=-5$$

$$3x-\frac{8}{3}x-4=-5$$

$$\frac{1}{3}x=-1$$

$$x=-3$$

$$y=\frac{2}{3}(-3)+1=-1$$

The solution is (–3, –1).

5. Solve the system of equations. [6.1]

$$\begin{cases} y=2x-5 & (1) \\ x=4y-1 & (2) \end{cases}$$

Substitute x from Eq. (2) into Eq. (1).

$$y=2(4y-1)-5$$

$$y=8y-2-5$$

$$-7y=-7$$

$$y=1$$

$$x=4(1)-1=3$$

The solution is (3, 1).

7. Solve the system of equations. [6.1]

$$\begin{cases} 6x+9y=15 & (1) \\ 10x+15y=25 & (2) \end{cases}$$

$$\begin{array}{rl} 2x+3y=5 & \frac{1}{3} \text{ times } (1) \\ \underline{2x+3y=5} & \frac{1}{5} \text{ times } (2) \\ 0=0 & \end{array}$$

Let $y=c$.

$$2x+3c=5$$

$$x=\frac{5-3c}{2}$$

The ordered-pair solutions are $\left(\frac{5-3c}{2},\ c\right)$.

9. Solve the system of equations. [6.2]

$$\begin{cases} 2x-3y+z=-9 & (1) \\ 2x+5y-2z=18 & (2) \\ 4x-y+3z=-4 & (3) \end{cases}$$

$$\begin{array}{rl} 2x-3y+z=-9 & (1) \\ -2x-5y+2z=-18 & -1 \text{ times } (2) \\ \hline 8y+3z=-27 & (4) \end{array}$$

$$\begin{array}{rl} -4x+6y-2z=18 & -2 \text{ times } (1) \\ 4x-y+3z=-4 & (3) \\ \hline 5y+z=14 & (5) \end{array}$$

$$\begin{cases} 2x-3y+z=-9 & (1) \\ -8y+3z=-27 & (4) \\ 5y+z=14 & (5) \end{cases}$$

$$\begin{array}{rl} -8y+3z=-27 & (4) \\ -15y-3z=-42 & -3 \text{ times } (5) \\ \hline -23y \quad =-69 & \\ y=3 & (6) \end{array}$$

$$\begin{cases} 2x-3y+z=-9 & (1) \\ -8y+3z=-27 & (4) \\ y=3 & (6) \end{cases}$$

$$-8(3)+3z=-27$$
$$3z=-3$$
$$z=-1$$

$$2x-3(3)-1=-9$$
$$2x=1$$
$$x=\frac{1}{2}$$

The ordered-triple solution is $\left(\frac{1}{2}, 3, -1\right)$.

11. Solve the system of equations. [6.2]

$$\begin{cases} x+3y-5z=-12 & (1) \\ 3x-2y+z=7 & (2) \\ 5x+4y-9z=-17 & (3) \end{cases}$$

$$\begin{array}{rl} -3x-9y+15z=36 & -3 \text{ times } (1) \\ 3x-2y+z=7 & (2) \\ \hline -11y+16z=43 & (4) \end{array}$$

$$\begin{array}{rl} -5x-15y+25z=60 & -5 \text{ times } (1) \\ 5x+4y-9z=-17 & (3) \\ \hline -11y+16z=43 & (5) \end{array}$$

$$\begin{cases} x+3y-5z=-12 & (1) \\ -11y+16z=43 & (4) \\ -11y+16z=43 & (5) \end{cases}$$

$$\begin{array}{rl} -11y+16z=43 & (4) \\ 11y-16z=-43 & -1 \text{ times } (5) \\ \hline 0=0 & (6) \end{array}$$

The system of equations is dependent.

Let $z = c$.

$$-11y+16c=43$$
$$y=\frac{16c-43}{11}$$
$$x+3\left(\frac{16c-43}{11}\right)-5c=-12$$
$$x=\frac{7c-3}{11}$$

The solution is $\left(\frac{7c-3}{11}, \frac{16c-43}{11}, c\right)$.

13. Solve the system of equations. [6.2]

$$\begin{cases} 3x+4y-6z=10 & (1) \\ 2x+2y-3z=6 & (2) \\ x-6y+9z=-4 & (3) \end{cases}$$

Rearrange the equations so that the equation with the 1 as the coefficient of x is in the first row.

$$\begin{cases} x-6y+9z=-4 & (3) \\ 2x+2y-3z=6 & (2) \\ 3x+4y-6z=10 & (1) \end{cases}$$

$$\begin{array}{rl} -2x+12y-18z=8 & -2 \text{ times } (3) \\ 2x+2y-3z=6 & (2) \\ \hline 14y-21z=14 & \\ 2y-3z=2 & (4) \end{array}$$

$$\begin{array}{rl} -3x+18y-27z=12 & -3 \text{ times } (3) \\ 3x+4y-6z=10 & (1) \\ \hline 22y-33z=22 & \\ 2y-3z=2 & (5) \end{array}$$

$$\begin{cases} x-6y+9z=-4 & (3) \\ 2y-3z=2 & (4) \\ 2y-3z=2 & (5) \end{cases}$$

$$\begin{array}{rl} 2y-3z=2 & (4) \\ -2y+3z=-2 & -1 \text{ times } (5) \\ \hline 0=0 & (6) \end{array}$$

$$\begin{cases} x-6y+9z=-4 & (3) \\ 2y-3z=2 & (4) \\ 0=0 & (6) \end{cases}$$

The system of equations is dependent.

Let $z = c$.

$$2y-3c=2$$
$$y=\frac{3c+2}{2}$$

$$x-6\left(\frac{3c+2}{2}\right)+9c=-4$$
$$x-9c-6+9c=-4$$
$$x=2$$

The solution is $\left(2, \frac{3c+2}{2}, c\right)$.

15. Solve the system of equations. [6.2]

$$\begin{cases} 2x+3y-2z=0 & (1) \\ 3x-y-4z=0 & (2) \\ 5x+13y-4z=0 & (3) \end{cases}$$

$$\begin{array}{rl} 6x+9y-6z=0 & 3 \text{ times } (1) \\ -6x+2y+8z=0 & -2 \text{ times } (2) \\ \hline 11y+2z=0 & (4) \end{array}$$

$$\begin{array}{rl} 15x-5y-20z=0 & 5 \text{ times } (2) \\ -15x-39y+12z=0 & -3 \text{ times } (3) \\ \hline -44y-8z=0 & \\ 11y+2z=0 & (5) \end{array}$$

$$\begin{cases} 2x+3y-2z=0 & (1) \\ 11y+2z=0 & (4) \\ 11y+2z=0 & (5) \end{cases}$$

$$\begin{array}{rl} 11y+2z=0 & (4) \\ -11y-2z=0 & -1 \text{ times } (5) \\ \hline 0=0 & (6) \end{array}$$

$$\begin{cases} 2x+3y-2z=0 & (1) \\ 11y+2z=0 & (4) \\ 0=0 & (6) \end{cases}$$

Let $z=c$ $\quad 11y+2c=0$

$$y=-\frac{2}{11}c$$

$$2x+3\left(-\frac{2c}{11}\right)-2c=0$$
$$2x=\frac{28c}{11}$$
$$x=\frac{14c}{11}$$

The solution is $\left(\frac{14}{11}c, -\frac{2}{11}c, c\right)$.

17. Solve the system of equations. [6.2]

$$\begin{cases} x-2y+z=1 & (1) \\ 3x+2y-3z=1 & (2) \end{cases}$$

$$\begin{array}{rl} -3x+6y-3z=-3 & -3 \text{ times } (1) \\ 3x+2y-3z=1 & (2) \\ \hline 8y-6z=-2 & \\ 4y-3z=-1 & (3) \end{array}$$

$$\begin{cases} x-2y+z=1 & (1) \\ 4y-3z=-1 & (3) \end{cases}$$

Let $z=c$. $\quad 4y-3c=-1$

$$y=\frac{3c-1}{4}$$

$$x-2\left(\frac{3c-1}{4}\right)+c=1$$
$$x=\frac{c+1}{2}$$

The solution is $\left(\frac{c+1}{2}, \frac{3c-1}{4}, c\right)$.

19. Solve the system of equations. [6.3]

$$\begin{cases} y=x^2-2x-3 \\ y=2x-7 \end{cases}$$

$$x^2-2x-3=2x-7$$
$$x^2-4x+4=0$$
$$(x-2)(x-2)=0$$
$$x=2$$

$$y=2(2)-7=-3$$

The solution is (2, –3).

21. Solve the system of equations. [6.3]

$$\begin{cases} y = 3x^2 - \ \ x + 1 \\ y = \ \ x^2 + 2x - 1 \end{cases}$$

$$3x^2 - x + 1 = x^2 + 2x - 1$$
$$2x^2 - 3x + 2 = 0$$
$$x = \frac{-(-3) \pm \sqrt{(-3)^2 - 4(2)(2)}}{2(2)}$$
$$x = \frac{3 \pm \sqrt{9-16}}{4} = \frac{3 \pm \sqrt{-7}}{4}$$

x has no real number solution. The system of equations is inconsistent. The graphs of the equations do not intersect. No real solution.

23. Solve the system of equations. [6.3]

$$(x+1)^2 + (y-2)^2 = 4 \quad (1)$$
$$2x + y = 4 \quad (2)$$

From Eq. (2), $y = -2x + 4$.

Substitute y in Eq. (1).

$$(x+1)^2 + (-2x+4-2)^2 = 4$$
$$(x+1)^2 + (-2x+2)^2 = 4$$
$$x^2 + 2x + 1 + 4x^2 - 8x + 4 = 4$$
$$5x^2 - 6x + 1 = 0$$
$$(5x-1)(x-1) = 0$$
$$x = \frac{1}{5} \text{ or } x = 1$$
$$y = -2\left(\frac{1}{5}\right) + 4 = \frac{18}{5}$$
$$y = -2(1) + 4 = 2$$

The solutions are $\left(\frac{1}{5}, \frac{18}{5}\right)$ and $(1, 2)$.

25. Solve the system of equations. [6.3]

$$\begin{cases} (x-2)^2 + (y+2)^2 = 4 & (1) \\ (x+2)^2 + (y+1)^2 = 17 & (2) \end{cases}$$

Expand the binomials. Then subtract.

$$x^2 - 4x + 4 + y^2 + 4y + 4 = 4$$
$$x^2 + 4x + 4 + y^2 + 2y + 1 = 17$$
$$-8x + 2y + 3 = -13$$
$$y = 4x - 8$$

Substitute y into Eq. (1).

$$(x-2)^2 + (4x-8+2)^2 = 4$$
$$(x-2)^2 + (4x-6)^2 = 4$$
$$x^2 - 4x + 4 + 16x^2 - 48x + 36 = 4$$
$$17x^2 - 52x + 36 = 0$$
$$(x-2)(17x-18) = 0$$
$$x = 2 \text{ or } x = \frac{18}{17}$$
$$y = 4(2) - 8 = 0$$
$$y = 4\left(\frac{18}{17}\right) - 8 = -\frac{64}{17}$$

The solutions are $(2, 0)$ and $\left(\frac{18}{17}, -\frac{64}{17}\right)$.

27. Solve the system of equations. [6.3]

$$\begin{cases} x^2 - 3xy + y^2 = -1 & (1) \\ 3x^2 - 5xy - 2y^2 = 0 & (2) \end{cases}$$

Factor Eq. (2).

$$(3x + y)(x - 2y) = 0$$
$$x = \frac{-y}{3} \text{ or } x = 2y$$

$x = -\frac{y}{3}$ implies $y = -3x$. Substitute for y in Eq. (1).

$$x^2 - 3x(-3x) + (-3x)^2 = -1$$
$$x^2 + 9x^2 + 9x^2 = -1$$
$$19x^2 = -1$$
$$x^2 = -\frac{1}{19}$$

This equation yields no real solutions. Substituting $x = 2y$ in Eq. (1) yields

$$(2y)^2 - 3(2y)y + y^2 = -1$$
$$4y^2 - 6y^2 + y^2 = -1$$
$$-y^2 = -1$$
$$y^2 = 1$$
$$y = \pm 1$$

The solutions are $(2, 1)$ and $(-2, -1)$.

29. Solve the system of equations. [6.3]

$$\begin{cases} 2x^2 - 5xy + 2y^2 = 56 & (1) \\ 14x^2 - 3xy - 2y^2 = 56 & (2) \end{cases}$$

Subtract Eq. (1) From Eq. (2)

$12x^2+2xy-4y^2=0$

Factor $2(3x+2y)(2x-y)=0$. Thus

$3x+2y=0$ or $2x-y=0$

$y=-\frac{3}{2}x$ $\quad y=2x$

Substituting $y=-\frac{3}{2}x$ into Eq. (1), we have

$$2x^2-5x\left(-\frac{3}{2}x\right)+2\left(-\frac{3}{2}x\right)^2=56$$
$$2x^2+\frac{15}{2}x^2+\frac{9}{2}x^2=56$$
$$14x^2=56$$
$$x^2=4$$
$$x=\pm 2$$

When $x=2, y=-\frac{3}{2}(2)=-3$;

When $x=-2, y=-\frac{3}{2}(-2)=3$.

Two solutions are (2, –3) and (–2, 3).

Substituting $y=2x$ into Eq. (1) yields 0 = 56. Thus the only solutions of the system are (2, –3), and (–2, 3).

31. Find the partial fraction decomposition. [6.4]

$$\frac{7x-5}{x^2-x-2}=\frac{7x-5}{(x-2)(x+1)}=\frac{A}{x-2}+\frac{B}{x+1}$$

$$7x-5=A(x+1)+B(x-2)$$
$$7x-5=Ax+A+Bx-2b$$
$$7x-5=(A+B)x+(A-2B)$$

$$\begin{cases}7=A+B\\-5=A-2B\end{cases}$$

$$\begin{array}{l}7=A+B\\ \underline{-5=A-2B}\\ 12=3B\\ 4=B\end{array}$$

$$A+4=7$$
$$A=3$$

$$\frac{7x-5}{x^2-x-2}=\frac{3}{x-2}+\frac{4}{x+1}$$

33. Find the partial fraction decomposition. [6.4]

$$\frac{2x-2}{(x^2+1)(x+2)}=\frac{Ax+B}{x^2+1}+\frac{C}{x+2}$$

$$2x-2=(Ax+B)(x+2)+C(x^2+1)$$
$$2x-2=Ax^2+2Ax+Bx+2B+Cx^2+C$$
$$2x-2=(A+C)x^2+(2A+B)x+(2B+C)$$

$$\begin{cases}0=A+C & (1)\\2=2A+B & (2)\\-2=2B+C & (3)\end{cases}$$

$$\begin{array}{lll}0=A & +C & (1)\\ \underline{2= \quad} & \underline{-2B-C} & -1 \text{ times } (3)\\ 2=A-2B & & (4)\end{array}$$

$$\begin{cases}0=A+C & (1)\\2=2A+B & (2)\\2=A-2B & (4)\end{cases}$$

$$\begin{array}{ll}4=4A+2B & 2 \text{ times } (2)\\ \underline{2=A-2B} & (4)\\ 6=5A & \\ \frac{6}{5}=A & \end{array}$$

$$2\left(\frac{6}{5}\right)+B=2$$
$$B=-\frac{2}{5}$$
$$\frac{6}{5}+C=0$$
$$C=-\frac{6}{5}$$

$$\frac{2x-2}{(x^2+1)(x-2)}=\frac{6x-2}{5(x^2+1)}+\frac{-6}{5(x+2)}$$

35. Find the partial fraction decomposition. [6.4]

$$\frac{11x^2-x-2}{x^3-x}=\frac{11x^2-x-2}{x(x-1)(x+1)}=\frac{A}{x}+\frac{B}{x-1}+\frac{C}{x+1}$$

$$11x^2-x-2=A(x-1)(x+1)+Bx(x+1)+Cx(x-1)$$
$$11x^2-x-2=Ax^2-A+Bx^2+Bx+Cx^2-Cx$$
$$11x^2-x-2=(A+B+C)x^2+(B-C)x+(-A)$$

$$\begin{cases}11=A+B+C & (1)\\-1=B-C & (2)\\-2=-A & (3)\end{cases}$$

$$\begin{cases} 11 = A+B+C & (1) \\ -1 = B-C & (2) \\ 2 = A & (4) \end{cases}$$

$$\begin{aligned} B+C &= 9 && \text{from (1) with } A=2 \\ B-C &= -1 && (2) \\ \hline 2B &= 8 \\ B &= 4 \end{aligned}$$

$$\begin{aligned} 4-C &= -1 \\ C &= 5 \end{aligned}$$

$$\frac{11x^2-x-2}{x^3-x} = \frac{2}{x} + \frac{4}{x-1} + \frac{5}{x+1}$$

37. Graph the solution set of the inequality. [6.5]

$4x-5y<20$

$y > \frac{4}{5}x - 4$ Use a dashed line.

Test point (0, 0): is in region

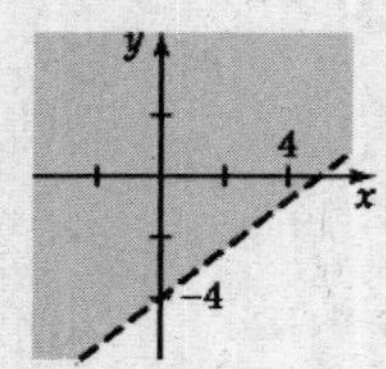

39. Graph the solution set of the inequality. [6.5]

$y \geq 2x^2 - x - 1$

vertex $\left(\frac{1}{4}, -\frac{9}{8}\right)$

Test point (0, 0): is in region

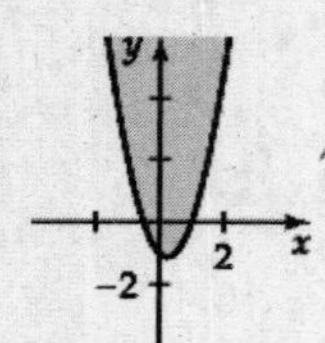

41. Graph the solution set of the inequality. [6.5]

$(x-2)^2 + (y-1)^2 > 4$

center (2, 1), $r = 2$

Test point (2, 0): not in region

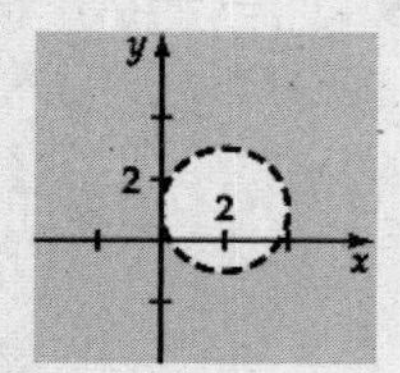

43. Graph the solution set of the inequality. [6.5]

$$\frac{(x-3)^2}{16} - \frac{(y+2)^2}{25} \leq 1$$

Test point (0, 0): is in region

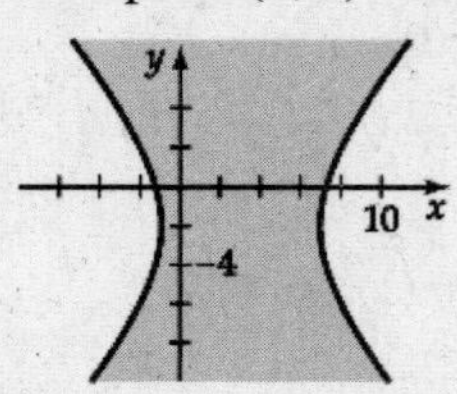

45. Graph the solution set of the inequality. [6.5]

$(2x-y+1)(2x-2y-2) > 0$

Test point (0, 0): not in region

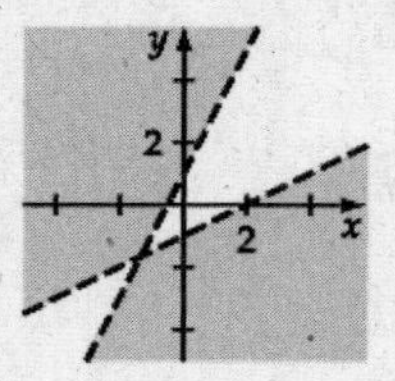

47. Graph the solution set of the inequality. [6.5]

$x^2y^2 < 1$

Test point (0, 0): is in region

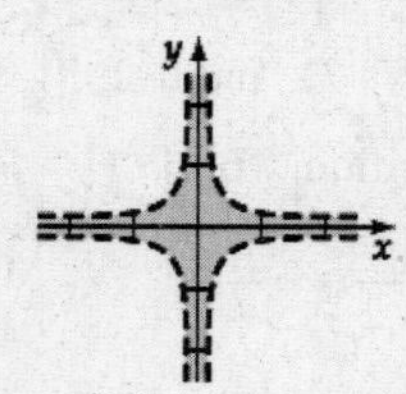

49. Sketch the graph of the solution set of the system of inequalities. [6.5]

$$\begin{cases} 2x-5y<9 \\ 3x+4y \geq 2 \end{cases}$$

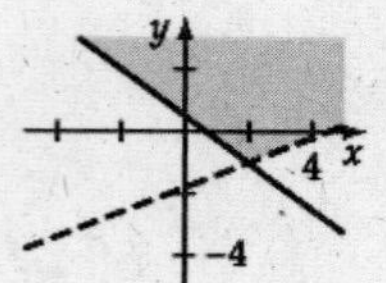

51. Sketch the graph of the solution set of the system of inequalities. [6.5]

$$\begin{cases} 2x+3y>6 \\ 2x-y>-2 \\ x\le 4 \end{cases}$$

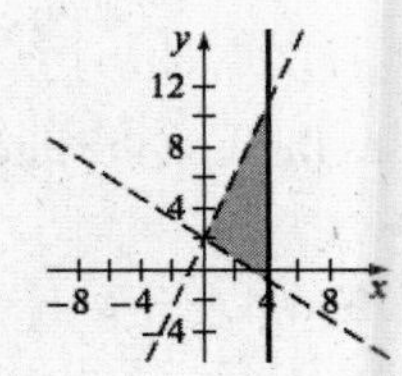

53. Sketch the graph of the solution set of the system of inequalities. [6.5]

$$\begin{cases} 2x+3y\le 18 \\ x+y\le 7 \\ x\ge 0,\ y\ge 0 \end{cases}$$

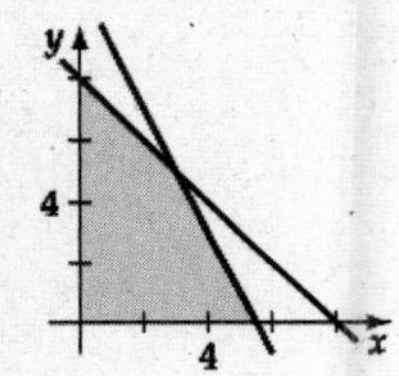

55. Sketch the graph of the solution set of the system of inequalities. [6.5]

$$\begin{cases} 3x+y\ge 6 \\ x+4y\ge 14 \\ 2x+3y\ge 16 \\ x\ge 0,\ y\ge 0 \end{cases}$$

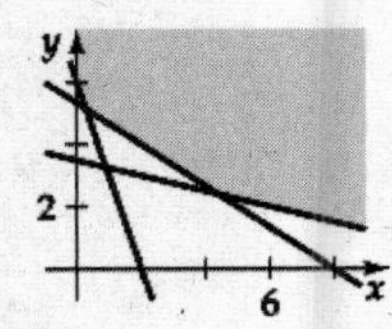

57. Sketch the graph of the solution set of the system of inequalities. [6.5]

$$\begin{cases} y<x^2-x-2 \\ y\ge 2x-4 \end{cases}$$

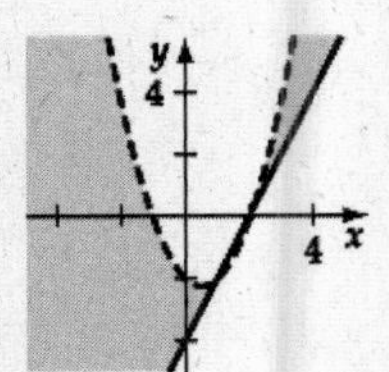

59. Sketch the graph of the solution set of the system of inequalities. [6.5]

$$\begin{cases} x^2+y^2-2x+4y>4 \\ y<2x^2-1 \end{cases}$$

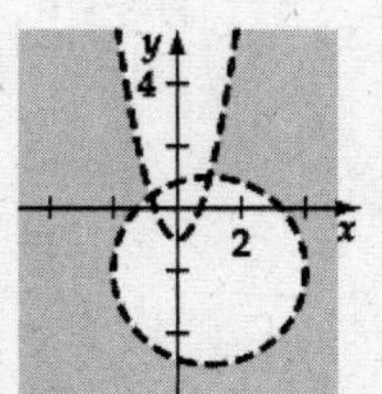

61. Solve the linear programming problem. [6.6]

$P=2x+2y$

$$\begin{cases} x+2y\le 14 \\ 5x+2y\le 30 \\ x\le 0,\ y\le 0 \end{cases}$$

	$P=2x+2y$	
(0,7)	14	
(6,0)	12	
(4,5)	18	maximum
(0,0)	0	

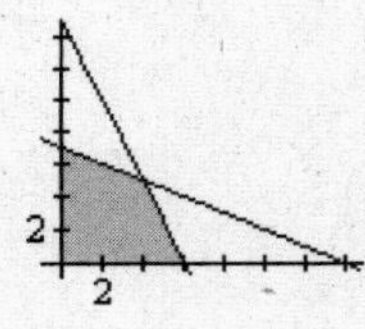

The maximum is 18 at (4, 5).

63. Solve the linear programming problem. [6.6]

$P=4x+y$

$$\begin{cases} 5x+2y\ge 16 \\ x+2y\ge 8 \\ 0\le x\le 20 \\ 0\le y\le 20 \end{cases}$$

	$P=4x+y$	
(0,8)	8	minimum
(2,3)	11	
(8,0)	32	
(20,0)	80	
(20,20)	100	
(0,20)	20	

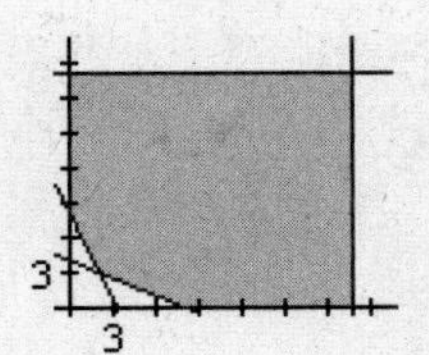

The minimum is 8 at (0, 8).

65. Solve the linear programming problem. [6.6]

$P = 6x + 3y$

$$\begin{cases} 5x + 2y \geq 20 \\ x + \ y \geq 7 \\ x + 2y \geq 10 \\ 0 \leq x \leq 15, \ 0 \leq y \leq 15 \end{cases}$$

$P = 6x + 3y$		
(0,10)	30	
(2,5)	27	minimum
(4,3)	33	
(10,0)	60	
(0,15)	45	
(15,15)	135	
(15,0)	90	

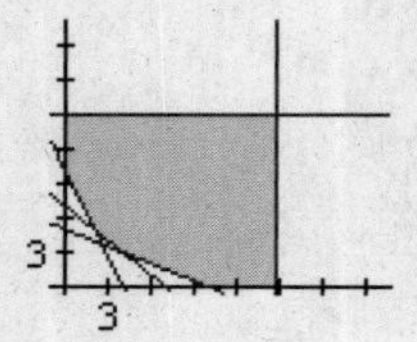

The minimum is 27 at (2, 5).

67. Find the equation. [6.2]

$y = ax^2 + bx + c$

$0 = a(1)^2 + b(1) + c$
$5 = a(-1)^2 + b(-1) + c$
$3 = a(2)^2 + b(2) + c$

$$\begin{cases} a + b + c = 0 & (1) \\ a - b + c = 5 & (2) \\ 4a + 2b + c = 3 & (3) \end{cases}$$

$$\begin{array}{rl} a + b + c = 0 & (1) \\ -a + b - c = -5 & -1 \text{ times } (2) \\ \hline 2b = -5 & (4) \end{array}$$

$$\begin{array}{rl} a - b + c = 5 & (2) \\ -4a - 2b - c = -3 & -1 \text{ times } (3) \\ \hline -3a - 3b = 2 & (5) \end{array}$$

$$\begin{cases} a + b + c = 0 & (1) \\ 2b = -5 & (4) \\ -3a - 3b = 2 & (5) \end{cases}$$

$2b = -5$

$b = -\frac{5}{2}$

$-3a - 3\left(-\frac{5}{2}\right) = 2$

$a = \frac{11}{6}$

$\frac{11}{6} - \frac{5}{2} + c = 0$

$c = \frac{2}{3}$

The equation of the graph that passes through the three points is $y = \frac{11}{6}x^2 - \frac{5}{2}x + \frac{2}{3}$.

69. Find the equation of the plane. [6.2]

$z = ax + by + c$

$2 = 2a + b + c$
$0 = 3a + b + c$
$-2 = -2a - 3b + c$

$$\begin{cases} 2a + b + c = 2 & (1) \\ 3a + b + c = 0 & (2) \\ -2a - 3b + c = -2 & (3) \end{cases}$$

$$\begin{array}{rl} 6a + 3b + 3c = 6 & 3 \text{ times } (1) \\ -6a - 2b - 2c = 0 & -2 \text{ times } (2) \\ \hline b + c = 6 & (4) \end{array}$$

$$\begin{array}{rl} 2a + b + c = 2 & (1) \\ -2a - 3b + c = -2 & (3) \\ \hline -2b + 2c = 0 & \\ -b + c = 0 & (5) \end{array}$$

$$\begin{array}{rl} 2a + b + \quad c = \ \ 2 & (1) \\ -2a - 3b + \quad c = -2 & (3) \\ \hline -2b + 2c = \ \ 0 & \\ -b + \quad c = \ \ 0 & (5) \end{array}$$

$$\begin{cases} 2a + b + c = 2 & (1) \\ b + c = 6 & (4) \\ -b + c = 0 & (5) \end{cases}$$

$$\begin{array}{rl} b + c = 6 & (4) \\ -b + c = 0 & (5) \\ \hline 2c = 6 & \\ c = 3 & (6) \end{array}$$

$$\begin{cases} 2a + b + c = 2 & (1) \\ b + c = 6 & (4) \\ c = 3 & (6) \end{cases}$$

$$b+3=6$$
$$b=3$$
$$2a+3+3=2$$
$$2a=-4$$
$$a=-2$$

The equation of the graph that passes through the three points is $z=-2x+3y+3$.

71. Find the rate of the wind and plane in calm air. [6.1]

Rate flying with the wind: $r+w$

Rate flying against the wind: $r-w$

$$\begin{cases}(r+w)5=855 & (1)\\ 5(r-w)=575 & (2)\end{cases}$$

$$r+w=171 \quad (1)$$
$$r-w=115 \quad (2)$$
$$2r=286$$
$$r=143$$

$$143+w=171$$
$$w=28$$

Rate of the wind is 28 mph.

Rate of the plane in calm air is 143 mph.

73. Find the values for a, b, and c. [6.2]

Given (a, b, c) with $ab=c$, $ac=b$.

If $ab=c$, then $b\cdot(ab)=a$ and $b^2=1$, $b=\pm1$, or $a=0$.

If $bc=a$, then $bc\cdot c=b$ and $c^2=1$, $c=\pm1$, or $b=0$.

If $ac=b$, then $a\cdot ac=c$ and $a^2=1$, $a\pm1$, or $c=0$.

The ordered triples are $(1, 1, 1)$, $(1, -1, -1)$, $(-1, -1, 1)$, $(-1, 1, -1)$, $(0, 0, 0)$.

Chapter 6 Test

1. Solve the system of equations. [6.1]

$$\begin{cases}3x+2y=-5 & (1)\\ 2x-5y=-16 & (2)\end{cases}$$

$$15x+10y=-25 \quad \text{5 times (1)}$$
$$4x-10y=-32 \quad \text{2 times (2)}$$
$$19x=-57$$
$$x=-3$$

$$3(-3)+2y=-5$$
$$2y=4$$
$$y=2$$

The solution is $(-3, 2)$.

3. Solve the system of equations. [6.2]

$$\begin{cases}x+3y-z=8 & (1)\\ 2x-7y+2z=1 & (2)\\ 4x-y+3z=13 & (3)\end{cases}$$

$$-2x-6y+2z=-16 \quad -2 \text{ times (1)}$$
$$2x-7y+2z=1 \quad (2)$$
$$-13y+4z=-15 \quad (4)$$

$$-4x-12y+4z=-32 \quad -4 \text{ times (1)}$$
$$4x-y+3z=13 \quad (3)$$
$$-13y+7z=-19 \quad (5)$$

$$\begin{cases}x+3y-z=8 & (1)\\ -13y+4z=-15 & (4)\\ -13y+7z=19 & (5)\end{cases}$$

$$-13y+4z=-15 \quad (4)$$
$$13y-7z=19 \quad -1 \text{ times (5)}$$
$$-3z=4$$
$$z=-\frac{4}{3} \quad (6)$$

$$\begin{cases}x+3y-z=8 & (1)\\ -13y+4z=-15 & (4)\\ z=-\frac{4}{3} & (6)\end{cases}$$

$$-13y+4\left(-\frac{4}{3}\right)=-15$$
$$y=\frac{29}{39}$$

$$x+3\left(\frac{29}{39}\right)-\left(-\frac{4}{3}\right)=8$$
$$x=\frac{173}{39}$$

The solution is $\left(\frac{173}{39}, \frac{29}{39}, -\frac{4}{3}\right)$.

5. Solve the system of equations. [6.2]

$$\begin{cases} 2x-3y+z=-1 & (1) \\ x+5y-2z=5 & (2) \end{cases}$$

$$\begin{array}{rl} 2x-3y+z=-1 & (1) \\ -2x-10y+4z=-10 & -2 \text{ times } (2) \\ \hline -13y+5z=-11 & (3) \end{array}$$

$$\begin{cases} 2x-3y+z=-1 & (1) \\ -13y+5z=-11 & (3) \end{cases}$$

The system of equations is dependent.

Let $z=c$.

$$-13y+5(c)=-11$$
$$y=\frac{5c+11}{13}$$

$$x+5\left(\frac{5c+11}{13}\right)-2c=5$$
$$x=\frac{c+10}{13}$$

The solution is $\left(\frac{c+10}{13}, \frac{5c+11}{13}, c\right)$.

7. Solve the system of equations. [6.3]

$$\begin{cases} y=x+3 & (1) \\ y=x^2+x-1 & (2) \end{cases}$$

Set the expressions equal to each other.

$$x^2+x-1=x+3$$
$$x^2=4$$
$$x=\pm 2$$

Substitute for x in Eq. (1).

$$y=2+3=5$$
$$y=-2+3=1$$

The solutions are (2, 5) and (–2, 1).

9. Graph the inequality. [6.5]

$$x^2+4y^2 \geq 16$$

Test point (0, 0): not in region

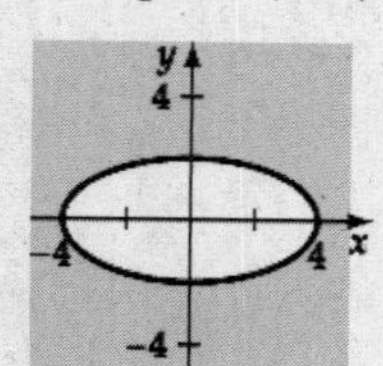

11. Graph the system of inequalities. [6.5]

$$\begin{cases} 2x-5y \leq 16 \\ x+3y \geq -3 \end{cases}$$

Test point (0, 0): is in region

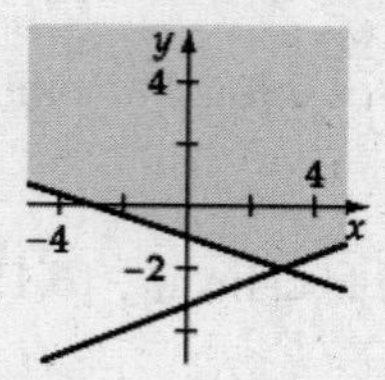

13. Graph the system of inequalities. [6.5]

$$\begin{cases} x+y \geq 8 \\ 2x+y \geq 11 \\ x \geq 0,\ y \geq 0 \end{cases}$$

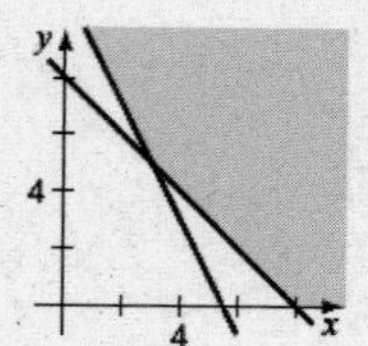

15. Find the partial fraction decomposition. [6.4]

$$\frac{3x-5}{x^2-3x-4}=\frac{3x-5}{(x-4)(x+1)}$$
$$=\frac{A}{x-4}+\frac{B}{x+1}$$

$$3x-5=A(x+1)+B(x-4)$$
$$3x-5=Ax+A+Bx-4B$$

$$\begin{cases} 3=A+B \\ -5=A-4B \end{cases}$$

$$\begin{array}{r} 3=A+B \\ 5=-A+4B \\ \hline 8=5B \end{array}$$

$$\frac{8}{5}=B$$

$$3=A+\frac{8}{5}$$
$$\frac{7}{5}=A$$

$$\frac{3x-5}{(x-4)(x+1)}=\frac{7}{5(x-4)}+\frac{8}{5(x+1)}$$

17. Find the length and width. [6.1]

$$x = 20 + y$$
$$2x + \pi y = 554.16$$

Using substitution,

$$2(20 + y) + \pi y = 554.16$$
$$40 + 2y + \pi y = 554.16$$
$$y(2 + \pi) = 514.16$$
$$y = \frac{514.16}{2 + \pi}$$
$$y \approx 100$$

$$x = 20 + y = 20 + 100 = 120$$

Length is 120 m and width is 100 m.

19. Find the amount of each to maximize profit. [6.6]

x = Acres of oats

y = Acres of barley

Constraints

$$\begin{cases} x + \quad y \le 160 \\ 15x + 13y \le 2200 \\ 15x + 20y \le 2600 \\ x \le 0, \quad y \ge 0 \end{cases}$$

Profit $= 120x + 150y$

$(0, 130)$	\$19,500	
$\left(\frac{680}{7}, \frac{400}{7}\right)$	\$20,228.57	maximum
$\left(146\frac{2}{3}, 0\right)$	\$17,600	
$(0, 0)$	0	

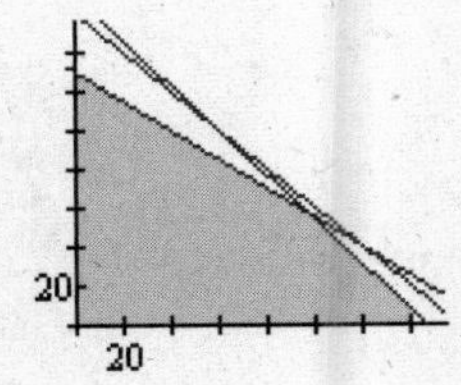

To maximize profit, $\frac{680}{7}$ acres of oats and $\frac{400}{7}$ acres of barley must be planted.

Cumulative Review Exercises

1. Find the slope. [2.3]

$$m = \frac{-\frac{1}{3} - 2}{4 - \left(-\frac{1}{2}\right)} = \frac{-\frac{7}{3}}{\frac{9}{2}} = -\frac{7}{3} \cdot \frac{2}{9} = -\frac{14}{27}$$

3. Evaluate. [P.1]

$$3(-2)^4 - 4(-2)^3 + 2(-2)^2 - (-2) + 1$$
$$= 3(16) - 4(-8) + 2(4) + 2 + 1$$
$$= 48 + 32 + 8 + 2 + 1$$
$$= 91$$

5. Find the equation of the parabola. [5.1]

Vertex = (4, 2), point (–1, 1), axis of symmetry $x = 4$

$$(x - 4)^2 = 4p(y - 2)$$
$$(-1 - 4)^2 = 4p(1 - 2)$$
$$25 = 4p(-1)$$
$$p = -\frac{25}{4}$$

$$(x - 4)^2 = 4\left(-\frac{25}{4}\right)(y - 2)$$
$$(x - 4)^2 = -25(y - 2)$$

7. Find the equation of the line. [2.3]

$$m = \frac{-1 - 2}{2 - (-4)} = \frac{-3}{6} = -\frac{1}{2}$$

$$y - (-1) = -\frac{1}{2}(x - 2)$$
$$y + 1 = -\frac{1}{2}x + 1$$
$$y = -\frac{1}{2}x$$

9. Solve. [4.5]

$$\log x - \log(2x - 3) = 2$$
$$\log \frac{x}{2x - 3} = 2$$
$$\frac{x}{2x - 3} = 10^2$$
$$x = 100(2x - 3)$$
$$x = 200x - 300$$
$$-199x = -300$$
$$x = \frac{300}{199}$$

11. Find $g\left(-\frac{1}{2}\right)$. [2.2]

$$g\left(-\frac{1}{2}\right)=\frac{-\frac{1}{2}-2}{-\frac{1}{2}}=\frac{-\frac{5}{2}}{-\frac{1}{2}}=5$$

13. Evaluate. [4.3]

$$\begin{aligned}\log_{0.25} 0.015625 &= \log_{0.25}(0.25)^3\\ &= 3\log_{0.25} 0.25\\ &= 3\end{aligned}$$

15. Find the polynomial from the given zeros. [3.4]

$$\begin{aligned}(x+2)(x-3i)(x+3i) &= (x+2)(x^2+9)\\ &= x^3+2x^2+9x+18\end{aligned}$$

17. Find the slant asymptote. [3.5]

$$\begin{array}{r} 2x+1 \\ x^2-x-1\overline{\big)\,2x^3 - x^2 \qquad -2} \\ \underline{2x^3-2x^2-2x} \\ x^2+2x-2 \\ \underline{x^2 - x - 1} \\ 3x-1 \end{array}$$

$$H(x)=\frac{2x^3-x^2-2}{x^2-x-1}=2x+1+\frac{3x-1}{x^2-x-1}$$

Slant asymptote: $y = 2x + 1$

19. Sketch the graph. [4.2]

$$F(x)=\frac{2^x-2^{-x}}{3}$$

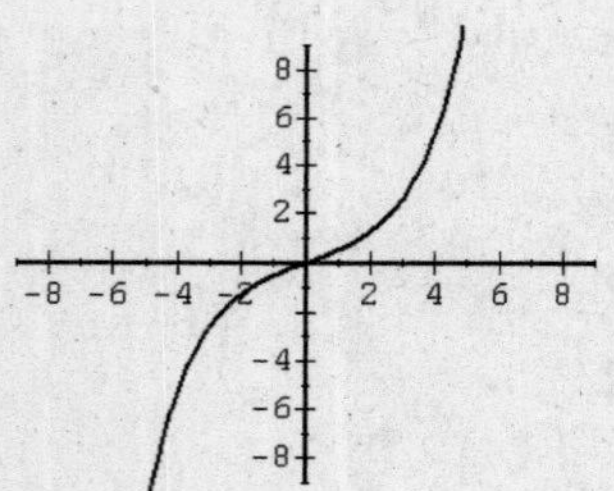

Chapter 7: Matrices

Section 7.1 Exercises

1. Writing the augmented, coefficient and constant matrices of the given system of equations.

$$\left[\begin{array}{rrr|r} 2 & -3 & 1 & 1 \\ 3 & -2 & 3 & 0 \\ 1 & 0 & 5 & 4 \end{array}\right], \begin{bmatrix} 2 & -3 & 1 \\ 3 & -2 & 3 \\ 1 & 0 & 5 \end{bmatrix}, \begin{bmatrix} 1 \\ 0 \\ 4 \end{bmatrix}$$

3. Writing the augmented, coefficient and constant matrices of the given system of equations.

$$\left[\begin{array}{rrrr|r} 2 & -3 & -4 & 1 & 2 \\ 0 & 2 & 1 & 0 & 2 \\ 1 & -1 & 2 & 0 & 4 \\ 3 & -3 & -2 & 0 & 1 \end{array}\right], \begin{bmatrix} 2 & -3 & -4 & 1 \\ 0 & 2 & 1 & 0 \\ 1 & -1 & 2 & 0 \\ 3 & -3 & -2 & 0 \end{bmatrix}, \begin{bmatrix} 2 \\ 2 \\ 4 \\ 1 \end{bmatrix}$$

The solutions and answers to Exercises 5 – 12 are not unique. The following solutions list the elementary row operations used to produce the row echelon forms shown.

5. Using elementary row operations to achieve row echelon form.

$$\begin{bmatrix} 2 & -1 & 3 & -2 \\ 1 & -1 & 2 & 2 \\ 3 & 2 & -1 & 3 \end{bmatrix} \xrightarrow{R_1 \longleftrightarrow R_2} \begin{bmatrix} 1 & -1 & 2 & 2 \\ 2 & -1 & 3 & -2 \\ 3 & 2 & -1 & 3 \end{bmatrix}$$

$$\begin{bmatrix} 1 & -1 & 2 & 2 \\ 2 & -1 & 3 & -2 \\ 3 & 2 & -1 & 3 \end{bmatrix} \xrightarrow[-3R_1+R_3]{-2R_1+R_2} \begin{bmatrix} 1 & -1 & 2 & 2 \\ 0 & 1 & -1 & -6 \\ 0 & 5 & -7 & -3 \end{bmatrix}$$

$$\begin{bmatrix} 1 & -1 & 2 & 2 \\ 0 & 1 & -1 & -6 \\ 0 & 5 & -7 & -3 \end{bmatrix} \xrightarrow{-5R_2+R_3} \begin{bmatrix} 1 & -1 & 2 & 2 \\ 0 & 1 & -1 & -6 \\ 0 & 0 & -2 & 27 \end{bmatrix}$$

$$\begin{bmatrix} 1 & -1 & 2 & 2 \\ 0 & 1 & -1 & -6 \\ 0 & 0 & -2 & 27 \end{bmatrix} \xrightarrow{-\frac{1}{2}R_3} \begin{bmatrix} 1 & -1 & 2 & 2 \\ 0 & 1 & -1 & -6 \\ 0 & 0 & 1 & -\frac{27}{2} \end{bmatrix}$$

Using a graphing calculator, we get the following equivalent answer.

$$\begin{bmatrix} 1 & \frac{2}{3} & -\frac{1}{3} & 1 \\ 0 & 1 & -\frac{11}{7} & \frac{12}{7} \\ 0 & 0 & 1 & -\frac{27}{2} \end{bmatrix}$$

7. Using elementary row operations to achieve row echelon form.

$$\begin{bmatrix} 4 & -5 & -1 & 2 \\ 3 & -4 & 1 & -2 \\ 1 & -2 & -1 & 3 \end{bmatrix} \xrightarrow{R_1 \longleftrightarrow R_3} \begin{bmatrix} 1 & -2 & -1 & 3 \\ 3 & -4 & 1 & -2 \\ 4 & -5 & -1 & 2 \end{bmatrix}$$

$$\xrightarrow[-4R_1+R_3]{-3R_1+R_2} \begin{bmatrix} 1 & -2 & -1 & 3 \\ 0 & 2 & 4 & -11 \\ 0 & 3 & 3 & -10 \end{bmatrix}$$

$$\xrightarrow{\frac{1}{2}R_2} \begin{bmatrix} 1 & -2 & -1 & 3 \\ 0 & 1 & 2 & -\frac{11}{2} \\ 0 & 3 & 3 & -10 \end{bmatrix}$$

$$\xrightarrow{-3R_2+R_3} \begin{bmatrix} 1 & -2 & -1 & 3 \\ 0 & 1 & 2 & -\frac{11}{2} \\ 0 & 0 & -3 & -\frac{13}{2} \end{bmatrix}$$

$$\xrightarrow{-\frac{1}{3}R_3} \begin{bmatrix} 1 & -2 & -1 & 3 \\ 0 & 1 & 2 & -\frac{11}{2} \\ 0 & 0 & 1 & -\frac{13}{6} \end{bmatrix}$$

Using a graphing calculator, we get the following equivalent answer.

$$\begin{bmatrix} 1 & -\frac{5}{4} & -\frac{1}{4} & \frac{1}{2} \\ 0 & 1 & 1 & -\frac{10}{3} \\ 0 & 0 & 1 & -\frac{13}{6} \end{bmatrix}$$

9. Using elementary row operations to achieve row echelon form.

$$\begin{bmatrix} 2 & 3 & -1 & 3 \\ 4 & 1 & 1 & -4 \\ 1 & -1 & 1 & 2 \end{bmatrix} \xrightarrow{R_1 \longleftrightarrow R_3} \begin{bmatrix} 1 & -1 & 1 & 2 \\ 4 & 1 & 1 & -4 \\ 2 & 3 & -1 & 3 \end{bmatrix}$$

$$\xrightarrow[-2R_1+R_3]{-4R_1+R_2} \begin{bmatrix} 1 & -1 & 1 & 2 \\ 0 & 5 & -3 & -12 \\ 0 & 5 & -3 & -1 \end{bmatrix}$$

$$\xrightarrow{\frac{1}{5}R_2} \begin{bmatrix} 1 & -1 & 1 & 2 \\ 0 & 1 & -\frac{3}{5} & -\frac{12}{5} \\ 0 & 5 & -3 & -1 \end{bmatrix}$$

$$\xrightarrow{-5R_2+R_3}\begin{bmatrix}1 & -1 & 1 & 2\\0 & 1 & -\frac{3}{5} & -\frac{12}{5}\\0 & 0 & 0 & 11\end{bmatrix}$$

Using a graphing calculator, we get the following equivalent answer.

$$\begin{bmatrix}1 & \frac{1}{4} & \frac{1}{4} & -1\\0 & 1 & -\frac{3}{5} & 2\\0 & 0 & 0 & 1\end{bmatrix}$$

11. Using elementary row operations to achieve row echelon form.

$$\begin{bmatrix}1 & -3 & 4 & 2 & 1\\2 & -3 & 5 & -2 & -1\\-1 & 2 & -3 & 1 & 3\end{bmatrix}$$

$$\xrightarrow[R_1+R_3]{-2R_1+R_2}\begin{bmatrix}1 & -3 & 4 & 2 & 1\\0 & 3 & -3 & -6 & -3\\0 & -1 & 1 & 3 & 4\end{bmatrix}$$

$$\xrightarrow{\frac{1}{3}R_2}\begin{bmatrix}1 & -3 & 4 & 2 & 1\\0 & 1 & -1 & -2 & -1\\0 & -1 & 1 & 3 & 4\end{bmatrix}$$

$$\xrightarrow{R_2+R_3}\begin{bmatrix}1 & -3 & 4 & 2 & 1\\0 & 1 & -1 & -2 & -1\\0 & 0 & 0 & 1 & 3\end{bmatrix}$$

Using a graphing calculator, we get the following equivalent answer.

$$\begin{bmatrix}1 & -\frac{3}{2} & \frac{5}{2} & -1 & -\frac{1}{2}\\0 & 1 & -1 & -2 & -1\\0 & 0 & 0 & 1 & 3\end{bmatrix}$$

13. Solve the system of equations using the Gaussian elimination method.

$$\begin{cases}x+5y=16\\4x+3y=13\end{cases}$$

$$\left[\begin{array}{cc|c}1 & 5 & 16\\4 & 3 & 13\end{array}\right]\xrightarrow{-4R_1+R_2}\left[\begin{array}{cc|c}1 & 5 & 16\\0 & -17 & -51\end{array}\right]$$

$$\xrightarrow{-\frac{1}{17}R_2}\left[\begin{array}{cc|c}1 & 5 & 16\\0 & 1 & 3\end{array}\right]$$

$$\begin{cases}x+5y=16\\y=3\end{cases}$$

$$x+5(3)=16$$
$$x=1$$

The solution is (1, 3).

15. Solve the system of equations using the Gaussian elimination method.

$$\begin{cases}2x-3y=13\\3x-4y=18\end{cases}$$

$$\left[\begin{array}{cc|c}2 & -3 & 13\\3 & -4 & 18\end{array}\right]\xrightarrow{\frac{1}{2}R_1}\left[\begin{array}{cc|c}1 & -\frac{3}{2} & \frac{13}{2}\\3 & -4 & 18\end{array}\right]$$

$$\xrightarrow{-3R_1+R_2}\left[\begin{array}{cc|c}1 & -\frac{3}{2} & \frac{13}{2}\\0 & \frac{1}{2} & -\frac{3}{2}\end{array}\right]\xrightarrow{2R_2}\left[\begin{array}{cc|c}1 & \frac{3}{2} & \frac{13}{2}\\0 & 1 & -3\end{array}\right]$$

$$\begin{cases}2x-3y=13\\y=-3\end{cases}$$

$$2x-3(-3)=13$$
$$2x=4$$
$$x=2$$

The solution is (2, –3).

17. Solve the system of equations using the Gaussian elimination method.

$$\begin{cases}x+2y-2z=-2\\5x+9y-4z=-3\\3x+4y-5z=-3\end{cases}$$

$$\left[\begin{array}{ccc|c}1 & 2 & -2 & -2\\5 & 9 & -4 & -3\\3 & 4 & -5 & -3\end{array}\right]\xrightarrow[-3R_1+R_3]{-5R_1+R_2}\left[\begin{array}{ccc|c}1 & 2 & -2 & -2\\0 & -1 & 6 & 7\\0 & -2 & 1 & 3\end{array}\right]$$

$$\xrightarrow{-1R_2}\left[\begin{array}{ccc|c}1 & 2 & -2 & -2\\0 & 1 & -6 & -7\\0 & -2 & 1 & 3\end{array}\right]$$

$$\xrightarrow{2R_2+R_3}\left[\begin{array}{ccc|c}1 & 2 & -2 & -2\\0 & 1 & -6 & -7\\0 & 0 & -11 & -11\end{array}\right]$$

$$\xrightarrow{-\frac{1}{11}R_3}\left[\begin{array}{ccc|c}1 & 2 & -2 & -2\\0 & 1 & -6 & -7\\0 & 0 & 1 & 1\end{array}\right]$$

$$\begin{cases} x+2y-2z=-2 \\ \quad y-6z=-7 \\ \qquad z=1 \end{cases}$$

$y-6(1)=-7$

$y=-1$

$x+2(-1)-2(1)=-2$

$x=2$

The solution is (2, –1, 1).

19. Solve the system of equations using the Gaussian elimination method.

$$\begin{cases} 3x+7y-7z=-4 \\ x+2y-3z=0 \\ 5x+6y+z=-8 \end{cases}$$

$$\left[\begin{array}{ccc|c} 3 & 7 & -7 & -4 \\ 1 & 2 & -3 & 0 \\ 5 & 6 & 1 & -8 \end{array}\right] \xrightarrow{R_1 \longleftrightarrow R_2} \left[\begin{array}{ccc|c} 1 & 2 & -3 & 0 \\ 3 & 7 & -7 & -4 \\ 5 & 6 & 1 & -8 \end{array}\right]$$

$$\xrightarrow[-5R_1+R_3]{-3R_1+R_2} \left[\begin{array}{ccc|c} 1 & 2 & -3 & 0 \\ 0 & 1 & 2 & -4 \\ 0 & -4 & 16 & -8 \end{array}\right]$$

$$\xrightarrow{4R_2+R_3} \left[\begin{array}{ccc|c} 1 & 2 & -3 & 0 \\ 0 & 1 & 2 & -4 \\ 0 & 0 & 24 & -24 \end{array}\right]$$

$$\xrightarrow{\frac{1}{24}R_3} \left[\begin{array}{ccc|c} 1 & 2 & -3 & 0 \\ 0 & 1 & 2 & -4 \\ 0 & 0 & 1 & -1 \end{array}\right]$$

$$\begin{cases} x+2y-3z=0 \\ \quad y+2z=-4 \\ \qquad z=-1 \end{cases}$$

$y+2(-1)=-4$

$y=-2$

$x+2(-2)-3(-1)=0$

$x=1$

The solution is (1, –2, –1).

21. Solve the system of equations using the Gaussian elimination method.

$$\begin{cases} x+2y-2z=3 \\ 5x+8y-6z=14 \\ 3x+4y-2z=8 \end{cases}$$

$$\left[\begin{array}{ccc|c} 1 & 2 & -2 & 3 \\ 5 & 8 & -6 & 14 \\ 3 & 4 & -2 & 8 \end{array}\right] \xrightarrow[-3R_1+R_3]{-5R_1+R_2} \left[\begin{array}{ccc|c} 1 & 2 & -2 & 3 \\ 0 & -2 & 4 & -1 \\ 0 & -2 & 4 & -1 \end{array}\right]$$

$$\xrightarrow{-\frac{1}{2}R_2} \left[\begin{array}{ccc|c} 1 & 2 & -2 & 3 \\ 0 & 1 & -2 & \frac{1}{2} \\ 0 & -2 & 4 & -1 \end{array}\right] \xrightarrow{2R_2+R_3} \left[\begin{array}{ccc|c} 1 & 2 & -2 & 3 \\ 0 & 1 & -2 & \frac{1}{2} \\ 0 & 0 & 0 & 0 \end{array}\right]$$

$$\begin{cases} x+2y-2z=3 \\ \quad y-2z=\frac{1}{2} \end{cases}$$

$y=2z+\frac{1}{2}$

$x+2\left(2z+\frac{1}{2}\right)-2z=3$

$x=2-2z$

Let z be any real number c.

The solution is $\left(2-2c,\ 2c+\frac{1}{2},\ c\right)$.

23. Solve the system of equations using the Gaussian elimination method.

$$\begin{cases} -x+5y-3z=4 \\ 3x+5y-z=3 \\ x+y=2 \end{cases}$$

$$\left[\begin{array}{ccc|c} -1 & 5 & -3 & 4 \\ 3 & 5 & -1 & 3 \\ 1 & 1 & 0 & 2 \end{array}\right] \xrightarrow{-1R_1} \left[\begin{array}{ccc|c} 1 & -5 & 3 & -4 \\ 3 & 5 & -1 & 3 \\ 1 & 1 & 0 & 2 \end{array}\right]$$

$$\xrightarrow[-1R_1+R_3]{-3R_1+R_2} \left[\begin{array}{ccc|c} 1 & -5 & 3 & -4 \\ 0 & 20 & -10 & 15 \\ 0 & 6 & -3 & 6 \end{array}\right]$$

$$\xrightarrow{\frac{1}{20}R_2} \left[\begin{array}{ccc|c} 1 & -5 & 3 & -4 \\ 0 & 1 & -\frac{1}{2} & \frac{3}{4} \\ 0 & 6 & -3 & 6 \end{array}\right]$$

$$\xrightarrow{-6R_2+R_3} \left[\begin{array}{ccc|c} 1 & -5 & 3 & -4 \\ 0 & 1 & -\frac{1}{2} & \frac{3}{4} \\ 0 & 0 & 0 & \frac{3}{2} \end{array}\right]$$

$$\begin{cases} x-5y+3z=-4 \\ y-\frac{1}{2}z=\frac{3}{4} \\ 0z=\frac{3}{2} \end{cases}$$

Because $0z=\frac{3}{2}$ has no solution, the system of equations has no solution.

25. Solve the system of equations using the Gaussian elimination method.

$$\begin{cases} x-3y+2z=0 \\ 2x-5y-2z=0 \\ 4x-11y+2z=0 \end{cases}$$

$$\left[\begin{array}{ccc|c} 1 & -3 & 2 & 0 \\ 2 & -5 & -2 & 0 \\ 4 & -11 & 2 & 0 \end{array}\right] \xrightarrow[-4R_1+R_3]{-2R_1+R_2} \left[\begin{array}{ccc|c} 1 & -3 & 2 & 0 \\ 0 & 1 & -6 & 0 \\ 0 & 1 & -6 & 0 \end{array}\right]$$

$$\xrightarrow{-1R_2+R_3} \left[\begin{array}{ccc|c} 1 & -3 & 2 & 0 \\ 0 & 1 & -6 & 0 \\ 0 & 0 & 0 & 0 \end{array}\right]$$

$$\begin{cases} x-3y+2z=0 \\ y-6z=0 \end{cases}$$

$y=6z$

$x-3(6z)+2z=0$

$x=16z$

Let z be any real number c. The solution is $(16c, 6c, c)$.

27. Solve the system of equations using the Gaussian elimination method.

$$\begin{cases} 2x+y-3z=4 \\ 3x+2y+z=2 \end{cases}$$

$$\left[\begin{array}{ccc|c} 2 & 1 & -3 & 4 \\ 3 & 2 & 1 & 2 \end{array}\right] \xrightarrow{-R_1+R_2} \left[\begin{array}{ccc|c} 2 & 1 & -3 & 4 \\ 1 & 1 & 4 & -2 \end{array}\right]$$

$$\xrightarrow{R_1 \longleftrightarrow R_2} \left[\begin{array}{ccc|c} 1 & 1 & 4 & -2 \\ 2 & 1 & -3 & 4 \end{array}\right]$$

$$\xrightarrow{-2R_1+R_2} \left[\begin{array}{ccc|c} 1 & 1 & 4 & -2 \\ 0 & -1 & -11 & 8 \end{array}\right]$$

$$\xrightarrow{-1R_2} \left[\begin{array}{ccc|c} 1 & 1 & 4 & -2 \\ 0 & 1 & 11 & -8 \end{array}\right]$$

$$\begin{cases} x+y+4z=-2 \\ y+11z=-8 \end{cases}$$

$y=-11z-8$

$x+(-11z-8)+4z=-2$

$x=7z+6$

Let z be any real number c. The solution is $(7c+6, -11c-8, c)$.

29. Solve the system of equations using the Gaussian elimination method.

$$\begin{cases} 2x+2y-4z=4 \\ 2x+3y-5z=4 \\ 4x+5y-9z=8 \end{cases}$$

$$\left[\begin{array}{ccc|c} 2 & 2 & -4 & 4 \\ 2 & 3 & -5 & 4 \\ 4 & 5 & -9 & 8 \end{array}\right] \xrightarrow{\frac{1}{2}R_1} \left[\begin{array}{ccc|c} 1 & 1 & -2 & 2 \\ 2 & 3 & -5 & 4 \\ 4 & 5 & -9 & 8 \end{array}\right]$$

$$\xrightarrow[-4R_1+R_3]{-2R_1+R_2} \left[\begin{array}{ccc|c} 1 & 1 & -2 & 2 \\ 0 & 1 & -1 & 0 \\ 0 & 1 & -1 & 0 \end{array}\right]$$

$$\xrightarrow{-1R_2+R_3} \left[\begin{array}{ccc|c} 1 & 1 & -2 & 2 \\ 0 & 1 & -1 & 0 \\ 0 & 0 & 0 & 0 \end{array}\right]$$

$$\begin{cases} x+y-2z=2 \\ y-z=0 \end{cases}$$

$y=z$

$x+z-2z=2$

$x=z+2$

Let z be any real number c. The solution is $(c+2, c, c)$.

31. Solve the system of equations using the Gaussian elimination method.

$$\begin{cases} x+3y+4z=11 \\ 2x+3y+2z=7 \\ 4x+9y+10z=20 \\ 3x-2y+z=1 \end{cases}$$

$$\left[\begin{array}{ccc|c} 1 & 3 & 4 & 11 \\ 2 & 3 & 2 & 7 \\ 4 & 9 & 10 & 20 \\ 3 & -2 & 1 & 1 \end{array}\right] \xrightarrow[\substack{-4R_1+R_3 \\ -3R_1+R_4}]{-2R_1+R_2} \left[\begin{array}{ccc|c} 1 & 3 & 4 & 11 \\ 0 & -3 & -6 & -15 \\ 0 & -3 & -6 & -24 \\ 0 & -11 & -11 & -32 \end{array}\right]$$

$$\xrightarrow{-\frac{1}{3}R_2}\left[\begin{array}{ccc|c}1 & 3 & 4 & 11 \\ 0 & 1 & 2 & 5 \\ 0 & -3 & -6 & -24 \\ 0 & -11 & -11 & -32\end{array}\right]$$

$$\xrightarrow{R_3 \longleftrightarrow R_4}\left[\begin{array}{ccc|c}1 & 3 & 4 & 11 \\ 0 & 1 & 2 & 5 \\ 0 & -11 & -11 & -32 \\ 0 & 0 & 0 & -9\end{array}\right]$$

$$\xrightarrow{-\frac{1}{11}R_3}\left[\begin{array}{ccc|c}1 & 3 & 4 & 11 \\ 0 & 1 & 2 & 5 \\ 0 & 1 & 1 & \frac{32}{11} \\ 0 & 0 & 0 & -9\end{array}\right]$$

$$\xrightarrow{-R_2+R_3}\left[\begin{array}{ccc|c}1 & 3 & 4 & 11 \\ 0 & 1 & 2 & 5 \\ 0 & 0 & -1 & -\frac{23}{11} \\ 0 & 0 & 0 & -9\end{array}\right]$$

$$\xrightarrow{-1R_3}\left[\begin{array}{ccc|c}1 & 3 & 4 & 11 \\ 0 & 1 & 2 & 5 \\ 0 & 0 & 1 & \frac{23}{11} \\ 0 & 0 & 0 & -9\end{array}\right]$$

$$\begin{cases} x+3y+4z=11 \\ y+2z=5 \\ z=\frac{23}{11} \\ 0z=-9 \end{cases}$$

Because $0z=-9$ has no solutions, the system of equations has no solution.

33. Solve the system of equations using the Gaussian elimination method.

$$\begin{cases} t+2u-3v+w=-7 \\ 3t+5u-8v+5w=-8 \\ 2t+3u-7v+3w=-11 \\ 4t+8u-10v+7w=-10 \end{cases}$$

$$\left[\begin{array}{cccc|c}1 & 2 & -3 & 1 & -7 \\ 3 & 5 & -8 & 5 & -8 \\ 2 & 3 & -7 & 3 & -11 \\ 4 & 8 & -10 & 7 & -10\end{array}\right]$$

$$\xrightarrow[\substack{-2R_1+R_3 \\ -4R_1+R_4}]{-3R_1+R_2}\left[\begin{array}{cccc|c}1 & 2 & -3 & 1 & -7 \\ 0 & -1 & 1 & 2 & 13 \\ 0 & -1 & -1 & 1 & 3 \\ 0 & 0 & 2 & 3 & 18\end{array}\right]$$

$$\xrightarrow{-1R_2}\left[\begin{array}{cccc|c}1 & 2 & -3 & 1 & -7 \\ 0 & 1 & -1 & -2 & -13 \\ 0 & -1 & -1 & 1 & 3 \\ 0 & 0 & 2 & 3 & 18\end{array}\right]$$

$$\xrightarrow{R_2+R_3}\left[\begin{array}{cccc|c}1 & 2 & -3 & 1 & -7 \\ 0 & 1 & -1 & -2 & -13 \\ 0 & 0 & -2 & -1 & -10 \\ 0 & 0 & 2 & 3 & 18\end{array}\right]$$

$$\xrightarrow{-\frac{1}{2}R_3}\left[\begin{array}{cccc|c}1 & 2 & -3 & 1 & -7 \\ 0 & 1 & -1 & -2 & -13 \\ 0 & 0 & 1 & \frac{1}{2} & 5 \\ 0 & 0 & 2 & 3 & 18\end{array}\right]$$

$$\xrightarrow{-2R_3+R_4}\left[\begin{array}{cccc|c}1 & 2 & -3 & 1 & -7 \\ 0 & 1 & -1 & -2 & -13 \\ 0 & 0 & 1 & \frac{1}{2} & 5 \\ 0 & 0 & 0 & 2 & 8\end{array}\right]$$

$$\begin{cases} t+2u-3v+w=-7 \\ u-v-2w=-13 \\ v+\frac{1}{2}w=5 \\ 2w=8 \end{cases}$$

$$v+\frac{1}{2}(4)=5$$
$$v=3$$
$$u-3-2(4)=-13$$
$$u=-2$$
$$t+2(-2)-3(3)+4=-7$$
$$t=2$$

The solution is (2, –2, 3, 4).

35. Solve the system of equations using the Gaussian elimination method.

$$\begin{cases} 2t-u+3v+2w=2 \\ t-u+2v+w=2 \\ 3t-2v-3w=13 \\ 2t+2u-2w=6 \end{cases}$$

$$\left[\begin{array}{cccc|c} 2 & -1 & 3 & 2 & 2 \\ 1 & -1 & 2 & 1 & 2 \\ 3 & 0 & -2 & -3 & 13 \\ 2 & 2 & 0 & -2 & 6 \end{array}\right] \xrightarrow{\frac{1}{2}R_1} \left[\begin{array}{cccc|c} 1 & -\frac{1}{2} & \frac{3}{2} & 1 & 1 \\ 1 & -1 & 2 & 1 & 2 \\ 3 & 0 & -2 & -3 & 13 \\ 2 & 2 & 0 & -2 & 6 \end{array}\right]$$

$$\xrightarrow[\substack{-3R_1+R_3 \\ -2R_1+R_4}]{-1R_1+R_2} \left[\begin{array}{cccc|c} 1 & -\frac{1}{2} & \frac{3}{2} & 1 & 1 \\ 0 & -\frac{1}{2} & \frac{1}{2} & 0 & 1 \\ 0 & \frac{3}{2} & -\frac{13}{2} & -6 & 10 \\ 0 & 3 & -3 & -4 & 4 \end{array}\right]$$

$$\xrightarrow{-2R_2} \left[\begin{array}{cccc|c} 1 & -\frac{1}{2} & \frac{3}{2} & 1 & 1 \\ 0 & 1 & -1 & 0 & -2 \\ 0 & \frac{3}{2} & -\frac{13}{2} & -6 & 10 \\ 0 & 3 & -3 & -4 & 4 \end{array}\right]$$

$$\xrightarrow[-3R_2+R_4]{-\frac{3}{2}R_2+R_3} \left[\begin{array}{cccc|c} 1 & -\frac{1}{2} & \frac{3}{2} & 1 & 1 \\ 0 & 1 & -1 & 0 & -2 \\ 0 & 0 & -5 & -6 & 13 \\ 0 & 0 & 0 & -4 & 10 \end{array}\right]$$

$$\xrightarrow{-\frac{1}{5}R_3} \left[\begin{array}{cccc|c} 1 & -\frac{1}{2} & \frac{3}{2} & 1 & 1 \\ 0 & 1 & -1 & 0 & -2 \\ 0 & 0 & 1 & \frac{6}{5} & -\frac{13}{5} \\ 0 & 0 & 0 & -4 & 10 \end{array}\right]$$

$$\xrightarrow{-\frac{1}{4}R_4} \left[\begin{array}{cccc|c} 1 & -\frac{1}{2} & \frac{3}{2} & 1 & 1 \\ 0 & 1 & -1 & 0 & -2 \\ 0 & 0 & 1 & \frac{6}{5} & -\frac{13}{5} \\ 0 & 0 & 0 & 1 & -\frac{5}{2} \end{array}\right]$$

$$\begin{cases} t-\frac{1}{2}u+\frac{3}{2}v+w=1 \\ u-v=-2 \\ v+\frac{6}{5}w=-\frac{13}{5} \\ w=-\frac{5}{2} \end{cases}$$

$$v+\frac{6}{5}\left(-\frac{5}{2}\right)=-\frac{13}{5}$$
$$v=\frac{2}{5}$$

$$u-\frac{2}{5}=-2$$
$$u=-\frac{8}{5}$$

$$t-\frac{1}{2}\left(-\frac{8}{5}\right)+\frac{3}{2}\left(\frac{2}{5}\right)-\frac{5}{2}=1$$
$$t=\frac{21}{10}$$

The solution is $\left(\frac{21}{10}, -\frac{8}{5}, \frac{2}{5}, -\frac{5}{2}\right)$.

37. Solve the system of equations using the Gaussian elimination method.

$$\begin{cases} 3t+10u+7v-6w=7 \\ 2t+8u+6v-5w=5 \\ t+4u+2v-3w=2 \\ 4t+14u+9v-8w=8 \end{cases}$$

$$\left[\begin{array}{cccc|c} 3 & 10 & 7 & -6 & 7 \\ 2 & 8 & 6 & -5 & 5 \\ 1 & 4 & 2 & -3 & 2 \\ 4 & 14 & 9 & -8 & 8 \end{array}\right] \xrightarrow{R_1 \longleftrightarrow R_3} \left[\begin{array}{cccc|c} 1 & 4 & 2 & -3 & 2 \\ 2 & 8 & 6 & -5 & 5 \\ 3 & 10 & 7 & -6 & 7 \\ 4 & 14 & 9 & -8 & 8 \end{array}\right]$$

$$\xrightarrow[\substack{-3R_1+R_3 \\ -4R_1+R_4}]{-2R_1+R_2} \left[\begin{array}{cccc|c} 1 & 4 & 2 & -3 & 2 \\ 0 & 0 & 2 & 1 & 1 \\ 0 & -2 & 1 & 3 & 1 \\ 0 & -2 & 1 & 4 & 0 \end{array}\right]$$

$$\xrightarrow{R_2 \longleftrightarrow R_3} \left[\begin{array}{cccc|c} 1 & 4 & 2 & -3 & 2 \\ 0 & -2 & 1 & 3 & 1 \\ 0 & 0 & 2 & 1 & 1 \\ 0 & -2 & 1 & 4 & 0 \end{array}\right]$$

$$\xrightarrow{-\frac{1}{2}R_2} \left[\begin{array}{cccc|c} 1 & 4 & 2 & -3 & 2 \\ 0 & 1 & -\frac{1}{2} & -\frac{3}{2} & -\frac{1}{2} \\ 0 & 0 & 2 & 1 & 1 \\ 0 & -2 & 1 & 4 & 0 \end{array}\right]$$

$$\xrightarrow{2R_2+R_4} \left[\begin{array}{cccc|c} 1 & 4 & 2 & -3 & 2 \\ 0 & 1 & -\frac{1}{2} & -\frac{3}{2} & -\frac{1}{2} \\ 0 & 0 & 2 & 1 & 1 \\ 0 & 0 & 0 & 1 & -1 \end{array}\right]$$

$$\begin{cases} t+4u+2v-3w=2 \\ u-\frac{1}{2}v-\frac{3}{2}w=-\frac{1}{2} \\ 2v+w=1 \\ w=-1 \end{cases}$$

$$2v+(-1)=1$$
$$v=1$$

$$u-\frac{1}{2}(1)-\frac{3}{2}(-1)=-\frac{1}{2}$$
$$u=-\frac{3}{2}$$

$$t+4\left(-\frac{3}{2}\right)+2(1)-3(-1)=2$$
$$t=3$$

The solution is $\left(3,\ -\frac{3}{2},\ 1,\ -1\right)$.

39. Solve the system of equations using the Gaussian elimination method.

$$\begin{cases} t-u+2v-3w=9 \\ 4t\ \ +11v-10w=46 \\ 3t-u+8v-6w=27 \end{cases}$$

$$\left[\begin{array}{cccc|c} 1 & -1 & 2 & -3 & 9 \\ 4 & 0 & 11 & -10 & 46 \\ 3 & -1 & 8 & -6 & 27 \end{array}\right]$$

$$\xrightarrow[-3R_1+R_3]{-4R_1+R_2}\left[\begin{array}{cccc|c} 1 & -1 & 2 & -3 & 9 \\ 0 & 4 & 3 & 2 & 10 \\ 0 & 2 & 2 & 3 & 0 \end{array}\right]$$

$$\xrightarrow{R_2\longleftrightarrow R_3}\left[\begin{array}{cccc|c} 1 & -1 & 2 & -3 & 9 \\ 0 & 2 & 2 & 3 & 0 \\ 0 & 4 & 3 & 2 & 10 \end{array}\right]$$

$$\xrightarrow{-2R_2+R_3}\left[\begin{array}{cccc|c} 1 & -1 & 2 & -3 & 9 \\ 0 & 2 & 2 & 3 & 0 \\ 0 & 0 & -1 & -4 & 10 \end{array}\right]$$

$$\begin{cases} t-u+\ 2v-3w=9 \\ 2u+\ 2v\ +3w\ =0 \\ -v-\ 4w=10 \end{cases}$$

$$v=-4w-10$$

$$2u+2(-4w-10)+3w=0$$
$$u=\frac{5}{2}w+10$$

$$t-\left(\frac{5}{2}w+10\right)+2(-4w-10)-3w=9$$
$$t=\frac{27}{2}w+39$$

Let w be any real number c. The solution is

$\left(\frac{27}{2}c+39,\ \frac{5}{2}c+10,\ -4c-10,\ c\right)$.

41. Solve the system of equations using the Gaussian elimination method.

$$\begin{cases} 3t-4u+v=2 \\ t+u-2v+3w=1 \end{cases}$$

$$\left[\begin{array}{cccc|c} 3 & -4 & 1 & 0 & 2 \\ 1 & 1 & -2 & 3 & 1 \end{array}\right]\xrightarrow{\frac{1}{3}R_1}\left[\begin{array}{cccc|c} 1 & -\frac{4}{3} & \frac{1}{3} & 0 & \frac{2}{3} \\ 1 & 1 & -2 & 3 & 1 \end{array}\right]$$

$$\xrightarrow{-R_1+R_2}\left[\begin{array}{cccc|c} 1 & -\frac{4}{3} & \frac{1}{3} & 0 & \frac{2}{3} \\ 0 & \frac{7}{3} & -\frac{7}{3} & 3 & \frac{1}{3} \end{array}\right]$$

$$\xrightarrow{\frac{3}{7}R_2}\left[\begin{array}{cccc|c} 1 & -\frac{4}{3} & \frac{1}{3} & 0 & \frac{2}{3} \\ 0 & 1 & -1 & \frac{9}{7} & \frac{1}{7} \end{array}\right]$$

$$\begin{cases} t-\frac{4}{3}u+\frac{1}{3}v\ \ \ =\frac{2}{3} \\ u-\ \ v+\frac{9}{7}w=\frac{1}{7} \end{cases}$$

$$u=v-\frac{9}{7}w+\frac{1}{7}$$

$$t-\frac{4}{3}\left(v-\frac{9}{7}w+\frac{1}{7}\right)+\frac{1}{3}v=\frac{2}{3}$$
$$t=v-\frac{12}{7}w+\frac{6}{7}$$

Let v be any real number c_1 and w be any real number c_2. The solution is

$\left(c_1-\frac{12}{7}c_2+\frac{6}{7},\ c_1-\frac{9}{7}c_2+\frac{1}{7},\ c_1,\ c_2\right)$.

43. Find the polynomial.

Because there are two points, the degree of the interpolating polynomial is at most 1. The form of the polynomial is $p(x)=a_1x+a_0$

Use this polynomial and the given points to find the system of equations.

$$p(-2)=a_1(-2)+a_0=-7$$
$$p(1)=a_1(1)+a_0=-1$$

The system of equations and the associated augmented matrix are $\begin{cases} -2a_1+a_0=-7 \\ a_1+a_0=-1 \end{cases}$ $\left[\begin{array}{cc|c} -2 & 1 & -7 \\ 1 & 1 & -1 \end{array}\right]$

The ref (**r**ow **e**chelon **f**orm) feature of a graphing calculator can be used to rewrite the augmented matrix in echelon form.

Consider using the function of your calculator that converts a decimal to a fraction.

The augmented matrix in echelon form and resulting system of equations are

$$\begin{bmatrix} 1 & -\frac{1}{2} & \vert & \frac{7}{2} \\ 0 & 1 & \vert & -3 \end{bmatrix} \qquad \begin{cases} a_1 - \frac{1}{2}a_0 = \frac{7}{2} \\ a_0 = -3 \end{cases}$$

Solving by back substitution yields

$a_0 = -3$ and $a_1 = 2$.

The interpolating polynomial is $p(x) = 2x - 3$.

45. Find the polynomial.

Because there are three points, the degree of the interpolating polynomial is at most 2.

The form of the polynomial is

$p(x) = a_2x^2 + a_1x + a_0$.

Use this polynomial and the given points to find the system of equations.

$$\begin{aligned} p(-1) &= a_2(-1)^2 + a_1(-1) + a_0 = 6 \\ p(1) &= a_2(1)^2 + a_1(1) + a_0 = 2 \\ p(2) &= a_2(2)^2 + a_1(2) + a_0 = 3 \end{aligned}$$

The system of equations and the associated augmented matrix are

$$\begin{cases} a_2 - a_1 + a_0 = 6 \\ a_2 + a_1 + a_0 = 2 \\ 4a_2 + 2a_1 + a_0 = 3 \end{cases} \qquad \begin{bmatrix} 1 & -1 & 1 & \vert & 6 \\ 1 & 1 & 1 & \vert & 2 \\ 4 & 2 & 1 & \vert & 3 \end{bmatrix}$$

The ref (**r**ow **e**chelon **f**orm) feature of a graphing calculator can be used to rewrite the augmented matrix in echelon form.

Consider using the function of your calculator that converts a decimal to a fraction.

The augmented matrix in echelon form and resulting system of equations are

$$\begin{bmatrix} 1 & \frac{1}{2} & \frac{1}{4} & \vert & \frac{3}{4} \\ 0 & 1 & -\frac{1}{2} & \vert & -\frac{7}{2} \\ 0 & 0 & 1 & \vert & 3 \end{bmatrix} \qquad \begin{cases} a_2 + \frac{1}{2}a_1 + \frac{1}{4}a_0 = \frac{3}{4} \\ a_1 - \frac{1}{2}a_0 = -\frac{7}{2} \\ a_0 = 3 \end{cases}$$

Solving by back substitution yields

$a_0 = 3$, $a_1 = -2$, and $a_2 = 1$.

The interpolating polynomial is $p(x) = x^2 - 2x + 3$.

47. Find the polynomial.

Because there are four points, the degree of the interpolating polynomial is at most 3.

The form of the polynomial is

$p(x) = a_3x^3 + a_2x^2 + a_1x + a_0$.

Use this polynomial and the given points to find the system of equations.

$$\begin{aligned} p(-2) &= a_3(-2)^3 + a_2(-2)^2 + a_1(-2) + a_0 = -12 \\ p(0) &= a_3(0)^3 + a_2(0)^2 + a_1(0) + a_0 = 2 \\ p(1) &= a_3(1)^3 + a_2(1)^2 + a_1(1) + a_0 = 0 \\ p(3) &= a_3(3)^3 + a_2(3)^2 + a_1(3) + a_0 = 8 \end{aligned}$$

The system of equations and the associated augmented matrix are

$$\begin{cases} -8a_3 + 4a_2 - 2a_1 + a_0 = -12 \\ a_0 = 2 \\ a_3 + a_2 + a_1 + a_0 = 0 \\ 27a_3 + 9a_2 + 3a_1 + a_0 = 8 \end{cases} \qquad \begin{bmatrix} -8 & 4 & -2 & 1 & \vert & -12 \\ 0 & 0 & 0 & 1 & \vert & 2 \\ 1 & 1 & 1 & 1 & \vert & 0 \\ 27 & 9 & 3 & 1 & \vert & 8 \end{bmatrix}$$

The ref (**r**ow **e**chelon **f**orm) feature of a graphing calculator can be used to rewrite the augmented matrix in echelon form.

Consider using the function of your calculator that converts a decimal to a fraction.

The augmented matrix in echelon form and resulting system of equations are

$$\begin{bmatrix} 1 & \frac{1}{3} & \frac{1}{9} & \frac{1}{27} & \vert & \frac{8}{27} \\ 0 & 1 & -\frac{1}{6} & \frac{7}{36} & \vert & -\frac{13}{9} \\ 0 & 0 & 1 & \frac{5}{6} & \vert & \frac{2}{3} \\ 0 & 0 & 0 & 1 & \vert & 2 \end{bmatrix}$$

$$\begin{cases} a_3 + \frac{1}{3}a_2 + \frac{1}{9}a_1 + \frac{1}{27}a_0 = \frac{8}{27} \\ a_2 - \frac{1}{6}a_1 + \frac{7}{36}a_0 = -\frac{13}{9} \\ a_1 + \frac{5}{6}a_0 = \frac{2}{3} \\ a_0 = 2 \end{cases}$$

Solving by back substitution yields

$a_0 = 2$, $a_1 = -1$, $a_2 = -2$ and $a_3 = 1$.

The interpolating polynomial is

$p(x) = x^3 - 2x^2 - x + 2$.

49. Find the polynomial.

Because there are three points, the degree of the interpolating polynomial is at most 2.

The form of the polynomial is

$p(x) = a_2x^2 + a_1x + a_0$.

Use this polynomial and the given points to find the system of equations.

$$p(-1) = a_2(-1)^2 + a_1(-1) + a_0 = 3$$
$$p(1) = a_2(1)^2 + a_1(1) + a_0 = 7$$
$$p(2) = a_2(2)^2 + a_1(2) + a_0 = 9$$

The system of equations and the associated augmented matrix are

$$\begin{cases} a_2 - a_1 + a_0 = 3 \\ a_2 + a_1 + a_0 = 7 \\ 4a_2 + 2a_1 + a_0 = 9 \end{cases} \qquad \left[\begin{array}{ccc|c} 1 & -1 & 1 & 3 \\ 1 & 1 & 1 & 7 \\ 4 & 2 & 1 & 9 \end{array}\right]$$

The ref (**r**ow **e**chelon **f**orm) feature of a graphing calculator can be used to rewrite the augmented matrix in echelon form.

Consider using the function of your calculator that converts a decimal to a fraction.

The augmented matrix in echelon form and resulting system of equations are

$$\left[\begin{array}{ccc|c} 1 & \frac{1}{2} & \frac{1}{4} & \frac{9}{4} \\ 0 & 1 & -\frac{1}{2} & -\frac{1}{2} \\ 0 & 0 & 1 & 5 \end{array}\right] \qquad \begin{cases} a_2 + \frac{1}{2}a_1 + \frac{1}{4}a_0 = \frac{9}{4} \\ a_1 - \frac{1}{2}a_0 = -\frac{1}{2} \\ a_0 = 5 \end{cases}$$

Solving by back substitution yields

$a_0 = 5$, $a_1 = 2$, and $a_2 = 0$.

The interpolating polynomial is $p(x) = 2x + 5$.

51. Find the equation of the plane.

The form of the polynomial is: $z = ax + by + c$.

Use this polynomial and the given points to find the system of equations.

$$-4 = a(-1) + b(0) + c$$
$$5 = a(2) + b(1) + c$$
$$-1 = a(-1) + b(1) + c$$

The system of equations and the associated augmented matrix are

$$\begin{cases} -a + c = -4 \\ 2a + b + c = 5 \\ -a + b + c = -1 \end{cases} \qquad \left[\begin{array}{ccc|c} -1 & 0 & 1 & -4 \\ 2 & 1 & 1 & 5 \\ -1 & 1 & 1 & -1 \end{array}\right]$$

The ref (**r**ow **e**chelon **f**orm) feature of a graphing calculator can be used to rewrite the augmented matrix in echelon form.

Consider using the function of your calculator that converts a decimal to a fraction.

The augmented matrix in echelon form and resulting system of equations are

$$\left[\begin{array}{ccc|c} 1 & \frac{1}{2} & \frac{1}{2} & \frac{5}{2} \\ 0 & 1 & 1 & 1 \\ 0 & 0 & 1 & -2 \end{array}\right] \qquad \begin{cases} a + \frac{1}{2}b + \frac{1}{2}c = \frac{5}{2} \\ b + c = 1 \\ c = -2 \end{cases}$$

Solving by back substitution yields

$a = 2$, $b = 3$, and $c = -2$.

The interpolating polynomial is $z = 2x + 3y - 2$.

53. Find the equation of the circle.

The form of the polynomial is: $x^2 + y^2 + ax + by = c$

Use this polynomial and the given points to find the system of equations.

$$(2)^2 + (6)^2 + a(2) + b(6) = c$$
$$(-4)^2 + (-2)^2 + a(-4) + b(-2) = c$$
$$(3)^2 + (-1)^2 + a(3) + b(-1) = c$$

The system of equations and the associated augmented matrix are

$$\begin{cases} 2a + 6b - c = -40 \\ -4a - 2b - c = -20 \\ 3a - b - c = -10 \end{cases} \qquad \left[\begin{array}{ccc|c} 2 & 6 & -1 & -40 \\ -4 & -2 & -1 & -20 \\ 3 & -1 & -1 & -10 \end{array}\right]$$

The ref (**r**ow **e**chelon **f**orm) feature of a graphing calculator can be used to rewrite the augmented matrix in echelon form.

Consider using the function of your calculator that converts a decimal to a fraction.

The augmented matrix in echelon form and resulting

system of equations are

$$\begin{bmatrix} 1 & \frac{1}{2} & \frac{1}{4} & 5 \\ 0 & 1 & -\frac{3}{10} & -10 \\ 0 & 0 & 1 & 20 \end{bmatrix} \quad \begin{cases} a-\frac{1}{2}b+\frac{1}{4}c=5 \\ b-\frac{3}{10}c=-10 \\ c=20 \end{cases}$$

Solving by back substitution yields

$a=2,\ b=-4,$ and $c=20$.

The interpolating polynomial is

$x^2+y^2+2x-4y=20$.

55. Use a graphing calculator to solve.

$$\left[\begin{array}{ccccc|c} 1 & 2 & -1 & 2 & 3 & 11 \\ 1 & -1 & 2 & -1 & 2 & 0 \\ 2 & 1 & -1 & 2 & -1 & 4 \\ 3 & 2 & -1 & 1 & -2 & 2 \\ 2 & 1 & -1 & -2 & 1 & 4 \end{array}\right]$$

$$\longrightarrow \left[\begin{array}{ccccc|c} 1 & 2 & -1 & 2 & 3 & 11 \\ 0 & 1 & -1 & 1 & \frac{1}{3} & \frac{11}{3} \\ 0 & 0 & 1 & -\frac{1}{2} & 3 & \frac{7}{2} \\ 0 & 0 & 0 & 1 & \frac{11}{6} & \frac{14}{3} \\ 0 & 0 & 0 & 0 & 1 & 2 \end{array}\right]$$

$$\begin{cases} x_1+2x_2-x_3+2x_4+3x_5=11 \\ x_2-x_3+x_4+\frac{1}{3}x_5=\frac{11}{3} \\ x_3-\frac{1}{2}x_4+3x_5=\frac{7}{2} \\ x_4+\frac{11}{6}x_5=\frac{14}{3} \\ x_5=2 \end{cases}$$

The solution is (1, 0, –2, 1, 2).

57. Use a graphing calculator to solve.

$$\left[\begin{array}{ccccc|c} 1 & 2 & -3 & -1 & 2 & -10 \\ -1 & -3 & 1 & 1 & -1 & 4 \\ 2 & 3 & -5 & 2 & 3 & -20 \\ 3 & 4 & -7 & 3 & -2 & -16 \\ 2 & 1 & -6 & 4 & -3 & -12 \end{array}\right]$$

$$\longrightarrow \left[\begin{array}{ccccc|c} 1 & 2 & -3 & -1 & 2 & -10 \\ 0 & 1 & 2 & 0 & -1 & 6 \\ 0 & 0 & 1 & \frac{4}{3} & -\frac{2}{3} & 2 \\ 0 & 0 & 0 & 1 & 3 & -7 \\ 0 & 0 & 0 & 0 & 0 & 0 \end{array}\right]$$

$$\begin{cases} x_1+2x_2-3x_3-x_4+2x_5=-10 \\ x_2+2x_3-x_5=6 \\ x_3+\frac{4}{3}x_4-\frac{2}{3}x_5=2 \\ x_4+3x_5=-7 \end{cases}$$

$x_4=-3x_5-7$

$x_3+\frac{4}{3}(-3x_5-7)-\frac{2}{3}x_5=2$

$x_3=\frac{14}{3}x_5+\frac{34}{3}$

$x_2+2\left(\frac{14}{3}x_5+\frac{34}{3}\right)-x_5=6$

$x_2=-\frac{25}{3}x_5-\frac{50}{3}$

$x_1+2\left(-\frac{25}{3}x_5-\frac{50}{3}\right)-3\left(\frac{14}{3}x_5+\frac{34}{3}\right)-(-3x_5-7)+2x_5=-10$

$x_1=\frac{77}{3}x_5+\frac{151}{3}$

Let x_5 be any real number c. The solution is

$$\left(\frac{77c+151}{3},\ \frac{-25c-50}{3},\ \frac{14c+34}{3},\ -3c-7,\ c\right).$$

59. Find the values for a for a unique solution.

$$\left[\begin{array}{ccc|c} 1 & 3 & -a^2 & a^2 \\ 3 & 4 & 2 & 3 \\ 2 & 3 & a & 2 \end{array}\right]$$

$$\xrightarrow[-2R_1+R_3]{-3R_1+R_2} \left[\begin{array}{ccc|c} 1 & 3 & -a^2 & a^2 \\ 0 & -5 & 3a^2+2 & -3a^2+3 \\ 0 & -3 & 2a^2+a & -2a^2+2 \end{array}\right]$$

$$\xrightarrow[5R_3]{-3R_2} \left[\begin{array}{ccc|c} 1 & 3 & -a^2 & a^2 \\ 0 & 15 & -9a^2-6 & 9a^2-9 \\ 0 & -15 & 10a^2+5a & -10a^2+10 \end{array}\right]$$

$$\xrightarrow{R_2+R_3} \left[\begin{array}{ccc|c} 1 & 3 & -a^2 & a^2 \\ 0 & 15 & -9a^2-6 & 9a^2-9 \\ 0 & 0 & a^2+5a-6 & -a^2+1 \end{array}\right]$$

$$\begin{cases} x+3y-a^2z=a^2 \\ 15y-(9a^2+6)z=9a^2-9 \\ \left(a^2+5a-6\right)z=-a^2+1 \end{cases}$$

For the system of equations to have a unique solution, a^2+5a-6 cannot be zero. Thus $a^2+5a-6\neq 0$, or $a\neq 1$ and $a\neq -6$.

The system of equations has a unique solution for all values of a except 1 and –6.

61. Find the values for a for no solution.

See the solution to exercise 59. For the system of equations to have no solution, a^2+5a-6 must be zero and $-a^2+1$ must not equal zero. Thus

$$a^2+5a-6=0 \quad \text{and} \quad -a^2+1\neq 0$$

$$(a+6)(a-1)=0 \qquad a^2\neq 1$$

$$a=-6 \text{ or } a=1 \qquad a\neq 1 \text{ or } a\neq -1$$

The system of equations will have no solution when $a=-6$.

Prepare for Section 7.2

P1. State the additive identity for real numbers.

0

P3. State the multiplicative identity for real numbers.

1

P5. State the order of the matrix.

3×1

Section 7.2 Exercises

1. Find $A+B$, $A-B$, $2B$, and $2A-3B$.

a. $A+B=\begin{bmatrix}2 & -1\\3 & 3\end{bmatrix}+\begin{bmatrix}-1 & 3\\2 & 1\end{bmatrix}=\begin{bmatrix}1 & 2\\5 & 4\end{bmatrix}$

b. $A-B=\begin{bmatrix}2 & -1\\3 & 3\end{bmatrix}-\begin{bmatrix}-1 & 3\\2 & 1\end{bmatrix}=\begin{bmatrix}3 & -4\\1 & 2\end{bmatrix}$

c. $2B=2\begin{bmatrix}-1 & 3\\2 & 1\end{bmatrix}=\begin{bmatrix}-2 & 6\\4 & 2\end{bmatrix}$

d. $2A-3B=2\begin{bmatrix}2 & -1\\3 & 3\end{bmatrix}-3\begin{bmatrix}-1 & 3\\2 & 1\end{bmatrix}=\begin{bmatrix}4 & -2\\6 & 6\end{bmatrix}-\begin{bmatrix}-3 & 9\\6 & 3\end{bmatrix}$

$=\begin{bmatrix}7 & -11\\0 & 3\end{bmatrix}$

3. Find $A+B$, $A-B$, $2B$, and $2A-3B$.

a. $A+B=\begin{bmatrix}0 & -1 & 3\\1 & 0 & -2\end{bmatrix}+\begin{bmatrix}-3 & 1 & 2\\2 & 5 & -3\end{bmatrix}=\begin{bmatrix}-3 & 0 & 5\\3 & 5 & -5\end{bmatrix}$

b. $A-B=\begin{bmatrix}0 & -1 & 3\\1 & 0 & -2\end{bmatrix}-\begin{bmatrix}-3 & 1 & 2\\2 & 5 & -3\end{bmatrix}=\begin{bmatrix}3 & -2 & 1\\-1 & -5 & 1\end{bmatrix}$

c. $2B=2\begin{bmatrix}-3 & 1 & 2\\2 & 5 & -3\end{bmatrix}=\begin{bmatrix}-6 & 2 & 4\\4 & 10 & -6\end{bmatrix}$

d. $2A-3B=2\begin{bmatrix}0 & -1 & 3\\1 & 0 & -2\end{bmatrix}-3\begin{bmatrix}-3 & 1 & 2\\2 & 5 & -3\end{bmatrix}$

$=\begin{bmatrix}0 & -2 & 6\\2 & 0 & -4\end{bmatrix}-\begin{bmatrix}-9 & 3 & 6\\6 & 15 & -9\end{bmatrix}$

$=\begin{bmatrix}9 & -5 & 0\\-4 & -15 & 5\end{bmatrix}$

5. Find $A+B$, $A-B$, $2B$, and $2A-3B$.

a. $A+B=\begin{bmatrix}-3 & 4\\2 & -3\\-1 & 0\end{bmatrix}+\begin{bmatrix}4 & 1\\1 & -2\\3 & -4\end{bmatrix}=\begin{bmatrix}1 & 5\\3 & -5\\2 & -4\end{bmatrix}$

b. $A-B=\begin{bmatrix}-3 & 4\\2 & -3\\-1 & 0\end{bmatrix}-\begin{bmatrix}4 & 1\\1 & -2\\3 & -4\end{bmatrix}=\begin{bmatrix}-7 & 3\\1 & -1\\-4 & 4\end{bmatrix}$

c. $2B=2\begin{bmatrix}4 & 1\\1 & -2\\3 & -4\end{bmatrix}=\begin{bmatrix}8 & 2\\2 & -4\\6 & -8\end{bmatrix}$

d. $2A-3B=2\begin{bmatrix}-3 & 4\\2 & -3\\-1 & 0\end{bmatrix}-3\begin{bmatrix}4 & 1\\1 & -2\\3 & -4\end{bmatrix}$

$=\begin{bmatrix}-6 & 8\\4 & -6\\-2 & 0\end{bmatrix}-\begin{bmatrix}12 & 3\\3 & -6\\9 & -12\end{bmatrix}$

$=\begin{bmatrix}-18 & 5\\1 & 0\\-11 & 12\end{bmatrix}$

7. Find $A+B$, $A-B$, $2B$, and $2A-3B$.

a. $A+B=\begin{bmatrix}-2 & 3 & -1\\0 & -1 & 2\\-4 & 3 & 3\end{bmatrix}+\begin{bmatrix}1 & -2 & 0\\2 & 3 & -1\\3 & -1 & 2\end{bmatrix}$

$=\begin{bmatrix}-1 & 1 & -1\\2 & 2 & 1\\-1 & 2 & 5\end{bmatrix}$

b. $A-B=\begin{bmatrix}-2 & 3 & -1\\0 & -1 & 2\\-4 & 3 & 3\end{bmatrix}-\begin{bmatrix}1 & -2 & 0\\2 & 3 & -1\\3 & -1 & 2\end{bmatrix}$

$=\begin{bmatrix}-3 & 5 & -1\\-2 & -4 & 3\\-7 & 4 & 1\end{bmatrix}$

c. $2B = 2\begin{bmatrix} 1 & -2 & 0 \\ 2 & 3 & -1 \\ 3 & -1 & 2 \end{bmatrix} = \begin{bmatrix} 2 & -4 & 0 \\ 4 & 6 & -2 \\ 6 & -2 & 4 \end{bmatrix}$

d. $2A - 3B = 2\begin{bmatrix} -2 & 3 & -1 \\ 0 & -1 & 2 \\ -4 & 3 & 3 \end{bmatrix} - 3\begin{bmatrix} 1 & -2 & 0 \\ 2 & 3 & -1 \\ 3 & -1 & 2 \end{bmatrix}$

$= \begin{bmatrix} -4 & 6 & -2 \\ 0 & -2 & 4 \\ -8 & 6 & 6 \end{bmatrix} - \begin{bmatrix} 3 & -6 & 0 \\ 6 & 9 & -3 \\ 9 & -3 & 6 \end{bmatrix}$

$= \begin{bmatrix} -7 & 12 & -2 \\ -6 & -11 & 7 \\ -17 & 9 & 0 \end{bmatrix}$

9. Find AB and BA, if possible.

$$AB = \begin{bmatrix} 2 & -3 \\ 1 & 4 \end{bmatrix}\begin{bmatrix} -2 & 4 \\ 2 & -3 \end{bmatrix}$$

$$= \begin{bmatrix} (2)(-2)+(-3)(2) & (2)(4)+(-3)(-3) \\ (1)(-2)+(4)(2) & (1)(4)+(4)(-3) \end{bmatrix}$$

$$= \begin{bmatrix} -10 & 17 \\ 6 & -8 \end{bmatrix}$$

$$BA = \begin{bmatrix} -2 & 4 \\ 2 & -3 \end{bmatrix}\begin{bmatrix} 2 & -3 \\ 1 & 4 \end{bmatrix}$$

$$= \begin{bmatrix} (-2)(2)+(4)(1) & (-2)(-3)+(4)(4) \\ (2)(2)+(-3)(1) & (2)(-3)+(-3)(4) \end{bmatrix}$$

$$= \begin{bmatrix} 0 & 22 \\ 1 & -18 \end{bmatrix}$$

11. Find AB and BA, if possible.

$$AB = \begin{bmatrix} 3 & -1 \\ 2 & 3 \end{bmatrix}\begin{bmatrix} 4 & 1 \\ 2 & -3 \end{bmatrix}$$

$$= \begin{bmatrix} (3)(4)+(-1)(2) & (3)(1)+(-1)(-3) \\ (2)(4)+(3)(2) & (2)(1)+(3)(-3) \end{bmatrix}$$

$$= \begin{bmatrix} 10 & 6 \\ 14 & -7 \end{bmatrix}$$

$$BA = \begin{bmatrix} 4 & 1 \\ 2 & -3 \end{bmatrix}\begin{bmatrix} 3 & -1 \\ 2 & -3 \end{bmatrix}$$

$$= \begin{bmatrix} (4)(3)+(1)(2) & (4)(-1)+(1)(3) \\ (2)(3)+(-3)(2) & (2)(-1)+(-3)(3) \end{bmatrix}$$

$$= \begin{bmatrix} 14 & -1 \\ 0 & -11 \end{bmatrix}$$

13. Find AB and BA, if possible.

$$AB = \begin{bmatrix} 2 & -1 \\ 0 & 3 \\ 1 & -2 \end{bmatrix}\begin{bmatrix} 1 & -2 & 3 \\ 2 & 0 & 1 \end{bmatrix}$$

$$= \begin{bmatrix} 2(1)+(-1)(2) & 2(-2)+(-1)(0) & 2(3)+(-1)(1) \\ (0)(1)+(3)(2) & 0(-2)+(3)(0) & 0(3)+(3)(1) \\ 1(1)+(-2)(2) & 1(-2)+(-2)(0) & 1(3)+(-2)(1) \end{bmatrix}$$

$$= \begin{bmatrix} 0 & -4 & 5 \\ 6 & 0 & 3 \\ -3 & -2 & 1 \end{bmatrix}$$

$$BA = \begin{bmatrix} 1 & -2 & 3 \\ 2 & 0 & 1 \end{bmatrix}\begin{bmatrix} 2 & -1 \\ 0 & 3 \\ 1 & -2 \end{bmatrix}$$

$$= \begin{bmatrix} 1(2)+(-2)(0)+3(1) & 1(-1)+(-2)(3)+3(-2) \\ 2(2)+(0)(0)+1(1) & 2(-1)+(0)(3)+(1)(-2) \end{bmatrix}$$

$$= \begin{bmatrix} 5 & -13 \\ 5 & -4 \end{bmatrix}$$

15. Find AB and BA, if possible.

$$AB = \begin{bmatrix} 2 & -1 & 3 \\ 0 & 2 & -1 \\ 0 & 0 & 2 \end{bmatrix}\begin{bmatrix} 2 & 0 & 0 \\ 1 & -1 & 0 \\ 2 & -1 & -2 \end{bmatrix}$$

$$= \begin{bmatrix} 4+(-1)+6 & 0+1+(-3) & 0+0+(-6) \\ 0+2+(-2) & 0+(-2)+1 & 0+0+2 \\ 0+0+4 & 0+0+(-2) & 0+0(0)+(-4) \end{bmatrix}$$

$$= \begin{bmatrix} 9 & -2 & -6 \\ 0 & -1 & 2 \\ 4 & -2 & -4 \end{bmatrix}$$

$$BA = \begin{bmatrix} 2 & 0 & 0 \\ 1 & -1 & 0 \\ 2 & -1 & -2 \end{bmatrix}\begin{bmatrix} 2 & -1 & 3 \\ 0 & 2 & -1 \\ 0 & 0 & 2 \end{bmatrix}$$

$$= \begin{bmatrix} 4+0+0 & -2+0+0 & 6+0+0 \\ 2+0+0 & -1+(-2)+0 & 3+1+0 \\ 4+0+0 & -2+(-2)+0 & 6+1+(-4) \end{bmatrix}$$

$$= \begin{bmatrix} 4 & -2 & 6 \\ 2 & -3 & 4 \\ 4 & -4 & 3 \end{bmatrix}$$

17. Find AB, if possible.

$$AB = \begin{bmatrix} 1 & -2 & 3 \end{bmatrix} \begin{bmatrix} 1 & 0 \\ 2 & -1 \\ 1 & 2 \end{bmatrix}$$
$$= \begin{bmatrix} 1(1)+(-2)(2)+3(1) & 1(0)+(-2)(-1)+3(2) \end{bmatrix}$$
$$= \begin{bmatrix} 0 & 8 \end{bmatrix}$$

19. Find AB, if possible.

The number of columns of the first matrix is not equal to the number of rows of the second matrix. The product is not possible.

21. Find AB, if possible.

$$AB = \begin{bmatrix} 2 & 3 \\ -4 & -6 \end{bmatrix} \begin{bmatrix} 3 & 6 \\ -2 & -4 \end{bmatrix}$$
$$= \begin{bmatrix} (2)(3)+(3)(-2) & (2)(6)+(3)(-4) \\ (-4)(3)+(-6)(-2) & (-4)(6)+(-6)(-4) \end{bmatrix}$$
$$= \begin{bmatrix} 0 & 0 \\ 0 & 0 \end{bmatrix}$$

23. Find AB, if possible.

The number of columns of the first matrix is not equal to the number of rows of the second matrix. The product is not possible.

25. Find the matrix X.

$$3X + A = B$$
$$3X + \begin{bmatrix} -1 & 3 \\ 2 & -1 \\ 3 & 1 \end{bmatrix} = \begin{bmatrix} 0 & -2 \\ 1 & 3 \\ 4 & -3 \end{bmatrix}$$
$$3X = \begin{bmatrix} 0 & -2 \\ 1 & 3 \\ 4 & -3 \end{bmatrix} - \begin{bmatrix} -1 & 3 \\ 2 & -1 \\ 3 & 1 \end{bmatrix}$$
$$3X = \begin{bmatrix} 1 & -5 \\ -1 & 4 \\ 1 & -4 \end{bmatrix}$$
$$X = \begin{bmatrix} \frac{1}{3} & -\frac{5}{3} \\ -\frac{1}{3} & \frac{4}{3} \\ \frac{1}{3} & -\frac{4}{3} \end{bmatrix}$$

27. Find the matrix X.

$$2X - A = X + B$$
$$2X - \begin{bmatrix} -1 & 3 \\ 2 & -1 \\ 3 & 1 \end{bmatrix} = X + \begin{bmatrix} 0 & -2 \\ 1 & 3 \\ 4 & -3 \end{bmatrix}$$
$$X - \begin{bmatrix} -1 & 3 \\ 2 & -1 \\ 3 & 1 \end{bmatrix} = \begin{bmatrix} 0 & -2 \\ 1 & 3 \\ 4 & -3 \end{bmatrix}$$
$$X = \begin{bmatrix} 0 & -2 \\ 1 & 3 \\ 4 & -3 \end{bmatrix} + \begin{bmatrix} -1 & 3 \\ 2 & -1 \\ 3 & 1 \end{bmatrix}$$
$$X = \begin{bmatrix} -1 & 1 \\ 3 & 2 \\ 7 & -2 \end{bmatrix}$$

29. Find A^2.

$$A^2 = A \cdot A = \begin{bmatrix} 2 & -3 \\ 1 & -1 \end{bmatrix} \begin{bmatrix} 2 & -3 \\ 1 & -1 \end{bmatrix}$$
$$= \begin{bmatrix} 2(2)+(-3)(1) & 2(-3)+(-3)(-1) \\ 1(2)+(-1)(1) & 1(-3)+(-1)(-1) \end{bmatrix} = \begin{bmatrix} 1 & -3 \\ 1 & -2 \end{bmatrix}$$

31. Find B^2.

$$B^2 = B \cdot B = \begin{bmatrix} 3 & -1 & 0 \\ 2 & -2 & -1 \\ 1 & 0 & 2 \end{bmatrix} \begin{bmatrix} 3 & -1 & 0 \\ 2 & -2 & -1 \\ 1 & 0 & 2 \end{bmatrix}$$
$$= \begin{bmatrix} 9+(-2)+0 & -3+2+0 & 0+1+0 \\ 6+(-4)+(-1) & -2+4+0 & 0+2+(-2) \\ 3+0+2 & -1+0+0) & 0+0+4 \end{bmatrix}$$
$$= \begin{bmatrix} 7 & -1 & 1 \\ 1 & 2 & 0 \\ 5 & -1 & 4 \end{bmatrix}$$

33. Write the system of equations.

$$\begin{bmatrix} 3 & -8 \\ 4 & 3 \end{bmatrix} \begin{bmatrix} x \\ y \end{bmatrix} = \begin{bmatrix} 11 \\ 1 \end{bmatrix}$$

$$\begin{cases} 3x - 8y = 11 \\ 4x + 3y = 1 \end{cases}$$

35. Write the system of equations.

$$\begin{bmatrix} 1 & -3 & -2 \\ 3 & 1 & 0 \\ 2 & -4 & 5 \end{bmatrix}\begin{bmatrix} x \\ y \\ z \end{bmatrix} = \begin{bmatrix} 6 \\ 2 \\ 1 \end{bmatrix}$$

$$\begin{cases} x - 3y - 2z = 6 \\ 3x + y = 2 \\ 2x - 4y + 5z = 1 \end{cases}$$

37. Write the system of equations.

$$\begin{bmatrix} 2 & -1 & 0 & 2 \\ 4 & 1 & 2 & -3 \\ 6 & 0 & 1 & -2 \\ 5 & 2 & -1 & -4 \end{bmatrix}\begin{bmatrix} x_1 \\ x_2 \\ x_3 \\ x_4 \end{bmatrix} = \begin{bmatrix} 5 \\ 6 \\ 10 \\ 8 \end{bmatrix}$$

$$\begin{cases} 2x_1 - x_2 + 2x_4 = 5 \\ 4x_1 + x_2 + 2x_3 - 3x_4 = 6 \\ 6x_1 + x_3 - 2x_4 = 10 \\ 5x_1 + 2x_2 - x_3 - 4x_4 = 8 \end{cases}$$

39. a. 3×4. There are three different fish in four different samples.

b. Fish A was caught in sample number 4.

c. Fish B. There are more 1's in this row than in any other row.

41. Compute the matrix.

$$0.98\begin{bmatrix} 2.0 & 1.4 & 3.0 & 1.4 \\ 0.8 & 1.1 & 2.0 & 0.9 \\ 3.6 & 1.2 & 4.5 & 1.5 \end{bmatrix}$$

$$= \begin{bmatrix} 0.98(2.0) & 0.98(1.4) & 0.98(3.0) & 0.98(1.4) \\ 0.98(0.8) & 0.98(1.1) & 0.98(2.0) & 0.98(0.9) \\ 0.98(3.6) & 0.98(1.2) & 0.98(4.5) & 0.98(1.5) \end{bmatrix}$$

$$= \begin{bmatrix} 1.96 & 1.37 & 2.94 & 1.37 \\ 0.78 & 1.08 & 1.96 & 0.88 \\ 3.53 & 1.18 & 4.41 & 1.47 \end{bmatrix}$$

43. a. $$H + A = \begin{bmatrix} 14 & 3 \\ 14 & 3 \\ 10 & 7 \end{bmatrix} + \begin{bmatrix} 12 & 5 \\ 7 & 10 \\ 8 & 9 \end{bmatrix} = \begin{bmatrix} 26 & 8 \\ 21 & 13 \\ 18 & 16 \end{bmatrix}$$

The matrix represents the total number of wins and losses for each team for the season.

b. $$H - A = \begin{bmatrix} 14 & 3 \\ 14 & 3 \\ 10 & 7 \end{bmatrix} - \begin{bmatrix} 12 & 5 \\ 7 & 10 \\ 8 & 9 \end{bmatrix} = \begin{bmatrix} 2 & -2 \\ 7 & -7 \\ 2 & -2 \end{bmatrix}$$

The matrix represents the difference between performances at home and performances away.

45. Find and interpret $A - B$.

$$A = \begin{bmatrix} 530 & 650 & 815 \\ 190 & 385 & 715 \\ 485 & 600 & 610 \\ 150 & 210 & 305 \end{bmatrix}, \quad B = \begin{bmatrix} 480 & 500 & 675 \\ 175 & 215 & 345 \\ 400 & 350 & 480 \\ 70 & 95 & 280 \end{bmatrix}$$

$$A - B = \begin{bmatrix} 50 & 150 & 140 \\ 15 & 170 & 370 \\ 85 & 250 & 130 \\ 80 & 115 & 25 \end{bmatrix}$$

$A - B$ is number sold of each item during the week.

47. Find CS and determine which company should be used.

$$C = \begin{bmatrix} 0.04 & 0.06 & 0.05 \\ 0.04 & 0.04 & 0.04 \\ 0.03 & 0.07 & 0.06 \end{bmatrix}, \quad S = \begin{bmatrix} 500 & 600 \\ 250 & 450 \\ 600 & 750 \end{bmatrix}$$

$$CS = \begin{bmatrix} 0.04 & 0.06 & 0.05 \\ 0.04 & 0.04 & 0.04 \\ 0.03 & 0.07 & 0.06 \end{bmatrix}\begin{bmatrix} 500 & 600 \\ 250 & 450 \\ 600 & 750 \end{bmatrix} = \begin{bmatrix} 65 & 88.5 \\ 54 & 72 \\ 68.5 & 94.5 \end{bmatrix}$$

To minimize commissions costs, customer S_1 should use company T_2.

49. Find the transformation.

$$R_x \cdot \begin{bmatrix} 2 & -3 \\ 5 & 6 \\ 1 & 1 \end{bmatrix} = \begin{bmatrix} 1 & 0 & 0 \\ 0 & -1 & 0 \\ 0 & 0 & 1 \end{bmatrix}\begin{bmatrix} 2 & -3 \\ 5 & 6 \\ 1 & 1 \end{bmatrix}$$

$$= \begin{bmatrix} 2 & -3 \\ -5 & -6 \\ 1 & 1 \end{bmatrix} \Rightarrow \begin{matrix} P'(2, -5) \\ Q'(-3, -6) \end{matrix}$$

51. Find the transformation.

$$R_{xy}\cdot\begin{bmatrix}-3 & -5\\ 1 & 3\\ 1 & 1\end{bmatrix} = \begin{bmatrix}0 & 1 & 0\\ 1 & 0 & 0\\ 0 & 0 & 1\end{bmatrix}\begin{bmatrix}-3 & -5\\ 1 & 3\\ 1 & 1\end{bmatrix}$$

$$= \begin{bmatrix}1 & 3\\ -3 & -5\\ 1 & 1\end{bmatrix} \Rightarrow \begin{matrix}P'(1,-3)\\ Q'(3,-5)\end{matrix}$$

53. Find the transformation.

$$R_{xy}\cdot T_{3,-1}\cdot\begin{bmatrix}-1 & 1 & 3\\ 5 & -2 & 4\\ 1 & 1 & 1\end{bmatrix}$$

$$= \begin{bmatrix}0 & 1 & 0\\ 1 & 0 & 0\\ 0 & 0 & 1\end{bmatrix}\begin{bmatrix}1 & 0 & 3\\ 0 & 1 & -1\\ 0 & 0 & 1\end{bmatrix}\begin{bmatrix}-1 & 1 & 3\\ 5 & -2 & 4\\ 1 & 1 & 1\end{bmatrix}$$

$$= \begin{bmatrix}0 & 1 & 0\\ 1 & 0 & 0\\ 0 & 0 & 1\end{bmatrix}\begin{bmatrix}2 & 4 & 6\\ 4 & -3 & 3\\ 1 & 1 & 1\end{bmatrix}$$

$$= \begin{bmatrix}4 & -3 & 3\\ 2 & 4 & 6\\ 1 & 1 & 1\end{bmatrix} \Rightarrow \begin{matrix}A'(4,2)\\ B'(-3,4)\\ C'(3,6)\end{matrix}$$

55. Find the transformation.

$$T_{1,-6}\cdot\begin{bmatrix}-1 & -1 & 4 & 4\\ 2 & 6 & 6 & 2\\ 1 & 1 & 1 & 1\end{bmatrix} = \begin{bmatrix}1 & 0 & 1\\ 0 & 1 & -6\\ 0 & 0 & 1\end{bmatrix}\begin{bmatrix}-1 & -1 & 4 & 4\\ 2 & 6 & 6 & 2\\ 1 & 1 & 1 & 1\end{bmatrix}$$

$$= \begin{bmatrix}0 & 0 & 5 & 5\\ -4 & 0 & 0 & -4\\ 1 & 1 & 1 & 1\end{bmatrix}$$

$$R_{180}\cdot\begin{bmatrix}0 & 0 & 5 & 5\\ -4 & 0 & 0 & -4\\ 1 & 1 & 1 & 1\end{bmatrix} = \begin{bmatrix}-1 & 0 & 0\\ 0 & -1 & 0\\ 0 & 0 & 1\end{bmatrix}\begin{bmatrix}0 & 0 & 5 & 5\\ -4 & 0 & 0 & -4\\ 1 & 1 & 1 & 1\end{bmatrix}$$

$$= \begin{bmatrix}0 & 0 & -5 & -5\\ 4 & 0 & 0 & 4\\ 1 & 1 & 1 & 1\end{bmatrix}$$

$$T_{-1,6}\cdot\begin{bmatrix}0 & 0 & -5 & -5\\ 4 & 0 & 0 & 4\\ 1 & 1 & 1 & 1\end{bmatrix}$$

$$= \begin{bmatrix}1 & 0 & -1\\ 0 & 1 & 6\\ 0 & 0 & 1\end{bmatrix}\begin{bmatrix}0 & 0 & -5 & -5\\ 4 & 0 & 0 & 4\\ 1 & 1 & 1 & 1\end{bmatrix} = \begin{bmatrix}-1 & -1 & -6 & -6\\ 10 & 6 & 6 & 10\\ 1 & 1 & 1 & 1\end{bmatrix}$$

$$\Rightarrow \begin{matrix}A'(-1,10) & C'(-6,6)\\ B'(-1,6) & D'(-6,10)\end{matrix}$$

57. a. Find the adjacency matrix.

$$A = \begin{bmatrix}0 & 1 & 1 & 0\\ 1 & 0 & 1 & 1\\ 1 & 1 & 0 & 1\\ 0 & 1 & 1 & 0\end{bmatrix}$$

b. Find A^2 and the number of walks.

$$A^2 = \begin{bmatrix}0 & 1 & 1 & 0\\ 1 & 0 & 1 & 1\\ 1 & 1 & 0 & 1\\ 0 & 1 & 1 & 0\end{bmatrix}\begin{bmatrix}0 & 1 & 1 & 0\\ 1 & 0 & 1 & 1\\ 1 & 1 & 0 & 1\\ 0 & 1 & 1 & 0\end{bmatrix} = \begin{bmatrix}2 & 1 & 1 & 2\\ 1 & 3 & 2 & 1\\ 1 & 2 & 3 & 1\\ 2 & 1 & 1 & 2\end{bmatrix}$$

A^2 gives the number of walks of length 2.

There are 2 walks of length 2 between vertex 4 and 1.

59. a. Find the adjacency matrix.

$$A = \begin{bmatrix}0 & 1 & 1 & 0\\ 1 & 0 & 1 & 1\\ 1 & 1 & 0 & 0\\ 0 & 1 & 0 & 0\end{bmatrix}$$

b. Find A^3 and the number of walks.

$$A^3 = \begin{bmatrix}0 & 1 & 1 & 0\\ 1 & 0 & 1 & 1\\ 1 & 1 & 0 & 0\\ 0 & 1 & 0 & 0\end{bmatrix}\begin{bmatrix}0 & 1 & 1 & 0\\ 1 & 0 & 1 & 1\\ 1 & 1 & 0 & 0\\ 0 & 1 & 0 & 0\end{bmatrix}\begin{bmatrix}0 & 1 & 1 & 0\\ 1 & 0 & 1 & 1\\ 1 & 1 & 0 & 0\\ 0 & 1 & 0 & 0\end{bmatrix}$$

$$= \begin{bmatrix}2 & 4 & 3 & 1\\ 4 & 2 & 4 & 3\\ 3 & 4 & 2 & 1\\ 1 & 3 & 1 & 0\end{bmatrix}$$

A^3 gives the number of walks of length 3.

There are 3 walks of length 3 between vertex 2 and 4.

61. a. Find the adjacency matrix.

$$A = \begin{bmatrix}0 & 1 & 1 & 0 & 0\\ 1 & 0 & 1 & 1 & 1\\ 1 & 1 & 0 & 1 & 1\\ 0 & 1 & 1 & 0 & 0\\ 0 & 1 & 1 & 0 & 0\end{bmatrix}$$

b. Find A^3 and the number of walks.

$$A^3 = \begin{bmatrix} 0 & 1 & 1 & 0 & 0 \\ 1 & 0 & 1 & 1 & 1 \\ 1 & 1 & 0 & 1 & 1 \\ 0 & 1 & 1 & 0 & 0 \\ 0 & 1 & 1 & 0 & 0 \end{bmatrix} \begin{bmatrix} 0 & 1 & 1 & 0 & 0 \\ 1 & 0 & 1 & 1 & 1 \\ 1 & 1 & 0 & 1 & 1 \\ 0 & 1 & 1 & 0 & 0 \\ 0 & 1 & 1 & 0 & 0 \end{bmatrix} \begin{bmatrix} 0 & 1 & 1 & 0 & 0 \\ 1 & 0 & 1 & 1 & 1 \\ 1 & 1 & 0 & 1 & 1 \\ 0 & 1 & 1 & 0 & 0 \\ 0 & 1 & 1 & 0 & 0 \end{bmatrix}$$

$$= \begin{bmatrix} 2 & 7 & 7 & 2 & 2 \\ 7 & 6 & 7 & 7 & 7 \\ 7 & 7 & 6 & 7 & 7 \\ 2 & 7 & 7 & 2 & 2 \\ 2 & 7 & 7 & 2 & 2 \end{bmatrix}$$

A^3 gives the number of walks of length 3.

There are 2 walks of length 3 between vertex 1 and 5.

63. a. Find the adjacency matrix.

$$A = \begin{bmatrix} 0 & 0 & 0 & 1 & 1 \\ 0 & 0 & 2 & 0 & 0 \\ 0 & 2 & 0 & 1 & 1 \\ 1 & 0 & 1 & 0 & 0 \\ 1 & 0 & 1 & 0 & 0 \end{bmatrix}$$

b. Find A^4 and the number of walks.

$$A^4 = \begin{bmatrix} 8 & 0 & 16 & 0 & 0 \\ 0 & 24 & 0 & 16 & 16 \\ 16 & 0 & 40 & 0 & 0 \\ 0 & 16 & 0 & 12 & 12 \\ 0 & 16 & 0 & 12 & 12 \end{bmatrix}$$

A^4 gives the number of walks of length 4.

There are 12 walks of length 4 between vertex 5 and 5.

65. a. Find the adjacency matrix.

$$A = \begin{bmatrix} 0 & 1 & 0 & 1 \\ 1 & 0 & 2 & 1 \\ 0 & 2 & 0 & 0 \\ 1 & 1 & 0 & 0 \end{bmatrix}$$

b. Find the matrix.

$$A^3 = \begin{bmatrix} 2 & 7 & 2 & 3 \\ 7 & 2 & 12 & 7 \\ 2 & 12 & 0 & 2 \\ 3 & 7 & 2 & 2 \end{bmatrix}$$

A^3 gives the number of walks of length 3.

67. a. Predict the percent drinking diet soda in 1 year.

For 1 year from now, $n = 2$.

$$\begin{bmatrix} 0.55 & 0.45 \end{bmatrix} \begin{bmatrix} 0.989 & 0.011 \\ 0.007 & 0.993 \end{bmatrix}^2 = \begin{bmatrix} 0.5443 & 0.4557 \end{bmatrix}$$

1 year from now 45.6% of the customers will be drinking diet soda.

b. Predict the percent drinking diet soda in 3 years.

For 3 years from now, $n = 6$.

$$\begin{bmatrix} 0.55 & 0.45 \end{bmatrix} \begin{bmatrix} 0.989 & 0.011 \\ 0.007 & 0.993 \end{bmatrix}^6 = \begin{bmatrix} 0.5334 & 0.4666 \end{bmatrix}$$

3 years from now 46.7% of the customers will be drinking diet soda.

69. a. Predict the percent of customers renting movies online 12 months from now.

For 12 months from now, $n = 12$.

$$\begin{bmatrix} 0.15 & 0.85 \end{bmatrix} \begin{bmatrix} 0.975 & 0.025 \\ 0.014 & 0.986 \end{bmatrix}^{12} = \begin{bmatrix} 0.2293 & 0.7707 \end{bmatrix}$$

12 months from now 22.9% of the customers will be renting movies online.

b. Predict the percent of customers renting movies online 24 months from now.

For 24 months from now, $n = 24$.

$$\begin{bmatrix} 0.15 & 0.85 \end{bmatrix} \begin{bmatrix} 0.975 & 0.025 \\ 0.014 & 0.986 \end{bmatrix}^{24} = \begin{bmatrix} 0.2785 & 0.7215 \end{bmatrix}$$

24 months from now 27.9% of the customers will be renting movies online.

71. Use Exercise 70 to find when store A has 50%.

When $n = 11$, $a > 0.5053$.

$$\begin{bmatrix} 0.25 & 0.75 \end{bmatrix} \begin{bmatrix} 0.98 & 0.02 \\ 0.05 & 0.95 \end{bmatrix}^{11} = \begin{bmatrix} 0.5053 & 0.4947 \end{bmatrix}$$

After 11 months, Store A will have 50% of the town's customers.

73. Using A and B as given and a calculator,

$$AB = \begin{bmatrix} 24 & 21 & -12 & 32 & 0 \\ -7 & -8 & 3 & 21 & 20 \\ 32 & 10 & -32 & 1 & 5 \\ 19 & -15 & -17 & 30 & 20 \\ 29 & 9 & -28 & 13 & -6 \end{bmatrix}$$

75. Using A as given and a calculator,

$$A^3 = \begin{bmatrix} 46 & -100 & 36 & 273 & 93 \\ 82 & -93 & 19 & 27 & 97 \\ 73 & -10 & -23 & 109 & 83 \\ 212 & -189 & 52 & 37 & 156 \\ 68 & -22 & 54 & 221 & 58 \end{bmatrix}$$

77. Using A and B as given and a calculator,

$$A^2 + B^2 = \begin{bmatrix} 76 & -8 & -25 & 30 & 6 \\ 14 & 16 & -10 & 14 & 2 \\ 39 & 0 & -45 & 22 & 27 \\ 0 & -4 & 23 & 83 & -16 \\ 56 & -20 & -22 & 7 & 5 \end{bmatrix}$$

79. a. $$R_{90} \cdot \begin{bmatrix} t \\ t+2 \\ 1 \end{bmatrix} = \begin{bmatrix} 0 & -1 & 0 \\ 1 & 0 & 0 \\ 0 & 0 & 1 \end{bmatrix} \begin{bmatrix} t \\ t+2 \\ 1 \end{bmatrix}$$

$$= \begin{bmatrix} -t-2 \\ t \\ 1 \end{bmatrix} \Rightarrow (-t-2,\ t)$$

$x = -t-2,\ y = t$

Using substitution, $x = -y-2$

$y = -x-2$

b. $$R_y \cdot \begin{bmatrix} t \\ 3t-1 \\ 1 \end{bmatrix} = \begin{bmatrix} -1 & 0 & 0 \\ 0 & 1 & 0 \\ 0 & 0 & 1 \end{bmatrix} \begin{bmatrix} t \\ 3t-1 \\ 1 \end{bmatrix}$$

$$= \begin{bmatrix} -t \\ 3t-1 \\ 1 \end{bmatrix} \Rightarrow (-t,\ 3t-1)$$

$x = -t,\ y = 3t-1$

$t = -x$

Using substitution, $y = 3(-x)-1$

$y = -3x-1$

c. $$T_{-1,-1} \cdot \begin{bmatrix} t \\ \frac{1}{t} \\ 1 \end{bmatrix} = \begin{bmatrix} 1 & 0 & -1 \\ 0 & 1 & -1 \\ 0 & 0 & 1 \end{bmatrix} \begin{bmatrix} t \\ \frac{1}{t} \\ 1 \end{bmatrix} = \begin{bmatrix} t-1 \\ \frac{1}{t}-1 \\ 1 \end{bmatrix}$$

$$R_{180} \cdot \begin{bmatrix} t-1 \\ \frac{1}{t}-1 \\ 1 \end{bmatrix} = \begin{bmatrix} -1 & 0 & 0 \\ 0 & -1 & 0 \\ 0 & 0 & 1 \end{bmatrix} \begin{bmatrix} t-1 \\ \frac{1}{t}-1 \\ 1 \end{bmatrix} = \begin{bmatrix} -t+1 \\ -\frac{1}{t}+1 \\ 1 \end{bmatrix}$$

$$T_{1,1} \cdot \begin{bmatrix} -t+1 \\ -\frac{1}{t}+1 \\ 1 \end{bmatrix} = \begin{bmatrix} 1 & 0 & 1 \\ 0 & 1 & 1 \\ 0 & 0 & 1 \end{bmatrix} \begin{bmatrix} -t+1 \\ -\frac{1}{t}+1 \\ 1 \end{bmatrix}$$

$$= \begin{bmatrix} -t+2 \\ -\frac{1}{t}+2 \\ 1 \end{bmatrix} \Rightarrow \left(-t+2,\ -\frac{1}{t}+2\right)$$

$x = -t+2,\ y = -\frac{1}{t}+2$

$t = -x+2$

Using substitution,

$$y = -\frac{1}{-x+2} + 2 = \frac{-1}{-x+2} + \frac{-2x+4}{-x+2} = \frac{-2x+3}{-x+2}$$

$$= \frac{2x-3}{x-2}$$

d. $$T_{2,-1} \cdot \begin{bmatrix} t \\ t^2 \\ 1 \end{bmatrix} = \begin{bmatrix} 1 & 0 & 2 \\ 0 & 1 & -1 \\ 0 & 0 & 1 \end{bmatrix} \begin{bmatrix} t \\ t^2 \\ 1 \end{bmatrix}$$

$$= \begin{bmatrix} t+2 \\ t^2-1 \\ 1 \end{bmatrix} \Rightarrow (t+2,\ t^2-1)$$

$x = t+2,\ y = t^2-1$

$t = x-2$

Using substitution,

$y = (x-2)^2 - 1$

$y = x^2 - 4x + 3$

e. $$R_{270} \cdot \begin{bmatrix} t \\ t^2 \\ 1 \end{bmatrix} = \begin{bmatrix} 0 & 1 & 0 \\ -1 & 0 & 0 \\ 0 & 0 & 1 \end{bmatrix} \begin{bmatrix} t \\ t^2 \\ 1 \end{bmatrix} = \begin{bmatrix} t^2 \\ -t \\ 1 \end{bmatrix} \Rightarrow (t^2,\ -t)$$

$x = t^2,\ y = -t$

Using substitution, $x = y^2$

f. $R_{90}\cdot\begin{bmatrix} t \\ t^2 \\ 1 \end{bmatrix}=\begin{bmatrix} 0 & -1 & 0 \\ 1 & 0 & 0 \\ 0 & 0 & 1 \end{bmatrix}\begin{bmatrix} t \\ t^2 \\ 1 \end{bmatrix}=\begin{bmatrix} -t^2 \\ t \\ 1 \end{bmatrix}$

$$T_{-2,1}\cdot\begin{bmatrix} -t^2 \\ t \\ 1 \end{bmatrix}=\begin{bmatrix} 1 & 0 & -2 \\ 0 & 1 & 1 \\ 0 & 0 & 1 \end{bmatrix}\begin{bmatrix} -t^2 \\ t \\ 1 \end{bmatrix}$$

$$=\begin{bmatrix} -t^2-2 \\ t+1 \\ 1 \end{bmatrix} \Rightarrow (-t^2-2,\ t+1)$$

$$x=-t^2-2,\quad y=t+1$$
$$t=y-1$$

Using substitution,

$$x=-(y-1)^2-2$$
$$x=-y^2+2y-3$$

Prepare for Section 7.3

P1. Find the multiplicative inverse of $-\frac{2}{3}$.

$-\frac{3}{2}$

P3. State the 3 elementary row operations for matrices.

1. Interchange any two rows.
2. Multiply each elements in a row by the same nonzero constant.
3. Replace a row by the sum of that row and a nonzero multiple of any other row.

P5. Solve for X.

$$AX=B$$
$$A^{-1}AX=A^{-1}B$$
$$X=A^{-1}B$$

Section 7.3 Exercises

1. Find the inverse of the matrix.

$$\left[\begin{array}{cc|cc} 1 & -3 & 1 & 0 \\ -2 & 5 & 0 & 1 \end{array}\right]$$

$$\xrightarrow{2R_1+R_2}\left[\begin{array}{cc|cc} 1 & -3 & 1 & 0 \\ 0 & -1 & 2 & 1 \end{array}\right]$$

$$\xrightarrow{-1R_2}\left[\begin{array}{cc|cc} 1 & -3 & 1 & 0 \\ 0 & 1 & -2 & -1 \end{array}\right]$$

$$\xrightarrow{3R_2+R_1}\left[\begin{array}{cc|cc} 1 & 0 & -5 & -3 \\ 0 & 1 & -2 & -1 \end{array}\right]$$

The inverse matrix is $\begin{bmatrix} -5 & -3 \\ -2 & -1 \end{bmatrix}$.

3. Find the inverse of the matrix.

$$\left[\begin{array}{cc|cc} 1 & 4 & 1 & 0 \\ 2 & 10 & 0 & 1 \end{array}\right]\xrightarrow{-2R_1+R_2}\left[\begin{array}{cc|cc} 1 & 4 & 1 & 0 \\ 0 & 2 & -2 & 1 \end{array}\right]$$

$$\xrightarrow{\frac{1}{2}R_2}\left[\begin{array}{cc|cc} 1 & 4 & 1 & 0 \\ 0 & 1 & -1 & \frac{1}{2} \end{array}\right]$$

$$\xrightarrow{-4R_2+R_1}\left[\begin{array}{cc|cc} 1 & 0 & 5 & -2 \\ 0 & 1 & -1 & \frac{1}{2} \end{array}\right]$$

The inverse matrix is $\begin{bmatrix} 5 & -2 \\ -1 & \frac{1}{2} \end{bmatrix}$.

5. Find the inverse of the matrix.

$$\left[\begin{array}{ccc|ccc} 1 & 2 & -1 & 1 & 0 & 0 \\ 2 & 5 & 1 & 0 & 1 & 0 \\ 3 & 6 & -2 & 0 & 0 & 1 \end{array}\right]$$

$$\xrightarrow[-3R_1+R_3]{-2R_1+R_2}\left[\begin{array}{ccc|ccc} 1 & 2 & -1 & 1 & 0 & 0 \\ 0 & 1 & 3 & -2 & 1 & 0 \\ 0 & 0 & 1 & -3 & 0 & 1 \end{array}\right]$$

$$\xrightarrow{-2R_2+R_1}\left[\begin{array}{ccc|ccc} 1 & 0 & -7 & 5 & -2 & 0 \\ 0 & 1 & 3 & -2 & 1 & 0 \\ 0 & 0 & 1 & -3 & 0 & 1 \end{array}\right]$$

$$\xrightarrow[-3R_3+R_2]{7R_3+R_1}\left[\begin{array}{ccc|ccc} 1 & 0 & 0 & -16 & -2 & 7 \\ 0 & 1 & 0 & 7 & 1 & -3 \\ 0 & 0 & 1 & -3 & 0 & 1 \end{array}\right]$$

The inverse matrix is $\begin{bmatrix} -16 & -2 & 7 \\ 7 & 1 & -3 \\ -3 & 0 & 1 \end{bmatrix}$.

7. Find the inverse of the matrix.

$$\left[\begin{array}{ccc|ccc} 1 & 2 & -1 & 1 & 0 & 0 \\ 2 & 6 & 1 & 0 & 1 & 0 \\ 3 & 6 & -4 & 0 & 0 & 1 \end{array}\right]$$

$$\xrightarrow[-3R_1+R_3]{-2R_1+R_2}\left[\begin{array}{ccc|ccc} 1 & 2 & -1 & 1 & 0 & 0 \\ 0 & 2 & 3 & -2 & 1 & 0 \\ 0 & 0 & -1 & -3 & 0 & 1 \end{array}\right]$$

$$\xrightarrow{\frac{1}{2}R_2}\left[\begin{array}{ccc|ccc} 1 & 2 & -1 & 1 & 0 & 0 \\ 0 & 1 & \frac{3}{2} & -1 & \frac{1}{2} & 0 \\ 0 & 0 & -1 & -3 & 0 & 1 \end{array}\right]$$

$$\xrightarrow{-1R_3}\left[\begin{array}{ccc|ccc} 1 & 2 & -1 & 1 & 0 & 0 \\ 0 & 1 & \frac{3}{2} & -1 & \frac{1}{2} & 0 \\ 0 & 0 & 1 & 3 & 0 & -1 \end{array}\right]$$

$$\xrightarrow{-2R_2+R_1}\left[\begin{array}{ccc|ccc} 1 & 0 & -4 & 3 & -1 & 0 \\ 0 & 1 & \frac{3}{2} & -1 & \frac{1}{2} & 0 \\ 0 & 0 & 1 & 3 & 0 & -1 \end{array}\right]$$

$$\xrightarrow[-\frac{3}{2}R_3+R_2]{4R_3+R_1}\left[\begin{array}{ccc|ccc} 1 & 0 & 0 & 15 & -1 & -4 \\ 0 & 1 & 0 & -\frac{11}{2} & \frac{1}{2} & \frac{3}{2} \\ 0 & 0 & 1 & 3 & 0 & -1 \end{array}\right]$$

The inverse matrix is $\begin{bmatrix} 15 & -1 & -4 \\ -\frac{11}{2} & \frac{1}{2} & \frac{3}{2} \\ 3 & 0 & -1 \end{bmatrix}$.

9. Find the inverse of the matrix.

$$\left[\begin{array}{ccc|ccc} 2 & 4 & -4 & 1 & 0 & 0 \\ 1 & 3 & -4 & 0 & 1 & 0 \\ 2 & 4 & -3 & 0 & 0 & 1 \end{array}\right]\xrightarrow{\frac{1}{2}R_1}\left[\begin{array}{ccc|ccc} 1 & 2 & -2 & \frac{1}{2} & 0 & 0 \\ 1 & 3 & -4 & 0 & 1 & 0 \\ 2 & 4 & -3 & 0 & 0 & 1 \end{array}\right]$$

$$\xrightarrow[-2R_1+R_3]{-1R_1+R_2}\left[\begin{array}{ccc|ccc} 1 & 2 & -2 & \frac{1}{2} & 0 & 0 \\ 0 & 1 & -2 & -\frac{1}{2} & 1 & 0 \\ 0 & 0 & 1 & -1 & 0 & 1 \end{array}\right]$$

$$\xrightarrow{-2R_2+R_1}\left[\begin{array}{ccc|ccc} 1 & 0 & 2 & \frac{3}{2} & -2 & 0 \\ 0 & 1 & -2 & -\frac{1}{2} & 1 & 0 \\ 0 & 0 & 1 & -1 & 0 & 1 \end{array}\right]$$

$$\xrightarrow[2R_3+R_2]{-2R_3+R_1}\left[\begin{array}{ccc|ccc} 1 & 0 & 0 & \frac{7}{2} & -2 & -2 \\ 0 & 1 & 0 & -\frac{5}{2} & 1 & 2 \\ 0 & 0 & 1 & -1 & 0 & 1 \end{array}\right]$$

The inverse matrix is $\begin{bmatrix} \frac{7}{2} & -2 & -2 \\ -\frac{5}{2} & 1 & 2 \\ -1 & 0 & 1 \end{bmatrix}$.

11. Use a graphing calculator to find the inverse.

$$\begin{bmatrix} 1 & -1 & 2 & 1 \\ 2 & -1 & 5 & 1 \\ 3 & -3 & 7 & 5 \\ -2 & 3 & -4 & -1 \end{bmatrix}$$

The inverse matrix is $\begin{bmatrix} \frac{19}{2} & -\frac{1}{2} & -\frac{3}{2} & \frac{3}{2} \\ \frac{7}{4} & \frac{1}{4} & -\frac{1}{4} & \frac{3}{4} \\ -\frac{7}{2} & \frac{1}{2} & \frac{1}{2} & -\frac{1}{2} \\ \frac{1}{4} & -\frac{1}{4} & \frac{1}{4} & \frac{1}{4} \end{bmatrix}$.

13. Use a graphing calculator to find the inverse.

$$\begin{bmatrix} 1 & -1 & 1 & 3 \\ 2 & -1 & 4 & 8 \\ 1 & 1 & 6 & 10 \\ -1 & 5 & 5 & 4 \end{bmatrix}$$

The inverse matrix is $\begin{bmatrix} 2 & \frac{3}{5} & -\frac{7}{5} & \frac{4}{5} \\ 4 & -\frac{7}{5} & -\frac{2}{5} & \frac{4}{5} \\ -6 & \frac{14}{5} & -\frac{1}{5} & -\frac{3}{5} \\ 3 & -\frac{8}{5} & \frac{2}{5} & \frac{1}{5} \end{bmatrix}$.

15. Solve the system using inverses.

$$\begin{bmatrix} 1 & 4 \\ 2 & 7 \end{bmatrix}\begin{bmatrix} x \\ y \end{bmatrix}=\begin{bmatrix} 6 \\ 11 \end{bmatrix} \quad (1)$$

Find the inverse of $\begin{bmatrix} 1 & 4 \\ 2 & 7 \end{bmatrix}$.

$$\left[\begin{array}{cc|cc} 1 & 4 & 1 & 0 \\ 2 & 7 & 0 & 1 \end{array}\right]\xrightarrow{-2R_1+R_2}\left[\begin{array}{cc|cc} 1 & 4 & 1 & 0 \\ 0 & -1 & -2 & 1 \end{array}\right]$$

$$\xrightarrow{-1R_2}\left[\begin{array}{cc|cc} 1 & 4 & 1 & 0 \\ 0 & 1 & 2 & -1 \end{array}\right]$$

$$\xrightarrow{-4R_2+R_1}\left[\begin{array}{cc|cc} 1 & 0 & -7 & 4 \\ 0 & 1 & 2 & -1 \end{array}\right]$$

The inverse of $\begin{bmatrix} 1 & 4 \\ 2 & 7 \end{bmatrix}$ is $\begin{bmatrix} -7 & 4 \\ 2 & -1 \end{bmatrix}$.

Multiply each side of Eq. (1) by the inverse matrix.

$$\begin{bmatrix} -7 & 4 \\ 2 & -1 \end{bmatrix}\begin{bmatrix} 1 & 4 \\ 2 & 7 \end{bmatrix}\begin{bmatrix} x \\ y \end{bmatrix}=\begin{bmatrix} -7 & 4 \\ 2 & -1 \end{bmatrix}\begin{bmatrix} 6 \\ 11 \end{bmatrix}$$

$$\begin{bmatrix} x \\ y \end{bmatrix}=\begin{bmatrix} 2 \\ 1 \end{bmatrix}$$

The solution is (2, 1).

17. Solve the system using inverses.

$$\begin{bmatrix}1 & -2\\3 & 2\end{bmatrix}\begin{bmatrix}x\\y\end{bmatrix}=\begin{bmatrix}8\\-1\end{bmatrix} \quad (1)$$

Find the inverse matrix of $\begin{bmatrix}1 & -2\\3 & 2\end{bmatrix}$.

$$\left[\begin{array}{cc|cc}1 & -2 & 1 & 0\\3 & 2 & 0 & 1\end{array}\right]\xrightarrow{-3R_1+R_2}\left[\begin{array}{cc|cc}1 & -2 & 1 & 0\\0 & 8 & -3 & 1\end{array}\right]$$

$$\xrightarrow{\frac{1}{8}R_2}\left[\begin{array}{cc|cc}1 & -2 & 1 & 0\\0 & 1 & -\frac{3}{8} & \frac{1}{8}\end{array}\right]$$

$$\xrightarrow{2R_2+R_1}\left[\begin{array}{cc|cc}1 & 0 & \frac{1}{4} & \frac{1}{4}\\0 & 1 & -\frac{3}{8} & \frac{1}{8}\end{array}\right]$$

The inverse matrix is $\begin{bmatrix}\frac{1}{4} & \frac{1}{4}\\-\frac{3}{8} & \frac{1}{8}\end{bmatrix}$.

Multiply each side of Eq. (1) by the inverse matrix.

$$\begin{bmatrix}\frac{1}{4} & \frac{1}{4}\\-\frac{3}{8} & \frac{1}{8}\end{bmatrix}\begin{bmatrix}1 & -2\\3 & 2\end{bmatrix}\begin{bmatrix}x\\y\end{bmatrix}=\begin{bmatrix}\frac{1}{4} & \frac{1}{4}\\-\frac{3}{8} & \frac{1}{8}\end{bmatrix}\begin{bmatrix}8\\-1\end{bmatrix}$$

$$\begin{bmatrix}x\\y\end{bmatrix}=\begin{bmatrix}\frac{7}{4}\\-\frac{25}{8}\end{bmatrix}$$

The solution is $\left(\frac{7}{4},-\frac{25}{8}\right)$.

19. Solve the system using inverses.

$$\begin{bmatrix}1 & 1 & 2\\2 & 3 & 3\\3 & 3 & 7\end{bmatrix}\begin{bmatrix}x\\y\\z\end{bmatrix}=\begin{bmatrix}4\\5\\14\end{bmatrix} \quad (1)$$

Find the inverse matrix of $\begin{bmatrix}1 & 1 & 2\\2 & 3 & 3\\3 & 3 & 7\end{bmatrix}$.

$$\left[\begin{array}{ccc|ccc}1 & 1 & 2 & 1 & 0 & 0\\2 & 3 & 3 & 0 & 1 & 0\\3 & 3 & 7 & 0 & 0 & 1\end{array}\right]\xrightarrow[-3R_1+R_3]{-2R_1+R_2}\left[\begin{array}{ccc|ccc}1 & 1 & 2 & 1 & 0 & 0\\0 & 1 & -1 & -2 & 1 & 0\\0 & 0 & 1 & -3 & 0 & 1\end{array}\right]$$

$$\xrightarrow{-R_2+R_1}\left[\begin{array}{ccc|ccc}1 & 0 & 3 & 3 & -1 & 0\\0 & 1 & -1 & -2 & 1 & 0\\0 & 0 & 1 & -3 & 0 & 1\end{array}\right]$$

$$\xrightarrow[R_3+R_2]{-3R_3+R_1}\left[\begin{array}{ccc|ccc}1 & 0 & 0 & 12 & -1 & -3\\0 & 1 & 0 & -5 & 1 & 1\\0 & 0 & 1 & -3 & 0 & 1\end{array}\right]$$

The inverse matrix is $\begin{bmatrix}12 & -1 & -3\\-5 & 1 & 1\\-3 & 0 & 1\end{bmatrix}$.

Multiply each side of Eq. (1) by the inverse matrix.

$$\begin{bmatrix}12 & -1 & -3\\-5 & 1 & 1\\-3 & 0 & 1\end{bmatrix}\begin{bmatrix}1 & 1 & 2\\2 & 3 & 3\\3 & 3 & 7\end{bmatrix}\begin{bmatrix}x\\y\\z\end{bmatrix}=\begin{bmatrix}12 & -1 & -3\\-5 & 1 & 1\\-3 & 0 & 1\end{bmatrix}\begin{bmatrix}4\\5\\14\end{bmatrix}$$

$$\begin{bmatrix}x\\y\\z\end{bmatrix}=\begin{bmatrix}1\\-1\\2\end{bmatrix}$$

The solution is (1, –1, 2).

21. Solve the system using inverses.

$$\begin{bmatrix}1 & 2 & 2\\-2 & -5 & -2\\2 & 4 & 7\end{bmatrix}\begin{bmatrix}x\\y\\z\end{bmatrix}=\begin{bmatrix}5\\8\\19\end{bmatrix} \quad (1)$$

Find the inverse matrix of $\begin{bmatrix}1 & 2 & 2\\-2 & -5 & -2\\2 & 4 & 7\end{bmatrix}$.

$$\left[\begin{array}{ccc|ccc}1 & 2 & 2 & 1 & 0 & 0\\-2 & -5 & -2 & 0 & 1 & 0\\2 & 4 & 7 & 0 & 0 & 1\end{array}\right]$$

$$\xrightarrow[-2R_1+R_3]{2R_1+R_2}\left[\begin{array}{ccc|ccc}1 & 2 & 2 & 1 & 0 & 0\\0 & -1 & 2 & 2 & 1 & 0\\0 & 0 & 3 & -2 & 0 & 1\end{array}\right]$$

$$\xrightarrow{-1R_2}\left[\begin{array}{ccc|ccc}1 & 2 & 2 & 1 & 0 & 0\\0 & 1 & -2 & -2 & -1 & 0\\0 & 0 & 3 & -2 & 0 & 1\end{array}\right]$$

$$\xrightarrow{\frac{1}{3}R_3}\left[\begin{array}{ccc|ccc}1 & 2 & 2 & 1 & 0 & 0\\0 & 1 & -2 & -2 & -1 & 0\\0 & 0 & 1 & -\frac{2}{3} & 0 & \frac{1}{3}\end{array}\right]$$

$$\xrightarrow{-2R_2+R_1}\left[\begin{array}{ccc|ccc}1 & 0 & 6 & 5 & 2 & 0\\0 & 1 & -2 & -2 & -1 & 0\\0 & 0 & 1 & -\frac{2}{3} & 0 & \frac{1}{3}\end{array}\right]$$

$$\xrightarrow[-6R_3+R_1]{2R_3+R_2}\left[\begin{array}{ccc|ccc}1&0&0&9&2&-2\\0&1&0&-\frac{10}{3}&-1&\frac{2}{3}\\0&0&1&-\frac{2}{3}&0&\frac{1}{3}\end{array}\right]$$

The inverse matrix is $\begin{bmatrix}9&2&-2\\-\frac{10}{3}&-1&\frac{2}{3}\\-\frac{2}{3}&0&\frac{1}{3}\end{bmatrix}$.

Multiply each side of Eq. (1) by the inverse matrix.

$$\begin{bmatrix}9&2&-2\\-\frac{10}{3}&-1&\frac{2}{3}\\-\frac{2}{3}&0&\frac{1}{3}\end{bmatrix}\begin{bmatrix}1&2&2\\-2&-5&-2\\2&4&7\end{bmatrix}\begin{bmatrix}x\\y\\z\end{bmatrix}=\begin{bmatrix}9&2&-2\\-\frac{10}{3}&-1&\frac{2}{3}\\-\frac{2}{3}&0&\frac{1}{3}\end{bmatrix}\begin{bmatrix}5\\8\\19\end{bmatrix}$$

$$\begin{bmatrix}x\\y\\z\end{bmatrix}=\begin{bmatrix}23\\-12\\3\end{bmatrix}$$

The solution is (23, –12, 3).

23. Solve the system using inverses.

$$\begin{bmatrix}1&2&0&1\\2&5&1&2\\2&4&1&1\\3&6&0&4\end{bmatrix}\begin{bmatrix}w\\x\\y\\z\end{bmatrix}=\begin{bmatrix}6\\10\\8\\16\end{bmatrix}\quad(1)$$

Find the inverse matrix of $\begin{bmatrix}1&2&0&1\\2&5&1&2\\2&4&1&1\\3&6&0&4\end{bmatrix}$.

$$\left[\begin{array}{cccc|cccc}1&2&0&1&1&0&0&0\\2&5&1&2&0&1&0&0\\2&4&1&1&0&0&1&0\\3&6&0&4&0&0&0&1\end{array}\right]$$

$$\xrightarrow[\substack{-2R_1+R_3\\-3R_1+R_4}]{-2R_1+R_2}\left[\begin{array}{cccc|cccc}1&2&0&1&1&0&0&0\\0&1&1&0&-2&1&0&0\\0&0&1&-1&-2&0&1&0\\0&0&0&1&-3&0&0&1\end{array}\right]$$

$$\xrightarrow{-2R_2+R_1}\left[\begin{array}{cccc|cccc}1&0&-2&1&5&-2&0&0\\0&1&1&0&-2&1&0&0\\0&0&1&-1&-2&0&1&0\\0&0&0&1&-3&0&0&1\end{array}\right]$$

$$\xrightarrow[-1R_3+R_2]{2R_3+R_1}\left[\begin{array}{cccc|cccc}1&0&0&-1&1&-2&2&0\\0&1&0&1&0&1&-1&0\\0&0&1&-1&-2&0&1&0\\0&0&0&1&-3&0&0&1\end{array}\right]$$

$$\xrightarrow[\substack{-R_4+R_2\\R_4+R_3}]{R_4+R_1}\left[\begin{array}{cccc|cccc}1&0&0&0&-2&-2&2&1\\0&1&0&0&3&1&-1&-1\\0&0&1&0&-5&0&1&1\\0&0&0&1&-3&0&0&1\end{array}\right]$$

The inverse matrix is $\begin{bmatrix}-2&-2&2&1\\3&1&-1&-1\\-5&0&1&1\\-3&0&0&1\end{bmatrix}$.

Multiply each side of Eq. (1) by the inverse matrix.

$$\begin{bmatrix}-2&-2&2&1\\-3&1&-1&-1\\-5&0&1&1\\-3&0&0&1\end{bmatrix}\begin{bmatrix}1&2&0&1\\2&5&1&2\\2&4&1&1\\3&6&0&4\end{bmatrix}\begin{bmatrix}w\\x\\y\\z\end{bmatrix}=\begin{bmatrix}-2&-2&2&1\\3&1&-1&-1\\-5&0&1&1\\-3&0&0&1\end{bmatrix}\begin{bmatrix}6\\10\\8\\16\end{bmatrix}$$

$$\begin{bmatrix}w\\x\\y\\z\end{bmatrix}=\begin{bmatrix}0\\4\\-6\\-2\end{bmatrix}$$

The solution is (0, 4, –6, –2).

25. Find the temperature at x_1 and x_2.

The average temperature for the two points,

$$x_1=\frac{35+50+x_2+60}{4}=\frac{145+x_2}{4}\text{ or }4x_1-x_2=145$$

$$x_2=\frac{x_1+50+55+60}{4}=\frac{165+x_1}{4}\text{ or }-x_1+4x_2=165$$

The system of equations and associated matrix equation are

$$\begin{cases}4x_1-x_2=145\\-x_1+4x_2=165\end{cases}\qquad\begin{bmatrix}4&-1\\-1&4\end{bmatrix}\begin{bmatrix}x_1\\x_2\end{bmatrix}=\begin{bmatrix}145\\165\end{bmatrix}$$

Solving the matrix equation by using an inverse matrix gives $\begin{bmatrix}x_1\\x_2\end{bmatrix}=\begin{bmatrix}49.7\\53.7\end{bmatrix}$

The temperatures are x_1 = 49.7°F, x_2 = 53.7°F.

27. Find the temperature at $x_1, x_2, x_3,$ and x_4.

The average temperature for the two points,

$$x_1 = \frac{50+60+x_2+x_3}{4} = \frac{110+x_2+x_3}{4}$$
or $4x_1 - x_2 - x_3 = 110$
$$x_2 = \frac{x_1+60+60+x_4}{4} = \frac{120+x_1+x_4}{4}$$
or $-x_1 + 4x_2 - x_4 = 120$
$$x_3 = \frac{50+x_1+x_4+50}{4} = \frac{100+x_1+x_4}{4}$$
or $-x_1 + 4x_3 - x_4 = 100$
$$x_4 = \frac{x_3+x_2+60+50}{4} = \frac{110+x_2+x_3}{4}$$
or $-x_2 - x_3 + 4x_4 = 110$

The system of equations and associated matrix equation are

$$\begin{cases} 4x_1-x_2-x_3=110 \\ -x_1+4x_2-x_4=120 \\ -x_1+4x_3-x_4=100 \\ -x_2-x_3+4x_4=110 \end{cases} \quad \begin{bmatrix} 4 & -1 & -1 & 0 \\ -1 & 4 & 0 & -1 \\ -1 & 0 & 4 & -1 \\ 0 & -1 & -1 & 4 \end{bmatrix}\begin{bmatrix} x_1 \\ x_2 \\ x_3 \\ x_4 \end{bmatrix} = \begin{bmatrix} 110 \\ 120 \\ 100 \\ 110 \end{bmatrix}$$

Solving the matrix equation by using an inverse matrix

gives $\begin{bmatrix} x_1 \\ x_2 \\ x_3 \\ x_4 \end{bmatrix} = \begin{bmatrix} 55 \\ 57.5 \\ 52.5 \\ 55 \end{bmatrix}$

The temperatures are $x_1 = 55°\text{F}$, $x_2 = 57.5°\text{F}$, $x_3 = 52.5°\text{F}$, $x_4 = 55°\text{F}$.

29. Find the number of children and adults.

A = number of adult tickets

C = number of child tickets

Saturday $\quad A + C = 100$

$\quad 20A + 15C = 1900$

$$\begin{bmatrix} 1 & 1 \\ 20 & 15 \end{bmatrix}\begin{bmatrix} A \\ C \end{bmatrix} = \begin{bmatrix} 100 \\ 1900 \end{bmatrix}$$

$$\begin{bmatrix} -3 & \frac{1}{5} \\ 4 & -\frac{1}{5} \end{bmatrix}\begin{bmatrix} 1 & 1 \\ 20 & 15 \end{bmatrix}\begin{bmatrix} A \\ C \end{bmatrix} = \begin{bmatrix} -3 & \frac{1}{5} \\ 4 & -\frac{1}{5} \end{bmatrix}\begin{bmatrix} 100 \\ 1900 \end{bmatrix}$$

$$\begin{bmatrix} A \\ C \end{bmatrix} = \begin{bmatrix} 80 \\ 20 \end{bmatrix}$$

On Saturday, 80 adults and 20 children took the tour.

Sunday $\quad A + C = 120$

$\quad 20A + 15C = 2275$

$$\begin{bmatrix} 1 & 1 \\ 20 & 15 \end{bmatrix}\begin{bmatrix} A \\ C \end{bmatrix} = \begin{bmatrix} 120 \\ 2275 \end{bmatrix}$$

$$\begin{bmatrix} -3 & \frac{1}{5} \\ 4 & -\frac{1}{5} \end{bmatrix}\begin{bmatrix} 1 & 1 \\ 20 & 15 \end{bmatrix}\begin{bmatrix} A \\ C \end{bmatrix} = \begin{bmatrix} -3 & \frac{1}{5} \\ 4 & -\frac{1}{5} \end{bmatrix}\begin{bmatrix} 120 \\ 2275 \end{bmatrix}$$

$$\begin{bmatrix} A \\ C \end{bmatrix} = \begin{bmatrix} 95 \\ 25 \end{bmatrix}$$

On Sunday, 95 adults and 25 children took the tour.

31. Find the amount of additive for each sample.

$x_1 =$ number of 100-gram portions of additive 1

$x_2 =$ number of 100-gram portions of additive 2

$x_3 =$ number of 100-gram portions of additive 3

Sample 1:
$$\begin{aligned} 30x_1 + 40x_2 + 50x_3 &= 380 \\ 10x_1 + 15x_2 + 5x_3 &= 95 \\ 10x_1 + 10x_2 + 5x_3 &= 85 \end{aligned}$$

$$\begin{bmatrix} 30 & 40 & 50 \\ 10 & 15 & 5 \\ 10 & 10 & 5 \end{bmatrix}\begin{bmatrix} x_1 \\ x_2 \\ x_3 \end{bmatrix} = \begin{bmatrix} 380 \\ 95 \\ 85 \end{bmatrix}$$

$$\begin{bmatrix} -\frac{1}{70} & -\frac{6}{35} & \frac{11}{35} \\ 0 & \frac{1}{5} & -\frac{1}{5} \\ \frac{1}{35} & -\frac{2}{35} & -\frac{1}{35} \end{bmatrix}\begin{bmatrix} 30 & 40 & 50 \\ 10 & 15 & 5 \\ 10 & 10 & 5 \end{bmatrix}\begin{bmatrix} x_1 \\ x_2 \\ x_3 \end{bmatrix} = \begin{bmatrix} -\frac{1}{70} & -\frac{6}{35} & \frac{11}{35} \\ 0 & \frac{1}{5} & -\frac{1}{5} \\ \frac{1}{35} & -\frac{2}{35} & -\frac{1}{35} \end{bmatrix}\begin{bmatrix} 380 \\ 95 \\ 85 \end{bmatrix}$$

$$\begin{bmatrix} x_1 \\ x_2 \\ x_3 \end{bmatrix} = \begin{bmatrix} 5 \\ 2 \\ 3 \end{bmatrix}$$

For Sample 1, 500 g of additive 1, 200 g of additive 2, and 300 g of additive 3 are required.

Sample 2:
$$\begin{aligned} 30x_1 + 40x_2 + 50x_3 &= 380 \\ 10x_1 + 15x_2 + 5x_3 &= 110 \\ 10x_1 + 10x_2 + 5x_3 &= 90 \end{aligned}$$

$$\begin{bmatrix} 30 & 40 & 50 \\ 10 & 15 & 5 \\ 10 & 10 & 5 \end{bmatrix}\begin{bmatrix} x_1 \\ x_2 \\ x_3 \end{bmatrix} = \begin{bmatrix} 380 \\ 110 \\ 90 \end{bmatrix}$$

$$\begin{bmatrix} -\frac{1}{70} & -\frac{6}{35} & \frac{11}{35} \\ 0 & \frac{1}{5} & -\frac{1}{5} \\ \frac{1}{35} & -\frac{2}{35} & -\frac{1}{35} \end{bmatrix}\begin{bmatrix} 30 & 40 & 50 \\ 10 & 15 & 5 \\ 10 & 10 & 5 \end{bmatrix}\begin{bmatrix} x_1 \\ x_2 \\ x_3 \end{bmatrix} = \begin{bmatrix} -\frac{1}{70} & -\frac{6}{35} & \frac{11}{35} \\ 0 & \frac{1}{5} & -\frac{1}{5} \\ \frac{1}{35} & -\frac{2}{35} & -\frac{1}{35} \end{bmatrix}\begin{bmatrix} 380 \\ 110 \\ 90 \end{bmatrix}$$

$$\begin{bmatrix} x_1 \\ x_2 \\ x_3 \end{bmatrix} = \begin{bmatrix} 4 \\ 4 \\ 2 \end{bmatrix}$$

For Sample 2, 400 g of additive 1, 400 g of additive 2, and 200 g of additive 3 are required.

33. Use a graphing calculator to find the inverse.

$$\begin{bmatrix} 2 & -2 & 3 & 1 \\ 5 & 2 & -2 & 3 \\ 6 & -1 & 2 & 3 \\ 2 & 3 & -1 & 5 \end{bmatrix}^{-1} \approx \begin{bmatrix} -5.667 & -3.667 & 5 & 0.333 \\ -27.667 & -18.667 & 24 & 2.333 \\ -19.333 & -13.333 & 17 & 1.667 \\ 15 & 10 & -13 & -1 \end{bmatrix}$$

35. Use a graphing calculator to find the inverse.

$$\begin{bmatrix} -\frac{2}{7} & 4 & -\frac{1}{6} \\ -2 & \sqrt{2} & -3 \\ \sqrt{3} & 3 & -\sqrt{5} \end{bmatrix}^{-1} \approx \begin{bmatrix} -0.150 & -0.217 & 0.302 \\ 0.248 & -0.024 & 0.013 \\ 0.217 & -0.200 & -0.195 \end{bmatrix}$$

37. Find the gross output for the given conditions.

$X = (I - A)^{-1}D$, where X is consumer demand, I is the identity matrix, A is the input-output matrix, and D is the final demand. Thus

$$X = \left(\begin{bmatrix} 1 & 0 & 0 \\ 0 & 1 & 0 \\ 0 & 0 & 1 \end{bmatrix} - \begin{bmatrix} 0.20 & 0.15 & 0.10 \\ 0.10 & 0.30 & 0.25 \\ 0.20 & 0.10 & 0.10 \end{bmatrix}\right)^{-1} \begin{bmatrix} 120 \\ 60 \\ 55 \end{bmatrix}$$

$$= \begin{bmatrix} 0.80 & -0.15 & -0.10 \\ -0.10 & 0.70 & -0.25 \\ -0.20 & -0.10 & 0.90 \end{bmatrix}^{-1} \begin{bmatrix} 120 \\ 60 \\ 55 \end{bmatrix} \approx \begin{bmatrix} 194.67 \\ 157.03 \\ 121.82 \end{bmatrix}$$

\$194.67 million worth of manufacturing, \$157.03 million worth of transportation, \$121.82 million worth of services.

39. Find the amount each industry should produce. The input-output matrix, A, is given by

$$A = \begin{bmatrix} 0.05 & 0.20 & 0.15 \\ 0.02 & 0.03 & 0.25 \\ 0.10 & 0.12 & 0.05 \end{bmatrix}$$

Consumer demand is given by

$$X = (I - A)^{-1}D$$

$$X = \left(\begin{bmatrix} 1 & 0 & 0 \\ 0 & 1 & 0 \\ 0 & 0 & 1 \end{bmatrix} - \begin{bmatrix} 0.05 & 0.20 & 0.15 \\ 0.02 & 0.03 & 0.25 \\ 0.10 & 0.12 & 0.05 \end{bmatrix}\right)^{-1} \begin{bmatrix} 30 \\ 5 \\ 25 \end{bmatrix}$$

$$= \begin{bmatrix} 0.95 & -0.20 & -0.15 \\ -0.02 & 0.97 & -0.25 \\ -0.10 & -0.12 & 0.95 \end{bmatrix}^{-1} \begin{bmatrix} 30 \\ 5 \\ 25 \end{bmatrix} \approx \begin{bmatrix} 39.69 \\ 14.30 \\ 32.30 \end{bmatrix}$$

\$39.69 million worth of coal, \$14.30 million worth of iron, \$32.30 million worth of steel.

41. Show that $AB = O$.

$$AB = \begin{bmatrix} 2 & -3 \\ -6 & 9 \end{bmatrix} \begin{bmatrix} -3 & 15 \\ -2 & 10 \end{bmatrix} = \begin{bmatrix} 2(-3)+(-3)(-2) & 2(15)+(-3)(10) \\ -6(-3)+9(-2) & -6(15)+9(10) \end{bmatrix} = \begin{bmatrix} 0 & 0 \\ 0 & 0 \end{bmatrix} = O$$

43. Show $AB = AC$.

$$AB = \begin{bmatrix} 2 & -1 \\ -4 & 2 \end{bmatrix} \begin{bmatrix} 3 & 4 \\ 1 & 5 \end{bmatrix} = \begin{bmatrix} 2(3)+(-1)(1) & 2(4)+(-1)(5) \\ -4(3)+2(1) & -4(4)+2(5) \end{bmatrix} = \begin{bmatrix} 5 & 3 \\ -10 & -6 \end{bmatrix}$$

$$AC = \begin{bmatrix} 2 & -1 \\ -4 & 2 \end{bmatrix} \begin{bmatrix} 4 & 7 \\ 3 & 11 \end{bmatrix} = \begin{bmatrix} 2(4)+(-1)(3) & 2(7)+(-1)(11) \\ -4(4)+2(3) & -4(7)+2(11) \end{bmatrix} = \begin{bmatrix} 5 & 3 \\ -10 & -6 \end{bmatrix}$$

Mid-Chapter 7 Quiz

1. Solve the system using Gaussian elimination method.

$$\begin{bmatrix} -2 & 1 & -4 & -4 \\ 0 & 1 & -3 & -2 \\ 5 & 1 & -1 & -4 \end{bmatrix} \xrightarrow{-\frac{1}{2}R_1} \begin{bmatrix} 1 & -\frac{1}{2} & 2 & 2 \\ 0 & 1 & -3 & -2 \\ 5 & 1 & -1 & -4 \end{bmatrix}$$

$$\xrightarrow{-5R_1+R_3} \begin{bmatrix} 1 & -\frac{1}{2} & 2 & 2 \\ 0 & 1 & -3 & -2 \\ 0 & \frac{7}{2} & -11 & -14 \end{bmatrix}$$

$$\xrightarrow{-\frac{7}{2}R_2+R_3}\begin{bmatrix}1 & -\frac{1}{2} & 2 & 2\\ 0 & 1 & -3 & -2\\ 0 & 0 & -\frac{1}{2} & -7\end{bmatrix}$$

$$\begin{cases} x-\frac{1}{2}y+2z=2\\ \quad y-3z=-2\\ \quad\quad -\frac{1}{2}z=-7\end{cases}$$

$z=14$

$y-3(14)=-2$

$y=40$

$x-\frac{1}{2}(40)+2(14)=2$

$x=-6$

The solution set is (–6, 40, 14)

Using a graphing calculator, we get the following equivalent answer.

$$\begin{bmatrix}1 & \frac{1}{5} & -\frac{1}{5} & -\frac{4}{5}\\ 0 & 1 & -\frac{22}{7} & -4\\ 0 & 0 & 1 & 14\end{bmatrix}$$

3. Evaluate.

$$A+C=\begin{bmatrix}5 & 1 & -4\\ -3 & 3 & -4\\ 1 & 5 & -2\end{bmatrix}+\begin{bmatrix}1 & -3 & 2\\ 5 & 0 & -4\\ 5 & -1 & -3\end{bmatrix}=\begin{bmatrix}6 & -2 & -2\\ 2 & 3 & -8\\ 6 & 4 & -5\end{bmatrix}$$

5. Evaluate.

$$AB=\begin{bmatrix}5 & 1 & -4\\ -3 & 3 & -4\\ 1 & 5 & -2\end{bmatrix}\begin{bmatrix}3 & -4\\ 3 & 2\\ 2 & -2\end{bmatrix}=\begin{bmatrix}10 & -10\\ -8 & 26\\ 14 & 10\end{bmatrix}$$

Prepare for Section 7.4

P1. Find the order of the matrix.

2

P3. Evaluate.

$(-1)^{1+1}(-3)+(-1)^{1+2}(-2)+(-1)^{1+3}(5)$

$=(-1)^2(-3)+(-1)^3(-2)+(-1)^4(5)$

$=-3+(-1)(-2)+5$

$=-3+2+5$

$=4$

P5. Simplify.

$$3\begin{bmatrix}-2 & 1\\ 3 & -5\end{bmatrix}=\begin{bmatrix}-6 & 3\\ 9 & -15\end{bmatrix}$$

Section 7.4 Exercises

1. Evaluate the determinant.

$$\begin{vmatrix}2 & -1\\ 3 & 5\end{vmatrix}=2(5)-(-1)(3)=10-(-3)=13$$

3. Evaluate the determinant.

$$\begin{vmatrix}5 & 0\\ 2 & -3\end{vmatrix}=5(-3)-(2)(0)=-15-0=-15$$

5. Evaluate the determinant.

$$\begin{vmatrix}4 & 6\\ 2 & 3\end{vmatrix}=4(3)-(2)(6)=12-12=0$$

7. Evaluate the determinant.

$$\begin{vmatrix}0 & 9\\ 0 & -2\end{vmatrix}=0(-2)-(0)(9)=0-0=0$$

9. Evaluate the minor and cofactor.

$$M_{11}=\begin{vmatrix}4 & -1\\ -5 & 6\end{vmatrix}=4(6)-(-5)(-1)=19$$

$$C_{11}=(-1)^{1+1}M_{11}=M_{11}=19$$

11. Evaluate the minor and cofactor.

$$M_{32}=\begin{vmatrix}5 & -3\\ 2 & -1\end{vmatrix}=5(-1)-2(-3)=1$$

$$C_{32}=(-1)^{3+2}M_{32}=-M_{32}=-1$$

13. Evaluate the minor and cofactor.

$$M_{22}=\begin{vmatrix}3 & 3\\ 6 & 3\end{vmatrix}=3(3)-6(3)=-9$$

$$C_{22}=(-1)^{2+2}M_{22}=M_{22}=-9$$

15. Evaluate the minor and cofactor.

$$M_{31}=\begin{vmatrix}-2 & 3\\ 3 & 0\end{vmatrix}=-2(0)-3(3)=-9$$

$$C_{31}=(-1)^{3+1}M_{31}=M_{31}=-9$$

17. Evaluate the determinant by expanding by cofactors.

$$\begin{vmatrix}2 & -3 & 1\\ 2 & 0 & 2\\ 3 & -2 & 4\end{vmatrix}=-2\begin{vmatrix}-3 & 1\\ -2 & 4\end{vmatrix}+0\begin{vmatrix}2 & 1\\ 3 & 4\end{vmatrix}-2\begin{vmatrix}2 & -3\\ 3 & -2\end{vmatrix}$$

$$=-2(-10)+0-2(5)=20-10$$

$$=10$$

19. Evaluate the determinant by expanding by cofactors.

$$\begin{vmatrix} -2 & 3 & 2 \\ 1 & 2 & -3 \\ -4 & -2 & 1 \end{vmatrix} = -2\begin{vmatrix} 2 & -3 \\ -2 & 1 \end{vmatrix} - 3\begin{vmatrix} 1 & -3 \\ -4 & 1 \end{vmatrix} + 2\begin{vmatrix} 1 & 2 \\ -4 & -2 \end{vmatrix}$$
$$= -2(-4) - 3(-11) + 2(6)$$
$$= 8 + 33 + 12$$
$$= 53$$

21. Evaluate the determinant by expanding by cofactors.

$$\begin{vmatrix} 2 & -3 & 10 \\ 0 & 2 & -3 \\ 0 & 0 & 5 \end{vmatrix} = 2\begin{vmatrix} 2 & -3 \\ 0 & 5 \end{vmatrix} - 0\begin{vmatrix} -3 & 10 \\ 0 & 5 \end{vmatrix} + 0\begin{vmatrix} -3 & 10 \\ 2 & -3 \end{vmatrix}$$
$$= 2(10) - 0 + 0$$
$$= 20$$

23. Evaluate the determinant by expanding by cofactors.

$$\begin{vmatrix} 0 & -2 & 4 \\ 1 & 0 & -7 \\ 5 & -6 & 0 \end{vmatrix} = 0\begin{vmatrix} 0 & -7 \\ -6 & 0 \end{vmatrix} - (-2)\begin{vmatrix} 1 & -7 \\ 5 & 0 \end{vmatrix} + 4\begin{vmatrix} 1 & 0 \\ 5 & -6 \end{vmatrix}$$
$$= 0 + 2(35) + 4(-6)$$
$$= 70 - 24$$
$$= 46$$

25. Evaluate the determinant by expanding by cofactors.

$$\begin{vmatrix} 4 & -3 & 3 \\ 2 & 1 & -4 \\ 6 & -2 & -1 \end{vmatrix} = 4\begin{vmatrix} 1 & -4 \\ -2 & -1 \end{vmatrix} - (-3)\begin{vmatrix} 2 & -4 \\ 6 & -1 \end{vmatrix} + 3\begin{vmatrix} 2 & 1 \\ 6 & -2 \end{vmatrix}$$
$$= 4(-9) + 3(22) + 3(-10)$$
$$= -36 + 66 - 30$$
$$= 0$$

27. Give a reason for each equality.

Row 2 consists entirely of zeros. Therefore, the determinant is zero.

29. Give a reason for each equality.

2 was factored from row 2.

31. Give a reason for each equality.

Row 1 was multiplied by –2 and added to row 2.

33. Give a reason for each equality.

2 was factored from column 1.

35. Give a reason for each equality.

The matrix is in triangular form.

The product of the elements on the main diagonal is –12.

Therefore, the value of the determinant is –12.

37. Give a reason for each equality.

Row 1 and row 3 were interchanged.

Therefore, the sign of the determinant was changed.

39. Give a reason for each equality.

Each row of the first determinant was multiplied by a to produce the second determinant.

41. Put the matrix in triangular form and evaluate.

$$\begin{vmatrix} 2 & 4 & 1 \\ 1 & 2 & -1 \\ 1 & 2 & 2 \end{vmatrix} = -\begin{vmatrix} 1 & 2 & -1 \\ 2 & 4 & -1 \\ 1 & 2 & 2 \end{vmatrix} R_1 \leftrightarrow R_2$$
$$= -\begin{vmatrix} 1 & 2 & -1 \\ 0 & 0 & 1 \\ 0 & 0 & 3 \end{vmatrix} \begin{matrix} -2R_1 + R_2 \\ -R_1 + R_3 \end{matrix}$$
$$= -(1)(0)(3) = 0$$

43. Put the matrix in triangular form and evaluate.

$$\begin{vmatrix} 1 & 2 & -1 \\ 2 & 3 & 1 \\ 3 & 4 & 3 \end{vmatrix} = \begin{vmatrix} 1 & 2 & -1 \\ 0 & -1 & 3 \\ 0 & -2 & 6 \end{vmatrix} \begin{matrix} -2R_1 + R_2 \\ -3R_1 + R_3 \end{matrix}$$
$$= \begin{vmatrix} 1 & 2 & -1 \\ 0 & -1 & 3 \\ 0 & 0 & 0 \end{vmatrix} -2R_2 + R_3$$
$$= (-1)(0) = 0$$

45. Put the matrix in triangular form and evaluate.

$$\begin{vmatrix} 0 & -1 & 1 \\ 1 & 0 & -2 \\ 2 & 2 & 0 \end{vmatrix} = -\begin{vmatrix} 1 & 0 & -2 \\ 0 & -1 & 1 \\ 2 & 2 & 0 \end{vmatrix} R_1 \leftrightarrow R_2$$
$$= -\begin{vmatrix} 1 & 0 & -2 \\ 0 & -1 & 1 \\ 0 & 2 & 4 \end{vmatrix} -2R_1 + R_3$$
$$= -\begin{vmatrix} 1 & 0 & -2 \\ 0 & -1 & 1 \\ 0 & 0 & 6 \end{vmatrix} 2R_2 + R_3$$
$$= -(1)(-1)(6) = 6$$

47. Put the matrix in triangular form and evaluate.

$$\begin{vmatrix} 1 & 2 & -1 & 2 \\ 1 & -2 & 0 & 3 \\ 3 & 0 & 1 & 5 \\ -2 & -4 & 1 & 6 \end{vmatrix} = \begin{vmatrix} 1 & 2 & -1 & 2 \\ 0 & -4 & 1 & 1 \\ 0 & -6 & 4 & -1 \\ 0 & 0 & -1 & 10 \end{vmatrix} \begin{matrix} -1R_1+R_2 \\ -3R_1+R_3 \\ 2R_1+R_4 \end{matrix}$$

$$= \begin{vmatrix} 1 & 2 & -1 & 2 \\ 0 & -4 & 1 & 1 \\ 0 & 0 & \frac{5}{2} & -\frac{5}{2} \\ 0 & 0 & -1 & 10 \end{vmatrix} -\frac{3}{2}R_2+R_3$$

$$= \begin{vmatrix} 1 & 2 & -1 & 2 \\ 0 & -4 & 1 & 1 \\ 0 & 0 & \frac{5}{2} & -\frac{5}{2} \\ 0 & 0 & 0 & 9 \end{vmatrix} \frac{2}{5}R_3+R_4$$

$$= 1(-4)\left(\frac{5}{2}\right)(9) = -90$$

49. Put the matrix in triangular form and evaluate.

$$\begin{vmatrix} 1 & 2 & 3 & -1 \\ 6 & 5 & 9 & 8 \\ 2 & 4 & 12 & -1 \\ 1 & 2 & 6 & -1 \end{vmatrix} = 3\begin{vmatrix} 1 & 2 & 1 & -1 \\ 6 & 5 & 3 & 8 \\ 2 & 4 & 4 & -1 \\ 1 & 2 & 2 & -1 \end{vmatrix} \quad \text{Factor 3 from } C_3$$

$$= 3\begin{vmatrix} 1 & 2 & 1 & -1 \\ 0 & -7 & -3 & 14 \\ 0 & 0 & 2 & 1 \\ 0 & 0 & 1 & 0 \end{vmatrix} \begin{matrix} -6R_1+R_2 \\ -2R_1+R_3 \\ -1R_1+R_4 \end{matrix}$$

$$= 3\begin{vmatrix} 1 & 2 & 1 & -1 \\ 0 & -7 & -3 & 14 \\ 0 & 0 & 2 & 1 \\ 0 & 0 & 0 & -\frac{1}{2} \end{vmatrix} -\frac{1}{2}R_3+R_4$$

$$= 3(1)(-7)(2)\left(-\frac{1}{2}\right) = 21$$

51. Use a graphing calculator to find the value of the determinant. Round to the nearest thousandth.

$$\begin{vmatrix} 2 & -2 & 3 & 1 \\ 5 & 2 & -2 & 3 \\ 6 & -1 & 2 & 3 \\ 2 & 3 & -1 & 5 \end{vmatrix} = 3$$

53. Use a graphing calculator to find the value of the determinant. Round to the nearest thousandth.

$$\begin{vmatrix} -\frac{2}{7} & 4 & -\frac{1}{6} \\ -2 & \sqrt{2} & -3 \\ \sqrt{3} & 3 & -\sqrt{5} \end{vmatrix} \approx -38.933$$

55. Find the area of the triangle.

$$\frac{1}{2}\begin{vmatrix} 2 & 3 & 1 \\ -1 & 0 & 1 \\ 4 & 8 & 1 \end{vmatrix} = \frac{1}{2}[3C_{12}+0C_{22}+8C_{32}]$$

$$= \frac{1}{2}\left[-3\begin{vmatrix} -1 & 1 \\ 4 & 1 \end{vmatrix} + 0 - 8\begin{vmatrix} 2 & 1 \\ -1 & 1 \end{vmatrix}\right]$$

$$= \frac{1}{2}[-3(-5)-8(3)] = \frac{1}{2}(15-24)$$

$$= \frac{1}{2}(-9) = -\frac{9}{2}$$

$$\left|-\frac{9}{2}\right| = \frac{9}{2}$$

The area of the triangle is $4\frac{1}{2}$ square units.

57. Find the area of the triangle.

$$\frac{1}{2}\begin{vmatrix} 4 & 9 & 1 \\ 8 & 2 & 1 \\ -3 & -2 & 1 \end{vmatrix} = \frac{1}{2}[4C_{11}+8C_{21}+(-3)C_{31}]$$

$$= \frac{1}{2}[4M_{11}-8M_{21}-3M_{31}]$$

$$= \frac{1}{2}\left[4\begin{vmatrix} 2 & 1 \\ -2 & 1 \end{vmatrix} - 8\begin{vmatrix} 9 & 1 \\ -2 & 1 \end{vmatrix} - 3\begin{vmatrix} 9 & 1 \\ 2 & 1 \end{vmatrix}\right]$$

$$= \frac{1}{2}[4(4)-8(11)-3(7)]$$

$$= \frac{1}{2}[16-88-21] = \frac{1}{2}(-93) = -\frac{93}{2}$$

$$\left|-\frac{93}{2}\right| = \frac{93}{2}$$

The area of the triangle is $46\frac{1}{2}$ square units.

59. Verify the statement.

$$\begin{vmatrix} x & y & 1 \\ x_1 & y_1 & 1 \\ x_2 & y_2 & 0 \end{vmatrix} = xC_{11}+yC_{12}+1C_{13} = 0$$

$$= xM_{11} - yM_{12} + 1M_{13} = 0$$

$$= x(0-y_2) - y(0-x_2) + (x_1y_2 - x_2y_1)$$

$$= 0$$

Since x_1, x_2, y_1 and y_2 are constants,

$x(0-y_2)-y(0-x_2)+(x_1y_2-x_2y_1)=0$ is a line in

the form $ax+by+c=0$, and (x_1, y_1) and (x_2, y_2) satisfy this equation.

61. Find the equation of the line.

$$\begin{vmatrix} x & y & 1 \\ -3 & 4 & 1 \\ 2 & -3 & 1 \end{vmatrix} = xC_{11}+yC_{12}+1C_{13}=0$$

$$= xM_{11}-yM_{12}+1M_{13}=0$$
$$= x(7)-y(-5)+1(1)=0$$
$$= 7x+5y+1=0$$

$7x+5y=-1$ is the equation of the line passing through the points (–3, 4) and (2,–3).

Prepare for Section 7.5

P1. Evaluate the determinant.

$$\begin{vmatrix} -5 & 2 \\ 3 & 1 \end{vmatrix} = -5(1)-3(2)=-5-6=-11$$

P3. Write the coefficient matrix.

$$\begin{bmatrix} 2 & -7 \\ 3 & 5 \end{bmatrix}$$

P5. Evaluate.

$$\begin{vmatrix} 3 & -1 \\ 2 & -3 \end{vmatrix} = -9+2=-7$$

$$\begin{vmatrix} 1 & 4 \\ -2 & 5 \end{vmatrix} = 5+8=13$$

$$\frac{\begin{vmatrix} 3 & -1 \\ 2 & -3 \end{vmatrix}}{\begin{vmatrix} 1 & 4 \\ -2 & 5 \end{vmatrix}} = -\frac{7}{13}$$

Section 7.5 Exercises

1. Solve using Cramer's Rule.

$$x_1 = \frac{\begin{vmatrix} 8 & 4 \\ 1 & -5 \end{vmatrix}}{\begin{vmatrix} 3 & 4 \\ 4 & -5 \end{vmatrix}} = \frac{-44}{-31} = \frac{44}{31}$$

$$x_2 = \frac{\begin{vmatrix} 3 & 8 \\ 4 & 1 \end{vmatrix}}{\begin{vmatrix} 3 & 4 \\ 4 & -5 \end{vmatrix}} = \frac{-29}{-31} = \frac{29}{31}$$

3. Solve using Cramer's Rule.

$$x_1 = \frac{\begin{vmatrix} -1 & 4 \\ 5 & -6 \end{vmatrix}}{\begin{vmatrix} 5 & 4 \\ 3 & -6 \end{vmatrix}} = \frac{-14}{-42} = \frac{1}{3}$$

$$x_2 = \frac{\begin{vmatrix} 5 & -1 \\ 3 & 5 \end{vmatrix}}{\begin{vmatrix} 5 & 4 \\ 3 & -6 \end{vmatrix}} = \frac{28}{-42} = -\frac{2}{3}$$

5. Solve using Cramer's Rule.

$$x_1 = \frac{\begin{vmatrix} 0 & 2 \\ -3 & 1 \end{vmatrix}}{\begin{vmatrix} 7 & 2 \\ 2 & 1 \end{vmatrix}} = \frac{6}{3} = 2$$

$$x_2 = \frac{\begin{vmatrix} 7 & 0 \\ 2 & -3 \end{vmatrix}}{\begin{vmatrix} 7 & 2 \\ 2 & 1 \end{vmatrix}} = \frac{-21}{3} = -7$$

7. Solve using Cramer's Rule.

$$x_1 = \frac{\begin{vmatrix} 0 & -7 \\ 0 & 4 \end{vmatrix}}{\begin{vmatrix} 3 & -7 \\ 2 & 4 \end{vmatrix}} = \frac{0}{26} = 0$$

$$x_2 = \frac{\begin{vmatrix} 3 & 0 \\ 2 & 0 \end{vmatrix}}{\begin{vmatrix} 3 & -7 \\ 2 & 4 \end{vmatrix}} = \frac{0}{26} = 0$$

9. Solve using Cramer's Rule.

$$x_1 = \frac{\begin{vmatrix} 2.1 & 0.3 \\ -1.6 & -1.4 \end{vmatrix}}{\begin{vmatrix} 1.2 & 0.3 \\ 0.8 & -1.4 \end{vmatrix}} = \frac{-2.46}{-1.92} = 1.28125$$

$$x_2 = \frac{\begin{vmatrix} 1.2 & 2.1 \\ 0.8 & -1.6 \end{vmatrix}}{\begin{vmatrix} 1.2 & 0.3 \\ 0.8 & -1.4 \end{vmatrix}} = \frac{-3.6}{-1.92} = 1.875$$

11. Solve using Cramer's Rule.

$$D = \begin{vmatrix} 3 & -4 & 2 \\ 1 & -1 & 2 \\ 2 & 2 & 3 \end{vmatrix} = -17$$

$$D_1 = \begin{vmatrix} 1 & -4 & 2 \\ -2 & -1 & 2 \\ -3 & 2 & 3 \end{vmatrix} = -21$$

$$D_2 = \begin{vmatrix} 3 & 1 & 2 \\ 1 & -2 & 2 \\ 2 & -3 & 3 \end{vmatrix} = 3$$

$$D_3 = \begin{vmatrix} 3 & -4 & 1 \\ 1 & -1 & -2 \\ 2 & 2 & -3 \end{vmatrix} = 29$$

$$x_1 = \frac{D_1}{D} = \frac{-21}{-17} = \frac{21}{17}$$

$$x_2 = \frac{D_2}{D} = \frac{3}{-17} = -\frac{3}{17}$$

$$x_3 = \frac{D_3}{D} = \frac{29}{-17} = -\frac{29}{17}$$

13. Solve using Cramer's Rule.

$$D = \begin{vmatrix} 1 & 4 & -2 \\ 3 & -2 & 3 \\ 2 & 1 & -3 \end{vmatrix} = 49$$

$$D_1 = \begin{vmatrix} 0 & 4 & -2 \\ 4 & -2 & 3 \\ -1 & 1 & -3 \end{vmatrix} = 32$$

$$D_2 = \begin{vmatrix} 1 & 0 & -2 \\ 3 & 4 & 3 \\ 2 & -1 & -3 \end{vmatrix} = 13$$

$$D_3 = \begin{vmatrix} 1 & 4 & 0 \\ 3 & -2 & 4 \\ 2 & 1 & -1 \end{vmatrix} = 42$$

$$x_1 = \frac{D_1}{D} = \frac{32}{49}$$

$$x_2 = \frac{D_2}{D} = \frac{13}{49}$$

$$x_3 = \frac{D_3}{D} = \frac{42}{49} = \frac{6}{7}$$

15. Solve using Cramer's Rule.

$$D = \begin{vmatrix} 0 & 2 & -3 \\ 3 & -5 & 1 \\ 4 & 0 & 2 \end{vmatrix} = -64$$

$$D_1 = \begin{vmatrix} 1 & 2 & -3 \\ 0 & -5 & 1 \\ -3 & 0 & 2 \end{vmatrix} = 29$$

$$D_2 = \begin{vmatrix} 0 & 1 & -3 \\ 3 & 0 & 1 \\ 4 & -3 & 2 \end{vmatrix} = 25$$

$$D_3 = \begin{vmatrix} 0 & 2 & 1 \\ 3 & -5 & 0 \\ 4 & 0 & -3 \end{vmatrix} = 38$$

$$x_1 = \frac{D_1}{D} = \frac{29}{-64} = -\frac{29}{64}$$

$$x_2 = \frac{D_2}{D} = \frac{25}{-64} = -\frac{25}{64}$$

$$x_3 = \frac{D_3}{D} = \frac{38}{-64} = -\frac{19}{32}$$

17. Solve using Cramer's Rule.

$$D = \begin{vmatrix} 4 & -5 & 1 \\ 3 & 1 & 0 \\ 1 & -1 & 3 \end{vmatrix} = 53$$

$$D_1 = \begin{vmatrix} -2 & -5 & 1 \\ 4 & 1 & 0 \\ 0 & -1 & 3 \end{vmatrix} = 50$$

$$D_2 = \begin{vmatrix} 4 & -2 & 1 \\ 3 & 4 & 0 \\ 1 & 0 & 3 \end{vmatrix} = 62$$

$$D_3 = \begin{vmatrix} 4 & -5 & -2 \\ 3 & 1 & 4 \\ 1 & -1 & 0 \end{vmatrix} = 4$$

$$x_1 = \frac{D_1}{D} = \frac{50}{53}$$

$$x_2 = \frac{D_2}{D} = \frac{62}{53}$$

$$x_3 = \frac{D_3}{D} = \frac{4}{53}$$

19. Solve using Cramer's Rule.

$$D=\begin{vmatrix}2 & 2 & -3\\ 1 & -3 & 2\\ 4 & -1 & 3\end{vmatrix}=-37$$

$$D_1=\begin{vmatrix}0 & 2 & -3\\ 0 & -3 & 2\\ 0 & -1 & 3\end{vmatrix}=0$$

$$D_2=\begin{vmatrix}2 & 0 & -3\\ 1 & 0 & 2\\ 4 & 0 & 3\end{vmatrix}=0$$

$$D_3=\begin{vmatrix}2 & 2 & 0\\ 1 & -3 & 0\\ 4 & -1 & 0\end{vmatrix}=0$$

$$x_1=\frac{D_1}{D}=\frac{0}{-37}=0$$

$$x_2=\frac{D_2}{D}=\frac{0}{-37}=0$$

$$x_3=\frac{D_3}{D}=\frac{0}{-37}=0$$

21. Solve for x_2.

$$D=\begin{vmatrix}2 & -3 & 4 & -1\\ 1 & 2 & 0 & 2\\ 3 & 1 & 0 & -2\\ 1 & -3 & 2 & -1\end{vmatrix}=-38$$

$$D_2=\begin{vmatrix}2 & 1 & 4 & -1\\ 1 & -1 & 0 & 2\\ 3 & 2 & 0 & -2\\ 1 & 3 & 2 & -1\end{vmatrix}=70$$

$$x_2=\frac{D_2}{D}=\frac{70}{-38}=-\frac{35}{19}$$

23. Solve for x_1.

$$D=\begin{vmatrix}1 & -3 & 2 & 4\\ 3 & 5 & -6 & 2\\ 2 & -1 & 9 & 8\\ 1 & 1 & 1 & -8\end{vmatrix}=-1310$$

$$D_1=\begin{vmatrix}0 & -3 & 2 & 4\\ -2 & 5 & -6 & 2\\ 0 & -1 & 9 & 8\\ -3 & 1 & 1 & -8\end{vmatrix}=1210$$

$$x_1=\frac{D_1}{D}=\frac{1210}{-1310}=-\frac{121}{131}$$

25. Solve for x_4.

$$D=\begin{vmatrix}0 & 3 & -1 & 2\\ 5 & 1 & 3 & -1\\ 1 & -2 & 0 & 9\\ 2 & 0 & 2 & 0\end{vmatrix}=120$$

$$D_4=\begin{vmatrix}0 & 3 & -1 & 1\\ 5 & 1 & 3 & -4\\ 1 & -2 & 0 & 5\\ 2 & 0 & 2 & 3\end{vmatrix}=160$$

$$x_4=\frac{D_4}{D}=\frac{160}{120}=\frac{4}{3}$$

27. Explain how to solve using Cramer's Rule.

$$D=\begin{vmatrix}2 & -3 & 1\\ 1 & 1 & -2\\ 4 & -1 & -3\end{vmatrix}=0$$

In order for us to use Cramer's Rule, the determinant of the coefficient matrix cannot be zero. The system of equations has infinitely many solutions.

29. Find k so that the system has a unique solution.

$$D=\begin{vmatrix}k & 3\\ k & -2\end{vmatrix}=-5k$$

For the system of equations to have a unique solution, the determinant of the coefficient matrix cannot be zero.

$$-5k=0$$
$$k=0$$

The system of equations has a unique solution for all values of k except $k=0$.

31. Find k so that the system has a unique solution.

$$D=\begin{vmatrix}1 & 2 & -3\\ 2 & k & -4\\ 1 & -2 & 1\end{vmatrix}=4k-8$$

For the system of equations to have a unique solution, the determinant of the coefficient matrix cannot be zero.

$$4k-8=0$$
$$4k=8$$
$$k=2$$

The system of equations has a unique solution for all values of k except $k=2$.

Exploring Concepts with Technology

$$XT=[0.428 \quad 0.572]$$
$$XT^2 \approx [0.45236 \quad 0.54764]$$
$$XT^3 \approx [0.47355 \quad 0.52645]$$
$$XT^{20} \approx [0.60209 \quad 0.39791]$$
$$XT^{40} \approx [0.61456 \quad 0.38544]$$
$$XT^{60} \approx [0.61533 \quad 0.38467]$$
$$XT^{100} \approx [0.61538 \quad 0.38462]$$

It appears that as the number of weeks increases, Super A will get slightly more than 61.5% of the neighborhood and Super B will get slightly less than 38.5% of the neighborhood.

Changing the market share *does not* affect the result (to 6 decimal places) after 100 weeks.

Three Department Stores: After 100 months, Super A will have 23.87% of the market share, Super B will have 33.65% of the market share, and Super C will have 42.48% of the market share.

Chapter 7 Review Exercises

1. Writing the augmented, coefficient and constant matrices of the given system of equations. [7.1]

$$\left[\begin{array}{cc|c} 3 & 5 & 6 \\ 2 & -7 & -1 \end{array}\right], \begin{bmatrix} 3 & 5 \\ 2 & -7 \end{bmatrix}, \begin{bmatrix} 6 \\ -1 \end{bmatrix}$$

3. Writing the augmented, coefficient and constant matrices of the given system of equations. [7.1]

$$\left[\begin{array}{cccc|c} 3 & 2 & -3 & 1 & 4 \\ 1 & 0 & -4 & 0 & -2 \\ -2 & -1 & 0 & -2 & 0 \end{array}\right], \begin{bmatrix} 3 & 2 & -3 & 1 \\ 1 & 0 & -4 & 0 \\ -2 & -1 & 0 & -2 \end{bmatrix}, \begin{bmatrix} 4 \\ -2 \\ 0 \end{bmatrix}$$

The solutions and answers to Exercises 5 – 12 are not unique. The following solutions list the elementary row operations used to produce the row echelon forms shown.

5. Using elementary row operations to achieve row echelon form. [7.1]

$$\begin{bmatrix} -1 & -2 & -8 \\ 3 & 7 & 3 \end{bmatrix} \xrightarrow{-R_1} \begin{bmatrix} 1 & 2 & 8 \\ 3 & 7 & 3 \end{bmatrix}$$
$$\xrightarrow{-3R_1+R_2} \begin{bmatrix} 1 & 2 & 8 \\ 0 & 1 & -21 \end{bmatrix}$$

Using a graphing calculator, we get the following equivalent answer.

$$\begin{bmatrix} 1 & \frac{7}{3} & 1 \\ 0 & 1 & -21 \end{bmatrix}$$

7. Using elementary row operations to achieve row echelon form. [7.1]

$$\begin{bmatrix} 3 & 3 & -4 \\ 3 & -1 & -4 \end{bmatrix} \xrightarrow{\frac{1}{3}R_1} \begin{bmatrix} 1 & 1 & -\frac{4}{3} \\ 3 & -1 & -4 \end{bmatrix}$$
$$\xrightarrow{-3R_1+R_2} \begin{bmatrix} 1 & 1 & -\frac{4}{3} \\ 0 & -4 & 0 \end{bmatrix} \xrightarrow{-\frac{1}{4}R_2} \begin{bmatrix} 1 & 1 & -\frac{4}{3} \\ 0 & 1 & 0 \end{bmatrix}$$

Using a graphing calculator, we get the same answer.

9. Using elementary row operations to achieve row echelon form. [7.1]

$$\begin{bmatrix} -5 & -3 & 6 & -14 \\ -10 & 7 & -6 & 3 \\ 1 & -1 & 1 & -1 \end{bmatrix}$$
$$\xrightarrow{R_1 \longleftrightarrow R_3} \begin{bmatrix} 1 & -1 & 1 & -1 \\ -10 & 7 & -6 & 3 \\ -5 & -3 & 6 & -14 \end{bmatrix}$$
$$\xrightarrow[5R_1+R_3]{10R_1+R_2} \begin{bmatrix} 1 & -1 & 1 & -1 \\ 0 & -3 & 4 & -7 \\ 0 & -8 & 11 & -19 \end{bmatrix}$$
$$\xrightarrow{-\frac{1}{3}R_2} \begin{bmatrix} 1 & -1 & 1 & -1 \\ 0 & 1 & -\frac{4}{3} & \frac{7}{3} \\ 0 & -8 & 11 & -19 \end{bmatrix}$$

$$\xrightarrow{8R_2+R_3}\begin{bmatrix}1 & -1 & 1 & -1\\ 0 & 1 & -\frac{4}{3} & \frac{7}{3}\\ 0 & 0 & \frac{1}{3} & -\frac{1}{3}\end{bmatrix}$$

$$\xrightarrow{3R_3}\begin{bmatrix}1 & -1 & 1 & -1\\ 0 & 1 & -\frac{4}{3} & \frac{7}{3}\\ 0 & 0 & 1 & -1\end{bmatrix}$$

Using a graphing calculator, we get the following equivalent answer.

$$\begin{bmatrix}1 & -\frac{7}{10} & \frac{3}{5} & -\frac{3}{10}\\ 0 & 1 & -\frac{18}{13} & \frac{31}{13}\\ 0 & 0 & 1 & -1\end{bmatrix}$$

11. Using elementary row operations to achieve row echelon form. [7.1]

$$\begin{bmatrix}3 & 3 & 4 & 2\\ 1 & 2 & 2 & 4\\ -4 & 4 & 0 & 2\end{bmatrix}\xrightarrow{R_1\longleftrightarrow R_2}\begin{bmatrix}1 & 2 & 2 & 4\\ 3 & 3 & 4 & 2\\ -4 & 4 & 0 & 2\end{bmatrix}$$

$$\xrightarrow[4R_1+R_3]{-3R_1+R_2}\begin{bmatrix}1 & 2 & 2 & 4\\ 0 & -3 & -2 & -10\\ 0 & 12 & 8 & 18\end{bmatrix}$$

$$\xrightarrow[\frac{1}{2}R_3]{-\frac{1}{3}R_2}\begin{bmatrix}1 & 2 & 2 & 4\\ 0 & 1 & \frac{2}{3} & \frac{10}{3}\\ 0 & 6 & 4 & 9\end{bmatrix}\xrightarrow{-6R_2+R_3}\begin{bmatrix}1 & 2 & 2 & 4\\ 0 & 1 & \frac{2}{3} & \frac{10}{3}\\ 0 & 0 & 0 & -11\end{bmatrix}$$

Using a graphing calculator, we get the following equivalent answer.

$$\begin{bmatrix}1 & -1 & 0 & -\frac{1}{2}\\ 0 & 1 & \frac{2}{3} & \frac{7}{12}\\ 0 & 0 & 0 & 1\end{bmatrix}$$

13. Solve the system using Gaussian elimination. [7.1]

$$\left[\begin{array}{cc|c}2 & -3 & 7\\ 3 & -4 & 10\end{array}\right]\xrightarrow{-1R_1+R_2}\left[\begin{array}{cc|c}2 & -3 & 7\\ 1 & -1 & 3\end{array}\right]$$

$$\xrightarrow{R_1\longleftrightarrow R_2}\left[\begin{array}{cc|c}1 & -1 & 3\\ 2 & -3 & 7\end{array}\right]$$

$$\xrightarrow{-2R_1+R_2}\left[\begin{array}{cc|c}1 & -1 & 3\\ 0 & -1 & 1\end{array}\right]$$

$$\xrightarrow{-1R_2}\left[\begin{array}{cc|c}1 & -1 & 3\\ 0 & 1 & -1\end{array}\right]$$

$$\begin{cases}x-y=3\\ y=-1\end{cases}\qquad \begin{aligned}x-(-1)&=3\\ x&=2\end{aligned}$$

The solution is (2, –1).

15. Solve the system using Gaussian elimination. [7.1]

$$\left[\begin{array}{cc|c}4 & -5 & 12\\ 3 & 1 & 9\end{array}\right]\xrightarrow{-1R_2+R_1}\left[\begin{array}{cc|c}1 & -6 & 3\\ 3 & 1 & 9\end{array}\right]$$

$$\xrightarrow{-3R_1+R_2}\left[\begin{array}{cc|c}1 & -6 & 3\\ 0 & 19 & 0\end{array}\right]$$

$$\xrightarrow{\frac{1}{19}R_2}\left[\begin{array}{cc|c}1 & -6 & 3\\ 0 & 1 & 0\end{array}\right]$$

$$\begin{cases}x-6y=3\\ y=0\end{cases}$$

$$\begin{aligned}x-6(0)&=3\\ x&=3\end{aligned}$$

The solution is (3, 0).

17. Solve the system using Gaussian elimination. [7.1]

$$\left[\begin{array}{ccc|c}1 & 2 & 3 & 5\\ 3 & 8 & 11 & 17\\ 2 & 6 & 7 & 12\end{array}\right]\xrightarrow[-2R_1+R_3]{-3R_1+R_2}\left[\begin{array}{ccc|c}1 & 2 & 3 & 5\\ 0 & 2 & 2 & 2\\ 0 & 2 & 1 & 2\end{array}\right]$$

$$\xrightarrow{\frac{1}{2}R_2}\left[\begin{array}{ccc|c}1 & 2 & 3 & 5\\ 0 & 1 & 1 & 1\\ 0 & 2 & 1 & 2\end{array}\right]$$

$$\xrightarrow{-2R_2+R_3}\left[\begin{array}{ccc|c}1 & 2 & 3 & 5\\ 0 & 1 & 1 & 1\\ 0 & 0 & -1 & 0\end{array}\right]$$

$$\xrightarrow{-1R_3}\left[\begin{array}{ccc|c}1 & 2 & 3 & 5\\ 0 & 1 & 1 & 1\\ 0 & 0 & 1 & 0\end{array}\right]$$

$$\begin{cases}x+2y+3z=5\\ y+z=1\\ z=0\end{cases}$$

$$\begin{aligned}y+0&=1\\ y&=1\end{aligned}$$

$$\begin{aligned}x+2(1)+3(0)&=5\\ x&=3\end{aligned}$$

The solution is (3, 1, 0).

19. Solve the system using Gaussian elimination. [7.1]

$$\left[\begin{array}{rrr|r} 2 & -1 & -1 & 4 \\ 1 & -2 & -2 & 5 \\ 3 & -3 & -8 & 19 \end{array}\right] \xrightarrow{R_1 \longleftrightarrow R_2} \left[\begin{array}{rrr|r} 1 & -2 & -2 & 5 \\ 2 & -1 & -1 & 4 \\ 3 & -3 & -8 & 19 \end{array}\right]$$

$$\xrightarrow[-3R_1+R_3]{-2R_1+R_2} \left[\begin{array}{rrr|r} 1 & -2 & -2 & 5 \\ 0 & 3 & 3 & -6 \\ 0 & 3 & -2 & 4 \end{array}\right]$$

$$\xrightarrow{\frac{1}{3}R_2} \left[\begin{array}{rrr|r} 1 & -2 & -2 & 5 \\ 0 & 1 & 1 & -2 \\ 0 & 3 & -2 & 4 \end{array}\right]$$

$$\xrightarrow{-3R_2+R_3} \left[\begin{array}{rrr|r} 1 & -2 & -2 & 5 \\ 0 & 1 & 1 & -2 \\ 0 & 0 & -5 & 10 \end{array}\right]$$

$$\xrightarrow{-\frac{1}{5}R_3} \left[\begin{array}{rrr|r} 1 & -2 & -2 & 5 \\ 0 & 1 & 1 & -2 \\ 0 & 0 & 1 & -2 \end{array}\right]$$

$$\begin{cases} x-2y-2z=5 \\ \quad y+\ z=-2 \\ \qquad z=-2 \end{cases}$$

$$y+(-2)=-2$$
$$y=0$$

$$x-2(0)-2(-2)=5$$
$$x=1$$

The solution is (1, 0, –2).

21. Solve the system using Gaussian elimination. [7.1]

$$\left[\begin{array}{rrr|r} 4 & -9 & 6 & 54 \\ 3 & -8 & 8 & 49 \\ 1 & -3 & 2 & 17 \end{array}\right] \xrightarrow{R_1 \longleftrightarrow R_3} \left[\begin{array}{rrr|r} 1 & -3 & 2 & 17 \\ 3 & -8 & 8 & 49 \\ 4 & -9 & 6 & 54 \end{array}\right]$$

$$\xrightarrow[-4R_1+R_3]{-3R_1+R_2} \left[\begin{array}{rrr|r} 1 & -3 & 2 & 17 \\ 0 & 1 & 2 & -2 \\ 0 & 3 & -2 & -14 \end{array}\right]$$

$$\xrightarrow{-3R_2+R_3} \left[\begin{array}{rrr|r} 1 & -3 & 2 & 17 \\ 0 & 1 & 2 & -2 \\ 0 & 0 & -8 & -8 \end{array}\right]$$

$$\xrightarrow{-\frac{1}{8}R_3} \left[\begin{array}{rrr|r} 1 & -3 & 2 & 17 \\ 0 & 1 & 2 & -2 \\ 0 & 0 & 1 & 1 \end{array}\right]$$

$$\begin{cases} x-3y+2z=17 \\ \quad y+\ 2z=-2 \\ \qquad z=\ 1 \end{cases}$$

$$y+2(1)=-2$$
$$y=-4$$

$$x-3(-4)+2(1)=17$$
$$x=3$$

The solution is (3, –4, 1).

23. Solve the system using Gaussian elimination. [7.1]

$$\left[\begin{array}{rrr|r} 1 & 4 & -3 & 2 \\ -2 & -8 & 6 & 1 \\ -1 & 4 & -6 & 3 \end{array}\right] \xrightarrow[R_1+R_3]{2R_1+R_2} \left[\begin{array}{rrr|r} 1 & 4 & -3 & 2 \\ 0 & 0 & 0 & 5 \\ 0 & 8 & -9 & 5 \end{array}\right]$$

$$\begin{cases} x+4y-3z=2 \\ \qquad 0z=5 \\ \quad 8y-9z=5 \end{cases}$$

Because $0z = 5$ has no solution, the system of equations has no solution.

25. Solve the system using Gaussian elimination. [7.1]

$$\left[\begin{array}{rrr|r} 2 & -5 & 1 & -2 \\ 2 & 3 & -3 & 2 \\ 2 & -1 & -1 & 2 \end{array}\right] \xrightarrow{\frac{1}{2}R_1} \left[\begin{array}{rrr|r} 1 & -\frac{5}{2} & \frac{1}{2} & -1 \\ 2 & 3 & -3 & 2 \\ 2 & -1 & -1 & 2 \end{array}\right]$$

$$\xrightarrow[-2R_1+R_3]{-2R_1+R_2} \left[\begin{array}{rrr|r} 1 & -\frac{5}{2} & \frac{1}{2} & -1 \\ 0 & 8 & -4 & 4 \\ 0 & 4 & -2 & 4 \end{array}\right]$$

$$\xrightarrow{\frac{1}{8}R_2} \left[\begin{array}{rrr|r} 1 & -\frac{5}{2} & \frac{1}{2} & -1 \\ 0 & 1 & -\frac{1}{2} & \frac{1}{2} \\ 0 & 4 & -2 & 4 \end{array}\right]$$

$$\xrightarrow{-4R_2+R_3} \left[\begin{array}{rrr|r} 1 & -\frac{5}{2} & \frac{1}{2} & -1 \\ 0 & 1 & -\frac{1}{2} & \frac{1}{2} \\ 0 & 0 & 0 & 2 \end{array}\right]$$

$$\begin{cases} x-\frac{5}{2}y+\frac{1}{2}z=-1 \\ \quad y-\frac{1}{2}z=\frac{1}{2} \\ \qquad 0z=2 \end{cases}$$

Because $0z = 2$ has no solution, the system of equations has no solution.

27. Solve the system using Gaussian elimination. [7.1]

$$\left[\begin{array}{rrr|r} 1 & -3 & 1 & -3 \\ 2 & 1 & -2 & 3 \end{array}\right] \xrightarrow{-2R_1+R_2} \left[\begin{array}{rrr|r} 1 & -3 & 1 & -3 \\ 0 & 7 & -4 & 9 \end{array}\right]$$

$$\xrightarrow{\frac{1}{7}R_2} \left[\begin{array}{rrr|r} 1 & -3 & 1 & -3 \\ 0 & 1 & -\frac{4}{7} & \frac{9}{7} \end{array}\right]$$

$$\begin{cases} x-3y+z=-3 \\ y-\frac{4}{7}z=\frac{9}{7} \end{cases}$$

$$y-\frac{4}{7}z=\frac{9}{7}$$
$$y=\frac{4}{7}z+\frac{9}{7}$$

$$x-3\left(\frac{4}{7}z+\frac{9}{7}\right)+z=-3$$
$$x=\frac{5}{7}z+\frac{6}{7}$$

Let z be any real number c.

The solution is $\left(\frac{5}{7}c+\frac{6}{7},\ \frac{4}{7}c+\frac{9}{7},\ c\right)$.

29. Solve the system using Gaussian elimination. [7.1]

$$\left[\begin{array}{rrr|r} 1 & 1 & 2 & -5 \\ 2 & 3 & 5 & -13 \\ 2 & 5 & 7 & -19 \end{array}\right] \xrightarrow[-2R_1+R_3]{-2R_1+R_2} \left[\begin{array}{rrr|r} 1 & 1 & 2 & -5 \\ 0 & 1 & 1 & -3 \\ 0 & 3 & 3 & -9 \end{array}\right]$$

$$\xrightarrow{-3R_2+R_3} \left[\begin{array}{rrr|r} 1 & 1 & 2 & -5 \\ 0 & 1 & 1 & -3 \\ 0 & 0 & 0 & 0 \end{array}\right]$$

$$\begin{cases} x+y+2z=-5 \\ y+\ z=-3 \end{cases}$$

$$y=-z-3$$
$$x+(-z-3)+2z=-5$$
$$x=-z-2$$

Let z be any real number c.

The solution is $(-c-2, -c-3, c)$.

31. Solve the system using Gaussian elimination. [7.1]

$$\left[\begin{array}{rrrr|r} 1 & 2 & -1 & 2 & 1 \\ 3 & 8 & 1 & 4 & 1 \\ 2 & 7 & 3 & 2 & 0 \\ 1 & 3 & -2 & 5 & 6 \end{array}\right] \xrightarrow[\substack{-2R_1+R_3 \\ -1R_1+R_4}]{-3R_1+R_2} \left[\begin{array}{rrrr|r} 1 & 2 & -1 & 2 & 1 \\ 0 & 2 & 4 & -2 & -2 \\ 0 & 3 & 5 & -2 & -2 \\ 0 & 1 & -1 & 3 & 5 \end{array}\right]$$

$$\xrightarrow{\frac{1}{2}R_2} \left[\begin{array}{rrrr|r} 1 & 2 & -1 & 2 & 1 \\ 0 & 1 & 2 & -1 & -1 \\ 0 & 3 & 5 & -2 & -2 \\ 0 & 1 & -1 & 3 & 5 \end{array}\right]$$

$$\xrightarrow[-1R_2+R_4]{-3R_2+R_3} \left[\begin{array}{rrrr|r} 1 & 2 & -1 & 2 & 1 \\ 0 & 1 & 2 & -1 & -1 \\ 0 & 0 & -1 & 1 & 1 \\ 0 & 0 & -3 & 4 & 6 \end{array}\right]$$

$$\xrightarrow{-1R_3} \left[\begin{array}{rrrr|r} 1 & 2 & -1 & 2 & 1 \\ 0 & 1 & 2 & -1 & -1 \\ 0 & 0 & 1 & -1 & -1 \\ 0 & 0 & -3 & 4 & 6 \end{array}\right]$$

$$\xrightarrow{3R_3+R_4} \left[\begin{array}{rrrr|r} 1 & 2 & -1 & 2 & 1 \\ 0 & 1 & 2 & -1 & -1 \\ 0 & 0 & 1 & -1 & -1 \\ 0 & 0 & 0 & 1 & 3 \end{array}\right]$$

$$\begin{cases} w+2x-\ y+2z=\ 1 \\ x+2y-\ z=-1 \\ y-\ z=-1 \\ z=\ 3 \end{cases}$$

$$y-3=-1$$
$$y=2$$
$$x+2(2)-3=-1$$
$$x=-2$$
$$w+2(-2)-2+2(3)=1$$
$$w=1$$

The solution is (1, –2, 2, 3).

33. Solve the system using Gaussian elimination. [7.1]

$$\left[\begin{array}{rrrr|r} 1 & 3 & 1 & -4 & 3 \\ 1 & 4 & 3 & -6 & 5 \\ 2 & 8 & 7 & -5 & 11 \\ 2 & 5 & 0 & -6 & 4 \end{array}\right] \xrightarrow[\substack{-2R_1+R_3 \\ -2R_1+R_4}]{-1R_1+R_2} \left[\begin{array}{rrrr|r} 1 & 3 & 1 & -4 & 3 \\ 0 & 1 & 2 & -2 & 2 \\ 0 & 2 & 5 & 3 & 5 \\ 0 & -1 & -2 & 2 & -2 \end{array}\right]$$

$$\xrightarrow[1R_2+R_4]{-2R_2+R_3} \left[\begin{array}{rrrr|r} 1 & 3 & 1 & -4 & 3 \\ 0 & 1 & 2 & -2 & 2 \\ 0 & 0 & 1 & 7 & 1 \\ 0 & 0 & 0 & 0 & 0 \end{array}\right]$$

$$\begin{cases} w+3x+\ y-4z=3 \\ x+2y-2z=2 \\ y+7z=1 \end{cases}$$

$$y=-7z+1$$

$$x+2(-7z+1)-2z=2$$
$$x=16z$$
$$w+3(16z)+(-7z+1)-4z=3$$
$$w=-37z+2$$

Let z be any real number c.

The solution is $(-37c+2,\ 16c,\ -7c+1,\ c)$.

35. Find the polynomial. [7.1]

Because there are three points, the degree of the interpolating polynomial is at most 2.

The form of the polynomial is

$$p(x)=a_2x^2+a_1x+a_0.$$

Use this polynomial and the given points to find the system of equations.

$$p(-1)=a_2(-1)^2+a_1(-1)+a_0=-4$$
$$p(2)=a_2(2)^2+a_1(2)+a_0=8$$
$$p(3)=a_2(3)^2+a_1(3)+a_0=16$$

The system of equations and the associated augmented matrix are

$$\begin{cases} a_2-\ a_1+a_0=-4 \\ 4a_2+2a_1+a_0=8 \\ 9a_2+3a_1+a_0=16 \end{cases} \qquad \left[\begin{array}{ccc|c} 1 & -1 & 1 & -4 \\ 4 & 2 & 1 & 8 \\ 9 & 3 & 1 & 16 \end{array}\right]$$

The ref (**r**ow **e**chelon **f**orm) feature of a graphing calculator can be used to rewrite the augmented matrix in echelon form.

Consider using the function of your calculator that converts a decimal to a fraction.

The augmented matrix in echelon form and resulting system of equations are

$$\left[\begin{array}{ccc|c} 1 & \frac{1}{3} & \frac{1}{9} & \frac{16}{9} \\ 0 & 1 & -\frac{2}{3} & \frac{13}{3} \\ 0 & 0 & 1 & -2 \end{array}\right] \qquad \begin{cases} a_2+\frac{1}{3}a_1+\frac{1}{9}a_0=\frac{16}{9} \\ a_1-\frac{2}{3}a_0=\frac{13}{2} \\ a_0=-2 \end{cases}$$

Solving by back substitution yields

$a_0=-2,\quad a_1=3,\quad$ and $\quad a_2=1.$

The interpolating polynomial is $p(x)=x^2+3x-2$.

37. Find $3A$. [7.2]

$$3A=3\begin{bmatrix} 2 & -1 & 3 \\ 3 & 2 & -1 \end{bmatrix}=\begin{bmatrix} 6 & -3 & 9 \\ 9 & 6 & -3 \end{bmatrix}$$

39. Find $-A+D$. [7.2]

$$-A+D=-\begin{bmatrix} 2 & -1 & 3 \\ 3 & 2 & -1 \end{bmatrix}+\begin{bmatrix} -3 & 4 & 2 \\ 4 & -2 & 5 \end{bmatrix}$$
$$=\begin{bmatrix} -2 & 1 & -3 \\ -3 & -2 & 1 \end{bmatrix}+\begin{bmatrix} -3 & 4 & 2 \\ 4 & -2 & 5 \end{bmatrix}$$
$$=\begin{bmatrix} -5 & 5 & -1 \\ 1 & -4 & 6 \end{bmatrix}$$

41. Find AB. [7.2]

$$AB=\begin{bmatrix} 2 & -1 & 3 \\ 3 & 2 & -1 \end{bmatrix}\begin{bmatrix} 0 & -2 \\ 4 & 2 \\ 1 & -3 \end{bmatrix}=\begin{bmatrix} -1 & -15 \\ 7 & 1 \end{bmatrix}$$

43. Find BA. [7.2]

$$BA=\begin{bmatrix} 0 & -2 \\ 4 & 2 \\ 1 & -3 \end{bmatrix}\begin{bmatrix} 2 & -1 & 3 \\ 3 & 2 & -1 \end{bmatrix}=\begin{bmatrix} -6 & -4 & 2 \\ 14 & 0 & 10 \\ -7 & -7 & 6 \end{bmatrix}$$

45. Find C^2. [7.2]

$$C^2=C\cdot C=\begin{bmatrix} 2 & 6 & 1 \\ 1 & 2 & -1 \\ 2 & 4 & -1 \end{bmatrix}\begin{bmatrix} 2 & 6 & 1 \\ 1 & 2 & -1 \\ 2 & 4 & -1 \end{bmatrix}$$
$$=\begin{bmatrix} 12 & 28 & -5 \\ 2 & 6 & 0 \\ 6 & 16 & -1 \end{bmatrix}$$

47. Find BAC. [7.2]

$$BAC=\begin{bmatrix} 0 & -2 \\ 4 & 2 \\ 1 & -3 \end{bmatrix}\begin{bmatrix} 2 & -1 & 3 \\ 3 & 2 & -1 \end{bmatrix}\begin{bmatrix} 2 & 6 & 1 \\ 1 & 2 & -1 \\ 2 & 4 & -1 \end{bmatrix}$$
$$=\begin{bmatrix} -6 & -4 & 2 \\ 14 & 0 & 10 \\ -7 & -7 & 6 \end{bmatrix}\begin{bmatrix} 2 & 6 & 1 \\ 1 & 2 & -1 \\ 2 & 4 & -1 \end{bmatrix}=\begin{bmatrix} -12 & -36 & -4 \\ 48 & 124 & 4 \\ -9 & -32 & -6 \end{bmatrix}$$

49. Find $AB-BA$. [7.2]

$$AB-BA=\begin{bmatrix} 2 & -1 & 3 \\ 3 & 2 & -1 \end{bmatrix}\begin{bmatrix} 0 & -2 \\ 4 & 2 \\ 1 & -3 \end{bmatrix}-\begin{bmatrix} 0 & -2 \\ 4 & 2 \\ 1 & -3 \end{bmatrix}\begin{bmatrix} 2 & -1 & 3 \\ 3 & 2 & -1 \end{bmatrix}$$

$$=\begin{bmatrix}-1 & -15\\ 7 & 1\end{bmatrix}-\begin{bmatrix}-6 & -4 & 2\\ 14 & 0 & 10\\ -7 & -7 & 6\end{bmatrix}$$

Not possible since AB is of order 2×2 and BA is of order 3×3.

51. Find $(A-D)C$. [7.2]

$(A-D)C$

$$=\left(\begin{bmatrix}2 & -1 & 3\\ 3 & 2 & -1\end{bmatrix}-\begin{bmatrix}-3 & 4 & 2\\ 4 & -2 & 5\end{bmatrix}\right)\begin{bmatrix}2 & 6 & 1\\ 1 & 2 & -1\\ 2 & 4 & -1\end{bmatrix}$$

$$=\begin{bmatrix}5 & -5 & 1\\ -1 & 4 & -6\end{bmatrix}\begin{bmatrix}2 & 6 & 1\\ 1 & 2 & -1\\ 2 & 4 & -1\end{bmatrix}=\begin{bmatrix}7 & 24 & 9\\ -10 & -22 & 1\end{bmatrix}$$

53. Find C^{-1}. [7.3]

$$\left[\begin{array}{ccc|ccc}2 & 6 & 1 & 1 & 0 & 0\\ 1 & 2 & -1 & 0 & 1 & 0\\ 2 & 4 & -1 & 0 & 0 & 1\end{array}\right]$$

$$\xrightarrow{\frac{1}{2}R_1}\left[\begin{array}{ccc|ccc}1 & 3 & \frac{1}{2} & \frac{1}{2} & 0 & 0\\ 1 & 2 & -1 & 0 & 1 & 0\\ 2 & 4 & -1 & 0 & 0 & 1\end{array}\right]$$

$$\xrightarrow[-2R_1+R_3]{-1R_1+R_2}\left[\begin{array}{ccc|ccc}1 & 3 & \frac{1}{2} & \frac{1}{2} & 0 & 0\\ 0 & -1 & -\frac{3}{2} & -\frac{1}{2} & 1 & 0\\ 0 & -2 & -2 & -1 & 0 & 1\end{array}\right]$$

$$\xrightarrow[\substack{-3R_2+R_1\\ 2R_2+R_3}]{-1R_2}\left[\begin{array}{ccc|ccc}1 & 0 & -4 & -1 & 3 & 0\\ 0 & 1 & \frac{3}{2} & \frac{1}{2} & -1 & 0\\ 0 & 0 & 1 & 0 & -2 & 1\end{array}\right]$$

$$\xrightarrow[-\frac{3}{2}R_3+R_2]{4R_3+R_1}\left[\begin{array}{ccc|ccc}1 & 0 & 0 & -1 & -5 & 4\\ 0 & 1 & 0 & \frac{1}{2} & 2 & -\frac{3}{2}\\ 0 & 0 & 1 & 0 & -2 & 1\end{array}\right]$$

$$C^{-1}=\begin{bmatrix}-1 & -5 & 4\\ \frac{1}{2} & 2 & -\frac{3}{2}\\ 0 & -2 & 1\end{bmatrix}$$

55. Write the system of equations. [7.2]

$$\begin{cases}2x-3y=5\\ 4x+5y=-1\end{cases}$$

57. Write the system of equations. [7.2]

$$\begin{cases}2x-y+3z=6\\ x-5y+4z=10\\ 2x+3y+7z=6\end{cases}$$

59. Find the adjacency matrix A and A^2. Determine the number of walks of length 2. [7.2]

$$A=\begin{bmatrix}0 & 1 & 0 & 1 & 0\\ 1 & 0 & 1 & 1 & 1\\ 0 & 1 & 0 & 0 & 0\\ 1 & 1 & 0 & 0 & 0\\ 0 & 1 & 0 & 0 & 0\end{bmatrix},\quad A^2=\begin{bmatrix}2 & 1 & 1 & 1 & 1\\ 1 & 4 & 0 & 1 & 0\\ 1 & 0 & 1 & 1 & 1\\ 1 & 1 & 1 & 2 & 1\\ 1 & 0 & 1 & 1 & 1\end{bmatrix}$$

A^2 gives the number of walks of length 2. There is 1 walk of length 2 from vertex 2 to vertex 4.

61. Find the inverse. [7.3]

$$\left[\begin{array}{cc|cc}2 & -2 & 1 & 0\\ 3 & -2 & 0 & 1\end{array}\right]\xrightarrow{\frac{1}{2}R_1}\left[\begin{array}{cc|cc}1 & -1 & \frac{1}{2} & 0\\ 3 & -2 & 0 & 1\end{array}\right]$$

$$\xrightarrow{-3R_1+R_2}\left[\begin{array}{cc|cc}1 & -1 & \frac{1}{2} & 0\\ 0 & 1 & -\frac{3}{2} & 1\end{array}\right]$$

$$\xrightarrow{R_2+R_1}\left[\begin{array}{cc|cc}1 & 0 & -1 & 1\\ 0 & 1 & -\frac{3}{2} & 1\end{array}\right]$$

The inverse matrix is $\begin{bmatrix}-1 & 1\\ -\frac{3}{2} & 1\end{bmatrix}$.

63. Find the inverse. [7.3]

$$\left[\begin{array}{cc|cc}-2 & 3 & 1 & 0\\ 2 & 4 & 0 & 1\end{array}\right]\xrightarrow{-\frac{1}{2}R_1}\left[\begin{array}{cc|cc}1 & -\frac{3}{2} & -\frac{1}{2} & 0\\ 2 & 4 & 0 & 1\end{array}\right]$$

$$\xrightarrow{-2R_1+R_2}\left[\begin{array}{cc|cc}1 & -\frac{3}{2} & -\frac{1}{2} & 0\\ 0 & 7 & 1 & 1\end{array}\right]$$

$$\xrightarrow{\frac{1}{7}R_2}\left[\begin{array}{cc|cc}1 & -\frac{3}{2} & -\frac{1}{2} & 0\\ 0 & 1 & \frac{1}{7} & \frac{1}{7}\end{array}\right]$$

$$\xrightarrow{\frac{3}{2}R_2+R_1}\left[\begin{array}{cc|cc}1 & 0 & -\frac{2}{7} & \frac{3}{14}\\ 0 & 1 & \frac{1}{7} & \frac{1}{7}\end{array}\right]$$

The inverse matrix is $\begin{bmatrix}-\frac{2}{7} & \frac{3}{14}\\ \frac{1}{7} & \frac{1}{7}\end{bmatrix}$.

65. Find the inverse. [7.3]

$$\left[\begin{array}{ccc|ccc} 1 & 2 & 1 & 1 & 0 & 0 \\ 2 & 6 & 4 & 0 & 1 & 0 \\ 3 & 8 & 6 & 0 & 0 & 1 \end{array}\right] \xrightarrow[-3R_1+R_3]{-2R_1+R_2} \left[\begin{array}{ccc|ccc} 1 & 2 & 1 & 1 & 0 & 0 \\ 0 & 2 & 2 & -2 & 1 & 0 \\ 0 & 2 & 3 & -3 & 0 & 1 \end{array}\right]$$

$$\xrightarrow{\frac{1}{2}R_2} \left[\begin{array}{ccc|ccc} 1 & 2 & 1 & 1 & 0 & 0 \\ 0 & 1 & 1 & -1 & \frac{1}{2} & 0 \\ 0 & 2 & 3 & -3 & 0 & 1 \end{array}\right]$$

$$\xrightarrow{-2R_2+R_3} \left[\begin{array}{ccc|ccc} 1 & 2 & 1 & 1 & 0 & 0 \\ 0 & 1 & 1 & -1 & \frac{1}{2} & 0 \\ 0 & 0 & 1 & -1 & -1 & 1 \end{array}\right]$$

$$\xrightarrow{-2R_2+R_1} \left[\begin{array}{ccc|ccc} 1 & 0 & -1 & 3 & -1 & 0 \\ 0 & 1 & 1 & -1 & \frac{1}{2} & 0 \\ 0 & 0 & 1 & -1 & -1 & 1 \end{array}\right]$$

$$\xrightarrow[-R_3+R_2]{R_3+R_1} \left[\begin{array}{ccc|ccc} 1 & 0 & 0 & 2 & -2 & 1 \\ 0 & 1 & 0 & 0 & \frac{3}{2} & -1 \\ 0 & 0 & 1 & -1 & -1 & 1 \end{array}\right]$$

The inverse matrix is $\begin{bmatrix} 2 & -2 & 1 \\ 0 & \frac{3}{2} & -1 \\ -1 & -1 & 1 \end{bmatrix}$.

67. Find the inverse. [7.3]

$$\left[\begin{array}{ccc|ccc} 3 & -2 & 7 & 1 & 0 & 0 \\ 2 & -1 & 5 & 0 & 1 & 0 \\ 3 & 0 & 10 & 0 & 0 & 1 \end{array}\right] \xrightarrow{\frac{1}{3}R_1} \left[\begin{array}{ccc|ccc} 1 & -\frac{2}{3} & \frac{7}{3} & \frac{1}{3} & 0 & 0 \\ 2 & -1 & 5 & 0 & 1 & 0 \\ 3 & 0 & 10 & 0 & 0 & 1 \end{array}\right]$$

$$\xrightarrow[-3R_1+R_3]{-2R_1+R_2} \left[\begin{array}{ccc|ccc} 1 & -\frac{2}{3} & \frac{7}{3} & \frac{1}{3} & 0 & 0 \\ 0 & \frac{1}{3} & \frac{1}{3} & -\frac{2}{3} & 1 & 0 \\ 0 & 2 & 3 & -1 & 0 & 1 \end{array}\right]$$

$$\xrightarrow{3R_2} \left[\begin{array}{ccc|ccc} 1 & -\frac{2}{3} & \frac{7}{3} & \frac{1}{3} & 0 & 0 \\ 0 & 1 & 1 & -2 & 3 & 0 \\ 0 & 2 & 3 & -1 & 0 & 1 \end{array}\right]$$

$$\xrightarrow{-2R_2+R_3} \left[\begin{array}{ccc|ccc} 1 & -\frac{2}{3} & \frac{7}{3} & \frac{1}{3} & 0 & 0 \\ 0 & 1 & 1 & -2 & 3 & 0 \\ 0 & 0 & 1 & 3 & -6 & 1 \end{array}\right]$$

$$\xrightarrow{\frac{2}{3}R_2+R_1} \left[\begin{array}{ccc|ccc} 1 & 0 & 3 & -1 & 2 & 0 \\ 0 & 1 & 1 & -2 & 3 & 0 \\ 0 & 0 & 1 & 3 & -6 & 1 \end{array}\right]$$

$$\xrightarrow[-1R_3+R_2]{-3R_3+R_1} \left[\begin{array}{ccc|ccc} 1 & 0 & 0 & -10 & 20 & -3 \\ 0 & 1 & 0 & -5 & 9 & -1 \\ 0 & 0 & 1 & 3 & -6 & 1 \end{array}\right]$$

The inverse matrix is $\begin{bmatrix} -10 & 20 & -3 \\ -5 & 9 & -1 \\ 3 & -6 & 1 \end{bmatrix}$.

69. Find the inverse. [7.3]

$$\left[\begin{array}{cccc|cccc} 1 & -1 & 2 & 3 & 1 & 0 & 0 & 0 \\ 2 & -1 & 6 & 5 & 0 & 1 & 0 & 0 \\ 3 & -1 & 9 & 6 & 0 & 0 & 1 & 0 \\ 2 & -2 & 4 & 7 & 0 & 0 & 0 & 1 \end{array}\right]$$

$$\xrightarrow[\substack{-3R_1+R_3 \\ -2R_1+R_4}]{-2R_1+R_2} \left[\begin{array}{cccc|cccc} 1 & -1 & 2 & 3 & 1 & 0 & 0 & 0 \\ 0 & 1 & 2 & -1 & -2 & 1 & 0 & 0 \\ 0 & 2 & 3 & -3 & -3 & 0 & 1 & 0 \\ 0 & 0 & 0 & 1 & -2 & 0 & 0 & 1 \end{array}\right]$$

$$\xrightarrow{-2R_2+R_3} \left[\begin{array}{cccc|cccc} 1 & -1 & 2 & 3 & 1 & 0 & 0 & 0 \\ 0 & 1 & 2 & -1 & -2 & 1 & 0 & 0 \\ 0 & 0 & -1 & -1 & 1 & -2 & 1 & 0 \\ 0 & 0 & 0 & 1 & -2 & 0 & 0 & 1 \end{array}\right]$$

$$\xrightarrow{-1R_3} \left[\begin{array}{cccc|cccc} 1 & -1 & 2 & 3 & 1 & 0 & 0 & 0 \\ 0 & 1 & 2 & -1 & -2 & 1 & 0 & 0 \\ 0 & 0 & 1 & 1 & -1 & 2 & -1 & 0 \\ 0 & 0 & 0 & 1 & -2 & 0 & 0 & 1 \end{array}\right]$$

$$\xrightarrow{R_2+R_1} \left[\begin{array}{cccc|cccc} 1 & 0 & 4 & 2 & -1 & 1 & 0 & 0 \\ 0 & 1 & 2 & -1 & -2 & 1 & 0 & 0 \\ 0 & 0 & 1 & 1 & -1 & 2 & -1 & 0 \\ 0 & 0 & 0 & 1 & -2 & 0 & 0 & 1 \end{array}\right]$$

$$\xrightarrow[-2R_3+R_2]{-4R_3+R_1} \left[\begin{array}{cccc|cccc} 1 & 0 & 0 & -2 & 3 & -7 & 4 & 0 \\ 0 & 1 & 0 & -3 & 0 & -3 & 2 & 0 \\ 0 & 0 & 1 & 1 & -1 & 2 & -1 & 0 \\ 0 & 0 & 0 & 1 & -2 & 0 & 0 & 1 \end{array}\right]$$

$$\xrightarrow[\substack{3R_4+R_2 \\ -1R_4+R_3}]{2R_4+R_1} \left[\begin{array}{cccc|cccc} 1 & 0 & 0 & 0 & -1 & -7 & 4 & 2 \\ 0 & 1 & 0 & 0 & -6 & -3 & 2 & 3 \\ 0 & 0 & 1 & 0 & 1 & 2 & -1 & -1 \\ 0 & 0 & 0 & 1 & -2 & 0 & 0 & 1 \end{array}\right]$$

The inverse matrix is $\begin{bmatrix} -1 & -7 & 4 & 2 \\ -6 & -3 & 2 & 3 \\ 1 & 2 & -1 & -1 \\ -2 & 0 & 0 & 1 \end{bmatrix}$.

71. Find the inverse. [7.3]

$$\left[\begin{array}{rrr|rrr} -1 & 8 & 1 & 1 & 0 & 0 \\ 1 & 3 & -1 & 0 & 1 & 0 \\ 6 & 5 & -6 & 0 & 0 & 1 \end{array}\right] \xrightarrow{-1R_1} \left[\begin{array}{rrr|rrr} 1 & -8 & -1 & -1 & 0 & 0 \\ 1 & 3 & -1 & 0 & 1 & 0 \\ 6 & 5 & -6 & 0 & 0 & 1 \end{array}\right]$$

$$\xrightarrow[-6R_1+R_3]{-R_1+R_2} \left[\begin{array}{rrr|rrr} 1 & -8 & -1 & -1 & 0 & 0 \\ 0 & 11 & 0 & 1 & 1 & 0 \\ 0 & 53 & 0 & 6 & 0 & 1 \end{array}\right]$$

$$\xrightarrow{\frac{1}{11}R_2} \left[\begin{array}{rrr|rrr} 1 & -8 & -1 & -1 & 0 & 0 \\ 0 & 1 & 0 & \frac{1}{11} & \frac{1}{11} & 0 \\ 0 & 53 & 0 & 6 & 0 & 1 \end{array}\right]$$

$$\xrightarrow{-53R_2+R_3} \left[\begin{array}{rrr|rrr} 1 & -8 & -1 & -1 & 0 & 0 \\ 0 & 1 & 0 & \frac{1}{11} & \frac{1}{11} & 0 \\ 0 & 0 & 0 & \frac{13}{11} & -\frac{53}{11} & 1 \end{array}\right]$$

The matrix does not have an inverse.

73. Find the inverse. [7.3]

$$\left[\begin{array}{rrrr|rrrr} 3 & 7 & -1 & 8 & 1 & 0 & 0 & 0 \\ 2 & 5 & 0 & 5 & 0 & 1 & 0 & 0 \\ 3 & 6 & -4 & 8 & 0 & 0 & 1 & 0 \\ 2 & 4 & -4 & 4 & 0 & 0 & 0 & 1 \end{array}\right]$$

$$\xrightarrow{\frac{1}{3}R_1} \left[\begin{array}{rrrr|rrrr} 1 & \frac{7}{3} & -\frac{1}{3} & \frac{8}{3} & \frac{1}{3} & 0 & 0 & 0 \\ 2 & 5 & 0 & 5 & 0 & 1 & 0 & 0 \\ 3 & 6 & -4 & 8 & 0 & 0 & 1 & 0 \\ 2 & 4 & -4 & 4 & 0 & 0 & 0 & 1 \end{array}\right]$$

$$\xrightarrow[\substack{-3R_1+R_3 \\ -2R_1+R_4}]{-2R_1+R_2} \left[\begin{array}{rrrr|rrrr} 1 & \frac{7}{3} & -\frac{1}{3} & \frac{8}{3} & \frac{1}{3} & 0 & 0 & 0 \\ 0 & \frac{1}{3} & \frac{2}{3} & -\frac{1}{3} & -\frac{2}{3} & 1 & 0 & 0 \\ 0 & -1 & -3 & 0 & -1 & 0 & 1 & 0 \\ 0 & -\frac{2}{3} & -\frac{10}{3} & -\frac{4}{3} & -\frac{2}{3} & 0 & 0 & 1 \end{array}\right]$$

$$\xrightarrow{3R_2} \left[\begin{array}{rrrr|rrrr} 1 & \frac{7}{3} & -\frac{1}{3} & \frac{8}{3} & \frac{1}{3} & 0 & 0 & 0 \\ 0 & 1 & 2 & -1 & -2 & 3 & 0 & 0 \\ 0 & -1 & -3 & 0 & -1 & 0 & 1 & 0 \\ 0 & -\frac{2}{3} & -\frac{10}{3} & -\frac{4}{3} & -\frac{2}{3} & 0 & 0 & 1 \end{array}\right]$$

$$\xrightarrow[\frac{2}{3}R_2+R_4]{1R_2+R_3} \left[\begin{array}{rrrr|rrrr} 1 & \frac{7}{3} & -\frac{1}{3} & \frac{8}{3} & \frac{1}{3} & 0 & 0 & 0 \\ 0 & 1 & 2 & -1 & -2 & 3 & 0 & 0 \\ 0 & 0 & -1 & -1 & -3 & 3 & 1 & 0 \\ 0 & 0 & -2 & -2 & -2 & 2 & 0 & 1 \end{array}\right]$$

$$\xrightarrow{-1R_3} \left[\begin{array}{rrrr|rrrr} 1 & \frac{7}{3} & -\frac{1}{3} & \frac{8}{3} & \frac{1}{3} & 0 & 0 & 0 \\ 0 & 1 & 2 & -1 & -2 & 3 & 0 & 0 \\ 0 & 0 & 1 & 1 & 3 & -3 & -1 & 0 \\ 0 & 0 & -2 & -2 & -2 & 2 & 0 & 1 \end{array}\right]$$

$$\xrightarrow{2R_3+R_4} \left[\begin{array}{rrrr|rrrr} 1 & \frac{7}{3} & -\frac{1}{3} & \frac{8}{3} & \frac{1}{3} & 0 & 0 & 0 \\ 0 & 1 & 2 & -1 & -2 & 3 & 0 & 0 \\ 0 & 0 & 1 & 1 & 3 & -3 & -1 & 0 \\ 0 & 0 & 0 & 0 & 4 & -4 & -2 & 1 \end{array}\right]$$

The matrix does not have an inverse.

75. Solve the system for the set of constants. [7.3]

a. $b_1 = 2,\ b_2 = -3$

$$\begin{bmatrix} 3 & 4 \\ 2 & 3 \end{bmatrix}\begin{bmatrix} x \\ y \end{bmatrix} = \begin{bmatrix} 2 \\ -3 \end{bmatrix}$$

$$\begin{bmatrix} 3 & -4 \\ -2 & 3 \end{bmatrix}\begin{bmatrix} 3 & 4 \\ 2 & 3 \end{bmatrix}\begin{bmatrix} x \\ y \end{bmatrix} = \begin{bmatrix} 3 & -4 \\ -2 & 3 \end{bmatrix}\begin{bmatrix} 2 \\ -3 \end{bmatrix}$$

$$\begin{bmatrix} x \\ y \end{bmatrix} = \begin{bmatrix} 18 \\ -13 \end{bmatrix}$$

The solution is (18, –13).

b. $b_1 = -2,\ b_2 = 4$

$$\begin{bmatrix} 3 & 4 \\ 2 & 3 \end{bmatrix}\begin{bmatrix} x \\ y \end{bmatrix} = \begin{bmatrix} -2 \\ 4 \end{bmatrix}$$

$$\begin{bmatrix} 3 & -4 \\ -2 & 3 \end{bmatrix}\begin{bmatrix} 3 & 4 \\ 2 & 3 \end{bmatrix}\begin{bmatrix} x \\ y \end{bmatrix} = \begin{bmatrix} 3 & -4 \\ -2 & 3 \end{bmatrix}\begin{bmatrix} -2 \\ 4 \end{bmatrix}$$

$$\begin{bmatrix} x \\ y \end{bmatrix} = \begin{bmatrix} -22 \\ 16 \end{bmatrix}$$

The solution is (–22, 16).

77. Solve the system for the set of constants. [7.3]

a. $b_1 = -1,\ b_2 = 2,\ b_3 = 4$

$$\begin{bmatrix} 2 & 1 & -1 \\ 4 & 4 & 1 \\ 2 & 2 & -3 \end{bmatrix}\begin{bmatrix} x \\ y \\ z \end{bmatrix} = \begin{bmatrix} -1 \\ 2 \\ 4 \end{bmatrix}$$

$$\begin{bmatrix} 1 & -\frac{1}{14} & -\frac{5}{14} \\ -1 & \frac{2}{7} & \frac{3}{7} \\ 0 & \frac{1}{7} & -\frac{2}{7} \end{bmatrix}\begin{bmatrix} 2 & 1 & -1 \\ 4 & 4 & 1 \\ 2 & 2 & -3 \end{bmatrix}\begin{bmatrix} x \\ y \\ z \end{bmatrix} = \begin{bmatrix} 1 & -\frac{1}{14} & -\frac{5}{14} \\ -1 & \frac{2}{7} & \frac{3}{7} \\ 0 & \frac{1}{7} & -\frac{2}{7} \end{bmatrix}\begin{bmatrix} -1 \\ 2 \\ 4 \end{bmatrix}$$

$$\begin{bmatrix} x \\ y \\ z \end{bmatrix} = \begin{bmatrix} -\frac{18}{7} \\ \frac{23}{7} \\ -\frac{6}{7} \end{bmatrix}$$

The solution is $\left(-\frac{18}{7}, \frac{23}{7}, -\frac{6}{7}\right)$.

b. $b_1 = -2,\ b_2 = 3,\ b_3 = 0$

$$\begin{bmatrix} 2 & 1 & -1 \\ 4 & 4 & 1 \\ 2 & 2 & -3 \end{bmatrix}\begin{bmatrix} x \\ y \\ z \end{bmatrix} = \begin{bmatrix} -2 \\ 3 \\ 0 \end{bmatrix}$$

$$\begin{bmatrix} 1 & -\frac{1}{14} & -\frac{5}{14} \\ -1 & \frac{2}{7} & \frac{3}{7} \\ 0 & \frac{1}{7} & -\frac{2}{7} \end{bmatrix}\begin{bmatrix} 2 & 1 & -1 \\ 4 & 4 & 1 \\ 2 & 2 & -3 \end{bmatrix}\begin{bmatrix} x \\ y \\ z \end{bmatrix} = \begin{bmatrix} 1 & -\frac{1}{14} & -\frac{5}{14} \\ -1 & \frac{2}{7} & \frac{3}{7} \\ 0 & \frac{1}{7} & -\frac{2}{7} \end{bmatrix}\begin{bmatrix} -2 \\ 3 \\ 0 \end{bmatrix}$$

$$\begin{bmatrix} x \\ y \\ z \end{bmatrix} = \begin{bmatrix} -\frac{31}{14} \\ \frac{20}{7} \\ \frac{3}{7} \end{bmatrix}$$

The solution is $\left(-\frac{31}{14}, \frac{20}{7}, \frac{3}{7}\right)$.

79. Evaluate the determinant. [7.4]

$$\begin{vmatrix} -2 & 3 \\ -5 & 1 \end{vmatrix} = -2(1) - (-5)(3) = -2 + 15 = 13$$

81. Find M_{11} and C_{11}. [7.4]

$$M_{11} = \begin{vmatrix} 0 & 3 \\ 2 & 5 \end{vmatrix} = 0(5) - 2(3) = -6$$

$$C_{11} = (-1)^{1+1} M_{11} = (1)(-6) = -6$$

83. Find M_{12} and C_{12}. [7.4]

$$M_{12} = \begin{vmatrix} 1 & 3 \\ -2 & 5 \end{vmatrix} = 1(5) - (-2)(3) = 11$$

$$C_{12} = (-1)^{1+2} M_{12} = (-1)(11) = -11$$

85. Evaluate the determinant by cofactors. [7.4]

$$\begin{vmatrix} 4 & 1 & 3 \\ -2 & 3 & -8 \\ -1 & 4 & -5 \end{vmatrix} = 4\begin{vmatrix} 3 & -8 \\ 4 & -5 \end{vmatrix} - 1\begin{vmatrix} -2 & -8 \\ -1 & -5 \end{vmatrix} + 3\begin{vmatrix} -2 & 3 \\ -1 & 4 \end{vmatrix}$$
$$= 4(17) - 1(2) + 3(-5)$$
$$= 51$$

87. Evaluate the determinant by cofactors. [7.4]

$$\begin{vmatrix} 8 & -3 & 0 \\ 10 & -2 & -5 \\ 3 & -4 & 0 \end{vmatrix} = 0\begin{vmatrix} 10 & -2 \\ 3 & -4 \end{vmatrix} - (-5)\begin{vmatrix} 8 & -3 \\ 3 & -4 \end{vmatrix} + 0\begin{vmatrix} 8 & -3 \\ 10 & -2 \end{vmatrix}$$
$$= 0 + 5(-23) + 0$$
$$= -115$$

89. Evaluate the determinant by cofactors. [7.4]

$$\begin{vmatrix} 1 & -1 & -4 & 2 \\ 0 & 1 & 5 & 0 \\ -2 & 4 & -4 & 0 \\ 3 & 4 & 5 & 0 \end{vmatrix} = -2\begin{vmatrix} 0 & 1 & 5 \\ -2 & 4 & -4 \\ 3 & 4 & 5 \end{vmatrix} + 0 - 0 + 0$$
$$= -2\left[0\begin{vmatrix} 4 & -4 \\ 4 & 5 \end{vmatrix} - 1\begin{vmatrix} -2 & -4 \\ 3 & 5 \end{vmatrix} + 5\begin{vmatrix} -2 & 4 \\ 3 & 4 \end{vmatrix}\right]$$
$$= -2[0 - 1(2) + 5(-20)]$$
$$= 204$$

91. Evaluate the determinant using elementary row or column operations. [7.4]

$$\begin{vmatrix} 2 & 6 & 4 \\ 1 & 2 & 1 \\ 3 & 8 & 6 \end{vmatrix} \xrightarrow{\text{Factor 2 from row 1}} 2\begin{vmatrix} 1 & 3 & 2 \\ 1 & 2 & 1 \\ 3 & 8 & 6 \end{vmatrix}$$

$$\xrightarrow[-3R_1 + R_3]{-1R_1 + R_2} 2\begin{vmatrix} 1 & 3 & 2 \\ 0 & -1 & -1 \\ 0 & -1 & 0 \end{vmatrix}$$

$$\xrightarrow{-R_2 + R_3} 2\begin{vmatrix} 1 & 3 & 2 \\ 0 & -1 & -1 \\ 0 & 0 & 1 \end{vmatrix}$$

$$= 2(1)(-1)(1) = -2$$

93. Evaluate the determinant using elementary row or column operations. [7.4]

$$\begin{vmatrix} 3 & -8 & 7 \\ 2 & -3 & 6 \\ 1 & -3 & 2 \end{vmatrix} \xrightarrow[-\frac{1}{3}R_1 + R_3]{-\frac{2}{3}R_1 + R_2} \begin{vmatrix} 3 & -8 & 7 \\ 0 & \frac{7}{3} & \frac{4}{3} \\ 0 & -\frac{1}{3} & -\frac{1}{3} \end{vmatrix}$$

$$\xrightarrow{\frac{1}{7}R_2 + R_3} \begin{vmatrix} 3 & -8 & 7 \\ 0 & \frac{7}{3} & \frac{4}{3} \\ 0 & 0 & -\frac{1}{7} \end{vmatrix}$$

$$= 3\frac{7}{3}\left(-\frac{1}{7}\right) = -1$$

95. Evaluate the determinant using elementary row or column operations. [7.4]

$$\begin{vmatrix} 1 & -1 & 2 & 1 \\ 2 & -1 & 6 & 3 \\ 3 & -1 & 8 & 7 \\ 3 & 0 & 9 & 9 \end{vmatrix} \xrightarrow{\text{Factor 3 from row 4}} 3\begin{vmatrix} 1 & -1 & 2 & 1 \\ 2 & -1 & 6 & 3 \\ 3 & -1 & 8 & 7 \\ 1 & 0 & 3 & 3 \end{vmatrix}$$

$$\begin{vmatrix} 1 & -1 & 2 & 1 \\ 2 & -1 & 6 & 3 \\ 3 & -1 & 8 & 7 \\ 3 & 0 & 9 & 9 \end{vmatrix} \xrightarrow{\text{Factor 3 from row 4}} 3\begin{vmatrix} 1 & -1 & 2 & 1 \\ 2 & -1 & 6 & 3 \\ 3 & -1 & 8 & 7 \\ 1 & 0 & 3 & 3 \end{vmatrix}$$

$$\xrightarrow[-1R_1+R_4]{\substack{-2R_1+R_2 \\ -3R_1+R_3}} 3\begin{vmatrix} 1 & -1 & 2 & 1 \\ 0 & 1 & 2 & 1 \\ 0 & 2 & 2 & 4 \\ 0 & 1 & 1 & 2 \end{vmatrix}$$

$$\xrightarrow[-1R_2+R_4]{-2R_2+R_3} 3\begin{vmatrix} 1 & -1 & 2 & 1 \\ 0 & 1 & 2 & 1 \\ 0 & 0 & -2 & 2 \\ 0 & 0 & -1 & 1 \end{vmatrix}$$

$$\xrightarrow[R_3+R_4]{\text{Factor } -2 \text{ from row 3}} -6\begin{vmatrix} 1 & -1 & 2 & 1 \\ 0 & 1 & 2 & 1 \\ 0 & 0 & 1 & -1 \\ 0 & 0 & 0 & 0 \end{vmatrix}$$

$$= -6(1)(1)(1)(0) = 0$$

97. Evaluate the determinant using elementary row or column operations. [7.4]

$$\begin{vmatrix} 1 & 2 & -2 & 1 \\ 2 & 5 & -3 & 1 \\ 2 & 0 & -10 & 1 \\ 3 & 8 & -4 & 1 \end{vmatrix} \xrightarrow[-R_4+R_3]{\substack{-R_4+R_1 \\ -R_4+R_2}} \begin{vmatrix} -2 & -6 & 2 & 0 \\ -1 & -3 & 1 & 0 \\ -1 & -8 & -6 & 0 \\ 3 & 8 & -4 & 1 \end{vmatrix}$$

$$\xrightarrow[\frac{1}{3}R_3+R_1]{\frac{1}{6}R_3+R_2} \begin{vmatrix} -\frac{7}{3} & -\frac{26}{3} & 0 & 0 \\ -\frac{7}{6} & -\frac{13}{3} & 0 & 0 \\ -1 & -8 & -6 & 0 \\ 3 & 8 & -4 & 1 \end{vmatrix}$$

$$\xrightarrow{-2R_2+R_1} \begin{vmatrix} 0 & 0 & 0 & 0 \\ -\frac{7}{6} & -\frac{13}{3} & 0 & 0 \\ -1 & -8 & -6 & 0 \\ 3 & 8 & -4 & 1 \end{vmatrix}$$

$$= 0\left(-\frac{13}{3}\right)(-6)(1) = 0$$

99. Solve the system using Cramer's Rule. [7.5]

$$x_1 = \frac{\begin{vmatrix} 2 & -3 \\ 2 & 5 \end{vmatrix}}{\begin{vmatrix} 2 & -3 \\ 3 & 5 \end{vmatrix}} = \frac{16}{19}$$

$$x_2 = \frac{\begin{vmatrix} 2 & 2 \\ 3 & 2 \end{vmatrix}}{\begin{vmatrix} 2 & -3 \\ 3 & 5 \end{vmatrix}} = \frac{-2}{19} = -\frac{2}{19}$$

101. Solve the system using Cramer's Rule. [7.5]

$$D = \begin{vmatrix} 2 & 1 & -3 \\ 3 & 2 & 1 \\ 1 & -3 & 4 \end{vmatrix} = 44, \quad D_1 = \begin{vmatrix} 2 & 1 & -3 \\ 1 & 2 & 1 \\ -2 & -3 & 4 \end{vmatrix} = 13$$

$$D_2 = \begin{vmatrix} 2 & 2 & -3 \\ 3 & 1 & 1 \\ 1 & -2 & 4 \end{vmatrix} = 11, \quad D_3 = \begin{vmatrix} 2 & 1 & 2 \\ 3 & 2 & 1 \\ 1 & -3 & -2 \end{vmatrix} = -17$$

$$x_1 = \frac{D_1}{D} = \frac{13}{44}$$

$$x_2 = \frac{D_2}{D} = \frac{11}{44} = \frac{1}{4}$$

$$x_3 = \frac{D_3}{D} = \frac{-17}{44} = -\frac{17}{44}$$

103. Solve the system using Cramer's Rule. [7.5]

$$D = \begin{vmatrix} 0 & 2 & 5 \\ 2 & -5 & 1 \\ 4 & 3 & 0 \end{vmatrix} = 138, \quad D_1 = \begin{vmatrix} 2 & 2 & 5 \\ 4 & -5 & 1 \\ 2 & 3 & 0 \end{vmatrix} = 108$$

$$D_2 = \begin{vmatrix} 0 & 2 & 5 \\ 2 & 4 & 1 \\ 4 & 2 & 0 \end{vmatrix} = -52, \quad D_3 = \begin{vmatrix} 0 & 2 & 2 \\ 2 & -5 & 4 \\ 4 & 3 & 2 \end{vmatrix} = 76$$

$$x_1 = \frac{D_1}{D} = \frac{108}{138} = \frac{54}{69} = \frac{18}{23}$$

$$x_2 = \frac{D_2}{D} = \frac{-52}{138} = -\frac{26}{69}$$

$$x_3 = \frac{D_3}{D} = \frac{76}{138} = \frac{38}{69}$$

105. Use Cramer's Rule to solve for x_3. [7.5]

$$D = \begin{vmatrix} 1 & -3 & 1 & 2 \\ 2 & 7 & -3 & 1 \\ -1 & 4 & 2 & -3 \\ 3 & 1 & -1 & -2 \end{vmatrix} = -252$$

$$D_3 = \begin{vmatrix} 1 & -3 & 3 & 2 \\ 2 & 7 & 2 & 1 \\ -1 & 4 & -1 & -3 \\ 3 & 1 & 0 & -2 \end{vmatrix} = -230$$

$$x_3 = \frac{D_3}{D} = \frac{-230}{-252} = \frac{115}{126}$$

107. Use transformations to find the endpoints. [7.2]

$$T_{3,-1}\cdot\begin{bmatrix}-5 & 4\\ 3 & -2\\ 1 & 1\end{bmatrix}=\begin{bmatrix}1 & 0 & 3\\ 0 & 1 & -1\\ 0 & 0 & 1\end{bmatrix}\begin{bmatrix}-5 & 4\\ 3 & -2\\ 1 & 1\end{bmatrix}=\begin{bmatrix}-2 & 7\\ 2 & -3\\ 1 & 1\end{bmatrix}$$

$$R_{xy}\cdot\begin{bmatrix}-2 & 7\\ 2 & -3\\ 1 & 1\end{bmatrix}=\begin{bmatrix}0 & 1 & 0\\ 1 & 0 & 0\\ 0 & 0 & 1\end{bmatrix}\begin{bmatrix}-2 & 7\\ 2 & -3\\ 1 & 1\end{bmatrix}=\begin{bmatrix}2 & -3\\ -2 & 7\\ 1 & 1\end{bmatrix}\Rightarrow \begin{matrix}A'(2,\ -2)\\ B'(-3,\ 7)\end{matrix}$$

109. Find the populations in 5 years. [7.2]

$$\begin{bmatrix}250{,}000 & 400{,}000\end{bmatrix}\begin{bmatrix}0.90 & 0.10\\ 0.15 & 0.85\end{bmatrix}^5 \approx\begin{bmatrix}356{,}777 & 293{,}223\end{bmatrix}$$

The city population will be about 357,000.

The suburb population will be about 293,000.

111. Find the temperatures at x_1 and x_2. [7.3]

The average temperature for the two points,

$$x_1=\frac{60+40+x_2+50}{4}=\frac{150+x_2}{4}\text{ or }4x_1-x_2=150$$

$$x_2=\frac{60+70+x_1+50}{4}=\frac{180+x_1}{4}\text{ or }-x_1+4x_2=180$$

The system of equations and associated matrix equation are

$$\begin{cases}4x_1-x_2=150\\ -x_1+4x_2=180\end{cases}\qquad \begin{bmatrix}4 & -1\\ -1 & 4\end{bmatrix}\begin{bmatrix}x_1\\ x_2\end{bmatrix}=\begin{bmatrix}150\\ 180\end{bmatrix}$$

Solving the matrix equation by using an inverse matrix gives $\begin{bmatrix}x_1\\ x_2\end{bmatrix}=\begin{bmatrix}52\\ 58\end{bmatrix}$

The temperatures are x_1 = 52°F, x_2 = 58°F.

113. The input-output matrix A is given by

$$A=\begin{bmatrix}0.05 & 0.06 & 0.08\\ 0.02 & 0.04 & 0.04\\ 0.03 & 0.03 & 0.05\end{bmatrix}$$

Consumer demand X is given by $X=(I-A)^{-1}D$.

$$X=\left(\begin{bmatrix}1 & 0 & 0\\ 0 & 1 & 0\\ 0 & 0 & 1\end{bmatrix}-\begin{bmatrix}0.05 & 0.06 & 0.08\\ 0.02 & 0.04 & 0.04\\ 0.03 & 0.03 & 0.05\end{bmatrix}\right)^{-1}\begin{bmatrix}30\\ 12\\ 21\end{bmatrix}$$

$$=\begin{bmatrix}0.95 & -0.06 & -0.08\\ -0.02 & 0.96 & -0.04\\ -0.03 & -0.03 & 0.95\end{bmatrix}^{-1}\begin{bmatrix}30\\ 12\\ 21\end{bmatrix}\approx\begin{bmatrix}34.47\\ 14.20\\ 23.64\end{bmatrix}$$

\$34.47 million computer division, \$14.20 million monitor division, \$23.64 million disk drive division.

Chapter 7 Test

1. Write the augmented matrix, coefficient matrix and constant matrix. [7.1]

$$\left[\begin{array}{ccc|c}2 & 3 & -3 & 4\\ 3 & 0 & 2 & -1\\ 4 & -4 & 2 & 3\end{array}\right],\ \begin{bmatrix}2 & 3 & -3\\ 3 & 0 & 2\\ 4 & -4 & 2\end{bmatrix},\ \begin{bmatrix}4\\ -1\\ 3\end{bmatrix}$$

3. Solve the system using Gaussian elimination. [7.1]

$$\left[\begin{array}{ccc|c}1 & -2 & 3 & 10\\ 2 & -3 & 8 & 23\\ -1 & 3 & -2 & -9\end{array}\right]\xrightarrow[R_1+R_3]{-2R_1+R_2}\left[\begin{array}{ccc|c}1 & -2 & 3 & 10\\ 0 & 1 & 2 & 3\\ 0 & 1 & 1 & 1\end{array}\right]$$

$$\xrightarrow{-R_2+R_3}\left[\begin{array}{ccc|c}1 & -2 & 3 & 10\\ 0 & 1 & 2 & 3\\ 0 & 0 & -1 & -2\end{array}\right]\xrightarrow{-R_3}\left[\begin{array}{ccc|c}1 & -2 & 3 & 10\\ 0 & 1 & 2 & 3\\ 0 & 0 & 1 & 2\end{array}\right]$$

$$\begin{cases}x-2y+3z=10\\ y+2z=3\\ z=2\end{cases}$$

$$y+2(2)=3$$
$$y=-1$$

$$x-2(-1)+3(2)=10$$
$$x=2$$

The solution is (2, –1, 2).

5. Solve the system using Gaussian elimination. [7.1]

$$\left[\begin{array}{ccc|c}4 & 3 & 5 & 3\\ 2 & -3 & 1 & -1\\ 2 & 0 & 2 & 3\end{array}\right]\xrightarrow{\frac{1}{4}R_1}\left[\begin{array}{ccc|c}1 & \frac{3}{4} & \frac{5}{4} & \frac{3}{4}\\ 2 & -3 & 1 & -1\\ 2 & 0 & 2 & 3\end{array}\right]$$

$$\xrightarrow[-2R_1+R_3]{-2R_1+R_2}\left[\begin{array}{ccc|c} 1 & \frac{3}{4} & \frac{5}{4} & \frac{3}{4} \\ 0 & -\frac{9}{2} & -\frac{3}{2} & -\frac{5}{2} \\ 0 & -\frac{3}{2} & -\frac{1}{2} & \frac{3}{2} \end{array}\right]$$

$$\xrightarrow{-\frac{2}{9}R_2}\left[\begin{array}{ccc|c} 1 & \frac{3}{4} & \frac{5}{4} & \frac{3}{4} \\ 0 & 1 & \frac{1}{3} & \frac{5}{9} \\ 0 & -\frac{3}{2} & -\frac{1}{2} & \frac{3}{2} \end{array}\right] \xrightarrow{\frac{3}{2}R_2+R_3}\left[\begin{array}{ccc|c} 1 & \frac{3}{4} & \frac{5}{4} & \frac{3}{4} \\ 0 & 1 & \frac{1}{3} & \frac{5}{9} \\ 0 & 0 & 0 & \frac{7}{3} \end{array}\right]$$

$$\begin{cases} x+\frac{3}{4}y+\frac{5}{4}z=\frac{3}{4} \\ y+\frac{1}{3}z=\frac{5}{9} \\ 0z=\frac{7}{3} \end{cases}$$

Because $0z=\frac{7}{3}$ has no solution, the system of equations has no solution.

7. Find $-3A$. [7.2]

$$-3A=-3\begin{bmatrix} -1 & 3 & 2 \\ 1 & 4 & -1 \end{bmatrix}=\begin{bmatrix} 3 & -9 & -6 \\ -3 & -12 & 3 \end{bmatrix}$$

9. Find $3B-2C$. [7.2]

$$3B-2C=3\begin{bmatrix} 2 & -1 & 3 \\ 4 & -2 & -1 \\ 3 & 2 & 2 \end{bmatrix}-2\begin{bmatrix} 1 & -2 & 3 \\ 2 & -3 & 8 \\ -1 & 3 & -2 \end{bmatrix}$$

$$=\begin{bmatrix} 6 & -3 & 9 \\ 12 & -6 & -3 \\ 9 & 6 & 6 \end{bmatrix}-\begin{bmatrix} 2 & -4 & 6 \\ 4 & -6 & 16 \\ -2 & 6 & -4 \end{bmatrix}=\begin{bmatrix} 4 & 1 & 3 \\ 8 & 0 & -19 \\ 11 & 0 & 10 \end{bmatrix}$$

11. Find $AB-A$. [7.2]

Use AB from Exercise 10. AB

$$-A=\begin{bmatrix} 16 & -1 & -2 \\ 15 & -11 & -3 \end{bmatrix}-\begin{bmatrix} -1 & 3 & 2 \\ 1 & 4 & -1 \end{bmatrix}=\begin{bmatrix} 17 & -4 & -4 \\ 14 & -15 & -2 \end{bmatrix}$$

13. Find $BC-CB$. [7.2]

$BC-CB$

$$=\begin{bmatrix} 2 & -1 & 3 \\ 4 & -2 & -1 \\ 3 & 2 & 2 \end{bmatrix}\begin{bmatrix} 1 & -2 & 3 \\ 2 & -3 & 8 \\ -1 & 3 & -2 \end{bmatrix}-\begin{bmatrix} 1 & -2 & 3 \\ 2 & -3 & 8 \\ -1 & 3 & -2 \end{bmatrix}\begin{bmatrix} 2 & -1 & 3 \\ 4 & -2 & -1 \\ 3 & 2 & 2 \end{bmatrix}$$

$$=\begin{bmatrix} -3 & 8 & -8 \\ 1 & -5 & -2 \\ 5 & -6 & 21 \end{bmatrix}-\begin{bmatrix} 3 & 9 & 11 \\ 16 & 20 & 25 \\ 4 & -9 & -10 \end{bmatrix}=\begin{bmatrix} -6 & -1 & -19 \\ -15 & -25 & -27 \\ 1 & 3 & 31 \end{bmatrix}$$

15. Find B^2. [7.2]

$$B^2=\begin{bmatrix} 2 & -1 & 3 \\ 4 & -2 & -1 \\ 3 & 2 & 2 \end{bmatrix}\begin{bmatrix} 2 & -1 & 3 \\ 4 & -2 & -1 \\ 3 & 2 & 2 \end{bmatrix}=\begin{bmatrix} 9 & 6 & 13 \\ -3 & -2 & 12 \\ 20 & -3 & 11 \end{bmatrix}$$

17. Find the minor and cofactor of b_{21}. [7.4]

$$M_{21}=\begin{vmatrix} -1 & 3 \\ 2 & 2 \end{vmatrix}=-2-6=-8$$

$$C_{21}=(-1)^{2+1}M_{21}=-(-8)=8$$

19. Find the determinant using elementary row operations. [7.4]

$$\begin{vmatrix} 1 & -2 & 3 \\ 2 & -3 & 8 \\ -1 & 3 & -2 \end{vmatrix}=\begin{matrix} -2R_1+R_2 \\ R_1+R_3 \end{matrix}\begin{vmatrix} 1 & -2 & 3 \\ 0 & 1 & 2 \\ 0 & 1 & 1 \end{vmatrix}$$

$$=-R_2+R_3\begin{vmatrix} 1 & -2 & 3 \\ 0 & 1 & 2 \\ 0 & 0 & -1 \end{vmatrix}$$

$$=(1)(1)(-1)$$

$$=-1$$

21. Use the inverse to solve the system of equations. [7.3]

$$\begin{bmatrix} 1 & -3 & 0 \\ -5 & -5 & 3 \\ 3 & 4 & -2 \end{bmatrix}\begin{bmatrix} x \\ y \\ z \end{bmatrix}=\begin{bmatrix} 1 \\ -2 \\ 0 \end{bmatrix}$$

$$\begin{bmatrix} -2 & -6 & -9 \\ -1 & -2 & -3 \\ -5 & -13 & -20 \end{bmatrix}\begin{bmatrix} 1 \\ -2 \\ 0 \end{bmatrix}=\begin{bmatrix} 10 \\ 3 \\ 21 \end{bmatrix}=\begin{bmatrix} x \\ y \\ z \end{bmatrix}$$

The solution is (10, 3, 21).

23. Find the vertices. [7.2]

$$\begin{bmatrix} 0 & -1 & 0 \\ 1 & 0 & 0 \\ 0 & 0 & 1 \end{bmatrix}\begin{bmatrix} -3 & 1 & 4 \\ -2 & 3 & 1 \\ 1 & 1 & 1 \end{bmatrix}=\begin{bmatrix} 2 & -3 & -1 \\ -3 & 1 & 4 \\ 1 & 1 & 1 \end{bmatrix}$$

$A^1(2,-3)$, $B^1(-3, 1)$, $C^1(-1, 4)$

25. Set up the matrix equation. [7.3]

$$X=(I-A)^{-1}D$$

$$\left(\begin{bmatrix} 1 & 0 & 0 \\ 0 & 1 & 0 \\ 0 & 0 & 1 \end{bmatrix}-\begin{bmatrix} 0.15 & 0.23 & 0.11 \\ 0.08 & 0.10 & 0.05 \\ 0.16 & 0.11 & 0.07 \end{bmatrix}\right)^{-1}\begin{bmatrix} 50 \\ 32 \\ 8 \end{bmatrix}$$

Cumulative Review Exercises

1. Find the equation of the circle.

$$(x+2)^2+(y-4)^2=25$$
$$x^2+4x+4+y^2-8y+16=25$$
$$x^2+y^2+4x-8y-5=0$$

3. Find the equation of the line. [2.3]

$$y-5=-\frac{1}{2}(x-(-4))$$
$$y-5=-\frac{1}{2}x-2$$
$$y=-\frac{1}{2}x+3$$

5. Find $h[k(0)]$. [4.2]

$$h[k(0)]=h[3^0]$$
$$=h[1]$$
$$=e^{-1}$$
$$\approx 0.3679$$

7. Solve the system of equations. [6.1]

$$\begin{cases} 3x-4y=4 & (1) \\ 2x-3y=1 & (2) \end{cases}$$

Solve (2) for x and substitute into (1).

$$2x-3y=1$$
$$2x=3y+1$$
$$x=\frac{3y+1}{2}$$

$$3\left(\frac{3y+1}{2}\right)-4y=4$$
$$9y+3-8y=8$$
$$y=5$$

$$x=\frac{3(5)+1}{2}=8$$

The solution is (8, 5).

9. Determine the domain. [2.2]

Domain: $\{x \mid -3 \le x \le 3\}$

11. Find the vertical asymptotes. [3.5]

$$x^2+4x-5=0$$
$$(x+5)(x-1)=0$$

$$x+5=0 \qquad x-1=0$$
$$x=-5 \qquad x=1$$

Vertical asymptotes: $x=-5$, $x=1$

13. Solve and express answer in interval notation. [1.5]

$$\frac{x+2}{x+1}>0$$

The quotient $\frac{x+2}{x+1}$ is positive.

The critical values are –2 and –1.

$\frac{x+2}{x+1}$ + + + + + + + + | - - | + + + + + + +

–4 –3 –2 –1 0 1

$$(-\infty,-2)\cup(-1,\infty)$$

15. Solve. [6.2]

$$\begin{cases} x-y+z=-1 & (1) \\ 2x+3y-z=1 & (2) \\ 3x-2y+3z=12 & (3) \end{cases}$$

$$\begin{array}{ll} x-y+z=-1 & (1) \\ 2x+3y-z=1 & (2) \\ \hline 3x+2y \quad =0 & (4) \end{array}$$

$$\begin{array}{ll} 6x+9y-3z=3 & \text{3 times (2)} \\ 3x-2y+3z=12 & (3) \\ \hline 9x+7y \quad =15 & (5) \end{array}$$

$$\begin{cases} x-y+z=-1 & (1) \\ 3x+2y=0 & (4) \\ 9x+7y=15 & (5) \end{cases}$$

$$\begin{array}{ll} -9x-6y=0 & -3 \text{ times (4)} \\ 9x+7y=15 & (5) \\ \hline y=15 & \end{array}$$

$$9x+7(15)=15 \quad \text{Substitute into Eq (5).}$$
$$9x+105=15$$
$$9x=-90$$
$$x=-10$$

$$-10-15+z=-1$$
$$z=24$$

The solution is (–10, 15, 24).

17. Solve. [4.5]

$$\begin{aligned} 125^x &= \frac{1}{25} \\ 5^{3x} &= 5^{-2} \\ 3x &= -2 \\ x &= -\frac{2}{3} \end{aligned}$$

19. Solve. Round to the nearest ten-thousandth. [4.5]

$$\begin{aligned} 10^x - 10^{-x} &= 2 \\ 10^x(10^x - 10^{-x}) &= 2(10^x) \\ (10^x)^2 - 2(10^x) - 1 &= 0 \end{aligned}$$

Let $u = 10^x$

$$u^2 - 2u - 1 = 0$$

$$\begin{aligned} u &= \frac{2 \pm \sqrt{4 - 4(1)(-1)}}{2} \\ &= \frac{2 \pm \sqrt{8}}{2} \\ &= 1 \pm \sqrt{2} \\ 10^x &= 1 + \sqrt{2} \\ \log 10^x &= \log(1 + \sqrt{2}) \\ x &= \log(1 + \sqrt{2}) \approx 0.3828 \end{aligned}$$

Chapter 8: Sequences, Series, and Probability

Section 8.1 Exercises

1. Find terms 1, 2, 3, and 8 of the sequence.

$$a_1 = 1(1-1) = 1\cdot 0 = 0$$
$$a_2 = 2(2-1) = 2\cdot 1 = 2$$
$$a_3 = 3(3-1) = 3\cdot 2 = 6$$
$$a_8 = 8(8-1) = 8\cdot 7 = 56$$

3. Find terms 1, 2, 3, and 8 of the sequence.

$$a_1 = 1 - \frac{1}{1} = 0$$
$$a_2 = 1 - \frac{1}{2} = \frac{1}{2}$$
$$a_3 = 1 - \frac{1}{3} = \frac{2}{3}$$
$$a_8 = 1 - \frac{1}{8} = \frac{7}{8}$$

5. Find terms 1, 2, 3, and 8 of the sequence.

$$a_1 = \frac{(-1)^{1+1}}{1^2} = \frac{(-1)^2}{1} = 1$$
$$a_2 = \frac{(-1)^{2+1}}{2^2} = \frac{(-1)^3}{4} = -\frac{1}{4}$$
$$a_3 = \frac{(-1)^{3+1}}{3^2} = \frac{(-1)^4}{9} = \frac{1}{9}$$
$$a_8 = \frac{(-1)^{8+1}}{8^2} = \frac{(-1)^9}{64} = -\frac{1}{64}$$

7. Find terms 1, 2, 3, and 8 of the sequence.

$$a_1 = \frac{(-1)^{2\cdot 1-1}}{3\cdot 1} = \frac{(-1)^{2-1}}{3} = -\frac{1}{3}$$
$$a_2 = \frac{(-1)^{2\cdot 2-1}}{3\cdot 2} = \frac{(-1)^{4-1}}{6} = -\frac{1}{6}$$
$$a_3 = \frac{(-1)^{2\cdot 3-1}}{3\cdot 3} = \frac{(-1)^{6-1}}{9} = -\frac{1}{9}$$
$$a_8 = \frac{(-1)^{2\cdot 8-1}}{3\cdot 8} = \frac{(-1)^{16-1}}{24} = -\frac{1}{24}$$

9. Find terms 1, 2, 3, and 8 of the sequence.

$$a_1 = \left(\frac{2}{3}\right)^1 = \frac{2}{3}$$
$$a_2 = \left(\frac{2}{3}\right)^2 = \frac{4}{9}$$
$$a_3 = \left(\frac{2}{3}\right)^3 = \frac{8}{27}$$
$$a_8 = \left(\frac{2}{3}\right)^8 = \frac{256}{6561}$$

11. Find terms 1, 2, 3, and 8 of the sequence.

$$a_1 = 1 + (-1)^1 = 1 + (-1) = 0$$
$$a_2 = 1 + (-1)^2 = 1 + 1 = 2$$
$$a_3 = 1 + (-1)^3 = 1 + (-1) = 0$$
$$a_8 = 1 + (-1)^8 = 1 + 1 = 2$$

13. Find terms 1, 2, 3, and 8 of the sequence.

$$a_1 = (1.1)^1 = 1.1$$
$$a_2 = (1.1)^2 = 1.21$$
$$a_3 = (1.1)^3 = 1.331$$
$$a_8 = (1.1)^8 = 2.14358881$$

15. Find terms 1, 2, 3, and 8 of the sequence.

$$a_1 = \frac{(-1)^{1+1}}{\sqrt{1}} = \frac{(-1)^2}{1} = 1$$
$$a_2 = \frac{(-1)^{2+1}}{\sqrt{2}} = \frac{(-1)^3}{\sqrt{2}} = -\frac{1}{\sqrt{2}} = -\frac{\sqrt{2}}{2}$$
$$a_3 = \frac{(-1)^{3+1}}{\sqrt{3}} = \frac{(-1)^4}{\sqrt{3}} = \frac{1}{\sqrt{3}} = \frac{\sqrt{3}}{3}$$
$$a_8 = \frac{(-1)^{8+1}}{\sqrt{8}} = \frac{(-1)^9}{2\sqrt{2}} = -\frac{1}{2\sqrt{2}} = -\frac{\sqrt{2}}{4}$$

17. Find terms 1, 2, 3, and 8 of the sequence.

$$a_1 = 1! = 1$$
$$a_2 = 2! = 2\cdot 1 = 2$$
$$a_3 = 3! = 3\cdot 2\cdot 1 = 6$$
$$a_8 = 8! = 8\cdot 7\cdot 6\cdot 5\cdot 4\cdot 3\cdot 2\cdot 1 = 40,320$$

19. Find terms 1, 2, 3, and 8 of the sequence.

$$a_1 = \log 1 = 0$$
$$a_2 = \log 2 \approx 0.3010$$
$$a_3 = \log 3 \approx 0.4771$$
$$a_8 = \log 8 \approx 0.9031$$

21. Find terms 1, 2, 3, and 8 of the sequence.

$$a_1 = 1 \qquad \frac{1}{7} = 0.\overline{142857}$$
$$a_2 = 4$$
$$a_3 = 2$$
$$a_8 = 4$$

23. Find terms 1, 2, 3, and 8 of the sequence.

$$a_1 = 3$$
$$a_2 = 3$$
$$a_3 = 3$$
$$a_8 = 3$$

25. Find the first three terms of the sequence.

$$a_1 = 5$$
$$a_2 = 2 \cdot a_1 = 2 \cdot 5 = 10$$
$$a_3 = 2 \cdot a_2 = 2 \cdot 10 = 20$$

27. Find the first three terms of the sequence.

$$a_1 = 2$$
$$a_2 = 2 \cdot a_1 = 2 \cdot 2 = 4$$
$$a_3 = 3 \cdot a_2 = 3 \cdot 4 = 12$$

29. Find the first three terms of the sequence.

$$a_1 = 2$$
$$a_2 = (a_1)^2 = (2)^2 = 4$$
$$a_3 = (a_2)^2 = (4)^2 = 16$$

31. Find the first three terms of the sequence.

$$a_1 = 2$$
$$a_2 = 2 \cdot 2 \cdot a_1 = 4 \cdot 2 = 8$$
$$a_3 = 2 \cdot 3 \cdot a_2 = 6 \cdot 8 = 48$$

33. Find the first three terms of the sequence.

$$a_1 = 3$$
$$a_2 = (a_1)^{1/2} = (3)^{1/2} = \sqrt{3}$$
$$a_3 = (a_2)^{1/3} = \left(3^{1/2}\right)^{1/3} = 3^{1/6} = \sqrt[6]{3}$$

35. Find a_3, a_4, and a_5.

$$a_1 = 1$$
$$a_2 = 3$$
$$a_3 = \frac{1}{2}(a_2 + a_1) = \frac{1}{2}(3+1) = \frac{1}{2}(4) = 2$$
$$a_4 = \frac{1}{2}(a_3 + a_2) = \frac{1}{2}(2+3) = \frac{1}{2}(5) = \frac{5}{2}$$
$$a_5 = \frac{1}{2}(a_4 + a_3) = \frac{1}{2}\left(\frac{5}{2} + 2\right) = \frac{1}{2}\left(\frac{9}{2}\right) = \frac{9}{4}$$

37. Find the next three terms of the Lucas sequence.

$$a_n = a_{n-1} + a_{n-2}$$
$$a_3 = 3 + 1 = 4$$
$$a_4 = 4 + 3 = 7$$
$$a_5 = 7 + 4 = 11$$

39. Evaluate the factorial expression.

$$7! - 6! = 7 \cdot 6! - 6! = 6!(7-1)$$
$$= 6! \cdot 6 = 6 \cdot 5 \cdot 4 \cdot 3 \cdot 2 \cdot 1 \cdot 6$$
$$= 4320$$

41. Evaluate the factorial expression.

$$\frac{9!}{7!} = \frac{9 \cdot 8 \cdot 7!}{7!} = 72$$

43. Evaluate the factorial expression.

$$\frac{8!}{3!5!} = \frac{8 \cdot 7 \cdot 6 \cdot 5!}{3 \cdot 2 \cdot 1 \cdot 5!} = 56$$

45. Evaluate the factorial expression.

$$\frac{100!}{99!} = \frac{100 \cdot 99!}{99!} = 100$$

47. Evaluate the series.

$$\sum_{i=1}^{5} i = 1 + 2 + 3 + 4 + 5 = 15$$

49. Evaluate the series.

$$\sum_{i=1}^{5} i(i-1)$$
$$= 1(1-1) + 2(2-1) + 3(3-1) + 4(4-1) + 5(5-1)$$
$$= 1 \cdot 0 + 2 \cdot 1 + 3 \cdot 2 + 4 \cdot 3 + 5 \cdot 4$$
$$= 0 + 2 + 6 + 12 + 20 = 40$$

51. Evaluate the series.

$$\sum_{k=1}^{4} \frac{1}{k} = \frac{1}{1} + \frac{1}{2} + \frac{1}{3} + \frac{1}{4} = \frac{12}{12} + \frac{6}{12} + \frac{4}{12} + \frac{3}{12} = \frac{25}{12}$$

53. Evaluate the series.

$$\sum_{j=1}^{8} 2j = 2\sum_{j=1}^{8} j$$
$$= 2(1+2+3+4+5+6+7+8)$$
$$= 2(36) = 72$$

55. Evaluate the series.

$$\sum_{i=1}^{5} (-1)^{i-1} 2^i$$
$$= (-1)^0 2^1 + (-1)^1 2^2 + (-1)^2 2^3 + (-1)^3 2^4 + (-1)^4 2^5$$
$$= 2 - 4 + 8 - 16 + 32$$
$$= 22$$

57. Evaluate the series.

$$\sum_{n=1}^{7}\log\left(\frac{n+1}{n}\right)$$
$$=\log\left(\frac{1+1}{1}\right)+\log\left(\frac{2+1}{2}\right)+\log\left(\frac{3+1}{3}\right)+\log\left(\frac{4+1}{4}\right)$$
$$+\log\left(\frac{5+1}{5}\right)+\log\left(\frac{6+1}{6}\right)+\log\left(\frac{7+1}{7}\right)$$
$$=\log\left(2\cdot\frac{3}{2}\cdot\frac{4}{3}\cdot\frac{5}{4}\cdot\frac{6}{5}\cdot\frac{7}{6}\cdot\frac{8}{7}\right)=\log 8$$
$$=3\log 2\approx 0.9031$$

59. Evaluate the series.

$$\sum_{k=0}^{8}\frac{8!}{k!(8-k)!}$$
$$=\frac{8!}{0!(8-0)!}+\frac{8!}{1!(8-1)!}+\frac{8!}{2!(8-2)!}+\frac{8!}{3!(8-3)!}$$
$$+\frac{8!}{4!(8-4)!}+\frac{8!}{5!(8-5)!}+\frac{8!}{6!(8-6)!}$$
$$+\frac{8!}{7!(8-7)!}+\frac{8!}{8!(8-8)!}$$
$$=\frac{8!}{0!8!}+\frac{8!}{1!7!}+\frac{8!}{2!6!}+\frac{8!}{3!5!}+\frac{8!}{4!4!}+\frac{8!}{5!3!}+\frac{8!}{6!2!}$$
$$+\frac{8!}{7!1!}+\frac{8!}{8!0!}$$
$$=1+8+\frac{8\cdot 7}{2}+\frac{8\cdot 7\cdot 6}{3\cdot 2}+\frac{8\cdot 7\cdot 6\cdot 5}{4\cdot 3\cdot 2}+\frac{8\cdot 7\cdot 6}{3\cdot 2\cdot 1}+\frac{8\cdot 7}{2}$$
$$+8+1$$
$$=1+8+28+56+70+56+28+8+1$$
$$=256$$

61. Write the series in summation notation.

$$\frac{1}{1}+\frac{1}{4}+\frac{1}{9}+\frac{1}{16}+\frac{1}{25}+\frac{1}{36}$$
$$=\frac{1}{1^2}+\frac{1}{2^2}+\frac{1}{3^2}+\frac{1}{4^2}+\frac{1}{5^2}+\frac{1}{6^2}$$
$$=\sum_{i=1}^{6}\frac{1}{i^2}$$

63. Write the series in summation notation.

$$2-4+8-16+32-64+128$$
$$=2^1(-1)^{1+1}+2^2(-1)^{2+1}+2^3(-1)^{3+1}+2^4(-1)^{4+1}$$
$$+2^5(-1)^{5+1}+2^6(-1)^{6+1}+2^7(-1)^{7+1}$$
$$=\sum_{i=1}^{7}2^i(-1)^{i+1}$$

65. Write the series in summation notation.

$$7+10+13+16+19$$
$$=7+(7+3)+(7+3\cdot 2)+(7+3\cdot 3)+(7+3\cdot 4)$$
$$=\sum_{i=0}^{4}(7+3i)$$

67. Write the series in summation notation.

$$\frac{1}{2}+\frac{1}{4}+\frac{1}{8}+\frac{1}{16}=\frac{1}{2}+\frac{1}{2^2}+\frac{1}{2^3}+\frac{1}{2^4}=\sum_{i=1}^{4}\frac{1}{2^i}$$

69. Approximate $\sqrt{7}$ by computing a_4 and compare to calculator value.

Let $N=7$.

$$a_1=\frac{7}{2}=3.5$$
$$a_2=\frac{1}{2}\left(3.5+\frac{7}{3.5}\right)=2.75$$
$$a_3=\frac{1}{2}\left(2.75+\frac{7}{2.75}\right)\approx 2.6477273$$
$$a_4\approx\frac{1}{2}\left(2.6477273+\frac{7}{2.6477273}\right)\approx 2.6457520$$

71. Determine the sum of the first 10 terms of the Fibonacci sequence.

$1+1=2$

$1+1+2=4$

$1+1+2+3=7$

The sum of the first n terms of a Fibonacci sequence equals the $(n+2)$ term – 1.

Therefore the sum of the first ten terms is (10+2) term – 1, i.e. 12th term – 1.

144 – 1 = 143

73. Find F_{10} and F_{15}.

$$F_{10}=\frac{\left(\frac{1+\sqrt{5}}{2}\right)^{10}-\left(\frac{1-\sqrt{5}}{2}\right)^{10}}{\sqrt{5}}=55$$

$$F_{15}=\frac{\left(\frac{1+\sqrt{5}}{2}\right)^{15}-\left(\frac{1-\sqrt{5}}{2}\right)^{15}}{\sqrt{5}}=610$$

Prepare for Section 8.2

P1. Solve for d.

$$-3 = 25 + (15-1)d$$
$$-3 = 25 + 14d$$
$$-28 = 14d$$
$$-2 = d$$

P3. Evaluate.

$$S = \frac{50\left[2(2) + (50-1)\frac{5}{4}\right]}{2} = \frac{50\left[4 + \frac{245}{4}\right]}{2}$$
$$= \frac{50\left[\frac{261}{4}\right]}{2} = \frac{6525}{4}$$

P5. Find the twentieth term.

$$a_{20} = 52 + (20-1)(-3) = -5$$

Section 8.2 Exercises

1. Find the 9th, 24th, and nth terms.

$$d = 10 - 6 = 4$$
$$a_n = 6 + (n-1)4 = 6 + 4n - 4 = 4n + 2$$
$$a_9 = 4\cdot 9 + 2 = 36 + 2 = 38$$
$$a_{24} = 4\cdot 24 + 2 = 98$$

3. Find the 9th, 24th, and nth terms.

$$d = 4 - 6 = -2$$
$$a_n = 6 + (n-1)(-2) = 6 - 2n + 2 = 8 - 2n$$
$$a_9 = 8 - 2\cdot 9 = 8 - 18 = -10$$
$$a_{24} = 8 - 2\cdot 24 = 8 - 48 = -40$$

5. Find the 9th, 24th, and nth terms.

$$d = -5 - (-8) = 3$$
$$a_n = -8 + (n-1)3 = -8 + 3n - 3$$
$$= 3n - 11$$
$$a_9 = 3\cdot 9 - 11 = 27 - 11 = 16$$
$$a_{24} = 3\cdot 24 - 11 = 72 - 11 = 61$$

7. Find the 9th, 24th, and nth terms.

$$d = 4 - 1 = 3$$
$$a_n = 1 + (n-1)3 = 1 + 3n - 3 = 3n - 2$$
$$a_9 = 3\cdot 9 - 2 = 27 - 2 = 25$$
$$a_{24} = 3\cdot 24 - 2 = 72 - 2 = 70$$

9. Find the 9th, 24th, and nth terms.

$$d = (a+2) - a = 2$$
$$a_n = a + (n-1)2 = a + 2n - 2$$
$$a_9 = a + 2\cdot 9 - 2 = a + 18 - 2 = a + 16$$
$$a_{24} = a + 2\cdot 24 - 2 = a + 48 - 2 = a + 46$$

11. Find the 9th, 24th, and nth terms.

$$d = \log 14 - \log 7 = \log\frac{14}{7} = \log 2$$
$$a_n = \log 7 + (n-1)\log 2$$
$$a_9 = \log 7 + 8\log 2$$
$$a_{24} = \log 7 + 23\log 2$$

13. Find the 9th, 24th, and nth terms.

$$d = \log a^2 - \log a = \log\frac{a^2}{a} = \log a$$
$$a_n = \log a + (n-1)\log a$$
$$= (1 + n - 1)\log a = n\log a$$
$$a_9 = 9\log a$$
$$a_{24} = 24\log a$$

15. Find the 20th term.

$$d = 15 - 13 = 2$$
$$a_4 = a_1 + (4-1)2 = 13$$
$$a_1 + 6 = 13$$
$$a_1 = 7$$
$$a_{20} = 7 + (20-1)2 = 7 + (19)2 = 7 + 38 = 45$$

17. Find the 17th term.

$$a_5 = -19 = a_1 + (5-1)d \qquad a_7 = -29 = a_1 + (7-1)d$$
$$-19 = a_1 + 4d \qquad -29 = a_1 + 6d$$
$$-19 - 4d = a_1 \qquad -29 = (-19 - 4d) + 6d$$
$$-29 = -19 + 2d$$
$$-10 = 2d$$
$$d = -5$$

$$a_1 = -4(-5) - 19 = 1$$
$$a_{17} = 1 + (17-1)(-5) = 1 + 16(-5) = 1 - 80 = -79$$

19. Find the nth partial sum.

$$a_1 = 3(1) + 2 = 5$$
$$a_{10} = 3(10) + 2 = 32$$
$$S_{10} = \frac{10}{2}(a_1 + a_{10}) = 5(5 + 32) = 185$$

21. Find the nth partial sum.

$$a_1 = 3-5(1) = -2$$
$$a_{15} = 3-5(15) = -72$$
$$S_{15} = \frac{15}{2}(a_1 + a_{15}) = \frac{15}{2}(-2+(-72)) = \frac{15}{2}(-74)$$
$$= -555$$

23. Find the nth partial sum.

$$a_1 = 6(1) = 6$$
$$a_{12} = 6(12) = 72$$
$$S_{12} = \frac{12}{2}(a_1 + a_{12}) = 6(6+72) = 6(78) = 468$$

25. Find the nth partial sum.

$$a_1 = 1+8 = 9$$
$$a_{25} = 25+8 = 33$$
$$S_{25} = \frac{25}{2}(a_1 + a_{25}) = \frac{25}{2}(9+33) = \frac{25}{2}(42) = 525$$

27. Find the nth partial sum.

$$a_1 = -1$$
$$a_{30} = -30$$
$$S_{30} = \frac{30}{2}(a_1 + a_{30}) = 15(-1+(-30))$$
$$= 15(-31)$$
$$= -465$$

29. Find the nth partial sum.

$$a_1 = 1+x$$
$$a_{12} = 12+x$$
$$S_{12} = \frac{12}{2}(a_1 + a_{12})$$
$$= 6(1+x+12+x)$$
$$= 78+12x$$

31. Find the nth partial sum.

$$a_1 = (1)x = x$$
$$a_{20} = (20)x = 20x$$
$$S_{20} = \frac{20}{2}(a_1 + a_{20}) = 10(x+20x) = 210x$$

33. Insert k arithmetic means between the numbers.

$-1,\ c_2,\ c_3,\ c_4,\ c_5,\ c_6,\ 23$

$$a_1 = -1$$
$$a_7 = a_1 + (n-1)d$$
$$23 = -1+(7-1)d$$
$$23 = -1+6d$$
$$24 = 6d$$
$$d = 4$$
$$c_2 = -1+(2-1)4 = -1+4 = 3$$
$$c_3 = -1+(3-1)4 = -1+(2)4 = 7$$
$$c_4 = -1+(4-1)4 = -1+(3)4 = 11$$
$$c_5 = -1+(5-1)4 = -1+(4)4 = 15$$
$$c_6 = -1+(6-1)4 = -1+(5)4 = 19$$

35. Insert k arithmetic means between the numbers.

$3,\ c_2,\ c_3,\ c_4,\ c_5,\ \frac{1}{2}$

$$a_1 = 3$$
$$a_6 = a_1 + (n-1)d$$
$$\frac{1}{2} = 3+(6-1)d$$
$$-\frac{5}{2} = 5d$$
$$d = -\frac{1}{2}$$
$$c_2 = 3+(2-1)\left(-\frac{1}{2}\right) = 3-\frac{1}{2} = \frac{5}{2}$$
$$c_3 = 3+(3-1)\left(-\frac{1}{2}\right) = 3-1 = 2$$
$$c_4 = 3+(4-1)\left(-\frac{1}{2}\right) = 3-\frac{3}{2} = \frac{3}{2}$$
$$c_5 = 3+(5-1)\left(-\frac{1}{2}\right) = 3-2 = 1$$

37. Show the sum.

$$a_1 = 1,\ d = 2$$
$$S_n = \frac{n[2(1)+(n-1)2]}{2} = \frac{n}{2}[2n] = n^2$$

39. Find the number of logs in the sixth row and the total number of logs for all six rows.

$$a_1 = 25,\ d = -1$$
$$a_6 = 25+(6-1)(-1) = 25-5 = 20$$
$$S_6 = \frac{6}{2}(25+20) = 3(45) = 135$$

20 logs stacked on sixth row,

135 logs in the six rows

41. Find the value of the 15th prize and the total amount awarded in prizes.

$a_1 = 5000, d = -250$
$a_{15} = 5000 + (15-1)(-250) = 5000 - 3500 = 1500$

The fifteenth prize is $1500.

$$S_{15} = \frac{15}{2}(5000+1500) = \frac{15}{2}(6500) = 48,750$$

The total amount of money distributed is $48,750.

43. Find the distance.

$a_1 = 16, d = 32$

$$S_7 = \frac{7}{2}[2(16)+(7-1)32] = \frac{7}{2}[32+192] = \frac{7}{2}[224] = 784$$

The total distance the object falls is 784 ft.

45. Find the number of cross threads.

$$\begin{aligned} a_n &= a_1 + (n-1)d \\ 46.9 &= 3.5 + (n-1)1.4 \\ 43.4 &= (n-1)1.4 \\ 31 &= n-1 \\ 32 &= n \end{aligned}$$

There are 32 cross threads.

47. Show $f(n)$ is an arithmetic sequence.

If $f(x)$ is a linear function, then $f(x) = mx + b$. To show that $f(n)$, where n is a positive integer, is an arithmetic sequence, we must show that $f(n+1) - f(n)$ is a constant. We have

$$\begin{aligned} f(n+1) - f(n) &= (m(n+1)+b) - (m(n)+b) \\ &= mn + m + b - mn - b \\ &= m \end{aligned}$$

Thus, the difference between any two successive terms is m, the slope of the linear function.

49. Find the formula for a_n in terms of a_1.

$a_1 = 4,\ a_n = a_{n-1} - 3$

Rewriting $a_n = a_{n-1} - 3$ as $a_n - a_{n-1} = -3$, we find that the difference between successive terms is the same constant –3.

Thus the sequence is an arithmetic sequence with $a_1 = 4$ and $d = -3$.

$a_n = a_1 + (n-1)d$

Substituting,

$a_n = 4 + (n-1)(-3) = 4 - 3n + 3 = 7 - 3n$

51. Show that a_n and b_n are arithmetic sequences and find a_{50}.

$a_1 = 1, a_n = b_{n-1} + 7; b_1 = -2, b_n = a_{n-1} + 1$

To show that a_n is an arithmetic sequence, we must show that $a_n - a_{n-1} = d$, where d is a constant. We begin by finding a relationship between a_n and a_{n-2}.

$$\begin{aligned} a_n &= b_{n-1} + 7 = a_{n-2} + 1 + 7 \\ a_n &= a_{n-2} + 8 \end{aligned}$$

This establishes a relationship between *alternate* successive terms. We now examine some terms of a_n.

$$\begin{aligned} a_1 &= 1 \\ a_2 &= b_1 + 7 = -2 + 7 = 5 \\ a_3 &= a_1 + 8 \qquad (a_n = a_{n-2} + 8) \\ a_4 &= a_2 + 8 \\ a_5 &= a_3 + 8 = (a_1 + 8) + 8 = a_1 + 2(8) \\ a_6 &= a_4 + 8 = (a_2 + 8) + 8 = a_2 + 2(8) \\ a_7 &= a_5 + 8 = (a_1 + 2(8)) + 8 = a_1 + 3(8) \\ a_8 &= a_6 + 8 = (a_2 + 2(8)) + 8 = a_2 + 3(8) \end{aligned}$$

Now consider two cases. First, n is an even integer, $n = 2k$.

$a_{2k} = a_2 + (k-1)8 \qquad k \geq 2$

When n is an odd integer, $n = 2k = 1$.

$a_{2k-1} = a_1 + (k-1)8 \qquad k \geq 2$

$$\begin{aligned} \text{Thus } a_{2k} - a_{2k-1} &= (a_2 + (k-1)8) - (a_1 + (k-1)8) \\ &= a_2 - a_1 = 5 - 1 = 4 \end{aligned}$$

Therefore, the difference between successive terms is the constant. To find a_{50}, use $a_n = a_1 + (n-1)d$.

$a_{50} = 1 + (49)(4) = 197$

To show that b_n is an arithmetic sequence, we have

$$b_n - b_{n-1} = (a_{n-1}+1)-(a_{n-2}+1) = a_{n-1}-a_{n-2}$$

Because a_n is an arithmetic sequence, $a_{n-1}-a_{n-2}$ is a constant. Thus b_n is an arithmetic sequence.

Prepare for Section 8.3

P1. Find the ratio of any two successive terms.

$$\frac{4}{2}=2,\quad \frac{8}{4}=2$$

The ratio is 2.

P3. Evaluate.

$$S=\frac{3(1-(-2)^5)}{1-(-2)}=33$$

P5. Write the first three terms of the sequence.

$$a_1 = 3\left(-\frac{1}{2}\right)^1 = -\frac{3}{2}$$

$$a_2 = 3\left(-\frac{1}{2}\right)^2 = \frac{3}{4}$$

$$a_3 = 3\left(-\frac{1}{2}\right)^3 = -\frac{3}{8}$$

Section 8.3 Exercises

1. Find the nth term of the geometric sequence.

$$r=\frac{8}{2}=4$$

$$a_n = 2\cdot 4^{n-1} = 2\cdot 2^{2(n-1)} = 2^{2n-1}$$

3. Find the nth term of the geometric sequence.

$$r=\frac{12}{-4}=-3$$

$$a_n=-4(-3)^{n-1}$$

5. Find the nth term of the geometric sequence.

$$r=\frac{4}{6}=\frac{2}{3}$$

$$a_n = 6\left(\frac{2}{3}\right)^{n-1}$$

7. Find the nth term of the geometric sequence.

$$r=-\frac{5}{6}$$

$$a_n=-6\left(-\frac{5}{6}\right)^{n-1}$$

9. Find the nth term of the geometric sequence.

$$r=\frac{-3}{9}=-\frac{1}{3}$$

$$a_n = 9\left(-\frac{1}{3}\right)^{n-1} = \left(-\frac{1}{3}\right)^{n-3}$$

11. Find the nth term of the geometric sequence.

$$r=\frac{-x}{1}=-x$$

$$a_n = 1(-x)^{n-1} = (-x)^{n-1}$$

13. Find the nth term of the geometric sequence.

$$r=\frac{c^5}{c^2}=c^3$$

$$a_n = c^2\left(c^3\right)^{n-1} = c^2c^{3n-3} = c^{3n-1}$$

15. Find the nth term of the geometric sequence.

$$r=\frac{\frac{3}{10{,}000}}{\frac{3}{100}}=\frac{1}{100}$$

$$a_n = \frac{3}{100}\left(\frac{1}{100}\right)^{n-1} = 3\left(\frac{1}{100}\right)^n$$

17. Find the nth term of the geometric sequence.

$$r=\frac{0.05}{0.5}=0.1$$

$$a_n = 0.5(0.1)^{n-1} = 5(0.1)^n$$

19. Find the nth term of the geometric sequence.

$$r=\frac{0.0045}{0.45}=0.01$$

$$a_n = 0.45(0.01)^{n-1} = 45(0.01)^n$$

21. Find the requested term.

$$a_1 = 2,\qquad a_5 = 162,\qquad a_n = a_1r^{n-1}$$

$$162 = 2r^{5-1}$$

$$r^4 = 81$$

$$r = 3$$

$$a_3 = 2(3)^{3-1} = 2\cdot 9 = 18$$

23. Find the requested term.

$$a_3 = \frac{4}{3}, \qquad a_6 = -\frac{32}{81}$$

$$\frac{\frac{4}{3}}{-\frac{32}{81}} = \frac{a_1(r)^{3-1}}{a_1(r)^{6-1}}$$

$$\frac{-27}{8} = \frac{1}{r^3}$$

$$r^3 = -\frac{8}{27}$$

$$r = \frac{-2}{3}$$

$$\frac{4}{3} = a_1\left(-\frac{2}{3}\right)^{3-1}$$

$$a_1 = \frac{4}{3}\left(\frac{9}{4}\right) = 3$$

$$a_2 = 3\left(\frac{-2}{3}\right) = -2$$

25. Classify the sequence $a_n = \frac{1}{n^2}$ as arithmetic, geometric or neither.

An arithmetic sequence has a common difference.

$$\frac{1}{1^2}, \frac{1}{2^2}, \frac{1}{3^2}, \frac{1}{4^2}, \ldots = 1, \frac{1}{2}, \frac{1}{9}, \frac{1}{16}, \ldots$$

No common difference; the sequence is not arithmetic.

A geometric sequence has a common ratio.

$$\frac{\frac{1}{2}}{1}, \frac{\frac{1}{9}}{\frac{1}{2}}, \frac{\frac{1}{16}}{\frac{1}{9}}, \ldots = \frac{1}{2}, \frac{2}{9}, \frac{9}{16}, \ldots$$

No common ratio; the sequence is not geometric.

Neither.

27. Classify the sequence $a_n = 2n - 7$ as arithmetic, geometric or neither.

An arithmetic sequence has a common difference.

$$2(1)-7,\ 2(2)-7,\ 2(3)-7,\ 2(4)-7, \ldots = -5, -3, -1, 1, \ldots$$

The common difference is $d = 2$.

Arithmetic

29. Classify the sequence $a_n = \left(-\frac{6}{5}\right)^n$ as arithmetic, geometric or neither.

A geometric sequence has a common ratio.

$$\text{common ratio: } r = \frac{\left(\frac{6}{5}\right)^{i+1}}{\left(-\frac{6}{5}\right)^{i}} = -\frac{6}{5}$$

Geometric

31. Classify the sequence $a_n = \frac{n}{2^n}$ as arithmetic, geometric or neither.

An arithmetic sequence has a common difference.

$$\frac{1}{2^1}, \frac{2}{2^2}, \frac{3}{2^3}, \frac{4}{2^4}, \ldots = \frac{1}{2}, \frac{1}{2}, \frac{3}{8}, \frac{1}{4}, \ldots$$

No common difference; the sequence is not arithmetic.

A geometric sequence has a common ratio.

$$\frac{\frac{1}{2}}{\frac{1}{2}}, \frac{\frac{3}{8}}{\frac{1}{2}}, \frac{\frac{1}{4}}{\frac{3}{8}}, \ldots = 1, \frac{3}{4}, \frac{2}{3}, \ldots$$

No common ratio; the sequence is not geometric.

Neither.

33. Classify the sequence $a_n = -3n$ as arithmetic, geometric or neither.

An arithmetic sequence has a common difference.

$$-3(1),\ -3(2),\ -3(3),\ -3(4), \ldots = -3, -6, -9, -12, \ldots$$

The common difference is $d = -3$.

Arithmetic

35. Classify the sequence $a_n = \frac{3^n}{2}$ as arithmetic, geometric or neither.

A geometric sequence has a common ratio.

$$\text{common ratio: } r = \frac{\frac{3^{i+1}}{2}}{\frac{3^i}{2}} = 3$$

Geometric

37. Find the sum of the finite geometric series.

$$a_1 = 3, \qquad a_2 = 9, \qquad r = \frac{9}{3} = 3$$

$$S_5 = \frac{3(1-3^5)}{1-3} = \frac{3(-242)}{-2} = 363$$

39. Find the sum of the finite geometric series.

$$a_1 = \frac{2}{3}, \quad a_2 = \frac{4}{9}, \quad r = \frac{\frac{4}{9}}{\frac{2}{3}} = \frac{2}{3}$$

$$S_6 = \frac{\frac{2}{3}\left[1-\left(\frac{2}{3}\right)^6\right]}{1-\frac{2}{3}} = \frac{\frac{2}{3}\left(\frac{665}{729}\right)}{\frac{1}{3}} = \frac{1330}{729}$$

41. Find the sum of the finite geometric series.

$$a_1 = 1,\ a_2 = -\frac{2}{5},\ r = -\frac{2}{5}$$

$$S_9 = \frac{1\left[1-\left(-\frac{2}{5}\right)^9\right]}{1-\left(-\frac{2}{5}\right)} = \frac{\frac{1,953,637}{1,953,125}}{\frac{7}{5}} = \frac{279,091}{390,625}$$

43. Find the sum of the finite geometric series.

$$a_1 = 1,\ a_2 = -2,\ r = -2$$

$$S_7 = \frac{1[1-(-2)^{10}]}{1-(-2)} = -341$$

45. Find the sum of the finite geometric series.

$$a_1 = 5,\ a_2 = 15,\ r = 3$$

$$S_{10} = \frac{5[1-3^{10}]}{1-3} = 147,620$$

47. Find the sum of the infinite geometric series.

$$a_1 = \frac{1}{3},\ a_2 = \frac{1}{9},\ r = \frac{1}{3}$$

$$S = \frac{\frac{1}{3}}{1-\frac{1}{3}} = \frac{\frac{1}{3}}{\frac{2}{3}} = \frac{1}{2}$$

49. Find the sum of the infinite geometric series.

$$a_1 = -\frac{2}{3},\ r = -\frac{2}{3}$$

$$S = \frac{-\frac{2}{3}}{1-\left(-\frac{2}{3}\right)} = -\frac{\frac{2}{3}}{\frac{5}{3}} = -\frac{2}{5}$$

51. Find the sum of the infinite geometric series.

$$a_1 = \frac{9}{100},\ r = \frac{9}{100}$$

$$S = \frac{\frac{9}{100}}{1-\frac{9}{100}} = \frac{\frac{9}{100}}{\frac{91}{100}} = \frac{9}{91}$$

53. Find the sum of the infinite geometric series.

$$a_1 = 0.1,\ r = 0.1$$

$$S = \frac{0.1}{1-0.1} = \frac{0.1}{0.9} = \frac{1}{9}$$

55. Find the sum of the infinite geometric series.

$$a_1 = 1,\ r = -0.4$$

$$S = \frac{1}{1-(-0.4)} = \frac{1}{1.4} = \frac{5}{7}$$

57. Write as a quotient of two integers in simplest form.

$$0.\overline{3} = \frac{3}{10} + \frac{3}{100} + \frac{3}{1000} + \cdots$$

$$a_1 = \frac{3}{10},\ r = \frac{\frac{3}{100}}{\frac{3}{10}} = \frac{1}{10}$$

$$0.\overline{3} = \frac{\frac{3}{10}}{1-\frac{1}{10}} = \frac{\frac{3}{10}}{\frac{9}{10}} = \frac{1}{3}$$

59. Write as a quotient of two integers in simplest form.

$$0.\overline{45} = \frac{45}{100} + \frac{45}{10,000} + \frac{45}{1,000,000} + \cdots$$

$$a_1 = \frac{45}{100},\ r = \frac{\frac{45}{10,000}}{\frac{45}{100}} = \frac{1}{100}$$

$$0.\overline{45} = \frac{\frac{45}{100}}{1-\frac{1}{100}} = \frac{\frac{45}{100}}{\frac{99}{100}} = \frac{5}{11}$$

61. Write as a quotient of two integers in simplest form.

$$0.\overline{123} = \frac{123}{1000} + \frac{123}{1,000,000} + \frac{123}{1,000,000,000} + \cdots$$

$$a_1 = \frac{123}{1000},\ r = \frac{1}{1000}$$

$$0.\overline{123} = \frac{\frac{123}{1000}}{1-\frac{1}{1000}} = \frac{123}{999} = \frac{41}{333}$$

63. Write as a quotient of two integers in simplest form.

$$0.\overline{422} = \frac{422}{1000} + \frac{422}{1,000,000} + \frac{422}{1,000,000,000} + \cdots$$

$$a_1 = \frac{422}{1000},\ r = \frac{1}{1000}$$

$$0.\overline{422} = \frac{\frac{422}{1000}}{1-\frac{1}{1000}} = \frac{422}{999}$$

65. Write as a quotient of two integers in simplest form.

$$0.25\overline{4} = \frac{25}{100} + \left[\frac{4}{1000} + \frac{4}{10,000} + \frac{4}{100,000} + \cdots\right]$$

$$a_1 = \frac{4}{1000},\ r = \frac{1}{10}$$

$$0.25\overline{4} = \frac{25}{100} + \frac{\frac{4}{1000}}{1 - \frac{1}{10}} = \frac{25}{100} + \frac{4}{900} = \frac{229}{900}$$

67. Write as a quotient of two integers in simplest form.

$$1.2\overline{084} = 1 + \frac{2}{10} + \left[\frac{84}{10,000} + \frac{84}{1,000,000} + \frac{84}{100,000,000} + \cdots\right]$$

$$a_1 = \frac{84}{10,000},\ r = \frac{1}{100}$$

$$1.2\overline{084} = 1 + \frac{2}{10} + \frac{\frac{84}{10,000}}{1 - \frac{1}{100}} = \frac{12}{10} + \frac{7}{825} = \frac{1994}{1650} = \frac{997}{825}$$

69. Find the future value in 8 years.

$P = 100,\ i = 0.09,\ n = 2,\ t = 8$

$$r = \frac{i}{n} = \frac{0.09}{2} = 0.045,\ m = nt = 2 \cdot 8 = 16$$

$$A = 100\frac{\left[(1+0.045)^{16} - 1\right]}{0.045} \approx 2271.93367$$

The future value is \$2271.93.

71. Find the amount each recipient would receive.

When a name was removed from the top of the list, the letter had been sent to

$5(5^5) = 15,625$ people

who sent 10 cents each for a total of

$0.10(15,625) = \$1562.50$ for each recipient.

73. Find the concentration to the nearest hundredth.

$A = 0.5,\ n = 4,\ k = -0.876,\ t = 4$

$$A + Ae^{kt} + Ae^{2kt} + Ae^{3kt}$$
$$= 0.5 + 0.5e^{-0.867(4)} + 0.5e^{2(-0.867)(4)} + 0.5e^{3(-0.867)(4)}$$
$$\approx 0.52 \text{ mg}$$

Or $A = 0.5,\ r = e^{kt},\ n = 4,\ k = -0.876,\ t = 4$

$$S_4 = \frac{0.5\left(1 - e^{-.867(4)(3)}\right)}{1 - e^{-.867(4)}} \approx 0.52 \text{ mg}$$

75. Find the dividend growth rate.

$$\text{Stock value} = \frac{D(1+g)}{i - g}$$

$$67 = \frac{1.32(1+g)}{0.20 - g} \quad \text{Solve for } g.$$

$$67(0.20 - g) = 1.32(1+g)$$
$$13.4 - 67g = 1.32 + 1.32g$$
$$12.08 = 68.32g$$
$$0.1768 = g$$

The dividend growth rather is 17.68%.

77. Find the price per share.

$$\text{Stock value (no dividend growth)} = \frac{D}{i} = \frac{2.94}{0.15} = \$19.60$$

79. Explain.

If g was **not** less than i in the Gordon model of stock valuation, the common ratio of the geometric sequence would be greater than 1 and the sum of the infinite geometric series would not be defined.

81. Find the net effect.

Using the multiplier effect,

$$\frac{25}{1 - 0.75} = 100$$

The net effect of \$25 million is \$100 million.

83. Find the number of grandparents.

The nth generation has $a_n = 2^n$ grandparents. Since this is a geometric sequence, the sum can be found by a formula.

$$S_n = \frac{a_1(1 - r^n)}{1 - r}$$

$$S_{10} = \frac{2(1 - 2^{10})}{1 - 2} = \frac{2(1 - 1024)}{-1} = 2046$$

When $n = 1$, $a_n = 2$ and there are no grandparents.

Therefore there are 2046 – 2 = 2044 grandparents by the 10th generation.

Mid-Chapter 8 Quiz

1. Find the fourth and eighth terms.

$$a_4 = \frac{4}{2^4} = \frac{4}{16} = \frac{1}{4}$$

$$a_8 = \frac{8}{2^8} = \frac{8}{256} = \frac{1}{32}$$

3. Evaluate.

$$\sum_{k=1}^{5}\frac{(-1)^{k-1}}{k^2}=\frac{(-1)^0}{1^2}+\frac{(-1)^1}{2^2}+\frac{(-1)^2}{3^2}+\frac{(-1)^3}{4^2}+\frac{(-1)^4}{5^2}$$
$$=1-\frac{1}{4}+\frac{1}{9}-\frac{1}{16}+\frac{1}{25}$$
$$=\frac{3019}{3600}$$

5. Find the sum of the first 25 terms.

$a_n=5-n$

$a_1=4,\ a_2=3,\ a_3=2$

$S_{25}=\frac{25}{2}[2(4)+(25-1)(-1)]=-200$

7. Find the sum of the first eight terms.

$a_n=(-3)^n$

$a_1=(-3)^1=-3$

$a_2=(-3)^2=9$

$r=\frac{9}{-3}=-3$

$S_8=\frac{-3\left(1-(-3)^8\right)}{1-(-3)}=4920$

Prepare for Section 8.4

P1. Show the statement is true for $n = 4$.

$$\sum_{i=1}^{4}\frac{1}{i(i+1)}=\frac{1}{1(1+1)}+\frac{1}{2(2+1)}+\frac{1}{3(3+1)}+\frac{1}{4(4+1)}$$
$$=\frac{1}{2}+\frac{1}{6}+\frac{1}{12}+\frac{1}{20}=\frac{4}{5}=\frac{4}{4+1}$$

P3. Simplify.

$$\frac{k}{k+1}+\frac{1}{(k+1)(k+2)}=\frac{k+2}{k+2}\cdot\frac{k}{k+1}+\frac{1}{(k+1)(k+2)}$$
$$=\frac{k^2+2k+1}{(k+1)(k+2)}$$
$$=\frac{(k+1)(k+1)}{(k+1)(k+2)}$$
$$=\frac{k+1}{k+2}$$

P5. Write in simplest form.

$$S_n+a_{n+1}=\frac{n(n+1)}{2}+n+1$$
$$=\frac{n^2+n}{2}+\frac{2n+2}{2}=\frac{n^2+3n+2}{2}$$
$$=\frac{(n+1)(n+2)}{2}$$

Section 8.4 Exercises

1. Prove the statement by mathematical induction.

1. Let $n = 1$. $S_1=3\cdot 1-2=1=\frac{1(3\cdot 1-1)}{2}$

2. Assume the statement is true for some positive integer k.

$$S_k=1+4+7+\cdots+(3k-2)=\frac{k(3k-1)}{2}$$

(Induction Hypothesis)

Verify that the statement is true when $n = k + 1$

$$S_{k+1}=\frac{(k+1)(3k+2)}{2}.$$

$a_k=3k-2,\ a_{k+1}=3k+1$

$$S_{k+1}=S_k+a_{k+1}$$
$$=\frac{k(3k-1)}{2}+3k+1=\frac{3k^2-k}{2}+\frac{6k+2}{2}$$
$$=\frac{3k^2+5k+2}{2}=\frac{(k+1)(3k+2)}{2}$$

By the Principle of Mathematical Induction, the statement is true for all positive integers.

3. Prove the statement by mathematical induction.

1. Let $n = 1$. $S_1=1^3=1=\frac{1^2(1+1)^2}{4}$

2. Assume $S_k=1+8+27+\cdots+k^3=\frac{k^2(k+1)^2}{4}$ is true for some positive integer k.(Induction Hypothesis).

Verify that $S_{k+1}=\frac{(k+1)^2(k+2)^2}{4}$.

$a_k=k^3,\ a_{k+1}=(k+1)^3$

$$S_{k+1}=S_k+a_{k+1}=\frac{k^2(k+1)^2}{4}+(k+1)^3$$
$$=\frac{k^2(k+1)^2+4(k+1)^3}{4}$$
$$=\frac{(k+1)^2(k^2+4k+4)}{4}=\frac{(k+1)^2(k+2)^2}{4}$$

By the Principle of Mathematical Induction, the statement is true for all positive integers.

5. Prove the statement by mathematical induction.

1. Let $n = 1$. $S_1 = 4\cdot1-1 = 3 = 1(2\cdot1+1)$

2. Assume that $S_k = 3+7+11+\cdots+(4k-1) = k(2k+1)$ is true for some positive integer k (Induction Hypothesis).

   Verify that $S_{k+1} + (k+1)(2k+3)$.

   $a_k = 4k-1, \quad a_{k+1} = 4k+3$

   $$\begin{aligned} S_{k+1} &= S_k + a_{k+1} \\ &= k(2k+1) + 4k + 3 \\ &= 2k^2 + 5k + 3 = (k+1)(2k+3) \end{aligned}$$

By the Principle of Mathematical Induction, the statement is true for all positive integers.

7. Prove the statement by mathematical induction.

1. Let $n = 1$. $S_1 = (2\cdot1-1)^3 = 1 = 1^2(2\cdot1^2-1)$

2. Assume that

   $$S_k = 1+27+125+\cdots+(2k-1)^3 = k^2(2k^2-1)$$

   is true for some positive integer k (Induction Hypothesis).

   Verify that $S_{k+1} = (k+1)^2(2k^2+4k+1)$.

   $a_k = (2k-1)^3, \quad a_{k+1} = (2k+1)^3$

   $$\begin{aligned} S_{k+1} = S_k + a_{k+1} &= k^2(2k^2-1) + (2k+1)^3 \\ &= 2k^4 - k^2 + 8k^3 + 12k^2 + 6k + 1 \\ &= 2k^4 + 8k^3 + 11k^2 + 6k + 1 \\ &= (k+1)(2k^3 + 6k^2 + 5k + 1) \\ &= (k+1)^2(2k^2 + 4k + 1) \end{aligned}$$

By the Principle of Mathematical Induction, the statement is true for all positive integers.

9. Prove the statement by mathematical induction.

1. Let $n = 1$. $S_1 = \dfrac{1}{(2\cdot1-1)(1\cdot1+1)} = \dfrac{1}{3} = \dfrac{1}{2\cdot1+1}$

2. Assume that

   $$S_k = \frac{1}{1\cdot3} + \frac{1}{3\cdot5} + \frac{1}{5\cdot7} + \cdots + \frac{1}{(2k-1)(2k+1)} = \frac{k}{2k+1}$$

   for some positive integer k (Induction Hypothesis).

   Verify that $S_{k+1} = \dfrac{k+1}{2k+3}$.

   $$a_k = \frac{1}{(2k-1)(2k+1)}, \quad a_{k+1} = \frac{1}{(2k+1)(2k+3)}$$

   $$\begin{aligned} S_{k+1} &= \frac{k}{2k+1} + \frac{1}{(2k+1)(2k+3)} \\ &= \frac{2k^2+3k+1}{(2k+1)(2k+3)} \\ &= \frac{(2k+1)(k+1)}{(2k+1)(2k+3)} = \frac{k+1}{2k+3} \end{aligned}$$

By the Principle of Mathematical Induction, the statement is true for all positive integers.

11. Prove the statement by mathematical induction.

1. Let $n = 1$.

   $$\begin{aligned} S_1 &= 1^4 = 1 \\ &= \frac{1(1+1)(2\cdot1+1)(3\cdot1^2+3\cdot1-1)}{30} = \frac{2\cdot3\cdot5}{30} = 1 \end{aligned}$$

2. Assume that

   $$\begin{aligned} S_k &= 1+16+81+\cdots+k^4 \\ &= \frac{k(k+1)(2k+1)(3k^2+3k-1)}{30} \end{aligned}$$

   for some positive integer k (Induction Hypothesis).

   Verify that

   $$S_{k+1} = \frac{(k+1)(k+2)(2k+3)(3k^2+9k+5)}{30}.$$

   $a_k = k^4, \quad a_{k+1} = (k+1)^4$

   $$\begin{aligned} S_{k+1} &= \frac{k(k+1)(2k+1)(3k^2+3k-1)}{30} + (k+1)^4 \\ &= \frac{(k+1)[k(2k+1)(3k^2+3k-1)+30(k+1)^3]}{30} \\ &= \frac{(k+1)[6k^4+39k^3+91k^2+89k+30]}{30} \\ &= \frac{(k+1)(k+2)(6k^3+27k^2+37k+15)}{30} \\ &= \frac{(k+1)(k+2)(2k+3)(3k^2+9k+5)}{30} \end{aligned}$$

By the Principle of Mathematical Induction, the statement is true for all positive integers.

13. Prove the inequality by mathematical induction.

1. Let $n = 4$. Then $\left(\frac{3}{2}\right)^4 = \frac{81}{16} = 5\frac{1}{16}$; $4+1=5$

   Thus, $\left(\frac{3}{2}\right)^n > n+1$ for $n = 4$.

2. Assume $\left(\frac{3}{2}\right)^k > k+1$ is true for some positive integer $k \geq 4$ (Induction Hypothesis).

   Verify that $\left(\frac{3}{2}\right)^{k+1} > k+2$.

$$\left(\frac{3}{2}\right)^{k+1} = \left(\frac{3}{2}\right)^k\left(\frac{3}{2}\right)$$
$$> (k+1)\left(\frac{3}{2}\right) = \frac{1}{2}(3k+3) = \frac{1}{2}(2k+k+3)$$
$$> \frac{1}{2}(2k+1+3) = k+2$$

Thus $\left(\frac{3}{2}\right)^{k+1} > k+2$. By the Principle of Mathematical Induction, $\left(\frac{3}{2}\right)^n > n+1$ for all $n \geq 4$.

15. Prove the inequality by mathematical induction.

1. Let $n = 1$.

$$0 < a < 1$$
$$0 < a \cdot a < a \cdot 1$$
$$a^{1+1} = a^2 < a$$

   Thus, if $0 < a < 1$, then $a^{1+1} < a^n$ for $n = 1$.

2. Assume $a^{k+1} < a^k$ is true for some positive integer k, if $0 < a < 1$ (Induction Hypothesis).

   Verify $a^{k+2} < a^{k+1}$.

$$0 < a < 1$$
$$0 < a \cdot a^{k+1} < 1 \cdot a^{k+1}$$
$$a^{k+2} < a^{k+1}$$

By the Principle of Mathematical Induction, if $0 < a < 1$, then $a^{n+1} < a^n$ for all positive integers.

17. Prove the inequality by mathematical induction.

1. Let $n = 4$. $1 \cdot 2 \cdot 3 \cdot 4 = 24$, $2^4 = 16$

   Thus, $1 \cdot 2 \cdot 3 \cdot 4 > 2^n$ for $n = 4$.

2. Assume $1 \cdot 2 \cdot 3 \cdot 4 \cdot \cdots \cdot k > 2^k$ is true for some positive integer $k \geq 4$ (Induction Hypothesis).

   Verify $1 \cdot 2 \cdot 3 \cdot \cdots \cdot k \cdot (k+1) > 2^{k+1}$.

$$1 \cdot 2 \cdot 3 \cdot \cdots \cdot k \cdot (k+1) > 2^k (k+1) > 2^k \cdot 2 = 2^{k+1}$$

Thus, $1 \cdot 2 \cdot 3 \cdot \cdots \cdot k \cdot (k+1) > 2^{k+1}$. By the Principle of Mathematical Induction, $1 \cdot 2 \cdot 3 \cdot \cdots \cdot n > 2^n$ for all $n \geq 4$.

19. Prove the inequality by mathematical induction.

1. Let $n = 1$ and $a > 0$. $(1+a)^1 = 1+a = 1+1 \cdot a$

   Thus $(1+a)^n \geq 1+na$ for $n = 1$.

2. Assume $(1+a)^k \geq 1+ka$ is true for some positive integer k (Induction Hypotheses).

   Verify $(1+a)^{k+1} \geq 1+(k+1)a$.

$$(1+a)^{k+1} = (1+a)^k(1+a)$$
$$\geq (1+ka)(1+a) = 1+(k+1)a+ka^2$$
$$\geq 1+(k+1)a$$

Thus $(1+a)^{k+1} \geq 1+(k+1)a$. By the Principle of Mathematical Induction, $(1+a)^n \geq 1+na$ for all positive integers n.

21. Prove the statement by mathematical induction.

1. Let $n = 1$. $1^2 + 1 = 2$, $2 = 2 \cdot 1$

   Thus 2 is a factor of $n^2 + n$ for $n = 1$.

2. Assume 2 is a factor of $k^2 + k$ for some positive integer k (Induction Hypothesis).

   Verify 2 is a factor of $(k+1)^2 + k + 1$.

$$(k+1)^2 + k + 1 = (k+1)(k+1+1)$$
$$= (k+1)(k+2)$$

   Since $k^2 + k = k(k+1)$, 2 is a factor of k or $k + 1$.

   If 2 is a factor of $k + 1$, then 2 is a factor of $(k+1)(k+2)$.

   If 2 is a factor of k, then 2 is a factor of $k + 2$.

Thus, 2 is a factor of $(k+1)^2+k+1$. By the Principle of Mathematical Induction, 2 is a factor of n^2+n for all positive integers.

23. Prove the statement by mathematical induction.

1. Let $n = 1$. $5^1-1=4, \quad 4=4\cdot 1$

   Thus, 4 is a factor of 5^n-1 for $n = 1$.

2. Assume 4 is a factor of 5^k-1 for some positive integer k (Induction Hypothesis).

   Verify 4 is a factor of $5^{k+1}-1$.

   Now $5^{k+1}-1=5\cdot 5^k-5+4=5\,(5^k-1)+4$ which is the sum of two multiples of 4.

Thus, 4 is a factor of $5^{k+1}-1$. By the Principle of Mathematical Induction, 4 is a factor of 5^n-1 for all positive integers.

25. Prove the statement by mathematical induction.

1. Let $n = 1$. $(xy)^1=xy=x^1y^1$

   Thus, $(xy)^n=x^ny^n$ for $n=1$.

2. Assume $(xy)^k=x^ky^k$ is true for some positive integer k (Induction Hypothesis).

   Verify $(xy)^{k+1}=x^{k+1}y^{k+1}$.

   $(xy)^{k+1}=(xy)^k(xy)^1=x^ky^k\cdot xy=x^{k+1}y^{k+1}$

Thus $(xy)^{k+1}=x^{k+1}y^{k+1}$. By the Principle of Mathematical Induction, $(xy)^n=x^ny^n$ for all positive integers.

27. Prove the statement by mathematical induction.

1. Let $n = 1$. $a^1-b^1=a-b$

   Thus $a-b$ is a factor of a^n-b^n for $n=1$.

2. Assume $a-b$ is a factor of a^k-b^k for some positive integer k (Induction Hypothesis).

   Verify $a-b$ is a factor of $a^{k+1}-b^{k+1}$.

   $$\begin{aligned}a^{k+1}-b^{k+1}&=(a\cdot a^k-ab^k)+(ab^k-b\cdot b^k)\\&=a(a^k-b^k)+b^k(a-b)\end{aligned}$$

   The sum of two multiples of $a-b$ is a multiple of $a-b$. Thus, $a-b$ is a factor of $a^{k+1}-b^{k+1}$.

By the Principle of Mathematical Induction, $a-b$ is a factor of a^n-b^n for all positive integers.

29. Prove the statement by mathematical induction.

1. Let $n = 1$. $ar^{1-1}=a\cdot 1=a=\dfrac{a(1-r^1)}{1-r}$

   Thus, the statement is true for $n=1$.

2. Assume $\displaystyle\sum_{k=1}^{j}ar^{k-1}=\frac{a(1-r^j)}{1-r}$ is true for some positive integer j.

   Verify $\displaystyle\sum_{k=1}^{j+1}ar^{k-1}=\frac{a(1-r^{j+1})}{1-r}$ is true for $n=j+1$.

   $$\begin{aligned}\sum_{k=1}^{j+1}ar^{k-1}&=\sum_{k=1}^{j}(ar^{k+1}-1)=\frac{a(1-r^j)}{1-r}+ar^j\\&=\frac{a(1-r^j)+ar^j(1-r)}{1-r}\\&=\frac{a[1-r^j+r^j-r^{j+1}]}{1-r}\\&=\frac{a(1-r^{j+1})}{1-r}\end{aligned}$$

By the Principle of Mathematical Induction,

$$\sum_{k=1}^{n}ar^{k-1}=\frac{a(1-r^n)}{1-r}$$

31. Prove the statement by mathematical induction.

1. If $N = 25$, then $\log 25!\approx 25.19>25$.

2. Assume $\log k!>k$ for $k>25$ (Induction Hypothesis).

   Prove $\log(k+1)\;!>k+1$.

   $$\begin{aligned}\log(k+1)!&=\log[(k+1)k!]\\&=\log(k+1)+\log k!\\&>\log(k+1)+k\end{aligned}$$

Because $k > 25$, $\log(k+1) > 1$.

Thus, $\log(k+1)! > k+1$.

Therefore, $\log n! > n$ for all $n > 25$.

33. Prove the statement by mathematical induction.

1. When $n = 1$, we have $(x^m)^1 = x^m$ and $x^{m \cdot 1} = x^m$.

   Therefore, the statement is true for $n = 1$.

2. Assume the statement is true for $n = k$. That is,

   assume $(x^m)^k = x^{mk}$ (Induction Hypothesis).

   Prove the statement is true for $n = k + 1$.

   $$x^{m(k+1)} = x^{mk+m} = x^{mk} \cdot x^m = (x^m)^k \cdot x^m = (x^m)^{k+1}$$

Thus, the statement is true for all positive integers n and m.

35. Prove the statement by mathematical induction.

1. When $n = 3$, we have $\left(\frac{3+1}{3}\right)^3 = \left(\frac{4}{3}\right)^3 = \frac{64}{27} < 3$.

   Thus the statement is true for $n = 3$.

2. Assume the statement is true for $n = k$. That is,

   assume $\left(\frac{k+1}{k}\right)^k < k$ (Induction Hypothesis).

   Prove the statement is true for $n = k + 1$. That is,

   prove $\left(\frac{k+2}{k+1}\right)^{k+1} < k+1$.

   We begin by noting that $\left(\frac{k+2}{k+1}\right) < \frac{k+1}{k}$.

   Therefore

   $$\left(\frac{k+2}{k+1}\right)^{k+1} < \left(\frac{k+1}{k}\right)^{k+1} = \left(\frac{k+1}{k}\right)^k \left(\frac{k+1}{k}\right)$$

   By the Induction Hypothesis, $\left(\frac{k+1}{k}\right)^k < k$; thus

   $$\left(\frac{k+1}{k}\right)^k \left(\frac{k+1}{k}\right) < k\left(\frac{k+1}{k}\right) = k+1$$

   We now have $\left(\frac{k+2}{k+1}\right)^{k+1} < k+1$. The induction is complete.

   Thus $\left(\frac{n+1}{n}\right)^n < n$ is true for all $n \geq 3$.

Prepare for Section 8.5

P1. Expand.

$$(a+b)^3 = (a+b)(a+b)(a+b) = a^3 + 3a^2b + 3ab^2 + b^3$$

P3. Evaluate.

$$0! = 1$$

P5. Evaluate.

$$\frac{7!}{3!(7-3)!} = \frac{5040}{6(24)} = 35$$

Section 8.5 Exercises

1. Evaluate the binomial coefficient.

$$\binom{7}{4} = \frac{7!}{4!(7-4)!} = \frac{7 \cdot 6 \cdot 5 \cdot 4!}{4!(3 \cdot 2 \cdot 1)} = 35$$

3. Evaluate the binomial coefficient.

$$\binom{9}{2} = \frac{9!}{2!(9-2)!} = \frac{9 \cdot 8 \cdot 7!}{2 \cdot 1 \cdot 7!} = 36$$

5. Evaluate the binomial coefficient.

$$\binom{12}{9} = \frac{12!}{9!(12-9)!} = \frac{12 \cdot 11 \cdot 10 \cdot 9!}{9! \cdot 3 \cdot 2 \cdot 1} = 220$$

7. Evaluate the binomial coefficient.

$$\binom{11}{0} = \frac{11!}{0!(11-0)!} = \frac{11!}{1 \cdot 11!} = 1$$

9. Expand the binomial.

$$(x+y)^5 = \binom{5}{0}x^5 + \binom{5}{1}x^4 \cdot y + \binom{5}{2}x^3 \cdot y^2 + \binom{5}{3}x^2 \cdot y^3 + \binom{5}{4}x \cdot y^4 + \binom{5}{5}y^5$$
$$= x^5 + 5x^4y + 10x^3y^2 + 10x^2y^3 + 5xy^4 + y^5$$

11. Expand the binomial.

$$(a-b)^4 = \binom{4}{0}a^4 + \binom{4}{1}a^3(-b) + \binom{4}{2}a^2(-b)^2 + \binom{4}{3}a(-b)^3 + \binom{4}{4}(-b)^4$$
$$= a^4 - 4a^3b + 6a^2b^2 - 4ab^3 + b^4$$

13. Expand the binomial.

$$(x+5)^4 = \binom{4}{0}x^4 + \binom{4}{1}x^3 5 + \binom{4}{2}x^2 5^2 + \binom{4}{3}x5^3 + \binom{4}{4}5^4$$
$$= x^4 + 20x^3 + 150x^2 + 500x + 625$$

15. Expand the binomial.

$$(a-3)^5 = \binom{5}{0}a^5 + \binom{5}{1}a^4(-3) + \binom{5}{2}a^3(-3)^2$$
$$+\binom{5}{3}a^2(-3)^3 + \binom{5}{4}a(-3)^4 + \binom{5}{5}(-3)^5$$
$$= a^5 - 15a^4 + 90a^3 - 270a^2 + 405ab^4 - 243$$

17. Expand the binomial.

$$(2x-1)^7 = \binom{7}{0}(2x)^7 + \binom{7}{1}(2x)^6(-1) + \binom{7}{2}(2x)^5(-1)^2$$
$$+\binom{7}{3}(2x)^4(-1)^3 + \binom{7}{4}(2x)^3(-1)^4$$
$$+\binom{7}{5}(2x)^2(-1)^5 + \binom{7}{6}(2x)(-1)^6 + \binom{7}{7}(-1)^7$$
$$= 128x^7 - 448x^6 + 672x^5 - 560x^4 + 280x^3$$
$$-84x^2 + 14x - 1$$

19. Expand the binomial.

$$(x+3y)^6 = \binom{6}{0}x^6 + \binom{6}{1}x^5(3y) + \binom{6}{2}x^4(3y)^2$$
$$+\binom{6}{3}x^3(3y)^3 + \binom{6}{4}x^2(3y)^4$$
$$+\binom{6}{5}x(3y)^5 + \binom{6}{6}(3y)^6$$
$$= x^6 + 18x^5y + 135x^4y^2 + 540x^3y^3$$
$$+1215x^2y^4 + 1458xy^5 + 729y^6$$

21. Expand the binomial.

$$(2x-5y)^4$$
$$= \binom{4}{0}(2x)^4 + \binom{4}{1}(2x)^3(-5y) + \binom{4}{2}(2x)^2(-5y)^2$$
$$+\binom{4}{3}(2x)(-5y)^3 + \binom{4}{4}(-5y)^4$$
$$= 16x^4 - 160x^3y + 600x^2y^2 - 1000xy^3 + 625y^4$$

23. Expand the binomial.

$$(x^2-4)^7$$
$$= \binom{7}{0}(x^2)^7 + \binom{7}{1}(x^2)^6(-4) + \binom{7}{2}(x^2)^5(-4)^2$$
$$+\binom{7}{3}(x^2)^4(-4)^3 + \binom{7}{4}(x^2)^3(-4)^4$$
$$+\binom{7}{5}(x^2)^2(-4)^5 + \binom{7}{6}(x^2)(-4)^6 + \binom{7}{7}(-4)^7$$
$$= x^{14} - 28x^{12} + 336x^{10} - 2240x^8 + 8960x^6$$
$$-21504x^4 + 28672x^2 - 16384$$

25. Expand the binomial.

$$(2x^2+y^3)^5 = \binom{5}{0}(2x^2)^5 + \binom{5}{1}(2x^2)^4(y^3)$$
$$+\binom{5}{2}(2x^2)^3(y^3)^2 + \binom{5}{3}(2x^2)^2(y^3)^3$$
$$+\binom{5}{4}(2x^2)(y^3)^4 + \binom{5}{5}(y^3)^5$$
$$= 32x^{10} + 80x^8y^3 + 80x^6y^6 + 40x^4y^9$$
$$+10x^2y^{12} + y^{15}$$

27. Expand the binomial.

$$(x+\sqrt{y})^5 = \binom{5}{0}x^5 + \binom{5}{1}x^4\cdot\sqrt{y} + \binom{5}{2}x^3\cdot(\sqrt{y})^2$$
$$+\binom{5}{3}x^2\cdot(\sqrt{y})^3 + \binom{5}{4}x\cdot(\sqrt{y})^4 + \binom{5}{5}(\sqrt{y})^5$$
$$= x^5 + 5x^4y^{1/2} + 10x^3y$$
$$+10x^2y^{3/2} + 5xy^2 + y^{5/2}$$

29. Expand the binomial.

$$\left(\frac{2}{x}-\frac{x}{2}\right)^4 = \binom{4}{0}\left(\frac{2}{x}\right)^4 + \binom{4}{1}\left(\frac{2}{x}\right)^3\left(-\frac{x}{2}\right) + \binom{4}{2}\left(\frac{2}{x}\right)^2\left(-\frac{x}{2}\right)^2$$
$$+\binom{4}{3}\left(\frac{2}{x}\right)\left(-\frac{x}{2}\right)^3 + \binom{4}{4}\left(-\frac{x}{2}\right)^4$$
$$= \frac{16}{x^4} - \frac{16}{x^2} + 6 - x^2 + \frac{x^4}{16}$$

31. Expand the binomial.

$$(s^{-2}+s^2)^6$$
$$=\binom{6}{0}(s^{-2})^6+\binom{6}{1}(s^{-2})^5(s^2)+\binom{6}{2}(s^{-2})^4(s^2)^2$$
$$+\binom{6}{3}(s^{-2})^3(s^2)^3+\binom{6}{4}(s^{-2})^2(s^2)^4+\binom{6}{5}(s^{-2})(s^2)^5$$
$$+\binom{6}{6}(s^2)^6$$
$$=s^{-12}+6s^{-8}+15s^{-4}+20+15s^4+6s^8+s^{12}$$

33. Find the term without expanding.

eighth term is $\binom{10}{7}(3x)^3(-y)^7=-3240x^3y^7$

35. Find the term without expanding.

third term is $\binom{12}{2}x^{10}(4y)^2=1056x^{10}y^2$

37. Find the term without expanding.

fifth term is $\binom{9}{4}(\sqrt{x})^5(-\sqrt{y})^4=126x^2y^2\sqrt{x}$

39. Find the term without expanding.

ninth term is $\binom{11}{8}\left(\frac{a}{b}\right)^3\left(\frac{b}{a}\right)^8=\frac{165b^5}{a^5}$

41. Find the term in the expansion.

$\binom{n}{i-1}a^{n-(i-1)}b^{i-1}$, if $b^{i-1}=b^8$, then $i=9$.

ninth term is $\binom{10}{8}(2a)^2(-b)^8=180a^2b^8$

43. Find the term in the expansion.

$(y^2)^{i-1}=y^8$, if $2i-2=8$, then $2i=10$ or $i=5$.

fifth term is $\binom{6}{4}(2x)^2(y^2)^4=60x^2y^8$

45. Find the term in the expansion.

sixth term is $\binom{10}{5}(3a)^5(-b)^5=-61,236a^5b^5$

47. Find the term in the expansion.

fifth term is $\binom{9}{4}(s^{-1})^5(s)^4=126s^{-1}$

sixth term is $\binom{9}{5}(s^{-1})^4(s)^5=126s$

49. Simplify the power of the complex number.

$$(2-i)^4=\binom{4}{0}(2^4)+\binom{4}{1}(2)^3(-i)^1+\binom{4}{2}(2)^2(-i)^2$$
$$+\binom{4}{3}2(-i)^3+\binom{4}{4}(-i)^4$$
$$=16+32(-i)+24(-1)+8(-i)^3+1$$
$$=16-32i-24+8i+1$$
$$=-7-24i$$

51. Simplify the power of the complex number.

$$(1+2i)^5=\binom{5}{0}(1)^5+\binom{5}{1}(1)^4(2i)^1+\binom{5}{2}(1)^3(2i)^2$$
$$+\binom{5}{3}(1)^2(2i)^3+\binom{5}{4}(1)(2i)^4+\binom{5}{5}(2i)^5$$
$$=1+10i-40-80i+80+32i$$
$$=41-38i$$

53. Simplify the power of the complex number.

$$\left(\frac{\sqrt{2}}{2}+i\frac{\sqrt{2}}{2}\right)^8$$
$$=\left(\frac{\sqrt{2}}{2}\right)^8(1+i)^8$$
$$+\frac{1}{16}(1+8i-28-56i+70+56i-28-8i+1)$$
$$=\frac{1}{16}(16)=1$$

Prepare for Section 8.6

P1. Evaluate.

$$7!=7\cdot6\cdot5\cdot4\cdot3\cdot2\cdot1=5040$$

P3. Evaluate.

$$\binom{7}{1}=\frac{7!}{1!(7-1)!}=\frac{7\cdot6!}{1(6!)}=7$$

P5. Evaluate.

$$\frac{10!}{(10-2)!}=\frac{10\cdot9\cdot8!}{8!}=90$$

Section 8.6 Exercises

1. Evaluate.

$$P(6,\ 2)=\frac{6!}{(6-2)!}=\frac{6\cdot 5\cdot 4!}{4!}=30$$

3. Evaluate.

$$C(8,\ 4)=\frac{8!}{4!(8-4)!}=\frac{8\cdot 7\cdot 6\cdot 5\cdot 4!}{4!\cdot 4\cdot 3\cdot 2\cdot 1}=70$$

5. Evaluate.

$$P(8,\ 0)=\frac{8!}{(8-0)!}=\frac{8!}{8!}=1$$

7. Evaluate.

$$C(7,\ 7)=\frac{7!}{7!(7-7)!}=\frac{7!}{7!\cdot 1}=1$$

9. Evaluate.

$$C(10,\ 4)=\frac{10!}{4!(10-4)!}=\frac{10\cdot 9\cdot 8\cdot 7\cdot 6!}{4\cdot 3\cdot 2\cdot 1\cdot 6!}=210$$

11. Find the number of ways.

Use the counting principle.

$3\cdot 2\cdot 2=12$

There are 12 different possible computer systems.

13. Find the number of ways.

Use the counting principle.

$2\cdot 2\cdot 2\cdot 2=16$

There are 16 possible light switch configurations.

15. Find the number of ways.

$P(6,6)=6!=720$

17. Find the number of ways.

$5\cdot 5\cdot 5=125$ ways

19. Find the number of tests.

Use the combination formula with $n=25$, $r=5$.

$$C(25,\ 5)=\frac{25!}{5!(20)!}=53{,}130$$

21. Explain.

There are 676 ways to arrange 26 letters taken two at a time ($26\cdot 26=676$). Now if there are more than 676 employees, then at least two employees have the same first and last initials.

23. Find the number of ways.

$C(6,\ 3)\cdot C(8,\ 3)=1120$

25. Find the number of ways.

$2^{10}=1024$

27. Find the number of ways.

$C(40,\ 6)=3{,}838{,}380$

29. **a.** Find the number of ways for 0 defective computers.

$C(7,\ 5)=21$

b. Find the number of ways for 1 defective computer.

$C(7,\ 4)\cdot C(3,\ 1)=35\cdot 3=105$

c. Find the number of ways for 3 defective computers.

$C(7,\ 2)\cdot C(3,\ 3)=21\cdot 1=21$

31. Find the possible number of serial numbers.

$3\cdot 12\cdot 5\cdot 10^7=1.8\times 10^9$

33. Find the number of ways to select at most 1 defective drive.

$C(10,\ 3)-C(8,\ 1)=120-8=112$

35. Find the number of ways.

$P(5,\ 5)=120$

37. Find the number of lines.

$C(7,\ 2)=21$

39. Find the number of ways.

$$\frac{16\cdot 14}{2}=112$$

41. Find the number of ways.

$C(20,\ 10)=184{,}756$

43. Find the number of ways.

$C(20,\ 12)\cdot C(12,\ 4)=125{,}970\cdot 495=62{,}355{,}150$

45. Find the number of possible cones.

A triple-decker cone could have all one flavor ice cream, or two different flavors with one scoop of the first flavor and two scoops of the second flavor (such as one scoop of vanilla and two scoops of chocolate), or two different flavors with two scoops of the first flavor and one scoop of the second flavor (such as two scoops of vanilla and one scoop of chocolate), or three different flavors.

$C(31, 1)+C(31, 2)+C(31, 2)+C(31, 3)$

$=\frac{31!}{1!30!}+\frac{31!}{2!29!}+\frac{31!}{2!29!}+\frac{31!}{3!28!}$

$= 31 + 465 + 465 + 4495$

$=5456$

47. Find the number of arrangements.

19!

49. **a.** Find the number of ways for no 2 people have the same birthday in a month.

$12\cdot11\cdot10\cdot9\cdot8\cdot7\cdot6=3,991,680$

b. Find the number of ways for 2 birthdays in a month.

$12^7-12\cdot11\cdot10\cdot9\cdot8\cdot7\cdot6=31,840,128$

51. Find the number of ways.

Let $a_1, a_2, ..., a_5$ be the long pieces and $b_1, b_2, ..., b_5$ be the short pieces. The pairs must have one a with one b. For a_1 there are 5 b's, for a_2 there are 4 b's, … . Thus there are $5\cdot4\cdot3\cdot2\cdot1=120$ pairs consisting of a long piece and a short piece.

53. Find the number of combinations.

To return to the original spot, the tourist must toss an equal number of heads and tails. This is $C(10, 5) =$ 252. There are 252 different toss combinations that return the tourist to the origin.

Prepare for Section 8.7

P1. Define the fundamental counting principle.

See Section 8.6.

P3. Evaluate.

$P(7,2)=\frac{7!}{(7-2)!}=\frac{7\cdot6\cdot5!}{5!}=42$

P5. Evaluate.

$\binom{8}{5}\left(\frac{1}{4}\right)^5\left(\frac{3}{4}\right)^{8-5}=\frac{8!}{5!(8-5)!}\left(\frac{1}{4}\right)^5\left(\frac{3}{4}\right)^3=\frac{189}{8192}$

Section 8.7 Exercises

1. List the elements in the sample space.

Label senators S_1, S_2 and representatives R_1, R_2, R_3.

The sample space is $S=\{S_1R_1, S_1R_2, S_1R_3, S_2R_1, S_2R_2, S_2R_3, R_1R_2, R_1R_3, R_2R_3, S_1S_2\}$

3. List the elements in the sample space.

Label coin H, T and integers 1, 2, 3, 4.

$S=\{H1, H2, H3, H4, T1, T2, T3, T4\}$

5. List the elements in the sample space.

Let the three cans be represented by A, B, and C and (x, y) represent the cans that balls 1 and 2 are placed in; e.g., (A, B) means ball 1 is in can A and ball 2 is in can B.

$S = \{(A, A), (A, B), (A, C), (B, A), (B, B), (B, C), (C, A), (C, B), (C, C)\}$

7. List the elements in the sample space.

$S=\{HSC, HSD, HCD, SCD\}$

9. List the elements in the sample space.

$S=\{$ae, ai, ao, au, ei, eo, eu, io, iu, ou$\}$

11. Express the event as a subset of the sample space.

$E = \{HHHH\}$

13. Express the event as a subset of the sample space.

$E=\{TTTT, HTTT, THTT, TTHT, TTTH, TTHH, THTH, HTHT, THHT, HTTH, HHTT\}$

15. Express the event as a subset of the sample space.

$E=\varnothing$

The sample space S for the events in Exercises 16—20 is

$S=\{(1,1), (1,2), (1,3), (1,4), (1,5), (1,6), (2,1), (2,2), (2,3), (2,4), (2,5), (2,6), (3,1), (3,2), (3,3), (3,4), (3,5), (3,6), (4,1), (4,2), (4,3), (4,4), (4,5), (4,6), (5,1), (5,2), (5,3), (5,4), (5,5), (5,6), (6,1), (6,2), (6,3), (6,4), (6,5), (6,6)\}$

17. Find the event: 2 numbers are the same.

$E=\{(1,1),(2,2),(3,3),(4,4),(5,5),(6,6)\}$

19. Find the event: second number is a 4.

$E=\{(1,4),(2,4),(3,4),(4,4),(5,4),(6,4)\}$

21. **a.** Find the probability: a king.

$P(\text{king})=\frac{4}{52}=\frac{1}{13}$

b. Find the probability: a spade.

$P(\text{spade}) = \frac{13}{52} = \frac{1}{4}$

23. Find the probabilities.

The sample space is

$S = \{HHHH, THHH, HTHH, HHHT, TTHH, THHT, HTTH, HHTT, THTH, HTHT, HHTH, HTTT, TTTH, THTT, TTHT, TTTT\}$

a. $E = \{TTTT\}$

$P(\text{all } T) = \frac{1}{16}$

b. $E = \{HHHH, THHH, HTHH, HHHT, TTHH, THHT, HTTH, HHTT, THTH, HTHT, HHTH, HTTT, TTTH, THTT, TTHT\}$

$P(\text{at least 1 } H) = \frac{15}{16}$

25. a. Yes, the events are mutually exclusive since $E_1 \cap E_2 = \varnothing$. That is, no two digit numbers that are perfect squares with the first digit the number 5.

b. $E_1 = \{50, 51, 52, 53, 54, 55, 56, 57, 58, 59\}$

$E_2 = \{16, 25, 36, 49, 64, 81\}$

$P(E_1) = \frac{10}{90}$ and $P(E_2) = \frac{6}{90}$

$P(E_1 \cup E_2) = \frac{10}{90} + \frac{6}{90} = \frac{16}{90} = \frac{8}{45}$

27. a. No, the events are not mutually exclusive, since the numbers 35 and 70 are divisible by both 5 and 7.

b. $E_1 = \{5, 10, 15, 20, 25, 30, 35, 40, 45, 50, 55, 60, 65, 70, 75, 80, 85, 90, 95, 100\}$

$E_2 = \{7, 14, 21, 28, 35, 42, 49, 56, 63, 70, 77, 84, 91, 98\}$

$P(E_1) = \frac{20}{100}$, $P(E_2) = \frac{14}{100}$ and $P(E_1 \cap E_2) = \frac{2}{100}$

$P(E_1 \cup E_2) = \frac{20}{100} + \frac{14}{100} - \frac{2}{100} = \frac{32}{100} = \frac{8}{25}$

29. Find the probability.

$P(\text{increase GNP}) + P(\text{increase inflation})$
$\quad - P(\text{increase GNP and inflation})$
$= 0.64 + 0.55 - 0.22 = 0.97$

31. Find the probability: at least one contract.

$P(\text{1st}) + P(\text{2nd}) - P(\text{1st and 2nd})$
$= \frac{1}{2} + \frac{1}{5} - \frac{1}{10} = \frac{6}{10} = \frac{3}{5}$

33. Find the probability: 0 does not occur.

Because sampling is with replacement, the events are independent. On one trial, the probability of not selecting a 0 is $\frac{9}{10}$. Therefore, the probability of not selecting a 0 on five trials is $\left(\frac{9}{10}\right)^5 = 0.59$.

35. Find the probability: at least \$50.

To receive at least \$50, an envelope with \$50 in cash or \$100 in cash must be selected. The probability is

$\frac{75}{500} + \frac{50}{500} = \frac{125}{500} = \frac{1}{4} = 0.25$.

37. Find the probability: alternating boys and girls.

There are $P(6, 6) = 720$ seating arrangements for the 6 children. There are $2 \cdot 3! \cdot 3!$ ways to have boys and girls alternate. Therefore the probability of boys and girls alternating is $\frac{2 \cdot 3! \cdot 3!}{720} = \frac{72}{720} = \frac{1}{10} = 0.1$.

39. Find the probability: 2 cards correct.

The subject can select $C(5, 2) = 10$ different sets of 2 cards. The magician must name the set the subject has drawn. Therefore, the probability that a magician can guess the answers is $\frac{1}{10}$ or 0.1.

41. Find the probability: Monday 8:00 A.M. chosen.

The probability of choosing Monday is $\frac{1}{5}$. The probability of choosing 8:00 A.M. is $\frac{1}{8}$. Therefore the probability of choosing Monday at 8:00 A.M. is

$\frac{1}{5} \cdot \frac{1}{8} = \frac{1}{40}$ or 0.025.

43. Find the probability: at least 1 unprofitable well.

The probability of at least one unprofitable

$= 1 -$ probability of all profitable

$= 1 - [(0.10)(0.10)(0.10)(0.10)]$

$= 1 - 0.0001 = 0.9999$

45. Find the probability: all 4 choose product *A*.

Assuming there is no preference, then the probability of choosing program *A* is $\frac{1}{2}$. The probability of all four companies choosing program *A* is $\left(\frac{1}{2}\right)^4 = \frac{1}{16}$.

47. Find the probability: at least one defective.

The probability of at least one defective equals 1 minus the probability of no defectives. The probability of no defectives is $(0.95)^5$. Therefore the probability of at least one defective is $1 - (0.95)^5 \approx 0.2262$

49. Find the probability 21 or more people show up.

This is a binomial probability with $p = \frac{3}{4}$, $q = \frac{1}{4}$, $n = 25$, and $k = 21, 22, 23, 24,$ and 25.

$$P = \binom{25}{21}\left(\frac{3}{4}\right)^{21}\left(\frac{1}{4}\right)^4 + \binom{25}{22}\left(\frac{3}{4}\right)^{22}\left(\frac{1}{4}\right)^3 + \binom{25}{23}\left(\frac{3}{4}\right)^{23}\left(\frac{1}{4}\right)^2 + \binom{25}{24}\left(\frac{3}{4}\right)^{24}\left(\frac{1}{4}\right)^1 + \binom{25}{25}\left(\frac{3}{4}\right)^{25}$$

The probability is approximately 0.2137.

51. Find the probability the rumor told 3 times without returning to originator.

Let the originator tell a member of the club. (Rumor told once.) This person repeats it to any of 7 remaining members with probability $\frac{7}{8}$. (Rumor told twice.) That person repeats it to any of 7 members with probability $\frac{7}{8}$. Probability of both events is $\left(\frac{7}{8}\right)^2 = \frac{49}{64}$.

53. Find the probability of winning a game presently tied.

Recall from Section 8.3 the Sum of a Geometric Series

$$S = \frac{a_1}{1-r}$$

$$S = \frac{p^2}{1-2p(1-p)} \qquad a_1 = p^2,\ r = 2p(1-p)$$

$$S = \frac{(0.55)^2}{1-2(0.55)(1-0.55)}$$

$$S \approx 0.599$$

The probability is 0.599.

Exploring Concepts with Technology

Casino 1

Mark	Catch	Win	Expectation
6	4	\$8	\$0.23
6	5	\$176	\$0.54
6	6	\$2960	\$0.38
			\$1.15

Casino 2

Mark	Catch	Win	Expectation
6	4	\$6	\$0.17
6	5	\$160	\$0.50
6	6	\$3900	\$0.50
			\$1.17

Casino 3

Mark	Catch	Win	Expectation
6	4	\$8	\$0.23
6	5	\$180	\$0.56
6	6	\$3000	\$0.39
			\$1.18

Casino 4

Mark	Catch	Win	Expectation
6	4	\$6	\$0.17
6	5	\$176	\$0.54
6	6	\$3000	\$0.39
			\$1.10

Each mathematical expectation was determined using the formula

$$\text{Mathematical expectation} = \frac{C(20,\ c)\cdot C(60,\ 6-c)}{C(80,\ 6)}W$$

W is the number of dollars you win for *c* catches.

Thus, casino 3 offers the greatest mathematical expectation. For each \$2 bet at casino 3, the gambler has a mathematical expectation of winning \$1.18.

Chapter 8 Review Exercises

1. Find the third and seventh terms. [8.1]

$$a_n = 3n+1$$
$$a_3 = 3(3)+1 = 10$$
$$a_7 = 3(7)+1 = 22$$

3. Find the third and seventh terms. [8.1]

$$a_n = n^2$$
$$a_3 = 3^2 = 9$$
$$a_7 = 7^2 = 49$$

5. Find the third and seventh terms. [8.1]

$$a_n = \frac{1}{n}$$
$$a_3 = \frac{1}{3}$$
$$a_7 = \frac{1}{7}$$

7. Find the third and seventh terms. [8.1]

$$a_n = 2^n$$
$$a_3 = 2^3 = 8$$
$$a_7 = 2^7 = 128$$

9. Find the third and seventh terms. [8.1]

$$a_n = \left(\frac{2}{3}\right)^n$$
$$a_3 = \left(\frac{2}{3}\right)^3 = \frac{8}{27}$$
$$a_7 = \left(\frac{2}{3}\right)^7 = \frac{128}{2187}$$

11. Find the third and seventh terms. [8.1]

$$a_1 = 2,\ a_n = 3a_{n-1}$$
$$a_2 = 3a_1 = 3\cdot 2 = 6$$
$$a_3 = 3a_2 = 3\cdot 6 = 18 \quad \bullet$$
$$a_4 = 3a_3 = 3\cdot 18 = 54$$
$$a_5 = 3a_4 = 3\cdot 54 = 162$$
$$a_6 = 3a_5 = 3\cdot 162 = 486$$
$$a_7 = 3a_6 = 3\cdot 486 = 1458 \quad \bullet$$

13. Find the third and seventh terms. [8.1]

$$a_1 = 1,\ a_n = -na_{n-1}$$
$$a_2 = -2a_1 = -2\cdot 1 = -2$$
$$a_3 = -3a_2 = -3\cdot(-2) = 6 \quad \bullet$$
$$a_4 = -4a_3 = -4\cdot 6 = -24$$
$$a_5 = -5a_4 = -5\cdot(-24) = 120$$
$$a_6 = -6a_5 = -6\cdot 120 = -720$$
$$a_7 = -7a_6 = -7\cdot(-720) = 5040 \quad \bullet$$

15. Find the third and seventh terms. [8.1]

$$a_1 = 1,\ a_2 = 2,\ a_n = a_{n-1}a_{n-2}$$
$$a_3 = a_2\cdot a_1 = 2\cdot 1 = 2 \quad \bullet$$
$$a_4 = a_3\cdot a_2 = 2\cdot 2 = 4$$
$$a_5 = a_4\cdot a_3 = 4\cdot 2 = 8$$
$$a_6 = a_5\cdot a_4 = 8\cdot 4 = 32$$
$$a_7 = a_6\cdot a_5 = 32\cdot 8 = 256 \quad \bullet$$

17. Find the third and seventh terms. [8.1]

$$a_1 = 1,\ a_2 = 2,\ a_n = 2a_{n-2} - a_{n-1}$$
$$a_3 = 2a_1 - a_2 = 2\cdot 1 - 2 = 0 \quad \bullet$$
$$a_4 = 2a_2 - a_3 = 2\cdot 2 - 0 = 4$$
$$a_5 = 2a_3 - a_4 = 2\cdot 0 - 4 = -4$$
$$a_6 = 2a_4 - a_5 = 2\cdot 4 - (-4) = 12$$
$$a_7 = 2a_5 - a_6 = 2\cdot(-4) - 12 = -20 \quad \bullet$$

19. Evaluate. [8.1]

$$5! + 3! = 5\cdot 4\cdot 3\cdot 2\cdot 1 + 3\cdot 2\cdot 1 = 120 + 6 = 126$$

21. Evaluate. [8.1]

$$\frac{10!}{6!} = \frac{10\cdot 9\cdot 8\cdot 7\cdot 6!}{6!} = 5040$$

23. Evaluate. [8.7]

$$\binom{12}{3} = \frac{12!}{3!(12-3)!} = \frac{12\cdot 11\cdot 10\cdot 9!}{3\cdot 2\cdot 1\cdot 9!} = 220$$

25. Evaluate. [8.1]

$$\sum_{k=1}^{5} k^2 = 1^2 + 2^2 + 3^2 + 4^2 + 5^2$$
$$= 1 + 4 + 9 + 16 + 25 = 55$$

27. Find the term for the arithmetic sequence. [8.2]

$$a_1 = 3,\ a_2 = 7,\ a_3 = 11,\ d = 7 - 3 = 4$$
$$a_n = 3 + (n-1)(4) = 3 + 4n - 4 = 4n - 1$$
$$a_{25} = 4(25) - 1 = 99$$

29. Find the term for the arithmetic sequence. [8.2]

$$a_1 = -2,\ a_{10} = 25$$
$$25 = -2 + (10-1)d$$
$$27 = 9d$$
$$d = 3$$
$$a_{15} = -2 + (15-1)(3) = 40$$

31. Find the term for the arithmetic sequence. [8.2]

$$a_n = 3n - 4$$
$$a_1 = 3(1) - 4 = -1$$
$$a_{20} = 3(20) - 4 = 56$$
$$S_{20} = \frac{20}{2}(-1 + 56) = 550$$

33. Find the sum for the arithmetic sequence. [8.2]

$$a_1 = 6,\ a_2 = 8,\ a_3 = 10,\ d = 8 - 6 = 2$$
$$S_{100} = \frac{100}{2}[2(6) + (100-1)(2)] = 10{,}500$$

35. Find the arithmetic means. [8.2]

$$13,\ c_2,\ c_3,\ c_4,\ c_5,\ 28$$

$$a_1 = 13$$
$$a_6 = a_1 + (n-1)d$$
$$28 = 13 + (6-1)d$$
$$15 = 5d$$
$$d = 3$$
$$c_2 = 13 + (2-1)(3) = 13 + 3 = 16$$
$$c_3 = 13 + (3-1)(3) = 13 + 6 = 19$$
$$c_4 = 13 + (4-1)(3) = 13 + 9 = 22$$
$$c_5 = 13 + (5-1)(3) = 13 + 12 = 25$$

37. Find the term for the geometric sequence. [8.3]

$$r = \frac{-2}{4} = -\frac{1}{2}$$
$$a_n = 4\left(-\frac{1}{2}\right)^{n-1}$$

39. Find the term for the geometric sequence. [8.3]

$$r = \frac{\frac{15}{4}}{5} = \frac{3}{4}$$
$$a_n = 5\left(\frac{3}{4}\right)^{n-1}$$

41. Find the sum for the geometric sequence. [8.3]

$$a_1 = 1,\ a_2 = 2,\ r = 2$$
$$S_8 = \frac{1(1-2^8)}{1-2} = 255$$

43. Find the sum for the geometric sequence. [8.3]

$$a_1 = 1,\ r = -\frac{1}{2}$$
$$S = \frac{1}{1-\left(-\frac{1}{2}\right)} = \frac{1}{\frac{3}{2}} = \frac{2}{3}$$

45. Evaluate the series. [8.1]

$$\sum_{k=1}^{5} 2\left(\frac{1}{5}\right)^{k-1} = 2\sum_{k=1}^{5}\left(\frac{1}{5}\right)^{k-1}$$
$$= 2\left(1 + \frac{1}{5} + \frac{1}{25} + \frac{1}{125} + \frac{1}{625}\right)$$
$$= 2\left(\frac{781}{625}\right) = \frac{1562}{625}$$

47. Evaluate the series. [8.3]

$$a_1 = 1,\ a_2 = -\frac{5}{6},\ r = -\frac{5}{6}$$
$$S = \frac{1}{1-\left(-\frac{5}{6}\right)} = \frac{1}{\frac{11}{6}} = \frac{6}{11}$$

49. Find the ratio of integers in lowest form. [8.3]

$$0.2\overline{3} = \frac{2}{10} + \left[\frac{3}{100} + \frac{3}{1000} + \frac{3}{10{,}000} \cdots\right]$$
$$a_1 = \frac{3}{100},\ r = \frac{1}{10}$$
$$0.2\overline{3} = \frac{2}{10} + \frac{\frac{3}{100}}{1-\frac{1}{10}} = \frac{2}{10} + \frac{1}{30} = \frac{7}{30}$$

51. Classify the sequence $a_n = n^2$ as arithmetic, geometric or neither. [8.3]

An arithmetic sequence has a common difference.

$$1^2,\ 2^2,\ 3^2,\ 4^2, \ldots = 1,\ 4,\ 9,\ 16, \ldots$$

No common difference; the sequence is not arithmetic.

A geometric sequence has a common ratio.

$$\frac{4}{1},\ \frac{9}{4},\ \frac{16}{9}, \ldots$$

No common ratio; the sequence is not geometric.

Neither.

53. Classify the sequence $a_n = (-2)^n$ as arithmetic, geometric or neither. [8.3]

A geometric sequence has a common ratio.

common ratio: $r=\frac{(-2)^{i+1}}{(-2)^i}=-2$

Geometric

55. Classify the sequence $a_n=\frac{n}{2}+1$ as arithmetic, geometric or neither. [8.3]

An arithmetic sequence has a common difference.

$\frac{1}{2}+1, \frac{2}{2}+1, \frac{3}{2}+1, \frac{4}{2}+1,...=\frac{3}{2}, 2, \frac{5}{2}, 3,...$

The common difference is $d=\frac{1}{2}$.

Arithmetic

57. Classify the sequence $a_n=n2^n$ as arithmetic, geometric or neither. [8.3]

An arithmetic sequence has a common difference.

$1(2)^1, 2(2)^2, 3(2)^3, 4(2)^4,...=2, 8, 24, 64,...$

No common difference; the sequence is not arithmetic.

A geometric sequence has a common ratio.

$\frac{8}{2}, \frac{24}{8}, \frac{64}{24},...$

No common ratio; the sequence is not geometric.

Neither.

59. Prove the statement $\sum_{i=1}^{n}(5i+1)=\frac{n(5n+7)}{2}$ by mathematical induction. [8.4]

1. For $n = 1$, we have $\sum_{i=1}^{1}(5i+1)=6$ and

$\frac{1(5+7)}{2}=6.$

Therefore that statement is true for $n = 1$.

2. Assume the statement is true for $n=k$.

$\sum_{i=1}^{k}(5i+1)=\frac{k(5k+7)}{2}$ Induction Hypothesis

Prove the statement is true for $n = k + 1$.

$$\sum_{i=1}^{k+1}(5i+1)$$
$$=\sum_{i=1}^{k}(5i+1)+5(k+1)+1=\sum_{i=1}^{k}(5i+1)+5k+6$$
$$=\frac{k(5k+7)}{2}+5k+6=\frac{k(5k+7)+10k+12}{2}$$
$$=\frac{5k^2+7k+10k+12}{2}=\frac{5k^2+17k+12}{2}$$
$$=\frac{(k+1)(5k+12)}{2}$$

Thus the statement is true for $n=k+1$. By the Induction Axiom, the statement is true for all positive integers.

61. Prove the statement $\sum_{i=0}^{n}\left(-\frac{1}{2}\right)^i=\frac{2\left(1-\left(-\frac{1}{2}\right)^{n+1}\right)}{3}$ by mathematical induction. [8.4]

1. This induction begins with $n=0$.

$$\sum_{i=0}^{0}\left(-\frac{1}{2}\right)^i=\left(-\frac{1}{2}\right)^0=1 \text{ and } \frac{2\left(1-\left(-\frac{1}{2}\right)^1\right)}{3}=1$$

Thus the statement is true when $n=0$.

2. Assume the statement is true for $n=k$.

$$\sum_{i=0}^{k}\left(-\frac{1}{2}\right)^i=\frac{2\left(1-\left(-\frac{1}{2}\right)^{k+1}\right)}{3} \quad \text{Induction Hypothesis}$$

Prove the statement is true for $n = k + 1$.

$$\sum_{i=0}^{k+1}\left(-\frac{1}{2}\right)^i=\sum_{i=0}^{k}\left(-\frac{1}{2}\right)^i+\left(-\frac{1}{2}\right)^{k+1}$$
$$=\frac{2\left(1-\left(-\frac{1}{2}\right)^{k+1}\right)}{3}+\left(-\frac{1}{2}\right)^{k+1}$$
$$=\frac{2\left(1-\left(-\frac{1}{2}\right)^{k+1}\right)+3\left(-\frac{1}{2}\right)^{k+1}}{3}$$
$$=\frac{2-2\left(-\frac{1}{2}\right)^{k+1}+3\left(-\frac{1}{2}\right)^{k+1}}{3}$$
$$=\frac{2+\left(-\frac{1}{2}\right)^{k+1}}{3}$$

Thus the statement is true for all integers $n\geq 0$.

63. Prove the statement $n^n \geq n!$ by mathematical induction. [8.4]

1. When $n = 1$, $1^1 = 1$ and $n! = 1$. The statement is true for $n = 1$.
2. Assume the statement is true for $n = k$.

$k^k \geq k!$ Induction Hypothesis

Prove the statement is true for $n = k + 1$. That is,

$(k+1)^{k+1} \geq (k+1)!$

$(k+1)^{k+1} = (k+1)(k+1)^k > (k+1)k^k$

By Induction Hypothesis $k^k \geq k!$. We have

$(k+1)k^k \geq (k+1)k! = (k+1)!$

Therefore $(k+1)^{k+1} \geq (k+1)!$

Therefore the statement is true for all integers $n \geq 1$.

65. Prove the statement 3 is a factor of $n^3 + 2n$ for all positive integers by mathematical induction. [8.4]

1.When $n = 1$, we have $1^3 + 2(1) = 3$. . Since 3 is a factor of 3, the statement is true for $n = 1$.

2. Assume the statement is true for $n = k$.

3 is a factor of $k^3 + 2k$ Induction Hypothesis

Prove the statement is true for $n = k + 1$. That is,

3 is a factor of $(k+1)^3 + 2(k+1)$.

$$(k+1)^3 + 2(k+1) = k^3 + 3k^2 + 3k + 1 + 2k + 2$$
$$= (k^3 + 2k) + 3(k^2 + k + 1)$$

By Induction Hypothesis, 3 is a factor of $k^3 + 2k$.

Three is also a factor of $3(k^2 + k + 1)$.

Thus 3 is a factor of $(k+1)^3 + 2(k+1)$.

The statement is true for all positive integers n.

67. Expand the binomial. [8.5]

$$(4a-b)^5 = \binom{5}{0}(4a)^5 + \binom{5}{1}(4a)^4(-b)^1 + \binom{5}{2}(4a)^3(-b)^2 + \binom{5}{3}(4a)^2(-b)^3 + \binom{5}{4}(4a)(-b)^4 + \binom{5}{5}(-b)^5$$
$$= 1024a^5 - 1280a^4b + 640a^3b^2 - 160a^2b^3 + 20ab^4 - b^5$$

69. Expand the binomial. [8.5]

$$(a-b)^7 = \binom{7}{0}a^7 + \binom{7}{1}a^6b + \binom{7}{2}a^5b^2 + \binom{7}{3}a^4b^3 + \binom{7}{4}a^3b^4 + \binom{7}{5}a^2b^5 + \binom{7}{6}ab^6 + \binom{7}{7}b^7$$
$$= a^7 - 7a^6b + 21a^5b^2 - 35a^4b^3 + 35a^3b^4 - 21a^2b^5 + 7ab^6 - b^7$$

71. Find the fifth term. [8.5]

$$\binom{7}{4}(3x)^3(-4y)^4 = 35(27x^3)(256y^4) = 241{,}920x^3y^4$$

73. Find the number of color selections. [8.6]

$12 \cdot 8 = 96$

75. Find the number of ways. [8.6]

There are 26 choices for each letter. By the Fundamental Counting Principle, there are 26^8 possible passwords.

77. Find the number of ways. [8.6]

This is a permutation with $n = 15$ and $r = 3$.

$$P(15,3) = \frac{15!}{(15-3)!} = \frac{15!}{12!} = 2730$$

79. Find the number of ways. [8.6]

There are $\binom{4}{1}$ ways to choose a supervisor and $\binom{12}{3}$ ways to choose 3 regular employees.

$\binom{4}{1}\binom{12}{3} = 4 \cdot 220 = 880$ shifts have 1 supervisor.

81. Find the number of ways. [8.6]

$$C(52,3)=\frac{52!}{3!(52-3)!}=\frac{52\cdot 51\cdot 50\cdot 49!}{3\cdot 2\cdot 1\cdot 49!}=22{,}100$$

83. a. Yes, the number 431 is in the sample space. [8.7]

b. Yes, with replacement, the number 313 is in the sample space.

85. List the elements: upward faces total 10. [8.7]

$\{(4, 6), (5, 5), (6, 4)\}$

87. a. Determine if mutually exclusive. [8.7]

The events are not mutually exclusive. For example, 61 is a prime number that is above 50.

b. Find the probability.

$$P(E_1)=\frac{25}{100},\ P(E_2)=\frac{50}{100},\ P(E_1\cap E_2)=\frac{10}{100}$$

$$P(E_1\cup E_2)=P(E_1)+P(E_2)-P(E_1\cap E_2)$$
$$=\frac{25}{100}+\frac{50}{100}-\frac{10}{100}=\frac{13}{20}$$

89. Find which one has the greater probability. [8.7]

The probability of drawing an ace and a 10 card from one regular deck of playing cards is

$$\frac{\binom{4}{1}\binom{16}{1}}{\binom{52}{2}}=\frac{4\cdot 16}{\frac{52\cdot 51}{2}}\approx 0.0483.$$

The probability of drawing an ace and a 10 card from two regular decks of playing cards is

$$\frac{\binom{8}{1}\binom{32}{1}}{\binom{104}{2}}=\frac{8\cdot 32}{\frac{104\cdot 103}{2}}\approx 0.0478.$$

Drawing an ace and a 10 card from *one* deck has the greater probability.

91. Find the probability. [8.7]

$$P=\binom{10}{8}(0.9)^8(0.1)^2\approx 0.19$$

93. Find the probability. [8.7]

There are $\binom{12}{3}$ ways of choosing 3 people from 12.

There are $\binom{11}{2}\cdot 1$ ways of choosing 2 people and the person with badge number 6.

$$\text{Probability}=\frac{\binom{11}{2}\cdot 1}{\binom{12}{3}}=\frac{\frac{11\cdot 10}{2}}{\frac{12\cdot 11\cdot 10}{3\cdot 2}}=\frac{1}{4}$$

95. Find the net effect. [8.3]

Using the multiplier effect,

$$\frac{15}{1-0.80}=75$$

The net effect of \$15 million is \$75 million.

Chapter 8 Test

1. Find the third and fifth terms. [8.1]

$$a_3=\frac{2^3}{3!}=\frac{8}{6}=\frac{4}{3}$$

$$a_5=\frac{2^5}{5!}=\frac{32}{120}=\frac{4}{15}$$

3. Classify the sequence as arithmetic, geometric or neither. [8.2]

$$a_{n+1}-a_n=[-2(n+1)+3]-(-2n+3)$$
$$=-2n-2+3+2n-3$$
$$=-2=\text{ constant}$$

arithmetic

5. Classify the sequence as arithmetic, geometric or neither. [8.2]

$$\frac{a_{n+1}}{a_n}=\frac{\frac{(-1)^{n+1-1}}{3^{n+1}}}{\frac{(-1)^{n-1}}{3^n}}=\frac{-1}{3}=\text{constant}$$

geometric

7. Find the sum of the series. [8.3]

$$\sum_{j=1}^{10}\frac{1}{2j}=\frac{1}{2}+\frac{1}{4}+\frac{1}{8}+\frac{1}{16}+\cdots\frac{1}{1024}$$

$$=\frac{\frac{1}{2}\left(1-\left(\frac{1}{2}\right)^{10}\right)}{1-\frac{1}{2}}=1-\left(\frac{1}{2}\right)^{10}$$

$$=1-\frac{1}{1024}=\frac{1023}{1024}$$

9. Find the twentieth term. [8.2]

$$a_3 = a_1 + (3-1)d = 7,$$
$$a_8 = a_1 + (8-1)d = 22$$

$$\begin{aligned} a_1 + 2d &= 7 \\ a_1 + 7d &= 22 \\ \hline -5d &= -15 \\ d &= 3 \end{aligned}$$

$$\begin{aligned} a_1 &= a_3 - 2(3) \\ &= 7-6 \\ &= 1 \end{aligned}$$

$$\begin{aligned} a_{20} &= a_1 + (20-1)d \\ &= 1 + (19)(3) \\ &= 58 \end{aligned}$$

11. Write as a quotient of integers in simplest form. [8.3]

$$\begin{aligned} 0.\overline{15} &= 0.15 + 0.0015 + 0.000015 + \ldots \\ &= \frac{0.15}{1-0.01} = \frac{0.15}{0.99} = \frac{15}{99} = \frac{5}{33} \end{aligned}$$

13. Prove by mathematical induction. [8.4]

1. Let $n = 7$

   $7! = 5040 \quad 3^7 = 2187$

   Thus $n! > 3^n$ for $n = 7$.

2. Assume $k! > 3^k$

   Verify $(k+1)! > 3^{k+1}$

$$\begin{aligned} k! &> 3^k \\ k+1 &> 3 \\ (k+1)k! &> 3 \cdot 3^k \\ (k+1)! &> 3^{k+1} \end{aligned}$$

Thus the formula has been established by the extended principle of mathematical induction.

15. Write the binomial expansion. [8.5]

$$\begin{aligned} \left(x+\frac{1}{x}\right)^6 &= x^6 + 6(x)^5\left(\frac{1}{x}\right) + 15(x)^4\left(\frac{1}{x}\right)^2 + 20(x)^3\left(\frac{1}{x}\right)^3 \\ &\quad + 15(x)^2\left(\frac{1}{x}\right)^4 + 6x\left(\frac{1}{x}\right)^5 + \left(\frac{1}{x}\right)^6 \\ &= x^6 + 6x^4 + 15x^2 + 20 + \frac{15}{x^2} + \frac{6}{x^4} + \frac{1}{x^6} \end{aligned}$$

17. Find the number of ways. [8.6]

$$52 \cdot 51 \cdot 50 = 132{,}600$$

19. Find the probability. [8.7]

$$\frac{C(8,\ 3)C(10,\ 2)}{C(18,\ 5)} = \frac{56 \cdot 45}{8568} = \frac{5}{17} \approx 0.294118$$

Cumulative Review Exercises

1. Solve, write in interval notation. [1.5]

$$\begin{aligned} |3-5x| &\le 4 \\ -4 \le 3-5x &\le 4 \\ -7 \le -5x &\le 1 \\ \frac{7}{5} \ge x &\ge -\frac{1}{5} \end{aligned}$$

The solution set is $\left[-\frac{1}{5}, \frac{7}{5}\right]$.

3. Graph $y = x^2 - x - 2$. [2.1]

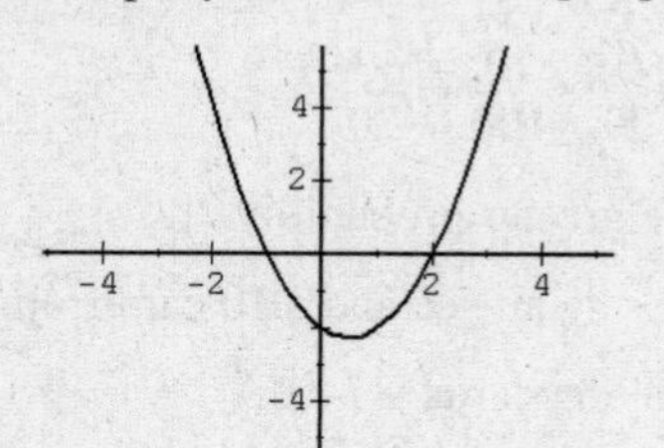

5. Divide. [3.1]

$$\begin{array}{r} x^2 - x + 1 \\ x+1 \overline{\smash{)}\, x^3 \qquad\qquad -1} \\ \underline{x^3 + x^2} \qquad\quad \\ -x^2 \qquad\quad \\ \underline{-x^2 - x} \quad \\ x - 1 \\ \underline{x+1} \\ -2 \end{array}$$

$$\frac{x^3-1}{x+1} = x^2 - x + 1 - \frac{2}{x+1}$$

7. Expand. [4.4]

$$\begin{aligned} \log_b\left(\frac{xy^2}{z^3}\right) &= \log_b x + \log_b y^2 - \log_b z^3 \\ &= \log_b x + 2\log_b y - 3\log_b z \end{aligned}$$

9. Solve the system of equations. [6.1]

$$\begin{cases} 2x-3y=8 & (1) \\ x+4y=-7 & (2) \end{cases}$$

Solve (2) for x and substitute into (1).

$x=-4y-7$

$$\begin{aligned} 2(-4y-7)-3y&=8 \\ -8y-14-3y&=8 \\ -11y&=22 \\ y&=-2 \end{aligned}$$

$x=-4(-2)-7=1$

The solution is (1, –2).

11. Simplify. [P.2]

$$-2\sqrt[4]{80}+3\sqrt[4]{405}=-4\sqrt[4]{5}+9\sqrt[4]{5}=5\sqrt[4]{5}$$

13. Find the coordinates of the vertex. [2.4]

$$-\frac{b}{2a}=-\frac{5}{2(-2)}=\frac{5}{4}$$

$$F\left(\frac{5}{4}\right)=-2\left(\frac{5}{4}\right)^2+5\left(\frac{5}{4}\right)-2=\frac{9}{8}$$

Vertex: $\left(\frac{5}{4},\frac{9}{8}\right)$

15. Find the horizontal asymptote. [3.5]

$y=0$

17. Solve. Round to the nearest tenth. [4.5]

$$\begin{aligned} 4^{2x+1}&=3^{x-2} \\ \ln 4^{2x+1}&=\ln 3^{x-2} \\ (2x+1)\ln 4&=(x-2)\ln 3 \\ 2x\ln 4+\ln 4&=x\ln 3-2\ln 3 \\ x\ln 4^2-x\ln 3&=-2\ln 3-\ln 4 \\ x(\ln 16-\ln 3)&=-2\ln 3-\ln 4 \\ x&=\frac{-2\ln 3-\ln 4}{\ln 16-\ln 3}\approx -2.1 \end{aligned}$$

19. Find the product. [7.2]

$$\begin{bmatrix} 3 & 2 \\ -2 & 1 \\ 1 & -4 \end{bmatrix}\begin{bmatrix} 2 & 3 & 1 & 1 \\ -2 & 0 & 4 & -3 \end{bmatrix}=\begin{bmatrix} 2 & 9 & 11 & -3 \\ -6 & -6 & 2 & -5 \\ 10 & 3 & -15 & 13 \end{bmatrix}$$